T0074944

Energy Systems

Energy Systems

A Project-Based Approach to Sustainability Thinking
for Energy Conversion Systems

Leon Liebenberg
University of Illinois
Urbana-Champaign

Library of Congress Cataloging-in-Publication Data applied for:

Hardback ISBN: 9781119869467

Cover Design: Wiley
Cover Image: Courtesy of the artist Alisa Singer

Set in 9.5/12.5pt STIXTwoText by Straive, Pondicherry, India

SKY10068634_022924

Contents

Preface

Energy is all-pervasive in the global economy. Every human activity involves the use of energy, either directly or indirectly. And every technological development depends on the availability of usable energy supplies. Unsurprisingly, energy conversion contributes to more than 70% of greenhouse-gas pollution. Even more than before, our engineers and policymakers, and indeed all citizens, need to be cognizant of the primary sources of energy. We also need to comprehend the processes that may be used to convert these fuels or resources into forms such as electricity and hydrogen that are easy to transport and convenient for the end-user.

This book emphasizes and probes what engineers and policymakers need to understand as they negotiate information overload and constantly changing energy scenarios. I summarize the basic equations of thermodynamics, fluid mechanics, and quantum mechanics in a just-in-time manner. I then go straight into applications, which include several case studies. The text is therefore multi-disciplinary in nature and forward looking because of its focus on what specialists will need to understand as we adopt new technologies or policies. Most chapters feature an up-to-date summary of salient energy technologies for quick reference by students and practitioners of energy engineering.

The book investigates energy supply and demand, the science of global warming, interpretations of sustainability, chemical fuels, carbon capture and storage, internal and external combustion engines, vapor power and refrigeration plants, and nuclear power. It further delves into the exciting world of direct energy conversion, with photovoltaic cells and fuel cells leading the way. Solar-heat, wind energy, and water energy are also covered, along with energy storage in its many forms. The book ends with a brief investigation into what we can do to decarbonize the transportation, industry, buildings, and electric power sectors, which together contribute nearly 95% of carbon emissions.

It is not possible to cover all energy conversion systems in a short book. I omit, or only mention in passing, some technologies that have not yet proven their viability or have small impact, such as liquid fuels derived from genetically modified algae, micro nuclear batteries, quantum glass batteries, ocean thermal energy conversion systems, and piezoelectric devices and other energy-scavenging systems. Instead, the book focuses on energy conversion systems that are already viable and recognized as such and are already having a large-scale impact.

Learning begins with learner engagement and learning leads to knowledge and understanding. Once learners understand, they become capable of effective action. They can reflect critically on their practice and engage in higher-order thinking. They are able to exercise judgment and create designs that reflect an appreciation of relevant constraints and uncertainties. Accordingly, this book employs multiple pedagogies of engagement, the aim of which is to arouse the curiosity and interest of the learner.

Each chapter begins with observations rather than merely stating general principles and deriving them. The intention is to provide context and to engage the reader as meaningfully as possible. The text captures important concepts first and then incrementally fills in finer details.

Energy technologies are contextualized in terms of their interrelated social, environmental, economic, and political impacts. Throughout the book, the reader is reminded that most sustainable energy systems rest on a platform or infrastructure that is inherently unsustainable. Instead of designing for sustainability, we should therefore perhaps be designing for perpetuity. This requires us to re-examine the deep, underlying structures that support us.

Our capacity to imagine holds the key to social, cultural, and technological development, and to effectively negotiating major contemporary challenges such as climate change. If we are to pursue different, better futures, we must be able to imagine what those futures might look, smell, and feel like. It is crucial, when we consider the case of climate change, to be able to anticipate what the impacts and implications of different energy scenarios might be. Accordingly, via many "big" and pressing questions, the reader is asked to "reimagine our future," particularly with a focus on sustainable energy. This kind of imaginative engagement is also important to generate new ideas and for grappling with advanced technologies. Can we envision where the linked solutions to sustainable energy might lead if we applied and extended what we have learned, and if we took responsibility for creating the world we want?

I believe that we will need to adopt the mindset of an entrepreneur to successfully negotiate an energy transition. I have therefore categorized the "reimagine our future" reflective questions under 10 attributes which I believe successful entrepreneurs have in common. These entrepreneurial mindsets and a summary of the reflective questions are provided on the book's companion website.

The book also includes interviews with young energy entrepreneurs. These interviews further serve as a reminder of the importance of an entrepreneurial mindset when negotiating climate change. Implicit in the mindset of these entrepreneurs is the creation of value, while being endlessly curious and making unusual connections. The reader is often asked to employ KEEN's "3-C" entrepreneurial mindset when investigating the potential of energy conversion systems. The book further features interviews with engineering students who share their thoughts about a sustainable energy future.

Problem-solving is another central theme in the book and numerous examples are provided with a view to describing problem-solving processes. Examples are provided in a guided, step-by-step manner that will facilitate independent study and reflection and help students develop and master the skill of solving problems in a systematic way. The examples provided are real-world, obtained from a diverse range of cultural contexts and drawing on diverse perspectives, to help ensure reader engagement. The focus is on case studies involving contemporary energy developments.

Quantitative and qualitative practice problems and questions are provided at the end of each chapter to help readers check and reinforce their understanding of energy fundamentals and related issues, some of which are complex. Many of the questions are open-ended and require readers to do independent research and think for themselves. Throughout, salient journal papers are cited to guide independent research. I mostly use SI units, which are summarized in Appendix A.

Importantly, the book also provides 10 guided mini projects to help the reader explore sustainability issues. These mini projects require a multidisciplinary approach, and often address open-ended design questions pertaining to the creation of optimal energy solutions based on best practices. The mini projects have been designed so that the reader can explore, analyze, and reflect on contemporary energy systems. The challenges related to energy production and consumption are – first and foremost – design problems, not technological problems. After all, the field of design is about imagining new possibilities that can transform the present. The reader is therefore prompted, throughout the book, to consider how we can reconceptualize our relationship with energy. This is a crucial step in resolving the global energy challenges.

Most examples, problems, and mini projects feature straightforward computational schemes. Some of these show how numerical analysis may be employed to provide more accurate results. I avoid the complexities of industrial design methodology with a view to making the book accessible to a broad audience. A reader with limited industrial experience or limited numerical skills will find here an accessible introduction to energy conversion systems that is nevertheless comprehensive and state of the art.

My intention is that when study of the text is combined with the mini projects, worked examples, practice problems, reflective exercises, and lectures, these elements will be mutually reinforcing to constitute a rich learning experience. The book's companion website contains instructor slides and solutions manuals for the end-of-chapter problems and for the 10 mini projects.

The text attempts to balance precision with conciseness. The 28 chapters are provided in "bite-sized" chunks, most of which can be covered in a traditional 50-min university lecture. Each of these chapters is presented in a self-contained or modular manner and can be presented in any sequence. The main book topics (which can typically be covered in about a week and hence are called "weeks") are each presented in a series of three related chapters, except for the final "week," which comprises a single chapter.

The choice of course material and the order in which it is presented should be tailored to the student body. This also applies to the use of mini projects. Based on experience, a series of three to six mini projects can form the bulk of course assessment in lieu of mid-term exams. Assessment of this sort, unlike the too-often used high-stake and stressful summative assessments like exams, makes for highly engaged students who enjoy learning at their own pace. The book's instructor resources feature, among others, lecture slides aimed at promoting active learning and guidelines on effective use of the book material.

The material in this book has been evaluated in several senior undergraduate- and junior graduate-level classes. I am indebted to my students whose creative ideas and incisive questions not only helped clarify my own explanations but led me to ask new questions. Several academics have influenced my thinking, and I am immensely grateful for their friendship and mentoring.

The traditional models of knowledge and design are no longer sufficient to effectively address our energy challenges. We need to think and operate differently. The complex and interconnected challenges we face require a new approach. It is my hope that this book will help to identify such new perspectives and approaches when tackling our sustainable energy challenge.

Leon Liebenberg
University of Illinois, Urbana-Champaign

Acknowledgments

I am deeply indebted to the Warren G. Lavey Family Charitable Fund for its grant, which initiated this book project. Thank you, Professors Warren Lavey and Holly Rosencranz, for your wonderful support. I would also like to extend a special word of thanks to the Kern Family Foundation for its generous grant. The related work for the Kern Entrepreneurial Engineering Network (KEEN) helped me to generate contextual content for this book. I am especially grateful to Dr. Doug Melton, Micheal Johnson, and Karen Wilken for their inspirational support and advice.

A huge thank you to Emeritus Professor Robert McKim (Religion), who read much of the manuscript and made incisive comments and suggestions. Thank you, Robert, for your constant inspiration. I also wish to thank my following colleagues from the University of Illinois at Urbana-Champaign (UIUC) for their critical review of several chapters: Professor Warren Lavey (Law), Emeritus Professor Bruce Hannon (Geography and Geographic Information Science), Emeritus Professor Don Wuebbles (Atmospheric Sciences), Emeritus Professor John Abelson (Materials Science and Engineering), Professor Daniel Andruczyk (Nuclear, Plasma, and Radiological Engineering), and Dr. Andrew Stumpf (Illinois Geothermal Coalition). Thank you also to UIUC colleagues for their enthusiastic encouragement and excellent advice: Emeritus Professor Robert McKim (Religion), Professor Warren Lavey (Law), Professor Gillen Wood (English), Professor Steve Marshak (Geology), Professor Bob Rauber (Atmospheric Sciences), Professors Tony Jacobi and Sanjiv Sinha (Mechanical Science and Engineering), and Professor Rizwan Uddin (Nuclear, Plasma, and Radiological Engineering). I am further grateful for the critical review of several chapters by Professor Andrew Chapman (Kyushu University), Dr. Johan Gouws (Consulting Engineer, Australia), and Uwe Franzmann (Hitachi Power).

I owe special gratitude to Professor Robert Jaffe (Massachusetts Institute of Technology) for his wonderful support. Also, his seminal book on energy systems was a constant inspiration.

I am eternally grateful for the mentoring of Professor Wesley Harris (MIT Aero & Astro Engineering) and the support from Professors Jenni Case (Engineering Education, Virginia Tech), Christine Hailey (College of Science and Engineering, Texas State University), Ian Jandrell (University of the Witwatersrand) and Emeritus Professors John Mitchell (College of Engineering, University of Wisconsin-Madison), Charles Tucker III, and Robert White (Mechanical Science and Engineering, UIUC), Frikkie van Niekerk (North-West University), André Nel (University of Johannesburg), Josua Meyer (Stellenbosch University), John Thome (EPFL), and Adrian Bejan (Duke).

I am indebted to Professor Michel Bellini (UIUC Center for Innovation in Teaching and Learning) and to Dr. Jay Mann and Dr. Chris Migotsky (UIUC Academy for Excellence in Engineering Education) for supporting my work in pedagogies of engagement. I also benefited immensely from the suggestions, critique, and fabulous teaching ideas from my team members in the Engagement in Engineering Education (or ENGINE) research group: Dr. Robert Baird, Dr. Ava Wolf, Dr. David

Favre, Jim Wentworth, Professor Chad Lane, Dr. Saad Shehab, Dr. Kate LaBore, Professor Shelly J. Schmidt, Cheelan Bo-Linn, Alex Pagano, Professor Tim Hale, Professor Justin Aronoff, Taylor Tucker, Esmee Vernooij, Jessica Mingee, Tierney Dufficy, and Professor Yuting Chen.

I wish to express my appreciation to the following individuals for their comments, ideas, or information pertaining to contemporary developments in energy engineering and sustainability: Patrick Forsythe (Generac), Dr. Sashin Anand and Marcos Guerrero (db|HMS), Dr. Arpit Dwivedi (Cache Energy), Dr. Andrew Stumpf (Illinois Geothermal Coalition), Dr. Nicolas Stauff (Argonne National Laboratory), Professor Robert Costanza (University College London), Mike Brewer (Abbott power plant), Mike Larson (UIUC Utilities Production), and Don Fournier (Energy Consultant, Illinois), and the following colleagues from UIUC. Emeritus Professor Ty Newell (Mechanical Science and Engineering), Emeritus Professor Clark Bullard (Mechanical Science and Engineering), and Frank Holcomb (Water Resources Engineering and Science).

I am furthermore deeply indebted to Dorothy Loudermilk for producing the superb illustrations and to Janine Smit for stellar editing. Thank you, too, to Shreyas Venkatarathinam (UIUC) and Peter Davis (University of St. Thomas) for their support, critique, and excellent suggestions. I also appreciate the expert assistance and advice of Andrew Eltzroth, Matthew Auston, Aaryaman Patel, Jonathan Lasso, Arya Haria, and Dimitri Kalinichenko with proofreading of the initial book notes, seeking permissions for copyrighted material, producing solutions manuals, and for their superb administrative assistance.

A big thank you to the students in my ENG 571 (Theory of Energy and Sustainability Engineering), NPRE 480 (Energy and Security), ENG 471 (Seminar in Energy and Sustainability Engineering), ME 400 (Energy Conversion Systems), TAM 335 (Fluid Mechanics), ME 370 (Mechanical Design I), ME 310 (Fluid Dynamics), ME 270 (Design for Manufacturability), and ME 200 (Thermodynamics) classes for testing ideas regarding the mini projects, worked examples, and reflective exercises and practice problems, many of which are featured in the book.

I am elated that so many energy entrepreneurs and engineering students took the time to share their experiences and views of a sustainable energy future, which are featured as short articles. Thank you to one and all.

The book has further benefited from the publication of the thought-provoking works of art of several artists and designers. I am especially grateful to Alisa Singer for allowing one of her seminal artworks to grace the cover of this book. Alisa's *environmental graphiti* [sic] merges climate science with art and helps foster the imagination. This ties in perfectly with the book's main objective, which is to help the reader reimagine a sustainable energy future. Thank you, Alisa.

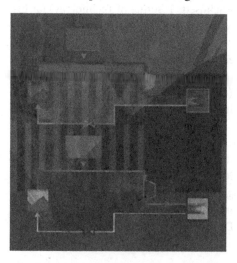

Alisa Singer's artwork, "Use of electric heat pumps to reduce emissions," which is featured on the book cover. *Source:* Environmental Graphiti.

A special word of thanks goes to Lauren Poplawski, commissioning editor at Wiley, who asked me to write this book. Lauren, I am most grateful for your insight, advice, organization, accommodating stance, and willingness to explore new financing options for this book. Thank you, too, to Jayashree Saishankar, managing editor at Wiley, Jeevaghan Devapal, Isabella Proietti and Sevanthi Vivekanandhan for fantastic editorial support.

For their endless encouragement, interest, and emotional support, I would like to express my gratitude to my brother, Terseus, and my in-laws, Henri and Marinette Favre. Finally, I would like to thank my wife, Salomé, for enduring, listening, understanding, and grounding the author and this book in sustainability thinking. (Thank you also, Suzie, for keeping Salomé company in my year-long "absence!")

Leon Liebenberg
University of Illinois at Urbana-Champaign

Notation

Symbol	Notation
a	Acceleration (m/s^2); amplitude (m)
A	Atomic mass number (number of nucleons in a nucleus); cross-sectional area (m^2); constant (−); aperture area (m^2)
α	Absorptivity (−); Seebeck coefficient (V/°C)
AFR	Air–fuel ratio (kg/kg)
B	Binding energy (MeV/nucleus)
$\bar{B}$	Magnetic field vector (T)
B_C	Coulomb repulsion (J)
b	Specific binding energy (MeV/nucleon)
β	Beta decay (years); tilt angle (°)
c	Speed of light (m/s); statistical function (−)
C	Capacitance (F)
c_d	Drag coefficient (−)
c_v	Specific heat capacity at constant volume (kJ/kg K)
c_p	Specific heat capacity at constant pressure (kJ/kg K)
$\bar{C}_v$	Molar heat capacity at constant volume (kJ/mol K)
$\bar{C}_p$	Molar heat capacity at constant pressure (kJ/mol K)
C_p	Power coefficient (−)
C_R	Concentration ratio (−)
COP	Coefficient of performance (−)
d	Distance (m)
D	Absorbed dose (Gy)
δ	Declination (°)
Δ	Atomic mass defect (MeV); difference
e	Specific energy (kJ/kg); electron; electronic charge (eV); exponential (−)
e	Specific exergy (kJ/kg)
$\bar{e}$	Molar specific exergy (kJ/mol)

(Continued)

Symbol	Notation
E	Energy (J); total energy (J); electric field (V/m); electromotive force V); effective dose (J)
$\mathbb{E}$	Exergy (J)
$\dot{E}$	Rate of energy transfer (W)
E_g	Energy gap (eV)
E_i	Initial energy (J)
E_F	Fermi energy (eV)
E_k	Kinetic energy (J)
E_p	Potential energy (J)
ε	Surface roughness height (m); emissivity (−); permittivity (F/m); fast fission factor (−); second-law efficiency (−)
ε_0	Vacuum permittivity (F/m)
η	First-law efficiency (−); neutron reproduction factor (−)
f	Friction factor (−); thermal neutron utilization factor (−)
ff	Fill factor (−)
F	Force (N); radiative forcing (W/m^2°C); Faraday constant (C/mol)
$\overline{F}$	Electromagnetic force (N)
g	Gravitational acceleration (m/s^2); specific Gibbs function (kJ/kg)
$\overline{g}$	Gibbs function of formation (kJ/kmol)
g_s	Geometric factor (−)
G	Gibbs energy (J)
γ	Adiabatic (isentropic) index (−)
Γ	Gamma function; lapse rate
h	Heat transfer coefficient (W/(m^{2}K)); Planck constant (Js); specific enthalpy (kJ/kg); head (m)
$\overline{h}$	Molar specific enthalpy (kJ/mol); molar enthalpy of formation (kJ/kmol)
H	Enthalpy (J); scale height (m); equivalent dose (Sv)
h_f	Specific enthalpy of liquid phase (kJ/kg)
h_g	Specific enthalpy of vapor phase (kJ/kg)
h_{fg}	Difference between specific enthalpy of vapor and liquid phases; enthalpy of vaporization or enthalpy of condensation (kJ/kg)
I	Moment of inertia (m^4); electric current (A); irradiance; insolation; irreversibility
i	Specific irreversibility (kJ/kg)
$\dot{I}$	Rate of irreversibility (kJ/kg s)
J	Kinetic energy flux (V/C)
k	Wave number (−); thermal conductivity (W/mK); neutron multiplication factor (−); absorption cross-section (barns)
k_B	Boltzmann constant (J/K)
K_T	Clearness index (−)
κ	Absorption coefficient (−)
l	Length (m)
L	Spectral irradiance (W/m^2 nm); latitude (°)
λ	Lambda fuel factor (−); wavelength (m); climate feedback parameter (W/m^2°C)

Symbol	Notation
λ_0	Planck feedback parameter (W/m^2 °C)
m	Mass (kg)
M	Molar mass (kg/kmol)
$\dot{m}$	Mass flow rate (kg/s)
m_e	Electron mass (kg)
m_n	Neutron mass (kg)
m_p	Proton mass (kg)
mep	Mean effective pressure (kPa)
μ	Dynamic viscosity (Pa s)
n	Number per unit length (−); number of moles (−); energy level (−)
$\hat{n}$	Unit normal vector
N	Number (−); particles (−); number of neutrons (−)
$\dot{N}$	Rate of moles (kmol/s)
N_A	Avogrado's number (/mol)
∇	Gradient (−); divergence
Nu	Nusselt number (−)
ω	Angular velocity (rad/s); angular frequency (rad/s)
p	Pressure (Pa); proton; resonance escape probability (−); photon momentum (J/(m/s))
P	Power (W); probability (−); irradiance (W/m^2)
Π	Peltier coefficient (W/A)
ϕ	Fuel-air ratio (kg/kg); humidity ratio (kg/kg); angular variable (°); phase
Φ	Flux (N m^2/C)
q	Specific heat (J/kg); electric charge (eV)
Q	Heat (J); electric charge (eV); fusion gain factor (−)
$\dot{Q}$	Heat rate (W)
r	Radial distance (m); compression ratio (−)
R	Radius; resistance (Ω); gas constant (kJ/kg K); thermal resistance (W/mK)
$\bar{R}$	Molar gas constant (kJ/kmol K)
R_f	Rate of fission reaction (cm^{-2}s^{-1})
r_c	Cut-off ratio (−)
Re	Reynolds number (−)
ρ	Mass density (kg/m^3); electrical resistivity (Ω^{-1}m^{-1}); reflectivity (−)
s	Displacement (m); specific entropy (kJ/kg K)
S	Entropy (kJ/K)
$\bar{s}$	Molar entropy (kJ/kmol K)
s_f	Specific entropy of liquid phase (kJ/kg K)
s_g	Specific entropy of vapor phase (kJ/kg K)
s_{fg}	Difference between specific entropy of vapor and liquid phases (kJ/kg K)
σ	Electrical conductivity (Ω m); Stefan–Boltzmann constant (W/(m^{2}K^4)); cross-section (barns); entropy production (kJ/K)

(Continued)

Symbol	Notation
$\dot{\sigma}$	Rate of entropy production (kJ/Ks)
Σ	Macroscopic cross-section (barn)
t	Time (s)
T	Temperature (K, °C); period (s); torque (Nm)
T_0	Temperature at dead state; reference temperature (K, °C)
T_H	High temperature (K, °C)
T_C	Low temperature (K, °C)
τ	Transmittance (−); time constant (s)
θ	Angular variable (°)
θ_{sub}	Angle subtended between the sun and the earth (°)
u	Specific internal energy (kJ/kg); atomic mass unit (Dalton) (u)
U	Internal energy (kJ); U-factor (thermal conductance) (W/m² K)
u_f	Specific internal energy of liquid phase (kJ/kg)
u_g	Specific internal energy of vapor phase (kJ/kg)
u_{fg}	Difference between specific internal energy of vapor and liquid phases (kJ/kg)
V	Speed (m/s); potential energy (J)
$\overline{V}$	Velocity vector (m/s)
V_e	Electrostatic potential (V)
V_n	Normal velocity (m/s)
ν	Specific volume (m³/kg); frequency (Hz)
ν_f	Specific volume of liquid phase (m³/kg)
ν_g	Specific volume of vapor phase (m³/kg)
ν_{fg}	Difference between specific volume of vapor and liquid phases (m³/kg)
$\mathcal{V}$	Volume (m³)
$\dot{\mathcal{V}}$	Volumetric flowrate (m³/s)
w	Specific work (kJ/kg); width (m)
W	Work (kJ)
$\dot{W}$	Work rate (power) (W)
ω	Rotational speed (rad/s); hour angle (°)
x	Spatial coordinate; dryness fraction (vapor quality) (kg/kg); mass fraction (kg/kg)
ξ	Logarithmic energy decrement (−)
y	Spatial coordinate (−); molar fraction (kmol/kmol)
z	Spatial coordinate (−)
z_0	Roughness length (m)
Z	Atomic number (number of protons) (−); thermoelectric figure of merit (−)

About the Companion Website

This book is accompanied by a companion website:

www.wiley.com/go/liebenberg/energy_systems

The website includes:

Abbreviations and acronyms
Glossary
Worked solutions for the 10 Mini Projects
Worked solutions for the chapter problems
Appendix B: Reflective Questions on "Reimagining our Future"

New York's daily carbon dioxide emissions
Buried in an inventory of New York City's greenhouse gas emissions was a single number: 54 million metric tons of CO_2 (equivalent) per year. **Adam Nieman** created Real World Visuals to explore ways to make numbers physically meaningful. What does a ton of carbon dioxide look like? The shown image places carbon emissions in a context that the audience could relate to on a personal level. The impact has been astonishing. Nieman's film (https://www.youtube.com/watch?v=JU6n9pjA-lg) and stills have appeared in media around the world. The film and stills illustrate New York City's daily average greenhouse gas emissions in 2010, equivalent to 148,000 tons of CO_2 to the atmosphere. In the shown image, each sphere represents 1 ton of CO_2.
Source: Courtesy of Real World Visuals.

Week 1 – What Is Energy?

The increasing use of energy has resulted in about 35 billion tons of carbon dioxide, produced by humans, being dumped into the atmosphere every year. This has undisputed adverse effects on Planet Earth and its inhabitants. To effectively address this urgent problem, we need to understand how to apply the general principles of energy conversion in specific and practical energy-related contexts.

Energy Systems: A Project-Based Approach to Sustainability Thinking for Energy Conversion Systems,
First Edition. Leon Liebenberg.
© 2024 John Wiley & Sons, Inc. Published 2024 by John Wiley & Sons, Inc.
Companion website: www.wiley.com/go/liebenberg/energy_systems

Chapter 1 – Introduction to energy: Measuring energy; total energy; kinetic energy and potential energy; internal energy; enthalpy; energy transfer by heat; energy transfer by work; energy transfer by mass; summary

Chapter 2 – Conservation of quantity of energy: First Law of Thermodynamics (conservation of energy) for a closed system and an open system; processes involving an ideal gas; thermal efficiency deduced from the First Law of Thermodynamics; summary

Chapter 3 – Destruction of quality of energy: Reversible and irreversible processes; entropy and the Second Law of Thermodynamics; entropy balance for closed systems and open systems; the *T*-d*s* relations; maximum attainable thermal efficiency; "Second-Law" efficiency: isentropic efficiencies; related fluid mechanics: head loss, fluid momentum; available work: exergy; exergy balance of a closed system and an open system; chemical exergy; exergetic efficiency; summary

Mini Project 1 – Energy and sustainability

1

Introduction to Energy

Energy is a measure of the capacity of an object or system to produce movement or change. It is the capacity to do work. Energy is not something physical. It is an abstract accounting concept, like money, which quantifies the amount of movement or change a system can produce (Feynman 1963; Schneider and Sagan 2005). Energy is therefore a quantitative measure of the condition (or state) of a system and can be assigned a numerical value. We call energy a *property* or characteristic of an object or system.

1.1 Measuring Energy

Energy can be expressed in many units, including the joule (J), kilowatt-hour (kWh), terawatt-year (TWy), British thermal unit (Btu), quadrillion Btu (quad), barrel of oil equivalent (BOE), ton of oil equivalent (TOE), therm, foot-pound (ft-lb), electron volt (eV), and calorie. Appendix A provides a summary of the units used to measure energy, including their prefixes and symbols.

Although this book mainly uses SI units (also known as metric units), it also refers to units that are commonly used today, such as the Btu and the quad. It may therefore be necessary to convert values between the various measurement systems. Such conversions are easily accomplished by *multiplying by unity*, expressed in an appropriate manner, such as the following:

$$1 = \frac{3600 \text{ s}}{\text{h}} \qquad 1 = \frac{1000 \text{ W}}{\text{kW}} \qquad 1 = \frac{1 \text{ J/s}}{\text{W}}$$

Example 1.1 Conversion Between Units
A convenient measure of energy is the flow of 1 W of power for an hour, which is denoted as 1 watt-hour (Wh). We know that energy (J) = power (W) × time (s), so 1 J = 1 Ws. A commonly used measure of electrical energy is the kilowatt-hour (kWh), which is equivalent to 1000 Wh. Since the joule (J) and watt-hour (Wh) are both units of energy, we will often convert between these two units. For example, the conversion from kWh to MJ can be accomplished as follows:

$$(1 \text{ kWh}) \left(\frac{1000 \text{ W}}{\text{kW}} \right) \left(\frac{1 \text{ J/s}}{\text{W}} \right) \left(\frac{3600 \text{ s}}{\text{h}} \right) \left(\frac{1 \text{ MJ}}{1 \times 10^6 \text{ J}} \right) = 3.6 \text{ MJ}$$

So, 1 kWh is equivalent to 3.6 MJ or 3600 kJ.

Energy Systems: A Project-Based Approach to Sustainability Thinking for Energy Conversion Systems,
First Edition. Leon Liebenberg.
© 2024 John Wiley & Sons, Inc. Published 2024 by John Wiley & Sons, Inc.
Companion website: www.wiley.com/go/liebenberg/energy_systems

Example 1.2   Conversion Between Units

In 2021, the USA consumed an average of 19.78 million barrels of petroleum a day, or a total of 7.22 billion barrels of petroleum over the entire year. The energy content of a barrel of oil is 6.2×10^9 J. Answer the following:

a) What is the average power (in TW) associated with this fuel usage?
b) What was the annual energy consumption of petroleum in 2021 in EJ?

a) What is the average power (in TW) associated with this fuel usage?

Solution strategy: Convert 19.78 million barrels of oil per day to energy/time to get power. Remember that 1 TW = 10^{12} W.

$$\left(\frac{19.78 \times 10^6 \text{ barrels}}{\text{day}}\right)\left(\frac{6.2 \times 10^9 \text{ J}}{\text{barrel}}\right)\left(\frac{1 \text{ day}}{24 \text{ h}}\right)\left(\frac{1 \text{ h}}{3600 \text{ s}}\right)\left(\frac{1 \text{ TW}}{10^{12} \text{ W}}\right) = 1.42 \text{ TW}$$

b) What was the annual energy consumption of petroleum in 2021 in EJ?

Solution strategy: Convert 1.42 TW to energy, where energy = power × time. Remember that 1 EJ = 10^{18} J.

$$(1.42 \text{ TW})\left(\frac{10^{12} \text{ J/s}}{\text{TW}}\right)\left(\frac{3600 \text{ s}}{\text{h}}\right)\left(\frac{24 \text{ h}}{\text{day}}\right)\left(\frac{365 \text{ days}}{\text{year}}\right)\left(\frac{1 \text{ EJ}}{10^{18} \text{ J}}\right) = 44.7 \text{ EJ}$$

When converting units, it is important to make sure that the mathematical expressions are dimensionally consistent, or that both sides of an equation have the same units. We should therefore cancel units as we progress with unit conversions. So, in Example 1.2, we see that [TW] and $\frac{1}{[\text{TW}]}$ cancel each other out, as do $\left[\frac{1}{\text{s}}\right]$ and [s], until only [EJ] remains, which matches the required unit on the right-hand side of the equation [EJ].

Table 1.1 illustrates the spectrum of energy quantities that this book covers.

Reimagine Our Future
How do you understand the terms "sustainable development," "sustainability," and "living sustainably"? What can be done to help inspire people to live more sustainably? Can you perhaps think of another word or phrase for "sustainability" that might get more attention and encourage greater action in this direction?

1.2   Total Energy

Energy comes in many forms, including mechanical (kinetic or potential-gravitational), electrical, magnetic, chemical, thermal, electromagnetic radiation, and thermonuclear (fission or fusion) energy. One form of energy can be converted to another (Figure 1.1), but energy always remains conserved. So, in every energy conversion event or process, there is the same total amount of energy at the end as there was in the beginning.

Table 1.1 Spectrum of energy quantities relative to 1 J.

Order of magnitude	Example
10^{-21} J	Energy of a hydrogen bond (0.04–0.13 eV)
10^{-19} J	Energy of ultraviolet light photons (1.6–3.1 eV)
10^{-18} J	Ground state ionization energy of hydrogen (13.6 eV)
10^{-12} J	Nuclear excitations, e.g., fission of a nucleus of ^{235}U (8 MeV per nucleon)
1 J	Kinetic energy of 0.1 kg falling 1 m against the Earth's gravity
10^2 J	Kinetic energy of an 800 g men's javelin thrown at 30 m/s
10^3 J	Vaporization of 1 g of water into steam at 100 kPa
10^6 J	Food energy in a Snickers candy bar (280 food calories)
10^7 J	One cubic meter of natural gas
10^8 J	One US gallon of gasoline
10^9 J	1 ton of TNT explosive
10^{12} J	1 ton of coal
10^{13} J	Energy equivalent of 1 g of matter, using $E = mc^2$
10^{19} J	Annual US energy consumption
10^{20} J	Annual world energy consumption in 2020
10^{21} J	Energy contained in the world's natural gas reserves
10^{22} J	Daily receipt by the Earth of solar energy

For small units, the equivalent joule values of eV are shown, with 1 eV = 1.60218×10^{-19} J.

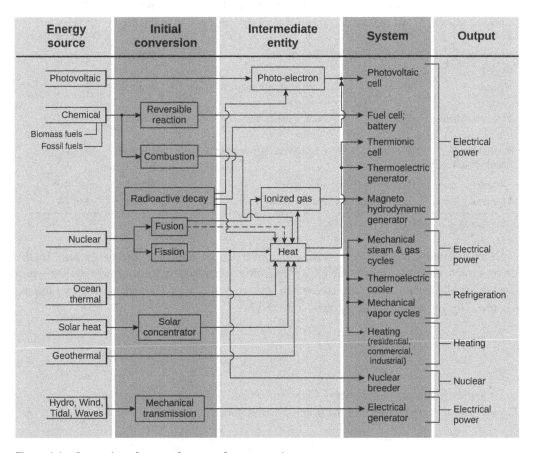

Figure 1.1 Conversion of energy from one form to another.

Table 1.2 Typical maximum conversion efficiencies for some salient energy forms.

Original energy form, E_{tot}	Converted energy form, E_i	Process	Conversion method	Maximum conversion efficiency, $\eta_{conversion}$ (%)
Electrical	Heat	Heat pump (in cooling mode)	Indirect	850 (coefficient of performance)
Electrical	Heat	Resistance heating	Direct	100
Electrical	Mechanical	Electrical motor	Direct	96
Mechanical	Electrical	Electrical generator	Direct	98
Chemical	Heat	Combustion furnace	Direct	93
Renewable (hydro)	Electrical	Hydroelectric turbine	Indirect	80
Chemical	Electrical	Combined gas turbine and steam turbine power plant	Indirect	65
Chemical	Electrical	Fuel cell	Direct	60
Chemical (coal, oil, gas)	Electrical	Combined heat, and power plant	Indirect	60
Chemical	Electrical	Battery	Direct	50
Chemical (oil, gas, coal)	Electrical	Fossil fuel-fired steam power plant	Indirect	47
Chemical	Electrical	Gas turbine power plant	Indirect	43
Chemical (diesel, gasoline)	Mechanical	Diesel or gasoline engine	Indirect	36 (gasoline) 55 (diesel)
Renewable (wind)	Electrical	Wind turbine	Indirect	45
Nuclear	Electrical	Nuclear power plant with steam turbines	Indirect	36
Electromagnetic (light)	Electrical	Photovoltaic cells	Direct	29
Renewable (geothermal)	Electrical	Geothermal power plant	Indirect	18
Electrical	Electromagnetic	LED lamps	Direct	110 lm/W
Electrical	Electromagnetic	Incandescent lamp	Direct	35 lm/W

Source: Liang et al. (2019) and Paoli and Cullen (2020).

Although energy remains conserved, the conversion of one form of energy into another usable form of energy is limited. Table 1.2 shows a few salient energy conversions and their typical conversion efficiencies, with the conversion efficiency $\eta_{conversion}$ defined as follows (also see Chapter 3 for the derivation of exergetic efficiency):

$$\eta_{conversion} = \frac{E_i}{E_{tot}} \tag{1.1}$$

with

E_i = the energy content of the energy form into which energy will be converted
E_{tot} = the total energy content of the original energy form

It is useful to classify the forms of energy into either microscopic or macroscopic forms of energy.

Microscopic forms of energy are related to the energy contained in molecules due to the movement of molecules or due to intermolecular forces – these are characterized by evaluating the *internal energy* of a system.

Macroscopic forms of energy are related to the gross characteristics of a substance on a large scale, compared to the mean free path of the molecules – these are characterized by evaluating the *kinetic energy* and *potential energy* of a system.

The *total energy* (*E*) is a property of a system and may be defined as the sum of all microscopic and macroscopic forms of energy (Black and Hartley 1985):

$$E = E_{microcopic} + E_{macroscopic} \qquad [\text{J}] \tag{1.2}$$

When one analyzes conventional energy conversion systems, it will be rare to contend with both microscopic and macroscopic forms of energy. For instance, when one is dealing with nuclear power, photovoltaics, or fuel combustion, the focus will be on the microscopic characteristics (internal energy). When one is dealing with wind or hydro power, the focus will be on the macroscopic aspects (kinetic energy or potential energy). Kinetic energy and potential energy are often also called *mechanical energy*. Mechanical energy is a form of energy that can be completely converted to work. With all energy conversion systems, we are interested in the transfer of *heat* (*Q*) and *work* (*W*), so we will also define these terms. It is important to acknowledge that energy can also be transferred by mass (*m*) or mass flow rate ($\dot{m}$).

1.3   Kinetic Energy and Potential Energy

The *kinetic energy* (E_k) of a quantity of mass (*m*) with velocity (*V*) is defined by:

$$E_k = \frac{1}{2} mV^2 \qquad [\text{J}] \tag{1.3}$$

The *specific kinetic energy* (or kinetic energy per unit mass) (e_k) is defined by:

$$e_k = \frac{E_k}{m} = \frac{V^2}{2} \qquad [\text{J/kg}] \tag{1.4}$$

Kinetic energy is a macroscopic property of a system, which means the change in kinetic energy is independent of the path followed between two end states of a process. The magnitude of the change in kinetic energy is only dependent on the mass and velocities at the beginning and end states.

A quantity of mass (*m*), which is elevated with height (*z*) above some arbitrary reference frame, possesses *potential energy* (E_p) in a gravitational field with acceleration of gravity (*g*):

$$E_p = mgz \qquad [\text{J}] \tag{1.5}$$

The *specific potential energy* (or potential energy per unit mass) (e_p) is:

$$e_p = \frac{E_p}{m} = gz \qquad [\text{J/kg}] \tag{1.6}$$

Like kinetic energy, potential energy is a macroscopic property of a system, which means the change in potential energy is independent of the path followed between two end states of a process. The magnitude of the change in potential energy is only dependent on the mass and positions at the beginning and end states.

1.4 Internal Energy

Internal energy (U) is a microscopic property of a system. At the molecular scale, molecules possess kinetic energy due to their masses and velocities. This is called translational kinetic energy. The molecules also possess vibrational and rotational energy. Molecules contain another form of energy due to intermolecular forces that operate over short distances. All these forms of microscopic energy are called the internal energy of a substance.

The average molecular velocity of a substance is proportional to its temperature. The higher the temperature, the greater the internal energy. In addition, the stronger the intermolecular forces, the greater the internal energy. In general, intermolecular forces are strongest for solids, which have small intermolecular distances; moderate for liquids, whose molecules are spaced further apart; and weakest for gases, whose molecules are furthest apart. Changing a substance from a solid to a liquid phase (or from a liquid to a gas phase) requires an increase in energy to overcome the intermolecular forces. The internal energy will therefore increase as the phase of a substance changes from a solid to a liquid to a gas (Tabor 1979).

Internal energy (U) cannot be measured directly. We can infer the value of U (in joules) based on measurements of a substance's temperature (T), pressure (p), and volume (V).

The specific internal energy may be defined as:

$$u = \frac{U}{m} \qquad [J/kg] \qquad (1.7)$$

For simple compressible systems (as will be dealt with in this book), where the changes in electrical, magnetic, chemical, and other macroscopic forms of energy are negligible compared to kinetic, potential, and internal energy, the total energy (E) of a system (Eq. 1.2) may be expressed as:

$$E = E_k + E_p + \cancel{E_{electric}} + \cancel{E_{magnetic}} + \cancel{E_{chemical}} + \cdots + U$$

$$= \frac{1}{2}mV^2 + mgz + mu \qquad [J] \qquad (1.8)$$

The total energy per unit mass is:

$$e = \frac{E}{m} = e_k + e_p + u$$

$$= \frac{V^2}{2} + gz + u \qquad [J/kg] \qquad (1.9)$$

1.5 Enthalpy

When one is dealing with energy conversion systems, it is convenient to define another thermodynamic property: *enthalpy* (H). For a system with pressure (p) and volume (V):

$$H = U + pV \qquad [J] \qquad (1.10)$$

The specific enthalpy (or enthalpy per unit mass) (h) for a system-specific volume ($v = \cancel{V}/m$) is:

$$h = u + pv \qquad\qquad [\text{J/kg}] \qquad\qquad\qquad (1.11)$$

As with internal energy, enthalpy cannot be measured directly, but it can be inferred from the measurement of a substance's temperature, pressure, and volume.

Reimagining Our Future, Together

Questioning ways of ensuring a sustainable future is something many of today's engineering students have in common. They are clearly alarmed by the rapid rate at which our non-renewable resources are dwindling and realize that innovative solutions are needed to safeguard our future. They are also aware of important environmental issues that affect and are affected by the way we live. Much of the concern expressed by engineering students (among others) revolves around climate change and alternative sources of energy, not just for the good of mankind, but for the good of the planet.

Juan David Campolargo (20) and **Aaryaman Patel** (20) are two engineering students who are already making a positive impact on the world around them. Their vision is driven by a collective curiosity, a desire to explore and understand the world around us, and an unwavering commitment to make a difference. "We don't do it alone, we bring people together, creating opportunities for collaboration and connection." They not only host the first-ever college talk show at the University of Illinois Urbana-Champaign but are tackling some of the most pressing problems of our time.

Their talk show features guests from all walks of life – from bus drivers to Nobel laureates, while some of their other initiatives include launching vertically landed rockets and spearheading energy sustainability campaigns. They have traveled the world, working on agricultural projects that make a real impact, have published hundreds of essays, and have captured thousands of pictures, all with one goal in mind – to bring people together on a mission to solve the challenges of today.

They share their thoughts about the future they crave and are striving to achieve. Their words are both inspirational and aspirational.

"We stand at the threshold of a brilliant future, where the world is our canvas, and the possibilities are endless. While some may be consumed by worry and doubt, we choose to see the boundless potential of what is to come. Together, we will tackle the most daunting challenges of our time, from climate change to information overload, and harness the power of technology to make the world get back to the future: a world where technology is about accomplishing more with less – fewer resources, less time, and less energy. This is the future we strive for, and it is waiting for us to shape it.

"Let us not forget the power of imagination. Imagine a future where the skies are filled with hypersonic planes, soaring faster than the speed of sound, powered by the cleanest, most sustainable sources. Picture advanced manufacturing facilities, where robots and humans work seamlessly together, all powered by the endless energy of the sun and wind. Envision high-speed transportation systems that whisk us from place to place without a single emission, and large-scale agriculture that feeds the world without depleting our natural resources.

"This is the sustainable energy future we strive for, and together, we will make it a reality. Let's shape the future, together!"

Juan David Campolargo (left) and Aaryaman Patel (right) are not only acing their engineering studies at the University of Illinois Urbana-Champaign (UIUC). These 20-year-old students also host and run the University's first talk show (The UIUC Talkshow), actively participate in humanitarian and engineering actions around the world, write and take photographs (AaryamanPatel.com/), share essays about important issues (JuanDavidCampolargo.com), and somehow also find time to run the UIUC Free Food initiative (@UIUCFreeFood) that provides food for the needy. *Source:* Juan David Campolargo and Aaryaman Patel.

1.6 Energy Transfer by Heat

Heat (Q) is a means of transferring energy due to the temperature difference between a system and its surroundings. Heat may be transferred across a boundary, even in the absence of mass flow. Only a temperature gradient is required to transfer heat.

Heat will *spontaneously* flow from a region of high temperature to a region of low temperature. In the process, work can be done. This happens in heat engines such as the internal combustion engine of a car. Heat can also flow from a region of low temperature to a region of high temperature, but work is required to achieve that. We call such systems refrigeration devices.

Throughout this text, we assume a sign convention when dealing with heat flow. Heat transfer into a system is assigned a positive value, while heat transfer from the system to its surroundings is assigned a negative value:

Q: $+$ for heat *added to* a system

     $-$ for heat *taken from* a system

In the absence of other forms of energy transfer across a system boundary, the transfer of heat into a system will increase the system's internal energy, while the transfer of heat out of a system will decrease its internal energy.

Unlike energy, heat is not a thermodynamic property. A quantity of heat can only be identified as it crosses a boundary. The amount of heat transferred in a process (say, from state 1 to state 2) depends on the path followed:

$$Q_{12} = \int_1^2 \delta Q \qquad \text{[J]} \qquad (1.12)$$

The specific heat (or heat per unit mass) (q) is:

$$q = \frac{Q}{m} \qquad \text{[J/kg]} \qquad (1.13)$$

The heat rate $(\dot{Q})$ can be calculated by evaluating the transfer of heat in a given period of time (Δt):

$$\dot{Q} = \frac{Q}{\Delta t} \qquad \text{[J/s] or [W]} \qquad (1.14)$$

If the mass of a substance changes with time, then the mass flow rate $(\dot{m})$ can be used to calculate the heat rate:

$$\dot{Q} = \dot{m}q \qquad \text{[J/s] or [W]} \qquad (1.15)$$

Example 1.3   Illustration of the Links Between Heat (Q), Heat rate $(\dot{Q})$, and Specific Heat (q)

If a system comprising 2 kg of a working fluid receives 100 kJ of heat from its environment in 20 s:

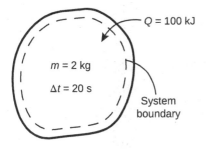

- The heat input will be $Q = 100$ kJ
- The heat rate will be $\dot{Q} = \dfrac{Q}{\Delta t} = \dfrac{100 \text{ kJ}}{20 \text{ s}} = 5$ kJ/s $= 5$ kW
- The specific heat will be $q = \dfrac{Q}{m} = \dfrac{100 \text{ kJ}}{2 \text{ kg}} = 50$ kJ/kg

Remember that heat flowing into a system has a positive value and heat flowing out of a system has a negative value. So, the heat rate would have been −5 kW if heat flowed out of the above system.

Heat can be transferred by three mechanisms: conduction, convection, and radiation.

1.6.1 Conduction

Conduction is the transfer of heat from more energetic particles of a solid, liquid, or gas to the adjacent, less energetic particles because of interaction between the particles. Figure 1.2 depicts heat conduction through a solid wall. Heat is transferred from the warm side at temperature T_H to the cooler side at temperature T_C. According to Fourier's Law of Heat Conduction, the rate of heat transfer for steady-state systems is proportional to the temperature gradient (dT/dx), the area perpendicular to the direction of heat transfer (A), and the thermal conductivity of the material (k):

$$\dot{Q} = kA\left(\frac{dT}{dx}\right) \equiv kA\left(\frac{T_H - T_C}{L}\right) \tag{1.16}$$

Table 1.3 lists some values of thermal conductivity. You will note that a material with a low thermal conductivity is a good thermal insulator, while a material with a high value of k is a good thermal conductor.

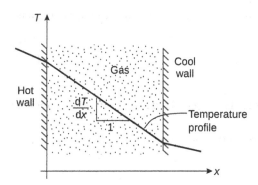

Figure 1.2 One-dimensional heat conduction through a planar wall of thickness Δx and surface area A.

Table 1.3 The thermal conductivity of some materials at 20 °C and 100 kPa.

Material	Thermal conductivity, k (W/mK)
Xenon	0.0057
Krypton	0.0094
Argon	0.018
Air	0.026
Peat soil (dry/saturated)	0.06/0.5
Softwood (white pine)	0.11
Hardwood (white oak)	0.18
Water	0.57
Brickwork	0.7
Stone concrete	0.9
Window glass	1.0
Rock	3
Sandy soil (dry/saturated)	3.2/2.2
Carbon steel	45
Aluminum	240

Source: Çengel (2003) and Lienhard and Lienhard (2005).

Example 1.4 Heat Loss Through a Room's Walls

Consider a room with a well-insulated ceiling and a nonconducting floor. Heat can only be transferred through brick walls with a total surface area of 450 m². The room's wall is 200 mm thick. The bricks and mortar have an average thermal conductivity of 0.72 W/mK. Calculate the heat conducted through the walls if the temperature inside the room is 20 °C and the temperature outside is 0 °C.

Analysis

From Eq. (1.16),

$$\dot{Q} = kA\left(\frac{T_H - T_C}{L}\right) = (0.72 \text{ W/mK})(450 \text{ m}^2)\left(\frac{[(20 + 273) - (0 + 273)]\text{K}}{0.2 \text{ m}}\right) = 32.4 \text{ kW}$$

Discussion

If the room loses 32.4 kW for the whole day (24 h), the total energy loss is 32.4 kW × 24 h = 778 kWh. If the room's temperature is to be maintained at 20 °C and an electrical heater is used to make up for the lost heat, the cost of providing that electrical energy (assuming an electrical cost of 10 ¢/kWh) will be 778 kWh × (10 ¢/kWh) = $77.80.

1.6.2 Convection

Convection involves the transfer of heat due to the combined effects of conduction and fluid motion. With *forced convection*, heat is transferred between a solid surface and an adjacent moving fluid by an external agent (like an electrical fan). With *free convection*, fluid motion is generated by buoyancy effects. For instance, if fluid density decreases with temperature, then a fluid near a hot surface will expand, rise, and convect heat away from the solid surface.

According to Newton's Law of Cooling, the rate of heat convection for steady-state systems is proportional to the temperature difference between the solid surface and the moving fluid ($T_s - T_f$), the surface area through which convection takes place (A), and the heat transfer coefficient (h):

$$\dot{Q} = hA(T_s - T_f) \tag{1.17}$$

The heat transfer coefficient (h) is dependent on the material type, geometry, and characteristics of the convective flow, among other things. The value of h is usually experimentally determined and may be presented as semi-empirical correlations. Table 1.4 lists some typical values for various materials and set-ups.

Table 1.4 The range of heat transfer coefficients for some types of convection.

Process	Heat transfer coefficient, h (W/m²K)
Free convection	
Gases	2–20
Liquids	50–1000
Forced convection	
Gases	25–300
Liquids	100–40,000
Convection with phase-change	
Boiling and condensation	2500–100,000

Source: Çengel (2003) and Lienhard and Lienhard (2005).

Example 1.5 Air Flowing in Pipes Heated by Cross-Flowing Steam

Consider a heat exchanger with air flowing inside 25.4-mm diameter tubes and steam being circulated across the tubes. The air is at 200 °C and the steam is at 220 °C. Very accurate measurements that consider the material type and geometry of the tubes, the velocity of air in the tubes, and the velocity of the steam outside reveal that the heat transfer coefficient is 120 W/m²K. If the heat exchanger is to transfer 50 kW$_{th}$, what is the required surface area of the tubes?

Analysis

From Eq. (1.17): $\dot{Q} = hA(T_s - T_f)$

So, $A = \dfrac{\dot{Q}}{h(T_s - T_f)} = \dfrac{50 \times 10^3 \text{ W}}{(120 \text{ W/m}^2\text{K})[(220 + 273) - (200 + 273)] \text{ K}} = 21 \text{ m}^2$

Discussion

Since the tubes are 25.4 mm in diameter, the required length of the tubes for this heat exchanger can be determined:

$$A = 21 \text{ m}^2 = \pi DL = \pi(25.4 \times 10^{-3} \text{ m})(L)$$

$$\therefore L = \frac{21 \text{ m}^2}{\pi(25.4 \times 10^{-3} \text{ m})} = 263 \text{ m}$$

1.6.3 Radiation

Radiation is the transfer of heat by the emission of electromagnetic waves. Unlike convection and conduction, the transfer of heat via radiation does not require the presence of a medium. For example, the Sun's electromagnetic rays reach the Earth despite traveling 150 million kilometers through a vacuum. Electromagnetic rays travel at the speed of light.

According to the *Stefan–Boltzmann Law*, the net rate of heat radiated between two surfaces is proportional to the difference of the fourth power of the temperature of the radiating body and the absorbing body $(T_s^4 - T_{surr}^4)$. In the Stefan–Boltzmann equation, A is the surface area, ε is the emissivity of the surface (with $\varepsilon = 1$ being a perfect emitter or absorber of electromagnetic waves, also called a "black body"), and σ is the *Stefan–Boltzmann constant*, with $\sigma = 5.67 \times 10^{-8} \text{ W/m}^2 \text{ K}^4$:

$$\dot{Q} = \varepsilon \sigma A(T_s^4 - T_{surr}^4) \tag{1.18}$$

Table 1.5 lists some values of emissivity.

Example 1.6 Radiation to a Metal Tube in a Furnace

Consider a 25.4-mm diameter 0.6-m long tube inside a furnace lined with fire bricks. The air inside the furnace is 1100 K and the surface temperature of the tube is 600 K. The emissivity of the steel tube is 0.7. Calculate the total heat transfer rate to the tube. You may disregard heat transfer by convection.

Table 1.5 Emissivity of some materials at 300 K.

Substance/object	Emissivity, ε
Sun	≈1.0
Ice, brick, marble, concrete	0.93–0.97
Aluminum	≈0.1
Gold	0.02
Silver	0.01

Source: Çengel (2003) and Lienhard and Lienhard (2005).

Analysis

From Eq. (1.18): $\dot{Q} = \varepsilon\sigma A\left(T_s^4 - T_{surr}^4\right)$

For the steel tube, $\varepsilon = 0.7$.

The surface area of the tube is $A = \pi D L = \pi(25.4 \times 10^{-3}\,\text{m})(0.6\,\text{m}) = 0.048\,\text{m}^2$, and $\sigma = 5.67 \times 10^{-8}\,\text{W/m}^2\text{K}^4$. Therefore:

$$\dot{Q} = \varepsilon\sigma A\left(T_s^4 - T_{surr}^4\right) = (0.7)\left(5.67 \times 10^{-8}\,\text{W/m}^2\text{K}^4\right)\left(0.048\,\text{m}^2\right)\left(1100^4 - 600^4\right)\text{K}^4$$

$$\dot{Q} \cong 2.5\,\text{kW}$$

Reimagine Our Future

Make a list of five things that demand a lot of electricity or heat. Flip that around and imagine those things requiring no electricity or heat. How could that change happen? How soon could it happen? What would that change mean for your life and for society?

1.7 Energy Transfer by Work

Work (W) is a means of transferring energy into coherent movement or displacement. Doing work requires the application of a force, which may be produced by a pressure difference, voltage difference, or concentration difference. Like heat, work can be transferred across a boundary in the absence of mass flow. There are several modes of work transfer. A few of these modes are shown in Table 1.6.

Like heat, work is not a thermodynamic property. It can only be identified as it crosses a boundary.

Throughout this text, we assume a sign convention when dealing with work transfer. Work transfer out of a system is assigned a positive value, while work transfer from the surroundings to a system is assigned a negative value.

W: + for work *done by* a system

         − for work *added to* a system

Table 1.6 Salient modes of work transfer.

Mode of work transfer	Generalized force (or "potential")	Generalized displacement (or "movement")	Work done on the surroundings	Example
Mechanical				
Expansion (also called "pressure-volume work")	Pressure (p)	Change in volume ($d\mathcal{V}$)	$\int_1^2 p\, d\mathcal{V}$	Expansion (or compression) of a gas against a piston in the cylinder of an internal combustion engine
Translation	Force (F)	Linear displacement (dx)	$\int_1^2 F\,dx$	Linear displacement of a mass by a linear force
Rotation	Torque (T)	Angular deformation ($d\theta$)	$\int_1^2 T\,d\theta$	A shaft being rotated by a motor
Electrical				
Resistance	Electrostatic potential (V_e) $= Q_e/C_e$	Electrical current (I) flowing in a period (dt)	$\int_1^2 V_e I\,dt$	Current flowing through a resistive heater
Capacitance	Electrostatic potential (V_e) $= Q_e/C_e$	Static electrical charge (dQ_e)	$\int_1^2 V_e\,dQ_e$	Discharge of an electrical capacitor Movement of an electron across an electrostatic potential
Inductance	Inductance (L_e)	$d\psi = d(L\,I_e)$	$\int_1^2 L_e\,d\psi$	Discharge of an electrical inductor

Source: Adapted from Karplus (1958) and Takahashi et al. (1970).

The amount of work transferred in a process (say, from state 1 to state 2) depends on the path followed:

$$W_{12} = \int_1^2 \delta W \qquad [J] \qquad (1.19)$$

The specific work (or work per unit mass) (w) is

$$w = \frac{W}{m} \qquad [\text{J/kg}] \qquad (1.20)$$

The work rate (or power) ($\dot{W}$) can be calculated by evaluating the transfer of work in each period of time (Δt):

$$\dot{W} = \frac{W}{\Delta t} \qquad [\text{J/s}] \text{ or } [\text{W}] \qquad (1.21)$$

If the mass of a substance changes with time, the mass flow rate can be used to calculate the work rate:

$$\dot{W} = \dot{m}w \qquad [\text{J/s}] \text{ or } [\text{W}] \qquad (1.22)$$

Example 1.7   Illustration of the Links Between Work (W), Work Rate ($\dot{W}$), and Specific Work (w)

If a system comprising 2 kg of a working fluid does 100 kJ of work on its environment in 20 s:

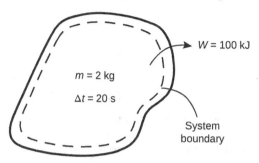

W = 100 kJ

m = 2 kg

Δt = 20 s

System boundary

- The work output will be $W = 100\,\text{kJ}$
- The work rate (or power) will be $\dot{W} = \dfrac{W}{\Delta t} = \dfrac{100\,\text{kJ}}{20\,\text{s}} = 5\,\text{kJ/s} = 5\,\text{kW}$
- The specific work will be $w = \dfrac{W}{m} = \dfrac{100\,\text{kJ}}{2\,\text{kg}} = 50\,\text{kJ/kg}$

Remember that work being done by a system on its environment has a positive value, and work being done on a system has a negative value.

So, the work rate would have been -5 kW if work were to be done on the above system.

In the absence of other forms of energy transfer across a system boundary, the transfer of work out of a system will decrease the internal energy of the system, and the transfer of work into a system will increase its internal energy.

Reimagine Our Future

Make an infographic or other illustration to show how the electricity you use is generated and how it is transmitted to your home. Share your insights on social media.

1.8   Energy Transfer by Mass

We have established that energy can be transferred via heat and work. As can be inferred from Eqs. (1.1) to (1.22), the quantity of energy in a system increases with the mass of the system, while the rate of energy transfer increases with the mass flow rate. This is because mass carries energy, as illustrated by Einstein's famous equation, $E = mc^2$. This is covered in more detail in Chapters 16 and 17.

If a system is *closed*, the mass of the substance within its boundaries remains constant during a process. If a system is *open*, the mass of the substance within its boundaries changes with time during a process.

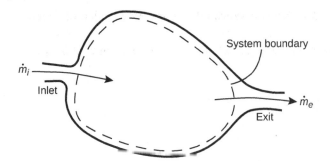

Figure 1.3 System used for the development of the conservation-of-mass equation.

Mass, like energy, is a conserved property that cannot be created or destroyed during a process. We can write a mass rate balance for the hypothetical system shown in Figure 1.3:

[Rate of change of mass contained within the system boundaries] = [Mass flow rate at inlet]
− [Mass flow rate at exit]

$$\text{(1.23)}$$

Or, in mathematical terms:

$$\frac{dm}{dt} = \dot{m}_i - \dot{m}_e \qquad \text{[kg/s]} \tag{1.24}$$

If the flow is *steady* (the mass of the system will not change):

$$\frac{dm}{dt} = 0 = \dot{m}_i - \dot{m}_e \text{ and } \dot{m}_i = \dot{m}_e \tag{1.25}$$

We know that:

$$m = \rho V \qquad \text{[kg]} \tag{1.26}$$

For an incompressible fluid (where the fluid's density remains constant), we have:

$$\dot{m} = \rho \dot{V} \qquad \text{[kg/s]} \tag{1.27}$$

For an incompressible fluid flowing in a duct of constant cross-section (A), the mass flow rate can be calculated as follows:

$$\dot{m} = \frac{dm}{dt} = \rho \frac{dV}{dt} = \rho \frac{d(Adx)}{dt} = \rho A \frac{dx}{dt} = \rho A V_n \tag{1.28}$$

where V_n Is the average velocity component, which is normal (perpendicular) to the surface of the duct.

Combining Eqs. (1.26) and (1.28) with Eq. (1.24), we can write the conservation of mass for an open system in general form:

$$\frac{dm}{dt} = \frac{d}{dt} \int_V \rho \, dV = \int_{A_i} \rho V_n \, dA - \int_{A_e} \rho V_n \, dA \tag{1.29}$$

Since the mass flow rate is often known in energy conversion system analysis, it is usually more convenient to use Eq. (1.24) than Eq. (1.29). Equation (1.29) is more useful when the velocity and density distributions across the inlet and exit areas are known.

Example 1.8 Illustration of the Conservation of Mass

The pressure vessel of a small nuclear reactor is filled with 185 m³ boiling water with density $\rho_w = 850 \text{ kg/m}^3$. Due to the failure of a pump needed to cool the reactor core, a pressure release valve at A opens and emits steam with density $\rho_{st} = 35 \text{ kg/m}^3$ and velocity $V = 200 \text{ m/s}$. If the steam passes through a 50-mm-diameter pipe at A, determine the time needed for all the water to escape. The temperature of the water and the velocity at A remain constant.

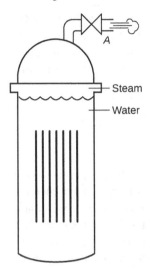

Required: We must find the time needed for 185 m³ of water to escape from the reactor core.

Solution strategy: Since steam exits the system, the system is open. To calculate the time rate of change of mass in the system, we can use the equation of conservation of mass for an open system, Eq. (1.26).

Assumptions: Assume the density of water in the reactor is constant since the temperature is constant. Also, assume that the tank contains only water in the liquid-only phase.

Analysis

Conservation of mass for an open system: $\dfrac{d}{dt}\displaystyle\int_{V} \rho \, dV = \int_{A_I} \rho V_n \, dA - \int_{A_e} \rho V_n \, dA$

There is only flow exiting the system, so the inlet flux term can be deleted:

$$\frac{d}{dt}\int_{V} \rho \, dV = \cancel{\int_{A_I} \rho V_n \, dA} - \int_{A_e} \rho V_n \, dA$$

From our assumptions, the equation reduces to the following:

$$\rho_w \frac{dV}{dt} + \rho_{st} VA = 0$$

So, $\dfrac{dV}{dt} = -\dfrac{\rho_{st}}{\rho_w} VA = -\left(\dfrac{35 \text{ kg/m}^3}{850 \text{ kg/m}^3}\right)(200 \text{ m/s})\left(\dfrac{\pi(0.05 \text{ m})^2}{4}\right) = -0.016 \text{ m}^3/\text{s}$

The negative sign indicates that the volume of water inside the vessel is decreasing. The time required for 185 m³ of water to escape from the vessel is:

$$t = \frac{V}{dV/dt} = \frac{185 \text{ m}^3}{0.016 \text{ m}^3/\text{s}} = 11{,}441 \text{ s}$$

This is equivalent to $(11{,}441 \text{ s}) \left(\dfrac{1 \text{ h}}{3600 \text{ s}} \right) = 3.2 \text{ h}$

Discussion: The water inside the reactor core will therefore leak out in 3.2 h.

Reimagine Our Future

Great social transformations – the end of slavery, the end of apartheid, women's and civil rights, the end of colonializations, the birth of environmentalism – all began with public awareness and engagement. Can you think of ways to enhance public awareness and accelerate engagement with a view to promoting *energy sustainability*? Can you reimagine the future in this respect?

1.9 Summary

Energy is a measure of the capacity of an object or system to do work. Although there are many kinds of work, we distinguish only a few forms of energy, such as electrical, electromagnetic, chemical, mechanical (kinetic or potential), nuclear or renewable (solar, wind, water, geothermal, or biomass).

The sum of all forms of microscopic and macroscopic energy is called the total energy. One form of energy can be transformed into another, and all forms of energy are measured in joules (J). Heat (Q) and work (W) are two of the main processes for transferring energy across a system boundary. Work (W) is a means of transferring energy into movement, while heat (Q) is a means of transferring energy via temperature gradients.

Problems

1.1 A quantity of 1.7 L of water is boiled for 3 min in an electric kettle fitted with a 1 kW electrical resistive element. Assume that 1.7 L of water has a mass of 1.7 kg. Heat is lost at a rate of 100 W via radiation and conduction through the kettle's transparent glass wall.

  a) What is the net heat transfer, in kilojoules?

  [Answer: 900 W; 162 kJ]

  b) What is the net specific heat transfer, in kilojoules per kilogram?

  [Answer: 95.3 kJ/kg]

1.2 A domestic gas-fired furnace has an output of 135,000 Btu/h. The homeowner wants to exchange it for an electrical heater. What is the equivalent electrical power in kilowatts? *Hint:* 1 Btu = 1055 J.

  [Answer: 39.6 kW]

1.3 An industrial boiler has an 800-kW heating capacity. The heat is produced by an electrical element, and the electricity is derived from a nearby photovoltaic plant. The thermal efficiency of the boiler is initially measured to be 90%. (Thermal efficiency of a boiler = actual heat delivered by the boiler divided by the theoretical amount of heat expected to be delivered.) After tuning the boiler, the maintenance engineer increases the combustion efficiency to 98%. The boiler operates for 4000 h a year on an intermittent basis. If the unit cost of electricity is 10 ¢/kWh, determine the annual energy and cost savings from tuning up the boiler.

[Answer: Before tuning the boiler: $32,000; after tuning the boiler: $6400; annual savings = $25,600]

1.4 The lighting needs in a university classroom in Illinois are met by light streaming in through large windows, and by 200 low-energy LED light bulbs (type LBRP6, manufactured by USAI Lighting). Each light bulb consumes 33 W of electricity. Conservatively, you may assume that the lightbulbs dissipate all electrical power as heat. The classroom operates with the lights on for 12 h a day, 250 days a year.

a) If the cost of electricity in Illinois is 10 ¢/kWh, determine the annual cost of the lighting in this room.

[Answer: $1980/year]

b) The classroom is filled with 170 very inquisitive and hard-thinking people. Each person radiates an average of 100 W of thermal energy to the surrounding air. The walls of the room are not well insulated, and heat also leaks through the large glass windows. It is estimated that around 50 kW of heat is lost via thermal radiation and conduction through the room's walls and large windows in the winter. Considering the heat emitted by the light bulbs and the people inside the room, and the heat lost through the windows, what is the net heat rate (in kW) gained in the room?

[Answer: −26.4 kW]

1.5 Energy use intensity (EUI) is an indicator of the energy efficiency of a building's design and/or operations. EUI can be thought of as the "miles per gallon" rating of the building industry. It is used in several different ways, including to set targets for energy performance before beginning design, to benchmark a building's designed or operational performance against others of the same building type, or to evaluate compliance with energy code requirements. It is important to remember that EUI varies with building type.

a) According to a survey conducted by the Energy Star accreditation organization, student apartment blocks and dormitories in Illinois have an EUI of 180 kBtu/ft^2.
    https://www.energystar.gov/sites/default/files/tools/DataTrends_Dormitory_20150129.pdf
    Convert 180 kBtu/ft^2 to SI units (GJ/m^2) showing all calculations.

[Answer: 2.04 GJ/m^2]

b) According to Friendly Power (https://esource.bizenergyadvisor.com/article/colleges-and-universities), a median university residence building has an area of 140,000 square feet. Calculate such a building's annual energy consumption (in joules) according to the Energy Star value of EUI. (Convert all values to SI units and report your answer in TJ.)

[Answer: 26.6 TJ]

c) This university building uses an average of 18.9 kWh of electricity and 17 cubic feet of natural gas per square foot of floor space each year. One cubic foot of gas typically contains about 1 kBtu of energy. Convert these energy values to the EUI for this building. State all your assumptions and show all calculations. Compare your answer from b with that of c. What can you deduce from this comparison?

> *Hint:* For electricity usage, convert the amount of kWh to kBtu/ft^2. For gas usage, calculate how many kBtu of energy are contained in 17 ft^3 of gas.

[Answer: Electricity: 65 kBtu/ft^2; Gas: 17 kBtu/ft^2]

What is the annual energy bill (for electricity and gas) for this building? Assume an electricity cost of 10 ¢/kWh and a gas cost of 30 ¢/therm.

[Answer: Electricity: $264,000 per year; Gas: $7082 per year]

d) The average carbon-dioxide emissions are 550 g/kWh and 5.5 kg/therm. If the university residence building reduces its consumption of electricity by 10%, what will the total emissions of carbon dioxide by this building per year be in tons (with 1 metric ton = 1000 kg)?

[Answer: Electricity: 1309 tons CO_2 per year; Gas: 135.7 tons CO_2 per year]

References

Black, W.Z. and Hartley, J.G. (1985). *Thermodynamics*. London: Harper & Row.

Çengel, Y.A. (2003). *Heat Transfer. A Practical Approach*. New York, NY: Mc Graw-Hill.

Feynman, R. (1963). *The Feynman Lectures on Physics. Book 1*. New York, NY: Addison-Wesley.

Karplus, W.J. (1958). *Analog Simulation*. New York, NY: McGraw-Hill.

Liang, T.S., Pravettoni, M., Deline, C., Stein, J.S. et al. (2019). A review of crystalline silicon bifacial photovoltaic performance characterisation and simulation. *Energy and Environmental Science* 12 (1): 116–148.

Lienhard, I.V. and Lienhard, J.H. (2005). *A Heat Transfer Textbook*. Cambridge, MA: Phlogiston Press.

Paoli, L. and Cullen, J. (2020). Technical limits for energy conversion efficiency. *Energy* 192: 116228.

Schneider, E.D. and Sagan, D. (2005). *Into the Cool: Energy Flow, Thermodynamics, and Life*. Chicago, IL: University of Chicago Press.

Tabor, D. (1979). *Gases, Liquids, and Solids*. Cambridge: Cambridge University Press.

Takahashi, Y., Rabins, M.J., and Auslander, D.M. (1970). *Control and Dynamic Systems*. Reading, MA: Addison-Wesley.

2

Conservation of Quantity of Energy

The principle of *conservation of energy* is the cornerstone of the analysis of energy conversion systems. This principle is also the foundation of the study of thermodynamics and is referred to as the *first law of thermodynamics*. The first law of thermodynamics establishes the relationships among heat, work, internal energy, and other forms of energy.

Reimagine Our Future

In the short term, mankind is only driven to real change in times of emergency or crisis. When that crisis or emergency occurs, the actions taken depend on the ideas that are lying around. Is it therefore not prudent to consider and experiment with alternative approaches and policies so that, when we confront a crisis, we have in our toolkit what we need? In the case of energy sustainability, what might this involve?

2.1 First Law of Thermodynamics (Conservation of Energy)

The *first law of thermodynamics* (or *principle of conservation of energy*) states that energy can be neither created nor destroyed during a process but only converted from one form to another.

Example 2.1 Illustration of Conservation of Energy: Rock Falling in Air

Consider a rock with a mass of 2 kg at an elevation of 40 m, which is released from rest to fall to the ground. We will disregard air friction. Initially, the rock has maximum potential energy and no kinetic energy. As it begins to fall, it loses potential energy and gains kinetic energy. The velocity of the rock can be calculated using $V^2 = V_0^2 + 2gs$, where s is the distance fallen. The table below shows that the sum of the rock's kinetic energy and potential energy remains constant at 784 J as the rock falls. Note that z is the height above the ground level.

Energy Systems: A Project-Based Approach to Sustainability Thinking for Energy Conversion Systems,
First Edition. Leon Liebenberg.
© 2024 John Wiley & Sons, Inc. Published 2024 by John Wiley & Sons, Inc.
Companion website: www.wiley.com/go/liebenberg/energy_systems

Elevation	Potential energy (A) $= mgz$	Velocity of rock $= \sqrt{V_0^2 + 2gz}$	Kinetic energy (B) $= \frac{1}{2}mV^2$	Total (A + B)
40 m	$= (2\,\text{kg})(9.8\,\text{ms}^{-2})(40\,\text{m})$ $= 784\,\text{J}$	$= \sqrt{0\,\text{ms}^{-1} + 2(9.8\,\text{ms}^{-2})(0\,\text{m})}$ $= 0\,\text{ms}^{-1}$	$= \frac{1}{2}(2\,\text{kg})(0\,\text{ms}^{-1})^2$ $= 0\,\text{J}$	784 J
30 m	$= (2\,\text{kg})(9.8\,\text{ms}^{-2})(30\,\text{m})$ $= 588\,\text{J}$	$= \sqrt{0\,\text{ms}^{-1} + 2(9.8\,\text{ms}^{-2})(10\,\text{m})}$ $= 14\,\text{ms}^{-1}$	$= \frac{1}{2}(2\,\text{kg})(14\,\text{ms}^{-1})^2$ $= 196\,\text{J}$	784 J
15 m	294 J	$= \sqrt{0\,\text{ms}^{-1} + 2(9.8\,\text{ms}^{-2})(25\,\text{m})}$ $= 22.14\,\text{ms}^{-1}$	$= \frac{1}{2}(?\,\text{kg})(??\,14\,\text{ms}^{-1})^2$ $= 490\,\text{J}$	784 J
0 m	0 J	$= \sqrt{0\,\text{ms}^{-1} + 2(9.8\,\text{ms}^{-2})(40\,\text{m})}$ $= 28\,\text{ms}^{-1}$	$= \frac{1}{2}(2\,\text{kg})(28\,\text{ms}^{-1})^2$ $= 784\,\text{J}$	784 J

If air friction were considered, the sum of potential and kinetic energy would not remain constant. It would grow less as air friction takes its toll. However, conservation of energy can still be observed by considering the heat energy dissipated to the air through friction.

Consider a hypothetical system with heat and work transfer across the system boundary. The system has an inlet and an exit, so there is a rate of energy transport due to mass crossing the system boundary.

The conservation of energy can be stated in words as follows:

$$\begin{bmatrix} \text{Rate of change of} \\ \text{energy contained within} \\ \text{the system boundaries} \end{bmatrix} = \begin{bmatrix} \text{Rate at which} \\ \text{energy enters} \\ \text{at the system} \\ \text{boundary} \end{bmatrix} - \begin{bmatrix} \text{Rate at which} \\ \text{energy leaves} \\ \text{at the system} \\ \text{boundary} \end{bmatrix} \qquad (2.1)$$

Or, in mathematical terms (Van Wylen and Sonntag 1976):

$$\dot{Q}_{net} - \dot{W}_{net} = \Delta \dot{E}_{net} = \underbrace{\int_{A_e} e(\rho V_n\,\mathrm{d}A)}_{\substack{\text{Rate of energy leaving} \\ \text{the system due to mass} \\ \text{flow across the boundary} \\ \text{at surface } A_e}} - \underbrace{\int_{A_i} e(\rho V_n\,\mathrm{d}A)}_{\substack{\text{Rate of energy entering} \\ \text{the system due to mass} \\ \text{flow across the boundary} \\ \text{at surface } A_i}} + \underbrace{\frac{\mathrm{d}}{\mathrm{d}t}\int_{\Psi} e\rho\,\mathrm{d}\Psi}_{\substack{\text{Rate of change of energy} \\ \text{within the system volume } \Psi}}$$

$$(2.2)$$

Equation (2.2) states that the *net change* (increase or decrease) in the total energy of the system during a process is equal to the difference between the total energy entering and the total energy leaving the system during that process. The net change in total energy of the system is also equal to the net heat transferred into the system minus the net work done by the system. Equation (2.2) is a *general* statement of the first law of thermodynamics.

When mass enters or leaves a system, work is required to maintain continuous flow. This work is referred to as *flow work* (or flow energy). The rate of flow work at the inlet or outlet is $\left(\dfrac{\delta W}{\mathrm{d}t}\right) = \int pv(\rho V_n \mathrm{d}A)$.

The total work rate can be obtained by evaluating energy flow at inlets and outlets:

$$\dot{Q}_{net} - \dot{W}_{net} = \Delta \dot{E}_{net} = \int_{A_e} (e + pv)\rho V_n\,\mathrm{d}A - \int_{A_i} (e + pv)\rho V_n\,\mathrm{d}A + \frac{\mathrm{d}}{\mathrm{d}t}\int_{\Psi} e\rho\,\mathrm{d}\Psi \qquad (2.3)$$

Combining Eq. (2.3) with Eqs. (1.9) and (1.11), we now have (Obert 1960; Van Wylen and Sonntag 1976):

$$\dot{Q}_{net} - \dot{W}_{net} = \Delta \dot{E}_{net} = \int_{A_e} \left(h + e_k + e_p\right) \rho V_n \, dA - \int_{A_i} \left(h + e_k + e_p\right) \rho V_n \, dA + \frac{d}{dt} \int_{\Psi} e\rho \, d\Psi$$

(2.4)

This equation can be simplified for use in closed and open systems, as shown in the next section.

2.2 First Law of Thermodynamics for a Closed System

In a *closed* system, mass cannot enter or leave the system, so Eq. (2.4) becomes:

$$\dot{Q}_{net} - \dot{W}_{net} = \frac{dE}{dt} \qquad \text{[J/s] or [W]}$$

(2.5)

Or, for a closed system undergoing a finite change between state points 1 and 2 over a time interval (Δt):

$$Q_{12} - W_{12} = \Delta E_{12} = \Delta U_{12} + \Delta E_k + \Delta E_p \qquad \text{[J]}$$

(2.6)

In words, the net heat flow minus the net work equals the change in energy of all forms.

Example 2.2   Illustration of the First Law of Thermodynamics for a Closed System
A rigid tank with a volume of 0.14 m^3 is filled with steam at 1 MPa and 250 °C. The tank is cooled so that the pressure falls to 350 kPa. Calculate (a) the final temperature and (b) the heat transferred, in kJ.

a) **Find the final temperature**
   Given: This is a constant-volume process as the tank is rigid. The mass of the steam inside the tank remains constant, so this is a closed system. Steam is changing from $p_1 = 1$ MPa and $T_1 = 250$ °C to $p_2 = 350$ kPa.
   Required: Find T_2
   Solution strategy: We will use the property tables of steam available in the Engineering Equation Solver (EES) software package, https://fchartsoftware.com/ees. (You could also use property tables for steam from many other sources, including the National Institute of Standards and Technology [NIST]'s Thermophysical Properties of Fluid Systems, or from any good textbook on thermodynamics.)
   Analysis: We must first establish whether state point 1 is situated in the superheated, saturated, or subcooled regime by comparing the temperature of the steam with its saturation temperature at 1 MPa. From EES, we note that:

Pressure (p), kPa	Saturation temperature (T_{sat}), °C
1000	179.88

$$T_{sat@1 \text{ MPa}} = 179.88 \,°C$$

and $T_1 = 250 \,°C > T_{sat @ 1 \text{ MPa}}$

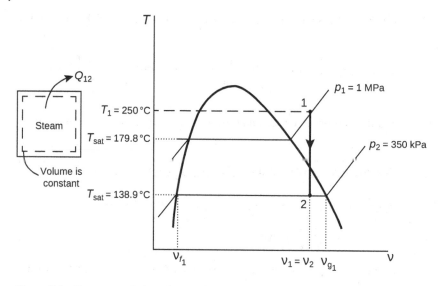

Figure 2.1 System consisting of steam inside a fixed-volume tank for Example 2.2 and the accompanying process diagram.

Following a line of constant pressure on a temperature volume diagram, we note that state point 1 is situated in the superheated regime (see Figure 2.1). We must therefore use the property tables of superheated steam to find properties associated with state point 1.

As going from 1 to 2 is a constant volume process, v_1 equals v_2. Once we have established the regime (superheated, saturated, subcooled) of state point 2, we can use v_2 to read the concomitant value of T_2:

	$p = 1.00$ MPa ($T_{sat} = 179.88$ °C)		
Temperature (T), °C	m³/kg (v)	kJ/kg (u)	kJ/kg (h)
250	0.23275	2710.4	2493.1

$$v_1 = 0.23275 \text{ m}^3/\text{kg} \equiv v_2$$

To find state point 2, take $v_2 = 0.23275$ m³/kg and $p_2 = 350$ kPa and compare the value of v_2 with that of $v_{f\,@\,350\,\text{kPa}}$ and $v_{g\,@\,350\,\text{kPa}}$. If $v_2 > v_{g\,@\,350\,\text{kPa}}$, state point 2 is superheated; if $v_2 < v_{f\,@\,350\,\text{kPa}}$, state point 2 is subcooled; if $v_{f\,@\,350\,\text{kPa}} < v_2 < v_{g\,@\,350\,\text{kPa}}$, state point 2 is saturated.

From the property tables of water, we have the following:

		Specific volume (m³/kg)		Internal energy (kJ/kg)		
Pressure (p), kPa	Saturation temperature (T_sat), °C	Saturated liquid (v_f)	Saturated vapor (v_g)	Saturated liquid (u_f)	Evaporation (u_fg)	Saturated vapor (u_g)
350	138.86	0.001079	0.52422	583.89	1964.6	2548.5

$$v_{g\,@\,350\,\text{kPa}} = 0.52422 \text{ m}^3/\text{kg} \text{ and } v_{f\,@\,350\,\text{kPa}} = 0.001079 \text{ m}^3/\text{kg}$$

$v_2 = 0.23275$ m³/kg lies between $v_{f @ 350 \text{ kPa}}$ and $v_{g @ 350 \text{ kPa}}$ and is therefore in the saturated regime.

We need to use the property tables for saturated steam to find T_2:

$$T_2 = T_{sat@350 \text{ kPa}} = 138.86\,°C$$

b) Find the heat transferred, in kJ

Solution strategy: As this is a closed system (all mass remains in the system), we will write down the first law of thermodynamics for a closed system. After simplifying the equation, we will use property tables for steam to find the required properties.

Assumptions: Disregard changes in kinetic energy, and the potential energy of steam.

Analysis:

First law of thermodynamics for an open system, Eq. (2.6):

$$Q_{12} - W_{12} = \Delta E_{12} = \Delta U_{12} + \Delta E_k + \Delta E_p$$

where $W_{12} = p\,dV$, but there is no change in volume, $dV = 0$, so $W_{12} = 0$.

As assumed, we will disregard ΔE_k and ΔE_p, so the first law becomes $Q_{12} = \Delta U_{12}$

To calculate the heat transferred (Q_{12}), we must find $\Delta U_{12} = m\,\Delta u_{12}$

where $m = \dfrac{V}{v} = \dfrac{0.14 \text{ m}^3}{0.23275 \text{ m}^3/\text{kg}} = 0.602$ kg

and $\Delta u_{12} = u_2 - u_1$

For superheated steam at 250 °C and 1 MPa, $u_1 = 2710.4$ kJ/kg (from the table).

For saturated steam at 138.86 °C and 350 kPa, $u_2 = u_f + x(u_g - u_f)$, where x is the dryness fraction of the steam.

$u_{g @ 350 \text{ kPa}} = 2548.5$ kJ/kg and $u_{f @ 350 \text{ kPa}} = 583.89$ kJ/kg

and $x = \dfrac{v_2 - v_{f2}}{v_{g2} - v_{f2}} = \dfrac{(0.23275 - 0.001079) \text{ m}^3/\text{kg}}{(0.52422 - 0.001079) \text{ m}^3/\text{kg}} = 0.442$

so, $u_2 = 583.89$ kJ/kg $+ 0.442(2548.5 - 583.89)$ kJ/kg $= 1452.2$ kJ/kg

Finally, $Q_{12} = m\,\Delta u_{12} = (0.602 \text{ kg})(1452.2 - 2710.4)$ kJ/kg $= -757.4$ kJ

Discussion: The negative sign in the answer indicates that the tank is losing 757.4 kJ of heat to the environment. This is in accord with our adopted sign convention for heat transfer (heat flowing into a system is positive; heat flowing out of a system is negative).

Reimagine Our Future

Does visiting your favorite natural places make you feel more connected to those places, to the other species that occupy them, and to other people who value them as you do?

2.3 First Law of Thermodynamics for an Open System

In an *open* system, mass can enter or leave the system. So, assuming uniform flow and using Eq. (1.25), Eq. (2.4) becomes (Obert 1960; Van Wylen and Sonntag 1976):

$$\dot{Q}_{net} - \dot{W}_{net} = \Delta \dot{E}_{net} = \sum_{exit} \dot{m}(h + e_k + e_p) - \sum_{inlet} \dot{m}(h + e_k + e_p) + \frac{dE}{dt} \qquad [\text{W}] \qquad (2.7)$$

where $\dot{m}$ is the mass flow rate (kg/s). For steady flow, $\dfrac{dE}{dt} = 0$. For a system comprising only one inlet and one outlet, with negligible changes in potential and kinetic energies, Eq. (2.7) simplifies to the following, commonly called the *steady flow energy equation*:

$$\dot{Q}_{net} - \dot{W}_{net} = (\dot{m}h)_{exit} - (\dot{m}h)_{inlet} \qquad \text{[W]} \qquad (2.8)$$

Example 2.3 Illustration of the First Law of Thermodynamics for an Open System

Steam enters a well-insulated industrial steam turbine at 4 MPa and 400 °C, and leaves the turbine at 75 kPa and 100 °C. The power output of the turbine is 2 MW, and changes in kinetic and potential energy are negligible. Calculate (a) the specific enthalpy in kJ/kg at the inlet and exit of the turbine, and (b) the mass flow rate of steam through the turbine, in kg/s.

Siemens SST-300 steam turbine, producing up to 45 MW. *Source:* © Siemens Energy.

a) **Find h_1 and h_2**

   Given: This problem involves mass flow rate, so the turbine is an open system. Steam is changing from $p_1 = 4$ MPa and $T_1 = 400\,°C$ to $p_2 = 75$ kPa and $T_2 = 100\,°C$, and $\dot{W} = 2$ MW.

   Required: We must find h_1 and h_2.

   Solution strategy: We will use the property tables of steam available in the EES software package to find the enthalpies.

   Analysis:

   We must first establish whether state point 1 is situated in the superheated, saturated, or subcooled regime by comparing the temperature of the steam with its saturation temperature at 4 MPa. From EES, we note that:

   $$T_{sat@4\ MPa} = 250.35\,°C$$

   and $T_1 = 400\,°C > T_{sat\,@\,4\ MPa}$

   Following a line of constant temperature on a temperature enthalpy diagram, we note that state point 1 is situated in the superheated regime (see Figure 2.2).

	p = 4.00 MPa (T_{sat} = 250.35 °C)		
Temperature (T), °C	m³/kg (v)	kJ/kg (u)	kJ/kg (h)
400	0.07343	2920.8	3214.5

We must therefore use the property tables of superheated steam to find properties associated with state point 1.

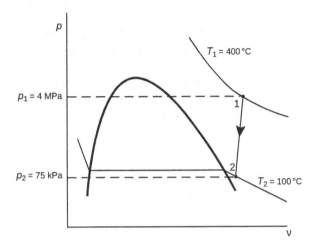

Figure 2.2 Process diagram for steam expanding in a turbine, Example 2.3.

$$h_1 = h_{@4\,\text{MPa},400\,°\text{C}} = 3214.5\,\text{kJ/kg}$$

To find the regime of state point 2 at 75 kPa and 100 °C, we note that

$$T_{sat@75\,\text{kPa}} = 91.76\,°\text{C}.$$

So, $T_2 = 100\,°\text{C} > T_{sat\,@\,75\,\text{kPa}}$ and state point 2 is situated in the superheated regime.

Pressure (p), kPa	Saturation temperature (T_{sat}), °C	Specific volume (m³/kg)		Internal energy (kJ/kg)			Enthalpy (kJ/kg)		
		Saturated liquid (v_f)	Saturated vapor (v_g)	Saturated liquid (u_f)	Evaporation (u_{fg})	Saturated vapor (u_g)	Saturated liquid (h_f)	Evaporation (h_{fg})	Saturated vapor (h_g)
75	91.76	0.001037	2.2172	384.36	2111.8	2496.1	384.44	2278.0	2662.4

From the property table for superheated steam, we note that we need to interpolate to find the enthalpy at 75 kPa (or 0.075 MPa):

Temperature (T) °C	p = 0.05 MPa (T_{sat} = 81.32 °C)			p = 0.10 MPa (T_{sat} = 99.61 °C)		
	m³/kg (v)	kJ/kg (u)	kJ/kg (h)	m³/kg (v)	kJ/kg (u)	kJ/kg (h)
100	3.4187	2511.5	2682.4	1.6959	2506.2	2675.8

Using linear interpolation:

$$\frac{h_2 - 2682.4}{2675.8 - 2682.4} = \frac{75 - 50}{100 - 50} \text{ and } h_2 = 2679.1\,\text{kJ/kg}$$

b) Find $\dot{m}$

Given: We are given the temperatures and pressures at the inlet and exit, and the power output of the turbine. We now also know the enthalpies at the inlet and exit.

Required: We must find the mass flow rate of the steam ($\dot{m}$).

Solution strategy: As this is an open system (with mass flowing through the system), we will start with the First Law of Thermodynamics for an open system, simplify the equation, and determine the mass flow rate of steam.

Assumptions: Changes in kinetic and potential energy of steam can be neglected as they usually make up a very small percentage of the enthalpy change. The turbine is well insulated, so we can disregard any heat losses to the environment. We may also assume that the turbine is in steady state, so $\dfrac{dE}{dt} = 0$.

Analysis:

First law of thermodynamics for an open system, from Eq. (2.11):

$$\dot{Q}_{net} - \dot{W}_{net} = \sum_{exit} \dot{m}\left(h + e_k + e_p\right) - \sum_{inlet} \dot{m}\left(h + e_k + e_p\right) + \frac{dE}{dt}$$

Based on our assumptions, the first law reduces to:

$-\dot{W}_{12} = \dot{m}(h_2 - h_1)$ or $\dot{W}_{12} = \dot{m}(h_1 - h_2)$

And because $\dot{W}_{12} = \dot{m}w_{12}$

we have $w_{12} = h_1 - h_2 = (3214.5 - 2679.1)$ kJ/kg $= 535.4$ kJ/kg

and finally, $\dot{m} = \dfrac{\dot{W}_{12}}{w_{12}} = \dfrac{2000 \text{ kJ/s}}{535.4 \text{ kJ/kg}} = 3.74$ kg/s

Discussion: Note that when we calculate the mass flow rate, the dimensions of the numerator were changed to kJ/s to cancel those in the denominator and give the final answer in kg/s. Converting to hourly flow rate, we see that the mass flow rate is 13.4 tons of steam per hour. To produce more power between the same inlet and exit conditions of the turbine, we would need to increase the mass flow rate.

Reimagine Our Future

Engineering emphasizes intellectual work. Is it however not time for engineers to also adopt emotional aspects, including empathizing with nature? How do we create the circumstances and places for engineers to "come to their senses"?

2.4   Processes Involving an Ideal Gas

When working with gases in this book, we assume that they behave as ideal gases. An ideal gas may be represented by its equation of state:

$$pv = RT \tag{2.9}$$

$$\text{or} \quad p\mathcal{V} = mRT \tag{2.10}$$

where m is the mass of gas. For a gas mixture, the total mass is the sum of all i components:

$$m = \sum_i m_i \tag{2.11}$$

The *mass fraction* (x_i) of any species is:

$$x_i = \frac{m_i}{m} \tag{2.12}$$

The *mole fraction* (y_i) of any species is:

$$y_i = \frac{N_i}{N} \tag{2.13}$$

Dalton's law requires that, in a mixture of perfect gases, each gas at N_i moles exerts a partial pressure (p_i) that is independent of the other components, and the total pressure (p) of a mixture of gases is equal to the sum of the partial pressures (Moran et al. 2018):

$$\frac{p_i}{p} = \frac{N_i}{N} = y_i \tag{2.14}$$

$$\text{and} \quad p = \sum_i p_i \tag{2.15}$$

Example 2.4 Molar Concentration of CO_2 in the Atmosphere

If the atmospheric concentration of CO_2 is 420 ppm, or 0.042%, then the partial pressure of CO_2 at $T_0 = 298$ K and $p_0 = 101.325$ kPa will be $p_{CO_2} = \left(\dfrac{0.042\, CO_2}{100\%}\right)(101.325\text{ kPa}) = 0.043$ kPa, and the

molar fraction will be $y_{CO_2} = \dfrac{p_{CO_2, 298\ K}}{p_0} = \dfrac{\left(\dfrac{0.042\, CO_2}{100\%}\right)101.325\text{ kPa}}{101.325\text{ kPa}} = 420 \times 10^{-6}$.

The *heat capacity* of a medium may be defined as the energy required to raise the temperature of the unit mass of a substance by one degree. That can be achieved in a constant volume or constant pressure process.

For a *constant volume* process, $\delta w = p\, dv = 0$ and the first law (Eq. 2.6) becomes $du = \delta q - \cancel{\delta w}$. So, all the transferred heat manifests as a change in internal energy. The specific heat at constant volume (c_v) is then:

$$c_v \doteq \left.\frac{\delta q}{dT}\right|_{\cancel{V}} = \left.\frac{du}{dT}\right|_{\cancel{V}} \qquad [\text{J/kg K}] \tag{2.16}$$

For a *constant pressure* process, the first law (Eq. 2.6) becomes $du = \delta q - p\, dv$. So, some of the transferred heat is used to change the internal energy and some of the transferred heat is used to bring about expansion or compression. Remembering that $h = u + p\, v$, the specific heat at constant pressure (c_p) is then:

$$c_p \doteq \left.\frac{\delta q}{dT}\right|_{p} = \left.\frac{dh}{dT}\right|_{p} \qquad [\text{J/kg K}] \tag{2.17}$$

The gas constant (R) is:

$$R = c_p - c_v \tag{2.18}$$

and the ratio of the specific heats is:

$$\gamma = \frac{c_p}{c_v} \tag{2.19}$$

The main thermodynamic processes involving an ideal gas are adiabatic ($q = 0$), isochoric ($V = $ constant), isobaric ($p = $ constant), and isothermal ($T = $ constant). The relationships summarized in Table 2.1 have been obtained by combining the first law (Eq. 2.6) with Eq. (2.9) and Eqs. (2.16) to (2.19) for an ideal gas, and simplifying the resulting expressions.

Table 2.1 Processes involving an ideal gas (1 indicates the begin state and 2 the end state of a process).

Adiabatic ($q = 0$)	Isochoric (V = constant)	Isobaric (p = constant)	Isothermal (T = constant)
$\left(\dfrac{pv}{T}\right)_1 = \left(\dfrac{pv}{T}\right)_2$	$\left(\dfrac{p\varkappa}{T}\right)_1 = \left(\dfrac{p\varkappa}{T}\right)_2$	$\left(\dfrac{\varkappa v}{T}\right)_1 = \left(\dfrac{\varkappa v}{T}\right)_2$	$\left(\dfrac{pv}{\varkappa}\right)_1 = \left(\dfrac{pv}{\varkappa}\right)_2$
$\dfrac{T_2}{T_1} = \left(\dfrac{v_1}{v_2}\right)^{\gamma-1}$			
$\dfrac{p_2}{p_1} = \left(\dfrac{v_1}{v_2}\right)^{\gamma}$			
$\dfrac{T_2}{T_1} = \left(\dfrac{p_2}{p_1}\right)^{\frac{\gamma-1}{\gamma}}$			
First law for a closed cycle: $\Delta u = \varkappa - w$ $\therefore w = -\Delta u \equiv -c_v \Delta T$	First law for a closed cycle: $\Delta u = q - w = q - p\,\varkappa v$ $\therefore q = \Delta u \equiv c_v \Delta T$	First law for a closed cycle: $\Delta u = q - w = q - p\Delta v$ $\therefore q = \Delta u + p\Delta v$ but $\Delta h = \Delta u + p\Delta v$ $\therefore q = \Delta h \equiv c_p \Delta T$	First law for a closed cycle: $\Delta u = q - w$ $= q - p\Delta v$ $\Delta u \equiv c_v \varkappa T = 0$ $\therefore q = w \equiv p\Delta v$
First law for an open cycle: $\Delta h = \varkappa - w$ $\therefore w = -\Delta h \equiv -c_p \Delta T$	First law for an open cycle: $\Delta h = q - w = q - p\,\varkappa v$ $\therefore q = \Delta h \equiv c_p \Delta T$	First law for an open cycle: $\Delta h = q - w = q - p\Delta v$ $\equiv c_p \Delta T$	First law for an open cycle: $\Delta h = q - w = q - p\Delta v$ $\equiv c_p \varkappa T$ $\therefore q = w \equiv p\Delta v$

2.5 Thermal Efficiency Deduced from the First Law of Thermodynamics

All thermal power cycles feature a heat rejection process, so the work done is always less than the heat supplied. From Eqs. (2.6) and (2.8), the work done is equal to the heat added minus the heat rejected.

The efficiency of a thermal power cycle may be defined as the fraction of heat supplied to a thermal cycle that is converted into work:

$$\eta_{th} = \frac{W_{net}}{Q_{in}} = \frac{W_{out} - W_{in}}{Q_{in}} \equiv \frac{Q_{in} - Q_{out}}{Q_{in}} = 1 - \frac{Q_{out}}{Q_{in}} \tag{2.20}$$

Note: For a cycle, there is no change in internal energy or no change in enthalpy (because the beginning and end points of a cycle are the same), so $W_{net} = Q_{net}$.

2.6 Summary

The first law of thermodynamics is based on energy conservation, i.e., on the conservation of a *quantity* of energy. In the first law, all forms of energy (Q, W, ΔU, ΔH) are equivalent in all respects. We will need another law, the second law of thermodynamics, to show that heat and work are, in fact, not equivalent in all respects. In Chapter 3, we will see that the second law also defines the *quality* of energy and the viability of processes or systems.

Problems

2.1 A desktop computer must be cooled by a fan (also known as a CPU fan) that maintains a volumetric flow rate of 0.34 m³/min.

    a) Determine the mass flow rate (in kg/s) of the air through the fan at an elevation of 3400 m where the air density is 0.7 kg/m³.

[Answer: 0.004 kg/s]

    b) If the average velocity of air must not exceed 110 m/min, determine the minimum required diameter of the fan (in mm).

[Answer: 63 mm]

2.2 An electrically driven pump increases water pressure from 100 kPa at the inlet to 900 kPa at the exit. Water enters the pump at 15 °C through a 10-mm diameter opening and exits through a 15-mm diameter opening. You may assume the density of water remains 1000 kg/m³ throughout. Determine the velocity (in m/s) of the water at the inlet and exit when the mass flow rate through the pump is 0.5 kg/s.

[Answer: Inlet velocity: 6.37 m/s; Exit velocity: 6.42 m/s]

2.3 A Refrigerant-134a (or R-134a) tank is located outdoors. The tank has a volume of 30 m³ and is filled with 1200 kg of R-134a at a pressure of 685.8 kPa. The tank is exposed to the sun, which causes the R-134a to reach a saturated vapor state. Calculate the following quantities:

    a) The initial temperature and state of the refrigerant prior to heating.
    b) The final temperature and pressure of the refrigerant after heating.
    c) The amount of heat transferred to the refrigerant, in MJ.

[Answers: (a) 26 °C, saturated steam; (b) 32 °C and 816 kPa; (c) 35.7 MJ]

Hint: Refer to the relevant extract of the thermodynamic properties of R-134a, below.

Temperature (T), °C	Saturated pressure (p_{sat}), kPa	Specific volume (m³/kg)		Internal energy (kJ/kg)			Enthalpy (kJ/kg)		
		Saturated liquid (v_f)	Saturated vapor (v_g)	Saturated liquid (u_f)	Evaporation (u_{fg})	Saturated vapor (u_g)	Saturated liquid (h_f)	Evaporation (h_{fg})	Saturated vapor (h_g)
26	685.84	0.0008313	0.029976	87.26	156.87	244.12	87.83	176.85	264.68
32	815.89	0.0008478	0.025108	95.79	151.35	247.14	96.48	171.14	267.62

2.4 Consider a Wilesco D18 model steam engine operating at 1 bar ambient pressure, while the boiler's pressure gauge reads 0.5 bar. The boiler boils off 150 mL water in 30 min. Assuming that the steam is in a saturated vapor state, estimate the heat rate (in watts) provided by the boiler.

[Answer: About 210 W]

Hint: When calculating energy values of steam inside the boiler, work with the *absolute pressure*, which is the *gauge pressure + atmospheric pressure*. You can access properties of steam in several Thermodynamics textbooks or from dedicated simulation packages such as Engineering Equation Solver (EES 2022).

The following is an extract of steam properties calculated using the EES package.

Saturated water

Pressure (*p*), kPa	Saturation temperature (*T*_{sat}), °C	Specific volume (m³/kg)		Enthalpy (kJ/kg)		
		Saturated liquid (v_f)	Saturated vapor (v_g)	Saturated liquid (h_f)	Evaporated (h_{fg})	Saturated vapor (h_g)
100	99.61	0.001043	1.6941	417.51	2257.5	2675.0
125	105.97	0.001048	1.3750	444.36	2240.6	2684.9
150	111.35	0.001053	1.1594	467.13	2226.0	2693.1

References

Borel, L. and Favrat, D. (2010). *Thermodynamics and Energy Systems Analysis. From Energy to Exergy.* Lausanne: EPFL Press/CRC Press.

EES (2022). *Engineering Equation Solver (EES)* developed by S.A. Klein and F.L. Alvarado. F Chart Software. https://fchartsoftware.com/ees (accessed 15 January 2022).

Moran, M.J., Shapiro, H.N., Boettner, D.D. et al. (2018). *Fundamentals of Engineering Thermodynamics.* New York, NY: Wiley.

Obert, E.F. (1960). *Concepts of Thermodynamics.* New York, NY: McGraw-Hill.

Van Wylen, G. and Sonntag, R.E. (1976). *Fundamentals of Classical Thermodynamics.* New York, NY: Wiley.

3

Destruction of Quality of Energy

The first law of thermodynamics is based on energy conservation, i.e., the conservation of a *quantity* of energy. In the first law, all forms of energy (Q, W, ΔU, ΔH) are equivalent in the accounting process.

The second law of thermodynamics recognizes that all forms of energy are *not* equivalent in their ability to do work (Obert 1960). The second law of thermodynamics will therefore help us assess the *quality* of energy. The second law of thermodynamics also identifies the direction of feasible processes and determines the theoretical performance limit of a process or cycle.

Reimagine Our Future
Sustainability is a goal that demands endless negotiation between humans and nature. Considering the uncertainty about the carrying capacity of Earth and the amount of damage that we have already incurred, what can we do to make up for lost time?

3.1 Reversible and Irreversible Processes

Real processes are never ideal. Real processes are accompanied by the dissipation of their ability to do work, or by the dissipation of available energy. Friction is a major cause of energy dissipation in real processes, as is the transfer of energy through finite temperature differences.

In thermodynamics, a system is *reversible* if it is an ideal system, if its direction can be reversed at any point in the process, and if both the system and its surroundings will recover to their original states (i.e., without leaving any permanent change on the surroundings or system). Reversibility can be attained by changing the properties of a frictionless system so that the process will comprise quasi-static – i.e., indefinitely slow – thermal equilibrium process steps.

In an *irreversible* process, the system and surroundings change in such a manner that they are unable to return to their original states.

A real thermal process can never be reversible as heat transfer can only occur through a finite temperature difference between a system and its surroundings. In thermodynamics, the concept of reversibility is useful if one is to conceptualize the most ideal process or cycle to determine the maximum work that the process or cycle (like a gas turbine or expanding gas in a piston

Energy Systems: A Project-Based Approach to Sustainability Thinking for Energy Conversion Systems,
First Edition. Leon Liebenberg.
© 2024 John Wiley & Sons, Inc. Published 2024 by John Wiley & Sons, Inc.
Companion website: www.wiley.com/go/liebenberg/energy_systems

cylinder) will be able to perform, or the minimum work that is required to operate a process or cycle (like a pump or a compressing gas in a piston cylinder).

Due to their assumed frictionless set-ups and quasi-static thermal equilibrium process steps, *reversible processes deliver the most work and consume the least.* Irreversible (real) processes will always have lower thermal efficiencies than reversible (ideal) processes. Reversible processes are also not spontaneous, while irreversible processes are.

Example 3.1   Reversible Work in Closed and Open Systems
In a *closed* system, the reversible work of a fluid of mass (m) at pressure (p) operating on a boundary area (A) over a differential distance (dL) is:

$$W_{rev,closed} = pA \, dL = p \, d\Psi = mp \, dv \tag{3.1}$$

As this is a closed system, the mass remains constant. Integrating Eq. (3.12) then gives:

$$W_{rev,closed} = m \int_{v_1}^{v_2} p \, dv \tag{3.2}$$

In an *open* system, disregarding the conversion of kinetic and potential energy, work is transferred due to compression or expansion minus flow work. So, using the expressions for flow work (contained in Eqs. (3.2) and (3.3)), Eq. (3.1) becomes:

$$W_{rev,open} = p \, d\Psi - d(p\Psi) = m(p \, dv - d(pv)) \tag{3.3}$$

$$W_{rev,open} = m \int_{v_1}^{v_2} p \, dv - m(p_2 v_2 - p_1 v_1) \tag{3.4}$$

Combining the first two terms on the right-hand side of the equation gives:

$$W_{rev,open} = -m \int_{p_1}^{p_2} v \, dp \tag{3.5}$$

3.2   Entropy and the Second Law of Thermodynamics

Available energy is that part of energy that, ideally, could be converted into work. In real (irreversible) processes, available energy will always decrease, while in ideal (reversible) processes, available energy will always be conserved. The concept of entropy is usually used to indicate changes in available energy.

For a *reversible* process, the following applies:

$$dS = \frac{\delta Q_{rev}}{T} \qquad [J/K] \tag{3.6}$$

where S is *entropy* (J/K) and s is specific entropy (in J/kg K). Entropy is a thermodynamic property, i.e., the change in entropy only depends on the beginning and end states of a process and not on the path followed between the states.

Clausius and others showed that, for an *irreversible* process, the following inequality holds:

$$dS > \frac{\delta Q_{irr}}{T} \tag{3.7}$$

Combining Eqs. (3.6) and (3.7) for *any* process:

$$dS \geq \frac{\delta Q}{T} \tag{3.8}$$

Equation (3.8) can be written in a differential format:

$$dS = \frac{\delta Q}{T} + \sigma \tag{3.9}$$

or in a finite format:

$$\Delta S = \sum \frac{Q}{T} + \sigma \tag{3.10}$$

$$\text{with} \quad \sigma \begin{cases} = 0 & \text{(no irreversibilities present within the system)} \\ > 0 & \text{(irreversibilities present within the system)} \\ < 0 & \text{(impossible)} \end{cases} \tag{3.11}$$

σ is sometimes called entropy production (measured in kJ/kg K).

Equations (3.9) and (3.10) are mathematical statements of the second law of thermodynamics. In Eqs. (3.9) and (3.10), the *equality* holds for reversible (ideal) processes, and the inequality for irreversible (real) processes. So, according to the second law of thermodynamics, *the entropy of an isolated system increases* in all real (irreversible) processes and is conserved in ideal (reversible) processes (Bejan et al. 1996):

$$\Delta S_{total} = \Delta S_{system} + \Delta S_{surroundings} \equiv \oint \frac{\delta Q}{T} + \sigma \geq 0 \tag{3.12}$$

Note that $Q = 0$ is an isolated system and its surroundings cannot exchange heat beyond their combined boundary (see Figure 3.1). Also, $\sigma \geq 0$ for reversible and irreversible processes. Further, ΔS_{system} can be negative. For instance, if a system discards heat to its surroundings, its entropy will be $\frac{-Q}{T}$ according to our adopted sign convention. If a system absorbs heat from its

Figure 3.1 An isolated system and its surroundings cannot exchange heat, work, or mass beyond the boundary of the isolated system.

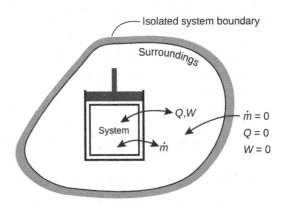

surroundings, its entropy will be $\dfrac{+Q}{T}$. When adding ΔS_{system} to $\Delta S_{surroundings}$, ΔS_{total} must always be greater than or equal to zero.

The fact that the second law of thermodynamics for a total system demands that $\Delta S_{total} \geq 0$ implies that entropy is not conserved in real (irreversible) systems and their surroundings but is always increasing. While the first law of thermodynamics informs that energy is always conserved, the second law informs that total entropy is never conserved.

3.3 Entropy Balance for Closed Systems

For a closed system (where mass remains constant), Eq. (3.10) applies. This can also be written in rate form:

$$\frac{dS}{dt} = \dot{S} = \sum \left(\frac{\dot{Q}}{T}\right) + \dot{\sigma} \qquad [\text{W/K}] \qquad (3.13)$$

where $\dot{S}$ is the entropy rate (W/K), $\dot{Q}$ is the rate of heat transfer (W), and $\dot{\sigma}$ is the rate of *entropy production* (W/K), with $\dot{\sigma} \geq 0$ for reversible and irreversible processes.

Example 3.2   Entropy Balance of a Closed System
A quantity of 1 kg of water vapor contained within a piston cylinder assembly, initially at 0.5 MPa and 300 °C, undergoes adiabatic expansion to a state where pressure is 0.1 MPa and the temperature is either (a) 150 °C, or (b) 100 °C. Using entropy balance, determine the viability of the process in each case.

Hint: Use the extract from the EES property table for water, shown below.

Temperature (T) °C	p = 0.50 MPa (T_{sat} = 151.83 °C) (s) kJ/kg K	p = 0.10 MPa (T_{sat} = 99.61 °C) (s) kJ/kg K
100	—	7.3611
150	—	7.6148
300	7.4614	—

Given: We are given the temperatures and pressures at the inlet and exit of two adiabatic processes.

Required: We must evaluate the viability of the two processes.

Solution strategy: We will use the entropy balance in Eqs. (3.9) and (3.11) to evaluate entropy production; from that, we will assess whether processes are real (irreversible), ideal (reversible), or impossible.

Analysis:

$$\text{Entropy balance}: \Delta S = \sum \left(\cancel{\frac{Q}{T}}\right) + \sigma \text{ because } Q = 0 \text{ (adiabatic process)} \qquad (3.14)$$

a) Let us first evaluate scenario (a): Inlet at 0.5 MPa and 300 °C; outlet at 0.1 MPa and 150 °C: From the property table, we note that:

$$T_{sat@0.5\,MPa} = 151.83\,°C$$

So, $T_1 = 300\,°C > T_{sat\,@\,0.5\,MPa}$, which means that state point 1 is in the superheated regime. We will use the superheated properties of water summarized in the above table.

$s_1 = 7.4614\,\text{kJ/kg K}$ and $s_2 = 7.6148\,\text{kJ/kg K}$

The entropy production is then, from Eq. (3.12):

$$\sigma = m(s_2 - s_1) \qquad\qquad (3.15)$$

$$\sigma = 1\,\text{kg}(7.6148 - 7.4614)\,\text{kJ/kg K}$$

$$\sigma = 0.153\,\text{kJ/K}$$

Discussion: As $\sigma > 0$, this adiabatic process is irreversible and will occur spontaneously.

b) Let us now evaluate the feasibility of scenario (b): Inlet at 0.5 MPa and 300 °C; outlet at 0.1 MPa and 100 °C:

From the property table, we note that:

$s_1 = 7.4614\,\text{kJ/kg K}$ and $s_2 = 7.3611\,\text{kJ/kg K}$

The entropy production is then, from Eq. (3.12):

$$\sigma = m(s_2 - s_1) \qquad\qquad (3.16)$$

$$\sigma = 1\,\text{kg}(7.3611 - 7.4614)\,\text{kJ/kg K}$$

$$\sigma = -0.1\,\text{kJ/K}$$

Discussion: As $\sigma < 0$, this adiabatic process is impossible. For this process to happen, the system would have to exchange heat with its surroundings.

3.4 Entropy Balance for Open Systems

Like energy, entropy can be transferred into or out of a control volume by mass. Since this is the principal difference between closed-system and open-system entropy rate balances, the open-system form of the balance equation can be obtained by modifying the closed-system form to account for such entropy transfer:

$$\frac{dS}{dt} = \sum \frac{\dot{Q}}{T} + \sum \dot{m}_i s_i + \sum \dot{m}_e s_e + \dot{\sigma} \qquad\qquad (3.17)$$

where $\dot{m}_i s_i$ and $\dot{m}_e s_e$ account, respectively, for the rates of entropy transfer accompanying mass flow at the inlets and exits.

With steady flow $\left(\dfrac{dS}{dt} = 0\right)$ and single inlets and exits, Eq. (3.17) simplifies to:

$$0 = \frac{\dot{Q}}{T} + \dot{m}_i s_i + \dot{m}_e s_e + \dot{\sigma} \qquad\qquad (3.18)$$

Example 3.3 Entropy Balance of an Open System

Consider an industrial steam turbine with a power output of 7 MW. The mass flow rate of steam is 8 kg/s. The turbine is not well insulated and loses 50 kW to the surroundings. The conditions at the inlet and exit are such that $s_1 = 7.0922$ kJ/kg K and $s_2 = 7.6148$ kJ/kg K. The steam is superheated at both the inlet and the exit. The temperature of the surrounding air is 25 °C. Calculate the entropy production rate in the turbine surroundings.

Given: We are given the steam mass flow rates, the entropies at the inlet and exit of the turbine, as well as the temperature of the surrounding air.

Required. We must calculate the entropy production rate of the turbine process.

Solution strategy: We will use the entropy rate balance of Eq. (3.16) to calculate the entropy production.

Analysis:

$$Entropy\ balance: 0 = \frac{\dot{Q}}{T} + \dot{m}_i s_i + \dot{m}_e s_e + \dot{\sigma}$$

$$\dot{\sigma} = \dot{m}(s_2 - s_1) - \frac{\dot{Q}}{T} \tag{3.19}$$

$$\dot{\sigma} = (8\ \text{kg/s})(7.6148 - 7.0922)\ \text{kJ/kg K} - \frac{(-50)\ \text{kJ/s}}{(25 + 273)\ \text{K}}$$

$$\dot{\sigma} = 4.180\ \text{kW/K} + 0.168\ \text{kW/K} = 4.35\ \text{kW/K}$$

Discussion: As $\dot{\sigma}$ is much larger than 0, the expansion process through the turbine is highly irreversible and happens spontaneously.

Reimagine Our Future

Imagine we were to collectively say "no" to purported "solutions" that only treat the symptoms of energy unsustainability and ignore or actively perpetuate the root causes. What would you say no to today?

3.5 The *T*-ds Relations

Using the previously derived equations for a reversible process or system, one can derive a set of equations that apply for all systems or processes. These are called the *T*-ds equations and are very useful when analyzing energy conversion systems.

Consider the first law of thermodynamics for a closed system:

$$\delta q - \delta w = du \tag{3.20}$$

For a reversible system, from Eqs. (3.2) and (3.6), we may now write Eq. (3.20) as follows:

$$T\,ds = du + p\,dv \qquad [\text{J}] \tag{3.21}$$

$$\text{But,}\ h = u + pv, \text{so}\ dh = du + p\,dv + v\,dp \tag{3.22}$$

Substituting Eq. (3.21) in (3.22) gives:

$$T\,ds = dh - v\,dp \qquad \text{[J]} \qquad (3.23)$$

Equations (3.21) and (3.23) were derived from reversible relations, but they now apply to *all* processes, reversible or irreversible; open or closed. Each of the quantities specified in these *T-ds* equations are thermodynamic properties and are independent of the path that a process follows.

3.5.1 Ideal Gas

Let us apply the *T-ds* relations to an ideal gas to arrive at equations that will often be used in this book. The ideal gas model assumes that pressure, specific volume, and temperature are related by:

$$pv = RT \qquad (3.24)$$

Specific internal energy and specific enthalpy also each depend solely on temperature: $u = u(T)$, $h = h(T)$, giving $du = c_v\,dT$ and $dh = c_p\,dT$, respectively (see Eqs. (2.9) and (2.10)).

Using these relationships, and integrating, the $T\,ds$ equations, Eqs. (3.21) and (3.23), give, respectively:

$$\Delta s = c_{v,avg} \ln \frac{T_2}{T_1} + R \ln \frac{v_2}{v_1} \qquad \text{[J/kg K]} \qquad (3.25)$$

$$\Delta s = c_{p,avg} \ln \frac{T_2}{T_1} - R \ln \frac{p_2}{p_1} \qquad \text{[J/kg K]} \qquad (3.26)$$

The values of $c_{v,avg}$ and $c_{p,avg}$ are evaluated at average temperatures $(T_1 + T_2)/2$. This assumption of average process temperatures is accurate enough for all the energy conversion systems considered in this book.

3.6 Maximum Attainable Thermal Efficiency (by Combining the First and Second Laws)

The thermal efficiency of a cycle was shown to be (Eq. 2.12):

$$\eta_{th} = \frac{W_{net}}{Q_{in}} = \frac{W_{out} - W_{in}}{Q_{in}} \equiv \frac{Q_{in} - Q_{out}}{Q_{in}} \qquad (3.27)$$

This definition of efficiency enables energy or heat losses to be estimated, but it gives only limited information about the optimal conversion of energy. The first law of thermodynamics indeed takes no account of an energy source in terms of its quality. We will develop a more comprehensive definition of thermal efficiency by incorporating the second law of thermodynamics.

Consider the simplest reversible cycle that can be devised to work between two temperatures, the *Carnot cycle* (Figure 3.2). This cycle comprises two isothermal processes (for heat addition and heat removal) and two isentropic processes (for expansion and compression).

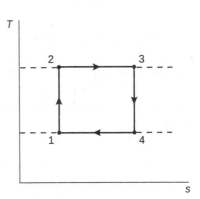

Figure 3.2 The Carnot power cycle.

As heat is added at T_H, the entropy increases by $\dfrac{+Q_H}{T_H}$, and as heat is removed at T_C, the entropy decreases by $\dfrac{-Q_C}{T_C}$.

Therefore, from the first law, Eq. (2.10), we have for the Carnot cycle (with $\Delta U_{cycle} = 0$):

$$Q_H - Q_C = W \tag{3.28}$$

and from the second law, Eq. (3.12), we have for the Carnot cycle (with $\Delta S_{cycle} = 0$):

$$0 = \left(\frac{Q_H}{T_H} + \frac{-Q_C}{T_C}\right) + \sigma \tag{3.29}$$

Combining Eqs. (3.28) and (3.29), and eliminating Q_C, we have (Black and Hartley 1985):

$$\frac{Q_H}{T_H} + \sigma = \frac{Q_H - W}{T_C} \tag{3.30}$$

So, the Carnot (reversible, ideal) cycle efficiency is:

$$\eta_{Carnot} = \frac{|W|}{Q_H} = \left(1 - \frac{T_C}{T_H}\right) - \frac{T_C \sigma}{Q_H} \tag{3.31}$$

The second law of thermodynamics, however, demands that $\sigma \geq 0$, and with $T_C > 0$, the last term is positive and reduces the efficiency. The *maximum efficiency*, also called *Carnot efficiency*, that can be obtained for a reversible heat engine ($\sigma = 0$) is therefore:

$$\eta_{rev} = \left(1 - \frac{T_C}{T_H}\right) \tag{3.32}$$

The efficiency of an irreversible (real) thermal machine must be smaller than or equal to that of a reversible (ideal) machine (see Section 3.1):

$$\eta_{irrev} \leq \eta_{rev} \tag{3.33}$$

Combining Eq. (3.33) with Eqs. (3.32) and (2.12) gives:

$$1 - \frac{Q_C}{Q_H} \leq 1 - \frac{T_C}{T_H} \tag{3.34}$$

$$\frac{Q_H}{T_H} + \frac{Q_C}{T_C} \leq 0 \tag{3.35}$$

Equation (3.35) shows how an actual (irreversible) process is limited by an ideal (reversible) one. This inequality is very useful for evaluating the viability of cycles, as illustrated in Example 3.4. When interpreting Eq. (3.35), remember our adopted convention that heat added to a system takes a positive sign, and heat discarded from a system takes a negative sign.

Equation (3.35) could also be written in a more general integral format, commonly called the *Clausius inequality*:

$$\oint \frac{Q}{T} \leq 0 \tag{3.36}$$

or $$\oint \frac{Q}{T} \leq -\sigma \tag{3.37}$$

with the entropy production (σ) interpreted as laid out in Eq. (3.11).

Example 3.4 Viability of a Cycle Based on the Combination of the First and Second Laws of Thermodynamics

Consider a heat engine that receives 300 kJ of heat at a temperature of 1000 K and discards 140 kJ of its heat at 300 K. Is this cycle possible?

Given: We are given the heat being transferred and the associated temperature.

Required: We must evaluate the viability of this cycle.

Solution strategy: We will use the inequality derived from the first and second laws of thermodynamics as stated in Eq. (3.34) to evaluate whether this cycle is possible. For the cycle to be possible, $\dfrac{Q_H}{T_H} + \dfrac{Q_C}{T_C}$ must be ≤ 0.

Analysis:

$$\frac{Q_H}{T_H} + \frac{Q_C}{T_C} = \frac{300 \text{ kJ}}{1000 \text{ K}} + \frac{-140 \text{ kJ}}{300 \text{ K}} = -0.167 \text{ kJ/K}$$

Remember, as 140 kJ of heat is being removed from the system, 140 kJ takes a *negative sign*, while the heat added (300 kJ) takes a *positive sign*.

Discussion: As $\dfrac{Q_H}{T_H} + \dfrac{Q_C}{T_C} < 0$, this cycle is possible. This cycle would be impossible if, for instance, the 140 kJ of heat were to be discarded at a higher temperature, say 600 K, which would make $\dfrac{Q_H}{T_H} + \dfrac{Q_C}{T_C} > 0$. The inequality $\dfrac{Q_H}{T_H} + \dfrac{Q_C}{T_C} \leq 0$ therefore provides a simple method to evaluate the viability of thermal cycles.

3.7 "Second Law" Efficiency: Isentropic Efficiencies

The thermal efficiency that we defined using the first law of thermodynamics, Eq. (2.12), can now be compared to the thermal efficiency of a reversible engine, Eq. (3.32), to give the *second law efficiency* (η_{II}):

$$\eta_{II} = \frac{\eta_{th}}{\eta_{rev}} \tag{3.38}$$

The second law efficiency is sometimes called the "isentropic efficiency" or the "adiabatic efficiency" when evaluating the performance of turbines, pumps, and compressors:

$$\text{Work} - \text{producing machine (e.g., a turbine): } \eta_s = \frac{\text{actual work}}{\text{isentropic work}} = \frac{W}{W_s} \tag{3.39}$$

$$\text{Work} - \text{consuming machine (e.g., a pump): } \eta_s = \frac{\text{isentropic work}}{\text{actual work}} = \frac{W_s}{W} \tag{3.40}$$

The isentropic efficiencies of work-producing machines are defined in a reciprocal manner relative to the isentropic efficiencies of work-consuming machines, so that the efficiencies will always be less than 100%, as per convention.

Example 3.5 Second Law Efficiency

Evaluate the performance of a heat engine with a thermal efficiency (as defined by Eq. (2.12)) of 25%, and which first (a) extracts heat from a source at 500 K and discards heat to a sink at 250 K, and

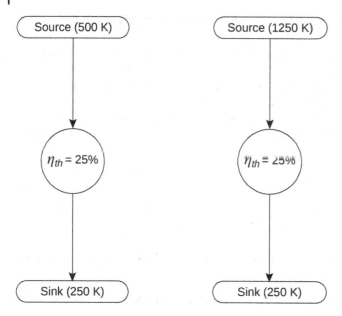

Figure 3.3 Set-up for evaluating Example 3.5.

then (b) extracts heat from a source at 1250 K and discards heat to the same sink at 250 K (Figure 3.3).

Given: We are given the source and sink temperatures of two heat engine scenarios.

Required: We must evaluate the performance of the engine relative to its performance under reversible (ideal) conditions for two scenarios.

Solution strategy: We will compare the second law efficiencies, Eq. (3.39), of the two scenarios. *Analysis:*

$$\eta_{II} = \frac{\eta_{th}}{\eta_{rev}}$$

a) For the first scenario, $T_H = 500$ K and $T_C = 250$ K, $\eta_{th} = 25\%$:

So, $\eta_{rev} = 1 - \frac{T_C}{T_H} = 1 - \frac{250}{500} = 1 - 0.5 = 0.5$ (or 50%)

Then, $\eta_{II} = \frac{\eta_{th}}{\eta_{rev}} = \frac{25\%}{50\%} = 50\%$

b) For the second scenario, $T_H = 1250$ K and $T_C = 250$ K, $\eta_{th} = 25\%$:

So, $\eta_{rev} = 1 - \frac{T_C}{T_H} = 1 - \frac{250}{1250} = 1 - 0.2 = 0.8$ (or 80%)

Then, $\eta_{II} = \frac{\eta_{th}}{\eta_{rev}} = \frac{25\%}{80\%} = 31.3\%$

Discussion: The engine operating in the second scenario has a much greater work potential (as it can achieve a maximum thermal efficiency of 80%) than the first one (which can only achieve a maximum efficiency of 50%), but the engine only has a thermal efficiency of 25%. So, the engine operating in the first scenario performs relatively better than the engine in the second scenario, i.e., $\eta_{II} = 50\%$ versus $\eta_{II} = 31.3\%$.

3.8   Related Fluid Mechanics: Head Loss, Fluid Momentum

Some of the systems investigated in this book will require fundamental thermal-fluid analysis. All thermal-fluid systems in this book will involve only a single inlet and a single exit. Following this assumption and using Eqs. (2.7) and (3.23), the energy balance can be written as follows for steady-state conditions:

$$w_{net} = \frac{\dot{W}_{net}}{\dot{m}} = -\int_1^2 v \, dp + \frac{1}{2}\left(V_1^2 - V_2^2\right) + g(z_1 - z_2) \tag{3.41}$$

3.8.1   Head Loss

Equation (3.41) can be rewritten to accommodate friction losses encountered by a viscous fluid flowing in a duct:

$$\frac{p_2}{\rho} + \frac{V_2^2}{2} + gz_2 = \frac{p_1}{\rho} + \frac{V_1^2}{2} + gz_1 + w_{net,in} - h_{loss} \tag{3.42}$$

Equation (3.42) is known as the *generalized Bernoulli equation* or the *energy equation*, where $w_{net,in}$ $\left(= \dfrac{\dot{W}_{net,in}}{\dot{m}}\right)$ is the work developed by the system per unit of mass flow rate, and h_{loss} is the *head loss* (also known as friction work or energy dissipation). If there is no friction (or other irreversibilities), Eq. (3.42) becomes Eq. (3.41).

The head loss (or loss in pressure head) can be calculated as follows for laminar and turbulent flow inside a pipe:

Laminar flow:
$$h_L = f\frac{L}{D}\frac{V^2}{2g} \tag{3.43}$$

where D is the pipe diameter, L is the pipe length, V is the fluid velocity, g is acceleration due to gravity, and f is the friction factor,

with $$f = \frac{64}{Re} \tag{3.44}$$

and $$Re = \frac{\rho VD}{\mu} \tag{3.45}$$

The Reynolds number (Re) is used to determine whether flow is laminar or turbulent, where ρ is the fluid density, V is the average fluid velocity, D is the characteristic length (such as pipe diameter), and μ is the dynamic viscosity.

Turbulent flow:
$$h_L = f\frac{L}{D}\frac{V^2}{2g} \tag{3.46}$$

Unlike laminar flow, there is no analytical way of determining the head loss in a pipe containing turbulent flow. It is therefore determined experimentally. The value of f can be read off diagrams such as the Moody diagram, or it can be calculated using a semi-empirical equation such as that of Haaland (Gerhart et al. 2020):

$$\frac{1}{\sqrt{f}} = -1.8\log\left(\left(\frac{\varepsilon/D}{3.7}\right)^{1.11} + \frac{6.9}{Re}\right) \tag{3.47}$$

Equation (3.47) shows that, with turbulent flow, the friction factor (f) is dependent on the surface roughness (ε) of the pipe (or duct) in which a fluid is flowing.

3.8.2 Momentum Equation

The design of many energy systems, including pumps and turbines, depends on the forces that a fluid flow exerts on them. These forces can be calculated using a linear momentum analysis, based on Newton's Second Law of Motion:

$$\sum \overline{F}_{sys} = \frac{\partial}{\partial t} \int_V \rho \overline{V} \, dV + \int_S \rho \overline{V} \, V_{f/s} \cdot \hat{n} \, dA \tag{3.48}$$

For steady flow, with the fluid entering through one port and exiting through one port, Eq. (3.48) becomes:

$$\overline{F}_{sys} = \dot{m}(\overline{V}_2 - \overline{V}_1) \tag{3.49}$$

Equation (3.49) reveals that the resultant external force acting on a fluid system equals the difference between the rates of momentum associated with mass flow exiting and entering the system. The resultant force includes forces due to pressure acting on the inlet (1) and exit (2), and the gravity force (Bejan et al. 1996).

3.9 Available Work: Exergy

The combination of the statements of the first and second laws of thermodynamics revealed that the maximum efficiency of a thermal machine is (Eq. (3.32)):

$$\eta_{rev} = \left(1 - \frac{T_C}{T_H}\right) \equiv \frac{W_{net}}{Q_{in}} \tag{3.50}$$

Since no temperature can be lower than 0 K, (T_C/T_H) will always be a finite value, which implies that the thermal efficiency, even of a reversible (ideal) engine, can never be 100%. So, according to the laws of thermodynamics, it is impossible to convert all heat input to useful work. However, we know from practice that work can be fully converted to heat, for example when electrical work is converted to heat when boiling water in a kettle. The laws of thermodynamics suggest the need to define parameters to assess the maximum amount of work achievable. Exergy accounting is used for this purpose, which considers both the first and second laws of thermodynamics (Borel and Favrat 2010).

Exergy (E) is the available energy for conversion from a source with reference to the ambient conditions. When evaluating exergy, the reference or ambient temperature (T_0) is usually taken to be 25 °C (298.15 K) and the reference or ambient pressure (p_0) is usually taken to be 1 atm (or 101.325 kPa). A system is said to be in the *dead state* when it is in thermodynamic equilibrium with its environment (at T_0 and p_0), i.e., if there is no temperature gradient or pressure gradient between a system and its environment, no useful work can be done.

Combining the energy and entropy balances, Eqs. (2.7) and (3.17), the following expression is obtained for the exergy (E) of a system at a specified state (Bejan et al. 1996):

$$E = (U - U_0) + p_0(V - V)_0 - T_0(S - S_0) + E_k + E_p \tag{3.51}$$

where U_0, p_0, V_0, T_0, and S_0 denote the internal energy, pressure, volume, temperature, and entropy of the system at a dead state, respectively. At the dead state, the kinetic energy (E_k) and potential energy (E_p) are each equal to 0.

The specific exergy, $e = \dfrac{E}{m}$, is:

$$e = (u - u_0) + p_0(v - v_0) - T_0(s - s_0) + V^2/2 + gz \tag{3.52}$$

Example 3.6 Exergy (or Availability) of Compressed Air

Lithium-ion battery packs used in a Tesla Model 3 electric vehicle can deliver 80 kWh of energy. Now, consider using a motor driven by compressed air, not electricity (e.g., Figure 3.4). If the air is compressed to 7 MPa at 25 °C, what volume air storage tank would be required to provide the compressed air with an exergy (or availability) of 80 kWh?

Given: We are given the required energy content of compressed air to propel a vehicle using an air motor.

Required: We must find the volume of compressed air that will provide an energy equivalent of 80 kWh. In doing so, we must evaluate the exergy of the compressed air.

Solution strategy: First, we will convert the energy content of the lithium-ion battery pack from kWh to J.

We will then assume that air is an ideal gas and use the ideal gas equation of state, $pV = mRT$.

To evaluate specific exergy, we will use Eq. (3.52): $e = (u - u_0) + p_0(v - v_0) - T_0(s - s_0) + V^2/2 + gz$.

The entropy change in the above expression will be evaluated using $\Delta s = s - s_0 = c_p \ln \dfrac{T}{T_0} - R \ln \dfrac{p}{p_0}$.

Assume the ambient temperature is 25 °C or 298.15 K and the ambient pressure is 1 bar or 101.325 kPa.

Analysis:

Convert 80 kWh to joules:

$$(80 \text{ kWh}) \left(\frac{1 \text{ J}}{1 \text{ Ws}} \right) \left(\frac{3600 \text{ s}}{h} \right) = 288 \times 10^3 \text{ kJ} = 288 \text{ MJ}$$

Figure 3.4 *AirPod* compressed air-propelled urban concept vehicle. *Source:* Courtesy of Wikimedia Commons, CC BY-SA 3.0.

The compressed air must store 288 MJ.

We can now find the various variables to calculate the exergy of air at 7 MPa at 25 °C:

$$p V = mRT \text{ or } pv = RT$$

$$\text{Compressed air}: v = \frac{RT}{p} = \frac{(0.287 \text{ kJ/kg K})(298.15 \text{ K})}{7000 \text{ kPa}} = 0.01222 \text{ m}^3/\text{kg}$$

$$\text{Ambient air}: v_0 = \frac{RT_0}{p_0} = \frac{(0.287 \text{ kJ/kg K})(298.15 \text{ K})}{101.325 \text{ kPa}} = 0.8445 \text{ m}^3/\text{kg}$$

$$\Delta s = s - s_0 = c_p \ln\frac{T}{T_0} - R\ln\frac{p}{p_0} = 0 - (0.287 \text{ kJ/kgK}) \ln\frac{7000 \text{ kPa}}{101.325 \text{ kPa}} = -1.2155 \text{ kJ/kg as } T = T_0$$

$$e = \left(u - u_0\right) + p_0(v - v_0) - T_0(s - s_0) + V^2/2 + gz$$

as $u = u_0$ because the internal energy of an ideal gas is only a function of temperature, and $T = T_0$; both the kinetic and potential energy are 0 as this is a stationary system evaluated with reference to its dead state. Therefore,

$$e = p_0(v - v_0) - T_0(s - s_0)$$

$$e = (101.325 \text{ kPa})(0.01222 - 0.8445) \text{ m}^3/\text{kg} - 298.15 \text{ K}(-1.2155 \text{ kJ/kg}) = 278.1 \text{ kJ/kg}$$

With $E = 288$ MJ, the mass of air is $m = \dfrac{E}{e} = \dfrac{288 \times 10^3 \text{ kJ}}{278.1 \text{ kJ/kg}} = 1035 \text{ kg}$

So, the required volume of air is $V = mv = (1035 \text{ kg})(0.01222 \text{ m}^3/\text{kg}) = 12.7 \text{ m}^3$

Discussion: For a car to be propelled by compressed air, and for the compressed air to have the same energy density as that of a Tesla EV battery pack (80 kWh), a massive air storage volume of 12.7 m³ would be required. This is not practical for a passenger car. There is just no space to accommodate 203 cylinders of 200-mm diameter and 2-m length, which will give the required volume of 12.7 m³. To remedy this, the air could be compressed to a much higher pressure, but this is associated with increased explosion danger if such a car were to be involved in an accident. Alternatively, such compressed air vehicles could only be used over very short ranges and at low speeds, such as the MDI concept car with a range of around 120 km and a maximum speed of 80 km/h. (In comparison, a Tesla Model 3 electric vehicle has a range of 400 km and a top speed of 250 km/h.)

3.10 Exergy Balance of a Closed System

Combining Eqs. (3.51) and (3.10), the exergy balance for a closed system is found (Moran et al. 2018):

$$\underbrace{E_2 - E_1}_{\text{exergy change}} = \underbrace{\int_1^2 \left(1 - \frac{T_0}{T_b}\right) \delta Q}_{\substack{\text{exergy transfer} \\ \text{accompanying} \\ \text{transfer, } E_q}} - \underbrace{[W - p_0(V_2 - V_1)]}_{\substack{\text{exergy transfer} \\ \text{accompanying} \\ \text{transfer, } E_w}} - \underbrace{T_0\sigma}_{\substack{\text{exergy} \\ \text{destruction,} \\ E_d}} \tag{3.53}$$

with $\qquad E_d = T_0\sigma$ $\begin{cases} = 0 & \text{(no irreversibilities present within the system)} \\ > 0 & \text{(irreversibilities present within the system)} \\ < 0 & \text{(impossible)} \end{cases}$ $\qquad$ (3.54)

For a closed system (i.e., no change in volume) at a steady state $\left(\dfrac{dE}{dt} = 0\right)$, Eq. (3.53) reduces to:

$$\cancel{\frac{d E}{dt}} = \sum_j \left(1 - \frac{T_0}{T_j}\right) \dot{Q}_j - \left[\dot{W} - p_0\left(\cancel{\dot{V}_2} - \cancel{\dot{V}_1}\right)\right] - T_0\dot{\sigma} \qquad (3.55)$$

3.11 Exergy Balance of an Open System

For an open system, the exergy balance must account for energy transfer accompanying mass transfer:

$$\frac{dE}{dt} = \sum_j \left(1 - \frac{T_0}{T_j}\right) \dot{Q}_j - \dot{W} + \sum_i \dot{m}_i e_{fi} - \sum_e \dot{m}_i e_{fe} - T_0\dot{\sigma} \qquad (3.56)$$

where e_{fi} accounts for the exergy per unit of mass entering at inlet i, and e_{fe} accounts for the exergy per unit of mass exiting at exit e. These terms, known as *specific flow exergy*, are derived using energy and entropy balances, and take the form:

$$e_f = (h - h_0) - T_0(s - s_0) + \frac{V^2}{2} + gz \qquad (3.57)$$

where h_0 and s_0 are, respectively, the specific enthalpies and specific entropies evaluated at T_0 and p_0.

For a one-inlet, one-exit open system at steady state, Eqs. (3.56) and (3.57) reduce to:

$$\frac{dE}{dt} = 0 = \sum_j \left(1 - \frac{T_0}{T_j}\right) \dot{Q}_j - \dot{W} + \dot{m}\left(e_{f1} - e_{f2}\right) - T_0\dot{\sigma} \qquad (3.58)$$

where $\quad e_{f1} - e_{f2} = (h_1 - h_2) - T_0(s_1 - s_2) + \dfrac{V_1^2 - V_2^2}{2} + g(z_1 - z_2) \qquad (3.59)$

3.12 Chemical Exergy

Equations (3.51) and (3.52) relate to *physical exergy* (or *thermo-mechanical exergy*), which is relevant for a wide range of practical energy conversion systems that do not involve chemical reactions. That is, there is no difference in chemical composition between the system and the environment. In such cases, the *chemical exergy* is the same at all states of interest and therefore cancels out when differences in exergy values between states are evaluated (Bejan et al. 1996).

Chemical exergy is important in energy conversion systems that feature chemical combustion, such as the burning of natural gas. The chemical exergy is the maximum work that a system can transfer at the pressure and temperature of the reference environment ($p_0 = 1$ bar $= 101.325$ kPa and $T_0 = 25\,°C = 298.15$ K, respectively) as the system changes to one with the same chemical composition, pressure, and temperature of the reference environment (Dincer and Rosen 2013).

3.12.1 Standard Chemical Exergy

For pure gases undergoing isothermal processes at reference states p_0 and T_0, the *standard chemical exergy* per mole $\left(\overline{e}_0^{ch}\right)$, using Eq. (3.26), is (Dincer and Rosen 2013):

$$\overline{e}_0^{ch} = \overline{R}T_0 \ln\left|\frac{p_0}{p_i}\right| \tag{3.60}$$

where $\overline{R}$ is the molar universal gas constant (8.314 kJ/kmol K), $p_0 = 1$ bar $= 101.325$ kPa, and p_i denotes the partial pressure of the chemical species.

Example 3.7 Standard Chemical Exergy of Water Vapor at Reference States p_0 and T_0
Calculate the standard chemical exergy of water vapor.
  Given: We know that $p_0 = 101.325$ kPa and $T_0 = (25 + 273.15)$ K $= 298.15$ K.
  Required: We must determine the standard chemical exergy, $\overline{e}^{ch}$.
  Solution strategy: We will assume that air is an ideal gas and use Eq. (3.60) to calculate the chemical exergy.

  Analysis:
  The chemical exergy is: $\overline{e}_0^{ch} = \overline{R}T_0 \ln\left[\frac{p_0}{p_i}\right]$
with the molar universal gas constant $\overline{R} = 8.314$ kJ/kmol K. From thermodynamic property tables, we find that the partial pressure of water is 2.2 kPa. Therefore:

$$\overline{e}_0^{ch} = \overline{R}T_0 \ln\left[\frac{p_0}{p_i}\right] = (8.314 \text{ kJ/kmol K})(298.15 \text{ K}) \ln\left(\frac{101.325 \text{ kPa}}{2.2 \text{ kPa}}\right)$$
$$\overline{e}_0^{ch} = 9.5\,\text{kJ/mol}$$

  The standard chemical exergies $\left(\overline{e}_0^{ch}\right)$ are used in calculations involving a mixture of gases. Table 3.1 shows a few chemical exergies of the main constituents of air, which will be used throughout this book.

Table 3.1 Standard chemical exergies (at reference states $p_0 = 101.325$ kPa and $T_0 = 298.15$ K) of the main constituents of atmospheric air.

Constituent component	$\overline{e}_0^{ch}$ (kJ/mol)	Mole fraction (%)	Partial pressure (p_i) (kPa)
N_2 (g)	0.69	75.67	75.78
O_2 (g)	3.97	20.34	20.39
CO_2 (g)	19.48	0.04	0.00335
H_2O (g)	9.5	3.34	2.2

Source: Szargut (1989) and Bejan et al. (1996).

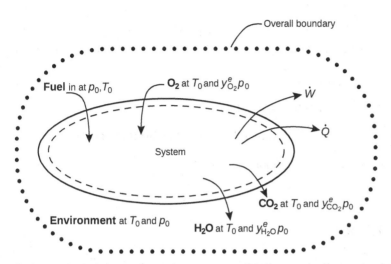

Figure 3.5 Conceptual system for evaluating the chemical exergy of fuel interacting with O_2.

3.12.2 Chemical Exergy of Gas Mixtures

The chemical exergy of a mixture of n gases, all of which are constituents of the atmosphere, can be deduced using a similar reasoning as used in Sections 3.9 and 3.10. Figure 3.5 illustrates the conceptual system for a hydrocarbon fuel combusting in oxygen at T_0 and p_0, and producing carbon dioxide, water, and heat as products. Assume that each of the n gases has a molar fraction of y_k and enters an isothermal chamber at T_0 with a partial pressure of $y_k^e p_0$. (The superscript e denotes the "environment" or ambient.) Each of the gases will then exit the isothermal chamber at T_0 with a partial pressure of y_k^e. The chemical exergy per mole of fuel $\left(\overline{e}^{ch}\right)$ is the maximum value of power that can be produced $\left(\dot{W}\right)$.

The molar chemical exergy per constituent at 101.325 kPa is (Dincer and Rosen 2013):

$$\overline{e}^{ch} = -\overline{R}T_0 \ln\left[\frac{y_k^e p_0}{p_0}\right] = -\overline{R}T_0 \ln y_k^e \tag{3.61}$$

The molar chemical exergy for all constituents is (Dincer and Rosen 2013):

$$\overline{e}^{ch} = -\overline{R}T_0 \sum y_k \ln\left(\frac{y_k^e}{y_k}\right)$$
$$\overline{e}^{ch} = \sum y_k \overline{e}_0^{ch} + \overline{R}T_0 \sum y_k \ln\left[y^k\right] \tag{3.62}$$

Example 3.8   Chemical Exergy of a Mixture of Water Vapor and Carbon Dioxide Vapor
Calculate the molar chemical exergy of a mixture of CO_2 vapor and H_2O vapor at reference states $p_0 = 1$ bar and $T_0 = 298$ K.

Given: Consider an isothermal chamber in which a fuel combusts and produces both CO_2 and H_2O in the vapor state.

Required: We must calculate the molar chemical exergy of the mixture of CO_2 and H_2O.

Solution strategy: We will use Eq. (3.62) to evaluate the molar chemical exergy.

$$\overline{e}^{ch} = \sum y_k \overline{e}_0^{ch} + \overline{R}T_0 \sum y_k \ln\left[y^k\right]$$

From Table 3.1, the standard chemical exergies $\left(\bar{e}_0^{ch}\right)$ for CO_2 and H_2O, respectively, are 19.48 kJ/mol and 9.5 kJ/mol. The mole fractions for CO_2 and H_2O, respectively, are 0.04% and 3.34%.

The chemical exergy of the mixture is then:

$$\bar{e}^{ch} = [0.04(19.48\,kJ/mol) + 0.0334(9.5\,kJ/mol)] +$$
$$\left(\frac{8.314\,kJ/kmol}{1000\,mol/kmol}\right)(298.15\,K)[0.04\ln(0.04\,kJ/mol) + 0.0334\ln(0.0334\,kJ/mol)]$$
$$\bar{e}^{ch} = 1.10\,kJ/mol + (-0.60) = 0.5\,kJ/mol$$

3.13 Exergetic Efficiency

To distinguish exergy-based and energy-based efficiencies, consider the steady-state system shown in Figure 3.6. The system represents a range of applications where the combustion of fuel provides heat, and where no work is being done. The system receives energy by heat transfer at the rate $\dot{Q}_s$ at the source temperature T_s, delivers $\dot{Q}_u$ at the use temperature T_u, and loses energy by heat transfer at the rate $\dot{Q}_\ell$ at temperature T_ℓ.

Applying closed-system *energy* and *exergy* rate balances at a steady state, from Eq. (3.56) (Moran et al. 2018):

$$\frac{dE}{dt} = 0 = \left[\left(1 - \frac{T_0}{T_s}\right)\dot{Q}_s - \left(1 - \frac{T_0}{T_u}\right)\dot{Q}_u - \left(1 - \frac{T_0}{T_\ell}\right)\dot{Q}_\ell\right] - T_0\dot{\sigma} \tag{3.63}$$

or alternatively:

$$\left(1 - \frac{T_0}{T_s}\right)\dot{Q}_s = \left(1 - \frac{T_0}{T_u}\right)\dot{Q}_u + \left(1 - \frac{T_0}{T_\ell}\right)\dot{Q}_\ell + T_0\dot{\sigma} \tag{3.64}$$

Equation (3.64) shows that the energy carried in by heat transfer $\dot{Q}_s$ is either used $\dot{Q}_u$ or lost $\dot{Q}_\ell$ by irreversibilities in the system. This can be described by an energy-based efficiency:

$$\eta = \frac{\dot{Q}_u}{\dot{Q}_s} \tag{3.65}$$

This definition of efficiency is commonly used to indicate the efficiency of energy conversion systems, as also used in Table 1.1.

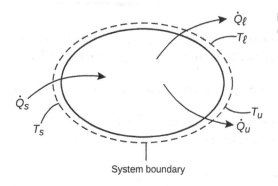

Figure 3.6 The system used to develop an expression for exergetic efficiency.

Equations (3.65) and (3.64) can be combined to provide an exergy-based efficiency (Moran et al. 2018):

$$\varepsilon = \frac{[1-(T_0/T_u)]\dot{Q}_u}{[1-(T_0/T_s)]\dot{Q}_s} = \eta\left[\frac{1-(T_0/T_u)}{1-(T_0/T_s)}\right] \tag{3.66}$$

This expression reveals that there are two ways to improve the exergetic efficiency: (a) increasing the energy conversion efficiency $\left(\eta = \dfrac{\dot{Q}_u}{\dot{Q}_s}\right)$, or (b) increasing the temperature of use (T_u) so that it better matches the source temperature (T_s).

The *higher* the temperature at which heat transfer occurs in cases where $T > T_0$, the more "valuable" the heat transfer and the greater the need to avoid direct heat transfer to the ambient environment, to cooling water, or to a refrigerated stream. The *lower* the temperature at which heat transfer occurs in cases where $T < T_0$, the more "valuable" the heat transfer and the greater the need to avoid direct heat transfer with the ambient environment or heated stream. In both cases, heat transfer across the T_0 boundary should be avoided (Bejan et al. 1996).

The following may be gleaned from Sections 3.9, 3.10, and 3.12:

- Mechanical work can be fully converted to heat, i.e., with 100% *energy* efficiency.
- Heat has a lower exergy than work. So, heat cannot be fully converted to work. Conversion of heat to work always has an *energy* efficiency less than 100%, but the conversion of heat to work can be achieved with 100% *exergy* efficiency.
- Exergy analysis helps to identify the upper limit for efficiency improvements.

Example 3.9 Exergy Change During an Expansion Process
Consider the steam turbine from Example 2.3. Steam enters the well-insulated industrial steam turbine at 4 MPa and 400 °C, and leaves the turbine at 75 kPa and 100 °C. The power output of the turbine is 2 MW and changes in kinetic and potential energy are negligible. Using an exergy analysis, determine the maximum work output in kJ/kg.

Given: We are given the entry and exit temperature and pressure of the steam. From Example 2.3, we also calculated the values of enthalpy at the turbine's inlet and exit.

Required: Using an exergy analysis, we must find the maximum work that this turbine can perform.

Solution strategy: As this is an open system, we will use Eq. (3.58) to calculate the exergy change through the turbine, which is equal to the maximum value of specific work output.

Analysis:
Maximum work output in kJ/kg (using an exergy analysis):

$$e_1 - e_2 = (h_1 - h_2) - T_0(s_1 - s_2) + \frac{V_1^2 - \cancel{V_2^2}}{2} + g(\cancel{z_1 - z_2})$$

From Example 2.3, we found the following:

$$h_1 = 3214.5 \text{ kJ/kg} \qquad h_2 = 2679.1 \text{ kJ/kg}$$

Using EES or the NIST Chemistry WebBook (https://webbook.nist.gov/chemistry), we can easily find s_1 and s_2:

$$s_1 = 6.7714 \text{ kJ/kg K} \qquad s_2 = 7.5282 \text{ kJ/kg K}$$

So,

$$e_1 - e_2 = (h_1 - h_2) - T_0(s_1 - s_2) = (3214.5 - 2679.1)\,\text{kJ/kg K} - 298.15\,\text{K}(6.7714 - 7.6953)\,\text{kJ/kg K}$$
$$= 810.86\,\text{kJ/kg}$$

Discussion: The exergy (available energy) of the steam is 810.86 kJ/kg. This is equal to the maximum work that the turbine can perform.

Reimagine Our Future

It appears most people would only practice energy efficiency (doing more with less) if everyone around them were doing the same. Is there a way around this obstacle?

3.14 Summary

The first law of thermodynamics applies to any cycle or process and states that heat and work are equivalent in all respects. However, we know from experience that, although work can be fully converted to heat (like when boiling water in an electric kettle), heat cannot be fully converted to work (like when using steam to produce work in a turbine). The second law of thermodynamics is therefore necessary as it explains that heat and work are not equivalent in all respects. The second law of thermodynamics also explains what the first law of thermodynamics does not: how the capacity for work is inevitably lost to heat, the most useless form of energy. Moreover, the second law of thermodynamics states that it is impossible to convert all heat into work – something is always lost in energy conversion.

When working with the second law of thermodynamics, a process or cycle can only be termed reversible (or ideal) in the absence of friction and if all process steps take place in a slow, quasi-isothermal manner. In such a reversible (or ideal) process, there would be no degradation of energy. On the other hand, if these conditions are not met, some inherent capacity to produce work in the system or its surroundings is permanently lost (Rogers and Mayhew 1967). According to the second law of thermodynamics, entropy must remain constant or increase for any isolated system. Entropy increases if the process is irreversible, as all real processes are. The degradation of energy therefore manifests as an increase in entropy and destruction of exergy (or destruction of available work).

We have also seen that electricity is a higher-quality form of energy than low-temperature heat. Exergy measures the quality of energy and informs us how far a system is from equilibrium and what the potential is of doing something useful with that energy. While exergy is about the potential to do something with energy, entropy tells us what happened to energy (Schneider and Sagan 2004).

Problems

3.1 A pump circulates water in a steam power plant. The water at the suction side of the pump is in a saturated liquid state at 45.81 °C. The enthalpy of the water here is 191.83 kJ/kg and the entropy is 0.6493 kJ/kg K. The water at the exit of the pump is in compressed liquid form at 45.81 °C. The enthalpy of the water here is 197.83 kJ/kg and the entropy is 0.6493 kJ/kg K. The ambient air is at 25 °C. Calculate the exergy destruction of this pumping process. Comment on your answer.

[Answer: 0 kJ/kg]

3.2 A boiler produces steam at 6 MPa. The gas furnace that feeds the boiler is operating at 1400 K. Compressed liquid water enters the boiler at 45.81 °C with enthalpy of 197.83 kJ/kg and entropy of 0.6493 kJ/kg K. Steam exits the boiler in the superheated state at 600 °C with enthalpy of 3658.4 kJ/kg and entropy of 7.1677 kJ/kg K. Calculate the exergy destruction of this boiling process. Comment on your answer.

[Answer: 1205.9 kJ/kg]

3.3 An adiabatic steam turbine receives superheated steam at 500 °C and 500 kPa. The steam has enthalpy of 3483.9 kJ/kg and entropy of 8.0873 kJ/kg K. The turbine does work on an electrical generator, after which the steam exits the turbine at 45.81 °C with enthalpy of 2565.5 kJ/kg and entropy of 8.0873 kJ/kg K. The ambient air is at 25 °C. Calculate the exergy destruction of this steam expansion process. Comment on your answer.

[Answer: 0 kJ/kg]

3.4 A steam condenser in a large power plant receives steam at 45.81 °C from a low-pressure turbine. The steam entering the condenser has enthalpy of 2565.5 kJ/kg and entropy of 8.0873 kJ/kg K. The steam condenses to saturated liquid water that exits the condenser at 45.81 °C with enthalpy of 191.83 kJ/kg and entropy of 0.6493 kJ/kg K. The ambient air is at 25 °C. Calculate the exergy destruction (or "irreversibility") of this condensing process. Comment on your answer.

[Answer: 0.92 kJ/kg]

References

Bejan, A., Tsatsaronis, G. and Moran, M. (1996). *Thermal Design and Optimization*. New York, NY: Wiley.

Black, W.Z. and Hartley, J.G. (1985). *Thermodynamics*. London: Harper & Row.

Borel, L. and Favrat, D. (2010). *Thermodynamics and Energy Systems Analysis. From Energy to Exergy*. Lausanne: EPFL Press/CRC Press.

Dincer, I. and Rosen, M.A. (2013). *Exergy: Energy, Environment, and Sustainable Development*. Elsevier.

Gerhart, A.L., Hochstein, J.I. and Gerhart, P.M. (2020). *Munson, Young and Okiishi's Fundamentals of Fluid Mechanics*. New York, NY: Wiley.

Moran, M.J., Shapiro, H.N., Boettner, D.D. et al. (2018). *Fundamentals of Engineering Thermodynamics*, 9th edition. New York, NY: Wiley.

Obert, E.F. (1960). *Concepts of Thermodynamics*. New York, NY: McGraw-Hill.

Rogers, G.F.C. and Mayhew, Y.R. (1967). *Engineering Thermodynamics – Work and Heat Transfer*. London: Longmans.

Schneider, E.D. and Sagan, D. (2004). *Into the Cool. Energy Flow, Thermodynamics, and Life*. Chicago, IL: University of Chicago Press.

Szargut, J. (1989). Chemical exergies of the elements. *Applied Energy*, 32(4): 269–286.

Mini Project 1

Energy and Sustainability

1) *Sustainable energy* may be defisned as a dynamic harmony between the equitable availability of energy-intensive goods and services to all people and the preservation of the Earth for future generations. Using this definition, discuss whether you think global energy use is sustainable. In your discussion, you should interpret at least five peer-reviewed journal papers from the last five years.

2) Do some research on the United Nations' Sustainable Development Goals (SDGs). You could start by studying the following material (some of the websites might have changed, in which case you could find your own sites that pertain to the SDGs):

"The 17 Goals", https://sdgs.un.org/goals

"We The People for The Global Goals", https://www.youtube.com/watch?v=RpqVmvMCmp0

"The Lazy Person's Guide to Saving the World", https://www.un.org/sustainabledevelopment/takeaction/

"SDG Academy", https://sdgacademy.org/

  i) What are the SDGs and why are they necessary (or not)?

  ii) Are the SDGs achievable by 2030? Why/Why not?

  iii) Briefly explain some important respects in which the 17 SDGs are interconnected. (Actions that are intended to promote one SDG may benefit or hinder progress towards other SDGs.) Please also use a clarifying diagram/s to illustrate typical interconnections. You might want to use online tools and articles as you prepare your answer, for example:

- Nilsson, M., Griggs, D. and Visbeck, M. (2016). Policy: Map the interactions between Sustainable Development Goals. *Nature* 534, pp. 320–322.
- https://www.nature.com/news/policy-map-the-interactions-between-sustainable-development-goals-1.20075
- "SDG Interlinkages Analysis & Visualisation Tool": https://sdginterlinkages.iges.jp/visualisationtool.html

3) List the various SDG targets related to energy (production, storage, distribution and management, and consumption).

4) How should policymakers weigh the cost of future damage from human-induced (or "anthropogenic") climate change against damages from current problems such as public health and safety?

5) If you were part of a start-up company, what would you do to ensure that your company's entrepreneurial activities were aligned with sustainable energy principles? How would you "sell" a business concept based on sustainable energy to venture capitalists? What would you show prospective investors to convince them that energy sustainability makes business sense apart from it being the right thing to do?

6) Summarize your findings pertaining to net-zero emissions in a high-impact manner using an online platform of your choice, for example:
   - Interactive infographic
   - ePortfolio (e.g., Wix, Reddit, Tumblr, Issuu, etc.)
   - Social media (e.g., Facebook, Instagram, Twitter/X)
   - Blogsite (e.g., Wix, Squarespace, WordPress)

Image of the future, from 1967
Panels from Athelstan Spilhaus's syndicated non-fiction comic strip, "Our New Age," which ran from 1958 to 1975 and raised awareness about environmental catastrophes. *Source:* The New York Times / Public domain.

Week 2 – Energy, Society, and Environment

Human activities currently result in about 35 billion tons of carbon dioxide being released into the atmosphere every year, with undisputed adverse effects on the Earth and its inhabitants. To effectively address this urgent problem, it is important to understand how the general principles of energy conversion apply in specific and practical energy-related contexts.

The increasing pace and intensity of global climate change is disturbing, as is the potential for unprecedented destructive impacts. Technology, infrastructure, global interconnectedness, the explosion of specialist knowledge, and the willingness of vast numbers of specialists and ordinary citizens in every country to make a difference will contribute to solving this problem. It will also enable us to explore bold and innovative solutions for climate action, while providing a basis for hope for a more sustainable future.

Energy Systems: A Project-Based Approach to Sustainability Thinking for Energy Conversion Systems,
First Edition. Leon Liebenberg.
© 2024 John Wiley & Sons, Inc. Published 2024 by John Wiley & Sons, Inc.
Companion website: www.wiley.com/go/liebenberg/energy_systems

Central to the idea of sustainability is clean, sustainable energy production. *Sustainable energy* may be defined as a dynamic harmony between the equitable availability of energy-intensive goods and services to all people, and the preservation of the Earth for future generations.

We are fortunate that the steps that are necessary to achieve this ideal of sustainable energy, and which have been deemed politically impossible, are quickly becoming technologically feasible and even politically inevitable. However, in addition to political measures, the practice of sound economic principles is vital to ensuring sustainable energy.

Chapter 4 – Energy usage and society: Energy demand is shifting to lower-carbon fuels; energy supply is evolving to meet demand projections; social transformation and energy transitions; images of the future; outlook

Chapter 5 – Energy usage and the environment: Our planet is rapidly warming due to the accumulation of greenhouse gases; the enhanced greenhouse effect; radiative forcing; surface warming due to cumulative emissions of CO_2; climate modeling and simulation; catastrophism versus techno-optimism; outlook

Chapter 6 – Interpretations of sustainability: Human Development Index (HDI), United Nations' Sustainable Development Goals (SDGs); the social, technological, environmental, economic, and political (STEEP) analysis of energy systems; the "doughnut" of social and planetary boundaries; Kaya identity; lifecycle analysis and embodied energy; economic metrics; outlook

Mini Project 2 – Energy scenarios for a sustainable world

4

Energy Usage and Society

Society's progress is intrinsically related to access to safe, reliable, and affordable energy. Economies, ecosystems, and chemical reactions arrange themselves around energy gradients. Those
energy gradients might emphasize differences in temperature, pressure, and chemical concentration that set up conditions for energy flow, and for producing work or movement (Schneider and
Sagan 2005).

As Figure 4.1 shows, there is a strong correlation between a country's gross domestic product
per person (also called "GDP per capita") and its energy consumption per person. Countries
also go through energy transitions as they develop. Figure 4.1 shows that as the poorest
countries (annual income <US$5000 per capita) begin to develop, they do not tend to use
significantly more energy. Demand only tends to accelerate once the average annual income
reaches around US$5000 per person (at purchasing power parity [PPP]). At around US
$15,000 per person, demand growth eases, as some uses (like domestic heating or cooling)
approach saturation and the economy diversifies from industrial to service sector activity
(Semieniuk et al. 2021; Shell 2022).

Reimagine Our Future

We want to end unsustainable energy production and use. And we want to end poor social
conditions. But will improvements in the social conditions of poor people cause their energy
consumption to increase and thereby contribute to unsustainability? Or might improvements
in social conditions actually result in reduced energy demand? Share your thoughts on social
media.

4.1 Energy Demand Is Shifting to Lower-Carbon Fuels

The global flow of energy begins with the supply of primary energy. That primary energy (oil, gas,
coal, nuclear energy, renewables, etc.) is converted into useful forms of heat and work (electricity)
using an array of devices and systems. The primary energy demanded by users depends on aspects
such as the service level, building occupant behavior, car driver behavior, and the characteristics of
the *passive* energy conversion device or system, such as the aerodynamic efficiency of a car or

Energy Systems: A Project-Based Approach to Sustainability Thinking for Energy Conversion Systems,
First Edition. Leon Liebenberg.
© 2024 John Wiley & Sons, Inc. Published 2024 by John Wiley & Sons, Inc.
Companion website: www.wiley.com/go/liebenberg/energy_systems

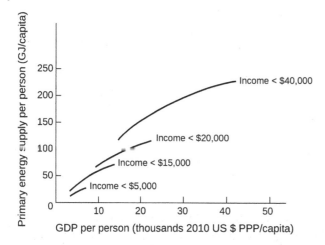

Figure 4.1 Annual gross domestic product (GDP) per capita in 2010 US Dollars (thousands) at purchasing power parity (PPP). *Source:* Adapted from Shell (2021) and Zhaoyuan and Ishwaran (2020).

insulation of a building. Energy demand is also determined by the primary-to-useful energy conversion efficiencies (such as those shown in Table 1.1).

In 2021, global *primary energy demand* was about 595 EJ (or 595×10^{18} J). This number could reach 712 EJ by 2040 (ExxonMobil 2019; BP 2022). However, if we implement demand-side management strategies, most notably through improved energy efficiency in industry, buildings, and transportation, we could *reduce* annual energy consumption to around 350 EJ by 2040 (Cullen and Allwood 2010). Every chapter in this book investigates ways in which better energy efficiency can be attained.

The global specific energy consumption in 2021 was 76 GJ/capita, and in the USA that figure was 280 GJ/capita (BP 2022). The USA consumes about 20% of the world's primary energy, despite only making up about 4% of the world's population.

Several *projections* estimate the world's energy demand over the next few decades. These projections are based on likely trends in technology, policy, consumer preferences, geopolitics, and economic development. Although these trends may vary over time, they help to evaluate society's progress toward meeting net zero objectives before 2050. These projections, such as the one by ExxonMobil (2019) shown in Figure 4.2, typically reveal the following:

- Oil will continue to play a leading role in the world's energy mix, with growing demand driven by commercial transportation and feedstocks for the chemicals industry.
- The demand for natural gas will grow the most of all energy types, reaching a quarter of all demand.
- Renewables and nuclear will see strong growth, contributing more than 40% of energy supplies to meet demand growth.
- Coal use will remain significant in parts of the developing world but will drop below 20% of the global share as China and the nations in the Organization for Economic Cooperation and Development (OECD) transition to lower-carbon sources like renewables, nuclear and natural gas.
- Electricity, an energy carrier and not an energy source, will grow three times faster than overall energy demand. This is supported by other projections, including those of the International Energy Agency (IEA 2020).

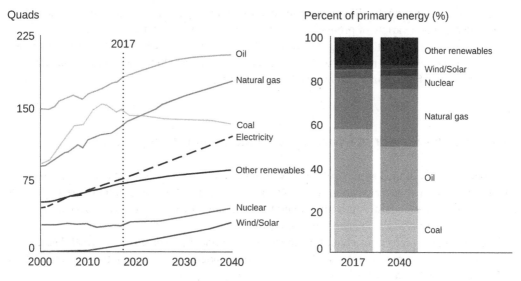

Figure 4.2 Projected global energy demand between 2017 and 2040. 1 quadrillion Btu = 1 quad; 1 quad = 1.06 EJ. *Source:* Adapted from ExxonMobil (2019).

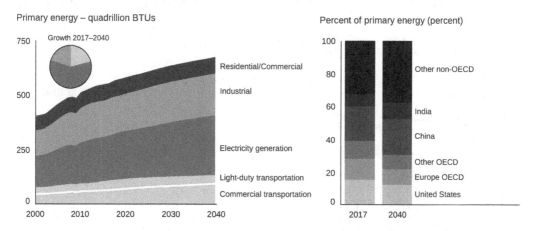

Figure 4.3 Projected global energy demand by sector between 2017 and 2040. 1 quadrillion Btu = 1 quad; 1 quad = 1.06 EJ. *Source:* Adapted from ExxonMobil (2019).

The global energy demand is also projected to vary by economic sector, as illustrated in Figure 4.3. In particular, the following well-accepted trends are expected (ExxonMobil 2019):

- Global energy demand will reach 675 quads or about 712 EJ in 2040, up 20% from 2017, reflecting a growing population and rising prosperity.
- Residential and commercial energy demand will grow continuously through 2040 as the energy needs of a growing population exceed efficiency improvements.
- Electricity generation will be the largest and fastest-growing sector, primarily reflecting expanding access to reliable electricity in developing countries.

- Industrial sector growth will support the construction of buildings and infrastructure, and the manufacture of a variety of products to meet the needs of the world's population.
- Commercial transportation will grow with expanding economies, which will increase the movement of goods. Personal mobility will also expand, but efficiency improvements and more electric vehicles will offset the increase in vehicle miles traveled.

4.2   Energy Supply Is Evolving to Meet Demand Projections

Fossil fuels are in limited supply. We will eventually run out of this precious resource. The same applies to fuel for nuclear power plants. "Renewable energy," however, as its name suggests, is available in endless quantities, thanks mainly to the sun. We can therefore expect much more use to be made of renewable energy in the future.

There are many projections about the global energy supply between 2017 and 2040. The ones by ExxonMobil (2019) are well accepted by analysts and academics. Their projections reveal the following (see also Figure 4.4):

- Technological improvements will help achieve more efficient fuel use and lower emissions intensity across all sources of supply.
- Oil will remain the primary energy source, essential for commercial transportation and chemicals.

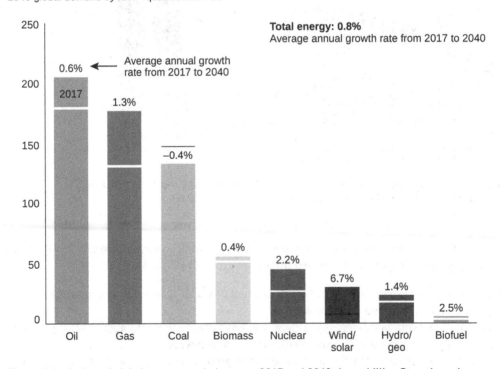

Figure 4.4   Projected global energy supply between 2017 and 2040. 1 quadrillion Btu = 1 quad; 1 quad = 1.06 EJ. *Source:* Adapted from ExxonMobil (2019).

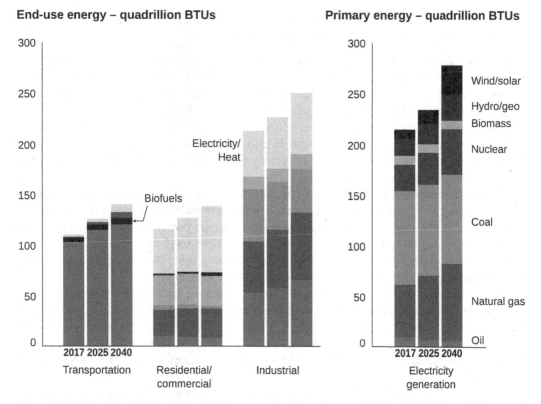

Figure 4.5 Projected global energy supply by economic sector between 2017 and 2040. 1 quadrillion Btu = 1 quad; 1 quad = 1.06 EJ. *Source:* Adapted from ExxonMobil (2019).

- The demand for natural gas will rise the most, largely to help meet increasing needs for electricity and lower-carbon industrial heat.
- Lower-carbon energy sources, including wind, solar, biofuels, and nuclear, will increase at the fastest pace.

Projections of global energy supplies show that the energy mix used to meet rising demand – while also addressing environmental impacts, including the risks of climate change – will vary by economic sector. Figure 4.5 illustrates that, between 2017 and 2040, the following trends can be expected (ExxonMobil 2019):

- Oil will remain essential for transportation, and growing commercial transportation will still rely on liquid fuels to meet more than 90% of demand.
- With a drive for cleaner and more efficient operations, the industrial sector will rely primarily on electricity and natural gas for growth; industrial demand for oil as a feedstock for chemicals, asphalt, lubricants, and other specialty products will grow.
- Electricity demand will rise in all end-use sectors, but the mix of fuel supply for electricity generation will shift to lower-carbon sources.

Experimenting, Not Just Dreaming About Sustainability Ideas

Plastics are found in many parts of the modern energy system, including solar panels, wind turbine blades, batteries, thermal insulation for buildings and electric vehicle parts. The global production of plastic is currently around 380 million tons per year. That number will increase to 1100 million tons per year by 2050 and account for about 15% of greenhouse-gas emissions.

Paul Cheek (30), a serial tech entrepreneur, entrepreneurship educator and software engineer, found himself pondering about what we can do to use less plastic, or to recycle more effectively. This led him to come up with a unique business-driven solution, marketed by the company Oceanworks, which he co-founded.

Oceanworks provides a global platform to trade in recycled plastic. The company purchases plastic recovered from the ocean, which is then resold to manufacturers to produce more sustainable products. Customers typically buy several shipping containers full of plastic every month under long-term contracts. The company has obtained US$3 million of funding to date with revenues in the order of US$15 million. In addition to reducing virgin plastic requirements, it is saving a lot of energy and reducing greenhouse-gas emissions.

Paul explains that the plastic produced today is derived mainly from petrochemicals, which are set to account for over a third of the growth in oil demand up to 2030, and nearly half up to 2050, ahead of trucks, aviation, and shipping. Using plastic also results in the use of a lot of energy. "Petrochemicals are poised to consume an additional 56 billion cubic meters of natural gas by 2030, equivalent to about half of Canada's total gas consumption in 2022."

In the past few years, Oceanworks has diverted thousands of tons of plastic from the ocean while serving hundreds of corporate customers in over 30 countries. "This has enabled the launch of a variety of high-profile sustainable products, such as Clorox ocean plastic trash bags, Sperry ocean plastic boat shoes and YKK ocean plastic zippers," says Paul.

Paul was recognized by Forbes 30 Under 30 in 2021. In this definitive list of young people changing the world, he was identified as 1 of 10 standouts who hold the future of clean tech in their hands.

Paul Cheek, co-founder and Chief Technical Officer of Oceanworks. Paul also teaches entrepreneurship at the MIT Sloan School of Management. He is especially passionate about experimental entrepreneurship, getting students to engage in real-world concept testing, market testing, pricing and business model development. *Source:* Paul Cheek, https://oceanworks.co.

4.3 Social Transformation and Energy Transitions

The social transformations that are especially relevant to energy transitions include the development of new systems of human-environmental interaction. Such changes occur when social, technological, economic, or political (governance) structures make the continuation of the existing system untenable. They involve multiple elements, including values and beliefs, practices, behaviors, incentives, institutions, and world views (Folke et al. 2021). We need to imagine the sustainable energy systems we wish to have and reflect on the various intertwined social-ecological interactions of technological change and social innovation.

4.3.1 Technological Aspects of Energy Transitions

We adore technology. It is certainly necessary for human betterment and has yielded huge benefits (Kelly 2011). But technological change alone will not lead to the needed transformations toward energy sustainability. The great power of technology has contributed to our acceptance of the idea of unlimited growth. Regrettably, technology has not helped us to align with nature's regenerative capacity (Arthur 2009).

Technologies can have beneficial and adverse impacts simultaneously. For example, electric vehicles whose batteries are charged from solar-derived electricity have zero local emissions and therefore promote well-being and a clean environment. On the other hand, the embodied energy in producing such an electric vehicle is accompanied by non-negligible CO_2 and other emissions, and lithium-ion batteries contain materials such as lithium, cobalt, manganese, nickel, and graphite (Section 27.3). The mining and processing of these materials impact negatively on local societies and ecosystems.

Indeed, the use of renewable technologies, including photovoltaic cells, wind turbines, "green" hydrogen fuel cells, or geothermal heating, does not produce greenhouse gases (like CO_2), but certainly emits greenhouse gases during their construction, maintenance, and decommissioning (see Chapters 20 and 24). The deployment of such technologies in a social-ecological context must be accompanied by suitable lifecycle analyses to evaluate their benefits and costs.

As sustainability pressures mount, so does the hope that artificial intelligence (AI) technologies, in combination with advanced sensors and robotics (including drones), will enhance society's capacity to detect, adapt, and effectively respond to climate and environmental change (Folke et al. 2021). To balance this view, one must of course also consider the apparent existential threat of AI (EmTech 2023).

Reimagine Our Future

Artificial intelligence (AI) has the potential of outstripping human intelligence by several orders of magnitude. Generative AI may also eventually not be well disposed to humans and develop consciousness beyond our understanding and control. This technology might not bode well for humans or the planet. Before we reach an irreversible situation, we would do well to ask if this is where we want to go. Is AI really needed and how might it change the way we supply and use energy? Discuss with your friends.

4.3.2 Techno-Economic Aspects of Energy Transitions

Focusing on techno-economic perspectives emphasizes the importance of physical energy flows, and matching energy conversion processes with energy supply, on the one hand, and the demand for energy, on the other, to run the economy and live our lives. The concept of a supply–demand balance means that a change in any type of energy supply or use must be balanced by corresponding changes in other types of supply or use. For example, the expansion of electricity generation from renewable sources must be accompanied by the increasing use of electricity, the phasing out of conventional sources, or increasing electricity exports. In the current economic paradigm of competitive markets, energy supply and use are in equilibrium if consumers are not prepared to pay a different price for energy, or if producers are not prepared to supply it at a different cost (Cherp et al. 2018).

Similarly, resource depletion leads to increasing extraction costs and may thus prompt shifts to other resources, more efficient equipment, or reduced consumption. Population growth often leads to increased demand and may trigger different supply options (Cherp et al. 2018).

A techno-economic perspective views technology as a method of extracting, converting, or using energy by means of equipment or infrastructure. A techno-economic focus therefore usually fails to provide an adequate account of inertia and innovation in energy technologies. Although the patterns of evolution of energy demand and use, infrastructure, and technologies can be derived from historical observations, such observations are rarely good predictors of the future. These limitations exist because technological innovation and diffusion, as well as policies, originate in systems that differ from energy-economy systems, which are the focus of the techno-economic perspective (Cherp et al. 2018).

4.3.3 Socio-Technical Aspects of Energy Transitions

Social scientists work hard to explain how technological innovations spread in society. Such socio-technological perspectives have, for instance, discovered that the adoption of a technology typically follows an S-curve, featuring initial development and rapid upscaling, followed by a plateau (Figure 4.6).

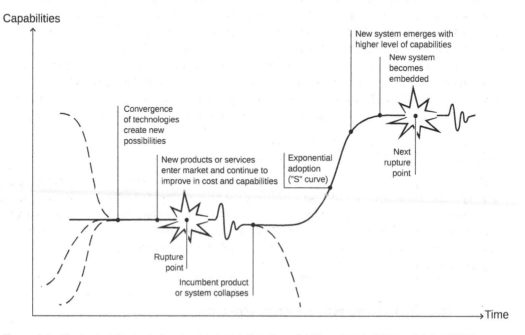

Figure 4.6 The typical S-curve of technology adoption. *Source:* Adapted from Arbib and Seba (2020).

In contrast to a techno-economic perspective, where technology is simply a method of extracting, converting, or using energy by means of equipment or infrastructure, the socio-technical perspective has a more complex and nuanced view of technology as a social phenomenon. The socio-technical perspective also considers the knowledge and practices embedded in infrastructure and other technical artifacts, shared by human actors, and circulating in social networks, collectively known as technological or socio-technical systems (Cherp et al. 2018).

Novel technologies that cannot initially compete within dominant regimes (e.g., because they are too costly or too complicated) emerge in protected niches, where they may mature and become competitive. The strategic niche management (SNM) approach emphasizes the need to foster such niches to facilitate technological innovation (Kemp et al. 1998).

An influential framework in socio-technical transition analysis, the multi-level perspective (MLP), points out that, because of regime resilience, niches do not automatically displace incumbent regimes, even when they become more effective in fulfilling a relevant social function. For a niche to replace an incumbent regime, the regime must first be destabilized, for example by external (landscape) pressures. Regime destabilization can occur along several distinct pathways, most of which represent nonlinear rapid change (Geels 2002).

Socio-technical perspectives traditionally have an excessive focus on novelty, which can overlook changes in the deployment of existing technologies that require only incremental innovation, e.g., natural gas replacing coal in steam power plants.

4.3.4 Political Aspects of Energy Transitions

Political perspectives on energy transitions focus on a change in policies that affect energy systems. Policy change is studied within several domains of political science. Because most energy policies are adopted and implemented by governments acting on behalf of nation states, the state is the main unit of analysis from a political perspective (Cherp et al. 2018).

Changes in global energy systems, in accord with sustainable development, will require the alignment of policy and regulatory actions, as well as other incentives. For new technologies to be accepted, they must be attractive (in terms of cost, convenience, performance, and more) to investors, purchasers, and users. Regulations and standards that target performance characteristics can help spur technological development and improve market attractiveness (Bierbaum and Matson 2013).

Techno-economic and political entities and processes are neither independent of nor subsumed into socio-technical systems. Rather, they make up semi-autonomous systems with their own dynamics, which co-evolve along with socio-technical systems. All three perspectives are therefore needed to explain national energy transitions (Cherp et al. 2018). Salient aspects of these three perspectives are summarized in Table 4.1.

4.3.5 Cultural Aspects of Energy Transitions

Culture may be defined as the norms, practices, and material artifacts in a society. Culture can operate at any scale, from individuals to groups (Stephenson et al. 2015). Using this definition, it is possible to evaluate how culture influences the transformation to low-carbon or no-carbon energies.

Table 4.1 Salient aspects of the techno-economic, socio-economic, and political perspectives of energy transitions.

Techno-economic perspective	Socio-technical perspective	Political perspective
Resources • Fossil fuel types, resources, reserves, extraction costs • Import and export of fuels and carriers • Type and potential of renewable resources • Cost of relevant technologies	**Innovation systems** • Presence and structure of national, sectoral, and technological innovation systems • Performance of innovation systems with respect to their functions, e.g., research and development activities, knowledge stock	**Goals of the state** • Type of state goals, e.g., energy security, access to modern energy, climate change mitigation, technological leadership • Factors affecting state goals, e.g., import dependence, international competition
Demand • Types and scale of energy uses • Energy intensity • Factors driving demand growth and decline, e.g., population and economic growth/decline; industrial restructuring	**Regimes and niches** • Structure, resources, and coordination of incumbent regimes • Structure and resources of newcomers' niches • Niche-regime interaction, including external support mechanisms	**Political interests** • Special interests (e.g., industrial lobbies) • Party ideologies and organized social movements • Voters' preferences
Infrastructure • Existing infrastructure for extraction, transportation, conversion, and use • Age of infrastructure • Manufacturing, import and export of equipment • Cost of operation and construction of infrastructure	**Technological diffusion** • Global maturity of relevant energy technologies • Location on core/periphery of technology • Possibilities for technology export	**Institutions and capacities** • State capacity, e.g., economic and other resources, political stability • Institutional arrangements, e.g., varieties of capitalism, party system, government system • International processes, e.g., policy diffusion, international agreements

Source: Adapted from Cherp et al. (2018).

Table 4.2 summarizes the cultural aspects that influence the adoption of three low-carbon transitional technologies and practices: ride-sharing, using automated vehicles, and whole-house retrofits. Ride-sharing is enabled by the proliferation of internet-connected mobile devices, but still resides in the technologically incremental category, as it involves cars. Automated vehicles feature artificial intelligence, advanced sensors, and robotics. The use of such high-technology vehicles demands a radical change in practices. Whole-house retrofits focus on redesigning homes to bring about reductions in energy demand, using high-efficiency insulation, heat pumps, and advanced ventilation, energy-efficient lighting, energy-efficient appliances, and obtaining electricity from renewable energy sources such as wind or solar.

Table 4.2 shows material artifacts associated with these cases, as well as the norms, values, and practices that may help or hinder their adoption. It follows from Table 4.2 that culture plays a key role in the adoption of changes in technology or practice, whether incremental or radical (Sovacool and Griffiths 2020).

Table 4.2 Culture as material artifacts, norms, and values, and practices in low-carbon transitions.

Cases	Ride-sharing	Automated vehicles	Whole-house retrofits
Technological change	Incremental	Radical	Radical
Behavioral change	Substantial	Substantial	Modest
Material artifacts	Smart phones, ride-hailing apps, and electric vehicles	Automated vehicles, electric vehicles, vehicle-to-vehicle (V2V) and vehicle-to-infrastructure (V2I) communication infrastructure	Homes, advanced insulation, heat pumps, ventilation with indoor air quality control, energy-efficient lighting, energy-efficient appliances, high-efficiency windows, integration with renewable energy (such as solar panels and energy storage), rainwater storage
Positive norms and values	Extraversion, sustainability, and dematerialization, mobility justice, religious restrictions, and aversions to driving	Enthusiasm for new technologies, safety, mobility justice, sustainability, and risk aversion	Collectivism and communal well-being, and responsibility
Negative norms and values	Introversion, personal safety, negative perception of professional drivers, convenience, independence, social status from car ownership, social awkwardness, and social exclusion	Aversion to new technologies, sensation seeking, and risk seeking from driving	Individualism and selfishness, convenience, conspicuous consumption, and adopting inefficiency as a symbol of social class
Positive practices	Sharing of personal space and belongings, and living in an urban environment	Safe and considerate driving styles, operating a car for utility, selecting vehicles based on safety, fuel economy, or sustainability	Attaching religious or spiritual significance to retrofits, and valuing the esthetics and style of energy efficiency
Negative practices	Driving as the default form of mobility, living in a rural environment, and communicating without a smartphone	Aggressive driving styles, operating a car for pleasure, selecting vehicles based on acceleration or top speed, and a threat to cyclists (given problems in detection)	Seeking to protect the heritage of inefficient buildings, and abandoning retrofits because they interfere with household lifestyles

Source: Adapted from Sovacool and Griffiths (2020).

Box 4.1 Equinox House, Urbana, Illinois

Equinox House is a 100% solar-powered home, built and owned by Professor Ty Newell and his family (Figure 4.7). This 2100 ft^2, four-bedroom, net-zero energy consumption house features automated fresh air control with distributed wireless CO_2 sensors, zoned air supply, and return dampers to ensure year-round excellent air quality (Build Equinox 2022).

The house further features energy-efficient appliances, a high-efficiency heat pump, energy-efficient LED lighting, and triple-paned windows. The windows are oriented to allow direct sunlight into the living space for the heat it provides during the cooler half of the year – beginning on the Fall equinox – and to exclude direct sunlight during the warmer half of the year – beginning on the Spring equinox – when it would increase the load on the cooling system. Equinox House was the first home in an Illinois municipality to be permitted to use rainwater harvesting.

The 8.2-kW photovoltaic array supplies 8000 kWh a year for use in the house, and 2000 kWh a year for charging the household's two electric vehicles, which each drive 8000 miles a year. The walls and roof of this beautiful house feature 12-in. (300-mm) thick structural insulation material with an equivalent R-value of 44, making this material more than four times more effective at preventing thermal transfer than the walls of a conventional house. The white north-facing roof reflects sunlight onto the photovoltaic array, while the white south-facing roof reflects sunlight in summer. The roof has an overhang that provides protection from the sun, but allows plentiful light to enter the skylight windows, making for comfortable indoor lighting all year round.

The home is characterized as a *net-zero* home. The table below compares the costs of this net-zero house with those of an equivalently sized conventional home without photovoltaic panels, a smart ventilation system, or insulation. The comparison is based on a 4.5% loan taken over 30 years, with no escalation in electrical utility costs (around 10 ¢/kWh). There is an insignificant

Figure 4.7 Equinox House (built in 2010), located in Urbana, Illinois. *Source:* Lyanne Alfaro / With permission of University of Illinois at Urbana-Champaign.

difference in monthly costs between the net-zero home and a conventional home. Once the mortgage has been repaid, the net-zero home features monthly savings of US$160 which increase with an escalation in utility costs.

	Equinox House "net-zero" home, cost	Conventional home, cost
House cost, US$	240,000	200,000
20% down-payment, US$	48,000	40,000
Mortgage, US$	192,000	160,000
Monthly bank payment, US$	973	811
Monthly utility cost, US$	0	160
Total monthly cost, US$	**973**	**971**

4.3.6 Social Media Aspects of Energy Transitions

The media (newspapers, television, and radio) play an important role in producing, constructing, and distributing representations of climate change and sustainability, especially in the journalistic mass media, but also in politics, science, and civil society. The participatory nature of social media plays a pivotal role in shaping individual attitudes, feelings, and behavior (Lazer et al. 2018). Social media changes our perceptions of the world and can generate large social mobilization and protests. The effective use of social media can have a beneficial effect on society's response to sustainability challenges.

To be sure, catastrophic climate scenarios and other dire warnings about the future may overwhelm some people. They can cause despair and inaction so that people continue with their unsustainable lifestyles, because it seems to them that any action they take will have little impact (De Tocqueville 1945; Leonard 2010). Dire warnings about our future should therefore be accompanied by feasible perspectives for action (Weber 2015).

Social media can raise awareness and elicit action (or non-action), in some cases shifting social norms and triggering reforms toward sustainability (Folke et al. 2021). Crowd-sourced fact-checking, when combined with computer-assisted analysis, may help curb disinformation or political bias in the quest for sustainability.

Passionate individuals using social media can have access to huge numbers of people and accomplish previously unthinkable things at a speed and on a scale previously unimaginable, for good or ill.

4.3.7 Social Innovation Aspects of Energy Transitions

Transformations toward sustainable energy cannot be achieved by adaptation alone, and certainly not only by incremental change; rather, what is needed is fundamental structural, systemic, and enabling transformations (Table 4.3). Such approaches are not necessarily mutually exclusive. In fact, they usually offer complementary analytical lenses on transformative change, as well as complementary approaches to understanding and trying to bring about real-world change.

Table 4.3 Structural transformation, systemic transformation, and enabling transformation.

Approach	Definition/emphasis	Advantages	Disadvantages	Example
Structural	Fundamental changes in the way societies govern, organize and practice, production, and consumption	Highlights the prevalent economic and political processes and associated interests that perpetuate current conditions	Lack of emphasis on environmental triggers and processes, individual agency, and the possibilities of incremental change; historical studies may downplay the role of complexity and serendipity	Emergent discourses on decarbonization or zero or degrowth economic structures; mass social mobilization around climate change and economic inequity
Systemic	Intentional change targeted at the interdependencies of specific institutions, technologies and constellations of actors to steer complex systems toward normative goals	Highlights interdependencies, connectivity across scale and geography, and the potential for non-linear shifts in system dynamics across scales; emphasizes the role of ecological dynamics in social change and vice versa	Critiqued for lack of emphasis on individual agency, power, and politics and/or its overly managerial approach, glossing over differences in capacities, governance structure, and politics	Low-carbon energy transitions, focusing on technology-centered developments, modulated by incentives and disincentives enacted in policy mixes
Enabling	Fostering the human agency, values, and capacities necessary to manage uncertainty, act collectively, identify, and enact pathways to desired futures	Recognizes the potential of human agents for collective action; explicitly addresses asymmetries in power and circumstances of social injustice	May neglect significant structural, political obstacles to social transformation; burdens those with the greatest vulnerability with the task of transformation	Community-led environmental action; hacker/maker spaces for grassroots innovation; common approaches to sustainable local economies

Source: Adapted from Scoones et al. (2020).

In some cases, change can be triggered by larger-scale ideological shifts and movements of capital, leading, in turn, to enhanced opportunity and agency for previously marginalized actors. In other cases, change may be more dispersed and fundamental in nature, cascading up from local innovations that disrupt system dynamics to create structural change (Scoones et al. 2020).

For *socially just and equitable transformations* to occur, necessary structural and systemic changes will, however, demand enabling and emancipatory change as well. The examples in Table 4.3 elucidate this point. A myriad of social innovations is currently underway in the field of renewable energy, all of them linking technologies with social, economic, and political sustainability aspirations. These changes seem to be unfolding as part of a cultural evolution, which is something that is clearly needed as urgently as decarbonization (Jørgensen et al. 2019).

The climate change crisis and its anticipated risks could inspire people to experiment with new ways of thinking. Social or moral entrepreneurs are mobilizing social networks, for instance, preparing for transformation.

In general, though, the resulting transformation goes beyond the adoption of a new technology or a local social innovation alone. Rather, it includes a portfolio of actions like investment in new infrastructure, the establishment of new markets, changes in incentives, the development of new social preferences, or the adjustment of user practices. Such transformations gain momentum when multiple innovations are linked, improving the functionality of each, and when there is collaboration to reconfigure systems (Folke et al. 2021).

Reimagine Our Future

We are not good at solving large system-based problems, especially if those occur over a prolonged period. Typically, we first deny the existence of the problem, then procrastinate, and when things get seriously bad, we tend to fiddle with small changes that often have unforeseeable and counterintuitive effects. Instead of trying to achieve a shift to renewable energy at a country level, would it therefore not make better sense for renewable energy to grow out of initiatives led by progressive neighborhoods, transition towns, forward-looking cities, agile companies, ethical investors, and imaginatively-led organizations of all kinds?

4.4 Images of the Future

Rapid transformation to sustainable energy will require a fundamental shift in narratives (the stories we tell ourselves to better understand our situation or to promote reflection and action). Such narratives will need to resonate, inspire, and provide hope. They will need to foster a collective recognition of our interdependence with nature and of the required social, political, and economic measures that are necessary to achieve our sustainability goals.

Although individuals might play important roles in such stories, as shapers or transmitters of the stories, such images of the future are usually not the creation of individuals. They are rather shaped by cultural intermediaries working in media, education, politics, or social movements, among others. The impact of networks and social structures therefore provides an important context for change to happen.

These images of the future (e.g., Figure 4.8) are more likely to be widely accepted if they offer possible courses of action; something to which reasonable people can aspire (Folke et al. 2021). Such images of the future will probably show the importance of technological change in societal transformation to attain energy sustainability. These stories of hope will also communicate the need for cultural transformation and the need for humanity to align with nature (Arthur 2009). Such models of the future are not intended to act as blueprints of the future, but rather as aids to the imagination (Gell-Mann 1994).

"Lasting and satisfying change can only be implemented through effective education, participation, a large measure of consensus, and the widespread perception by individuals that they have personal stake in the outcome" of our sustainability actions (Gell-Mann 1994). No single scientist or engineer is going to step in and fix this crisis. A transformation will only occur if large masses of people start thinking critically and creatively about our interlinked sustainability challenges, and if they start implementing their ideas, like rebels with a just and ethical cause.

Figure 4.8 Image of the future: A future forest growing in Qatar, made possible by evaporation-cooled greenhouses, desalinated water, and concentrated solar power to run the whole system (Porritt 2013). *Source:* Sahara Forest Project.

4.5 Outlook

The world faces the challenge of meeting the needs of a constantly growing human population, and of doing so sustainably – i.e., without affecting the ability of future generations to meet their needs. Energy plays a pivotal role in this challenge, not only because of its importance to economic development, but also because of the myriad interactions it has on other critical sustainability issues, including water, food, and clean air.

The global energy transition continues to evolve and gain momentum across sectors – threatening to leave behind businesses that fail to decarbonize fast enough. As the world pivots to low-carbon resources, the global energy demand for electricity could triple to 50% by 2050 if net-zero carbon emissions are being sought by then. Fossil fuel demand could also peak as soon as 2024 – driven primarily by increased electric vehicle uptake and policy changes but could be delayed by geopolitical factors such as the war in Ukraine. From the oil industry to nascent technologies and renewables, all sectors of the economy are involved in making the net-zero transition. We have most of the technology to achieve this goal. The question is how quickly our societies, economies, and policies can change – i.e., do we have the resolve to bring about the urgent transformation to sustainable energy?

Problems

4.1 What is the equivalent mass (in billions of tons, or gigatons) of CO_2 in the atmosphere corresponding to a CO_2 concentration of 425 ppm (by volume)?

[Answer: About 3400 Gt]

Hint: Estimate the mass of air in the atmosphere, assuming the average atmospheric pressure is 101.3 kPa and the Earth's radius is 6378 km. The downward force of the air on the Earth typically remains constant. Remember that force = pressure/area, with force = mass ×

g. Air typically comprises 21% oxygen and 79% nitrogen, so that 1 mol of air has a mass of 29 g. Carbon dioxide has a molar mass of 44 g.

4.2 Estimate your total energy consumption per year in barrels of oil equivalent (BOE). State and substantiate your assumptions and show all calculations. Also calculate and show your energy footprint using a free online energy calculator, e.g., https://www.nature.org/en-us/get-involved/how-to-help/carbon-footprint-calculator or https://www.lifestylecalculator.com/doconomy.

Do another energy calculation for the lifestyle that you would like to lead in 10 years' time. Compare that figure with the one from your present lifestyle on campus. How realistic are these energy calculations, and what would make them more realistic or accurate?

[Answer: Open-ended question.]

Hint: 1 BOE = 6.1 GJ

4.3 Consider a domestic back-up power generator that runs on natural gas instead of diesel (e.g., Generac Guardian). Develop a high-impact fact sheet of no fewer than 300 words (with diagrams and photographs) that elucidate how the widespread adoption of this technology can help reduce US greenhouse-gas emissions compared to the use of traditional diesel-fuel power generators. Your discussion must make it clear how this technology targets SDG 7 (clean energy) and SDG 13 (climate action), as well as any of the other Sustainable Development Goals (SDGs).

[Answer: Open-ended question; independent research required.]

4.4 Dar-Lon Chang lives in the Geos neighborhood, an all-electric, solar-powered community in Arvada, Colorado, with 28 net-zero homes that each generate as much electricity from rooftop solar panels as they use over a year. The original 28 homes do not use fossil gas, as the homes have heat pump water heaters, electric or induction stoves, and space heating and cooling provided by geothermal ("ground-source") heat pumps for the larger homes and air-source heat pumps for the town homes. A new developer installed gas pipelines adjacent to the Geos net-zero homes, shattering the net-zero promise of Geos for the next phase of 91 homes and the dreams of the former developer and the neighborhood's residents.

Dar-Lon and his neighbors have lodged fierce protests against the new developer and the City of Arvada. However, fighting the new developer in court over breaking its promise to continue building net-zero in a press release would be expensive for the residents with no guarantee of prevailing. The new developer claims that "net-zero" can be taken to mean that solar panels only provide enough electricity over a year for use of the electric appliances, while ignoring the energy used by the fossil gas appliances. Given his experience with the broken promise of Geos, Dar-Lon believes that affordable all-electric, net-zero homes in new neighborhoods will not become widely available soon without city codes or state laws that ban fossil gas or heavily incentivize all-electric building in new developments. While Dar-Lon works with advocacy groups to influence policy and legislation to make new net-zero neighborhoods popular, he is also working to transition oil and gas workers into new careers decarbonizing and weatherizing existing neighborhoods with district heating using networked geothermal heat pumps and solar-powered microgrids.

Dar-Lon Chang and other Geos residents protest the installation of natural gas lines in October 2021. *Source:* Angie Olivero.

a) Guided by Table 4.1, summarize the interrelated techno-economic, socio-technical, and political aspects at play in the Geos neighborhood in Arvada, Colorado. Briefly discuss the ramifications of your findings.

[Answer: Open-ended question.]

b) Guided by Table 4.2, summarize the cultural aspects that influence the adoption of the all-electric, solar-powered, net-zero homes of the Geos neighborhood.

[Answer: Open-ended question.]

c) Considering the required structural transformation, systemic transformation, and enabling transformation (Table 4.3), what can Dar-Lon and his fellow Geos residents do to bring this situation to the attention of people in other neighborhoods in Colorado and around the country? What can Dar-Lon do to motivate developers, builders, and city government to design and successfully implement communities with net-zero energy consumption? What "story of hope" can he share?

[Answer: Open-ended question.]

4.5 "The transformation to sustainable energy requires advanced technology and a new organizing system, which encompasses our values and beliefs, institutions, and reward systems that will enable an optimal decision to be taken across society, and the structures that manage, control, govern, and influence its population. The best combination of technology and an organizing system that is available dictates the winners." (Arbib and Seba 2020)

Discuss this statement in no fewer than 300 words using the Tesla Megapack organization as a case study (https://www.tesla.com/megapack). In your discussion, be sure to indicate how Tesla Megapack's competitors are hamstrung by protecting their existing product suite and how those competitors are locked into existing business models, thought processes, cultures, and incentive structures that favor incremental progress rather than disruptive innovation. How did Tesla Megapack employ the "3-C entrepreneurial mindset" (https://orchard-prod. azurewebsites.net/media/Framework/KEEN_framework_new.pdf): being curious, making connections between apparently disparate aspects, and creating value? How can that 3-C entrepreneurial mindset be replicated in other renewable energy industries?

[Answer: Open-ended question; independent research required.]

References

Arbib, J. and Seba, T. (2020). Rethinking humanity. Five foundational sector disruptions, the lifecycle of civilizations, and the coming age of freedom. www.rethinkx.com

Arthur, W.B. (2009). *The Nature of Technology: What It Is and How It Evolves*. London: Allen Lane/ Penguin Books.

Bierbaum, R.M. and Matson, P.A. (2013). Energy in the context of sustainability. *Daedalus* 142 (1): 146–161.

BP (2022). BP statistical review of world energy, 2022. https://www.bp.com/content/dam/bp/business-sites/en/global/corporate/pdfs/energy-economics/statistical-review/bp-stats-review-2022-full-report.pdf

Build Equinox (2022). Conditioning Energy Recovery Ventilation (CERV) system. http://www. buildequinox.com

Cherp, A., Vinichenko, V., Jewell, J., Brutschin, E. et al. (2018). Integrating techno-economic, socio-technical and political perspectives on national energy transitions: a meta-theoretical framework. *Energy Research and Social Science* 37: 175–190.

Cullen, J.M. and Allwood, J.M. (2010). The efficient use of energy: tracing the global flow of energy from fuel to service. *Energy Policy* 38 (1): 75–81.

De Tocqueville, A. (1945). *Democracy in America*. New York, NY: Vintage Books. (Originally published in 1835).

EmTech (2023). Geoffrey Hinton talks about the "existential threat" of AI. https://www. technologyreview.com/2023/05/03/1072589/video-geoffrey-hinton-google-ai-risk-ethics

ExxonMobil (2019). *Outlook for Energy: A Perspective to 2040*. Irving, TX: Exxon Mobil.

Folke, C., Polasky, S., Rockström, J., Galaz, V. et al. (2021). Our future in the Anthropocene biosphere. *Ambio* 50 (4): 834–869.

Geels, F.W. (2002). Technological transitions as evolutionary reconfiguration processes: a multi-level perspective and a case-study. *Research Policy* 31 (8–9): 1257–1274.

Gell-Mann, M. (1994). *The Quark and the Jaguar*. London: Little, Brown and Company.

IEA (2020). *Electricity Market Report December 2020*. International Energy Agency.

Jørgensen, P.S., Folke, C. and Carroll, S.P. (2019). Evolution in the Anthropocene: informing governance and policy. *Annual Review of Ecology, Evolution and Systematics* 50: 527–546.

Kelly, K. (2011). *What Technology Wants*. New York: Penguin.

Kemp, R., Schot, J. and Hoogma, R. (1998). Regime shifts to sustainability through processes of niche formation: the approach of strategic niche management. *Technology Analysis and Strategic Management* 10 (2): 175–198.

Lazer, D.M., Baum, M.A., Benkler, Y., Berinsky, A.J. et al. (2018). The science of fake news. *Science* 359 (6380): 1094–1096.

Leonard, A. (2010). *The Story of Stuff: How Our Obsession with Stuff Is Trashing the Planet, Our Communities, and Our Health – and a Vision for Change*. New York: Free Press/Simon and Schuster.

Porritt, J. (2013). *The World we Made: Alex McKay's Story from 2050*. London: Phaidon Press.

Schneider, E.D. and Sagan, D. (2005). *Into the Cool. Energy Flow, Thermodynamics, and Life*. Chicago: University of Chicago Press.

Scoones, I., Stirling, A., Abrol, D., Atela, J. et al. (2020). Transformations to sustainability: combining structural, systemic and enabling approaches. *Current Opinion in Environmental Sustainability* 42: 65–75.

Semieniuk, G., Taylor, L., Rezai, A. and Foley, D.K. (2021). Plausible energy demand patterns in a growing global economy with climate policy. *Nature Climate Change* 11 (4): 313–318.

Shell (2022). *Shell World Energy Model. A View to 2100*. The Hague, Netherlands: Shell Oil Corporation. https://www.shell.com/energy-and-innovation/the-energy-future/scenarios/shell-scenarios-energy-models/world-energy-model.html

Sovacool, B.K. and Griffiths, S. (2020). Culture and low-carbon energy transitions. *Nature Sustainability* 3 (9): 685–693.

Stephenson, J., Barton, B., Carrington, G., Doering, A. et al. (2015). The energy cultures framework: exploring the role of norms, practices and material culture in shaping energy behaviour in New Zealand. *Energy Research and Social Science* 7: 117–123.

Weber, E.U. (2015). Climate change demands behavioral change: what are the challenges? *Social Research* 82: 561–581.

Zhaoyuan, X. and Ishwaran, M. (2020). Overview: high-quality energy for high-quality growth: China's energy revolution in the new era. In: *China's Energy Revolution in the Context of the Global Energy Transition*, pp.1–39. Shell International BV.

5

Energy Usage and the Environment

The statistics presented in Section 4.3 suggest that the primary global energy consumption is on course to increase by more than 30% within the two decades leading up to 2040. Electricity makes up only 17% of final global energy consumption and accounts for less than a third of global fossil fuel use. Globally, the industrial sector is the largest user of energy and is heavily reliant on direct fossil fuels (see also Chapter 28). Fossil fuels also account for around 95% of energy consumption in transport (IEA 2022).

To make matters worse, over the last 25 years, the developed world has shifted much of its carbon-intensive manufacture of steel, cement, ammonia, and plastics to the developing world. The unfortunate result is that the developed world's adoption of wind, solar, and other forms of renewable energy will end up being very large sources of carbon emissions in the developing world.

Europe and Japan have, respectively, reduced their primary energy use by 8% and 20% between 2010 and 2020 (IEA 2021; Eurostat 2022). China's and India's energy use are, however, still soaring, and Africa's energy use is rising from per capita levels seen in Europe in the nineteenth century (Cembalest 2021; BP 2022).

Burning such massive amounts of fossil fuels is harming air quality across the globe, and permanently changing the world's climate. The growing accumulation of heat trapping gases from fossil fuel combustion is resulting in rising temperatures, longer and more severe droughts, more intense storms, rising sea levels, acidifying oceans, and the loss of many species. This is accompanied by increasing hunger and worsening geopolitical tensions in many parts of the world.

The choice is no longer whether we should transform. Rather, the choice is between choosing transformations that we like, or, being transformed by the Earth due to our actions and inactions (Rockström et al. 2021; Plumer and Zhong 2022). The dangers from climate change are bigger and unfolding faster than previously expected, and humanity may struggle to adapt to the consequences unless greenhouse-gas emissions are significantly reduced in the next few years (Folke et al. 2021; Pörtner et al. 2022). To help us achieve this massive task requires focus, courage, and an entrepreneurial spirit like never before.

5.1 Our Planet Is Rapidly Warming Due to an Accumulation of Greenhouse Gases

If the Earth had no atmosphere, the planet would absorb and reradiate incoming solar energy directly back into space. A steady-state energy balance between incoming and outgoing (mainly infrared) energy predicts that the Earth's mean surface temperature would be 255 K (−18 °C) with

Energy Systems: A Project-Based Approach to Sustainability Thinking for Energy Conversion Systems, First Edition. Leon Liebenberg.
© 2024 John Wiley & Sons, Inc. Published 2024 by John Wiley & Sons, Inc.
Companion website: www.wiley.com/go/liebenberg/energy_systems

no atmosphere, which would make our planet frozen and basically uninhabitable in terms of life as we know it (Pierrehumbert 2011). Fortunately, the presence of natural "greenhouse" gases in the atmosphere importantly affects this energy balance and ensures an average temperature on Earth of about 14 °C, making life as we know it possible.

However, human activities are producing excessive greenhouse gases (especially carbon dioxide, methane, nitrous oxide, and halocarbons) that result in excessive heat trapping, which is rapidly heating the planet (Pörtner et al. 2022). Professor Don Wuebbles has spent several decades modeling our atmosphere and explains that the greenhouse effect is like having a blanket covering the Earth and that human activities are like adding another blanket, rapidly warming the Earth (Wuebbles 2022).

5.1.1 Carbon Dioxide

Excessive carbon dioxide is emitted during the burning of fossil fuels, with other significant human-related contributions coming from cement and steel production, and from land-use change (Chapter 28). By burning fossil fuels, humans are accelerating the part of the natural geologic carbon cycle that transfers carbon in rocks and sediments to the atmosphere, processes with time scales of tens to hundreds of millennia (Bruhwiler et al. 2021).

The world might be getting more energy efficient every year, but levels of emissions keep rising (see Figure 5.1). That is why most *deep decarbonization* strategies rely on replacing fossil fuels rather than "merely" reducing fossil fuel consumption per capita or per unit of performance (Cembalest 2021). Most statistical projections reveal that global energy-related carbon levels should begin tapering off after 2040 (Figure 5.1).

CO_2 released from combustion of fossil fuels equilibrates among the various carbon reservoirs of the atmosphere, the ocean, and the terrestrial biosphere on timescales of a few centuries. A sizeable fraction of the CO_2 remains in the atmosphere, awaiting a return to the solid Earth by much slower

CO2 emissions (billions of tons)

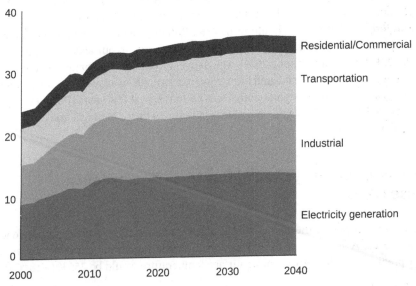

Figure 5.1 Global energy-related CO_2 emissions in billions of tons. *Source:* Adapted from ExxonMobil (2019).

weathering processes and deposition of $CaCO_3$. Common measures of the atmospheric lifetime of CO_2, including the *e-folding* time scale, disregard the "long tail." (*e*-Folding is the time interval in which an exponentially growing quantity increases by a factor of *e*; it is therefore the base-*e* analog of doubling time.) Leaving this out in the calculation of global warming potential leads many to underestimate the longevity of anthropogenic global warming. Contemporary atmospheric models show that 20–35% of CO_2 remains in the atmosphere after equilibration with the ocean for 200–2000 years. Neutralization by $CaCO_3$ draws the airborne fraction of CO_2 down further in timescales of 3000–7000 years (Archer et al. 2009; Buis 2019).

In the atmosphere, CO_2 absorbs long-wavelength infrared radiation emitted from the Earth that would otherwise go to space, and reradiates this energy, in part, back to the Earth. The Earth's atmosphere is therefore getting concomitantly warmer, based on this well-documented *enhanced greenhouse effect* (see Figure 5.2). Disturbingly, the rate of warming over the past 15 years is well above the long-term trend, indicative of the importance of human-caused change (Kramer et al. 2021).

Figure 5.2 plots the increase in the Earth's surface temperature as a function of accumulated CO_2 since 1876. The indicated 1.5 °C average warming temperature is the widely accepted temperature limit to minimize climate-related damage (Scientific American 2019; Pörtner et al. 2022; UNFCCC 2022).

The average concentration of CO_2 in the atmosphere was about 422 ppm in May 2022, steadily rising at about 2.35 ppm per year, as shown in Figure 5.3. The CO_2 level in preindustrial atmosphere

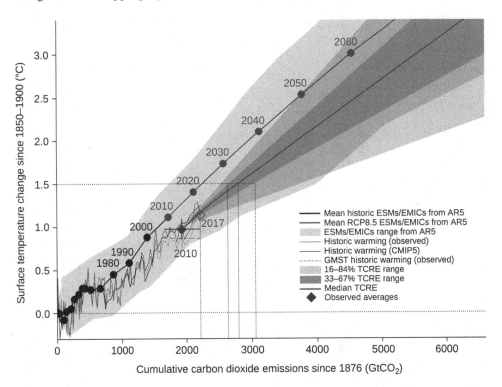

Figure 5.2 Temperature changes from 1850 to 1900 versus cumulative CO_2 emissions since 1 January 1876, based on various emission scenarios. The dotted black lines illustrate the remaining carbon budget estimates for a 1.5 °C temperature rise. Note these remaining budgets exclude possible Earth system feedbacks that could reduce the budget, such as CO_2 and CH_4 release from permafrost thawing and tropical wetlands. *Source:* Pörtner et al. (2022) / Reproduced with permission from The Intergovernmental Panel on Climate Change.

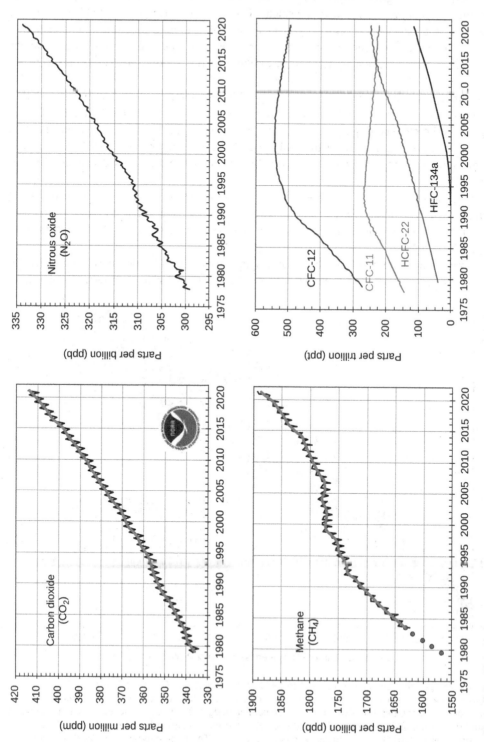

Figure 5.3 Global average abundance of the major greenhouse gases: carbon dioxide, methane, nitrous oxide, and halocarbons (CFC-12, CFC-11, HCFC-22, HFC-134a). *Source:* National Oceanic and Atmospheric Administration (NOAA).

was about 300 ppm and the current levels of around 420 ppm have not been seen on Earth for two to three million years. The high CO_2 concentrations were then due to the release of large, trapped bubbles of CO_2 by volcanic eruptions (Wuebbles 2022).

The concentration of CO_2 in the atmosphere is a million times greater than that of CH_4 and N_2O, and a billion times greater than that of halocarbons. Carbon dioxide is therefore the primary greenhouse gas. Its accumulation in the atmosphere correlates strongly with rises in the Earth's surface temperature, as illustrated in Figures 5.2 and 5.3.

5.1.2 Methane

The current CH_4 abundance in the atmosphere is unprecedented over at least the last 800,000 years, having increased from 700 ppb in preindustrial times to a concentration of around 1900 ppb in 2022, mostly due to human activities (Wuebbles 2022). Methane has a much shorter residence time in the atmosphere than CO_2, only around 10 years. Emissions of CH_4 are, however, 27 times more effective at trapping heat in the climate system than an equivalent emission of CO_2 over a 100-year time horizon (Wuebbles and Hayhoe 2002; Bruhwiler et al. 2021).

The production of fossil fuels accounts for 30–35% of all anthropogenic CH_4 emissions. Livestock, agriculture, swamps and marshes, landfills, and sewage account for another ~60%, with the remainder due to biomass and biofuel burning. Natural emissions, mainly from wetlands and terrestrial aquatic systems, are thought to make up ~40% of global emissions (Bruhwiler et al. 2021).

5.1.3 Nitrous Oxide

The concentration of nitrous oxide in our atmosphere was around 335 ppb in 2022 (see Figure 5.3), again showing an increasing trend compared to preindustrial times, in large part due to human activities (Wuebbles 2022). Nitrous oxide has an atmospheric lifetime of ~120 years and a global warming potential 298 times larger than that of CO_2 (Bruhwiler et al. 2021).

About 66% of anthropogenic nitrous-oxide emissions are likely due to agriculture, mostly from the use of industrially produced fertilizer and from the cultivation of nitrogen-fixing crops. Another 15% of total N_2O emissions come from fossil-fuel combustion. Biomass burning produces 11% of global N_2O emissions, with the remaining emissions coming from other sources such as waste waters. Population growth nearly guarantees that N_2O will continue to increase in the atmosphere, since producing enough food to feed the growing human population will require fertilizer use (Hansen et al. 2017).

5.1.4 Halocarbons

Halogenated gases (or halocarbons) are carbon-based molecules that contain chlorine, fluorine, or bromine. Halocarbons have been produced industrially in significant quantities since the mid-twentieth century to meet societal needs for refrigeration, air conditioning, insulation, solvents, and fire protection. Many halogenated gases are resistant to destruction by natural processes. Once these chemicals escape from refrigerators or foams to the atmosphere, they persist for decades to centuries. They are powerful greenhouse gases, with a global warming potential up to 14,000 times larger than that of CO_2. Atmospheric concentrations of halogenated gases are fortuitously 6 orders of magnitude lower than those of CO_2 (see Figure 5.3), so their contribution to global warming is only about 17% as much as that of CO_2.

5.1.5 Global Change in Surface Temperature and Climate Change

The rise in average global temperature (Figure 5.2) is causing global climate change, and vice versa (Pörtner et al. 2022). The knowledge base on the observed and projected impacts and risks generated by climate hazards, exposure, and vulnerability has increased as we have observed impacts attributed to climate change (Voosen 2022). These impacts and risks may be expressed in terms of their damage, harm, and economic and non-economic losses. The impacts and risks are illustrated in Figure 5.4 as a function of the rise in global temperature.

Figure 5.4 indicates that global temperatures have already increased by an average of 1.1 °C (or 2 °F) since the nineteenth century as we have pumped heat-trapping gases into the atmosphere by burning coal, oil, and gas for energy, and worsened the problem by cutting down forests (Pörtner et al. 2022). Many world leaders have pledged to implement measures that could limit total global warming to no more than 1.5 °C compared with preindustrial levels. Considering the slow implementation of these pledges, we would be fortunate if we could limit global warming to 2 °C by 2050.

Figure 5.4 further suggests that only immediate emission reductions across all sectors might limit global warming to 1.5 °C. If average global warming passes 1.5 °C, which it soon will, even humanity's best efforts to adapt could falter. The cost of defending coastal communities against rising sea levels could exceed what many nations can afford. Coral reefs, which buffer coastlines against storms and sustain fisheries for millions of people, will face more frequent bleaching from ocean heat waves and will decline by 70–90%. The number of people around the world exposed to severe coastal flooding could increase by more than one-fifth without new protections (Pörtner et al. 2022).

When coupled with population growth and land use change, future climate variability is predicted to have profound impacts on global food security (Molotoks et al. 2021; Lin et al. 2022). If global warming reaches 1.5 °C – as is now likely within the next few years – roughly 10% of the world's farmland could become unsuitable for growing food (Wheeler and Von Braun 2013). Livestock and outdoor workers could also face rising levels of heat stress that make farming increasingly more difficult (Pörtner et al. 2022).

At 2 °C of warming, between 800 million and 3 billion people globally could face chronic water scarcity because of drought, including more than one-third of the population in southern Europe. Crop yields and fish harvests could start declining in many places. An additional 1.4 million children in Africa could face severe malnutrition, stunting their growth (Hoegh-Guldberg et al. 2019; Pörtner et al. 2022).

At 3 °C of warming, the risk of extreme weather events could increase fivefold before the end of the century. Flooding from sea-level rises and heavier rainstorms could cause four times as much economic damage worldwide as it does today. As much as 30% of known plant and animal species on land could face a high risk of extinction (Pörtner et al. 2022).

Clearly, the increasing amounts of greenhouse gases in our atmosphere is detrimental to our well-being and to the well-being of much of life on Earth.

Reimagine Our Future

The anodyne term "global warming" does not begin to describe the anticipated frequency and severity of looming climate disasters. Perhaps "planetary destabilization" is a more accurate description?

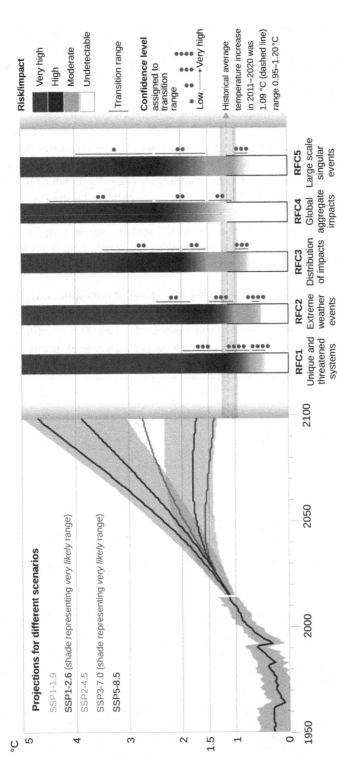

Figure 5.4 (a) Global surface temperature changes in °C relative to 1850–1900 based on a variety of scenarios for future emissions and concentrations of greenhouse gases; very likely ranges are shown for SSP1-2.6 and SSP3-7.0, where "SSP" is a certain "shared socioeconomic pathway." (b) Reason for concern (RFC) framework, which communicates scientific understanding about the accrual of risk for five broad categories. Diagrams are shown for each RFC, assuming low to no adaptation (i.e., adaptation is fragmented, localized, and comprises incremental adjustments to existing practices). The transition to a very high-risk level has an emphasis on irreversibility and adaptation limits. The horizontal broken line at around 1.09 °C denotes present global warming, which is used to separate the observed, past impacts below the line from the future projected risks above it. *Source:* Pörtner et al. (2022) / Reproduced with permission from The Intergovernmental Panel on Climate Change.

5.2   The Enhanced Greenhouse Effect

Solar radiation falls on the Earth with an energy distribution characteristic of a "blackbody" (i.e., a perfect emitter and absorber of radiative energy) radiating at the sun's surface temperature of roughly 6000 K. Most of this incoming radiation falls in the 0.2–4 μm range of wavelengths, i.e., "short wavelength radiation." The Earth absorbs this energy and re-emits it at longer wavelengths characteristic of a much cooler blackbody. Most of the re-emitted radiation falls in the 5–30 μm range (i.e., "long wavelength infrared radiation"), as shown in Figure 5.5.

Most gases in the atmosphere are poor absorbers of short wavelength solar radiation (see Figure 5.5), but some of them – especially water vapor – absorb some of the longer wavelength radiation re-emitted by the Earth (Mitchell 1989). Carbon dioxide is not as strong a greenhouse gas as water vapor, but it absorbs energy in longer infrared wavelengths (12–15 μm), which water vapor does not do, partially closing the "window" through which heat radiated by the Earth's surface would normally escape to space (see Figure 5.5). The absorbed long wavelength or infrared energy is partially re-radiated to space and partially back to the Earth's surface, thereby contributing to net warming of the Earth (Figure 5.6). The increase in the Earth's surface temperature resulting from the absorptive behavior of the atmosphere is what is known as the "greenhouse effect" (Corti and Peter 2009).

The greenhouse effect occurs when chemical species that absorb significantly in the 8–12 μm wavelength window are introduced into the atmosphere and affect the radiative balance of the Earth: more energy is absorbed, and the mean global temperature rises. The quantitative effect of chemical species like CO_2, CH_4, and N_2O will, of course, depend on several factors, including

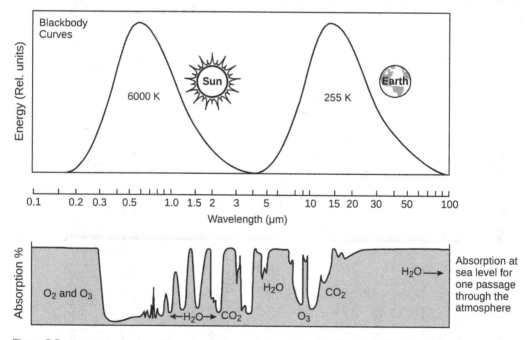

Figure 5.5   Atmospheric absorption of radiation emissions from blackbodies. The upper panel shows the spectral distribution of emissions from blackbodies at 6000 K (the sun) and 255 K (the Earth). The lower panel shows the percentage of absorption for radiation passing through the atmosphere, indicating the region of strong infrared (heat) absorption, 12–15 μm, of the thermal spectrum re-emitted by the Earth. Much of this heat is reradiated to the Earth and is the primary cause of human-induced (anthropogenic) global warming, and climate change. *Source:* Mitchell (1989) / Reproduced with permission from John Wiley & Sons.

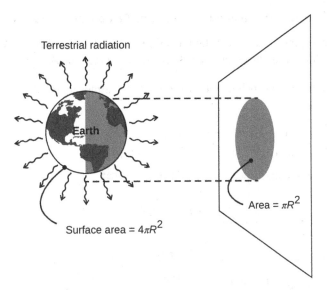

Figure 5.6 The Earth intercepts the incoming solar radiation with its cross-sectional area πR^2, but emits long-wave terrestrial radiation over its entire spherical surface area $4\pi R^2$.

the atmospheric concentrations of the species and their absorptive strengths. The radiative balance will also be strongly affected by atmospheric and surface phenomena (Emanuel 2018), including the albedo effect discussed below in a simple model of the Earth's radiation processes.

5.2.1 Albedo Effect

The solar radiation incident on the Earth's upper atmosphere has an intensity of $I_0 = 1.37 \, \text{kW/m}^2$; this value is known as the *solar constant* (Jaffe and Taylor 2018). A fraction of this incoming electromagnetic radiation is reflected by the Earth and is known as *albedo, α*. The atmosphere reflects radiation with $\alpha \approx 0.06$, clouds with $\alpha \approx 0.14$, and the Earth's surface with $\alpha \approx 0.10$, which gives a total albedo of $\alpha \approx 0.30$.

The radiation power received by an area of the Earth's surface facing the sun is:

$$P = (1 - \alpha)I_0 \, \pi R^2 \tag{5.1}$$

where R is the radius of the Earth (6380 km) (Stephens 2005).

This incoming radiation is distributed over the surface of the Earth ($4\pi R^2$). The thermal radiation *emitted* from the full surface of the Earth can then be estimated using Boltzmann's equation for electromagnetic radiation and treating the Earth as a blackbody emitter:

$$P = \left(4\pi R^2\right)\sigma T_s^4 \tag{5.2}$$

where σ is Boltzmann's constant $= 5.67 \times 10^{-8} \, \text{W/m}^2\text{K}^4$ and T_s is the surface temperature of the Earth in kelvin.

For radiative balance, set Eq. (5.1) equal to Eq. (5.2):

$$\left(4\pi R^2\right)\sigma T_s^4 = (1 - \alpha)I_0 \, \pi R^2 \tag{5.3}$$

Using $\alpha = 0.30$ and $I_0 = 1.37 \, \text{kW/m}^2$, we obtain $T_s = 267 \, \text{K}$ (or $-6 \, °\text{C}$). This is much lower than the average surface temperature of around 287 K (14 °C) because we are not yet accounting for the "greenhouse effect" of the gases trapped in our atmosphere. We will account for that next.

5.2.2 Absorption and Re-Radiation by Carbon Dioxide and Other Greenhouse Gases

Using Figure 5.7 to model the incoming and outgoing radiation, we can write an energy balance assuming a full albedo of $\alpha = 0.30$ (Corti and Peter 2009):

Radiative energy balance at the top of the atmosphere, where T_a is the temperature of the atmosphere:

$$I_0(1-0.3)\left(\pi R^2\right) = I_{earth}\left(4\pi R^2\right)$$
$$I_{earth} = 0.7\, I_0/4 \approx 240 \text{ W/m}^2 \equiv \sigma T_a^4 \tag{5.4}$$

Radiative energy balance at the Earth's surface:

$$\sigma T_s^4 = \underbrace{(0.7\, I_0/4)}_{\sigma T_a^4} + \sigma T_a^4 = 2\sigma T_a^4 \tag{5.5}$$

Solving for Eqs. (5.3) and (5.5) yields:

$$T_a = 255\,\text{K} \quad \text{and} \quad T_s = 303\,\text{K}$$

This simple model (Figure 5.7) yields an average surface temperature of 303 K (30 °C). Therefore, by including an "enhanced greenhouse effect" by accounting for the absorption (especially by atmospheric carbon dioxide and water vapor) and reradiation of infrared energy to the Earth, we have obtained a much higher surface temperature than the 14 °C obtained from Eq. (5.3). Earth would not support life as we know it without its naturally occurring greenhouse gases

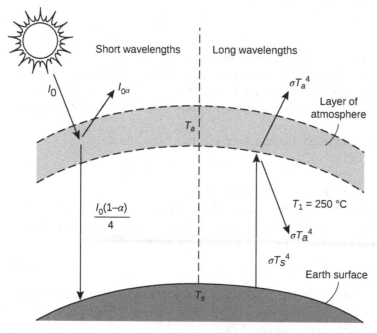

Figure 5.7 Simplified model of the "enhanced greenhouse effect." A single-layer atmosphere at T_a is perfectly transparent to solar (shortwave) radiation and perfectly opaque to longwave (infrared) radiation. About one-third of the sun's radiation is reflected to space (by albedo, α) or absorbed by the atmosphere. Naturally occurring greenhouse gases do not absorb most of the incoming shortwave energy from the sun, but they do absorb the longwave energy re-radiated from the Earth's surface, and some of that infrared energy is reflected to the Earth, just as happens in a greenhouse that grows flowers or vegetables in cold, but sunny areas. This energy is then re-emitted in all directions, keeping the surface of the planet much warmer than it would otherwise be. *Source:* Adapted from Mitchell (1989), Stephens (2005), and Taylor (2005).

(e.g., from water vapor, the decomposition of organic matter, or from volcanic eruptions), but the "enhanced greenhouse effect" brought on by anthropogenic carbon-dioxide emissions has dramatically changed the Earth's energy balance.

More accurate predictions of the enhanced greenhouse effect can be obtained by considering more atmospheric layers beyond the simple example shown here (Pierrehumbert 2011). Computational models must also consider the absorption spectra of each of the greenhouse gases and account for the effect of convective transport, as well as the variation of temperature ("lapse rate") and pressure in the various layers of the atmosphere, among other factors. Regardless, the climate's temperature response to additional greenhouse gases is still a source of uncertainty. Different models sometimes produce significant differences in projections of global average surface temperature; these differences largely result from different evaluations of how atmospheric dynamical processes (weather) and cloud processes will be affected by changes in climate. Scientists therefore often use multiple models to account for the variability and represent this as a range of outcomes (e.g., Figures 5.2 and 5.4).

Figure 5.8 shows measurements of the mean global energy balance. As illustrated, the Earth's surface receives almost twice as much radiation from the atmosphere (342 W/m^2) as it does from

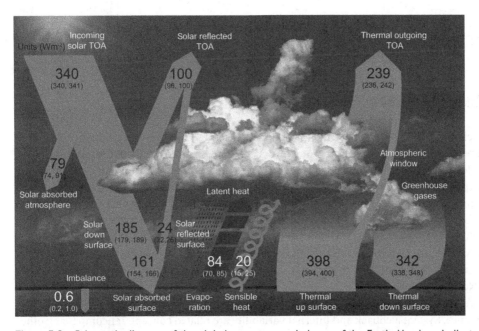

Figure 5.8 Schematic diagram of the global mean energy balance of the Earth. Numbers indicate the best estimates for the magnitudes of the globally averaged energy balance components, together with their uncertainty ranges, representing present-day climate conditions at the beginning of the twenty-first century.

Incoming solar radiation is $\dfrac{(1370 \, \text{Wm}^{-2}) \, \pi R^2}{4 \pi R^2} \approx 340 \, \text{Wm}^{-2}$, as explained in Eq. (5.4). Since albedo is assumed to

be 0.3, the amount of incoming radiation reflected to space is $\dfrac{(1370 \, \text{Wm}^{-2}) \, \pi R^2 \alpha}{4 \pi R^2} \approx 100 \, \text{Wm}^{-2}$. The energy

re-radiated from the Earth at (288 K) is $\dfrac{\sigma (4\pi R^2) T_s^4}{4\pi R^2} = (5.67 \times 10^{-8} \, \text{W/m}^2\text{K}^4)(288 \, \text{K})^4 \approx 397 \, \text{Wm}^{-2}$. A small

amount of radiation passes through the atmosphere, with the remaining about 350 Wm^{-2} absorbed by greenhouse gases in the atmosphere. The atmosphere then radiates about

$\dfrac{2\sigma (4\pi R^2) T_a^4}{4\pi R^2} = 2(5.67 \times 10^{-8} \, \text{W/m}^2\text{K}^4)(234 \, \text{K})^4 \approx 342 \, \text{Wm}^{-2}$ back to the Earth. Units: 2014 Wm^{-2}. *Source:*

Pörtner et al. (2022) / Reproduced with permission from The Intergovernmental Panel on Climate Change.

the sun (161 W/m^2). To balance this extra input of radiation – the radiation emitted by atmospheric greenhouse gases and clouds – the Earth's surface warms up and thereby emits more radiation itself. This is the essence of the *enhanced greenhouse effect*.

Example 5.1   Carbon Content of the Atmosphere

Find a relationship between parts per million of CO_2 in the atmosphere and the amount of gigatons (Gt) of carbon in the atmosphere. Assume that the atmosphere has a mass of 5.12×10^{21} g.

Given: We are given the constituents of air, and the total mass of the atmosphere.

Required: We must find how many tons of carbon will result in 1 ppm of CO_2.

Solution strategy: We will assume that the whole atmosphere is at standard temperature and pressure and has a homogenous density of 1291.8 g/m^3.

Analysis:

$$1 \text{ ppm } CO_2 = \left(\frac{1 \text{ m}^3 \text{ CO}_2}{10^6 \text{ m}^3 \text{ air}}\right)\left(\frac{44.61 \text{ mol}}{\text{m}^3 \text{ CO}_2}\right)\left(\frac{12 \text{ g C}}{\text{mol C}}\right)\left(\frac{5.12 \times 10^{21} \text{ g air}}{1291.8 \dfrac{\text{g air}}{\text{m}^3 \text{ air}}}\right)\left(\frac{10^{-15} \text{ Gt C}}{\text{g C}}\right) = 2.12 \text{ Gt C}$$

$$(5.6)$$

Discussion: We have found that $1 \text{ ppm } CO_2 \equiv 2.12 \text{ Gt C}$. The CO_2 concentration in 2022 was about 422 ppm. The total amount of carbon in the atmosphere in 2022 was: (422 ppm CO_2) × (2.12 Gt C/ppm) = 894.6 Gt C

5.3   Radiative Forcing

Climate scientists have developed a metric that more readily expresses the heating (or cooling) effect of greenhouse and other "forcings" onto the climate system, called radiative forcing. *Radiative forcing* is defined as the net difference in absorbed and radiated energy per unit area, W/m^2, at the top of the troposphere (Rohrschneider et al. 2019). The *troposphere* is the first and lowest layer of the Earth's atmosphere and contains 75% of the total mass of the planetary atmosphere, 99% of the total mass of water vapor and aerosols, and is where most weather phenomena occur (Douglass and Clader 2002).

A positive radiative forcing will lead to increased downward radiation to the Earth's surface, which would lead to an increase in surface temperature. Negative radiative forcing will result in cooling (North et al. 1981).

Radiative forcing is used to quantify and compare the external drivers of change to the Earth's energy balance. System feedback and internal variability are related concepts, encompassing other factors that also influence the direction and magnitude of imbalance.

A popular framework to analyze the radiative response to external radiative forcing is the *linear forcing-feedback framework*. It is based on a linear relationship between global net top of the atmosphere radiative flux (N), radiative forcing (F), and surface temperature perturbation (ΔT), which, in turn, actuates the radiative feedback parameter (λ) (Stephens 2005; Rohrschneider et al. 2019):

$$N = F + \lambda \, \Delta T \tag{5.7}$$

where

$$\lambda = F/\Delta T \qquad [(W/m^2)/^\circ C] \tag{5.8}$$

and ΔT is the mean surface response to the radiative forcing F.

If $N = 0$, as when equilibrium has been reached, the radiative forcing $F_0 = -\lambda_0 \Delta T$ can be used to estimate the change in global surface temperature ΔT. If the Earth's temperature increases, with all other parameters remaining constant, the upward flux of energy will increase and the radiative forcing will decrease, giving a negative value of λ_0. The value of λ_0 has been accurately determined and verified (Jaffe and Taylor 2018):

$$\lambda_0 \approx -3.2 \pm 0.1 \text{ W/m}^2 \text{ °C} \tag{5.9}$$

If there were no climate feedback, the radiative forcing would give rise to a change in temperature ΔT_0, which, in turn, would give rise to a uniform temperature response that cancels the original radiative forcing (Jaffe and Taylor 2018):

$$0 = F + \lambda_0 \Delta T_0$$

or

$$\Delta T_0 = -F/\lambda_0 \tag{5.10}$$

Radiative forcing has been steadily increasing, especially due to increasing greenhouse gas concentrations, as shown in Figure 5.9. The forcing function for CO_2 is $F_{CO_2} \approx 1.8 \pm 0.2 \text{ W/m}^2$; for methane it is $F_{CH_4} \approx 0.5 \text{ W/m}^2$; for nitrous oxide it is $F_{N_2O} \approx 0.17 \pm 0.03 \text{ W/m}^2$; and for halocarbons it is $F_{\text{halo-C}} \approx 0.29 \pm 0.04 \text{ W/m}^2$ (Pörtner et al. 2022).

Small particles in the Earth's atmosphere (called aerosols) like sulfur oxides could have a negative forcing, as shown in Figure 5.9 and discussed in Box 5.1. Scientists and engineers are experimenting with *solar geoengineering*, which involves spraying aerosols into the stratosphere to bring about negative forcing. The aerosols either scatter incoming solar radiation directly or nucleate reflective clouds. However, the aerosols have short residence times, around weeks to a few months.

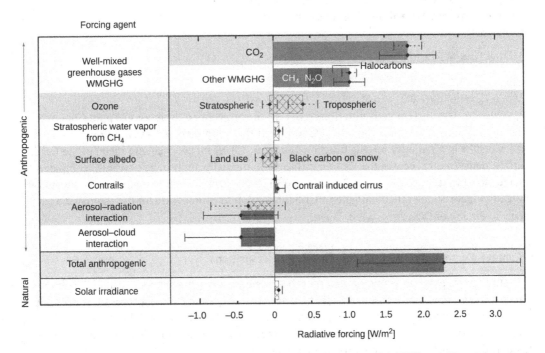

Figure 5.9 Radiative forcings. Box 5.2 discusses the global warming potential (GWP) and CO_2-equivalent of greenhouse gases. *Source:* Pörtner et al. (2022) / Reproduced with permission from The Intergovernmental Panel on Climate Change.

Box 5.1 Volcanic Eruptions and Global Cooling

The massive eruption of Mount Pinatubo in 1991 spewed some 20 million tons of sulfur dioxide into the sky, mostly above the troposphere, into the stratosphere. By reflecting sunlight back into space, the particles in the stratosphere helped push global temperatures down about 0.5 °C over the next two years (Dartevelle et al. 2002). The explosion of Mount Tambora in Indonesia in 1815 was famously followed by the "Year Without a Summer" in 1816 (Wood 2014), a gloomy period that helped inspire the creation of two of literature's most enduring horror creatures; vampires, and Frankenstein's monster (Hannah 2018). Scientists and engineers are now considering mimicking the climatic cooling effects of volcanic eruptions by artificially introducing sulfuric aerosols (among others) into the stratosphere. However, uncertainties abound regarding the use of such geoengineering experiments, including their short-term and long-term consequences on climate.

Consequently, massive amounts of such aerosols would need to be continuously introduced into the atmosphere to bring about a global cooling effect.

Example 5.2   The Temperature Response Generated by Current Levels of Atmospheric CO_2

Estimate the Earth's surface temperature change due to only the effect of current levels of atmospheric CO_2.

Given: Without any feedback effects, we must calculate the change in surface temperature due to only the effect of current levels of carbon dioxide. From Figure 5.9, $F_{CO_2} \approx 1.8 \pm 0.2\,\text{W/m}^2$.

Required: The change in global surface temperature ΔT.

Solution strategy: We will use Eqs. (5.9) and (5.10).

Analysis:

$$\Delta T_0 = -F/\lambda_0$$

with $\lambda_0 \approx -3.2 \pm 0.1\ \text{W/m}^2\,°\text{C}$ and $F = 1.8\ \text{W/m}^2$

So, $\Delta T_0 = -1.8/(-3.2) = 0.56\ °\text{C}$

Discussion: The current concentration of CO_2 in the atmosphere is about 420 ppm (Figure 5.3). This high concentration of carbon dioxide results in the trapping of longwave infrared radiation, which results in a radiative forcing function of around 1.8 W for every square meter of the Earth's surface. This, in turn, causes the Earth's surface to increase by an average of 0.56 °C compared to preindustrial times (around 1750). Therefore, 0.56 °C is the change in temperature that would be needed to increase the rate of outgoing thermal radiation to compensate for the estimated anthropogenic forcing of CO_2, $F_{CO_2} = 1.8\ \text{W/m}^2$.

Context, Community, and Collaboration for Just Transitions

There appears to be no viable future other than the one in which we adopt sustainable energy technologies. Most people will agree that we have the technology to meet zero-emission goals, and that it is currently a matter of lackadaisical policy change that is inhibiting wide-scale implementation. These are the views of **Amelia Keating** (22), who majors in Biological Engineering at the University of Illinois Urbana-Champaign (UIUC).

Amelia believes that the importance of how policies are enacted with respect to social context cannot be overstated. Amelia says that "technological advancement is a tool that humans have used to create and amplify divides between communities. This looks as if it is also affecting access to resources."

To ensure the just transition to a sustainable future, Amelia explains that policies will need to explicitly focus on the equitable and sustainable adoption of technology. "Diversity benefits progress. As an engineering student in 2023, I see an assembly of determined and inquisitive people – partly driven by fear, partly by fascination. This cohort's motivation is different from that of past generations; the consequences of technologies that lacked environmental forethought are becoming increasingly rapid and destructive." Amelia is of the opinion that one cannot approach innovation without being acutely aware of how our ideas will impact the world.

"I'm inspired by processes like the fungal rot pretreatment of biomass for cellulosic ethanol production, combined heat and power generation, and contextual engineering methodology. I think collaboration is key to solving the impending crisis. This applies to all engineering domains, but more importantly, varied vocations and walks of life." Amelia further believes that nurturing strong communities is the underlying basis of supporting ethical progress. "I am excited to see what is to come."

Amelia Keating (22) at the Saguaro National Park in Arizona. Amelia is a senior student at the University of Illinois at Urbana-Champaign and majors in Biological Engineering with a minor in Horticulture.
Source: Amelia Keating.

5.4 Surface Warming Due to Cumulative Emissions of CO$_2$

Due to the global warming potential, long residence time, and vast quantities of anthropogenic CO$_2$ in the atmosphere, it is important to quantify the cumulative emissions of CO$_2$. The data presented in Table 5.1 shows that, between 1850 and 2019, approximately 2390 Gt of CO$_2$ was spewed into the atmosphere from human activities. According to projections by the Intergovernmental Panel on

Table 5.1 Estimates of historical CO_2 emissions and remaining carbon budgets.

Global warming between 1850–1900 and 2010–2019 (°C)		Historical CO_2 emissions from 1850 to 2019 (Gt of CO_2)					
1.07		2390					
Approximate global warming relative to 1850–1900 until temperature limit (°C)	Additional global warming relative to 2010–2019 until temperature limit (°C)	Estimated carbon budgets from the beginning of 2020 (Gt of CO_2) Likelihood of limiting global warming to temperature limit					Variations in reductions in non-CO_2 emissions
		17%	33%	50%	67%	83%	
1.5	0.43	900	650	500	400	300	Higher or lower reductions in non-CO_2 emissions can increase or decrease the values on the left by 220 Gt CO_2 or more
1.7	0.63	1450	1050	850	700	550	
2.0	0.93	2300	1700	1350	1150	900	

Estimated remaining carbon budgets are calculated from the beginning of 2020 and extend until global net-zero CO_2 emissions are reached. They refer to CO_2 emissions, while accounting for the global warming effect of non-CO_2 emissions. Global warming in this table refers to the human-induced global surface temperature increase, which excludes the impact of natural variability on global temperatures in individual years.
Source: Adapted from Pörtner et al. (2022).

Climate Change (IPCC), if humanity wants a 50% chance of limiting global warming to only 1.5 °C by 2050, we have approximately a remaining 500 Gt of CO_2 that can be emitted (Table 5.1). With current annual CO_2 emissions reaching 35 Gt or more, we therefore have only about $500/35 \approx$ 14 years left before we exceed the carbon budget. For an 83% chance of keeping global warming to below 1.5 °C, we cannot produce more than about 300 Gt of CO_2, which means that we have 8.5 years left at the current annual rate of emissions before we exceed the carbon budget.

As can be deduced from Figure 5.5 (upper panel), the absorption cross-section of CO_2 peaks at about 15 μm and decays approximately exponentially on wings on either side of the peak. So, the radiative forcing from atmospheric CO_2 grows logarithmically with the concentration of CO_2.

The radiative forcing due to a change in concentration of CO_2 can be approximated as (Pörtner et al. 2022):

$$F = 5.35 \ln \left(\frac{\text{ending } CO_2 \text{ concentration (ppm)}}{\text{starting } CO_2 \text{ concentration (ppm)}} \right) \qquad [\text{W/m}^2] \qquad (5.11)$$

Reimagine Our Future

The odds of staving off the worst of climate disasters are about 50–50. For perspective, would you get in a car with those odds of a fatal accident?

Example 5.3 Estimations of Radiative Functions Based on Cumulative Levels of CO_2 in the Atmosphere

Assuming that the preindustrial concentration of CO_2 was 280 ppm, what is (a) the current radiative forcing function CO_2, and (b) the radiative forcing function for a doubling of CO_2 concentration compared to the preindustrial time, $F_{2\times}^{(RF)}$?

Given: The preindustrial concentration of CO_2 is 280 ppm.

Required: Without any feedback effects, we must estimate (a) the radiative forcing function in, say, 2022, compared to preindustrial times, and (b) the forcing function that coincides with a doubling of CO_2 concentration compared to preindustrial times.

Solution strategy: We can use Eq. (5.10) to estimate the forcing function.

Analysis:

$$F = 5.35 \ \ln\left(\frac{\text{ending } CO_2 \text{ concentration (ppm)}}{\text{starting } CO_2 \text{ concentration (ppm)}}\right)$$

a) The starting CO_2 concentrationis 280 ppm and the ending CO_2 concentration can be read off Figure 5.3 (carbon dioxide) as 420 ppm. So:

$$F = 5.35 \ln\left(\frac{420 \text{ ppm}}{280 \text{ ppm}}\right) = 2.17 \text{ W/m}^2$$

b) $F_{2\times} = 5.35 \ln\left(\frac{2 \times \text{ending } CO_2 \text{ concentration (ppm)}}{\text{starting } CO_2 \text{ concentration (ppm)}}\right) = 5.35 \ln(2) = 3.71 \text{ W/m}^2$

Box 5.2 CO_2 Equivalent and Global Warming Potential

Due to the pivotal effect of atmospheric CO_2 on climate change, the warming potential of greenhouse gases and other chemical species is usually expressed in terms of *carbon dioxide equivalent* (CO_2-eq) values. CO_2-eq is the amount of CO_2 that would give the same change in absorbed infrared radiation in the atmosphere over a period (usually 100 years).

Another metric often used is the *global warming potential* (GWP), which is the heat absorbed by any greenhouse gas in the atmosphere as a multiple of the heat that would be absorbed by the same mass of carbon dioxide (Wuebbles 2022). So, the GWP of CO_2 is 1. Methane has a GWP (over 100 years) of 27, meaning that, for example, a leak of one tonne of CH_4 is equivalent to emitting 27 t of CO_2 (Table 5.2) (EPA 2022).

Table 5.2   Comparison of 100-year GWP values.

Greenhouse gas	GWP over 100 years
Carbon dioxide, CO_2	1
Methane, CH_4	27
Nitrous oxide, N_2O	273
Refrigerant-134a, 1,1,1,2-Tetrafluoroethane	1526
Hexafluoroethane, C_2F_6	12,400
Sulfur hexafluoride, SF_6	25,200

Source: Adapted from EPA (2022).

5.5 Climate Modeling and Simulation

The previously discussed surface warming due to radiative forcing can be expanded to account for other nonlinear interactions that can produce more intense extreme weather. So, in addition to accounting for temperature effects, more advanced climate models also consider feedback effects, couplings, time delays, self-regulation, and tipping points.

The climate is a complex system that comprises distinct parts that co-exist dynamically. Intervening in one climatic aspect will probably affect every other part of the system (Hannon and Ruth 2001). The whole is indeed more than the sum of its parts.

Considering the immense complexity of the greenhouse effect, climatic consequences do not usually follow immediately upon the causes and are often indirectly linked to those causes. This can result in inaction or sluggish action from decision-makers. So, to grasp the reality (as a whole), it is not sufficient to perceive only the details; the details must be connected to include the all-important effect of interconnections (Solé 2011).

Feedback effects and other interconnections are considered in Earth system models (ESMs) such as general circulation models (GCMs) that focus on the atmosphere. ESMs model atmospheric dynamics (including the atmospheric lapse rate and cloud dynamics), ocean dynamics (including marine biota), the cryosphere (ice packs, sea ice, glaciers), and the biosphere (including the effects of albedo, evaporation and plant growth rates, and the Earth's surface roughness). Such models require an advanced scientific understanding of mitigation pathways. Most of these models are not readily accessible to policymakers or engineers. However, policymakers and engineers often only need an imprecise but true ("rough but right") simulation of reality, as provided by integrated assessment models (IAMs) (Fork and Koningstein 2021). The En-ROADS simulation package is one such tool.

The Energy Rapid Overview and Decision Support (En-ROADS) tool is a publicly available, widely used, online policy simulation model designed to complement IAMs in public outreach. En-ROADS represents the energy-economy-climate system at a globally aggregated level. It has been grounded in the best available science and is calibrated to fit historical data and projections from multiple IAMs across most climate change scenarios. Through an intuitive interface, users choose assumptions, policies, and actions to model the mitigation of greenhouse-gas emissions, receiving immediate feedback on likely energy, emissions, and climate pathways until 2100, enabling users to explore policies and their implications, scenarios, and uncertainties for themselves (Rooney-Varga et al. 2020; Kapmeier et al. 2021).

Figure 5.10 shows the salient elements of the En-ROADS simulation tool.

Like all complex, dynamic systems, the climate-energy system includes many "stocks," "flows," time delays, and feedback processes. For example, the primary driver of human-caused climate change is the accumulation of atmospheric CO_2 (a "stock") in response to emissions (its "inflow") that is roughly double current sinks (its "outflows") (Sterman et al. 2012).

The required energy transition requires that existing stocks of fossil fuel-based infrastructure decline over time, due to outflows such as retirement, and that the services they provide gradually be replaced (via inflows such as production) by stocks of energy-efficient, low-carbon capital. While climate and energy systems also include many complex interactions and feedback processes, even the simple process of a stock accumulating in response to its flows is poorly understood by most people (Sterman et al. 2012).

People also tend to underestimate the time required for a stock to accumulate or decline in response to its flows (Hannon and Ruth 2001). These misconceptions, when directed to the

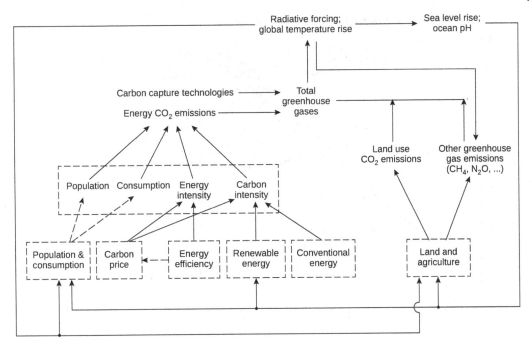

Figure 5.10 Energy-economy-climate interactions considered in a systems-based simulation tool such as En-ROADS. *Source:* Adapted from Sterman et al. (2012). Arrows indicate a causal relationship, which may be immediate or delayed, large or small, positive (reinforcing) or negative (depleting).

climate-energy system, can lead to grave underestimations of the scale and urgency of the action needed to meet international climate goals (Meadows et al. 1992).

For example, correlating current CO_2 emissions with current CO_2 concentrations leads to the false assumption that reducing greenhouse-gas emissions would be sufficient for stabilizing CO_2 concentrations at 500 ppm. If emissions exceed removals, CO_2 concentrations will continue to rise. Similarly, many people tend to underestimate the time needed to transition to a low-carbon economy, which would require the existing stock of fossil fuel-based infrastructure to be drained through gradual retirement, as new stocks of low-carbon infrastructure are gradually built (Kapmeier et al. 2021). This could take decades. The En-ROADS simulation tool, among others, can help decision-makers and designers better comprehend the stock-flow dynamics of both energy infrastructure and atmospheric CO_2 concentrations.

Regardless of the sensitivity of climate models and their capacity to capture all the relevant variables, it is abundantly clear that we will not be able to stop climate change overnight, or even by 2100. Nonetheless, we can and should limit the amount of climate change by reducing human-caused emissions of greenhouse gases, switching to renewable and non-carbon-based fuel, and increasing energy efficiency. Figure 5.11 illustrates the possible effect of such climate strategies.

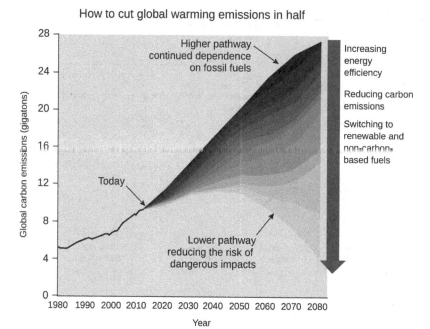

How to cut global warming emissions in half

Figure 5.11 Reducing carbon emissions from a higher global warming scenario (RCP8.5, or "business as usual") to a lower global warming scenario (RCP4.5, a moderate scenario in which emissions peak around 2040 and then decline) can be accomplished with a combination of many technologies and policies. In this example, these emission-reduction "wedges" could include increasing the energy efficiency of appliances, vehicles, buildings, electronics, and electricity generation, reducing carbon emissions from fossil fuels by switching to lower-carbon fuels or capturing and storing carbon, and switching to renewable and non-carbon-emitting sources of energy, including solar, wind, wave, biomass, tidal, and geothermal power. The shapes and sizes of the wedges shown here are illustrative only. *Source:* Walsh et al. (2014) / U.S. Global Change Research Program.

Box 5.3   "Net-zero" or "Carbon Neutral"

Nowadays, one often reads about companies employing "net-zero" or "carbon-neutral" strategies (IEA 2021; McKinsey 2022). What does that mean?

Becoming *carbon neutral* or achieving *net zero* involves two components: Companies need to decarbonize their business by lowering their emissions and then compensate for unavoidable emissions through carbon offset programs like reforestation projects and carbon removal technologies.

For now, compensating for emissions is often a gamble. Forests form the basis of *carbon offsets* but using them is complicated. They can burn, releasing their stored carbon into the air and undermining the offset they provide. Carbon offset projects based on forest conservation are also often criticized for failing to provide real carbon savings and simply shifting deforestation elsewhere. They are also criticized for riding roughshod over local forest communities and for allowing Western companies to put off cutting their emissions. Most technologies to remove carbon already in the atmosphere are prohibitively expensive and not available on a commercial scale.

These uncertainties and concerns mean that carbon offsets should be marginal in corporate plans, according to the standards set by Science Based Targets, a non-profit group that assesses corporate goals (https://sciencebasedtargets.org). Most companies will only be able to rely on these tools to offset 10% of their emissions, at most.

Some companies, like those in the fossil fuel sector, will need to radically change the core of their business models. There is no workaround.

For further information, see the "Carbon Offset Guide," https://www.offsetguide.org/understanding-carbon-offsets (Table 5.3).

Table 5.3 Examples of major carbon offset programs.

"Compliance" carbon offset programs (run by governmental bodies)	Geographic coverage	Label used for offset credits
Clean Development Mechanism (CDM)	Low- and middle-income countries	Certified Emission Reduction (CER)
California Compliance Offset Program	USA	Air Resources Board Offset Credit (ARBOC)
Joint Implementation (JI)	High-income countries	Emission Reduction Unit (ERU)
Regional Greenhouse Gas Initiative (RGGI)	Northeast USA	RGGI CO_2 Offset Allowance (ROA)
Alberta Emission Offset Program (AEOP)	Alberta, Canada	Alberta Emissions Offset Credit (AEOC)

"Voluntary" carbon offset programs (run by NGOs)	Geographic coverage	Label used for offset credits
American Carbon Registry	USA, some international	Emission Reduction Tonne (ERT)
Climate Action Reserve (CAR)	USA, Mexico	Climate Reserve Tonne (CRT)
The Gold Standard	International	Verified Emission Reduction (VER)
Plan Vivo	International	Plan Vivo Certificate (PVC)
The Verified Carbon Standard	International	Verified Carbon Unit (VCU)

Source: Adapted from Broekhoff et al. (2019).

Reimagine Our Future

Could carbon-offset programs save Miami from being flooded?

5.6 Catastrophism Versus Techno-Optimism

It will not benefit us to assume a dystopian or catastrophic view of the future. Nor would it be help-ful if we were to rely on the false magic of techno-optimism. *Catastrophists* cannot imagine how a rapidly growing global human population would meet its food, water, and energy needs; but, in the past three generations, we have done exactly that, despite a tripling of the global population since 1950. *Techno-optimists*, who promise seemingly endless, near-miraculous technological solutions, must also contend with a poor track record. For example, the promise of boundless energy from nuclear fusion or from the safe use of nuclear fission has yet to be met (as discussed in Chapter 17).

We should, instead, objectively view our current situation. It is crucial to understand the facts if the sustainable energy challenge is to be tackled meaningfully (Wuebbles 2021). We must acknowl-edge that we will not be able to sever our ties with fossil fuels quickly or cheaply. Complete decar-bonization of the global economy by 2050, as proposed by the IPCC and others, is only conceivable at the cost of unimaginable global economic retreat, or because of extraordinarily rapid transfor-mations that rely on near-miraculous technical advances (see also Chapter 28). And how do we effectively address the difficult challenge of convincing populations of the need for significant expenditures when those benefits will not be seen for decades to come?

To help decision-makers, engineers, and energy entrepreneurs, it is important to view matters from varying perspectives. The well-known *3-C entrepreneurial mindset* (Figure 5.12) can be helpful in questioning current practices or technologies and envisioning new ones. The 3-C entrepreneurial

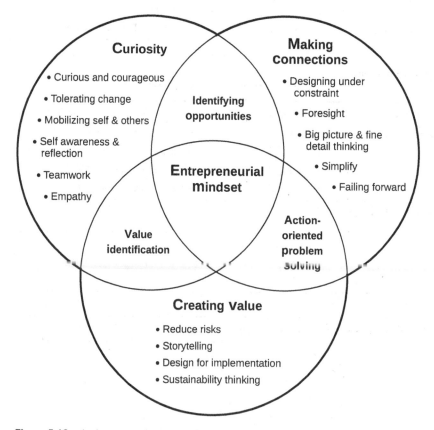

Figure 5.12 An interpretation of the *3-C entrepreneurial mindset*.

mindset revolves around being *c*urious, making *c*onnections between apparently disparate aspects, and *c*reating value (KEEN 2022, n.d.). We will use the 3-C entrepreneurial mindset throughout the book to assess energy technologies and policies.

Reimagine Our Future

Might focusing our attention on environmental crises lead us to ignore, or pay less attention to, how these crises fuel (or are fueled by) social, economic, or political crises? How many examples can you and your friends come up with in which social, economic, and political factors resulted in (or from) adverse energy scenarios, including increased CO_2 emissions resulting from coal combustion?

5.7 Outlook

Our energy choices directly affect the Earth's capacity to support life. Our choices, including choices we make that result in the emission of greenhouse gases, principally carbon dioxide, methane, nitrous oxide, and particulates from combusting carbon-based fossil fuels, make a difference to the atmosphere, to climate, and to all ecosystems on Earth. Climate change, in turn, affects all components of human and natural systems, adding both complexity and urgency to the search for sustainable energy solutions (Wuebbles 2021).

The concept of sustainability implicitly recognizes system feedback and interdependencies among social, technological, ecological, economic, and political systems. Mitigating climate change will require transitions in the energy sector which will require a systemic approach. This transition will demand a substantial *reduction in fossil fuel use*, widespread *electrification* (using renewable energy sources and nuclear power), *improved energy efficiency* (doing more with less), and the use of *alternative fuels* such as hydrogen and biofuels.

Importantly, the transition to a decarbonized future must be *just and equitable*. Developed nations will have to play a larger role in assisting developing nations to implement low-carbon or no-carbon energy strategies (Lubchenco and Kerry 2021). Fortunately, many of the easiest and cheapest opportunities to reduce energy use, produce clean energy, and reduce climate and other environmental changes can be found in developing countries, where infrastructure has yet to be built, where there is potential to greatly improve the efficiency of energy use, and where land-use practices can decrease greenhouse-gas emissions.

In theory, developing countries can *leapfrog* to available clean-energy technologies. But low-income countries face significant market barriers to technology adoption. International support and investment must be ramped up.

Regardless, high-income economies – the world's major carbon (and other) emitters – must replace their stock of high-carbon-emitting technologies with no-carbon or low-carbon alternatives. They must also invest in emerging technologies.

Humanity appears to be at a crossroads. The decisions we make now can secure a livable future (Ceballos and Ehrlich 2018; Voosen 2022). We have most of the technology and know-how required to limit global warming. But we lack a sense of urgency and the resolve to implement drastic measures to bring about global decarbonization. The reality is that any sufficiently effective steps in *decarbonizing will be decidedly non-magical, gradual, uncomfortable, and costly* (Chapter 28). We will need a realistic grasp of our past, present, and uncertain future when imagining sustainable energy scenarios. In Chapter 6, we will consider some perspectives on energy sustainability.

Problems

5.1 Explain qualitatively what the "enhanced greenhouse effect" is. Why is it crucial for our wellbeing? What can we do to stabilize or reverse the effects of the "enhanced greenhouse effect"?

[Answer: Open-ended question. Refer to Section 5.3.]

5.2 What is the resultant change in the global surface temperature of surface albedo, which has a radiative forcing of $-2\,W/m^2$?

[Answer: $-0.6\,°C$]

5.3 Disregarding any climate feedback effects, calculate (a) the estimated change in global surface temperature since preindustrial (1750) times, and (b) the rise in temperature if atmospheric CO_2 levels were to double compared to the levels in 1750. You may assume that radiative forcing was $1.1\,W/m^2$ in 1750 and $3.3\,W/m^2$ in 2022, and that $\lambda_0 \approx -3.2\,W/m^2\,°C$.

[Answers: (a) $0.7\,°C$; (b) $1.2\,°C$]

5.4 Using the total anthropogenic radiative forcing of about $2.3\,W/m^2$ (see Figure 5.9) and an equilibrium temperature increase of $0.8\,°C$ increased by those forcings, calculate the following: (a) The value of λ; (b) The equilibrium temperature change for a doubling of atmospheric CO_2.

[Answers: (a) $-1.84\,W/m^2\,°C$; (b) $2\,°C$]

5.5 Explain qualitatively what is meant by "global warming potential." Also, explain how cumulative emissions of CO_2 give rise to global surface temperature change. How are the global warming potentials of other greenhouse gases (such as CH_4 or N_2O) accounted for? Which greenhouse gas currently has the largest effect on global warming of the Earth's surface?

[Answer: Open-ended question. Refer to Section 5.4.]

5.6 In addition to increasing the Earth's surface temperature, what other adverse environmental and societal effects result from burning fossil fuels (coal, oil, and gas) or when fissioning nuclear fuels to produce energy? Briefly discuss each.

[Answer: Open-ended question that requires independent research.]

5.7 *Net-zero emissions* may be defined as the balancing of emissions by absorbing equivalent amounts of CO_2 from the atmosphere. The concept of net-zero emissions is the defining approach of international climate efforts. But some scientists contend that this strategy simply allows the perpetuation of the status quo (i.e., "business-as-usual") and is certain to fail. Do online research and discuss, in no fewer than 300 words, the purported facts and fallacies surrounding "net-zero emissions." Cite all your references.

[Answer: Open-ended question that requires independent research. Also read Box 5.3.]

5.8 Discuss, in no fewer than 300 words, how policymakers should weigh the cost of future damages from human-induced (or "anthropogenic") climate change against damages from current problems such as public health and safety. How can policymakers employ the "3-C entrepreneurial mindset" in their work? The 3-C mindset revolves around being curious, making connections between apparently disparate aspects, and creating value (https://orchard-prod.azurewebsites.net/media/Framework/KEEN_framework_new.pdf).

[Answer: Open-ended question that requires independent research.]

5.9 The Regional Greenhouse Gas Initiative (RGGI) is a cooperative, market-based effort among the states of Connecticut, Delaware, Maine, Maryland, Massachusetts, New Hampshire, New Jersey, New York, Rhode Island, Vermont, and Virginia to cap and reduce CO_2 emissions from the power sector. It represents the first cap-and-trade regional initiative implemented in the USA (RGGI 2022).

According to this plan, the total amount of annual CO_2 emissions is "capped" at around 90 million tons. Power plant operators purchase allowances (or carbon credits) that represent a permit to emit a specific amount of CO_2. Proceeds from the sale of allowances are used to support efforts in the region to foster energy efficiency and promote renewable energy. A power utility that emits less than its projected CO_2 allotment can sell unneeded allowances to utilities that are unable to meet their obligations. This is called "trade."

While RGGI auction prices are relatively low compared to estimates of the social cost of carbon, the policy represents a direct increase in the cost of production and should have behavioral impacts on producers. Using plant-level data and several identification strategies, researchers have found that there are reductions in emissions from coal-fired plants in RGGI states, but there is mixed evidence at natural gas-fired plants.

Perform an investigation into *cap-and-trade systems* and produce a fact sheet of no less than 300 words (with diagrams) to document your findings. The fact sheet must summarize the characteristics of cap-and-trade programs. It should discuss the RGGI and compare it to cap-and-trade programs employed in other countries.

[Answer: Open-ended question that requires independent research.]

5.10 Sustainable energy may be defined as a dynamic harmony between the equitable availability of energy-intensive goods and services to all people, and the preservation of the Earth for future generations.

If you were the founder of a start-up company, what would you do to ensure that your company's entrepreneurial activities are aligned with such-defined *sustainable energy* principles? How would you "sell" a business concept based on sustainable energy to venture capitalists? How would you employ the "3-C entrepreneurial mindset" in your business plan? The "3-C" mindset revolves around being curious, making connections between apparently disparate aspects, and creating value (https://orchard-prod.azurewebsites.net/media/Framework/KEEN_framework_new.pdf). What would you show prospective investors to convince them that energy sustainability strategies make business sense apart from being the ethically right thing to do? Summarize your thoughts in a high-impact, two-page fact sheet of no fewer than 300 words (with diagrams or photographs).

[Answer: Open-ended question that requires independent research.]

References

Archer, D., Eby, M., Brovkin, V., Ridgwell, A. et al. (2009). Atmospheric lifetime of fossil fuel carbon dioxide. *Annual Review of Earth and Planetary Sciences* 37: 117–134.

BP (2022). bp statistical review of world energy, 2022, 71st edition.

Broekhoff, D., Gillenwater, M., Colbert-Sangree, T. and Cage, P. (2019). Securing climate benefit: a guide to using carbon offsets. Stockholm Environment Institute and Greenhouse Gas Management Institute. https://www.offsetguide.org/wp-content/uploads/2020/03/Carbon-Offset-Guide_3122020.pdf.

Bruhwiler, L., Basu, S., Dutlcr, J.H., Chatterjee, A. et al. (2021). Observations of greenhouse gases as climate indicators. *Climatic Change* 165 (1): 1–18.

Buis, A. (2019). The atmosphere: getting a handle on carbon dioxide. NASA, Global Climate Change. https://climate.nasa.gov/news/2915/the-atmosphere-getting-a-handle-on-carbon-dioxide.

Ceballos, G. and Ehrlich, P.R. (2018). The misunderstood sixth mass extinction. *Science* 360 (6393): 1080–1081.

Cembalest, M. (2021). Eye on the market. *2021 Annual Energy Paper*. JP Morgan Private Bank. https://am.jpmorgan.com/content/dam/jpm-am-aem/global/en/insights/eye-on-the-market/future-shock-amv.pdf.

Corti, T. and Peter, T. (2009). A simple model for cloud radiative forcing. *Atmospheric Chemistry and Physics* 9 (15): 5751–5758.

Dartevelle, S., Ernst, G.G., Stix, J. and Bernard, A. (2002). Origin of the Mount Pinatubo climactic eruption cloud: Implications for volcanic hazards and atmospheric impacts. *Geology* 30 (7): 663–666.

Douglass, D.H. and Clader, B.D. (2002). Climate sensitivity of the earth to solar irradiance. *Geophysical Research Letters* 29 (16): 33-1–33-4.

Emanuel, K. (2018). *What We Know About Climate Change*. Boston, MA: MIT Press.

EPA (2022). Inventory of US greenhouse gas emissions and sinks, 1990–2020. https://www.epa.gov/system/files/documents/2022-04/us-ghg-inventory-2022-main-text.pdf.

Eurostat (2022). Energy statistics – an overview. Eurostat. https://ec.europa.eu/eurostat/statistics-explained/index.php?title=Energy_statistics_-_an_overview.

ExxonMobil, E. (2019). *Outlook for Energy: A Perspective to 2040*. Irving, TX: Exxon Mobil.

Folke, C., Polasky, S., Rockström, J., Galaz, V. et al. (2021). Our future in the Anthropocene biosphere. *Ambio* 50 (4): 834–869.

Fork, D. and Koningstein, R. (2021). How engineers can disrupt climate change. *IEEE Spectrum* 58 (7): 24–29.

Hannah, D. (2018). *A Year Without a Winter*. New York: Columbia University Press.

Hannon, B. and Ruth, M. (2001). *Dynamic Modeling*. New York: Springer Science and Business Media.

Hansen, J., Sato, M., Kharecha, P., Von Schuckmann, K. et al. (2017). Young people's burden: requirement of negative CO_2 emissions. *Earth System Dynamics* 8 (3): 577–616.

Hoegh-Guldberg, O., Jacob, D., Taylor, M., Guillén Bolaños, T. et al. (2019). The human imperative of stabilizing global climate change at 1.5 °C. *Science* 365 (6459): 1263.

IEA (2021). Net zero by 2050. A roadmap for the global energy sector. International Energy Agency. www.iea.org.

IEA (2022). *World Energy Outlook 2022*. International Energy Agency. https://www.iea.org/reports/world-energy-outlook-2022.

Jaffe, R.L. and Taylor, W. (2018). *The Physics of Energy*. Cambridge, UK: Cambridge University Press.

Kapmeier, F., Greenspan, A., Jones, A. and Sterman, J. (2021). Science-based analysis for climate action: how HSBC Bank uses the En-ROADS climate policy simulation. *System Dynamics Review: The Journal of the System Dynamics Society* 37 (4): 333–352.

KEEN (2022). Engineering unleashed. Transforming engineering education. The Kern Family Foundation http://www.engineeringunleashed.com.

KEEN (n.d.). The KEEN Framework. A guide for entrepreneurial mindset. Engineering Unleashed, The Kern Family Foundation. https://orchard-prod.azurewebsites.net/media/Framework/KEEN_framework_new.pdf.

Kramer, R.J., He, H., Soden, B.J., Oreopoulos, L. et al. (2021). Observational evidence of increasing global radiative forcing. *Geophysical Research Letters* 48 (7): 2020GL091585.

Lin, H.I., Yu, Y.Y., Wen, F.I. and Liu, P.T. (2022). Status of food security in East and Southeast Asia and challenges of climate change. *Climate* 10 (3): 40.

Lubchenco, J. and Kerry, J.F. (2021). Climate science speaks: "act now". *Science* 373 (6561): 1285–1285.

Masters, G.M. and Ela, W.P. (2014). *Introduction to Environmental Engineering and Science.* New York: Pearson.

McKinsey (2022). *The Net-Zero Transition. What It Would Cost, What It Could Bring.* Boston, MA: McKinsey Global Institute.

Meadows, D.H., Meadows, D.L. and Randers, J. (1992). *Beyond the Limits: Confronting Global Collapse Envisioning a Sustainable Future.* Chelsea Green Publishing Company.

Mitchell, J.F. (1989). The "greenhouse" effect and climate change. *Reviews of Geophysics* 27 (1): 115–139.

Molotoks, A., Smith, P. and Dawson, T.P. (2021). Impacts of land use, population, and climate change on global food security. *Food and Energy Security* 10 (1): e261.

North, G.R., Cahalan, R.F. and Coakley Jr, J.A. (1981). Energy balance climate models. *Reviews of Geophysics* 19 (1): 91–121.

Pierrehumbert, R.T. (2011). Infrared radiation and planetary temperature. In: *AIP Conference Proceedings* 1401 (1): 232–244. American Institute of Physics.

Plumer, B. and Zhong, R. (2022). Climate change is harming the planet faster than we can adapt, UN. *The New York Times*, 28 February 2022.

Pörtner, H.O., Roberts, D.C., Adams, H., Adler, C. et al. (2022). IPCC Sixth Assessment Report. Climate Change 2022: Impacts, adaptation and vulnerability. Intergovernmental Panel on Climate Change. https://www.ipcc.ch/report/ar6/wg2/downloads/report/IPCC_AR6_WGII_SummaryForPolicymakers.pdf and https://www.ipcc.ch/report/ar6/wg1/downloads/report/IPCC_AR6_WGI_SPM.pdf.

RGGI (2022). Fact sheet: Regional Greenhouse Gas Initiative (RGGI). https://www.rggi.org/sites/default/files/Uploads/Fact%20Sheets/RGGI_101_Factsheet.pdf.

Rockström, J., Gupta, J., Lenton, T.M., Qin, D. et al. (2021). Identifying a safe and just corridor for people and the planet. *Earth's Future* 9 (4): 2020EF001866.

Rohrschneider, T., Stevens, B. and Mauritsen, T. (2019). On simple representations of the climate response to external radiative forcing. *Climate Dynamics* 53 (5): 3131–3145.

Rooney-Varga, J.N., Kapmeier, F., Sterman, J.D., Jones, A.P. et al. (2020). The climate action simulation. *Simulation and Gaming* 51 (2): 114–140.

Scientific American (2019). Climate Clincher. The argument that global warming is part of a natural cycle is dead. *Scientific American* 86. Nov. 1, https://www.scientificamerican.com/article/global-warming-is-not-part-of-natural-climatevariability/.

Solé, R.V. (2011). *Phase Transitions.* Princeton, New York: Princeton University Press.

Stephens, G.L. (2005). Cloud feedbacks in the climate system: a critical review. *Journal of Climate* 18 (2): 237–273.

Sterman, J., Fiddaman, T., Franck, T.R., Jones, A. et al. (2012). Climate interactive: the C-ROADS climate policy model. *System Dynamics Review* 28 (3): 295–305.

Taylor, F.W. (2005). *Elementary Climate Physics.* Oxford, UK: Oxford University Press.

UNFCCC (2022). Key aspects of the Paris Agreement. United Nations Framework Convention on Climate Change (UNFCCC). https://unfccc.int/most-requested/key-aspects-of-the-paris-agreement#:~:text=The%20Paris%20Agreement's%20central%20aim,further%20to%201.5%20degrees%20Celsius.

Voosen, P. (2022). UN panel warns of warming's toll and an 'adaptation gap'. *Science* 375 (6584): 948–948.

Walsh, J.D., Wuebbles, D.J., Hayhoe, K., Kossin, J. et al. (2014). Appendix 4: Frequently asked questions. In: *Climate Change Impacts in the United States: The Third National Climate Assessment*, J.M. Melillo, T.C. Richmond and G.W. Yohe. Washington, DC: US Global Change Research Program, US Government Printing Office, pp. 790–820.

Wheeler, T. and Von Braun, J. (2013). Climate change impacts on global food security. *Science* 341 (6145): 508–513.

Wood, G.D.A. (2014). *Tambora: The Eruption that Changed the World*. Princeton, NJ: Princeton University Press.

Wuebbles, D.J. (2021). Ethics in climate change: a climate scientist's perspective. *Geological Society Special Publication* 508 (1): 285–296.

Wuebbles, D.J. (2022). Personal communication, 8/5/2022. Professor Don Wuebbles, School of Atmospheric Sciences, University of Illinois at Urbana-Champaign.

Wuebbles, D.J. and Hayhoe, K. (2002). Atmospheric methane and global change. *Earth-Science Reviews* 57 (3–4): 177–210.

6

Interpretations of Sustainability

Our globally interconnected systems and challenges suggest that we need to take a multidisciplinary and systemic approach to help us reimagine our societies, technologies, economies, environments, and politics. To effectively do this, we need to shift perspectives, open new lines of investigation, stimulate dialog and the exchange of ideas, and instigate new ways of seeing and being (Ryan 2020). We need to do whatever we can to minimize the devastating consequences of a *business-as-usual* approach.

How can we think systemically about what the essential components are of the sustainable use of energy? The matter is contested and there are competing theories, some of which are outlined in this chapter. It is assumed that any framework that will provide adequate guidance for the future should require *rapid and deep decarbonization*. Rapid and deep decarbonization requires the transformation of sociotechnical systems – the interlinked mix of technologies, infrastructure, organizations, markets, regulations, and user practices that together deliver societal functions such as personal mobility. These systems have developed over many decades, and the alignment and coevolution of their elements make them resistant to change (Geels et al. 2017).

To assist us in assessing the various contesting theories and frameworks presented here, it might be helpful to also have access to indicators and interpretations of sustainability. *Sustainability* is a societal goal or guiding ideal that relates to the ability of people to safely co-exist for the foreseeable future. (See "Glossary" in the book's companion website for a more comprehensive definition.) *Sustainable energy* may be defined as a dynamic harmony between the equitable availability of energy-intensive goods and services to all people, and the preservation of the Earth for future generations.

Indicators are widely used as a tool for communicating energy issues to policymakers and the public. A properly designed indicator or set of indicators transforms the basic statistical information that is available, which can provide a deeper understanding of an issue or dimension and help develop a clearer picture of the whole system, including its interlinkages and trade-offs (Mayer 2008). Such indicators should be relevant to the system, easily understandable, reliable, and based on information that is accessible to others. We will review some commonly used social, technological, economic, environmental, and political (or governance) metrics.

Energy Systems: A Project-Based Approach to Sustainability Thinking for Energy Conversion Systems,
First Edition. Leon Liebenberg.
© 2024 John Wiley & Sons, Inc. Published 2024 by John Wiley & Sons, Inc.
Companion website: www.wiley.com/go/liebenberg/energy_systems

6.1 Human Development Index (HDI)

A well-known measure of a country's standard of living is the Human Development Index (HDI). This index combines indicators of education, life expectancy, and income. As Figure 6.1 shows, there is a strong correlation between a country's HDI and its energy consumption per person (or "per capita"). The figure also shows that the HDI saturates above 5 kWh per person per day. This suggests that economic growth beyond an energy consumption of 5 kWh per person per day does not make people any happier.

However, a significant portion of the global population still faces serious challenges in accessing energy that is sufficient for them to have an adequate standard of living. This impacts negatively on their well-being. The promotion of accessible energy is expected to become increasingly challenging, considering that the global population is projected to grow from 8 billion in 2022 to 9.2 billion by 2040 (UN 2022).

Reimagine Our Future

Imagine we were to respond to Pope Francis's calls to go beyond *efficiency* (doing more with less) to *sufficiency* (consuming less). Imagine we also redefined our notion of progress to include more emphasis on people's quality of life. How might we implement these changes in our thinking? Share your thoughts on social media.

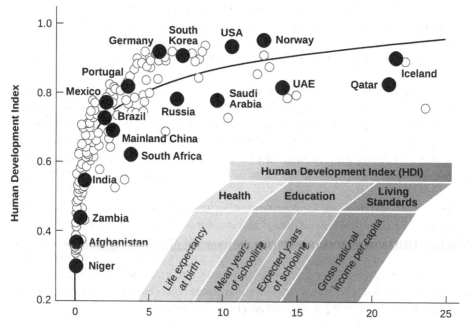

Per Capita Primary Energy Consumption (kilowatt-hours per person per day)

Figure 6.1 Energy is essential for society's progress. *Source:* Adapted from Neumayer (2001) and ExxonMobil (2019).

6.2   United Nations' Sustainable Development Goals

The United Nations has proposed 17 interlinked objectives and 169 targets designed to serve as a "shared blueprint for peace and prosperity for people and the planet now and into the future" (UN 2022). The 17 Sustainable Development Goals (SDGs) are the following, with SDG 7 pertaining directly to our use of energy (UN 2022): SDG 1: No Poverty; SDG 2: Zero Hunger; SDG 3: Good Health and Wellbeing; SDG 4: Quality Education; SDG 5: Gender Equality; SDG 6: Clean Water and Sanitation; *SDG 7: Affordable and Clean Energy;* SDG 8: Decent Work and Economic Growth; SDG 9: Industry, Innovation and Infrastructure; SDG 10: Reduced Inequality; SDG 11: Sustainable Cities and Communities; SDG 12: Responsible Consumption and Production; SDG 13: Climate Action; SDG 14: Life Below Water; SDG 15: Life on Land; SDG 16: Peace and Justice Strong Institutions; and SDG 17: Partnerships to achieve the Goal.

Each SDG features several targets. Importantly, the targets for SDG 7 (Affordable and Clean Energy) are:

- *Target 7.1:* By 2030, ensure universal access to affordable, reliable, and modern energy services.
- *Target 7.2:* By 2030, increase substantially the share of renewable energy in the global energy mix.
- *Target 7.3:* By 2030, double the global rate of improvement in energy efficiency.
- *Target 7.a:* By 2030, enhance international cooperation to facilitate access to clean energy research and technology, including renewable energy, energy efficiency and advanced and cleaner fossil-fuel technology, and promote investment in energy infrastructure and clean energy technology.
- *Target 7.b:* By 2030, expand infrastructure and upgrade technology for supplying modern and sustainable energy services for all in developing countries, in particular the least developed countries, small island developing states, and land-locked developing countries, in accordance with their respective programs of support.

Figure 6.2 depicts the interactions of the 16 non-energy SDGs as they pertain to achieving the goal of clean energy (SDG 7). The positive interactions between SDG 7 (Affordable and Clean Energy) and the other SDGs clearly outweigh the negative ones, both in number and magnitude. This suggests that efforts to ensure access to modern energy forms for the world's poorest, and to deploy renewables rapidly and accelerate the pace of energy efficiency improvements in all countries are very likely to promote the broader Sustainable Development Agenda.

Figure 6.2 illustrates that replacing coal and natural gas in electricity generation with solar, wind and most other renewables, and subsequently using that electricity to power end-use processes in the transport, buildings, and industrial sectors will help improve the air quality of cities throughout the world (SDG 3). Cleaner air, in turn, means healthier populations that can contribute more productively to the economy. There is high certainty regarding these interconnections, hence the high score of +2 (McCollum et al. 2018).

There are many other interesting connections between the SDGs that are relevant here. For example, if an expansion of renewables leads to large-scale bioenergy production globally, there will be a risk of competition with land for food production (SDG 2) and water for multiple uses (SDG 6). Increased food prices could potentially result from such a scenario, which would be to the detriment of the poor worldwide. There is, however, lower certainty regarding these interconnections, hence the range of scores between −1 and +1 (McCollum et al. 2018).

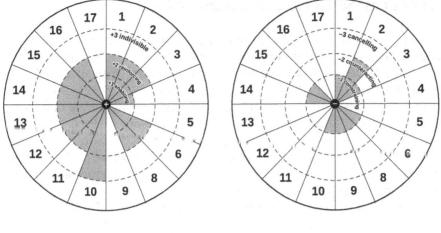

SDG 1: No Poverty
SDG 2: Zero Hunger
SDG 3: Good Health and Wellbeing
SDG 4: Quality Education
SDG 5: Gender Equality
SDG 6: Clean Water and Sanitation

SDG 7: Affordable and Clean Energy
SDG 8: Decent Work and Economic Growth
SDG 9: Industry, Innovation and Infrastructure
SDG 10: Reduced Inequality
SDG 11: Sustainable Cities and Communities
SDG 12: Responsible Consumption and Production

SDG 13: Climate Action
SDG 14: Life Below Water
SDG 15: Life on Land
SDG 16: Peace and Justice,
　　　　Strong Institutions
SDG 17: Partnerships to Achieve
　　　　the Goal

Figure 6.2 Connecting the United Nations' Sustainable Development Goal number 7 with the 16 non-energy SDGs. The scale used to depict the interconnections between the SDGs are: +3: indivisible; +2: reinforcing; +1: enabling; 0: consistent; −1: constraining; −2: counteracting; −3: canceling. The absence of a colored wedge in either the left or right panels indicates a lack of positive or negative interactions, respectively; if wedges are absent in both panels for a given SDG, that indicates a score of 0 ("consistent"). Only one positive or negative score is shown per SDG; in instances where multiple interactions are present at the underlying target level (positive and negative are treated separately), the individual score with the greatest magnitude is shown. Some SDG linkages may involve more than simple two-way interactions (e.g., the energy-water-land "nexus"), but that is not illustrated. *Source:* McCollum et al. (2018) / IOP Publishing Ltd / Licensed under CC BY 3.0.

These linkages are, however, context dependent and case specific as they hinge on where, when, and how the pursuit of targets occurs. So, when assessing interactions for the purposes of policy implementation, it is important to clearly articulate such dependencies, and where they exist (Singh et al. 2018; Fonseca et al. 2020).

Many of the interactions between the targets of SDG 7 (clean energy) and those of the other SDGs are either unidirectional (i.e., an impact of SDG 7 on the others, or vice versa) or bidirectional (simultaneously impacting on one another, although often not in complete symmetry). For example, consider the unidirectional impact of access to energy on education (SDG 4), employment (SDG 8), and quality healthcare (SDG 3). An example of a bidirectional impact pertains to the large-scale utilization of renewable bioenergy, which can have negative effects on food production, and thus on the goal of ending hunger (SDG 2), while in the reverse direction, ending hunger and protecting terrestrial ecosystems (SDG 15) may impose limits on how much cropland is available for bioenergy production altogether (McCollum et al. 2018).

Reimagine Our Future

Consider what you can do personally to bring about energy sustainability. Are there ways in which you can exert and expand your influence to help those around you reframe their thinking about energy sustainability? Can you lead by example? Can you lead by teaching, facilitating, advising, advocating, encouraging, informing, mentoring, or networking? Can you inspire others? Can you think more innovatively and empathize more deeply, challenging perceptions and inviting others to reimagine the future?

A black kite (*Milvus migrans*), flying over a polluted Hong Kong, as seen from Victoria Peak. *Source:* Prof Ida Kubiszewski / University College London (UCL).

6.3 The Social, Technological, Environmental, Economic, and Political (STEEP) Analysis of Energy Systems

Sustainable energy systems may be assessed using the so-called STEEP framework, which analyses social, technological, economic, environmental, and political/governance paradigms:

- *Social* paradigms: How can we ensure healthy and just societies? (*Goal:* Meet the needs of all people now and in the future, prioritizing those in most need.)
- *Technological* paradigms: How can we use our technology responsibly? (*Goal:* Effective policies on appropriate technology, precautionary principles, and appropriate values.)
- *Environmental* paradigms: How can we live within nature's limits? (*Goal:* Respect and improve the environment so that resources are unimpaired for the future.)
- *Economic paradigms:* How can we achieve a sustainable economy? (*Goal:* Provide opportunity for all. Environmental and social costs to fall on those who impose them.)
- *Political/governance* challenges: How do we ensure good governance? (*Goal:* Promote participation in all forms and levels of governance to engage people's energy, creativity, and diversity.)

Table 6.1 summarizes some contemporary STEEP challenges or developments related to energy.

Table 6.1 Some contemporary STEEP challenges or developments related to energy.

Social	Technological	Environmental	Economic	Political
Increasing energy usage as wealth increases	Production of heat and electricity on a small scale using low-carbon (or no-carbon) sources	Treatment of flue gases at fossil-fuel power plants	Employing degrowth measures	Weaponization of energy by energy-controlling countries or companies
Excessive reliance on personal transportation	Improved efficiencies of power stations fired by fossil fuels (gas, coal, oil)	Biofuels	Increased investment in renewable energies	Reduction and eventual elimination of petroleum for transportation
Rising energy prices, leading to fuel poverty for many people		Radioactive waste from nuclear power plants	Improved energy return on investment (EROI) in renewable energy systems	
Lack of electricity, especially in developing countries	Deployment of a hydrogen economy	Carbon capture, storage, and utilization	Hike in energy prices due to carbon sequestration at source (such as fossil-fuel power plants)	Subsidies for technology or infrastructure related to renewable energy
Immense fatalities among workers in the fossil-fuel mining (or drilling) industries	Electricity demand-side management	Offshore wind farms	Increasing ratio of fuel use to gross domestic product (GDP)	Carbon taxation
	Retrofitting buildings and homes with photovoltaics	Small hydropower		Stranded capital assets, such as coal-fired power stations
	Microgrids			
	Waste to energy			

The global energy system is currently facing several challenges, most notably high consumption levels, lack of energy access, environmental concerns like climate change and air pollution, energy security concerns, and the need for a long-term sustainability focus. Addressing these urgent issues simultaneously will require a fundamental transformation of the global energy system.

This book investigates some technological solutions as they relate to societal, environmental, economic, and political (or governance) aspects. It should become clear that an integrated approach needs to be followed, based on renewable and some non-renewable energy sources and systems. Unfortunately, there is no "magic silver bullet" or single solution that will solve our sustainable energy challenge in its entirety.

6.4 The "Doughnut" of Social and Planetary Boundaries

The doughnut-shaped *safe and just space* framework (Raworth 2017) is regarded as a holistic tool for envisioning human development on a stable and resilient planet (Fanning et al. 2022). The doughnut combines two core concepts: *biophysical boundaries* that avoid critical planetary degradation, and a sufficient *social foundation* that avoids critical human deprivation, and which closely aligns with the social priorities of the UN's SDGs (Fanning et al. 2022).

In the doughnut framework, the social thresholds include self-reported life satisfaction and life expectancy, nutrition, sanitation, income poverty, access to energy, education, social support, democratic quality, equality, and employment. The biophysical thresholds pertain to carbon-dioxide

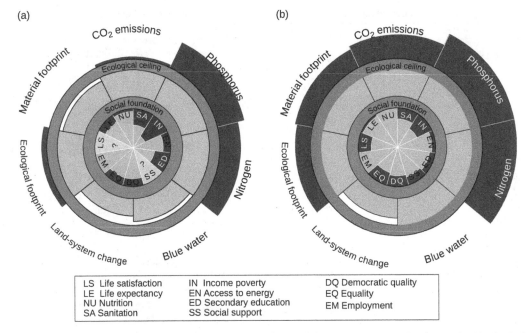

LS Life satisfaction	IN Income poverty	DQ Democratic quality
LE Life expectancy	EN Access to energy	EQ Equality
NU Nutrition	ED Secondary education	EM Employment
SA Sanitation	SS Social support	

Figure 6.3 Global performance relative to the doughnut's safe and just space, on the basis of the biophysical boundaries and social thresholds measured in (a) 1992 and (b) 2015. The dark circles show the ecological ceiling and social foundation, which encompass the doughnut of social and planetary boundaries. The small wedges show the average population-weighted social performance relative to each social threshold. The larger wedges show total resource use relative to each global biophysical boundary, starting from the outer edge of the social foundation. The outer wedges show shortfalls below social thresholds or overshoot beyond biophysical boundaries. Light-colored wedges show indicators with missing data. *Source:* Fanning et al. (2022) / Reproduced with permission from Springer Nature.

emissions, the consumption of phosphorous and nitrogen, land-system change, the ecological footprint, and the material footprint. Figure 6.3 illustrates the doughnut framework applied to the world in 1992, compared to 2015. Although many countries have apparently improved their social conditions since 1992, that has come at the price of transgressing the planet's biophysical boundaries.

When applying the doughnut framework to countries, Figure 6.4 reveals that high-income countries with high levels of social achievement have levels of resource use far beyond anything that could sustainably be extended to all people. Although low-income countries have shown progress in reducing social shortfalls, they have generally been transgressing biophysical boundaries (such as CO_2 emissions) faster than they have been achieving social thresholds.

Figure 6.4 also shows that the slow rate of social progress (between 1992 and 2015) is coupled with *ecological overshoot* at the global scale, which is already overwhelming the regenerative capacity of the biosphere and exposing humanity to a high risk of destabilizing the Earth's system (Fanning et al. 2022). It is apparent that deep transformations are needed to safeguard human and planetary health.

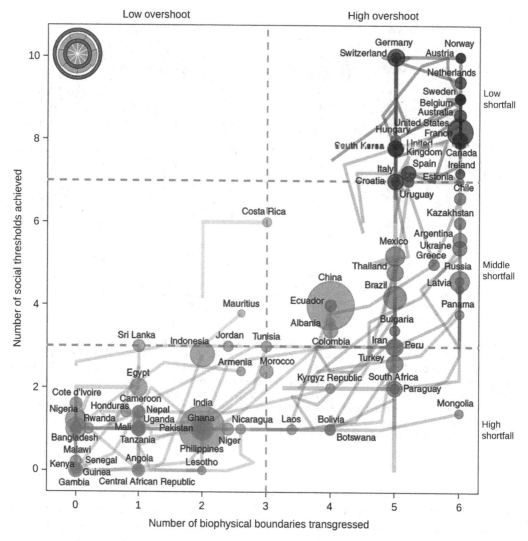

Figure 6.4 The number of social thresholds achieved versus the number of biophysical boundaries transgressed by countries over time, 1992–2015. Circles are sized according to population. Ideally, countries should be situated in the circular safe-and-just "doughnut" space located in the top-left corner. Country trajectories between 1992 and 2015 are shown with solid lines. Costa Rica comes closest to meeting social and biophysical objectives. *Source:* Fanning et al. (2022) / Reproduced with permission from Springer Nature.

6.5   Kaya Identity

The responses to the environmental concerns of our actions can be neatly summarized with the following formulation, which is sometimes called the *Kaya identity* (Kaya and Yokoburi 1997):

$$CO_2 \text{ emissions} = \text{Population} \times \frac{\text{GDP}}{\text{Population}} \times \frac{\text{Energy}}{\text{GDP}} \times \underbrace{\frac{CO_2 \text{ emissions}}{\text{Energy}}}_{\text{Carbon intensity}} \tag{6.1}$$

The first two terms in this equation imply that a reduction in human-caused CO_2 emissions could be achieved by controlling population growth or reducing wealth (as measured by GDP). Presently, the global focus is on the third and fourth terms: energy use per GDP and CO_2 emitted per unit of energy consumed. Our energy usage can be reduced by employing energy efficiency and other demand-side management strategies. The CO_2 emissions of energy conversion systems can be reduced using more renewable and low-carbon energy sources (e.g., solar, wind, geothermal, bio-fuel) and nuclear power. The CO_2 emissions can also be mitigated by employing carbon capture, sequestration, and usage; or by planting more trees that absorb CO_2.

Reimagine Our Future

What could we do to reduce hyperconsumption and improve the circular production of what we consume? Do you think that such a *circular economy* might decrease our energy usage? Discuss with your friends.

6.6 Lifecyle Analysis and Embodied Energy

Lifecycle analysis is a quantitative methodology for assessing the environmental, economic, and social impacts of a product, service, or system, related to its function and accounting for its overall lifecycle. Part of an LCA comprises calculating the sum of all energy spent to produce a product – also called the "embodied" energy – to extract and process the initial materials, use the product, and dispose of the waste (see Table 6.2). Box 6.1 summarizes the embodies energy of a typical 3 MW wind turbine.

It is generally easier and cheaper to bring about changes to a product or system at the design stage. This underscores the importance of performing the early assessment of technologies to facil-itate the effective integration of measures for environmental impact mitigation or to identify proc-ess alternatives that minimize lifecycle impact (Gerber 2014). When producing LCAs, extensive databases are available, among which the most widely used are the GaBi (2022) and ecoinvent (2022) tools.

Table 6.2 Embodied energy values of some common materials.

Material	Embodied energy (MJ/kg)
Concrete	0.95
Fiberglass	28
Hardboard	16
Galvanized sheet steel	13.1
Stainless steel	56.7

Source: Adapted from Hammond and Jones (2008); GaBi (2022).

Box 6.1   Embodied Energy and Lifecycle Cost of a 3-MW Wind Turbine

As wind turbines get bigger, more energy is required for their manufacture. It is important to consider whether these increases in wind turbine size, and thus embodied energy, can be adequately justified by equivalent increases in the energy yield of such systems. Table 6.3 shows the materials used in the construction and maintenance of a 3-MW wind turbine for a lifetime of 20 years, as well as the results of an LCA.

The net energy produced over the lifecycle of the wind turbine is equal to the gross energy output of the turbine minus the initial embodied energy, the energy associated with the necessary replacement of components during the 20-year period and the energy required for its operation, maintenance, repair, and eventual disposal. The ratio of the embodied energy (E_{in}) to the net energy produced (E_{out}) by the turbine over its lifetime (L) gives the energy yield ratio (EYR):

$$EYR = \frac{E_{out} \times L}{E_{in}} \qquad (6.2)$$

The net lifecycle energy produced over a 20-year period was 2049 TJ for the 3-MW wind turbine, which gives an EYR of 23. This shows that the turbine produces a significantly larger amount of energy than is required for its manufacture, operation, and maintenance during its effective life.

The total net (avoided) greenhouse-gas emissions of the wind turbine may be calculated by subtracting the initial and recurring embodied emissions from the net emissions that would have been released from traditional fossil fuel-based energy production for the equivalent quantity of energy produced by the wind turbine. The total net avoided greenhouse-gas emissions equate to 122,961 t of greenhouse gases over a 20-year service life. The net annual avoided emissions associated with the 3-MW turbine of 6148 t CO_2-eq are equivalent to the annual emissions from about 520 typical Western households.

Source: Adapted from Crawford (2009).

Table 6.3   The material usage, embodied energy, and embodied greenhouse-gas emissions of a 3-MW wind turbine.

Embodied energy of wind turbine (GJ)		Embodied greenhouse-gas emissions of wind turbine (t CO_2-eq)	
Initial embodied energy	84,237	Initial embodied emissions	5054
Recurring embodied energy	7939	Recurring embodied emissions	476
Embodied energy/MW-rated output	30,725	Embodied emissions/MW rated output	1844

Materials used in a typical 3-MW wind turbine include the following: 1140 t concrete, 240 t steel, 4 t copper, 2.4 t aluminum, 0.7 t plastic, 12 t fiber glass, 8 t epoxy, and 1.2 t paint. The embodied energy and greenhouse-gas emissions of these materials are summarized below.
Source: Adapted from Crawford (2009).

Considering the Adverse Effects of Renewable Energy

We have known about the energy crisis for decades. We also understand that fossil fuels will eventually run out, and that we should adopt renewable energy technologies on a large scale. Unfortunately, the various forms of renewable energy are not without their drawbacks.

Simon Zhang (25), a doctoral engineering graduate, works in a university laboratory that aims to produce biofuels from organic waste using a high-pressure and high-temperature process called hydrothermal liquefaction. Although this technology is far from commercialization, Simon enthuses that it would be ideal for converting wet biomass to oil. "We will be able to produce green oil. In the interim, though, we need to get things right with the roll-out of existent renewable energy systems," he remarks.

He explains the drawbacks of the currently implemented forms of renewable energy, such as wind, hydro-electric, solar, and nuclear power. "Large wind farms extract momentum from the air and change the meteorology downwind," Simon observes. "There is also concern about noise, bird strikes, and aesthetics, with many people adopting the attitude of 'not in my backyard'." Hydropower also changes the features of waterbodies and ecosystems.

Although solar electricity is an impressive initiative, solar-electricity farms take up valuable farmland. "The typical solar farm requires constant maintenance of the moving components," he observes. "Solar farms in deserts require constant cleaning due to sand deposition, and the solar cells only last around 25 years."

Considering the introduction of electric vehicles, Simon reminds us that electric vehicles primarily use lithium-ion batteries. "Lithium mining requires large amounts of water and releases toxic waste." Lithium, like many other minerals and metals used in renewable technologies, has limited reserves. To make matters worse, lithium-ion batteries are not recycled to the extent that they should be.

Simon observes that nuclear power might be an important part of the energy mix. This form of power is currently being generated through nuclear fission. However, research into nuclear fusion techniques is delivering an alternative that does not have the radioactivity dangers that nuclear fission has. "If only we could scale up the fusion ignition that was recently successfully demonstrated at the National Ignition Facility, we would be set for ages."

"In the meantime," remarks Simon, "we need a diverse energy portfolio."

What we also need is brilliant engineers like Simon Zhang to help imagine new technologies.

Doctoral engineering student, **Simon Zhang** (25), in his office at the University of Illinois, Urbana-Champaign. *Source:* Simon Zhang.

6.7 Economic Metrics

The practice of sound economic principles is vital to ensuring sustainable energy. A bankrupt system does not lead to sustainability. Economic indicators provide quantitative information to evaluate the financial viability of a new technology or process. This section probes how economic measurement can be incorporated into the process of evaluating whether the use of energy is sustainable. Of practical importance is *lifecycle costing*, which includes the initial cost of an energy system and the operating cost over the life of the system.

6.7.1 Present Value

The value of money changes with time due to inflation and other factors. The concept of present worth or *present value* (P) is based on the fact that a dollar today is worth more than a dollar one year from now, because a dollar can today be invested to earn more money during the time one would be waiting to receive the future dollar. The *future value* (F) of an invested sum of money (P) after n years with the *interest* compounded at i (% per year) is

$$F = P(1 + i)^n. \tag{6.3}$$

Note that the interest rate is sometimes also called the *discount rate*. Future costs are discounted to their present value according to the discount rate. This makes it possible to compare, for instance, a photovoltaic park that has a high initial cost, but low operating costs to a natural gas-fired combined cycle power plant that may have relatively low initial costs, but a high fuel cost over its lifetime.

Equation (6.3) can be rewritten to indicate the present value (P) of a future (F) sum of money:

$$P = \frac{F}{(1 + i)^n} \tag{6.4}$$

For instance, at 5% interest rate ($i = 0.05$), a $1 of savings one year from now is currently worth $\frac{\$1}{(1 + 0.05)^1} = \$0.95$. If invested over two years, its current value is $\frac{\$1}{(1 + 0.05)^2} = \$0.907$, etc.

6.7.2 Levelized Cost

Levelizing is a method of expressing costs or revenues that occur once (or at irregular intervals) as equivalent equal payments (or annuities) at regular intervals (Bejan et al. 1996). For instance, consider a loan of present value P on a solar-electric system that must be repaid as a series of equal annual payments A over n years. The present value may be calculated as follows:

$$P = A \frac{[1 - (1 + i)^{-n}]}{i} \tag{6.5}$$

Note that when $i = 0$, $P = An$.

The *capital recovery factor* (*CRF*) is the fraction of the capital cost per year that gives the annual revenue required to pay off the capital loan and all interest. The CRF is defined as:

$$CRF = \begin{cases} \dfrac{i}{[1 - (1 + i)^{-n}]} & \text{if } i \neq 0 \\ \dfrac{1}{n} & \text{if } i = 0 \end{cases} \tag{6.6}$$

6.7.3 Net Present Value

The *net present value* (*NPV*) may be used as a measure of economic value when comparing different investment options in a lifecycle cost analysis. The present value of a future cost *C* evaluated at year *j* can be calculated with Eq. (6.4):

$$P = \frac{C}{(1 + i)^j} \tag{6.7}$$

The net present value of the cost *C* to be paid annually for *n* years is:

$$NPV = \sum_{j=1}^{n} \frac{C}{(1 + i)^j} \tag{6.8}$$

If the cost is inflated at an annual *inflation rate* of *r*, the cost in year *j* is $C_j = C(1 + r)^j$ so that the net present value becomes:

$$NPV = \sum_{j=1}^{n} C\left(\frac{1 + r}{1 + i}\right)^j \tag{6.9}$$

6.7.4 Levelized Cost of Electricity (or Levelized Cost of Energy)

The *levelized cost of electricity* (*LCOE*), also sometimes called the levelized cost of energy, is the constant-dollar electricity (or energy) price that would be required over the life of the power plant to cover all operating expenses, the payment of debt and accrued interest on initial project expenses, and the payment of an acceptable return to investors (EIA 2022). *LCOE* is typically measured in ¢/kWh or $/MWh.

The *LCOE* may be calculated as the sum of annual levelized costs for an energy conversion system divided by annual energy production (E_{annual}):

$$LCOE = \frac{(NPV)(CRF)}{E_{annual}} \tag{6.10}$$

A simplified version of *LCOE* considers the interest rate (or "discount rate") (*i*) on the capital ($C_{capital}$) loaned to build a power plant, the inflation rate (*r*), the number of years in operation (*n*), the cost of operations and maintenance (*O & M*), and the cost of fuel.

To break even, a power plant should accomplish the following objectives:

- Pay off the capital loan and all the interest that the capital would have earned during the period of operation of the plant.
- Pay the annual operations and maintenance costs.

Equation (6.7) can be simplified as follows (Short et al. 1995):

$$LCOE = \frac{\text{Total lifecycle cost}}{\sum_{j=1}^{n} \dfrac{E_{annual}}{(1 + i)^n}}$$

$$= \frac{\left[CRF + (O \& M \text{ ratio})_{fixed}\right] (C_{capital})}{(8760 \times CF)} + (O \& M)_{variable} + (\text{Fuel}) \tag{6.11}$$

where *CRF* is defined in Eq. (6.6); *CF* is the capacity factor of the power plant, which is the average annual power output of the plant as a fraction of the rated power output; 8760 is the number of hours in a year; $C_{capital}$ is the capital cost in \$/kW; $(O \& M \text{ ratio})_{fixed}$ is the fixed operations and maintenance cost taken as a fraction of the capital cost; *(Fuel)* is the annual fuel cost; and $(O \& M)_{variable}$ is the variable operations and maintenance cost, which depends on the amount of electricity produced.

Example 6.1   Levelized Cost of Electricity

In 2023, the "rule of thumb" was that \$2 million is required to install a 1-MW solar-photovoltaic plant with battery storage (Lazard 2022). Calculate the levelized cost of electricity of a 300 MW$_e$ photovoltaic power plant in the Mojave Desert with an annualized capacity factor of 45%. The power plant has a lifetime of 30 years, and the loan is repaid over this period. The annual operations and maintenance cost is 3% of the capital cost. The interest rate (cost of capital) is 8%.

Required: We must find the *LCOE* of a 300 MW$_e$ photovoltaic plant.

Solution strategy: After determining the cost of the power plant and the *O&M* cost, we will determine the *LCOE* using Eq. (6.3) and the capital recovery factor using Eq. (6.4).

Assumptions: The variable operations and maintenance costs $(O \& M)_{variable}$ are assumed to be zero. Also, assume that land lease costs are incorporated into the fixed *O&M* costs.

Analysis:

$$\text{Levelized cost of electricity} = LCOE = \frac{\left[CRF + (O \& M \text{ ratio})_{fixed}\right](C_{capital})}{(8760 \times CF)} + \cancel{(O \& M)_{variable}} + \cancel{(Fuel)}$$

There is no fuel cost with a photovoltaic plant, so we immediately cancel that term in our LCOE formula. We also neglect $(O \& M)_{variable}$ as stated in the assumptions.

$$CRF = \frac{i}{\left[1 - (1 + i)^{-n}\right]} = \frac{0.08}{\left[1 - (1 + 0.08)^{-30}\right]} = 0.089$$

The capital needed to build the power plant, $C_{capital}$, is \$2 million for 1 MW, or \$2000/kW.
The annual O&M cost ratio is $(O \& M \text{ ratio})_{fixed} = 0.03$.
The capacity factor is $CF = 0.45$.
Therefore,

$$LCOE = \frac{\left[CRF + (O \& M \text{ ratio})_{fixed}\right](C_{capital})}{(8760 \times CF)} = \frac{[0.089 + 0.03](\$2000/kW)}{(8760 \text{ h} \times 0.45)} = \$0.06/kWh$$

The *LCOE* is therefore 6 ¢/kWh.

Discussion: The *LCOE* of 6 ¢/kWh is a conservative estimate based on the assumptions. This problem also disregards any financial or tax incentives offered to developers of renewable energy technologies. In some states in the United States, investment tax credits of up to 30% are available.

6.7.5   Economies of Scale and Learning Curve Estimation

Economy of scale (or the "cost-capacity relation," "bigger is cheaper") pertains to the non-linear reduction of capital equipment costs when the number or size of the equipment being purchased is increased. The economy of scale can be expressed as follows (Dieter and Schmidt 2021):

$$\frac{C_1}{C_0} = \left(\frac{L_1}{L_0}\right)^x \tag{6.12}$$

or $$\frac{C_1}{L_1} = \left(\frac{C_0}{L_0}\right)\left(\frac{L_1}{L_0}\right)^{x-1} \tag{6.13}$$

where C_0 is the cost of equipment at size or capacity, L_0. The exponent x varies from about 0.4 to 0.8 (Dieter and Schmidt 2021).

Example 6.2 Economy of Scale for Gas Turbine Plants

A 1-MW_e gas turbine electrical plant costs $2500/$kW_e$. Using an economy of scale calculation, estimate what a 100-MW_e gas turbine would cost. A scaling exponent of 0.67 is typical for this application.

From Eq. (6.13): $\frac{C_{100\,MW_e}}{L_{100\,MW_e}} = 2500\,\$/kW_e\left(\frac{100\,MW_e}{1\,MW_e}\right)^{0.67-1} = 547\,\$/kW_e$, which represents a substantial saving compared to the smaller gas turbine.

Related to economy of scale is the concept of *learning curves*, which refers to products or systems becoming cheaper over time. This is related to the fact that workers and managers, for instance, gain experience ("learn") in their manufacturing jobs. This implies that more products can be produced in a period, which decreases manufacturing costs (Dieter and Schmidt 2021). This improvement phenomenon can be expressed by a learning curve, such as the one suggested in Box 6.2 and illustrated in Box 6.3.

Box 6.2 Lazard's Levelized Cost of Electricity

Lazard (2022) has calculated the *LCOE* of all modern power plants since 2008. Figure 6.5 provides a comparison of several power conversion technologies. As the cost of renewable energy continues to decline, certain technologies (e.g., onshore wind and utility-scale solar), which became cost-competitive with conventional generation several years ago on a new-build basis, continue to maintain their competitiveness with the marginal cost of selected existing conventional generation technologies.

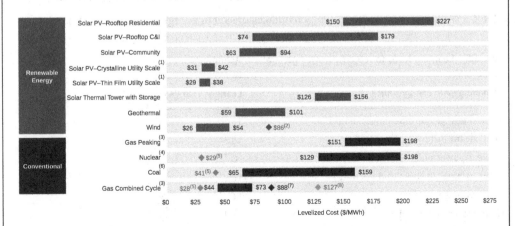

Figure 6.5 Comparison of the levelized costs of energy. The *LCOE* of utility-scale solar-photovoltaics (~$35/$MWh_e$) is cheap compared to that of large gas power plants (~$170/$MWh_e$) and large nuclear power plants (~$165/$MWh_e$). *Source:* Lazard (2022) / Reproduced with permission of Lazard.

Table 6.4 Exponent values for typical learning curve factors (P), where $P = 2^n$.

Learning curve factor, P (%)	n
25	-2
45	-0.868
65	-0.621
85	-0.234

The learning curve can be expressed by (Dieter and Schmidt 2021):

$$y = kx^n \tag{6.14}$$

where y is the production effort, expressed in hours per unit or \$ per unit; k is the effort to manufacture the first unit of production; x is the quantity; and n is the negative slope of the learning curve, expressed as a decimal. A factor of 20%, for example, means that, for each cumulative doubling of the product's sales or installations, the price falls by about 20%.

Table 6.4 shows some typical learning curve percentages and their n exponents. These exponents can easily be calculated, as illustrated in Example 6.3

Example 6.3 Calculate the Learning Curve Exponent for a Technology on a 65% Learning Curve

We will use Eq. (6.14) for this analysis: $y = kx^n$

For a 65% learning curve, $y_2 = 0.65y_1$ for $x_2 = 2x_1$, then:

$$\frac{y_2}{y_1} = \left(\frac{x_2}{x_1}\right)^n$$

$$\frac{0.65y_1}{y_1} = \left(\frac{2x_1}{x_1}\right)^n$$

$$n \log 2 = \log 0.65$$

$$n = -0.621$$

This value is also shown in Table 6.4, which shows some common learning curve factors.

Table 6.5 shows some learning curve percentages of salient energy conversion technologies.

Table 6.5 Learning rates for electric power generation technologies.

Technology and energy source	Learning rate (%)
Natural gas, gas turbine power plant	15
Coal, steam turbine power plant	8.3
Nuclear power plant	0
Wind turbine (onshore and offshore)	12
Biomass power plant	11
Hydroelectric power plant	1.4
Photovoltaic power plant	24.1

Source: McDonald and Schrattenholzer (2001); Rubin et al. (2015); ITRPV (2022).

Box 6.3   Learning Curve to Produce Photovoltaic Cells

As the price of photovoltaics drops (Figure 6.6), the number of installations skyrockets (Figure 6.7). The two are linked – the price decline both causes and is caused by the increasing number of solar installations. Figure 6.6 shows a learning factor of 24.1% from 1976 through 2021. Short-term fluctuations are partly due to supply chain issues. The "learning curve" is the biggest story in solar power.

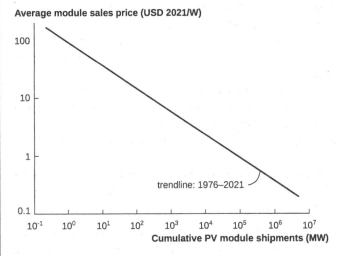

Figure 6.6   A 24.1% learning curve between 1976 and 2021 for the global production of photovoltaic power. *Source:* Adapted from ITRPV (2022).

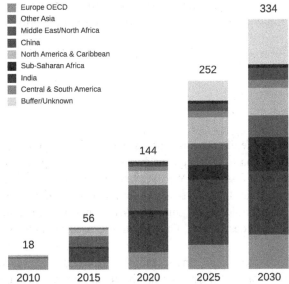

Figure 6.7   Global photovoltaic installation estimate and forecast (in GW), January 2022. *Source:* Adapted from BloombergNEF (2022).

New technologies, like photovoltaics, are first deployed in small and relatively cost-insensitive niches where their performance attributes are valued. The learning effects reduce costs and enable the technology to succeed in a broader range of applications. Increased deployment leads to further cost reductions, which lead to further deployment. This trend follows the S-curve, illustrated in Figure 4.8.

There are four principal ways in which costs come down over time (National Academies 2016):

1) *Research and development*: This process is uncertain in its outcomes, often creates significant spillover benefits for other sectors, and may feature long time horizons.
2) *Learning by doing*: This occurs as a byproduct of manufacturing and deployment. Companies develop (generally incremental) improvements to industrial operations, installation procedures, and sales and financing processes.
3) *Economies of scale*: These result from companies or industries getting larger, spreading some relatively fixed costs over a larger volume of product sales. At the industry level, one way in which economies of scale might reduce costs is greater specialization, as companies emerge to focus on one component of the product or supply chain rather than all companies being required to be generalists.
4) *Learning by waiting*: This involves harnessing the spillover effects from other industries, technologies, or countries. This is "the result primarily of innovation that occurs elsewhere and not of accelerating technology deployment."

The drawback of learning curves pertains to *complacency*. If people know that costs are going to fall, why should they buy solar panels now? However, costs do not fall just due to the passage of time. The fourth item, "learning by waiting," depends on somebody else doing the research and development, deployment, and learning by doing. It means being a free-rider on the efforts of others.

Photovoltaics did not become more affordable just because 1980 (or 2008) turned into 2015, but because, during that time, millions of people around the world installed solar, even when it was not the most cost-effective option. Now, in many cases, photovoltaics cost less than grid power – thanks to those pioneers who brought the industry to its current state of maturity.

Source: Adapted from O'Connor (2016).

6.8 Outlook

Our energy choices have a profound impact on the life support capacity of the Earth, including on our atmosphere, ecosystems, and on climate – mainly through the emission of human-produced greenhouse gases from combusting carbon-based fossil fuels. We have known this for a long time. Climate change affects all components of human and natural systems (see Figure 6.8), adding both complexity and urgency to the search for sustainable energy solutions. We do not need to accept another century of resource depletion, exergetic destruction, and pollution caused by inefficient energy conversion systems.

If *design* has always been about looking forward – and doing so with the hope that what was to come would be better than what happened before – it must now also be about looking back and reflecting on how our lives (and those of future generations) might not have been improved by

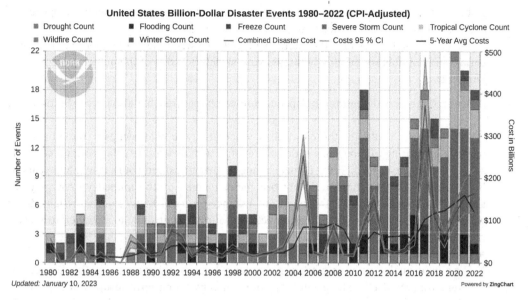

Figure 6.8 US *billion-dollar event* frequency, 1980–2021. The USA has sustained 323 separate weather and climate disasters since 1980, where overall damages/costs reached or exceeded $1 billion. Total direct costs since 2018 exceed $2.2 trillion (CPI-adjusted to 2022). Between 1980 and 2021, there were 7.7 weather and climate-related billion-dollar events in the United States. Between 2017 and 2021, there were 17.8 events. The billion-dollar events are becoming more frequent, and as shown, they are also leading to greater costs. Central to this increased frequency and severity of weather and climate events is CO_2 emissions from fossil-fuel energy conversion. *Source:* U.S. Department of Commerce / Public Domain.

the prevailing growth economy. To secure a safe, reliable, and low-carbon energy future, we must alter both *technologies and human behavior* (Dietz et al. 2009; Byrne et al. 2022). We have focused a lot on technical fixes rather than ways to alter lifestyles and social norms (Sovacool and Dworkin 2014).

Our search for sustainable energy will bear fruit if we conserve and regenerate nature. That implies simultaneously addressing the interlinked social, technological, environmental, economic, and political challenges.

Problems

6.1 List all the UN's SDGs related to energy (production, storage, distribution and management, and consumption). Briefly explain your selection. *Source:* https://sdgs.un.org/goals.

[Answer: Open-ended question. Independent research required, but also refer to Section 6.2.]

6.2 "Nations remain dependent on fossil fuels and are struggling to build up supplies precisely at a moment when scientists say the world must slash its use of oil, gas, and coal to avert irrevocable damage to the planet. This dependency on fossil fuels is exacerbated by countries coping with energy shocks following widespread geopolitical instability, including wars. Governments are therefore focused on alleviating near-term energy shocks and boosting global oil production to help ensure energy security. This is not in line with the agenda to decarbonize society as a matter of urgency" (Plumer et al. 2022).

Produce a sketch that shows the interdependencies between the UN's various SDGs as they pertain to fossil fuel usage. Briefly discuss your sketch in terms of the quoted paragraph by Plumer et al.

[Answer: Open-ended question. Independent research required, but also refer to Section 6.2.]

6.3 "While countries could greatly reduce their vulnerability to wild swings in the oil and gas markets by shifting to renewable sources of energy such as wind or solar power – which is also in the playbook for fighting climate change – that transition could take many years. It is therefore important that we do not become consumed by the immediate energy crisis and neglect longer-term policies to cut our reliance on fossil fuels. Such shortsightedness would surely lead to more oil and gas shocks and further contribute to a dangerously over-heated Earth" (Rockström et al. 2021).

a) Discuss the above statement by virtue of a STEEP analysis as elucidated in Table 6.1.

b) As a policymaker at an engineering company, how would you employ the "3-C entre-preneurial mindset" to help you form your company's long-term sustainability policies? The "3-C" mindset entails you being curious, making connections between apparently disparate aspects, and creating value (https://orchard-prod.azurewebsites.net/media/Framework/KEEN_framework_new.pdf).

[Answer: Open-ended question. Independent research required, but also refer to Section 6.3.]

6.4 Study the fact sheet of the National Renewable Energy Laboratory (NREL) pertaining to "Lifecycle greenhouse gas emissions from electricity generation: Update," https://www.nrel.gov/docs/fy21osti/80580.pdf

a) Explain the use of the "g CO_2-equivalent/kWh" metric to depict "carbon intensity." (Also refer to the discussion in Box 5.2.)

b) Briefly discuss the electricity-producing technologies with the highest and lowest carbon emissions.

c) Briefly discuss the energy storage technologies with the highest and lowest emissions.

[Answer: Open-ended question. Independent research required.]

6.5 Using the online Kaya Identity Scenario Prognosticator tool, (a) find the best- and worst-case scenarios for business-as-usual assuming a global population of 11 billion people in 2100, and the best- and worst-case scenarios for business-as-usual. Also, (b) find the amount of carbon-free energy needed (in terawatts) to stabilize atmospheric CO_2 at 500 ppm and at 350 ppm. Summarize your simulation results in a one-page fact sheet.

[Answer: Open-ended question. Independent research required, but also refer to Section 6.5.]

6.6 Study the interactive world map "A good life for all within planetary boundaries," https://goodlife.leeds.ac.uk/national-snapshots/world-map. (a) Investigate the number of biophysi-cal boundaries that China, India, South Africa, and the United States transgress in compar-ison to the number of social thresholds they achieve. Ideally, countries should achieve the social thresholds without transgressing the biophysical boundaries. (b) Then, go to https://goodlife.leeds.ac.uk/national-snapshots/explore-relationships/and apply the sliders to each of the social indicators to raise or lower the minimum thresholds that you choose for a "good life." Briefly comment on what your choices would mean for sustainability if they were extended to all people.

[Answer: Open-ended question. Independent research required, but also refer to Section 6.6.]

6.7 A 500-MW$_e$ combined cycle (gas turbine and steam turbine) power plant costs $350 million to build. If the power plant has an 85% capacity factor, fuel cost is 2 ¢/kWh, it has fixed operations and maintenance costs of 4% of the plant capital cost, the cost of variable operations and maintenance is $12/MWh, and money is loaned at an interest rate of 8% with a 30-year payback period, what is the levelized cost of electricity?

[Answer: 5.5 ¢/kWh.]

6.8 One of the gas boilers at an old steam power plant must be replaced at a cost of $4 million. If this money is loaned at an interest rate of 7% with a 20-year payback period, what total amount will the power plant be paying back to the bank after 20 years?

[Answer: around $7.5 million.]

References

Bejan, A., Tsatsaronis, G. and Moran, M. (1996). *Thermal Design and Optimization*. New York, NY: Wiley.

BloombergNEF (2022). Solar – 10 predictions for 2022. https://about.bnef.com/blog/solar-10-predictions-for-2022.

Byrne, J., Lund, P. and Taminiau, J. (2022). Rapid climate transformation requires transformative policy and science thinking – an editorial essay. *Wiley Interdisciplinary Reviews: Energy and Environment* 11 (1): e428.

Crawford, R.H. (2009). Life cycle energy and greenhouse emissions analysis of wind turbines and the effect of size on energy yield. *Renewable and Sustainable Energy Reviews* 13 (9): 2653–2660.

Dieter, G.E. and Schmidt, L. (2021). *Engineering Design*. New York, NY: McGraw-Hill.

Dietz, T., Gardner, G.T., Gilligan, J., Stern, P.C. and Vandenbergh, M.P. (2009). Household actions can provide a behavioral wedge to rapidly reduce US carbon emissions. *Proceedings of the National Academy of Sciences of United States of America* 106 (44): 18452–18456.

ecoinvent (2022). ecoinvent® LCA database. http://www.ecoinvent.org

EIA (2022). Energy outlook 2022. US Energy Information Administration, Washington, DC. 3 March 2022. https://www.eia.gov/outlooks/aeo (accessed 12 December 2021).

ExxonMobil (2019). *Outlook for Energy: A Perspective to 2040*. Irving, TX: ExxonMobil.

Fanning, A.L., O'Neill, D.W., Hickel, J. and Roux, N. (2022). The social shortfall and ecological overshoot of nations. *Nature Sustainability* 5 (1): 26–36.

Fonseca, L.M., Domingues, J.P. and Dima, A.M. (2020). Mapping the sustainable development goals relationships. *Sustainability* 12 (8): 3359.

GaBi (2022). GaBi LCI® and MCI configuration tool. https://gabi.sphera.com/america/index.

Geels, F.W., Sovacool, B.K., Schwanen, T. and Sorrell, S. (2017). Sociotechnical transitions for deep decarbonization. *Science* 357 (6357): 1242–1244.

Gerber, L. (2014). *Designing Renewable Energy Systems: A Life Cycle Assessment Approach*. Boca Raton, FL: CRC Press.

Hammond, G.P. and Jones, C.I. (2008). Embodied energy and carbon in construction materials. *Proceedings of the Institution of Civil Engineers-Energy* 161 (2): 87–98.

ITRPV (2022). International Technology Roadmap for Photovoltaic (ITRPV): 2021 Results. https://www.vdma.org/international-technology-roadmap-photovoltaic

Kaya, Y. and Yokoburi, K. (1997). *Environment, Energy, and Economy: Strategies for Sustainability*. Tokyo, Japan: United Nations University Press.

Lazard (2022). Lazard's levelized cost of energy analysis – version 14.0. https://www.lazard.com/perspective/lcoe2020.

Mayer, A.L. (2008). Strengths and weaknesses of common sustainability indices for multidimensional systems. *Environment International* 34 (2): 277–291.

McCollum, D.L., Echeverri, L.G., Busch, S., Pachauri, S. et al. (2018). Connecting the sustainable development goals by their energy inter-linkages. *Environmental Research Letters* 13 (3): 033006.

McDonald, A. and Schrattenholzer, L. (2001). Learning rates for energy technologies. *Energy Policy* 29 (4): 255–261.

Neumayer, E. (2001). The human development index and sustainability – a constructive proposal. *Ecological Economics* 39(1): 101–114.

O'Connor P. (2016). What is the learning curve—and what does it mean for solar power and for electric vehicles? https://blog.ucsusa.org/peter-oconnor/what-is-the-learning-curve.

Plumer, B., Friedman, L. and Gelles, D. (2022). As war rages, a struggle to balance energy crunch and climate crisis. *The New York Times*, 10 March 2022. https://www.nytimes.com/2022/03/10/climate/climate-oil-crisis-global.html.

Raworth, K. (2017). A doughnut for the Anthropocene: humanity's compass in the 21st century. *The Lancet Planetary Health* 1 (2): e48–e49.

Rockström, J., Gupta, J., Lenton, T.M., Qin, D. et al. (2021). Identifying a safe and just corridor for people and the planet *Earth's Future* 9 (4): e2020EF001866.

Rubin, E.S., Azevedo, I.M., Jaramillo, P. and Yeh, S. (2015). A review of learning rates for electricity supply technologies. *Energy Policy* 86: 198–218.

Ryan, Z. (2020). The design imagination. In: Hiesinger, K.B. and Millar, M. (eds.) *Designs for Different Futures*. 41–47. New Haven, CT: Yale University Press.

Short, W., Packey, D.J. and Holt, T. (1995). *A Manual for the Economic Evaluation of Energy Efficiency and Renewable Energy Technologies* (No. NREL/TP-462-5173). Golden, CO: National Renewable Energy Lab (NREL).

Singh, G.G., Cisneros-Montemayor, A.M., Swartz, W., Cheung, W. et al. (2018). A rapid assessment of co-benefits and trade-offs among Sustainable Development Goals. *Marine Policy* 93 (July): 223–231.

Sovacool, B.K. and Dworkin, M.H. (2014). *Global Energy Justice*. Cambridge, UK: Cambridge University Press.

UN (2022). United Nations' 17 Sustainable Development Goals. https://sdgs.un.org/goals

Mini Project 2

Energy Scenarios for a Sustainable World

In this climate action simulation (adapted from Rooney-Varga et al. 2020), you will learn about the dynamics of the climate-energy system by simulating the climate and energy outcomes of your own decisions with the interactive computer model En-ROADS. You will also engage in the social dynamics of climate and energy decision-making in a role-play simulation. Please note that some of the terminology used in the En-ROADS simulation might differ from that used in the UN's SDGs.

In this mini project, you will work with the En-ROADS simulation model to create a scenario that can mitigate global warming to the internationally agreed target of less than 2 °C compared to preindustrial times. In addition, you will consider the economic, political, and social issues relevant to the successful implementation of your policies.

Your Steps

1) *Read the following paper:* Grubler et al. (2018).
2) *Access En-ROADS* at https://en-roads.climateinteractive.org/. Review these materials to help you understand how to use this model:
   - A 20-min introductory video to En-ROADS: https://youtu.be/7Muh-eoPd3g
   - Quick guide to the En-ROADS policy levers: https://www.climateinteractive.org/wp-content/uploads/2019/09/EnROADS-one-page-guide-to-control-panel-v9-new-layout.pdf
3) *Develop a scenario to meet your goals:* Use En-ROADS to develop your vision of how to successfully limit global warming to less than 2 °C as per the Paris Agreement. Ensure your scenario factors are in the goals listed below.
4) *Submit a write-up expressing your vision*: After developing your preferred scenario, your team must write one thoughtful and concise response to the stated goals below. Develop your vision and approach on the issues. Instructions on how to format your write-up are at the end of this assignment.

Your Goals

Your project mission is to recommend a set of global policies, investments, and actions that meet the following five objectives as fully as possible. When addressing these objectives, you should revisit the UN's SDG targets and check for parity between the SDG goals and the five stated

Energy Systems: A Project-Based Approach to Sustainability Thinking for Energy Conversion Systems,
First Edition. Leon Liebenberg.
© 2024 John Wiley & Sons, Inc. Published 2024 by John Wiley & Sons, Inc.
Companion website: www.wiley.com/go/liebenberg/energy_systems

objectives. Also comment on disparities between the stated objectives and see if you could rather achieve the SDG targets.

Objective 1: Limit global warming. Global warming above 2 °C will cause dangerous and irreversible impacts that will harm our prosperity, security, health, and lives. (Warming of less than 2 °C will also have destructive consequences – some of which we are already seeing today – especially as it approaches 2 °C.) However, scientists believe that if warming is kept below the 2 °C threshold, the risks will be more manageable.

Objective 2: Preserve and create a healthy economy. Your policies should strive to lead a global energy transition that would preserve and create a healthy global economy. You must decide how to balance the short-term costs of climate actions with the long-term costs of damages from climate inaction. Note that financial costs are not explicitly predicted in the model as they are uncertain and controversial. You may offer your own hypotheses on the financial impacts of different decisions.

While assessing the economics, also consider the potential to offset short-term costs with additional co-benefits, e.g., benefits to the economy, public health, national security, and other areas that could provide benefits on top of the direct benefits of mitigating global warming.

Objective 3: Promote equity, well-being, and a just transition. Consider the impacts of your policies on both high-income and low-income nations, and on the rich and poor people within nations. Consider whether your policies will disproportionately harm certain groups and how to mitigate such harms (e.g., if you favor policies that would reduce or shut down fossil-fuel use, how will your policies address the resulting unemployment of people employed by the fossil-fuel industry?). Also consider how opportunities for the new green economy can be shared more equitably and not leave marginalized groups behind. Consider the social SDGs 1–12.

Objective 4: Protect the environment. Many environmental challenges besides climate change threaten human welfare – e.g., water shortage, air and water pollution (smog, particulates), soil loss, plastic pollution, anoxic zones (dead zones) in rivers and oceans, the extinction of species, etc. Your proposals should minimize other harmful effects on the environment. Consider the environmental SDGs 13–15.

Objective 5: Be realistic, but not cynical. Imagine a scenario of what could be possible if human civilization operates at its best. Consider the governance SDGs 16–17.

Your Tool: *En-ROADS*

You will be testing your recommendations using the En-ROADS simulation model. En-ROADS is an interactive tool for simulating the long-term impacts of policy actions available to mitigate global warming, e.g., policies affecting energy supply, energy efficiency, carbon emission prices, land use, and other crucial factors that can mitigate greenhouse-gas emissions. En-ROADS is grounded in the best available peer-reviewed science about climate impacts, solutions and the complex interactions of the climate, energy, land, population, and economic systems.

Please note: En-ROADS is a global model, which means that the policy levers simulate if the action were to be applied to the whole world. En-ROADS does not attempt to address the complex nuances of how different countries and political groups might respond to each policy. En-ROADS is focused on the physical science of the feasibility and impacts of each solution, as well as on the application of assumed societal values.

More information on En-ROADS is available at http://www.climateinteractive.org/simulations/en-roads

Your Write-Up

Write a memo describing your proposals. Please respond to all the questions and organize your write-up into the following sections. There is no minimum or maximum length for your write-up. Create a compelling analysis with clear and focused writing.

Section A: Plan

1) *Policies*: Summarize your team's plan using the **template provided at the end of this assignment**. You may choose to share screenshots of specific graphs that caught your attention and are worth noting.

Section B: Meeting the Goals

2) *Climate*: How well do you think your proposal does in meeting the Paris Agreement's climate goals? If it does not meet the goals, why is this acceptable to you?
3) *Economy*: If the world followed your recommendations, how would the economy be different at different points in the future, e.g., 2030, 2050, and 2100? In what ways would it be better? In what ways would it be worse?
4) *Equity and wellbeing*: How does your proposal strive to increase equity across nations and different peoples?
5) *Environment*: To what extent might your proposal mitigate other environmental challenges (e.g., biodiversity, pollution, water and air quality)? To what extent might your proposals cause or worsen other environmental problems?
6) *Realism without cynicism*: What would it take for your proposal to be realized? What barriers might arise in the implementation of your proposals, and how might they be addressed? To get started, what actions and priorities are needed as soon as possible from businesses, civil society, governments, or the public?

Section C: Reflections

7) *Winners/losers*: Who would be the biggest winners and losers globally in your proposed future? Create a table with two columns: one for winners and one for losers.
8) *Surprises from En-ROADS:* What surprised you about the behavior of the energy and climate system as captured in the simulation? For example, what actions had a bigger or smaller effect than you thought they would have? Did you discover why that might be?
9) *Feelings*: How did your insights from the model and this assignment make you feel (your feelings, not thoughts)?
10) *Hope and personal action*: What trends in the world give you hope that your proposals are possible? What can you personally do to help create the necessary changes?

Use this template to present your plan in Section A of your write-up:

1) Provide a short, memorable name for your plan.
2) Paste a screenshot of the main En-ROADS interface showing your results.
3) Present bullet points summarizing your most important policies and outcomes.

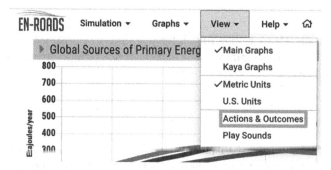

Figure MP 2.1 Actions and outcomes.

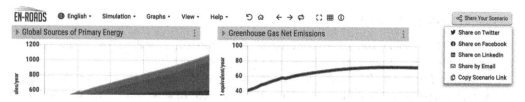

Figure MP 2.2 Copy scenario link.

4) Paste the text from "Actions and outcomes" to document all the assumptions and policy settings you choose in your scenario. Access the "Actions and outcomes" feature from the "View" menu (see Figure MP 2.1).
5) Paste your URL from the "Copy scenario link" option in the "Share your scenario" feature on the top tool bar (see Figure MP 2.2).
6) Optional – You may choose to share screenshots of up to three specific graphs that caught your attention and are worth noting. For each graph you share, please add a brief note about why it is worth noting.

References

Grubler, A., Wilson, C., Bento, N., Boza-Kiss, B. et al. (2018). A low energy demand scenario for meeting the 1.5 °C target and sustainable development goals without negative emission technologies *Nature Energy* 3 (6): 515–527.

Rooney-Varga, J.N., Kapmeier, F., Sterman, J.D. and Jones, A.P. (2020). The climate action simulation. *Simulation and Gaming* 51 (2): 114–140.

Source: Elements by Visual Capitalist.

Energy Systems: A Project-Based Approach to Sustainability Thinking for Energy Conversion Systems,
First Edition. Leon Liebenberg.
© 2024 John Wiley & Sons, Inc. Published 2024 by John Wiley & Sons, Inc.
Companion website: www.wiley.com/go/liebenberg/energy_systems

Week 3 – Fuels

The combustion or "burning" of chemical fuels is important in contemporary energy conversion systems. Coal, oil (petroleum), and natural gas will continue to play important roles in our energy future. Coal, the most CO_2-polluting of fossil fuels, will remain prominent in China and in some other countries that are not members of the Organization for Economic Cooperation and Development (OECD).

Although global consumption appears to have peaked in 2013, oil is predicted to remain the largest source of primary energy until 2050, as it is essential for commercial transportation and the chemicals industry. Natural gas demand will probably continue to rise, largely to help meet increasing needs for electricity and lower-CO_2-emission industrial heat. Reducing global CO_2 emissions to net-zero by 2050 will, however, require energy combustion systems to rapidly convert to low-carbon fuels or, preferably, no-carbon fuels. Hydrogen (H_2) and ammonia (NH_3) top the list of candidate fuels in this regard, but innovations are required to ensure net-zero carbon emissions in their life-cycles. Biofuels such as biodiesel and biogas will therefore also become increasingly important, especially if their production is paired with carbon capture systems.

Chapter 7 – Chemical fuels: Combustion of fuel in air; enthalpy of formation; enthalpy of combustion, and heating values; entropy production and exergy destruction during combustion; the Gibbs function and spontaneous reactions; outlook

Chapter 8 – Coal: Coal classification, consumption, and reserves; combustion of coal; the future of coal

Chapter 9 – Oil and natural gas, and hydrogen and biofuels: Oil classification, consumption, reserves, and combustion of oil; diesel, and gasoline; natural gas classification, consumption, reserves, and combustion; hydrogen and ammonia; biofuels; the future of oil, gas, hydrogen and ammonia, and biofuel

Mini Project 3 – Combustion of fossil fuels

7

Chemical Fuels

Figure 7.1 shows that coal, oil, and natural gas are projected to continue to play dominant roles as fuels until 2040. Solar and wind are projected to contribute 6.7% and nuclear power 6% to the total global energy mix in 2040; but oil is predicted to contribute 30%, natural gas 28%, and coal 14%. The battle for decarbonization will clearly be a struggle and we will have to contend with fossil fuels for many years to come. It is therefore important to understand the chemistry involved in converting coal, oil, and gas to useful energy.

A *fuel* is a substance that will spontaneously participate in an exothermic reaction. For power plant applications, the fuel must be readily available, and its use must be economically and environmentally justifiable.

Fuels can be classified as nuclear fuels or chemical fuels, as shown in Table 7.1. Chapters 7, 8, and 9 only deal with chemical fuels, specifically coal, oil, gas, hydrogen, ammonia, and biofuels. Chapters 19–21 deal with nuclear fuels and Chapter 24 deals with the electrochemical (not chemical or combustion) reaction of hydrogen in fuel cells.

7.1 Combustion of Fuel in Air

Combustion is a chemical reaction during which the *fuel* is oxidized (or "burned") and a large quantity of energy is released. With combustion, the weak molecular bonds of the reactants (which usually comprise C, H, or O atoms) are broken and replaced by stronger bonds in the products (usually comprising CO_2 and H_2O, among others) while liberating the excess bond energy as heat. Combustion is therefore characterized by a fuel that loses its electrons in an uncontrolled manner to an oxidizer, usually air.

Chemical fuel is a combustible material that may come in many different forms, but the emphasis in this book is on fossil fuels (or "hydrocarbons"), where the principal combustible elements are carbon and hydrogen. Sulfur, present in small quantities in most fossil fuels, is also combustible and should be accounted for in combustion calculations.

In a combustion process, mass is conserved. So, the mass of the products is equal to the mass of the reactants. For instance, when hydrogen is completely oxidized in oxygen to form water, conservation of mass requires the chemical reaction to be balanced as follows (Glassman et al. 2014):

$$H_2 + \frac{1}{2}O_2 \rightarrow H_2O \tag{7.1}$$

Energy Systems: A Project-Based Approach to Sustainability Thinking for Energy Conversion Systems, First Edition. Leon Liebenberg.
© 2024 John Wiley & Sons, Inc. Published 2024 by John Wiley & Sons, Inc.
Companion website: www.wiley.com/go/liebenberg/energy_systems

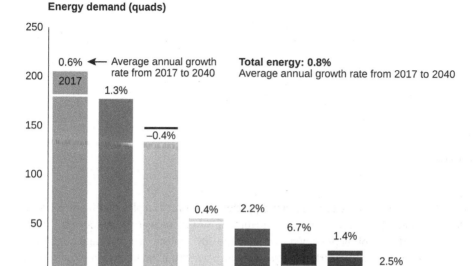

Figure 7.1 Global demand by fuel between 2017 and 2040. *Note:* Units are in quadrillion Btus (or "quads"), with 1 quad = 1.055 EJ. *Source:* ExxonMobil (2021) / Reproduced with permission of Exxon Mobil Corporation.

Table 7.1 Classification of fuels.

Fuel	Classification	Types and some salient examples
Chemical fuel	Fossil	*Solids:* Coal
		Liquids: Oil (petroleum), shale oil, oil extracted from tar sands
		Gases: Natural gas
	Manufactured (synthetic)	*Solids:* Coke, charcoal
		Liquids: Diesel, gasoline, kerosene, liquefied petroleum gas (LPG), coal tar, naphtha, ethanol
		Gases: Hydrogen, propane, methane, coal gas, water gas, blast furnace gas, coke oven gas, compressed natural gas (CNG)
	Biomass	Wood and wood processing wastes – firewood, wood pellets and wood chips, lumber and furniture mill sawdust and waste, and black liquor from pulp and paper mills.
		Agricultural crops and waste materials, including corn, soybeans, sugar cane, switchgrass, woody plants, and algae, and crop and food processing residue.
		Biogenic material in municipal solid waste, including paper, cotton and wool products, and food, yard and wood waste.
		Animal manure and human sewage
		Algae (microalgae and macroalgae)
	Waste material	Plastic
Nuclear fuel	Fission	Uranium, plutonium, thorium
	Fusion	Deuterium, tritium

Source: Adapted from EIA (2021).

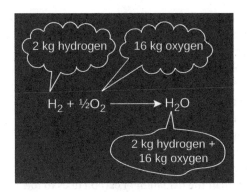

In this chemical reaction, 1 kmol of hydrogen combines with $\frac{1}{2}$ kmol of oxygen to form 1 kmol of water. From a mass perspective, 2 kg of hydrogen (i.e., 2(1 kg/kmol H) = 2 kg) combines with 16 kg of oxygen (i.e., $\frac{1}{2}$(2 × 16 kg/kmol O) = 16 kg) to form 18 kg of water (i.e., 2(1 kg/kmol H) + 1(16 kg/kmol O) = 18 kg). The molecular weights are 1 for a hydrogen atom and 16 for an oxygen atom. These molecular weights are obtained from periodic tables, of which the online version of the *NIST Chemistry WebBook* is especially useful: https://webbook.nist.gov/chemistry. In the combustion process, the mass of the products equals that of the reactants. The mass of each element is also conserved.

Most combustion will not be with oxygen, but rather with *air*. The composition of dry air on a volume basis is 78.1% nitrogen (N_2), 20.9% oxygen (O_2), and 0.9% argon (Ar), with lesser amounts of carbon dioxide, helium, hydrogen, and neon. The volume proportion of air is frequently approximated in combustion calculations by lumping the noncombustibles (argon and others) with nitrogen, giving *79% nitrogen* and *21% oxygen*. This is done as nitrogen is typically inert at normal temperatures. (Although nitrogen dissociates at high temperatures to form pollutants such as nitric oxide, which reacts with sunlight to form photochemical smog, we will not consider that complex chemistry when dealing with nitrogen.) Dry air may therefore be approximated as follows on a volume basis:

$$0.21 \, mol \, O_2 + 0.79 \, mol \, N_2 \rightarrow 1 \, mol \, air \tag{7.2}$$

For 1 mol of oxygen, the composition of dry air is:

$$1 \, mol \, O_2 + \frac{0.79}{0.21} \, mol \, N_2 \rightarrow \frac{1}{0.21} \, mol \, air$$
$$1 \, mol \, O_2 + 3.76 \, mol \, N_2 \rightarrow 4.76 \, mol \, air \tag{7.3}$$

Table 7.2 explains how the molecular weight of air is approximated as 29. This means that 1 mol of air weighs 29 g, or 1 kmol of air weighs 29 kg.

$$1 \, kmol \, O_2 + 3.76 \, kmol \, N_2 \rightarrow 4.76 \, kmol \, air$$
$$1 \, kmol \, air \approx 29 \, kg$$

Table 7.2 Molecular weight of air (atomic masses: N = 14; O = 16).

	Mole fraction (y_i) (kmol/kmol)	Molar mass (kg/kmol)	Mass (kg)	Mass fraction (x_i) (kmol/kmol)
O_2	0.21	2 × 16 = 32	6.72	0.23
N_2	0.79	2 × 14 = 28	22.12	0.77
		Mass of 1 kmol of air:	28.84 ≈ 29	

The fuel must be brought above its *ignition temperature* to start the combustion. The minimum *ignition temperatures* in atmospheric air are approximately 260 °C for gasoline, 400 °C for carbon, 580 °C for hydrogen, and 630 °C for methane (Glassman et al. 2014). Proportions of the fuel and air must be in the proper range for combustion to begin. For example, natural gas does not burn in air in concentrations less than 5% or greater than about 15%.

For *complete combustion* to occur, the oxidizer (air) must be available in sufficient quantity so that no unoxidized carbon (in the form of C or CO) and no unoxidized hydrogen (in the form of H_2 or OH) can exist in the products. *Incomplete combustion* implies a deficiency in the amount of oxidizer, resulting in the production of unburned fuel and amounts of C, CO, H_2, or OH. With insufficient oxygen, the hydrogen in the fuel still completely reacts with the oxygen to form water; the oxidation of hydrogen proceeds to completion much more readily than the oxidation of carbon does to form carbon dioxide.

If there is just enough oxygen (or other oxidizer) available to ensure complete combustion, the reaction is termed *stoichiometric*. The stoichiometric proportion of air is therefore the minimum amount of air necessary to oxidize all C in the fuel to CO_2, all H in the fuel to H_2O, and all S (if any) to SO_2 and SO_3. In normal practice, SO_2 and SO_3 cannot be cooled to room temperature, or else highly corrosive H_2SO_4 might form. This results in a loss of heat, even when the combustion is stoichiometric.

If there is less air (*excess fuel*) available than is required for stoichiometric combustion, the fuel-air mixture is called *rich*. Operating a boiler with a rich fuel–air mixture could lead to incomplete combustion and excessive pollution. In practice, excess air is always supplied to ensure complete combustion (Severns et al. 1948). If there is more air (*excess oxygen*) available than is required for stoichiometric combustion, the fuel–air mixture is called *lean*. When excess air is supplied, it must be heated, thus removing useful energy from the boiler process.

The ratio of mass of air to mass of fuel is called the *air–fuel ratio (AFR)*:

$$AFR = \frac{m_{air}}{m_{fuel}} = \frac{N_{air}M_{air}}{N_{fuel}M_{fuel}} \tag{7.4}$$

where N is the number of moles (mol) and M is the molar mass (kg/mol).

When analyzing pollutants, it is often necessary to calculate the partial pressure of products. *Dalton's Law* is invoked to do these calculations. This law states that, in a mixture of perfect gases, each gas – at N_i moles – exerts a partial pressure (p_i) that is independent of the other components. The total pressure (p) of a mixture of gases is equal to the sum of the partial pressures (Moran et al. 2018):

$$\frac{p_i}{p} = \frac{N_i}{N} = y_i \tag{7.5}$$

and $$p = \sum_i p_i \tag{7.6}$$

Example 7.1 Stoichiometric Combustion of Iso-Octane with Air

Iso-octane (C_8H_{18}) is burned in air at stoichiometric conditions. Calculate (a) the mass of each of the products and reactants for a kilomole of fuel, (b) the air–fuel ratio, and (c) the partial pressure of the water vapor in the exhaust.

Required: We must find masses of reactants and products, and then determine the *AFR* for C_8H_{18} burning stoichiometrically in air. We must then determine the partial pressure of H_2O in the exhaust.

Solution strategy: We will balance the stoichiometric equation for C_8H_{18} combusting in air. We will then calculate the masses of reactants and products, using:

$$mass(kg) = (number\ of\ moles,\ mol) \times (molar\ mass, kg/mol)$$

Then we can determine the $AFR = \dfrac{m_{air}}{m_{fuel}}$.

Dalton's Law, Eqs. (7.5) and (7.6) will then be used to calculate the partial pressure of H_2O.

Assumptions: Assume that N_2 is inert. So, the same amount of N_2 appears before and after combustion. Also, assume that the atmospheric pressure is 100 kPa.

Analysis:

Stoichiometric combustion of C_8H_{18} in air:

$$C_8H_{18} + a\underbrace{[O_2 + 3.76\,N_2]}_{air} \rightarrow b\,CO_2 + c\,H_2O + d\,3.76\,N_2$$

We then balance the equation to find a, b, c, and d:

C: $8 = b$

H: $18 = 2c$ $\qquad\qquad c = 9$

O: $2a = 2b + c$ $\qquad a = 12.5$

N: $12.5(3.76) = 3.76d$ $\qquad d = 12.5$

The balanced equation is: $C_8H_{18} + 12.5[O_2 + 3.76\,N_2] \rightarrow 8\,CO_2 + 9\,H_2O + 47N_2$

a) We can now find the masses of each of the reactants and products. The molar mass for C is 12 kg/kmol, for H it is 1 kg/kmol, for O it is 16 kg/kmol, and for N it is 14 kg/kmol. The molar mass of air is assumed to be 29 kg/kmol.

$$m_{air} = N_{air}M_{air} = [12.5\,kmol\,O_2 + (12.5 \times 3.76)\,kmol\,N_2] \times 29\,kg/kmol\,air = 1725.5\,kg$$

$$m_{C_8H_{18}} = N_{C_8H_{18}}M_{C_8H_{18}} = (8\,kmol \times 12\,kg/kmol) + (18\,kmol \times 1\,kg/kmol) = 114\,kg$$

$$m_{CO_2} = N_{CO_2}M_{CO_2} = 8\,kmol \times (12 + 2(16))\,kg/kmol = 352\,kg$$

$$m_{H_2O} = N_{H_2O}M_{H_2O} = 9\,kmol \times (2(1) + 16)\,kg/kmol = 162\,kg$$

$$m_{O_2} = N_{O_2}M_{O_2} = 12.5\,kmol \times (2(16))\,kg/kmol = 400\,kg$$

$$m_{N_2} = N_{N_2}M_{N_2} = 47\,kmol \times (2(14))\,kg/kmol = 1316\,kg$$

b) The air–fuel ratio can be determined as follows:

$$AFR = \frac{m_{air}}{m_{fuel}} = \frac{1725.5\,kg}{114\,kg} = 15.1\,kg\,air/kg\,fuel$$

c) The partial pressure of H_2O vapor can be determined by Dalton's Law, $\dfrac{p_i}{p} = \dfrac{N_i}{N} = y_i$

$$p_{H_2O} = \frac{N_{H_2O}}{N}p = \frac{9\,mol}{(8+9+47)\,mol}(100\,kPa) = 14\,kPa$$

Discussion: For each kilogram of this gasoline, 15.1 kg of air is required for combustion. Importantly, for every 1 kg of fuel that is burned, nearly 3 kg of CO_2 is emitted. Unsurprisingly, therefore, the burning of gasoline (and diesel) for transportation purposes contributes to around 15% of anthropogenic CO_2 emissions.

Combustion often involves either an excess or an insufficient amount of air relative to the theoretic (stoichiometric) amount. If, for instance, there was 25% excess air, then the reaction would balance as follows:

$$C_8H_{18} + \underbrace{1.25}_{\substack{25\% excess \\ air}} \times \{12.5[O_2 + 3.76\,N_2]\} \rightarrow 8\,CO_2 + 9\,H_2O + 3.15\,O_2 + \underbrace{1.25}_{\substack{25\% excess \\ air}} [47\,N_2]$$

Reimagine Our Future

Are the young people of affluent countries ready to put distant benefits of decarbonization ahead of their more immediate goals? Would they be willing to sustain this course for more than 50 years? What will it take to get older people to acknowledge that their collective ignorance is not only putting their retirements at risk, but also the planet they are leaving behind for their children? Are older people willing to do the ethical thing and reinvest their retirement funds in carbon-free funds? Can we all, young and old, act wisely, with foresight, and with an understanding of what is at stake? Or will we act only when deteriorating conditions compel us to act?

7.2 Enthalpy of Formation, Enthalpy of Combustion, and Heating Values

For chemically reacting substances, the change in chemical composition and the change in thermodynamic state must be known before changes in properties can be determined. With combustion, the products and reactants are chemically different. The system will release heat during the (exothermic) reaction so that the products can return to the same state (pressure and temperature) of the reactants. Disregarding changes in kinetic and potential energy, the conservation of energy demands that heat transfer from a combustion chamber during the reaction should be:

$$Q = \Delta h \tag{7.7}$$

7.2.1 Enthalpy of Formation

It is customary to assign a zero value of enthalpy to a state at a chosen temperature. The thermodynamic reference states, where enthalpies are 0, are $p = 1$ atm and $T = 298$ K. The *enthalpy of formation* (or heat of formation) of a compound is the difference between the enthalpy of 1 mol of the compound at the standard reference state and the enthalpy of the stable elements that form the compound evaluated at the standard reference state. The enthalpy of formation is given the symbol $\bar{h}_f^0$ and is mathematically (Black and Hartley 1985):

$$\bar{h}_f^0 = \bar{h}_f^{compound} - \sum \left(N_i \bar{h}_i^0\right)_{elements} \tag{7.8}$$

where N_i is the number of moles in the balanced chemical equation and $\bar{h}$ is the molar enthalpy of the element. The zero superscript $\left(^0_f\right)$, which is sometimes also shown in other books with a degree symbol ($^\circ$), denotes that the enthalpies are evaluated at the standard reference state.

As the enthalpy of all stable elements (such as gaseous H_2, N_2, O_2, and solid C) is defined to be zero at the standard reference state, Eq. (7.8) reduces to:

$$\bar{h}_f^0 = \bar{h}_f^{compound} \tag{7.9}$$

Equation (7.9) states that the enthalpy of a compound at the standard reference state is numerically equal to the enthalpy of formation of that compound at the standard reference state.

Table 7.3 gives some useful enthalpies of formation. According to the sign convention adopted in Chapter 1, the enthalpy (or heat) of formation is negative if the process is exothermic (it gives off heat to the environment); the enthalpy (or heat) of formation is positive if the process is endothermic (it receives heat from the environment). Iso-octane, for instance, has a heat of formation of $\bar{h}_f^0 = -250$ kJ/mol, indicating the vast quantities of heat liberated when C_8H_{18} is formed from its elements, C and H (Law 2010).

The enthalpy of any state other than the standard state (1 atm, 298 K) can be found by adding the change in enthalpy relative to the standard case (Law 2010):

$$\bar{h}_{T,p}^0 = \left(\bar{h}_f^0\right)_{298K,1\,atm} + \left(\Delta\bar{h}\right)_{298K,1atm\rightarrow T,p} \tag{7.10}$$

If the enthalpy values for $\Delta\bar{h}$ cannot be found from standard enthalpy tables, the chemical species could be assumed to be an ideal gas, and the change in enthalpy could be approximated as:

$$\Delta\bar{h} = \bar{c}_p\Delta T \tag{7.11}$$

Table 7.3 Some enthalpies of formation at standard reference conditions, 1 atm and 298 K; (*s*) indicates a solid phase, (*ℓ*) indicates a liquid phase, and (*g*) indicates a gaseous phase.

Substance	Formula	$\bar{h}_f^0$ (kJ/mol)
Solid-carbon; diatomic hydrogen, nitrogen, oxygen	C (*s*), H_2 (*g*), N_2 (*g*), O_2 (*g*)	0
Methane gas	CH_4 (*g*)	−75
Water, vapor	H_2O (*g*)	−242
Water, liquid	H_2O (*ℓ*)	−286
Carbon dioxide	CO_2 (*g*)	−394
Iso-octane	C_8H_{18} (*ℓ*)	−250
Propane, gas	C_3H_8 (*g*)	−104
Ethanol, liquid	C_2H_5OH (*ℓ*)	−278
Hydrogen, gas	H (*g*)	+218
Oxygen, gas	O (*g*)	+249
Nitrogen, gas	N (*g*)	+473

Source: Stull et al. (1969), NIST-JANAF (1998), and Moran et al. (2018).

Table 7.4 Some common enthalpies of combustion and Gibbs free energies at standard reference conditions, 1 atm and 298 K.

Substance	Formula	$\Delta \bar{h}_c^0$ (kJ/mol)
Hydrogen	H_2 (g)	−285.8
Methane gas	CH_4 (g)	−890
Propane gas	C_3H_8 (g)	−2220
Iso-octane	C_8H_{18} (ℓ)	−5471
Ethanol, liquid	C_2H_5OH (ℓ)	−1368
Oxygen, gas	O (g)	+249
Nitrogen, gas	N (g)	+473

Source: Stull et al. (1969), NIST-JANAF (1998), and Glassman et al. (2014).

7.2.2 Enthalpy of Combustion

According to Hess's Law, the *enthalpy of combustion* (or heat of combustion) may be defined as the difference between the enthalpy of the *products* at the reference state (1 atm, 298 K), and the enthalpy of the *reactants* at the same reference state when the reaction consists of the complete combustion of 1 mol of fuel in stoichiometric proportions of air (Moran et al. 2018):

$$Q = \Delta \bar{h}_c^0 = \sum_{products} N_i \left(\bar{h}_i^0 + \Delta \bar{h}_{298 \text{ K,1 atm} \rightarrow T,p} \right) - \sum_{reactants} N_i \left(\bar{h}_i^0 + \Delta \bar{h}_{298 \text{ K,1 atm} \rightarrow T,p} \right) \qquad (7.12)$$

Or, in words:

Net heat produced = (heat contents of products) less (heat contents of reactants) (7.13)

Table 7.4 lists the enthalpies of combustion of some common fuels.

7.2.3 Heating Values of Fuels

The heating value of a fuel may then be defined as the amount of heat transferred from the fuel when it is burned stoichiometrically such that the reactants are at the standard reference state and the products are returned to the standard reference state (1 atm, 298 K). The heating value is numerically equal to the enthalpy of combustion, but it has the opposite sign:

$$\Delta \bar{h}_c = -(\text{heating value}) \qquad (7.14)$$

The *higher heating value (HHV)* is the heating value of the fuel when the water in the products is condensed to liquid. The *lower heating value (LHV)* is the heating value of the fuel when the water in the products is in the vapor phase. The *HHV* is therefore greater than the *LHV* by the latent heat of condensation of water (around 44 kJ/kmol or 2257 kJ/kg at 1 atm and 298 K):

$$HHV = LHV + \left(N\bar{h}_{fg} \right)_{H_2O} \qquad (7.15)$$

Table 7.5 provides the *HHV*s and *LHV*s of some common fuels in kJ/kg. To obtain the heating values in kJ/mol, multiply the kJ/kg value by the molar mass (kg/kmol).

Table 7.5 Higher heating values and lower heating values of some fuels and their constituent components at 1 atm and 298 K.

Fuel	Formula (showing main combustible elements, where relevant)	Molar mass (M) (kg/kmol)	Density (kg/L)	Enthalpy of vaporization/condensation (h_{fg}) (kJ/kg)	Specific heat (c_p) (kJ/kgK)	Enthalpy of formation ($\bar{h}_f^0$) (kJ/kmol)	Gibbs function of formation ($\bar{g}_f^0$) (kJ/kmol)	Absolute entropy ($\bar{s}^0$) (kJ/kmol K)	HHV (kJ/kg)	LHV (kJ/kg)
Water vapor	H_2O (g)	18	0.008	2257	1.996	−241,820	−228,500	188.72	—	—
Oxygen	O_2 (g)	32	0.0014	213	0.916	0	0	205.03	—	—
Nitrogen	N_2 (g)	28	0.00165	199	1.04	0	0	191.5	—	—
Carbon dioxide	CO_2 (g)	44	0.00198	179.5	0.84	−393,520	−394,380	213.69	—	—
Carbon monoxide	CO (g)	28	1.25	216	1.04	−110,530	−137,150	197.54	10,160	—
Hydrogen	H (g)	1	0.000082	220	14.31	218,000	203,290	114.72	141,700	120,000
Carbon	C (s)	12	2	—	0.708	0	0	5.74	32,800	32,800
Sulfur	S (s)	32	1.98	—	0.732	0	0	32.05	9200	9200
Liquid fuels										
Crude oil	82–87% C, 10–14% H_2	Variable	0.856	287	1.69	—	—	—	43,050	45,300
Iso-octane	C_8H_{18} (ℓ)	114	0.703	363	2.23	−249,910	6610	360.79	47,890	44,430
Gasoline	$C_nH_{1.87n}$ (ℓ)	105	0.75	350	2.4	−208,450	17,320	463.67	47,300	44,000
Light diesel	$C_nH_{1.8n}$ (ℓ)	170	0.81	270	2.2	−291,010	50,150	622.83	46,100	43,200
Heavy fuel oil	82–87% C, 10–14% H_2	Variable	0.959	222–352	2.09	—	—	—	43,260	40,870
LPG, propane or butane	C_3H_8 (ℓ) or C_4H_{10} (ℓ)	44–58	0.533	386	1.68–1.73	−104,700	—	—	49,840	46,280

(Continued)

Table 7.5 (Continued)

Fuel	Formula (showing main combustible elements, where relevant)	Molar mass (M) (kg/kmol)	Density (kg/L)	Enthalpy of vaporization/condensation (h_{fg}) (kJ/kg)	Specific heat (c_p) (kJ/kgK)	Enthalpy of formation ($\bar{h}_f^0$) (kJ/kmol)	Gibbs function of formation ($\bar{g}_f^0$) (kJ/kmol)	Absolute entropy ($\bar{s}^0$) (kJ/kmol K)	HHV (kJ/kg)	LHV (kJ/kg)
Kerosene (paraffin)	$C_{12}H_{26}$ (ℓ) to $C_{15}H_{32}$ (ℓ)	170	0.78–0.81	251	2.01	—	—	—	46,200	43,000
Acetylene	C_2H_2 (ℓ)	26	0.91	164	1.69	226,730	209,170	200.85	49,900	—
Methanol	CH_3OH (ℓ)	32	0.79	1168	2.53	−238,810	−166,290	126.8	22,660	19,920
Ethanol	C_2H_5OH(ℓ)	46	0.79	919	2.44	−277,690	−174,890	160.7	29,670	26,810
Gaseous fuels										
Methane	CH_4 (g)	16	0.67	511	2.2	−74,873	−50,751	186.3	55,530	50,050
Propane	C_3H_8 (g)	44	0.5	335	2.77	−103,850	−23,490	269.91	50,330	46,340
Natural gas	CH_4 (g), C_2H_6 (g), C_3H_8 (g), C_4H_{10} (g), C_6H_{14} (g), H_2S (g)	CH_4: 16, C_2H_6: 30, C_3H_8: 44, C_4H_{10}: 58, C_6H_{14}: 86, H_2S: 34	0.78	CH_4: 511, C_2H_6: 488, C_3H_8: 428, C_4H_{10}: 386, C_6H_{14}: 336, H_2S: 353	2.34				52,200	47,100
Ammonia	NH_3 (g)	17	0.69	1372	2.18	−46,190	−16,590	192.33	22,500	18,600
Hydrogen	H_2 (g)	2	0.09	447	14.4	0	0	130.68	141,800	120,000

Source: Heywood (1988), NIST-JANAF (1998), Glassman et al. (2014), and Moran et al. (2018).

Example 7.2   Enthalpy of Formation and Enthalpy of Combustion for Methane Burning in Air

Calculate the enthalpy of combustion (in kJ/kg) of methane burning in stoichiometric air. All products of combustion are in a gaseous phase.

Required: We must find the enthalpy change of reactants and products to calculate the enthalpy of combustion.

Solution strategy: We will balance the stoichiometric equation for CH_4 combusting in air. We will read off the enthalpies of formation from Table 7.3 and calculate the enthalpies of combustion using Eq. (7.12).

Assumptions: Combustion occurs at 101.325 kPa and 298 K.

Analysis:

$$CH_4(g) + 2(O_2 + 3.76N_2)(g) \rightarrow CO_2(g) + 2H_2O(g) + 7.52N_2(g)$$

Enthalpy of combustion: $Q = \Delta \bar{h}_c^0 = \sum_{products} N_i \left(\bar{h}_i^0 + \Delta \bar{h}_{298\,K,1\,atm \rightarrow T,p} \right) - \sum_{reactants} N_i \left(\bar{h}_i^0 + \Delta \bar{h}_{298\,K,1\,atm \rightarrow T,p} \right)$

Reactant	N_i	$\bar{h}_i^0$ (kJ/mol)	$N_i \bar{h}_i^0$ (kJ/mol)	Product	N_i	$\bar{h}_i^0$ (kJ/mol)	$N_i \bar{h}_i^0$ (kJ/mol)
CH_4	1	−75	−75	CO_2	1	−394	−394
O_2	2	0	0	H_2O	2	−242	−484
N_2	7.52	0	0	N_2	7.52	0	0
Σ:			−75	Σ:			−878

$$\therefore Q = \Delta \bar{h}_c^0 = \sum_{products} N_i \bar{h}_i^0 - \sum_{reactants} N_i \bar{h}_i^0 = -878 \text{ kJ/mol} - (-75 \text{ kJ/mol}) = -803 \text{ kJ/mol}$$

or, −803,000 kJ/kmol

The molar mass of CH_4 is 1(12 kg/kmol) + 4 (1 kg/kmol) = 16 kg/kmol

The heating value of CH_4 is $-\Delta \bar{h}_c = \left(\dfrac{803{,}000 \text{ kJ}}{\text{kmol}} \right) \left(\dfrac{1 \text{ kmol}}{16 \text{ kg}} \right) \approx 50{,}188 \text{ kJ/kg}$

Discussion: As expected, this heating value corresponds with the lower heating value of methane in Table 7.5, as all products of combustion are in the gaseous phase.

7.3   Entropy Production and Exergy Destruction During Combustion

7.3.1   Entropy

The entropy balance developed in Eq. (3.17) for a steady-state, steady-flow process can be rewritten for a *combustion* process (which has no heat losses, adiabatic) as follows:

$$\frac{dS}{dS} = 0 = \sum \frac{\dot{Q}}{T} + \sum \dot{m}_i s_i - \sum \dot{m}_e s_e + \dot{\sigma} \tag{7.16}$$

or $\sigma = S_{products} - S_{reactants} = m_p s_{products} - m_r s_{reactants} = N_p \bar{s}_{products} - N_r \bar{s}_{reactants}$ \hfill (7.17)

where σ is the entropy production and S is the total entropy, N is the number of moles, and $\bar{s}$ is the molar absolute entropy.

The entropy values used in Eq. (7.17) are absolute entropies, measured with respect to 0 K. Table 7.5 lists the absolute molar entropy $(\bar{s}^0)$ of several substances at standard temperature (298 K) and pressure (101.325 kPa). The value of molar entropy at any temperature can be found in standard thermodynamics tables, like those available on the *National Institute of Standards and Technology-Joint Army-Navy-Air Force* (NIST-JANAF) website, https://janaf.nist.gov

When evaluating the entropy of a component of an ideal gas mixture, the temperature and partial pressure of the component must be used. The temperature of the component is taken to be the same as the mixture, and the partial pressure of a component is equal to the mixture pressure multiplied by the mole fraction of the component (Bejan et al. 1996). For the *i*-th component of an ideal gas mixture and with a mole fraction of y_i, the absolute entropy at pressures *other than* $p_0 = 101.325$ kPa can be calculated *for any temperature* (T) (using Eq. [3.60]):

$$\bar{s}_i(T,p) = \bar{s}_i^0(T,p_0) - \bar{R} \ln \frac{y_i p}{p_0} \tag{7.18}$$

with $\bar{R} = 8.314$ kJ/kmol K, and:

$$S_i = N_i \bar{s}_i(T,p_i) = N_i \left(\bar{s}_i^0(T,p_0) - \bar{R} \ln \frac{y_i p}{p_0} \right). \tag{7.19}$$

The values of molar fraction may be obtained from Table 2.1.

Reimagine Our Future

University education is currently geared toward preparing young people for jobs and careers in an economy designed to expand without limits. Is the idea of *expansion without limits* compatible with the goal of sustainability? What sort of university courses would help promote sustainability? Can you imagine courses in which there is more emphasis on fostering wonder, gratitude, and a sense of limits?

7.3.2 Exergy

For a one-inlet, one-exit open system at steady state, Eqs. (3.58) and (3.59) for calculating total exergy rate and specific exergy change can be rewritten for a combustion process, and by disregarding changes in kinetic and potential energy changes:

$$\frac{dE}{dt} = 0 = \sum_j \left(1 - \frac{T_0}{T_j} \right) \dot{Q}_j - \dot{W} + \dot{m}(e_{f1} - e_{f2}) - T_0 \dot{\sigma} \tag{7.20}$$

where $e_{f1} - e_{f2} = (h_1 - h_2) - T_0(s_1 - s_2) + \frac{V_1^2 - V_2^2}{2} + g(z_1 - z_2)$ \tag{7.21}

The *maximum available work* can then be determined as follows for a generalized system with several inlets and exits:

$$W_{max} = \sum m_i(h_i - T_0 s_i) - \sum m_e(h_e - T_0 s_e) \tag{7.22}$$

The Gibbs function (g, or G) is a significant variable in chemical reactions and is conveniently defined as follows:

$$g = h - T\Delta s \tag{7.23}$$

The *Gibbs free energy* is therefore the difference between the enthalpy and heat and measures the energy in a system that is available to do work. A system seeks to minimize its free energy by both reducing its enthalpy and increasing the entropy.

Values of Gibbs formation energy are shown in Table 7.5 in convenient molar form, $\bar{g}_f^0$. The Gibbs formation energy is calculated in a similar manner as the enthalpy of formation (i.e., with respect to 25 °C and 101.325 kPa). So, values of $\bar{g}_f^0$ can be read from Table 7.5 or calculated using:

$$\bar{g}_f^0 = \bar{h}_f^0 - T\left(s_i - \bar{s}^0\right) \tag{7.24}$$

For a combustion process, Eq. (7.22) could be rewritten as follows in molar format (Bejan et al. 1996):

$$W_{max} = \sum_{reactants} N_i\left(\bar{h}_f^0 + \Delta\bar{h} - T_0\bar{s}\right)_i - \sum_{products} N_e\left(\bar{h}_f^0 + \Delta\bar{h} - T_0\bar{s}\right)_e \tag{7.25}$$

Or, using the Gibbs function

$$W_{max} = \sum_{reactants} N_i\bar{g}_i - \sum_{products} N_e\bar{g}_e. \tag{7.26}$$

Performing exergy analyses of combustion processes are cumbersome due to the number of different species of matter that are involved. Eq. (7.25) or (7.26) must be solved for each species involved in the combustion process. The total exergy is obtained by summing the component values.

Using Eqs. (3.58) and (3.59), the *irreversibility* of a combustion process can be calculated as follows (Bejan et al. 1996):

$$\text{Irreversibility} = I = \sum_{products} N_e T_0\bar{s}_e - \sum_{reactants} N_i T_0\bar{s}_i - Q \tag{7.27}$$

and
$$\text{Irreversibility rate} = \dot{I} = \sum_{products} \dot{N}_e T_0\bar{s}_e - \sum_{reactants} \dot{N}_i T_0\bar{s}_i - \dot{Q} \tag{7.28}$$

The *exergy destroyed* during a combustion process is:

$$E_{destroyed} = T_0\sigma \tag{7.29}$$

and the rate of exergy destruction is:

$$\dot{E}_{destroyed} = T_0\dot{\sigma} \tag{7.30}$$

Example 7.3 explains how to calculate the heat obtained from lean burning of iso-octane, while Box 7.1 discusses the energy efficiency of natural gas-fired condensing boilers.

Example 7.3   Heat Obtained from the Lean Burning of Iso-Octane

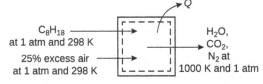

An industrial gas turbine uses liquid iso-octane (C_8H_{18}) for fuel. The iso-octane has an enthalpy of formation of −249,910 kJ/kmol. The fuel is mixed with 25% excess air to bring about the required combustion characteristics. The air and the fuel enter at 298 K (25 °C) and 1 atm, and the products of combustion leave at 1000 K in a vapor state. Calculate (a) the heat transfer from the combustor

---

Box 7.1 Natural Gas-Fired Condensing Boiler

When a boiler burns natural gas, CO_2 vapor, and H_2O vapor are created as products of the combustion process. The exit flue gas temperature can be between 150 °C and 200 °C. In a conventional boiler, this useful energy is wasted as the flue gases exit the chimney. In total, about 10% of the fuel's energy is tied up as H_2O vapor and lost to the environment (Che et al. 2004; Day et al. 2008).

With a condensing boiler (Figure 7.2), the exit flue gas temperature is reduced to below its dew point temperature by a cooling medium, usually water, which recovers both sensible and latent heat (about 2.4 MJ/kg) from the flue gas, increasing the boiler's efficiency by around 10%.

Although natural gas usually comprises mostly CH_4, it may also contain other substances. This means that the products of combustion might also contain SO_x, NO_x, dust, and soot. In a condensing boiler, these mainly acidic constituents can be totally dissolved in the condensing water and drained from the system, thus eliminating harmful emissions. This water along with acidic byproducts of the fuel combustion process are drained out of the boiler into a neutralizer.

Despite the high energy conversion efficiencies that can be attained by condensing boilers, the fact remains that natural gas is a finite (non-renewable) resource. Its combustion releases vast amounts of the greenhouse gas CO_2. (Also see Problem 7.4.)

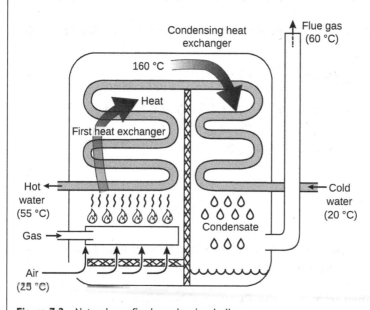

Figure 7.2 Natural gas-fired condensing boiler.

---

per kilomole of fuel, (b) the heat transfer rate of the combustor if the fuel flow rate is 0.3 kg/s, (c) the entropy generation during the process, and (d) the rate of irreversibility.

Required: We must find the heat transfer from the combustor per kilomole of C_8H_{18} burning lean fuel with 25% excess air.

Solution strategy: (a) We will subtract the enthalpies of formation of the reactants from those of the products to determine the *LHV*. We will use the enthalpies for formation listed in Table 7.3 when calculating Q:

$$Q = \Delta \overline{h}_c^0 = \sum_{products} N_i \left(\overline{h}_i^0 + \Delta \overline{h}_{298K,1\ atm \to T,p} \right) - \sum_{reactants} N_i \left(\overline{h}_i^0 + \Delta \overline{h}_{298K,1\ atm \to T,p} \right)$$

Assumptions: Steady-state and steady-flow conditions prevail.
Analysis:

a) Heat transfer

From Example 7.1, the balanced equation for C_8H_{18} burning stoichiometrically is:

$$C_8H_{18} + 12.5[O_2 + 3.76\ N_2] \to 8\ CO_2 + 9\ H_2O + 47N_2$$

So, for 25% excess air, the balanced equation becomes the following:

$$C_8H_{18} + 1.25\{12.5[O_2 + 3.76N_2]\} \to 8CO_2 + 9H_2O + 3.13O_2 + 1.25(47N_2)$$
$$\therefore \quad C_8H_{18} + 15.63O_2 + 58.75N_2 \to 8CO_2 + 9H_2O + 3.13O_2 + 58.75N_2$$

Heat of combustion: $Q = \Delta \overline{h}_c^0 = \sum_{products} N_i \left(\overline{h}_i^0 + \Delta \overline{h}_{298K,1\ atm \to T,p} \right) - \sum_{reactants} N_i \left(\overline{h}_i^0 + \Delta \overline{h}_{298K,1\ atm \to T,p} \right)$

Reactants: Since the air is composed of stable elements (N_2 and O_2) and enters at standard reference conditions (1 atm and 298 K), the enthalpy of the reactants is simply equal to that of the fuel:

$$\sum_{reactants} N_i \left(\overline{h}_i^0 + \Delta \overline{h}_{298\ K \to 298\ K} \right) = N_{C_8H_{18}} \left(\overline{h}_f^0 + \Delta \overline{h} \right)_{C_8H_{18}} + N_{O_2} \left(\overline{h}_f^{\emptyset} + \cancel{\Delta h} \right)_{O_2} + N_{N_2} \left(\overline{h}_f^{\emptyset} + \cancel{\Delta h} \right)_{N_2}$$

$$\sum_{reactants} N_i \left(\overline{h}_i^0 + \Delta \overline{h}_{298K \to 298K} \right) = 1(-249,910)\ kJ/kmol = -249,910\ kJ/kmol$$

Products (using values from Tables 7.3 and 7.4):

$$\sum_{products} N_i \left(\overline{h}_i^0 + \Delta \overline{h}_{298K \to 1000K} \right) = N_{CO_2} \left(\overline{h}_f^0 + \Delta \overline{h} \right)_{CO_2} + N_{H_2O} \left(\overline{h}_f^0 + \Delta \overline{h} \right)_{H_2O} + N_{O_2} \left(\overline{h}_f^{\emptyset} + \Delta \overline{h} \right)_{O_2}$$
$$+ N_{N_2} \left(\overline{h}_f^{\emptyset} + \Delta \overline{h} \right)_{N_2}$$

with $\Delta \overline{h}_{CO_2} = \left(c_p \Delta T \right)_{CO_2} M_{CO_2} = (0.84\ kJ/kg\ K)(1000\ K - 298\ K)(44\ kg/kmol) = 25,946\ kJ/kmol$

with $\Delta \overline{h}_{H_2O} = \left(c_p \Delta T \right)_{H_2O} M_{H_2O} = (1.996\ kJ/kg\ K)(1000\ K - 298\ K)(18\ kg/kmol) = 25,221\ kJ/kmol$

with $\Delta \overline{h}_{O_2} = \left(c_p \Delta T \right)_{O_2} M_{O_2} = (0.916\ kJ/kg\ K)(1000\ K - 298\ K)(32\ kg/kmol) = 20,577\ kJ/kmol$

with $\Delta \overline{h}_{N_2} = \left(c_p \Delta T \right)_{N_2} M_{N_2} = (1.04\ kJ/kg\ K)(1000\ K - 298\ K)(28\ kg/kmol) = 20,442\ kJ/kmol$

$$\sum_{products} N_i \left(\overline{h}_i^0 + \Delta \overline{h}_{298K \to 1000K} \right) = 8\underbrace{(-394,000 + 25,946)}_{\left(\overline{h}_f^0 + \Delta \overline{h} \right)_{CO_2}}\ kJ/kmol + 9\underbrace{(-242,000 + 25,221)}_{\left(\overline{h}_f^0 + \Delta \overline{h} \right)_{H_2O}}\ kJ/kmol$$

$$\sum_{products} N_i \left(\overline{h}_i^0 + \Delta \overline{h}_{298K \to 900K} \right) = +3.13\underbrace{(20,577\ kJ/kmol)}_{\left(\overline{h}_f^{\emptyset} + \Delta \overline{h} \right)_{O_2}} + 58.75\underbrace{(20,442\ kJ/kmol)}_{\left(c_p \Delta T \right)_{N_2}}$$

$$= -3.63 \times 10^6\ kJ/kmol\ fuel$$

Then $Q = \Delta \overline{h}_c^0 = -3.63 \times 10^6\ kJ/kmol - (-249,910)\ kJ/kmol = -3.4 \times 10^6\ kJ/kmol\ fuel$

b) Heat rate of combustor

The heat rate is $\dot{Q} = \dot{m}q$

The specific heat is $q = \dfrac{3.4 \times 10^6 \text{ kJ/kmol}}{114 \text{ kg/kmol}} = 29{,}651 \text{ kJ/kg}$

$\dot{Q} = \dot{m}q = (0.3 \text{ kg/s})(29{,}651 \text{ kJ/kg}) \simeq 8.9 \text{ MW}$

c) Entropy production

From Eq. (7.19): $S_i = N_i \bar{s}_i(T, p_i) = N_i\left(\bar{s}_i^0(T, p_0) - \bar{R}\ln\dfrac{y_i p_i}{p_0}\right)$, and $p_i = y_i p_{total}$. Using the NIST-JANAF thermochemical data tables to obtain $\bar{s}_i^0$, the following calculations can be made:

Component	Moles (N_i)	Mole fraction (y_i)	$\bar{s}_i^0$ (kJ/kmol K)	$-\bar{R}\ln\frac{y_i p}{p_0}$	$N_i\bar{s}_i$
Reactants					
$C_8H_{18,\ 298K}$	1	$= 1/75.38$ $= 0.01$	360.79	$-8.314 \ln (0.01) = 38.29$	322.5
$O_{2,\ 298K}$	15.63	$= 15.63/75.38$ $= 0.21$	205.03	$-8.314 \ln (0.21) = 12.97$	3001.89
$N_{2,\ 298K}$	58.75	$= 58.75/75.38$ $= 0.78$	191.5	$-8.314 \ln (0.78) = 2.07$	11,129
Σ	75.38	1.0			14,453
				$S_{reactants} = 14{,}453$ kJ/kmol K	
Products					
$CO_{2,\ 1000K}$	8	$= 8/78.88$ $= 0.101$	269.33	$-8.314 \ln (0.101) = 19.06$	2002.2
$H_2O_{,\ 1000K}$	9	$= 9/78.88$ $= 0.114$	232.71	$-8.314 \ln (0.114) = 18.05$	1931.9
$O_{2,\ 1000K}$	3.13	$= 3.13/78.88$ $= 0.039$	243.59	$-8.314 \ln (0.039) = 26.97$	678.02
$N_{2,\ 1000K}$	58.75	$= 58.75/78.88$ $= 0.745$	228.2	$-8.314 \ln (0.745) = 2.44$	13,263
Σ	78.88	1.0			17,876
				$S_{products} = 17{,}876$ kJ/kmol K	

Therefore, the entropy produced is $\sigma = (17{,}876 - 14{,}453)$ kJ/kmol K $= 3423$ kJ/kmol K.

d) Rate of irreversibility

The rate of irreversibility of the combustion process $= \sum_{products} \dot{N}_e T_0 \bar{s}_e - \sum_{reactants} \dot{N}_i T_0 \bar{s}_i - \dot{Q}$

$= T_0 \dot{\sigma}$, with $\dot{N} = \dfrac{\dot{m}_{C_8H_{18}}}{M_{C_8H_{18}}} = \dfrac{0.3 \text{ kg/s}}{(114 \text{ kg/kmol})} = 2.63 \times 10^{-3}$ kmol/s

So, the rate of irreversibility (or rate of destruction of exergy) $= T_0 \dot{\sigma} = (298 \text{ K})$ $(2.63 \times 10^{-3} \text{ kmol/s})(3423 \text{ kJ/kmol K}) = 2.68$ MW.

Discussion: The heat obtained from iso-octane is -3.4×10^6 kJ/kmol fuel. This is 38% less than the lower heating value of iso-octane (about -5.1×10^6 kJ/kmol fuel or 44,430 kJ/kg). The rate of heat transfer is less than the *LHV* as the combustion reaction is *not* under stoichiometric conditions and the products of combustion *do not* leave the combustor at the standard reference state.

The combustor's exergy is 8.9 MW, but its irreversibility (thus, exergy destruction) is 2.68 MW. The exergy destruction is therefore about 30%. The large exergy destruction in any combustor is due to the presence of the irreversibilities associated with chemical reaction, heat transfer, and friction. The inherent high temperature, high flow rate, and reaction chemistry of a combustor or furnace will always ensure that it is a major site of exergy loss.

7.4 The Gibbs Function and Spontaneous Reactions

From Chapter 3, we have

$$T \, dS = dU + p \, d\cancel{V} \tag{7.31}$$

From the second law of thermodynamics (Chapter 3), we know that, for irreversible systems:

$$dS_{total} = dS_{system} + dS_{surroundings} > 0 \tag{7.32}$$

Combining Eqs. (7.31) and (7.32), we have:

$$dH - T \, dS \leq 0 \tag{7.33}$$

The Gibbs function may be defined as:

$$G = H - TS \tag{7.34}$$

and represents the maximum amount of non-expansion work (like electric work) that can be extracted from a reaction.

If the Gibbs function is differentiated for a constant-pressure process:

$$dG = dH - T \, dS \tag{7.35}$$

Substituting Eq. (7.35) in Eq. (7.33) gives:

$$(dG)_{T,p} \leq 0 \tag{7.36}$$

Equation (7.36) provides a criterion for *spontaneous reactions,* which demands that the *Gibbs function must decrease*. For instance, in Section 9.3, we will show that when a reaction is exothermic ($\Delta H < 0$) and the entropy of the products is higher than that of the reactants ($\Delta S > 0$), the reaction will be spontaneous as $\Delta G < 0$. If the Gibbs function is equal to zero, however, chemical equilibrium has occurred. If the Gibbs function increases, the reaction will not proceed spontaneously and will need a heat input (endothermic).

7.5 Outlook

During the combustion of chemical fuels, exergy losses (i.e., destruction of thermodynamic quality) are large, typically around 30% of the energy input. Making improvements on the combustion side of energy conversion processes are therefore important and will have a knock-on benefit as they will also give rise to higher "first-law" efficiencies (Hammond 2004). It is difficult, however, to reduce the irreversibilities associated with uncontrolled electron exchange between reactants and the high-temperature gradients inherent to combustion reactions. The recovery of waste heat from flue gases is an important step in the right direction, as is improving the metallurgical limit of combustor materials to facilitate higher combustion temperatures. It is also crucial to only use the combustion of chemical fuels in applications where high-grade heating is of absolute necessity, and where renewable energy sources (and nuclear power) cannot be used.

Problems

7.1 (a) Write down the combustion reaction for diesel fuel ($C_{12}H_{23}$) and balance this equation. Then, (b) determine the air–fuel ratio when the engine operates under stoichiometric conditions; (c) A city-bus is equipped with a diesel engine that consumes 11.05 kg of diesel per hour. The bus drives the same route for 4 hours a day, 5 days a week, 50 weeks a year. Using the balanced stoichiometric equation, calculate the amount of CO_2 that the bus emits per year.

[Answers: (a) $C_{12}H_{23} + 17.75(O_2 + 3.76N_2) \rightarrow 11.5H_2O + 12CO_2 + 66.74N_2$]; (b) $AFR = 14.7$; (c) about 35 tons CO_2 per year.]

7.2 An accurate flue gas analyzer is used to determine the composition of the flue gas emanating from a furnace that is combusting a hydrocarbon fuel. The flue gas analyzer reports the following results: 11.9% CO_2, 4.1% O_2, 1% CO, and 83% N_2. (a) Find the composition of the hydrocarbon fuel being burned, (b) the air–fuel ratio, and (c) the amount of excess air.

Hint: With the chemical reaction equation, insert the percentages of the products, and solve for the unknown number of C and H atoms in the fuel, i.e., assume: $C_aH_b + c(O_2 + 3.76\,N_2) \rightarrow d\,H_2O + 11.9\,CO_2 + 4.1\,O_2 + 1.0\,CO + 83\,N_2$, and solve for a, b, c, and d.

[Answers: (a) $C_{12.9}H_{22} + 22(O_2 + 3.76\,N_2) \rightarrow 11\,H_2O + 11.9\,CO_2 + 4.1\,O_2 + 1.0\,CO + 83\,N_2$]; (b) $AFR = 17.2$; (c) 19.6% excess air.]

7.3 A Generac home generator uses gaseous propane (C_3H_8) as fuel, with gaseous H_2O in the products of combustion. (a) Write down the chemical reaction of C_3H_8 burning stoichiometrically in air and balance the equation. Then, (b) calculate the enthalpy of combustion (in MJ/kg) of propane at 25 °C, and (c) at 500 K.

[Answers: (a) $C_3H_8 + 5(O_2 + 3.76\,N_2) \rightarrow 3CO_2 + 4H_2O + 18.8N_2$; (b) -46.5 MJ/kg; (c) -42.8 MJ/kg.]

7.4 A Viessmann Vitocrossal 200 kW condensing boiler uses natural gas (in the vapor phase) as fuel (see also Box 7.1). The natural gas can be assumed to comprise only methane (CH_4). The methane and air enter the gas burner at 25 °C and 1 atm. The products of combustion leave the furnace at 60 °C in a liquid state. (a) Write down the chemical reaction of CH_4 if it burns with 5% excess air. Then, (b) calculate the air–fuel ratio, (c) heat transfer from the furnace, (d) mass flow rate of methane (in kg/s), and (e) mass flow rate of air. (f) Also calculate the irreversibility (or exergy destruction) of the process.

[Answers: (a) $CH_4 + 2.1(O_2 + 3.76\ N_2) \rightarrow CO_2 + 2H_2O + 0.1O_2 + 7.9\ N_2$; (b) $AFR = 18.1$; (c) $-876{,}227$ kJ/kg mol fuel; (d) 3.6 g/s; (e) 65.3 g/s air; (f) 754,940 kJ/kmol fuel.]

References

Bejan, A., Tsatsaronis, G. and Moran, M. (1996). *Thermal Design and Optimization*. New York: John Wiley and Sons Inc.

Black, W.Z. and Hartley, J.G. (1985). *Thermodynamics*. London: Harper & Row.

Che, D., Liu, Y. and Gao, C. (2004). Evaluation of retrofitting a conventional natural gas fired boiler into a condensing boiler. *Energy Conversion and Management* 45 (20): 3251–3266.

Day, A.R., Ratcliffe, M.S. and Shepherd, K. (2008). *Heating Systems, Plant and Control*. New York: Wiley.

EIA (2021). *Biomass Explained*. https://www.eia.gov/energyexplained/biomass.

ExxonMobil (2021). *Outlook for Energy, 2021*. ExxonMobil. https://corporate.exxonmobil.com/Energy-and-innovation/Outlook-for-Energy/Energy-supply.

Glassman, I., Yetter, R.A. and Glumac, N.G. (2014). *Combustion*. Cambridge, MA: Academic Press.

Hammond, G.P. (2004). Engineering sustainability: thermodynamics, energy systems, and the environment. *International Journal of Energy Research* 28 (7): 613–639.

Heywood, J.B. (1988). *Internal Combustion Engine Fundamentals*. New York, NY: McGraw-Hill.

Law, C.K. (2010). *Combustion Physics*. Cambridge, UK: Cambridge University Press.

Moran, M. J., Shapiro, H. N., Boettner, D. D. and Bailey, M. B. (2018). *Fundamentals of Engineering Thermodynamics*, 9th edition. New York, NY: Wiley.

NIST-JANAF (1998). *NIST-JANAF Thermochemical Tables*. http://www.janaf.nist.gov.

Severns, W.H., Degler, H.E. and Miles, J.C. (1948). *Steam, Air, and Gas Power*. New York, NY: Wiley.

Stull, D.R., Westrum, E.F. and Sinke, G.C. (1969). *The Chemical Thermodynamics of Organic Compounds*. New York, NY: Wiley.

8

Coal

As shown earlier in Figure 7.1, coal currently accounts for about a quarter of the world's energy needs. In 2050, coal is expected to meet about 15% of global energy demand and to produce 30% of energy-related CO_2 emissions (Jakob et al. 2020; IEA 2021a; ExxonMobil 2022), despite the devastating impacts of coal combustion on the climate, the environment (e.g., Figure 8.1), and on society (Palmer et al. 2010; Lindberg et al. 2011; EIA 2021; Finkelman et al. 2021).

Reimagine Our Future
Imagine if people took fewer or shorter trips or chose to walk or bike or take public transportation on those trips. Compare this scenario with one in which we turn to electric vehicles to solve our transportation woes. Which scenario is preferable?

8.1 Coal

Due to the critical role of coal in the past 200 years, it would not be an understatement to say that industrialization pivoted on the following chemical reaction:

$$\underbrace{2\,CH}_{\substack{\text{simplified} \\ \text{representation} \\ \text{of coal}}} + \frac{3}{2}O_2 \rightarrow 2\,CO + H_2O \tag{8.1}$$

Coal is a combustible black or brownish-black sedimentary rock, formed over millions of years as rock strata called coal seams. Coal comprises mostly carbon, with variable amounts of other elements, chiefly hydrogen, sulfur, oxygen, and nitrogen. As a fossil fuel that is burned for heat, coal supplies about a quarter of the world's primary energy and two-fifths of its electricity (IEA 2021a).

Most of the energy content in coal is associated with its *carbon content*. Higher-carbon coals usually have a higher energy content, are more valued in the marketplace, and are more suited for electricity generation. Power plants designed for high-carbon content fuels have a higher generating efficiency and lower capital cost and could be more effectively designed for CO_2 capture (Deutsche et al. 2007).

Energy Systems: A Project-Based Approach to Sustainability Thinking for Energy Conversion Systems,
First Edition. Leon Liebenberg.
© 2024 John Wiley & Sons, Inc. Published 2024 by John Wiley & Sons, Inc.
Companion website: www.wiley.com/go/liebenberg/energy_systems

Figure 8.1 Coal strip-mining in West Virginia. Such coal-mining operations "bring about changes in surface water and groundwater movement as a result of mining and waste disposal, the infiltration of groundwater and surface water by hazardous contaminants resulting from solid waste disposal, acid rain, and increased mobilization of trace elements from coal. Undesirable health effects may result for those segments of the occupational workforce near certain components of the fuel system, and for the general population because of the release of oxides of sulfur and nitrogen, particulates, and toxic organic compounds during production, transportation, or utilization of coal and its products". *Source:* Elliott (1981) / With permission from The Ohio Valley Environmental Coalition (OVEC).

The *sulfur* in coal tends to decrease the efficiency of pulverized-coal boilers, due to the need to maintain higher boiler outlet temperatures to avoid the condensation of sulfuric acid in the flue gas, which results in corrosion problems in downstream equipment. Sulfuric acid is formed from sulfur that is oxidized to SO_3 and which reacts with water to form H_2SO_4 (Macrae 1966). High boiler outlet temperatures are needed to carry heat out of the boiler, as not all of it is used to convert liquid-water to steam. Coal with a high sulfur content also increases the power requirements for flue gas desulfurization units and operating costs. Further, for CO_2 capture, high-sulfur coal has a detrimental effect on carbon capture technologies (Deutsche et al. 2007).

Coal *ash* is the residue of inorganic (mineral) matter that forms after combustion. Coal ash typically comprises silicon, aluminum, iron, magnesium, calcium, titanium, sodium, and potassium (Rose and Cooper 1977). Coal ash is incombustible and is thus termed inert. Ash content and properties affect the boiler design and operation. High-ash coals cause increased erosion and reduce efficiency. They may be more effectively handled in circulating fluidized bed boilers. Boilers are designed for the ash to exit the boiler, either as molten slag ("wet bottom boilers"), particularly for low fusion-temperature ash, or as fly ash ("dry bottom boilers"). Most large boilers are dry-ash designs. For integrated gas combined cycle (IGCC) plants, coal ash consumes heat energy to melt it, requires more water per unit of carbon in the slurry, increases the cost per kilowatt-electric (kW_e), and reduces the overall generating efficiency (Deutsche et al. 2007).

Coal also contains *moisture* (H_2O), which is incombustible and thus inert. Coals with a higher moisture content reduce the generating efficiency in pulverized-coal combustion plants and reduce gasifier efficiency in IGCC plants, increasing the cost per kW_e. Boiler size and cost also increase with higher-moisture coals.

Acid corrosion due to the sulfur contents of coal is also dependent on the water content in the flue gas. The water content of the flue gas is pivotal to determining the *acid dew point* of the flue gases. The following relationship makes accurate predictions (Pierce 1977):

$$\frac{1000}{t_{adp} + 273.15} = 1.7846 + 0.02671 \log_{10} p_{H_2O} - 0.1031 \log_{10} p_{SO_3}$$
$$+ 0.0329 \left(\log_{10} p_{H_2O}\right)\left(\log_{10} p_{SO_3}\right)$$

(8.2)

where t_{adp} is the acid dew point in °C, and where the partial pressures p_{H_2O} and p_{SO_3} are measured in bar (with 1 bar = 100 kPa).

Acid corrosion can be counteracted by using steam to preheat the boiler feedwater to a few degrees above the acid dew point; a 30 °C "buffer" temperature is typically applied. So, if t_{adp} is calculated to be 150 °C, the combustion process is designed for $t_{adp} = 180$ °C to be safe. The chimney stack should also be well insulated to avoid cold spots, where condensation can occur. A more expensive method is to neutralize the acid *in situ* by spraying magnesium hydroxide ($Mg(OH)_2$) into the flue gases. This forms H_2O and $MgSO_4$; the latter being much less acidic than H_2SO_4.

8.1.1 Coal Classification

Coal can be broadly classified as follows:

Anthracite and bituminous coal are located up to a few thousand meters beneath the Earth's surface, although conventional mining only goes a few hundred meters deep. Lignite is located close to the Earth's surface and can therefore be more easily mined.

The term *total coal* refers to the sum of hard coal and brown coal after conversion to a common energy unit, *tonne of coal equivalent* – TCE. The conversion is done by multiplying the heating value of the coal by the total volume of hard and brown coal used, measured in physical units, i.e., in tonnes. One TCE has an energy content of 29.3 GJ and corresponds to 0.7 t of oil equivalent (TOE).

8.1.2 Coal Consumption

Coal is used in four possible ways (IEA 2022):

- As *a primary input to produce electricity,* or a secondary or tertiary fuel that is used elsewhere or sold. This is referred to as *use in conversion* processes, e.g., coking coal used to produce coke in a coke oven or steam coal used to produce electricity.
- As a fuel used to support a conversion process. This is referred to as *energy industry own use*, e.g., a coke oven gas used to heat the coke oven or steam coal used to operate the power plant.
- As a fuel consumed in manufacturing, industry, mining and construction, in transport, in agriculture, in commercial and public services, and in households. This is referred to as *use in the final consumption sectors*, e.g., steam coal used to produce heat in cement kilns, and steam coal used to produce industrial process steam.
- As a raw material. This is referred to as *non-energy use*, e.g., coal tar used as a chemical feedstock.

Between 2000 and 2022, coal consumption has declined in the United States (from 6000 to 2000 TWh), Germany (1000 to less than 500 TWh), Japan, and the United Kingdom. In this period, coal consumption in China has however skyrocketed (7500 to 25,000 TWh) and in India (2500 to 5000 TWh). During Russia's war in Ukraine and the linked global energy crisis, most countries in the developing world in 2021 began burning cheap coal again, resulting in a global production surpassing eight billion tonnes in a single year for the first time and eclipsing the previous record set in 2013.

Reimagine Our Future

Will energy sustainability be achieved by better managing our energy sources (like oil or uranium) or rather by learning to manage ourselves to live within ecological limits?

8.1.3 Coal Reserves

Coal is certain to play a major role in the world's energy future for two reasons. First, it is the lowest-cost fossil source for base-load electricity generation. Second, in contrast to oil and natural gas, coal resources are widely distributed around the world (Deutsche et al. 2007).

Coal reserves are spread between developed and developing countries (see also Box 8.1). The estimated total recoverable global coal reserves are about 1000 Gt: 754 Gt of anthracite and bituminous coal, and 321 Gt of sub-bituminous coal and lignite (IEA 2021a). The *global reserve-to-production (R/P) ratio* in 2020 shows that, at current production levels, the United States has about 470 years of coal left (BP 2022). Globally, however, we have about 154 years of coal left in the ground.

The current oil crisis has spurred many countries to ramp up their production of anthracite and bituminous coal. Due to lignite's high moisture content and non-agglomerating nature, its specific energy content is low, and its transportation is limited to the area surrounding the coal mine. There is therefore no global market for lignite; it must be produced and used close to source.

Box 8.1   Proved (or "Proven") Reserves Versus Resources

Unlike *resources*, which is the amount that could technically be extracted, according to BP (2022), total *proved (or 'proven') reserves* of coal is generally taken to be those quantities that geological and engineering information indicates with reasonable certainty can be recovered in the future from known reservoirs under existing economic and operating conditions. Coal reserves can therefore vary with coal and carbon prices.

8.2 Combustion of Coal

It follows from the previous discussion that coal has no precise chemical structure. The heat content of coal will vary depending on the amount of carbon and other components.

DuLong's formula is often used to estimate (within 1.5% accuracy) the *gross calorific value (GCV)* or higher heating value (*HHV*) in terms of the mass fractions of carbon, hydrogen, oxygen, and sulfur from the *ultimate (or elementary) analysis* of coal (Macrae 1966; Elliott 1981):

$$GCV = \left\{ 33{,}830(\%C) + 144{,}300\left(\%H - \frac{\%O}{8}\right) + 9420(\%S) \right\} \frac{1}{100} \qquad [kJ/kg] \qquad (8.3)$$

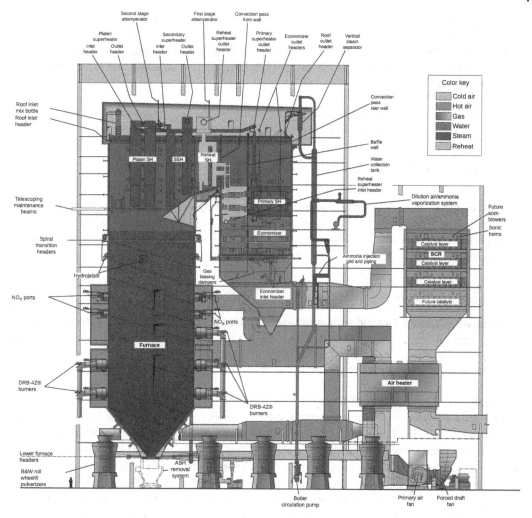

Figure 8.2 B&W supercritical (248 bar, 568 °C) 420 MW$_{th}$ boiler. *Source:* Image reproduced with permission from B&W.

In a coal-fired steam power plant, the coal is pulverized or gasified to ensure good combustion. Figure 8.2 illustrates the scale and complexity of a utility-size boiler that uses pulverized coal as fuel.

Coal combustion calculations are laborious, but straightforward. Each constituent of coal must be separately balanced with the amount of air. The products of combustion can then be calculated as illustrated in Example 8.1.

The main emissions arising from coal combustion are (Rose and Cooper 1977; Elliott 1981):

- The sulfur in the coal is converted to sulfur oxides (SO$_x$), the bulk (usually well over 90%) is sulfur dioxide (SO$_2$) and the remainder is sulfur trioxide (SO$_3$).
- Nitrogen oxides (NO$_x$) are formed from nitrogen compounds in the coal and from the nitrogen in the combustion air.
- Carbon monoxide (CO)
- Unburned hydrocarbons (HC)

- Trace elements such as mercury, selenium, and boron, among many others.
- Chlorides and fluorides are emitted in gaseous form, but in insignificantly small proportions.
- Particulates, which are made up of smoke (unburned solid combustible matter) and some ash. Ash that does not fly off remains molten in the boiler, where it solidifies and forms bottom ash that is tapped out.

The combustion efficiency of a coal furnace (or steam boiler) can be calculated by accounting for the energy lost in the flue gases and the coal ash:

$$\text{Combustion efficiency} = \frac{\text{Energy input} - \text{Energy lost in flue gases and ash}}{\text{Energy input}} \tag{8.4}$$

The flue gas enthalpy comprises *sensible* (with accompanying temperature change) and *latent* (with no temperature change) components, which must be accounted for, as shown in Example 8.1. A modern boiler (such as the one shown in Figure 8.2) can achieve energy conversion efficiencies close to 90%.

Reimagine Our Future

Energy utilities produce solid and liquid waste, as well as air emissions. Where do they dump their waste? (If you are in the United States, check out https://www.epa.gov/superfund to help answer this question.) How would you feel if you lived close to a dumping site?

Example 8.1 Combustion of Coal

An ultimate analysis of a bituminous coal sample provides the following mass fraction of constituents: 66% C, 4.1% H_2, 7.2% O_2, 1.3% N_2, 1.7% S, 12% moisture, 7.7% ash, 5% of the sulfur oxidizes to SO_3, and the remainder (95%) to SO_2. The ash contains 5% carbon, which will not participate in combustion. The fuel's higher heating value (*HHV*) (or gross calorific value) is measured to be 27 MJ/kg. To ensure complete combustion, the coal will be burned with 20% excess air at 100 kPa. Calculate (a) the air–fuel ratio, (b) the mass of CO_2, H_2O, SO_2, SO_3, O_2, and N_2 emitted per kg of coal burned, (c) the acid dew point temperature, and (d) the combustion efficiency.

Required: We must find the air–fuel ratio when combusting this coal in 20% excess air. We must also find the mass of products for each kilogram of coal burned, as well as the acid dew point temperature.

Solution strategy: (a) We will first balance the reaction of the combustible elements (C, H, S) with O_2. After determining the amount of O_2 required for stoichiometric combustion, we will calculate the amount of required air, using the composition of air given in Table 7.2. The air–fuel ratio can then be calculated. (b) Each of the products of combustion will then be separately analyzed, including S, of which 5% oxidizes to SO_3 and 95% to SO_2. (c) That will provide the molar information required to use Eq. (8.2) to calculate the acid dew point of the flue gas. (d) To calculate the combustion efficiency, we must do an enthalpy balance for the boiler with the datum state for water as a liquid (corresponding to the *HHV* of coal). Values of c_p can be obtained from the NIST-JANAF web page.

Assumptions: The coal combusts totally so that its heating value of 27 MJ/kg is available. We will disregard any radiation losses. Assume that the ambient air is at 100 kPa and 25 °C. We will also assume that the coal ash leaves the boiler at the flue gas exit temperature. Take the specific heat capacity of coal and ash to be equal to 0.84 kJ/kgK (Rose and Cooper 1977).

Analysis:

a) For C:

$$C + O_2 \rightarrow CO_2$$
$$12 \text{ kg C} + 32 \text{ kg O}_2 \rightarrow 44 \text{ kg CO}_2$$

$$\frac{\text{mass O}_2}{\text{mass C}} = \frac{32 \text{ kg}}{12 \text{ kg}} = 2.67 \text{ kg O}_2/\text{kg C}$$

$$\frac{\text{mass CO}_2}{\text{mass C}} = \frac{44 \text{ kg}}{12 \text{ kg}} = 3.67 \text{ kg CO}_2/\text{kg C}$$

For H:

$$H_2 + \frac{1}{2}O_2 \rightarrow H_2O$$
$$2 \text{ kg H}_2 + 16 \text{ kg O}_2 \rightarrow 18 \text{ kg H}_2O$$

$$\frac{\text{mass O}_2}{\text{mass H}_2} = \frac{16 \text{ kg}}{2 \text{ kg}} = 8 \text{ kg O}_2/\text{kgH}_2$$

$$\frac{\text{mass H}_2O}{\text{mass H}_2} = \frac{18 \text{ kg}}{2 \text{ kg}} = 9 \text{ kg H}_2O/\text{kg H}_2$$

For S:

$$S + O_2 \rightarrow SO_2$$
$$32 \text{ kg S} + 32 \text{ kg O}_2 \rightarrow 64 \text{ kg SO}_2$$

$$\frac{\text{mass O}_2}{\text{mass S}} = \frac{32 \text{ kg}}{32 \text{ kg}} = 1 \text{ kg O}_2/\text{kg S}$$

$$\frac{\text{mass SO}_2}{\text{mass S}} = \frac{64 \text{ kg}}{32 \text{ kg}} = 2 \text{ kg SO}_2/\text{kg S}$$

Oxygen required to burn combustible elements:
For C:
The coal sample contains 66% carbon and 7.7% ash. Of that ash, 5% contains C, which means that there will be 95% ash in the residue leaving the boiler.

So, for 100 kg of this coal, $(66 \text{ kg C}) - \left(7.7 \text{ kg ash} \frac{0.05 \text{ C}}{1.00 \text{ ash}}\right) = 65.6 \text{ kg C}$ will participate in combustion.

The mass of oxygen required to combust 65.6 kg of C is:

$$\frac{O_2}{C} \times C = (2.67 \text{ kg O}_2/\text{kg C})(65.6 \text{ kg C}) = 175.15 \text{ kg O}_2/100 \text{ kg coal}$$

For H:
The mass of oxygen required to combust 4.1 kg of H_2 is:

$$\frac{O_2}{H_2} \times H_2 = (8 \text{ kg O}_2/\text{kg H}_2)(4.1 \text{ kg H}_2) = 32.8 \text{ kg O}_2/100 \text{ kg coal}$$

For S:
The mass of oxygen required to combust 1.7 kg of S is:

$$\frac{O_2}{S} \times S = (1 \text{ kg O}_2/\text{kg S})(1.7 \text{ kg S}) = 1.7 \text{ kg O}_2/100 \text{ kg coal}$$

The *gross* mass of oxygen required to burn 100 kg of coal is 175.15 kg + 32.8 kg + 1.7 kg = 209.65 kg

The *net* amount of oxygen required = gross amount of O_2 required to combust 100 kg of coal less than the O_2 in the fuel:

Net oxygen = 209.65 kg − 7.1 kg = 202.6 kg O_2/100 kg coal

Required air
From Table 7.2 there is 23 kg O_2 per 100 kg air.
For 202.6 kg of O_2

the required stoichiometric air = $(202.6 \text{ kg } O_2)\left(\dfrac{100 \text{ kg air}}{23 \text{ kg } O_2}\right)$ = 880.1 kg air/100 kg coal

This 880.1 kg of air comprises the following, on a mass basis:

$(0.76 \text{ kg } N_2/\text{kg air})(880.1 \text{ kg air})$ = 668.9 kg N_2

$(0.0005 \text{ kg } CO_2/\text{kg air})(880.1 \text{ kg air})$ = 0.44 kg CO_2

20% excess air = $(880.1 \text{ kg air})(0.2)$ = 176 kg excess air

The air–fuel ratio for this coal is therefore = $\dfrac{(880.1 \text{ kg air} + 176 \text{ kg excess air})}{100 \text{ kg fuel}}$ = 10.6.

b) Mass of CO_2 per kg of coal burned:

$$\left(\frac{\text{mass } CO_2}{\text{mass C}}\right) \times (\text{C in } 100 \text{ kg coal}) = (3.67 \text{ kg } CO_2/\text{kg C})(65.6 \text{ kg C}) = 240.8 \text{ kg } CO_2/100 \text{ kg coal}$$

Mass of H_2O per kg of coal burned:

$$\left(\frac{\text{mass } H_2O}{\text{mass } H_2}\right)(H_2 \text{ in } 100 \text{ kg coal}) = (9 \text{ kg } H_2O/\text{kg } H_2)(4.1 \text{ kg } H_2) = 36.9 \text{ kg } H_2O/100 \text{ kg coal}$$

Mass of moisture (H_2O) per kg of coal burned:
The coal comprises 12% moisture, so there is 12 kg H_2O per 100 kg of coal.
Total mass of H_2O produced = $(36.9 + 12)$ kg H_2O/100 kg coal = 48.9 kg H_2O/100 kg coal
Mass of S converted to SO_3 and SO_2 per kg of coal burned:
5% of the S oxidizes to SO_3 = $(1.7 \text{ kg S}/100 \text{ kg coal})(0.05 \text{ kg S}/\text{kg } SO_3)$ = 0.085 kg S/100 kg coal goes to SO_3
95% of the S oxidizes to SO_2 = $(1.7 \text{ kg S}/100 \text{ kg coal}) - (0.085 \text{ kg S}/100 \text{ kg coal})$ = 1.615 kg S/100 kg coal goes to SO_2
Mass of SO_2 per kg of coal burned:

$$S + O_2 \rightarrow SO_2$$
$$32 \text{ kg S} + 32 \text{ kg } O_2 \rightarrow 64 \text{ kg } SO_2$$

So, the mass of SO_2 produced is

$$\underbrace{\left(\frac{1.615 \text{ kg S}}{100 \text{ kg coal}}\right)}_{S} + \underbrace{\left(\frac{32 \text{ kg } O_2}{32 \text{ kg S}} \times 1.615 \text{ kg S}\right)}_{O_2} = \underbrace{3.23 \text{ kg } SO_2/100 \text{ kg fuel}}_{SO_2}$$

Mass of SO_3 per kg of coal burned:

$$S + 1.5 \, O_2 \rightarrow SO_3$$
$$32 \text{ kg S} + 48 \text{ kg } O_2 \rightarrow 80 \text{ kg } SO_3$$

So, the mass of SO_3 produced is

$$\underbrace{\left(\frac{0.085 \text{ kg S}}{100 \text{ kg coal}}\right)}_{S} + \underbrace{\left(\frac{48 \text{ kg O}_2}{32 \text{ kg S}} \times 0.085 \text{ kg S}\right)}_{O_2} = \underbrace{0.2125 \text{ kg SO}_3/100 \text{ kg fuel}}_{SO_3}$$

Mass of N_2 per kg of coal burned:

Nitrogen is inert and the same amount will appear in the reactants as in the products. From Table 7.2, there is 79% of N_2 in air on a molar basis, which is equivalent to 76% on a mass basis. We previously determined that 808.9 kg + 176 kg = 1056 kg of air is required for every 100 kg of coal. So, the amount of nitrogen required (on a mass basis) is:

$$(0.76 \text{ kg N}_2/\text{kg air})(1056 \text{ kg air}/100 \text{ kg coal}) = 802.6 \text{ kg N}_2/100 \text{ kg coal}$$

Mass of O_2 per kg of coal burned:

From Table 7.2, there is 21% of O_2 in air. The amount of oxygen supplied is:

$$(0.21 \text{ kg O}_2/\text{kg air})(1056 \text{ kg air}/100 \text{ kg coal}) = 221.8 \text{ kg O}_2/100 \text{ kg coal}$$

We previously determined that the net amount of oxygen is 202.6 kg O_2/100 kg coal. The excess oxygen is therefore the oxygen supplied less than the net oxygen required = (221.8 − 202.6) kg O_2/100 kg coal = 19.2 kg excess O_2/100 kg coal

The results are summarized in the table below, which also confirm that mass is conserved.

Fuel component	Mass (kg)	Oxygen required per 100 kg coal (kg)	Product	Stoichiometric product mass per 100 kg coal (kg)	Excess air per 100 kg coal (kg)	Total product mass per 100 kg coal (kg)	Specific heat (c_p) (kJ/kgK)	Enthalpy (kJ/K)
C	65.6	175.15	CO_2	240.8	0.44(0.2) = 0.088	240.9	0.84	202.36
H_2	4.1	32.8	H_2O	36.9	—	36.9	1.996	73.65
O_2	7.1	−7.1	O_2	0	202.6(0.2) = 40.52	40.52	0.92	37.28
N_2	1.3	0	N_2	668.9	668.9(0.2) = 133.78	802.7	1.04	834.7
S to SO_2	1.615	1.615	SO_2	3.23	—	3.23	0.61	1.97
S to SO_3	0.085	0.1275	SO_3	0.2125	—	0.2125	Approx. 0	0
H_2O (moisture)	12.0	0	H_2O	12.0	—	12.0	1.996	23.95

Flue gas (sensible) enthalpy: 1174

c) The acid dew point temperature, t_{adp}:

From Eq. (8.2):

$$\frac{1000}{t_{adp} + 273.15} = 1.7846 + 0.02671 \log_{10} p_{H_2O} - 0.1031 \log_{10} p_{SO_3} + 0.0329 \left(\log_{10} p_{H_2O}\right)\left(\log_{10} p_{SO_3}\right)$$

with t_{adp} in °C and p in bar.

We can determine partial pressure using Dalton's Law:

$$p_i = \frac{N_i}{N} p$$

We can calculate the total number of moles as show in the table below:

Product	Total product per 100 kg coal (kg) ÷	Molar mass (kg/kmol) =	N (kmol)
CO_2	240.9	44	5.47
H_2O	48.9	18	2.72
O_2	40.52	32	1.26
N_2	902.75	28	22.48
SO_2	3.23	64	0.05
SO_3	0.2125	80	0.0027
		Total:	32

The partial pressure of H_2O is:

$$p_{H_2O} = \frac{N_{H_2O}}{N} p = \left(\frac{2.72 \text{ kmol}}{32 \text{ kmol}}\right) 1 \text{ bar} = 85 \times 10^{-3} \text{ bar}$$

The partial pressure of SO_3 is:

$$p_{SO_3} = \frac{N_{SO_3}}{N} p = \left(\frac{0.0027 \text{ kmol}}{32 \text{ kmol}}\right) 1 \text{ bar} = 84.38 \times 10^{-6} \text{ bar}$$

Therefore,

$$\frac{1000}{t_{adp} + 273.15} = 1.7846 + 0.02671 \, \log_{10}\left(85 \times 10^{-3}\right) - 0.1031 \, \log_{10}\left(84.3 \times 10^{-6}\right)$$

$$+ 0.0329 \left(\log_{10} 85 \times 10^{-3}\right)\left(\log_{10} 84.3 \times 10^{-6}\right) \frac{1000}{t_{adp} + 273.15}$$

$$= 1.7846 + (-0.029) - (-0.420) + 0.144$$

$$\therefore t_{adp} = 158 \,°C$$

At 158 °C, the flue gas temperature will be "cool" enough to form sulfuric acid, which condenses on metal surfaces to cause acid corrosion. To prevent this from happening, engineers usually work with a 30 °C margin. In the case of the current coal being combusted, the flue gas temperature should therefore not be lower than 188 °C.

d) Combustion efficiency:

$$\text{Combustion efficiency} = \frac{\text{Energy input} - \text{Energy lost in flue gases and ash}}{\text{Energy input}}$$

Energy input: For 100 kg of coal, the energy input is $(100 \text{ kg})(27{,}000 \text{ kJ/kg}) = 2.7 \times 10^6 \text{ kJ}$

Energy lost in flue gases is equal to the flue gas enthalpy + enthalpy of ash and carbon lost in residue. We need to account for the fact that ash contains 5% of carbon, which will not participate in combustion.

We will calculate the energy flows relative to a datum of 25 °C, and use the acid dew point temperature calculated in (c) as the flue exit temperature (184 °C):

Flue gas enthalpy = sensible heat loss + latent heat loss

$$= \underbrace{(1174.1 \text{ kJ/K})(184 - 25) \text{ K}}_{\text{sensible enthalpy}} + (36.9 \text{ kg } H_2O + 12 \text{ kg } H_2O)\underbrace{(2442.5 \text{ kJ/kg})}_{\substack{\text{latent heat of} \\ \text{water at 25 °C}}}$$

$$= 186{,}680 \text{ kJ} + 119{,}438 \text{ kJ} = 306{,}119 \text{ kJ}$$

Ash enthalpy = ash + unburned carbon

Total mass of ash and carbon $= 7.7 \text{ kg ash} + 7.7 \text{ kg ash} \underbrace{\left(\dfrac{0.05 \text{ kg C}}{1 \text{ kg ash}}\right)}_{\substack{\text{ratio of unburned} \\ \text{carbon in ash}}} = 8.1 \text{ kg}$

$\therefore$ Ash enthalpy $= (8.1 \text{ kg})(0.84 \text{ kJ/kgK})(185 - 25)\text{K} + (0.4 \text{ kg})(32,800 \text{ kJ/kgK})$

$\therefore$ Ash enthalpy $= 14,209 \text{ kJ}$

$\therefore$ Combustion efficiency $= \dfrac{2,700,000 \text{ kJ} - (306,119 + 14,209) \text{ kJ}}{2,700,000 \text{ kJ}} = 88\%$

Discussion: The boiler is 88% efficient in converting the chemical energy available in coal to heat. Box 8.2 illustrates the typical carbon dioxide emissions per unit of energy produced. Box 8.3 explains how coal can be used to produce synthetic fuel.

Box 8.2   Carbon-Dioxide Emission from the Combustion of Coal

As illustrated in Example 8.1, bituminous coal with 66% carbon content produces 240.8 kg of CO_2 for every 100 kg of coal combusted, or 2.41 kg of CO_2 for every kg of coal combusted. The gross calorific (heating) value of this coal is 27 MJ/kg. If we assume that the coal will be burned in a boiler at a steam power plant, and that the power plant has an overall energy conversion efficiency of 30%, then the amount of CO_2 produced per kWh_e is:

$$\left(\frac{27 \text{ MJ}}{\text{kg coal}}\right)(0.3)\left(\frac{1 \text{ kg coal}}{2.41 \text{ kg } CO_2}\right)\left(\frac{1 \text{ kWh}}{3.6 \text{ MJ}}\right) = 0.93 \text{ kWh/kg } CO_2 = 1.07 \text{ kg } CO_2/\text{kWh}$$

Coal is "dirty". The combustion of common coal produces around 1 kg of CO_2 for every kg of coal burned in a typical steam power plant. Modern coal combustion incorporates technologies to capture CO_2 in the flue gas, as discussed in Chapter 14.

Box 8.3   Production of Synthetic Liquid Fuel from Coal

The syngas generated by coal gasification can also be used to produce synthetic fuels, such as with the *Fischer-Tropsch process*. For the Fischer–Tropsch reaction to form diesel fuel (C_nH_{2n+2}), the syngas needs to have an H_2 to CO ratio of about 2 : 1. The chemical reaction takes the following form (Sørensen 2017):

Fischer-Tropsch diesel: $(2n + 1)H_2 + n\,CO \rightarrow (C_nH_{2n+2}) + n\,H_2O$ (8.5)

Comparing Eqs. (8.4) and (8.5), one CO_2 molecule is produced for each C atom incorporated in the Fischer–Tropsch product. However, due to the required high temperatures and system irreversibilities, as well as other inefficiencies, producing synthetic fuel from coal can produce 3.5 times the CO_2 produced by burning conventional hydrocarbons. To its advantage, the CO_2 produced in the Fischer–Tropsch process is an almost pure stream at an intermediate pressure, which makes it ideal, although expensive, for carbon capture (Deutsche et al. 2007).

In South Africa, Sasol (https://www.sasol.com) produces about 200,000 barrels of liquid fuel per day using the Fischer–Tropsch process (see Figure 8.3). The high capital and operational costs have, however, kept this technology uncompetitive.

Figure 8.3 Part of the Sasol coal-to-gas plant in Secunda, South Africa, rated to be the "most polluting plant in the world." *Source:* AntimatterMachine / Wikimedia Common.

Interconnected Sustainability Solutions Are Required, Also in University Education

Renewable energy generation and clean fuels are effective solutions to counteract climate change when widely implemented. However, their strategic placement within larger electric grids, as well as energy storage, resources, and supporting policy, must also be considered.

Jessica Nicholson (24), a BS in Mechanical Engineering graduate from the University of Illinois at Urbana-Champaign (UIUC), believes that consistent reliability and cost innovations are necessary for renewables to remain competitive in the energy sector.

In 2022, she completed a research project to define reliability standards for proposed US Army Corps of Engineers microgrid systems with high proportions of renewable energy. In that work, she helped project high variable renewable energy (VRE) system service availability and optimize its parameters for reliability. "I will continue to focus on innovations in renewable energy reliability, cost, and infrastructure," says Jessica.

In her years as a student, it was already clear that Jessica wanted to help bring about rapid change. As a Student Senator, she reminded the student leadership that "excessive pollution, resource overuse, and environmental degradation have progressed climate change to a critical point, where ensuring the well-being of our present and future generations has become a significant challenge." She observed that rapid, detrimental environmental changes are becoming increasingly evident and require work within all technical, social, and economic circles of influence to change. "We need to update our university programs to reflect these urgent needs." With her passion about sustainability, she went on to champion a proposal for all UIUC undergraduates to complete general education courses in sustainability.

When she was only nine years old, Jessica wrote letters to President Obama, senators, and several other influential people, imploring them to bring about urgent changes to help establish a more sustainable way of life. Most of them, including President Obama, responded with kind letters of support. "After all this time," says Jessica, "investment in several forms of renewable energy—such as wind, solar, hydroelectric, and geothermal power—is imperative, more than ever before, to ensure that fossil-free energy is possible in every community." Jessica's spirited approach has clearly not dwindled.

We are living in times when we are reminded that the limited capacities of the present grid power transmission and energy storage must increase to account for the variability of added renewable energy. "Not only is the improved maintenance of the present electrical grids necessary, but so is the development of new grid infrastructure to successfully integrate new renewable energy generation," she adds.

With clear-minded, techno-savvy, brilliant engineers such as Jessica, we have a good chance of bringing about these urgent changes.

Jessica Nicholson, a mechanical engineer on a mission.
Source: Jessica Nicholson, https://www.jessicanicholson.info/.

8.3 The Future of Coal

Coal-fired electricity generation peaked in 2020–2022. The prevailing energy crisis will probably ensure that coal-fired power plants will be around for many more years. This makes the goal of achieving net zero carbon emissions before 2050 even more difficult to achieve. To keep that goal alive, coal-fired power plants must be retrofitted with pollution control (and carbon capture) technologies, and they should be completely phased out by 2040. This would require shutting around 870 GW of existing coal-fired power plants. In this scenario, only around 470 million tons of coal-equivalent will be used in 2050, and most of those facilities will be equipped with carbon capture technologies (IEA 2021b).

International collaboration will be of importance to facilitate coal substitutes. Retraining and regional revitalization programs will be essential to reduce the social impact of job losses at the local level and to enable workers and communities to find alternative livelihoods.

Problems

8.1 An ultimate analysis of a bituminous coal sample provides the following mass fraction of constituents: 72.8% C, 4.8% H_2, 6.2% O_2, 1.5% N_2, 2.2% S, 3.5% moisture, and 9% ash. To ensure complete combustion, the coal will be burned with 25% excess air at 100 kPa. Determine the air–fuel ratio.

[Answer: 12.4 : 1]

8.2 An ultimate analysis of a bituminous coal sample provides the following mass fraction of constituents: 75% C, 4% H_2, 7% O_2, 1% N_2, 2% S, 4% moisture, 7% ash, 5% of the sulfur oxidizes to SO_3, and the remainder (95%) to SO_2. The ash contains 5% carbon, which will not participate in combustion. To ensure complete combustion, the coal will be burned with 40% excess air at 100 kPa. (a) Estimate the gross calorific value of this coal using DuLong's formula. Determine (b) the air–fuel ratio, (c) the kg of CO_2 emitted per kg of coal burned, and (d) the amount of coal used per day in a 400 MW power plant with a 75% capacity factor and 40% thermal efficiency.

[Answers: (a) 30 MJ/kg; (b) 14 : 1; (c) 2.75 kg CO_2/kg coal; (d) 2150 tons of coal per day]

8.3 Bituminous coal with a gross calorific value of 25 MJ/kg is used in a boiler that produces steam for a turbine that generates 100 MW_e. The thermal efficiency of the power plant (i.e., the ratio of work done to the amount of heat provided by the boiler) is 39%. The plant has a capacity factor of 80%. Calculate the amount of CO_2 that a carbon capture system with 97% efficiency will need to process every day. Clearly state your assumptions.

[Answer: 610 tons/day]

8.4 Consider bituminous coal with a gross calorific value of 27 MJ/kg and the following ultimate analysis: 78.5% C, 4.7% H_2, 1.7% N_2, 1.9% S, 4.2% O_2, 4% moisture, and 5% ash. Construct graphs of the CO_2 and O_2 in the flue gas plotted against the percentage of excess air. Vary the percentage of excess air from 0% (stoichiometric) to 100% in steps of 20%. Comment on the nature of the CO_2 and O_2 graphs.

[Answer: At 0% excess air, there will be about 25.8% CO_2 and 0% O_2. At 20% excess air, there will be about 21.8% CO_2 and 3.7% O_2. At 40% excess air, there will be about 18.8% CO_2 and 6.4% O_2. At 60% excess air, there will be about 16.5% CO_2 and 8.4% O_2. At 80% excess air, there will be about 14.7% CO_2 and 10% O_2. At 100% excess air, there will be about 13.3% CO_2 and 11.3% O_2.]

8.5 The schematic diagram in Figure 8.4 shows a 500 MW_e IGCC unit with CO_2 capture. Using the metrics provided in Table 8.1, together with the data provided in Figure 8.4, produce a fact sheet of no less than 300 words (with accompanying diagrams, where necessary) about IGCC

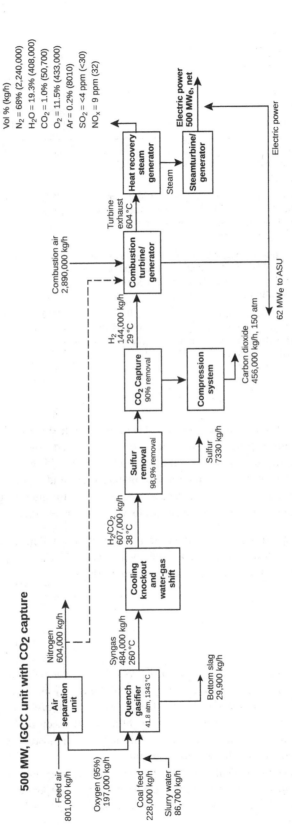

Figure 8.4 Schematic of an ultra-supercritical pulverized coal unit that provides steam for a 500 MWe steam turbine or generator.
Source: Adapted from Deutsche et al. (2007).

Table 8.1 Classification of coal.

Type	Description	Carbon content (%)	Lower heating value (MJ/kg)	Higher heating value (MJ/kg)
Anthracite ("hard coal" or "black coal")	Anthracite is a high-rank, non-agglomerating coal, usually only used for metallurgical applications	86–97	24	31
Bituminous	Bituminous coal is an agglomerating coal, about 100–300 million years old, and is mainly used for electricity generation and in the iron and steel industry	45–86	24	27
Sub-bituminous	Sub-bituminous coal is a non-agglomerating ("crumbly") coal, which is at least 100 million years old. It is mainly used for electricity generation	35–45	17	24
Lignite ("brown coal")	Lignite is a non-agglomerating ("soft" and "crumbly") coal, mainly used for electricity generation. It may contain various toxic heavy metals, including naturally occurring radioactive materials	25–35	15	17

Source: Adapted from IEA (2022).

technology and how it addresses the United Nations' Sustainable Development Goals (SDGs). Your fact sheet must specifically illustrate to decision-makers how IGCC technology helps to address the SDGs, and how it enhances and benchmarks sustainability performance, improves risk management, enhances data management and reporting practices, improves resource allocation and reduces expenses, improves environmental performance, reduces social impact, and improves communications with the stakeholder.

[Answer: This question is open-ended and requires independent research. The fact sheet should at least illustrate how IGCC technology impacts on SDGs 3, 7, 8, 9, 11, 12, and 13. Refer to Mini Project 1 at the end of Chapter 3 for more tips.]

References

BP (2022). BP Energy Outlook. 2022 edition. https://www.bp.com/content/dam/bp/business-sites/en/global/corporate/pdfs/energy-economics/energy-outlook/bp-energy-outlook-2022.pdf

Deutsche, J., Moniz, E.J., Ansolabehere, S., Beer, J. et al. (2007). The future of coal - an interdisciplinary MIT study. Massachusetts Institute of Technology. http://web.mit.edu/coal

EIA (2021). Coal and the environment. US Energy Information Administration. https://www.eia.gov/energyexplained/coal/coal-and-the-environment.php

Elliott, M.A. (1981). *Chemistry of Coal Utilization*. Second supplementary volume. New York, NY: Wiley.

ExxonMobil (2022). Outlook for Energy - 2021. ExxonMobil. https://corporate.exxonmobil.com/Energy-and-innovation/Outlook-for-Energy/Energy-supply#Liquids

Finkelman, R.B., Wolfe, A. and Hendryx, M.S. (2021). The future environmental and health impacts of coal. *Energy Geoscience* 2 (2): 99–112.

IEA (2021a) World energy outlook, 2021. International Energy Agency (IEA). https://iea.blob.core.windows.net/assets/4ed140c1-c3f3-4fd9-acae-789a4e14a23c/WorldEnergyOutlook2021.pdf

IEA (2021b). Net Zero by 2050. A roadmap for the global energy sector. International Energy Agency. https://iea.blob.core.windows.net/assets/deebef5d-0c34-4539-9d0c-10b13d840027/NetZeroby2050-ARoadmapfortheGlobalEnergySector_CORR.pdf

IEA (2022). Coal information. International Energy Agency (IEA). https://www.iea.org/data-and-statistics/data-product/coal-information-2#coal-statistics-oecd-and-selected-countries

Jakob, M., Steckel, J.C., Jotzo, F., Sovacool, B.K. et al. (2020). The future of coal in a carbon-constrained climate. *Nature Climate Change* 10 (8): 704–707.

Lindberg, T.T., Bernhardt, E.S., Bier, R., Helton, A.M. et al. (2011). Cumulative impacts of mountaintop mining on an Appalachian watershed. *Proceedings of the National Academy of Sciences of Sciences of the United States of America* 108 (52): 20929–20934.

Macrae, J.C. (1966). *An Introduction to the Study of Fuel*. Amsterdam, Netherlands: Elsevier Publishing Company.

Palmer, M.A., Bernhardt, E.S., Schlesinger, W.H., Eshleman, K.N. et al. (2010). Mountaintop mining consequences. *Science* 327 (5962): 148–149.

Pierce, R.R. (1977). Estimating acid dew points in stack gases. *Chemical Engineering (New York)* 84 (8): 125–128.

Rose, J.W. and Cooper, T.R. (1977). *Technical Data on Fuel*. New York, NY: Wiley.

Sørensen, H. (2017). *Renewable Energy. Physics, Engineering, Environmental Impacts, Economics, and Planning*. Cambridge, MA: Academic Press/Elsevier.

9

Oil and Natural Gas, and Hydrogen and Biofuels

The various scenarios to achieve net-zero carbon emission by 2050 indicate a broad range of potential outcomes for the future energy mix and show that no single energy source or technology solution is sufficient to achieve a "lower 2 °C" pathway (IEA 2021a; Pörtner et al. 2022). All lower-emission sources play important roles. The scenarios include unprecedented deployment of solar, wind, carbon capture and storage, hydrogen, and biofuels such as biodiesel and biomethane.

However, oil and natural gas will remain a significant part of the energy mix for decades in these scenarios, reinforcing the need for continued innovation to bring about cleaner oil and gas combustion (Deloitte 2022). According to recent projections to achieve net-zero emissions by 2050, the demand for coal must fall by 95%, oil by 75%, and natural gas by 55% compared to 2020 levels (IEA 2021b) (see Figure 9.1).

While the term *petroleum* can be applied to all hydrocarbons that occur in the Earth, both liquid and gaseous, it is more generally applied to the liquid that is recovered from *oil* wells. It is also sometimes called *crude oil*. The gas component of petroleum is called *natural gas* (or just "gas"). Natural gas can, however, also come from a gas field, instead of being separated via distillation from crude oil (Macrae 1966). Together, oil and gas constitute the principal prime energy sources of Earth in current use, together supplying nearly 53% of the world's energy needs in 2021 (IEA 2021b).

Reimagine Our Future
Sweden taxes polluters around $120 per ton of CO_2 emitted. Sweden's carbon tax generates considerable revenue. These funds may be used for purposes linked to the carbon tax, such as addressing the undesirable distributional consequences of carbon tax or financing other climate-related measures. Should your country impose a tax on carbon rather than (or as well as) on income? Discuss this with your friends.

9.1 Oil

Oil is chiefly found in sedimentary rocks such as shale. However, tar sands (also called "bituminous sands" or "oil sands") hold large quantities of bitumen that can be processed to yield 70% synthetic crude oil (Harder 1982).

Energy Systems: A Project-Based Approach to Sustainability Thinking for Energy Conversion Systems, First Edition. Leon Liebenberg.
© 2024 John Wiley & Sons, Inc. Published 2024 by John Wiley & Sons, Inc.
Companion website: www.wiley.com/go/liebenberg/energy_systems

Production (EJ)

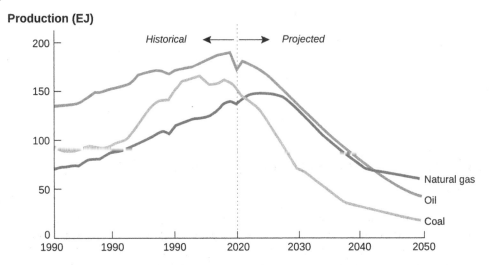

Figure 9.1 Coal, oil, and natural gas production to achieve net-zero emissions by 2050. *Source:* Adapted from IEA (2021a).

Oil is regarded as a fossil fuel due to its suggested animal or vegetable origins, probably due to the anaerobic bacterial decay of marine life. It is eventually compacted into limestone and shale. Due to the action of intense heat and pressure over millions of years, this organic material decomposed under the rock cover into various types of hydrocarbons. Compared to another fossil fuel – coal – oil is simpler to transport due to its liquid form.

9.1.1 Oil Classification

Oil comprises mainly hydrocarbons, together with sulfur, nitrogen, oxygen, and possibly also some ash. The ash in heavy fuel oil contains traces of silicon, aluminum, iron, calcium, sodium, vanadium, nickel, lead, copper, and tin. Although the types of hydrocarbons in oil vary widely in their proportions, depending on their origins, the ultimate analysis of crude oil tends to be in the following ranges (Rose and Cooper 1977):

$$C: 82 - 87\% \qquad H_2: 10 - 14\% \qquad O_2 + N: 1 - 3\% \qquad S: 0.5 - 3\% \qquad Ash: 0 - 0.1\%$$

Like coal, oil contains significant amounts of sulfur that can cause acid corrosion in boilers and downstream equipment (see Section 8.1 for a comparison with coal). The acid dew point of flue gas can be determined using Eq. (8.2), as was the case for coal.

Crude oil is usually classified according to the type of residue that remains after distillation:

- *Paraffins* ("alkanes"), C_nH_{2n+2}: Examples include methane (CH_4), ethane (C_2H_6), propane (C_3H_8), *n*-octane, and iso-octane (C_8H_{18}). Note that "iso-octane" usually refers to 2,2,4-trimethyl-pentane, suggesting five carbon atoms in a straight chain (hence, pentane) with three methyl (CH_3) branches located at C-atoms 2, 2, and 4, respectively.
- *Naphthenes* ("cycloalkanes" or "cycloparaffins"), C_nH_{2n}: Examples include cyclopropane (C_3H_6), cyclobutane (C_4H_8), and cyclopentane (C_5H_{10}).
- *Olefins* ("alkenes"), C_nH_{2n}: Examples include ethene or ethylene (C_2H_4), propene or propylene (C_3H_6), butene or butylene (C_4H_8).
- *Acetylenes* ("alkynes"), C_nH_{2n-2}: Acetylene (C_2H_2) is the prime member of this group.

- *Aromatics*, C_nH_{2n-6}: Examples include benzene (C_6H_6), toluene (C_7H_8), and xylene (C_8H_{10}).
  Hydrocarbon fuels can be blended with other constituents after distillation to provide fuels with unique characteristics, such as alcohols, which present another fuel category:
- *Alcohols*, $C_nH_{2n+1}OH$: Examples include methanol (CH_3OH) and ethanol (C_2H_5OH).

The fractional distillation of crude oil produces, in order of increasing boiling points: gas, gasoline (or "petrol"), naphtha, kerosene, diesel, lubricating oil, heavy fuel oil, and residue such as asphalt (or "bitumen"). Table 7.5 summarizes the salient properties of crude oil and its distillation products.

9.1.2 Oil Consumption

Global oil demand in 2020 amounted to 90 million barrels per day (IEA 2021a). Oil consumption is especially increasing in emerging economies, driven by increasing prosperity and rising living standards, including the ownership of cars with internal combustion engines. In 2020, emerging economies consumed around 55% of oil. By 2050, emerging economies will consume an estimated 80% of global oil (BP 2022). In contrast, oil demand in the developed world has been following a downward trend between 2015 and 2020 but picked up in 2021 due to the energy crisis which accompanied Russia's war in Ukraine.

Apart from its use as a fuel for all types of vehicles (cars, trucks, trains, airplanes, ships) with internal combustion engines, oil is also used as a feedstock for petrochemicals, including plastic. However, as the global economies decarbonize, the pace of growth of oil as a feedstock will also slow down. This implies that plastic recycling will become increasingly important, and the use of single-use plastic will become increasingly rare.

Reimagine Our Future

Virtually everything we make, use, eat, wear, build, refrigerate, light, and transport depends on using fossil fuels, a finite resource, with its disastrous emission of CO_2. It appears that modern man has become "vaccinated" against actions to live within ecological constraints. Do you agree?

9.1.3 Oil Reserves

There are about 6.05 trillion barrels of proved oil shale reserves, and more than 2 trillion barrels of oil from tar sands. In 2020, the estimated total recoverable global oil reserves were 1732 billion barrels of oil (BP 2021). The Organization of the Petroleum Exporting Countries (OPEC) holds 70.2% of the global reserves. The top countries in terms of reserves are Venezuela (17.5% of global reserves), closely followed by Saudi Arabia (17.2%), and Canada (9.7%). The global reserve-to-production (R/P) ratio reveals that oil reserves in 2020 accounted for just over 50 years of current production (BP 2021). This means that, at the current level of production, global oil reserves will last approximately another 50 years before being depleted.

9.1.4 Combustion of Oil, Diesel, and Gasoline

As *heavy fuel oil* has a similar chemical composition to coal, albeit with a larger percentage of carbon and hydrogen, oil combustion is treated identically to coal combustion. Example 8.1 covers the combustion of coal. The problem-solving procedure will be the same for oil. No additional example will therefore be provided regarding the combustion of oil.

The combustion of *diesel* is, however, illustrated in Example 9.1. A similar problem-solving procedure would be followed to analyze the combustion of *gasoline* ("petrol") and other liquid and gaseous hydrocarbons. The average chemical formula for gasoline is C_8H_{15}, ranging from around C_7H_{13} to C_9H_{17}. For simplicity, though, gasoline is often approximated as iso-octane, C_8H_{18}.

The average chemical formula for common diesel fuel is $C_{12}H_{23}$, ranging from around $C_{10}H_{20}$ to $C_{15}H_{28}$ (also see Table 7.5 for the classification of fuels).

As discussed in Chapters 7 and 8, fuel can be burned with more or less than the stoichiometric air requirement.

With *excess air (or fuel-lean) combustion*, the excess air appears in the products in unchanged form. A lean mixture helps reduce CO, HC, and NO_x emissions, unless the mixture becomes so lean that misfiring occurs. In Example 7.1, we burned C_8H_{18} (iso-octane) in stoichiometric air. This gave the following reaction:

$$C_8H_{18} + 12.5[O_2 + 3.76\ N_2] \rightarrow 8\ CO_2 + 9\ H_2O + 47\ N_2$$

Then, in Example 7.2, the same fuel was burned with 400% excess air, which gave the following reaction, where the excess air (oxygen and nitrogen) remains unchanged:

$$C_8H_{18} + 4\{12.5[O_2 + 3.76\ N_2]\} \rightarrow 8\ CO_2 + 9\ H_2O + 37.5\ O_2 + 4(47\ N_2)$$
$$\therefore \quad C_8H_{18} + 62\ O_2 + 188\ N_2 \rightarrow 8\ CO_2 + 9\ H_2O + 37.5\ O_2 + 188\ N_2$$

If the fuel is burned with less than the stoichiometric air requirement, i.e., with *fuel-rich combustion*, there is insufficient O_2 to fully oxidize the fuel C and H_2 to CO_2 and H_2O. The products then include a mixture of CO_2 and H_2O, along with CO, H_2, and N_2.

The ratio of the actual fuel–air ratio $(F/A)_{actual}$ to the stoichiometric ratio $(F/A)_{stoich}$ is called the fuel/air equivalence ratio ϕ:

$$\phi = \frac{(F/A)_{actual}}{(F/A)_{stoich}} \tag{9.1}$$

For *fuel-lean mixtures*: $\phi < 1$ and O_2 also appears in the products of combustion.
For *stoichiometric mixtures*: $\phi = 1$.
For *fuel-rich mixtures*: $\phi > 1$ and CO and H_2 also appear in the products of combustion.
For *all air-to-fuel mixtures*, the products of combustion will include N_2, CO_2, H_2O.

Sometimes, it is more convenient to work with the air–fuel ratios, in which case the lambda (λ) factor may be defined:

$$\lambda = \frac{1}{\phi} = \frac{(A/F)_{actual}}{(A/F)_{stoich}} \tag{9.2}$$

If the fuel is reformulated to contain oxygen (such as with alcohols), the procedure for balancing the combustion equation is the same as with gasoline and diesel, except that fuel–oxygen must be accounted for. For instance, for *methanol* (CH_3OH), the balanced stoichiometric equation is:

$$CH_3\ OH + 1.5[O_2 + 3.76\ N_2] \rightarrow CO_2 + 2\ H_2O + 1.5\ (3.76)\ N_2 \tag{9.3}$$

and for the widely used *ethanol* (C_2H_5OH), the balanced stoichiometric equation is:

$$C_2H_5\ OH + 3[O_2 + 3.76\ N_2] \rightarrow 2\ CO_2 + 3\ H_2O + 3\ (3.76)\ N_2 \tag{9.4}$$

Whereas the air–fuel ratio is 6.5 for methanol and 9 for ethanol, it is approximately 14.7 for gasoline and diesel.

As explained in Box 9.1, gasoline is often blended with other fuels (such as ethanol) to reduce emissions. When fuels are blended in certain mass fractions (x_i), the effective air–fuel ratio can be calculated as follows:

$$AFR_{blend} = \sum (AFR)_i (x_i) \tag{9.5}$$

From Table 7.2, mass fractions (x_i) can be obtained from volume fractions (y_i):

$$y_i = \frac{V_i}{V} \tag{9.6}$$

$$x_i = \frac{m_i}{m} = \frac{\rho_i V_i}{\sum(\rho_i V_i)} \equiv \frac{\rho_i \dfrac{V_i}{V}}{\sum\left(\rho_i \dfrac{V}{V_i}\right)} = \frac{\rho_i y_i}{\sum(\rho_i y_i)} \tag{9.7}$$

The heat of combustion of a blended fuel can be calculated by considering the mass fractions of the constituent fuel components:

$$Q_{blend} = \sum (Q_i x_i) \tag{9.8}$$

Box 9.1   Some Alternative Fuels: Oxygenated Fuels, Reformulated Gasoline, FlexFuel, LPG

In the USA, the US Clean Air Act of 1990 mandated an *oxygenated fuels program* and a *reformulated gasoline program*, which requires changes in the composition of gasoline to reduce tailpipe emissions. The oxygenation program demands that gasoline contains at least 2.7% oxygen (by mass) to reduce the emission of HC and CO. The reformulated gasoline program requires gasoline in certain parts of the country to have a minimum oxygen content of 2% (by mass) and a maximum benzene content of 1%. The main oxygenation component is ethanol. The US Energy Policy Act of 2005 amended the 1990 Clean Air Act, which requires gasoline to contain 10% renewable fuels such as ethanol ("E10 gasoline"). Depending on the time of year, certain US states sell gasoline containing 15% ethanol ("E15 gasoline") and 20% ethanol ("E20 gasoline").

Some vehicles have a *FlexFuel* capability, according to which they can operate on "E85" ethanol-gasoline blends containing between 51 and 83% ethanol. These flexible-fuel or dual-fuel vehicle engines can effectively burn any proportion of the fuel blend as fuel injection and ignition timing are automatically adjusted according to the blend detected by a fuel composition sensor.

If an "E85" fuel comprises 60% ethanol (HHV = 29.67 MJ/kg; density = 790 kg/m³) and 40% iso-octane (HHV = 47.3 MJ/kg; density = 750 kg/m³), then the higher heating value (HHV) of the fuel blend can be calculated using Eqs. (9.4)–(9.7):

$$X_{ethanol} = \frac{\rho_i y_i}{\sum(\rho_i y_i)} = \frac{(790)0.6}{(790)0.6 + (750)0.4} = 0.61$$

$$\therefore X_{gasoline} = 1 - 0.61 = 0.39$$

The stoichiometric air–fuel ratios can easily be determined for both methanol and gasoline (see Examples 9.1 and 9.2 for the procedure). The stoichiometric air–fuel ratios for these fuel components are:

$$AFR_{stoich,ethanol} = 4.5 \text{ and } AFR_{stoich,gasoline} = 15.1$$

The heating value of the fuel blend is then:

$$Q_{blend} = \sum(Q_i x_i) = (29.67 \text{ MJ/kg})0.61 + (47.3 \text{ MJ/kg})0.39 = 36.5 \text{ MJ/kg}$$

Liquefied petroleum gas is a mixture of propane and butane (Table 7.5), usually pressurized to 200 kPa to 2 MPa, depending on the ratio of propane to butane, and the operating temperature. If it is used as a fuel for vehicles or stationary applications such as electricity generators, it is called *Autogas* in the United States. Autogas produces about 15% less CO_2 emissions than its gasoline counterpart and is used in about nine million vehicles worldwide (WI PGA 2021).

Example 9.1 Combustion of Diesel

Diesel ($C_{12}H_{23}$), with a density of 0.835 kg/L and a heating value of 42.5 MJ/kg, is burned stoichiometrically in air. Calculate (a) the air–fuel ratio, and (b) the kg of CO_2 emitted per kg of diesel burned.

Required: We must find the air–fuel ratio when combusting diesel stoichiometrically in air. We must also find the mass of CO_2 for each kg of diesel burned.

Solution strategy: (a) We will first balance the combustion reaction and then calculate the amount of required air and fuel, as well as the amount of CO_2 produced per kg of diesel burned.

Assumptions: Assume that N_2 is inert. So, the same amount of N_2 appears before and after combustion.

Analysis:

a) Air–fuel ratio

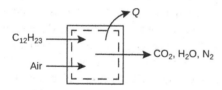

Combustion of diesel in air: $C_{12}H_{23} + a\underbrace{[O_2 + 3.76 \, N_2]}_{air} \rightarrow b \, CO_2 + c \, H_2O + d \, 3.76 \, N_2$

We then balance the equation to find a, b, c, and d:

C: $12 = b$

H: $23 = 2c$ $\qquad\qquad c = 11.5$

O: $2a = 2b + c$ $\qquad\quad a = 17.75$

N: $2a(3.76) = 3.76(2d)$ $\quad d = 17.75$

The balanced equation is then:

$$C_{12}H_{23} + 17.75\underbrace{[O_2 + 3.76 \, N_2]}_{air} \rightarrow 12 \, CO_2 + 11.5 \, H_2O + 66.74 N_2$$

We can now find the masses of each of the reactants and products. The molar masses for C is 12 kg/kmol, for H it is 1 kg/kmol, for O it is 16 kg/kmol, and for N it is 14 kg/kmol. The molar mass of air is assumed to be 29 kg/kmol.

$$m_{air} = N_{air}M_{air} = (17.75 \text{ kmol O}_2 + (17.75 \times 3.76) \text{ kmol N}_2) \times 29 \text{ kg/kmol air} = 2450.2 \text{ kg}$$

$$m_{C_{12}H_{23}} = N_{C_{12}H_{23}}M_{C_{12}H_{23}} = (12 \text{ kmol} \times 12 \text{ kg/kmol}) + (23 \text{ kmol} \times 1 \text{ kg/kmol}) = 167 \text{ kg}$$

$$m_{CO_2} = N_{CO_2}M_{CO_2} = 12 \text{ kmol}(12 + (16)) \text{ kg/kmol} = 528 \text{ kg}$$

$$m_{H_2O} = N_{H_2O}M_{H_2O} = 11.5 \text{ kmol} \times (2(1) + 16) \text{ kg/kmol} = 207 \text{ kg}$$

$$m_{O_2} = N_{O_2}M_{O_2} = 17.75 \text{ kmol} \times (2(16)) \text{ kg/kmol} = 568 \text{ kg}$$

$$m_{N_2} = N_{N_2}M_{N_2} = 66.74 \text{ kmol} \times (2(14)) \text{ kg/kmol} = 1868.7 \text{ kg}$$

Then, the air–fuel ratio can be determined:

$$AFR = \frac{m_{air}}{m_{fuel}} = \frac{2450.2 \text{ kg}}{167 \text{ kg}} = 14.7 \text{ kg air/kg fuel}$$

b) Mass of CO_2 per unit mass of fuel

$$\frac{m_{CO_2}}{m_{C_{12}H_{23}}} = \frac{528 \text{ kg}}{167 \text{ kg}} = 3.16 \text{ kg CO}_2/\text{kg diesel}$$

Discussion: The combustion of diesel in air under stoichiometric conditions requires an air–fuel ratio of 14.7: 1 and produces 3.16 kg of CO_2 for every kg of diesel burned. A diesel-powered city bus is typically equipped with a 12.9-L diesel engine. If the bus operates on a typical urban driving cycle, the fuel consumption will be about 11.05 kg/h. If the bus operates for 4 h a day, 5 days a week, and 500 weeks a year, the diesel engine will emit 35 t of CO_2 per year.

9.2 Natural Gas

Natural gas refers to any naturally occurring combination of gaseous hydrocarbons. Its origins are like that of petroleum. But, among fossil fuels, natural gas has the lowest carbon content, emits less CO_2 per unit of energy, and burns relatively cleanly and efficiently with very few non-carbon emissions (Harder 1982). Unlike oil, natural gas requires limited processing to prepare it for end-use. However, natural gas has a low energy density and is difficult and expensive to transport and store. Most natural gas supplies are delivered by pipeline, and delivery costs represent a large fraction of the total cost (MIT 2011).

Unlike other fossil fuels, natural gas plays a major role in most sectors of the modern economy – from electrical power generation (accounting for about 24% of global electricity generation in 2022), to industrial, commercial, and residential use. In the US market, the price of natural gas is set regionally rather than globally, which is the case with oil. This is important in determining market share when there is an opportunity for substituting an incumbent fuel (MIT 2011).

9.2.1 Classification and Storage of Natural Gas

Natural gas can be separated ("stripped") from crude oil, in which case it is called *associated natural gas*. Natural gas can also come from coal beds or from a gas field. Large fields of natural gas have recently been developed by the hydraulic fracturing (or "fracking") of the shale rock formations in which the *shale gas* is found (Jaffe and Taylor 2018).

Natural gas comprises more than 80% methane (CH_4) and is either stored in gas form at a pressure of around 20 MPa as *compressed natural gas (CNG)* or as *liquefied natural gas (LNG)* at around $-162\,°C$ in a thermally insulated ("cryogenic") tank. LNG requires about a third the storage volume of CNG, but large amounts of energy are required to liquefy and store it. CNG is therefore the preferred choice for the storage of natural gas at refueling stations.

9.2.2 Consumption of Natural Gas

Gas demand in 2020 amounted to around 3800 billion m^3 (BP 2021; IEA 2021a, 2021b). Industry in emerging markets and developing economies is the key driver of growth in natural gas demand. Natural gas is also replacing coal (and biomass) in a range of industrial applications, including cement and steel production (IEA 2021b). The global trade of LNG is steadily increasing and will soon overtake CNG as the dominant form of natural gas.

Natural gas is increasingly being used as a feedstock to produce low-carbon hydrogen until "green" hydrogen (produced using solar or wind) becomes financially viable (see Box 9.3).

9.2.3 Reserves of Natural Gas

The current global reserve-to-production ratio shows that natural gas reserves in 2020 accounted for 48.8 years of current production. The Middle East (110.4 years) and Commonwealth of Independent States (70.5 years) are the regions with the highest R/P ratio (BP 2021). Combined, at the current level of production, global natural gas reserves will last approximately another 50 years before being depleted.

9.2.4 Combustion of Natural Gas

Natural gas is combusted in furnaces at residential, commercial, and industrial levels. The natural gas is pre-treated to remove any abrasive material, liquid components, and sulfur compounds like the highly toxic hydrogen sulfide (H_2S), if present.

Natural gas is primarily (80–99%) composed of methane (CH_4) and may contain smaller amounts of ethane (C_2H_6), propane (C_3H_8), and butane (C_4H_{10}), and trace amounts of pentane and hexane, nitrogen (N_2), CO_2, H_2O, O_2, H_2S, and He. From Table 7.5, natural gas has an average higher heating value of 52.2 MJ/kg and a lower heating value of 47.1 MJ/kg.

Example 9.2 Combustion of Methane in a Gas Turbine
An 80-MW combustor of a modern gas turbine is fueled by gaseous methane (CH_4). Methane is supplied at 25 °C and 1 atm and is burned in 40% excess air. The air is pre-heated and enters the combustor at 177 °C (450 K). The combustion products exit the combustor as gases at 1500 °C (1773 K) and 1 atm. (a) How much heat is emitted (in kJ/kmol CH_4)? (b) What is the mass flow rate of the fuel (in kg/s)? (c) What is the mass flow rate of air (in kg/s)? (d) Determine the entropy generation (in kJ/kmol K), and (g) the rate of irreversibility.

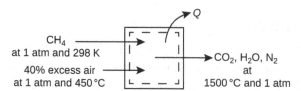

Required: We must first calculate the heat emitted by the combustor, which burns CH_4 as a lean mixture. Then, the mass flow rates of the CH_4 and air must be determined, followed by a calculation of the entropies of the products and reactants, and a determination of the entropy generation and rate of irreversibility.

Solution strategy: We first need to balance the combustion equation. Knowing the heat capacities of each of the reactants and products (from Table 7.5), and the entry and exit temperatures of the reactants and products, we can calculate the change in enthalpies ($\Delta h = c_p \Delta T$) and the heat produced by the reactants, as well as that absorbed by the reactants. The difference will provide the combustor's net heat output. We know the heat rate (80 MW) and have just determined the heat. The ratio of the two values will give the mass flow rate of CH_4. The air–fuel ratio can easily be determined from the balanced reaction, from which we can calculate the required mass flow rate of air. The absolute entropies can be calculated at their appropriate temperatures and pressure, either analytically, or by using the NIST-JANAF tables of thermochemical properties, https://janaf.nist.gov.

Assumptions: Neglect heat and power losses, and any changes in gas velocity (kinetic energy). Assume that all gases can be treated as perfect gases.

Analysis:

a) Heat emitted in kJ/kmol

The balanced equation for CH_4 burning stoichiometrically is:

$$CH_4 + 2[O_2 + 3.76\,N_2] \rightarrow CO_2 + 2\,H_2O + 7.52\,N_2$$

So, for 40% excess air, the balanced equation becomes the following:

$$CH_4 + 1.4\{2[O_2 + 3.76\,N_2]\} \rightarrow CO_2 + 2\,H_2O + 0.8\,O_2 + \{(1.4)7.52\,N_2\}$$
$$\therefore \quad CH_4 + 2.8\,O_2 + 10.53\,N_2 \rightarrow CO_2 + 2\,H_2O + 0.8\,O_2 + 10.53\,N_2$$

The heat of combustion is

$$Q = \Delta \bar{h}_c^0 = \sum_{products} N_i\left(\bar{h}_i^0 + \Delta \bar{h}_{298K,1atm \rightarrow T,p}\right) - \sum_{reactants} N_i\left(\bar{h}_i^0 + \Delta \bar{h}_{298K,1atm \rightarrow T,p}\right)$$

Reactants:

CH_4 is supplied at standard reference conditions, so the enthalpy of formation can be read from Table 7.5. The enthalpy of formation of the stable components of air is 0, as noted in Table 7.5. However, the air enters at 450 K, so we will have to calculate the additional enthalpy compared to air at a standard reference temperature of 298 K. We will use Eq. (7.10) for that:

$$\sum_{reactants} N_i(\bar{h}_i^0 + \Delta \bar{h}_{298K \rightarrow T}) = N_{CH_4}\left(\bar{h}_f^0 + \cancel{\Delta \bar{h}}\right)_{CH_4} + N_{O_2}\left(\cancel{\bar{h}_f^0} + \Delta \bar{h}\right)_{O_2} + N_{N_2}\left(\cancel{\bar{h}_f^0} + \Delta \bar{h}\right)_{N_2}$$

with $\bar{h}_{CH_4}^0 = -74,873$ kJ/kmol

with $\Delta \bar{h}_{CH_4} = \left(c_p \Delta T\right)_{CH_4} M_{CH_4} = 0$ kJ/kmol

with $\Delta \bar{h}_{O_2} = \left(c_p \Delta T\right)_{O_2} M_{O_2} = (0.916\text{ kJ/kgK})(450 - 298)\text{ K }(32\text{ kg/kmol}) = 4455.4$ kJ/kmol

with $\Delta \bar{h}_{N_2} = \left(c_p \Delta T\right)_{N_2} M_{N_2} = (1.04\text{ kJ/kgK})(450 - 298)\text{ K }(28\text{ kg/kmol}) = 4426.2$ kJ/kmol

Therefore:

$$\sum_{reactants} N_i\left(\overline{h}_i^0 + \Delta\overline{h}_{298K\to T}\right) = 1\underbrace{\left(-74,873 \text{ kJ/kmol}\right)}_{\left(\overline{h}_f^0 + \Delta\overline{h}\right)_{CH_4}} + 2.8\underbrace{\left(4455.4 \text{ kJ/kmol}\right)}_{\left(\overline{h}_f^0 + \Delta\overline{h}\right)_{O_2}}$$

$$+ 10.53 \underbrace{\left(4426.2 \text{ kJ/kmol}\right)}_{\left(\overline{h}_f^0 + \Delta\overline{h}\right)_{N_2}} = -15,790 \text{ kJ/kmol fuel}$$

Products (using values from Tables 7.3 and 7.4, and the NIST-JANAF tables):

$$\sum_{products} N_i\left(\overline{h}_i^0 + \Delta\overline{h}_{298K\to 1773K}\right) = N_{CO_2}\left(\overline{h}_f^0 + \Delta\overline{h}\right)_{CO_2} + N_{H_2O}\left(\overline{h}_f^0 + \Delta\overline{h}\right)_{H_2O} + N_{O_2}\left(\overline{h}_f^0 + \Delta\overline{h}\right)_{O_2}$$

$$+ N_{N_2}\left(\overline{h}_f^0 + \Delta\overline{h}\right)_{N_2}$$

with $\Delta\overline{h}_{CO_2} = \left(c_p\Delta T\right)_{CO_2} M_{CO_2} = (0.84 \text{ kJ/kgK})(1773 - 298) \text{ K } (44 \text{ kg/kmol}) = 54,516 \text{ kJ/kmol}$

with $\Delta\overline{h}_{H_2O} = \left(c_p\Delta T\right)_{H_2O} M_{H_2O} = (1.996 \text{ kJ/kgK})(1773 - 298) \text{ K } (18 \text{ kg/kmol}) = 52,994 \text{ kJ/kmol}$

with $\Delta\overline{h}_{O_2} = \left(c_p\Delta T\right)_{O_2} M_{O_2} = (0.916 \text{ kJ/kgK})(1773 - 298) \text{ K } (32 \text{ kg/kmol}) = 43,235 \text{ kJ/kmol}$

with $\Delta\overline{h}_{N_2} = \left(c_p\Delta T\right)_{N_2} M_{N_2} = (1.04 \text{ kJ/kgK})(1773 - 298) \text{ K } (28 \text{ kg/kmol}) = 49,952 \text{ kJ/kmol}$

$$\sum_{products} N_i\left(\overline{h}_i^0 + \Delta\overline{h}_{298K\to 1773K}\right) = 1\underbrace{\left(-394,000 + 54,516\right)}_{\left(\overline{h}_f^0 + \Delta\overline{h}\right)_{CO_2}} \text{kJ/kmol} + 2\underbrace{\left(-242,000 + 52,994\right)}_{\left(\overline{h}_f^0 + \Delta\overline{h}\right)_{H_2O}} \text{kJ/kmol}$$

$$\sum_{products} N_i\left(\overline{h}_i^0 + \Delta\overline{h}_{298K\to 900K}\right) = + 0.8\underbrace{\left(43,235 \text{ kJ/kmol}\right)}_{\left(\overline{h}_f^0 + \Delta\overline{h}\right)_{O_2}} + 10.53 \underbrace{\left(49,952 \text{ kJ/kmol}\right)}_{\left(\overline{h}_f^0 + \Delta\overline{h}\right)_{N_2}}$$

$$= -230.6 \times 10^3 \text{ kJ/kmol fuel}$$

The combustor produces

$$Q = \Delta\overline{h}_c^0 = -230.6 \times 10^3 \text{ kJ/kmol} - (-15,790) \text{ kJ/kmol} = -246,390 \text{ kJ/kmol fuel}.$$

b) Fuel flow rate

$$\dot{N}_{CH_4} = \frac{\dot{Q}}{Q} = \frac{-80,000 \text{ kJ/s}}{-246,390 \text{ kJ/kmol}} = 0.325 \text{ kmol CH}_4/\text{s}$$

The mass flow rate of air is then:

$$\dot{m}_{CH_4} = M_{CH_4}\dot{N}_{CH_4} = \left(\frac{16 \text{ kg CH}_4}{\text{kmol CH}_4}\right)(0.325 \text{ kmol CH}_4/\text{s}) = 5.2 \text{ kg/s}$$

c) Air flow rate

The air–fuel ratio is $AFR = \dfrac{N_{air}}{N_{total}} = \dfrac{(2.8 + 10.53) \text{ kmol air}}{1 \text{ kmol fuel}} = 13.33 \text{ kmol air/kmol fuel}$

The molar flow rate of air is then:

$\dot{N}_{air} = M_{CH_4} = (AFR)\dot{N}_{CH_4} = (13.33 \text{ kmol air/kmol CH}_4)(0.325 \text{ kmol CH}_4/\text{s}) = 4.33 \text{ kmol air/s}$

The air flow rate (in kg/s) is: $\dot{m}_{air} = M_{air}\dot{N}_{air} = \left(\dfrac{29 \text{ kg air}}{\text{kmol air}}\right)(4.33 \text{ kmol air/s}) = 125.6 \text{ kg air/s}$

d) Entropy production

From Eq. (7.19): $S_i = N_i \bar{s}_i(T, p_i) = N_i \left(\bar{s}_i^0(T, p_0) - \bar{R} \ln \dfrac{y_i p_i}{p_0} \right)$, and $p_i = y_i p_{total}$; therefore:

Component	Moles (N_i)	Mole fraction (y_i)	(kJ/kmol K)	$-\bar{R} \ln \dfrac{y_i p}{p_0}$	$N_i \bar{s}_i$
Reactants					
$CH_{4@298K}$	1	1.00	360.79	0	360.79
$O_{2@450K}$	2.8	0.21	217.3	$-8.314 \ln (0.21) = 12.978$	644.78
$N_{2@450K}$	10.53	0.79	203.5	$-8.314 \ln (0.79) = 1.958$	2163.5
				$S_{reactants} = 3169$ kJ/kmol K	
Products					
$CO_{2@1773K}$	1	$= 1/14.33$ $= 0.0697$	298.6	$-8.314 \ln (0.0697) = 22.15$	320.75
$H_2O_{@1773K}$	2		258	$-8.314 \ln (0.1396) = 16.37$	548.74
$O_{2@1773K}$	0.8	$= 2/14.33 = 0.8/14.33$ $= 0.1396 = 0.056$	264.2	$-8.314 \ln (0.056) = 23.96$	230.53
$N_{2@1773K}$	10.53	$= 10.53/14.33$ $= 0.735$	247.8	$-8.314 \ln (0.735) = 2.56$	2636.3
Σ	14.33	1.0			3736.3
				$S_{products} = 3736.3$ kJ/kmol K	

Therefore, the entropy produced is $\sigma = (3736.3 - 3169)$ kJ/kmol K $= 567.3$ kJ/kmol K.

e) Rate of irreversibility

The rate of irreversibility of the combustion process

$$= \sum_{products} \dot{N}_e T_0 \bar{s}_e - \sum_{reactants} \dot{N}_i T_0 \bar{s}_i - \dot{Q} \equiv T_0 \dot{\sigma}$$

with $\dot{N}_{CH_4} = 0.325$ kmol CH_4/s

So, the rate of irreversibility is

$$= T_0 \dot{\sigma} = (298 \text{ K})(0.325 \text{ kmol } CH_4/s)(567.3 \text{ kJ/kmol K}) = 54.9 \text{ MW}$$

Discussion:
The combustor's exergy is 80 MW, and its irreversibility is 54.9 MW, resulting in nearly 69% exergy destruction. The high level of exergy destruction is due to the presence of irreversibilities associated with the chemical reaction and heat transfer at the very high exit temperature (1500 °C). A combustor's function is to provide high-temperature gases for the downstream turbine unit. This means that there is little opportunity to improve on this combustor's design, unless the exit temperature can somehow be reduced, and if the combustor could burn fuel with less excess air. Box 9.2 explains how much carbon dioxide is emitted per unit of energy.

Box 9.2   Carbon-Dioxide Emission from the Combustion of Natural Gas

As illustrated in Example 9.2, methane (the major component of natural gas) produces 44 kg of CO_2 for every 16 kg of CH_4 burned, thus 2.75 kg of CO_2 for every kg of natural gas combusted. The lower heating value of natural gas is 50.05 MJ/kg. If we assume that the natural gas will be burned at a gas turbine power plant with 40% thermal efficiency, then the amount of CO_2 produced per kWh of electricity produced is:

$$\left(\frac{50.05 \text{ MJ}}{\text{kg gas}}\right)(0.4)\left(\frac{1 \text{ kg gas}}{2.75 \text{ kg } CO_2}\right)\left(\frac{1 \text{ kWh}}{3.6 \text{ MJ}}\right) = 2.02 \text{ kWh/kg } CO_2 = 0.49 \text{ kg } CO_2/\text{kWh}$$

From Box 8.2, an electric power plant fired by coal produces about 1.07 kg CO_2/kWh. If the power plant is fired by natural gas, it produces "only" 0.49 kg CO_2/kWh. Burning natural gas is therefore about 50% less polluting than coal. Unsurprisingly, therefore, most coal-fired power plants have been (or are in the process of being) converted to plants that can burn natural gas. More about this in Chapter 14.

Reimagine Our Future

Replacing fossil fuels with renewable energy will diminish our carbon footprint. In doing so, we would be wise to also live differently. Instead of maintaining our current standard of living based on hyperconsumption, greed, waste, ecological ruin, wars, and injustice and inequality, would *now* not be the perfect time to adopt a less frenetic society built on the precepts of efficiency ("doing more with less"), social justice, frugality, and with a long-term view?

9.3   Hydrogen and Ammonia

9.3.1   Hydrogen

Hydrogen (H_2) use in the energy sector is largely confined to oil refining and the production of ammonia (NH_3) and methanol (CH_3OH). In 2020, the global hydrogen demand was around 90 million tonnes. This hydrogen was mainly produced from fossil fuels (mostly natural gas), which emitted about 900 million tonnes of CO_2 (IEA 2021b).

To reach net-zero carbon emissions by 2050, 13% of the global energy demand in 2050 will have to be met by hydrogen-based fuels, which include hydrogen (H_2) and ammonia (NH_3), as well as synthetic hydrocarbon fuels produced from *syngas* (H_2 and CO), (IEA 2021a). By 2050, most of that H_2 will have to be produced using renewable energy (solar and wind) for the electrolysis of water, rather than being derived from fossil fuels such as natural gas and coal.

If H_2 is produced by *steam methane reforming (SMR)* using a liquid-H_2O input, the following endothermic reaction holds (Sørensen and Spazzafumo 2018):

$$CH_4 + 2H_2O \; (\ell) + \Delta h_c^0 \rightarrow CO_2 + 4H_2 \qquad \Delta h_c^0 = 257.3 \text{ kJ/mol} \tag{9.9}$$

where Δh_c^0 is the enthalpy of combustion.

If H_2 is produced by *electrolysis of liquid-H_2O* using an electrolyte, the following endothermic reaction pertains (Sørensen and Spazzafumo 2018):

$$2H_2O \ (\ell) + \Delta h_c^0 \rightarrow 2H_2 + O_2 \qquad \Delta h_c^0 = 285.8 \ kJ/mol \qquad (9.10)$$

From Eq. (7.35), the enthalpy of combustion $\left(\Delta h_c^0\right)$ can be related to the Gibbs function as follows:

$$\Delta g = \Delta h - T\Delta s \qquad (9.11)$$

For electrolysis, the Gibbs function at standard conditions (ΔG^0) is the amount of *electricity* required for the reaction. For liquid-H_2O, $\Delta G^0 = 237.1 \ kJ/mol$. This means that the *heat* $(T\Delta S)$ required for the reaction is: 285.8 kJ/mol less 237.1 kJ/mol, which equals 48.7 kJ/mol. We can write the electrolysis reaction as follows:

$$2H_2O \ (\ell) + 237.1 \ kJ/mol \ electricity + 48.7 \ kJ/mol \ heat \rightarrow 2H_2 + O_2 \qquad (9.12)$$

Comparing Eqs. (9.9) and (9.12) shows that producing H_2 using electrolysis powered by renewable sources (or nuclear power) produces no CO_2, which is obviously favored for a net-zero economy. But, in 2022, the cost of producing H_2 via electrolysis was about \$5 per kg of H_2, whereas it was only \$2 per kg of H_2 for reforming CH_4 (ETC 2021).

Example 9.3 Production of Carbon Dioxide and Consumption of Methane in the Steam Methane Reforming (SMR) Process

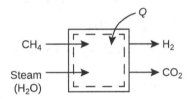

a) Calculate the amount of CO_2 produced and the amount of methane consumed in the SMR process, with the reaction given by Eq. (9.4).
b) Calculate the amount of CO_2 produced and methane used *for producing the heat* required to bring about the endothermic reaction given by Eq. (9.5). It has been experimentally determined that the endothermic reaction requires an input of 2.25 kWh per m³ of hydrogen (T-raissi and Block 2004).

Analysis:

a) CO_2 produced and CH_4 consumed for the SMR process

From Table 7.5, the *HHV* of H_2 is 141.8 MJ/kg or $(141.8 \ MJ/kg) \times \left(\dfrac{2 \ kg \ H_2}{kmol}\right) = 283.6 \ MJ/kmol$, or inverting this, 0.0035 kmol H_2/MJ. The CO_2 produced during the SMR process is therefore:

$$\left(\frac{0.0035 \ kmol \ H_2}{MJ}\right)\left(\frac{1 \ kmol \ CO_2}{4 \ kmol \ H_2}\right) = 874 \times 10^{-6} \ kmol \ CO_2/MJ \ .$$ With a molecular weight of

44 kg per kmol, the amount of CO_2 produced during the SMR process is:
$$\left(\frac{874 \times 10^{-6} \text{ kmol } CO_2}{\text{MJ}}\right)\left(\frac{44 \text{ kg}}{\text{kmol } CO_2}\right) \simeq 0.04 \text{ kg } CO_2/\text{MJ}.$$

The amount of CH_4 produced can be calculated as follows:
$$\left(\frac{0.0035 \text{ kmol } H_2}{\text{MJ}}\right)\left(\frac{1 \text{ kmol } CH_4}{4 \text{ kmol } H_2}\right) = 875 \times 10^{-6} \text{ kmol } CH_4/\text{MJ}.$$

With a molecular weight of 16 kg per kmol, the amount of CH_4 produced during the SMR process is: $\left(\frac{875 \times 10^{-6} \text{ kmol } CO_2}{\text{MJ}}\right)\left(\frac{16 \text{ kg}}{\text{kmol } CH_4}\right) = 0.014 \text{ kg } CH_4/\text{MJ}.$

b) CO_2 produced and CH_4 consumed for producing heat

The energy in CH_4 required to produce a kilomole of H_2 can be calculated as follows:
$$\left(\frac{2.25 \text{ kWh}}{m^3 H_2}\right)\left(\frac{3.6 \text{ MJ}}{\text{kWh}}\right)\left(\frac{1 \text{ m}^3}{1000 \text{ L}}\right)\left(\frac{22{,}400 \text{ L}}{\text{mol}}\right) = 181.4 \text{ MJ/kmol } H_2$$

In Example 9.2 and in Box 9.2, we calculated that 2.75 kg CO_2 is produced for every kg of CH_4 burned. The amount of CO_2 produced per MJ is therefore: $\left(\frac{50.05 \text{ MJ}}{\text{kg gas}}\right)\left(\frac{1 \text{ kg gas}}{2.75 \text{ kg } CO_2}\right) =$ 18.2 MJ/kg $CO_2 \equiv 0.05$ kg $CO_2/$MJ

Therefore, $\left(\frac{181.4 \text{ MJ}}{\text{kmol } H_2}\right)\left(\frac{0.05 \text{ kg } CO_2}{\text{MJ}}\right) = 9.07 \text{ kg } CO_2/\text{kmol } H_2$

In (a) we calculated that the *HHV* of H_2 is equivalent to 0.035 kmol $H_2/$MJ. Therefore, the amount of CO_2 produced per MJ is $\left(\frac{9.07 \text{ kg } CO_2}{\text{kmol } H_2}\right)\left(\frac{0.0035 \text{ kmol } H_2}{\text{MJ}}\right) = 0.03 \text{ kg } CO_2/\text{MJ}.$

Then, as 1 kmol of CH_4 is burned to produce 1 kmol of CO_2, we can calculate the CH_4 consumed:
$$\left(\frac{0.03 \text{ kg } CO_2}{\text{MJ}}\right)\left(\frac{1 \text{ kmol } CO_2}{44 \text{ kg } CO_2}\right)\left(\frac{16 \text{ kg } CH_4}{\text{kmol } CH_4}\right)\left(\frac{1 \text{ mol } CH_4}{1 \text{ mol } CO_2}\right) = 0.01\text{kg } CH_4/\text{MJ}.$$

Discussion:
The sum of the CO_2 from the SMR process (0.04 kg $CO_2/$MJ) and from the energy used to generate heat and for the SMR process (0.05 kg $CO_2/$MJ) is 0.09 kg $CO_2/$MJ. Box 9.3 explains that hydrogen produced in this manner is called "grey hydrogen" as it is produced from a fossil fuel, and the CO_2 emitted is not captured. If the CO_2 is captured, the produced hydrogen is called "blue hydrogen."

The total quantity of CH_4 consumed to generate H_2 is the sum of that used in the SMR process (0.014 kg $CH_4/$MJ). The quantity burned to generate the heat and high pressure needed for the process (0.01kg $CH_4/$MJ) is 0.024 kg $CH_4/$MJ.

Table 7.5 shows that H_2 is about three times as energy-dense as natural gas (principally methane), and more than twice as energy-dense as gasoline. Better still, burning hydrogen is much cleaner than burning hydrocarbons. When mixed with pure oxygen, it only produces water as a byproduct of combustion.

Producing, storing, and transporting hydrogen on an industrial scale is challenging. Although hydrogen has a lower heating value (*LHV*) of 120 MJ/kg, it has a volumetric heating value of 10 MJ/m^3 at an atmospheric pressure of 100 kPa. If hydrogen is compressed to 30 MPa, its volumetric energy density is 2.7 MJ/m^3. Table 9.1 compares the energy densities of hydrogen and some hydrocarbon fuels, which suggests that hydrogen needs to be compressed to high pressures if it

Table 9.1 Average energy density by weight and volume for various forms of hydrogen storage and some hydrocarbon fuels.

	Energy density		Density
Storage form	MJ/kg	MJ/m³	kg/m³
Hydrogen gas at 100 kPa	120	10	0.09
Hydrogen gas at 20 MPa	120	1900	15.9
Hydrogen gas at 30 MPa	120	2700	22.5
Hydrogen liquid	120	8700	71.9
Hydrogen in metal hydrides	2.1	11,450	5480
Methane at 100 kPa	56	37.4	0.67
Methanol	21	17,000	0.79
Ethanol	28	22,000	0.79

Source: Adapted from Sørensen and Spazzafumo (2018).

is to become comparable to methanol, ethanol, and natural gas (methane) as a practical combustion fuel. Today, hydrogen is either stored as a gas at 35–70 MPa, or as a liquid in cryogenic tanks at a temperature of around −253 °C.

As a fuel for internal combustion engines (i.e., not for gas turbines that are external combustion engines), due to the lower energy density of H_2 at the pressures suitable for piston cylinders, the displacement volume must be two to three times as large as for gasoline engines. This, of course, has practical and economic ramifications (Sørensen and Spazzafumo 2018).

A key hurdle to implementing hydrogen in combustion engines is that its *flame temperatures* are almost 300 °C higher than methane, for instance (Lefebvre and Ballal 2010). Hydrogen's *flame speed* (around 3.5 m/s) is also more than seven times that of methane (0.45 m/s), and the autoignition delay time of hydrogen is more than three times lower than methane. These result in challenging combustion and safety concerns due to hydrogen's wide flammability range (Sørensen and Spazzafumo 2018). Gas turbine manufacturers are working to control the flame of the highly reactive H_2 while maintaining the integrity of the combustion system.

The widely touted *hydrogen economy* (Figure 9.2) could generate $140 billion in annual revenue and support 700,000 jobs (Hydrogen Council 2021). The infrastructure requirements of a hydrogen

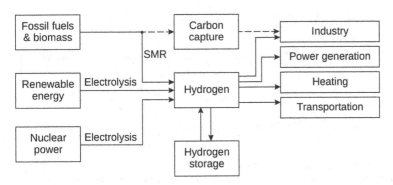

Figure 9.2 Depiction of a hydrogen economy. "SMR" refers to steam methane reforming, where methane reacts with steam under 3–25 bar pressure in the presence of a catalyst to produce hydrogen, carbon monoxide, and a relatively small amount of carbon dioxide. *Source:* Adapted from Chaube et al. (2021).

economy will shift from small-scale H_2 refueling stations and trailer trucks in 2022, to pipeline usage in 2050, and underground H_2 storage in large-scale depleted fields and salt caverns by 2100 (Chaube et al. 2021).

Some projections indicate that around 3% of the global energy need could be met in 2050 by around 268 million tons of H_2. In Japan, however, H_2 could underpin up to 25% of that country's energy needs by 2050. That H_2 will mostly be imported and will initially be derived from fossil fuels (such as CH_4), and carbon capture and storage pathways. The H_2 will also be used to synthesize methanol, which will mostly be used for transportation (Chaube et al. 2021).

In the United States, it is projected that most of the H_2 will be imported in 2050, possibly produced via biomass and some fossil fuels (including coal). The main uses of H_2 will be for use as city gas and for transportation. It is anticipated that H_2 will provide for about 7.6% of the United States' energy needs in 2050 (Chaube et al. 2021) and could provide as much as 6.5% of global energy (McKinsey 2022).

Box 9.3   The Future of Hydrogen Is Multi-colored

Depending on how hydrogen is manufactured, it is assigned a certain color, as shown in Figure 9.3. Hydrogen will play a major role in decarbonizing the energy system, beginning with blue hydrogen, and eventually shifting to green hydrogen.

- *Green hydrogen* is produced by splitting H_2O by electrolysis, through which the electricity is provided by renewable energy (solar or wind or a combination of both).
- *Blue hydrogen* is obtained when natural gas is split into H_2 and CO_2, either by steam methane reforming or auto thermal reforming (ATR), through which the CO_2 is captured and stored. Blue hydrogen is not a zero-emission fuel. Examining the lifecycle greenhouse-gas emissions of blue hydrogen, especially considering that around 3.5% of CH_4 is leaked during its drilling and transportation, the greenhouse gas footprint of blue hydrogen is more than 20% greater than burning natural gas or coal for heat (Howarth and Jacobson 2021).

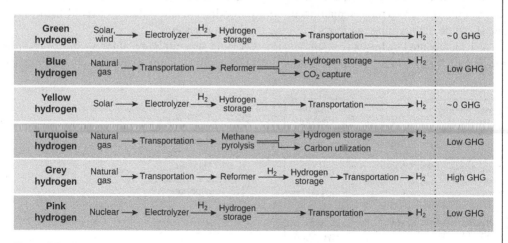

Figure 9.3   Six methods for producing hydrogen and their concomitant "colors" or labels. One could also identify "gold hydrogen", which is tapped from natural subsurface accumulations, and "orange hydrogen,", which is stimulated by pumping water into deep-source rocks.

- *Yellow hydrogen* is like green hydrogen, except that electrolysis is achieved solely through solar power.
- *Turquoise hydrogen* is produced through the pyrolysis of methane (CH_4). This method generally produces solid carbon, but CO_2 is produced in the high-temperature process.
- *Gray* hydrogen is produced in a similar manner as blue hydrogen, except that CO_2 is not captured.
- *Pink* hydrogen is like green hydrogen, except that electricity for electrolysis is obtained from nuclear power, a nonrenewable energy source.

Example 9.4 Combustion of Hydrogen in an Industrial Gas Turbine

A 20-MW gas turbine is fueled by gaseous hydrogen (H_2) that has an overall energy conversion efficiency of 25% based on the lower heating value of the fuel. Hydrogen is supplied at 25 °C and is burned in oxygen at a stoichiometric ratio in the main portion of the combustor. Oxygen (from air) is compressed in the gas turbine and enters the combustor at 450 °C. The combustion gases exit the combustor at 1100 °C. (a) What is the *gross* heat rate (in kW) of the gas turbine's combustor? (b) What is the required mass flow rate of the fuel (in kg/s)? (c) What is the *net* heat rate of the combustor, accounting for the additional heat obtained from the reactant O_2 and the heat losses to the products CO_2 and H_2O? (d) Calculate the total air flow required for stoichiometric combustion. (e) What is the required flow rate of O_2?

Hint: See Eq. (7.12) and Example 9.2 for important background theory.

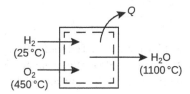

Required: We must first calculate the gross heat rate of the combustor, assuming a 25% thermal efficiency and using the *LHV* of H_2. We must then find the mass flow rate of the reactants and products for H_2 burning in stoichiometric O_2. Knowing that, we can calculate the additional heat provided by the reactant O_2 and the losses of heat in the products CO_2 and H_2O. Knowing those values, we can calculate the net heat rate of the combustor using Eq. (7.12). The required air flow for stoichiometric combustion can then finally be determined.

Solutions strategy: We will use Table 7.5 to find the *LHV* of methane, from which we can calculate the gross heat rate of the combustor. After balancing the combustion equation, we can find the required mass flow rates for each of the reactants and products. Knowing the heat capacities of each of the reactants and products (from Table 7.5), and the entry and exit temperatures of the reactants and products, we can calculate the enthalpies and the extra heat delivered by O_2, as well as the heat lost in the products. The net heat rate can then be calculated as the gross heat rate plus the heat rate of O_2 less the heat rates lost to the products CO_2 and H_2O using Eq. (7.12).

Assumptions: Steady-state and steady-flow conditions prevail. Neglect heat and power losses and any changes in gas velocity (kinetic energy). Assume that all gases can be treated as perfect gases.

Analysis:

a) Gross heat rate of combustor

With a 25% thermal efficiency based on the *LHV* of H_2, the 20-MW turbine has a gross heat rate of:

$$\dot{Q} = \frac{\dot{W}}{\eta_{th}} = \frac{20 \text{ MW}}{0.25} = 80 \text{ MW}.$$

b) Fuel flow rate

The heat rate $\dot{Q}$ (kJ/s) is also equal to the mass flow rate (kg/s) multiplied by the *LHV* (kJ/kg) of methane:

$$\dot{Q} = 80 \text{ MW} = \dot{m}_{fuel}(LHV).$$

We find the *LHV* of CH_4 from Table 7.5 to be 120,000 kJ/kg, so:

$$\dot{m}_{fuel} = \frac{80 \times 10^3 \text{ kJ/s}}{120,000 \text{ kJ/kg}} = 0.67 \text{ kg/s}$$

c) Net heat rate of combustor

From Eq. (7.14): Net heat rate = gross heat rate of H_2 + (additional heat from O_2) − (heat loss to CO_2 and H_2O)

with $\quad \underbrace{H_2}_{\substack{\text{molecular mass} \\ = 2}} \quad + \quad \underbrace{0.5 \, O_2}_{\substack{\text{molecular mass} \\ = 0.5(32) = 16}} \quad \rightarrow \quad \underbrace{H_2O}_{\substack{\text{molecular mass} \\ = (2(1) + 16) = 18}}$

$\qquad\qquad\quad \underbrace{}_{\text{flowrate} = 0.2 \text{ kg/s}} \qquad \underbrace{}_{\text{flowrate} = 1.6 \text{ kg/s}} \qquad \underbrace{}_{\text{flowrate} = 1.8 \text{ kg/s}}$

The fuel flow rate of 1 mol of H_2 is 0.2 kg/s, and H_2 has a molecular mass of 2. The molar flow rates of the other components can then be determined from a mass (and mass flow rate) balance of products and reactants, which requires 18 kg (and 1.8 kg/s) on the right- and left-hand side of the combustion equation, which gives:

Flow rate of 0.5 mol O_2 = 1.6 kg/s
Flow rate of 1 mol H_2O = 1.8 kg/s

The additional heat input from $O_2 = \dot{m}_{O_2} c_{p,O_2} \Delta T$
From Table 7.5, $c_{p,O_2} = 0.916$ kJ/kg K
So, the additional heat *input* from $O_2 = \dot{m}_{O_2} c_{p,O_2} \Delta T = (1.6 \text{ kg/s})(0.916 \text{ kJ/kg K})(450 - 25)\text{K} = 622.8 \text{ kJ/s}$
The heat *output* (loss) from $H_2O = \dot{m}_{H_2O} c_{p,H_2O} \Delta T$
From Table 7.5, $c_{p,H_2O} = 1.996$ kJ/kg K
So, the heat output from H_2O is $(1.8 \text{ kg/s})(1.996 \text{ kJ/kg K})(1100 - 25) \text{ K} = 3862.3 \text{ kJ/s}$
Therefore, the heat content of the product H_2O between 1100 and 25 °C is 3862.3 kJ/s, and the heat content of the reactant O_2 between 25 and 450 °C is 622.8 kJ/s. Thus, the net heat rate of the fuel = $(80,000 + 622.8 - 3862.3)$ kJ/s = 76,761 kJ/s $\simeq$ 77.8 MW.

d) Flow rate of O_2

The required flow rate of O_2 to produce 76,761 kJ/s is:

$$\dot{m}_{O_2} = \frac{\dot{Q}_{net}}{\Delta h_{O_2}} = \frac{76,761 \text{ kJ/s}}{(0.916 \text{ kJ/kg K})(1100 - 450) \text{ K}} = 128.9 \text{ kg/s}$$

The total flow rate of O_2 is $(128.9 + 1.6) = 130.5$ kg/s

Discussion: To produce a work rate of 20 MW, the gas turbine with its 25% thermal efficiency must combust hydrogen with a gross heating value of 80 MW or a net heat rate of 77.8 MW. The gas turbine's combustor therefore has a conversion efficiency of $\dfrac{77.8\,\text{MW}}{80\,\text{MW}} \times 100\% \simeq 97.3\%$. From a combustion viewpoint, hydrogen is probably the nearest thing to an ideal fuel. It is characterized by high flame speeds, wide burning limits, easy ignition, and freedom from soot formation. The main drawbacks of hydrogen lie in its very low density and low boiling point, which necessitate the use of large, heavily insulated storage tanks. It is also costly to produce (Lefebvre and Ballal 2010).

9.3.2 Ammonia

Around 180 million tons of *ammonia* (NH_3) was produced in 2021, mainly as feedstock in the chemicals industry and for producing fertilizers (IEA 2021b). Recently, however, NH_3 has been promoted as a carbon-free energy carrier (Sørensen and Spazzafumo 2018; MacFarlane et al. 2020). It is cheaper and easier to transport and store NH_3 than H_2. But, converting H_2 to NH_3 or other energy carriers adds additional cost.

Currently, NH_3 is produced from nitrogen and hydrogen through the thermally catalyzed Haber-Bosch process, which operates under harsh conditions (450 °C and 2 MPa), requiring large, centralized plants and high capital investment. NH_3 production in 2021 accounted for about 1% of global energy consumption and produced about 1.4% of global CO_2 emissions, mainly due to its use of steam methane reforming (IEA 2021b).

Figure 9.4 summarizes the main routes for producing H_2 and NH_3 from biomass, steam methane reforming, electrolysis of water, and coal gasification. NH_3, as an energy carrier, is attractive because the cycle is carbon-green. Methane is attractive because the locally produced CO_2 can

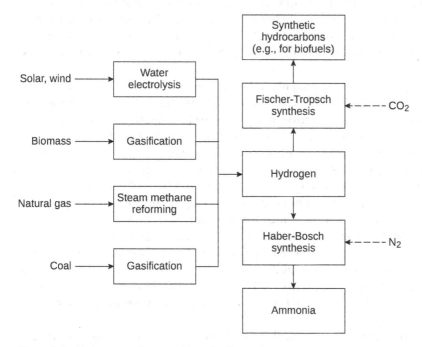

Figure 9.4 Hydrogen and ammonia production pathways.

potentially be sequestered. Biomass is attractive because, if deployed alongside sequestration, it results in a carbon-negative cycle (Sørensen and Spazzafumo 2018).

Although NH_3 might be easier to handle than H_2, Table 7.5 shows that NH_3 has an *LHV* of only 18.6 MJ/g compared to 120 MJ/kg for H_2. High compression ratios, low speeds, and high loads have been found preferable for NH_3-fueled engines, which is primarily related to ammonia's low flame speed. This also means that engines with large displacement volumes (such as large and low-speed ship engines) are better suited for the combustion of NH_3. However, the combustion of NH_3 must deal with high fuel-NO_x emissions, thus warranting exhaust after-treatment or multi-stage combustion (Dimitriou and Javaid 2020).

9.4 Biofuels

Biomass, or current growth (like plants, wood, and algae), comprises all recent (not fossil) vegetable and animal growth and its residues (like discarded vegetable oil, food waste, manure or municipal sewage, landfill gas, or agricultural wastes). *Biofuels* are therefore hydrocarbon-based and renewable. However, the combustion of biofuels produces CO, CO_2, NO_x, and other gases and particulates, sometimes at levels exceeding those from fossil fuels like oil and gas. Biofuels can be used directly or indirectly.

With *direct-fired systems*, the biomass is burned in a furnace (or boiler) to typically produce high-pressure steam. Typical feedstock is bark, wood chips, sawdust, and a wide variety of other elements. The most common types of boilers are fixed-bed boilers (or "stokers") and "fluidized bed boilers" and can be up to 300 MW in size. With indirect systems, also called *gasification systems*, the biomass is heated and anaerobically digested to produce a gaseous fuel, called syngas, of which the main combustible products are H_2 and CO. Non-combustible CO_2 is also produced (Lieuwen et al. 2010). Gasification can take place in fixed-bed gasifiers or in fluidized-bed gasifiers. The feedstock could comprise algae (see Box 9.4), wood, rice hulls, sugarcane peel, sewage sludge, or wood residues, among others. The capacity of these gasification systems is usually below 30 MW (IEA 2021b).

In 2020, 4 EJ of energy was produced from new biofuels (such as biodiesel), compared to 193 EJ from crude oil, 170 EJ from coal, 144 EJ from natural gas, 30 EJ from nuclear, 37 EJ from hydro, 13 EJ from wind, and 6.5 EJ from solar. Approximately 86% of modern bioenergy is used for heating applications, with 9% used for transport and 5% for electricity (BP 2021; IEA 2021a).

Most biofuels are combusted and produce greenhouse-gas emissions. Only a small percentage of biofuels (such as bio-hydrogen) is used for non-combustion applications, such as hydrogen fuel cells that rely on electrochemical reactions. (Hydrogen-powered fuel cells will be covered in Chapter 24.) The use of biomass derived from agricultural crops raises social issues pertaining to the "fuel or food" dilemma, as well as the use of land (Mason et al. 2015).

Notable examples of *liquid biofuels* are biodiesel, methanol, ethanol, bio-oil, and biojet fuel. Liquid biofuels are expensive and there is a limited supply of affordable and sustainable feedstocks. In 2022, it cost $70 to $130 per barrel of oil equivalent (BOE) to produce conventional biofuels, and $85–160/BOE to produce advanced biofuels (Rystad Energy 2022). To achieve net-zero emissions by 2050, it is predicted that total biofuel demand would need to increase by more than 3.5 thousand BOE/day up to 2030 (IEA 2021b).

Every part of the world has significant scope to produce *biogases* – biosyngas, biomethane, and bio-hydrogen. Biogas can provide access to clean cooking, but the relatively high upfront cost of biodigesters and the challenges of the continuous maintenance of dispersed and small-scale units have slowed their deployment. Biomethane faces challenges of costs and availability, and several non-economic barriers need to be overcome, including sourcing adequate volumes of feedstocks of consistent quality. Nonetheless, the global potential is significant (IEA 2021b).

An assessment of sustainable feedstocks indicates that around 25 EJ of biomethane could be developed globally, and a further 7 EJ of biomethane could be produced from biomass gasification. With today's high prices for natural gas, around a quarter of the global sustainable potential is cost competitive (IEA 2021b).

Around 70% of the biogas developed today is used for power and heat, 20% for cooking purposes, and 10% upgraded to biomethane. More than half of biogas production today takes place in Asia, which also has the largest growth potential, given the availability of significant volumes of organic feedstocks such as crop residues and rising levels of municipal solid waste. Well-managed biogas projects not only help reduce emissions but also provide co-benefits such as rural development and local job creation. Some potential biogas feedstocks would produce methane emissions if left untreated, so converting them to biogas can prevent such emissions. If methane leaks occur during biogas production, however, they would add to the overall greenhouse-gas emissions. Presently, there is some uncertainty about the extent to which this happens (IEA 2021b).

Box 9.4   Biodiesel and Biojet Fuel from Microalgae

Microalgae grown in a nutritionally complete medium without a starvation period generally comprise (by dry mass) 30–50% protein, 20–40% carbohydrate, and 8–15% lipids. Of these components, lipids are especially rich in energy. The lipid content of the biomass can be increased by manipulating the conditions under which an alga is grown.

Some microalgae are rich in oils or can be induced to accumulate copious amounts of oils. Algal oils include triglyceride oils, hydrocarbons, xanthophylls (i.e., oxygen-containing carotenoids), sterols, and others. Algal chlorophyll is a chloroform-soluble oil. It has a 16-carbon hydrocarbon side chain, which can be cleaved to provide combustible fuel equivalent to diesel. For comparison, petroleum diesel is a mixture of hydrocarbons with between eight and 21 carbons per molecule. Around 75% of the diesel hydrocarbons are saturated, and the rest are aromatic. Algal crude oil has an energy content of around 35.8 MJ/kg, or nearly 80% of the average energy content of petroleum.

Algal crude oil is extracted from dried biomass, but drying requires a lot of energy and is expensive. The solvent-based extraction of biomass paste has therefore been developed as an alternative oil recovery method and is as effective as extraction from dry biomass. The triglyceride component of algal oil can be used to produce biodiesel. This involves transesterification, a catalyzed reaction of a triglyceride with an alcohol that produces fatty esters and glycerol. In the production of biodiesel, triglyceride is reacted with methanol. The resulting methyl esters are called biodiesel.

The calorific value of biodiesel ($C_{19}H_{35}O_2$) is around 39 MJ/kg, or about 10% less than conventional diesel ($C_{12}H_{23}$). Unlike conventional diesel, biodiesel contains around 10% oxygen by mass. Pure biodiesel can be used directly in a diesel engine, or it may be blended with conventional diesel before use. Alternatively, biodiesel can be deoxygenated to form various hydrocarbon fuels, including biojet fuel. Compared to petroleum-derived diesel, biodiesel combustion produces a much lower emission of unburned hydrocarbons, CO, sulfates, polycyclic aromatic hydrocarbons, and particulate matter. However, NO_x emissions are higher than those of conventional diesel (DoE 2021).

Algae is fast-growing compared with traditional terrestrial feed crops. One kilogram of algae consumes roughly 1.8 kg of CO_2 to grow. In comparison, a mature tree absorbs around 22 kg of CO_2 a year. This makes algal biofuel systems ideal if paired with carbon capture systems. Additionally, the photosynthesizing organisms do not need fresh water or arable land to grow, lessening the demand on increasingly scarce natural resources. But before they can produce biofuels, the tiny droplets of oil found in algae need to be harvested. These cells are about 10 μm or 0.001 mm wide. Collecting and processing oil from microalgae is expensive, resulting in the cost of algae biodiesel being around $8 a gallon in 2021 (Kruger et al. 2022).

Source: Christi (2018) / With permission of Elsevier.

9.5 The Future of Oil, Gas, Hydrogen and Ammonia, and Biofuel

Oil and gas have been the lifeblood of the global economy and underpin the geopolitical order. There will therefore be major upheavals as the world transitions to clean energy (Bordoff and O'Sullivan 2022). Even when the world achieves net-zero carbon emissions, it will hardly mean the end of fossil fuels. If the world reached net-zero emissions by 2050, we would still be using nearly half as much natural gas as we do today, and about a third as much oil (IEA 2021a).

9.5.1 The Future of Oil

Oil literally drove the twentieth century, but its eclipse now seems increasingly assured. In around 2000, British Petroleum (BP) briefly rebranded itself "Beyond Petroleum." Soon that may be more than a slogan. BP and Shell, along with other oil companies like Saudi Aramco and ExxonMobil, are investing heavily in renewable energy projects. Petroleum companies are rebranding themselves as energy companies.

The coronavirus pandemic (2019–2023), followed by Russia's war in Ukraine (2021, onward), resulted in high oil prices, and hence also high gasoline prices. It may be that 2019 or 2020 was the year of "peak oil", after which the demand for oil will consistently decrease. The world's taste for oil is also being driven down by pressure to decarbonize global economies. The game-changer for oil is set to be the rise of electric vehicles (IEA 2021a; Deloitte 2022; Rystad Energy 2022). As Figure 9.1 suggests, average oil demand to achieve net-zero emissions by 2050 must fall by 4% per year between 2020 and 2050. That is a tall order and meeting it will result in significant stranded capital and value.

Many high-cost oil producers (such as those in Canada and Russia's Arctic territory) will probably be priced out of the market as demand and oil prices fall. The Gulf states, which produce cheap, low-carbon oil, could see their market share increase. Other oil-producing countries that seek to be leaders in the fight against climate change (such as Norway, the United Kingdom, and the United States) will, in future, constrain their domestic output in response to rising public pressure and hasten the transition away from fossil fuels (Bordoff and O'Sullivan 2022).

However, none of these proposed urgent transitions to renewable energy will occur if national security concerns come into conflict with climate change ambitions. It will therefore be crucial to establish and maintain global peace and international collaboration on an unprecedented scale.

9.5.2 The Future of Natural Gas

As many countries are retrofitting or shutting down coal-fired power plants, electrical companies have been adopting natural gas-fired power plants, as they emit nearly 60% less CO_2 than their coal counterparts. Natural gas-fired power plants provide nearly 40% of the United States' electricity, and electrical companies are planning to build more than 200 new ones in the United States. If these gas-fired power plants are built, along with the gas pipelines to support them, they are likely to run for 30 or 40 years – long past the point that carbon emissions from electricity production need to approach zero if we are to have a reasonable climate future.

As the world begins to use less natural gas (see Figure 9.1), the market share of the small number of players that can produce it most cheaply and most cleanly will rise. For instance, it will be difficult for Europe to decrease its dependence on Russian gas. Regardless, declines in demand for natural gas will slow after 2040, and more than half the natural gas used globally in 2050 will be used to produce hydrogen in facilities with carbon capture and utilization technologies (IEA 2021a).

9.5.3 The Future of Hydrogen and Ammonia

After 2030, oil companies will focus on the production and export of low-carbon fuels, especially hydrogen and ammonia. These fuels will be critical to the transition to net-zero emissions given their potential to decarbonize hard-to-electrify sectors, such as steel production, fueling trucks, ships, and other heavy vehicles, and balancing grids supplied primarily by renewable sources of energy that can experience intermittent disruptions (also see Chapter 28). For countries that lack natural gas but have the capacity to store CO_2 underground, the cheapest way to get hydrogen – which is hard to transport over long distances – may well be to import natural gas and then convert it into hydrogen close to where it will be used (Bordoff and O'Sullivan 2022).

To reach net-zero carbon emissions by 2050, hydrogen demand is projected to increase to 530 million tons, more than six times that produced in 2020 (IEA 2021a). A projected 50% of the hydrogen will be used in steel manufacturing, chemicals production, and transportation; 30% will be converted to hydrogen-based fuels, mainly NH_3 for ship propulsion and electricity generation, synthetic kerosene for aviation, and synthetic methane blended into gas networks; and around 17% will be used in gas-fired power plants to balance electricity generation from solar photovoltaics and wind. Overall, hydrogen-based fuels will account for about 13% of global energy demand in 2050 (IEA 2021a).

9.5.4 The Future of Biofuels

Urgent measures are needed to incentivize the rapid development and deployment of advanced liquid and gas biofuel technologies. Measures that could boost the scaling up of advanced biofuels by 2030 include the following (IEA 2021b): incentives for co-processing bio-oil in existing oil refineries or fully converting oil refineries to biorefineries; retrofitting ethanol plants with carbon capture and storage systems; new infrastructure to provide for the increased injection of biomethane into gas networks; minimizing fugitive biomethane emissions from the supply chain; and promoting the deployment of biofuel plants close to existing industrial hubs where carbon capture and utilization projects are planned.

Problems

9.1 Determine how many kilograms of CO_2 are emitted for each kWh of electricity produced when burning either natural gas (assuming it comprises only methane) or coal (assuming it comprises 100% carbon) in a Rankine cycle with 36% energy conversion efficiency. From Table 7.5, the lower heating value of methane is 50,050 kJ/kg and that of carbon (coal) is 32,800 kJ/kg.

[Answer: Coal combustion produces 1.1 kg CO_2/kWh$_e$ and natural gas combustion produces about 0.6 kg CO_2/kWh$_e$.]

9.2 Calculate (a) the air–fuel ratio, and (b) the mass of CO_2 per kg of fuel for butane (C_4H_{10}), which burns in 20% excess air.

[Answers: (a) 18.56; (b) 3.03 kg CO_2 per kg C_4H_{10}.]

9.3 The 2022 Ford Explorer FFV AWD features a 3.3-L engine that can run on E85 FlexFuel. The car is refueling at a gas station that offers E85 fuel comprising a blend of 80% ethanol (C_2H_5OH) and 20% gasoline (assumed to be C_8H_{18}) by volume. Assume that the engine runs lean with an equivalence ratio of $\phi = 0.95$. What is (a) the required air–fuel ratio, and (b) the heat of combustion of the E85 fuel blend?

Hint: Calculate the stoichiometric AFRs for ethanol and gasoline independently. Then convert the volume fractions of gasoline and ethanol to their respective mass fractions using Eq. (9.7). Equation (9.5) can then be used to determine the AFR of the blended fuel, and Eq. (9.8) can be used to determine the effective heat value of the blended fuel. The example in Box 9.1 shows the problem-solving procedure. You must, however, remember to implement the equivalence ratio in this problem.

[Answers: (a) 6.86; (b) 33,020 kJ/kg.]

9.4 An ultimate analysis of a sample of heavy fuel oil provides the following mass fraction of constituents: 85% C, 11% H_2, 0% O_2, 0% N_2, 3% S 1.7%, 1% moisture, and 0% ash. Assuming that the only products of combustion are CO_2, H_2O, and CO_2, calculate the following: (a) the mass yields of each of the products of combustion when 100 kg of oil is burned with 5% excess air, (b) the mass of air required for the combustion of 100 kg of oil with 5% excess air, and (c) the combustion efficiency. *Hint:* Study Example 8.1 for a similar problem.

[Answers: (a) 317.7 kg O_2; (b) 1437.7 kg air; (c) 85.5%.]

9.5 The emission of CO_2 and oxides of sulfur (which can be assumed to be entirely SO_2) from two types of power plants are of interest for their impact on global warming and on environmental pollution. Estimate the emissions of these two pollutants in kg/MWh of electricity generated from the following combination of power plant and fuel: (a) and (b). State all assumptions and show all calculations. *Hint:* Study Example 8.1 for a similar problem.

a) A conventional pulverized coal-fired power plant with no flue gas desulfurization. The thermal efficiency of the plant is 38%. Coal comprises 66% C and 1.7% S. The gross calorific value of the bituminous coal is 27 MJ/kg.

[Answers: 850 kg CO_2/MWh$_e$; 12 kg SO_2/MWh$_e$.]

b) A conventional oil-fired power station with no control of nitrous oxides and no flue gas desulfurization. The plant has a thermal efficiency of 40%, and the fuel oil comprises 85.4% C and 2% S. The oil has a gross calorific value of 42.9 MJ/kg.

[Answers: 658 kg CO_2/MWh$_e$; 11.7 kg SO_2/MWh$_e$.]

9.6 The Nicor Gas Company supplies natural gas to users in Illinois. An ultimate analysis of a sample of the provided natural gas gives the following mass fraction of constituents: 96% CH_4, 2% C_2H_6, 0.1% C_3H_8, 0.1% C_4H_{10}, 0.05% C_6H_{14}, 0.03% C_5H_{12}, and 1.12% N_2. The products of combustion comprise 6% CO_2, 0.2% CO, 7% O_2, and 81% N_2. The natural gas is burned in stoichiometric air. Calculate the air–fuel ratio on a molar basis (i.e., how many kilomoles of air is required per kilomole of natural gas). *Hint:* Study Example 8.1 for a similar problem. Data source: https://www.nicorgas.com/content/dam/southern-co-gas/shared/pdf/spanish/AGL-Natural-Gas-SDA103015.pdf

[Answer: 12.4 kmol air/kmol fuel.]

9.7 "As the energy sector continues to transition into a cleaner one, each of its industries is embracing change. In this sense, downstream operations — which include all the processes involved in converting oil and gas into the finished product — are likewise undergoing a significant transformation to remain competitive and address sustainability. Transformations, including transitioning traditional refineries into biofuel hubs, developing new fueling solutions such as hydrogen fuel cells, renewable natural gas, and renewable fuels, or advancing petrochemical processes to comply with the economy's circularity, are underway and are rapidly evolving into more advanced technologies"

Produce a two-page fact sheet of no fewer than 400 words (with accompanying graphics) in which you interpret the above statement and explain how you think the use of Big Data, data analytics, and machine learning will help downstream actors in the energy sector to sustainably grow, reduce their environmental footprint, and prepare for the energy needs of the future.

[Answer: This question is open-ended and requires independent research.]

9.8 For many years, small biogas systems have collected methane from landfills, sewage plants, and farms. Now, in the USA and Europe, the growth of this renewable form of natural gas is taking off as businesses capture large amounts of methane from manure, food waste, and other sources.

An energy entrepreneur is proposing converting manure from several hog (pig) farms to produce biogas (bio-methane), also known as renewable natural gas. The methane will be transported through a 30-mi pipeline and will be processed in an anaerobic digester before being injected into an existing natural gas pipeline. (a) If the hog facility processes 10 billion gallons of waste every year, how much biogas could be produced? Clearly state your assumptions and show all your calculations, including relevant chemical calculations. (b) If this biogas were to be burned in a gas turbine with 40% thermal efficiency, how much power could be produced? How many homes would typically be powered by such means? (c) How can the energy entrepreneur use the "3-C" entrepreneurial mindset to streamline their energy innovation? The "3-C" mindset comprises the entrepreneur being curious, making connections between apparently disparate aspects, and creating value. (https://orchard-prod.azureweb-sites.net/media/Framework/KEEN_framework_new.pdf)

[Answers: (a) 35–50 billion ft^3 of biogas per year; (b) 280 MW; around 220,000 homes; (c) open-ended, independent research required.]

9.9 "Bioenergy is widely seen as being in competition with food for land resources. For instance, plants that use the mode of photosynthesis known as crassulacean acid metabolism (CAM) have the potential to generate globally significant quantities of renewable electricity without displacing productive agriculture and perhaps even increasing food supply. CAM plants require in the order of 10-fold less water per unit of dry biomass produced than common C3 and C4 crops do, and because of their succulence, are endowed with substantial water-storage capacities that help buffer intermittent water availability. This allows them to thrive in areas where traditional agriculture struggles, either because of low rainfall, or because the seasonality or unpredictability of rainfall is too great to allow profitable arable farming" (Mason et al. 2015).

(a) Describe how the anaerobic digestion of a CAM crop such as *Opuntia ficus-indica* (Figure 9.5), coupled with the Fischer-Tropsch process (see Box 8.3), could be used to produce aviation fuel. (b) Discuss the technical feasibility and economic viability of such an energy conversion process in the form of a high-impact fact sheet or infographic of no more than one page.

[Answer: Open-ended question that requires independent research.]

Figure 9.5 *Opuntia ficus-indica* biofuel crops. *Source:* Courtesy of Elqui global energy.

9.10 The typical energy content of municipal solid waste (MSW) in the developed world is 10 MJ/kg. Globally, fewer than 10% of MSW facilities use waste-to-energy plants. Most MSW stays in the landfill where it emits harmful methane, among other products. However, the combustion of MSW in combined heat and power (CHP) plants represents a useful means of disposing of this waste. See Figure 9.6 for a modern example of a waste-to-energy plant. European waste-to-energy plants incinerate waste that is approximately 50% biogenic (i.e., originating from renewable biomass). Therefore, energy from waste-to-energy plants have a lower carbon intensity than that produced through fossil fuel-based power plants. When combined with carbon capture technologies, waste-to-energy plants would enable energy production with negative CO_2 emissions, and thus help mitigate climate change. (Adapted from Magnanelli et al. 2021.)

Estimate the annual reduction in CO_2 emissions if a 55 MW_e natural gas-fired power plant were to be replaced by a 55 MW_e municipal solid waste-to-energy plant. Also estimate how much waste the MSW plant would consume per year. You may assume the MSW plant has an energy conversion efficiency of 38%.

[Answer: Around 100,000 tons of CO_2 per year.]

Figure 9.6 The Copenhill waste-to-energy plant in Copenhagen, Denmark, featuring an artificial ski-slope. When running at full capacity, the plant can burn and convert 440,000 t of municipal solid waste into clean energy annually, enough to provide over 30,000 homes with electricity, and 72,000 homes with heating. *Source:* Copenhill Amager Bakke, BIG – Bjarke Ingels Group, With permission Rasmus Hjortshøj.

References

Bordoff, J. and O'Sullivan, M.L. (2022). Green upheaval: the new geopolitics of energy. *Foreign Affairs* 101: 68.

BP (2021). Statistical review of world energy 2021. 70th edition. BP, https://www.bp.com/content/dam/bp/business-sites/en/global/corporate/pdfs/energy-economics/statistical-review/bp-stats-review-2021-full-report.pdf.

BP (2022). BP energy outlook. 2022 edition. https://www.bp.com/content/dam/bp/business-sites/en/global/corporate/pdfs/energy-economics/energy-outlook/bp-energy-outlook-2022.pdf.

Chaube, A., Chapman, A., Minami, A., Stubbins, J. et al. (2021). The role of current and emerging technologies in meeting Japan's mid-to long-term carbon reduction goals. *Applied Energy* 304 (December): 117669.

Christi, Y. (2018). Introduction to algal biofuels. In: *Biomass, Biofuels, Biochemicals: Biofuels from Algae.* (A. Pandey, D.J. Lee, J.S. Chang, Y. Chisti et al.). Amsterdam, Netherlands: Elsevier.

Deloitte (2022). 2022 Oil and gas industry outlook. https://www2.deloitte.com/us/en/pages/energy-and-resources/articles/oil-and-gas-industry-outlook.html.

Dimitriou, P. and Javaid, R. (2020). A review of ammonia as a compression ignition engine fuel. *International Journal of Hydrogen Energy* 45 (11): 7098–7118.

DoE (2021). Biodiesel vehicle emissions. Alternative Fuels Data Center. https://afdc.energy.gov/vehicles/diesels_emissions.html.

ETC (2021). *Making the Hydrogen Economy Possible: Accelerating Clean Hydrogen in an Electrified Economy.* London, UK: Energy Transitions Commission.

Harder, E.L. (1982). *Fundamentals of Energy Production.* New York, NY: Wiley.

Howarth, R.W. and Jacobson, M.Z. (2021). How green is blue hydrogen? *Energy Science & Engineering* 9 (10): 1676–1687.

Hydrogen Council (2021). Hydrogen insights. A perspective on hydrogen investment, market development and cost competitiveness. *The Hydrogen Council* https://hydrogencouncil.com/wp-content/uploads/2021/02/Hydrogen-Insights-2021-Report.pdf.

IEA (2021a). Net Zero by 2050. A roadmap for the global energy sector. International Energy Agency. https://iea.blob.core.windows.net/assets/deebef5d-0c34-4539-9d0c-10b13d840027/NetZeroby2050-ARoadmapfortheGlobalEnergySector_CORR.pdf.

IEA (2021b). World energy outlook, 2021. International Energy Agency (IEA), https://iea.blob.core.windows.net/assets/4ed140c1-c3f3-4fd9-acae-789a4e14a23c/WorldEnergyOutlook2021.pdf.

Jaffe, R.L. and Taylor, W. (2018). *The Physics of Energy.* Cambridge, UKI: Cambridge University Press.

Kruger, J.S., Wiatrowski, M., Davis, R.E., Dong, T. et al. (2022). Enabling production of algal biofuels by techno-economic optimization of co-product suites. *Frontiers in Chemical Engineering* 3 803513.

Lefebvre, A.H. and Ballal, D.R. (2010). *Gas Turbine Combustion. Alternative Fuels and Emissions.* Boca Raton, FL: CRC Press.

Lieuwen, T., Yang, V. and Yetter, R. (2010). *Synthesis Gas Combustion. Fundamentals and Applications.* Boca Raton, FL: CRC Press.

MacFarlane, D.R., Choi, J., Suryanto, B.H., Jalili, R. et al. (2020). Liquefied sunshine: transforming renewables into fertilizers and energy carriers with electromaterials. *Advanced Materials* 32 (18): 1904804.

Macrae, J.C. (1966). *An Introduction to the Study of Fuel.* Amsterdam, Netherlands: Elsevier.

Magnanelli, E., Mosby, J. and Becidan, M. (2021). Scenarios for carbon capture integration in a waste-to-energy plant. *Energy* 227: 120407.

Mason, P.M., Glover, K., Smith, J.A.C., Willis, K.J. et al. (2015). The potential of CAM crops as a globally significant bioenergy resource: moving from 'fuel or food' to 'fuel and more food'. *Energy and Environmental Science* 8 (8): 2320–2329.

McKinsey (2022). *Global Energy Perspective*. New York: McKinsey. https://www.mckinsey.com/~/media/McKinsey/Industries/Oil%20and%20Gas/Our%20Insights/Global%20Energy%20Perspective%202022/Global-Energy-Perspective-2022-Executive-Summary.pdf.

MIT (2011). *The Future of Natural Gas. An Interdisciplinary MIT Study*. Boston, MA: MIT.

Pörtner, H.O., Roberts, D.C., Adams, H., Adler, C. et al. (2022). IPCC Sixth Assessment Report. Climate change 2022: Impacts, adaptation and vulnerability. Intergovernmental Panel on Climate Change. https://www.ipcc.ch/report/ar6/wg2/downloads/report/IPCC_AR6_WGII_SummaryForPolicymakers.pdf.

Rose, J.W. and Cooper, J.R. (1977). *Technical Data on Fuel*. Edinburgh: Scottish Academic Press.

Rystad Energy (2022). *Oil and gas*. Oslo, Norway: Rystad Energy. https://www.rystadenergy.com/energy-themes/oil--gas.

Sørensen, B. and Spazzafumo, G. (2018). *Hydrogen and Fuel Cells: Emerging Technologies and Applications*. New York, NY: Elsevier.

T-raissi, A. and Block, D.L. (2004). Hydrogen: automotive fuel of the future. *IEEE Power and Energy Magazine* 2 (6): 40–45.

WLPGA (2021). *Guide to new autogas markets 2021*. World LPG Association. https://online.fliphtml5.com/addge/whce/#p=1.

Mini Project 3

Combustion of Fossil Fuels

Fossil fuels make up 80% of the current global primary energy demand. The global energy system is the source of approximately two-thirds of global CO_2 emissions. Inasmuch as methane (CH_4) and other short-lived climate pollutant emissions are believed to be severely underestimated, it is likely that energy production and use are the source of an even greater share of emissions than previously considered.

Fossil fuels (coal, oil, and gas) are derived from natural materials produced from photosynthesis (over millions of years). These fuels have high heat contents, are easy to combust, readily available, consistent in their molecular composition, but contain "unwanted" impurities when burned (i.e., they give off atmospheric pollutants and corrosion-causing materials). You are tasked with evaluating three candidate fuels for a steam power plant's boiler (or steam generator) – coal, oil, and natural gas, with the constituent properties given below:

i) Coal
  - *Higher heating value*: 27 MJ/kg
  - *Chemical composition*: 66% C, 4.1% H, 7.2% O, 1.3% N, 1.7% S, 12% moisture, 7.7% ash
  - *Typical composition of ash*: 40% Si, 26.2% Al, 20% Fe, 1.5% Mg, 3.5% Ca, 1.1% Ti, 2.9% Na, 4.8% K
ii) Heavy fuel oil
  - *Higher heating value*: 42.9 MJ/kg
  - *Chemical composition*: 85.4% C, 11.4% H, 0.1% O, 0.1% N, 2.8% S, 0.1% moisture, 0.1% ash
  - *Typical composition of ash*: 3.6% Si, 0.3% Al, 1.8% Fe, 0% Mg, 0.1% Ca, 0% Ti, 0% K, 7.4% Na, 68.9% V, 17.1% Ni, 0.8% Pb, 0% Cu, 0% Sn
iii) Natural gas
  - *Higher heating value*: 52.2 MJ/kg
  - *Chemical composition*: 98.43% methane, 0.44% ethane, 0.19% propane, 0.0275% *i*-butane, 0.0275% *n*-butane, 0.0275% *i*-pentane, 0.0275% *n*-pentane, 0.415% CO_2, 0.415% N

When evaluating these three fuels, you must answer the following questions:

1) Perform combustion analyses for each of these three fuels: (i) coal, (ii) heavy fuel oil, and (iii) natural gas. Evaluate how many kilograms of each product of combustion will be produced for each kilogram of fuel burned. Assume that each of these three fuels will be burned in 20% excess air.

Energy Systems: A Project-Based Approach to Sustainability Thinking for Energy Conversion Systems, First Edition. Leon Liebenberg.
© 2024 John Wiley & Sons, Inc. Published 2024 by John Wiley & Sons, Inc.
Companion website: www.wiley.com/go/liebenberg/energy_systems

2) Considering SDG 7 (clean and affordable energy) of the United Nations' Sustainable Development Goals, briefly explain which of these three fuels is the "best"? In your discussion, you should also elaborate on how extracting, transporting, and burning these three fuels will impact on other SDGs, including SDG 3 (good health and well-being), SDG 6 (clean water and sanitation), SDG 9 (industry, innovation, and infrastructure), SDG 10 (reduced inequalities), SDG 11 (sustainable cities and communities), SDG 12 (responsible production and consumption), SDG 13 (climate action), SDG 14 (life below water), and SDG 15 (life on land).

3) When performing the above evaluation, you will note the strengths and shortcomings of each of these boiler fuels. Briefly discuss alternative boiler fuels that might solve many of the issues you have identified, and which might also be more sustainable alternatives (like left-over biomass or discarded plastic). Your discussion should be linked to the targets stated to achieve SDG 7. You should also briefly explore the implications for achieving other SDGs. Further, you should discuss the interconnected social, technological, economic, environmental, and governance issues when shifting from one fuel to another.

Images of our future climate, from 1962
Panels from Athelstan Spilhaus's syndicated non-fiction comic strip, "Our New Age," which ran from 1958 to 1975 and raised awareness about environmental catastrophes. *Source:* The Fabulous Fifties.

Week 4 – Gas Cycles

Internal and external combustion engines use a working fluid (such as air), which remains in the gaseous phase throughout the cycle. These engine cycles are referred to as *gas cycles*. This is different than the *vapor cycles* discussed in chapters 13 and 14, which feature working fluids (such as water) that are alternatively vaporized and condensed.

Internal combustion engines that operate on fossil fuels consume around 25% of the world's energy and produce about 10% of the world's greenhouse-gas emissions. Gasoline-fueled internal combustion engines are the dominant prime mover for passenger cars, while diesel fuel is preferred for larger vehicles, due to diesel engines' better fuel efficiency. Although new fuels such as biofuels could be produced more sustainably than their fossil fuel counterparts, biofuels have a low to negligible influence on the lifecycle greenhouse-gas emissions of current and future gasoline- or diesel-fueled cars. It is therefore crucial for vehicles fueled by fossil fuels to use advanced emission control systems.

External combustion engines such as gas turbines are ubiquitous in the power plant industry. Gas turbines have a high efficiency, are reliable, operationally flexible, fuel flexible, and have low-emission power options. However, if gas turbines are to remain relevant in a decarbonizing global society, these engines must be able to burn hydrogen and low-carbon fuels. Gas turbines must also use advanced emission control technologies, especially to curtail their high thermal nitric oxide (NO_x) emissions. Stirling engines are external combustion engines that are uniquely positioned to use solar concentrators to provide high-temperature heat rather than burning fuel.

Chapter 10 – Internal combustion gas engines: Engine performance parameters; the Otto cycle for spark-ignition engines; the Diesel cycle for compression-ignition engines; alternative fuel-injection technologies; supercharging and turbo-charging; outlook

Chapter 11 – External combustion gas engines: Gas turbines: classification, combustion, engine performance parameters; Brayton cycle for gas turbines; Stirling engines: the Stirling cycle; outlook

Chapter 12 – Emission control of internal and external combustion engines: Emissions from the combustion of gasoline and diesel; controlling the emissions of gasoline and diesel combustion; reducing emissions from natural gas combustion; outlook

Mini Project 4 – Domestic back-up generator using natural gas or propane

10

Internal Combustion Gas Engines

Despite their massive use of fossil fuels and polluting emissions, internal combustion engines represent one of the most significant inventions of modern times, especially in terms of their primary application as a *portable* power source. The main applications are in vehicles, trains, ships, and aircraft. Internal combustion engines are also used for *stationary* applications, such as emergency power generators.

Internal combustion gas engines (or just "internal combustion engines") comprise primarily a cylinder, almost always stationary, and a piston. Together, these form a combustion chamber of variable volume. A fuel-air mixture is burned in the combustion chamber and the expanding gases operate on the piston, which performs work. It is therefore called an *internal* combustion engine. (In contrast with an *external* combustion engine, such as a gas turbine, the combustion process occurs externally of the prime mover.)

Air is brought to the cylinder. Depending on the cycle, fuel is introduced into the air before, during, or after compression, and mixed with it. Upon ignition of the fuel-air mixture, the heat developed raises the pressure of the products of combustion during one of the strokes of the piston. The pressure imbalance on the piston causes the piston to move, with the motion transferred to a connecting rod and crankshaft, which is linked to the vehicle's wheels via a transmission (or "gearbox").

The events of the cycle (such as an Otto cycle or a Diesel cycle) on which an internal combustion engine operates are mainly controlled by valves located in ports leading to and from a cylinder. In general, a cylinder comprises one or more inlet valves and one or more exhaust valves. These valves are usually mechanically operated by linking with the crankshaft via the valve gear train (Figure 10.1). As the internal combustion engine develops power in a cyclical manner, the power fluctuations are lessened using a heavy flywheel and with overlapping power impulses by employing multi-cylinder arrangements.

An internal combustion engine cycle comprises either two or four strokes. A *two-stroke cycle* has an intake stroke and a power stroke. A *four-stroke cycle* generally has an intake stroke, a compression stroke, an expansion (or "power") stroke, and an exhaust stroke (Figure 10.2). Therefore, a two-stroke engine produces power on every downward movement of the piston, while a four-stroke engine produces power with every other downward stroke. Although two-stroke engines are mechanically simpler and have greater power-to-weight ratios (kW per kg engine mass) than their four-stroke engine counterparts, two-stroke engines have higher emissions due to them simultaneously taking in fresh air–fuel while exhausting burned and unburned combustion products.

Energy Systems: A Project-Based Approach to Sustainability Thinking for Energy Conversion Systems,
First Edition. Leon Liebenberg.
© 2024 John Wiley & Sons, Inc. Published 2024 by John Wiley & Sons, Inc.
Companion website: www.wiley.com/go/liebenberg/energy_systems

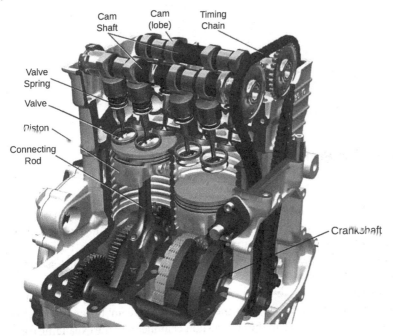

Figure 10.1 The piston-cylinder arrangement of an internal combustion engine, showing valves and cams, and the crankshaft. *Source:* Machine Design.

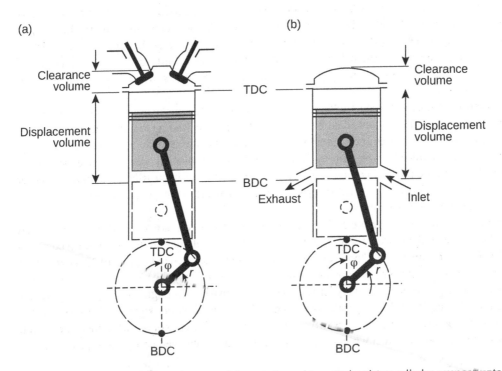

Figure 10.2 The main geometric features of four-stroke and two-stroke piston-cylinder arrangements, (a) four-stroke engine and (b) two-stroke engine. TDC = top dead-center; BDC = bottom dead-center.

Throughout the power cycle, the working fluid (air, or combusted fuel/air) remains in the gaseous phase. These cycles are therefore also referred to as *gas cycles*.

Reimagine Our Future

We will not be motivated to respond to energy unsustainability unless we feel personally connected to the issue. Create a blog entry or share a picture online that pertains to our energy unsustainability and that will, in your view, help people feel this personal connection. Ask your social network to comment on the feelings, thoughts, or questions that your words or photo(s) evoke.

10.1 Engine Performance Parameters

Figure 10.2 illustrates the main geometrical features of two-stroke and four-stroke piston-cylinder arrangements. The length of the piston's movement is called the stroke (s), which is twice the radius of the crankshaft. The diameter of the cylinder is called the bore (b). The *displacement volume* (V) is the volume through which the piston sweeps between the top-dead-center (TDC) and the bottom-dead-center (BDC) positions of the piston. If there is more than one piston, the total displacement is the displacement of one cylinder multiplied by the number of cylinders (n):

$$V = \left(\frac{\pi}{4}b^2 s\right)n \tag{10.1}$$

The engine's *compression ratio* (r) is the ratio of maximum to minimum volume:

$$r = \frac{V_{max}}{V_{min}} = \frac{V_{BDC}}{V_{TDC}} \tag{10.2}$$

From Table 1.6, *expansion work* (W) can be calculated using the prevailing pressure and change in volume:

$$W = \int p \, dV \tag{10.3}$$

The engine's *power* (or work rate) depends on the work (W) measured in joules, rotational speed of the crankshaft (N) measured in revolutions per minute (or rpm), number of cylinders (n), and the number of power strokes per crankshaft revolution:

$$4\text{-stroke engine (two revolutions of crankshaft per power stroke)}: \dot{W} = \frac{nW(N/60)}{2}$$

$$2\text{-stroke engine (one revolution of crankshaft per power stroke)}: \dot{W} = nW(N/60) \tag{10.4}$$

Power (measured in watts) is also equal to the product of torque (T), measured in newton-meters, and rotational velocity (ω), measured in radians per second:

$$\dot{W} = T\omega \tag{10.5}$$

To convert between rotational speed N (rpm) and rotational velocity ω (rad/s), use:

$$\omega = \frac{2\pi N}{60} \qquad [\text{rad/s}] \tag{10.6}$$

Engines are often compared based on their *mean (or average) effective pressure* (*mep*), usually specified in MPa, and defined as:

$$mep = \frac{W}{V} \qquad [\text{MPa}] \tag{10.7}$$

Using Eqs. (10.1) and (10.4), power output can also be written in terms of *mep*:

$$\dot{W} = \frac{(mep)\,V\,(N/60)}{2} \tag{10.8}$$

The *specific fuel consumption* (*sfc*) of an internal combustion engine is the ratio of fuel consumption rate $(\dot{m}_{fuel})$ to power $(\dot{W})$ and is often specified in grams of fuel per kWh produced:

$$sfc = \frac{\dot{m}_{fuel}}{\dot{W}} \qquad [\text{g/kWh}] \tag{10.9}$$

The thermal efficiency of an internal combustion engine is the ratio of power output to input rate of heat. The power (heat rate) input is equal to the product of the fuel flow rate and the specific heat of combustion (q_c):

$$\eta_{th} = \frac{\dot{W}}{\dot{Q}} = \frac{\dot{W}}{\dot{m}_{fuel}\,q_c} \equiv \frac{1}{(sfc)q_c} \tag{10.10}$$

The specific heat of combustion (q_c) can be determined using the values of higher heating value (*HHV*) or lower heating value (*LHV*), specified in Table 7.5. We will again assume that air is an ideal gas, in which case we can write down the following heat and air–fuel relationships:

$$\dot{Q} = \dot{m}_{fuel}\,q_c \qquad [\text{kW}] \tag{10.11}$$

with

$$\dot{m}_{fuel} = \dot{m} - \dot{m}_{air} \tag{10.12}$$

The air–fuel ratio (*AFR*) is:

$$AFR = \frac{\dot{m}_{air}}{\dot{m}_{fuel}} \equiv \frac{m_{air}}{m_{fuel}} \tag{10.13}$$

Therefore:

$$\dot{m}_{fuel} = \frac{\dot{m}}{AFR + 1} \tag{10.14}$$

and

$$m_{fuel} = \frac{m}{AFR + 1} \tag{10.15}$$

From our knowledge of an ideal gas (see Section 2.4), we can determine the mass of gas in a fixed volume at a certain absolute temperature and pressure:

$$m = \frac{p\,V}{RT} \tag{10.16}$$

where R is the gas constant of the air–fuel mixture and m is the mass of air and fuel.

Also:

$$R = c_p - c_v \tag{10.17}$$

and

$$\gamma = \frac{c_p}{c_v} \tag{10.18}$$

Substituting Eqs. (10.14) to (10.18) in (10.11), we obtain the following handy relationships for heat produced (kJ) or the heat rate of fuel (kW) (Kirkpatrick 2020):

$$Q = \underbrace{\left(\frac{p\mathcal{V}}{RT}\right)}_{\text{mass}} \left(\frac{q_c}{AFR + 1}\right) \tag{10.19}$$

$$\text{or} \quad \dot{Q} = \left(\frac{p\dot{\mathcal{V}}}{RT}\right) \left(\frac{q_c}{AFR + 1}\right) \tag{10.20}$$

The air–fuel ratios are calculated as explained in Chapters 7–9.

Example 10.1 Characteristics of a Gasoline Engine

Consider a modern four-stroke, four-cylinder gasoline engine that runs at 3000 rpm with ambient pressure equal to 100 kPa. Gasoline is accurately represented by iso-octane (C_8H_{18}). The engine operates stoichiometrically with an air–fuel ratio of 15. All products of combustion leave the engine in gaseous form. The engine bore is 80 mm and the stroke is 100 mm, while the compression ratio is 11. If the engine's thermal efficiency is 30%, calculate (a) the engine's power output, (b) the specific fuel consumption, and (c) the mean effective pressure.

Required: We must find (a) the engine's power with knowledge of the type of fuel, engine displacement, engine speed, and engine efficiency. Once the power has been calculated, we can determine the specific fuel consumption and the mean effective pressure.

Solution strategy: The power can be calculated with $\dot{W} = \dfrac{(mep)\,\mathcal{V}\,(N/60)}{2}$. We do, however, not know the *mep*. Another way of calculating the power output is from the heat rate and the engine's thermal efficiency: $\dot{W} = \dfrac{(mep)\,\mathcal{V}\,(N/60)}{2} \equiv \eta_{th}\,\dot{m}_{fuel}\,q_c$

The thermal efficiency is given as 30%. The fuel flow rate is $\dot{m}_{fuel} = \dfrac{\dot{m}_{air}}{AFR}$ with $\dot{m}_{air} = \dfrac{\dot{\mathcal{V}}}{v}$. If the gas is assumed to be air, then its density at 100 kPa can be determined. The specific volume of air is the inverse of density: $v = \dfrac{1}{\rho}$. Lastly, the specific heat of combustion can be read from Table 7.5.

Assumptions: Assume that the engine has no mechanical losses and that the combustion process is ideal.

Analysis:

a) Power output

The engine displacement is $\mathcal{V} = \left(\frac{\pi}{4}b^2 s\right)n = \left[\frac{\pi}{4}(0.08\text{ m})^2 0.1\text{ m}\right](4\text{ cylinders}) = 2 \times 10^{-3}\text{ m}^3$. This is therefore a two-liter engine.

The volumetric flow rate of air in the cylinders is:

$$\dot{\mathcal{V}} = \mathcal{V}\,(N/60) = (2 \times 10^{-3}\text{ m}^3)\left(\frac{3000\text{ rpm}}{60\text{ s/ min}}\right) = 0.1\text{ m}^3/\text{s}.$$

The density of air at 100 kPa is 1.2 kg/m³, so its specific volume is

$$v = \frac{1}{\rho} = \frac{1}{1.2\text{ kg/m}^3} = 0.83\text{ m}^3/\text{kg}.$$

The mass flow rate of air is: $\dot{m}_{air} = \dfrac{\dot{V}}{v} = \dfrac{0.1 \text{ m}^3/\text{s}}{0.83 \text{ m}^3/\text{kg}} = 0.12 \text{ kg/s}$

The fuel flow rate is: $\dot{m}_{fuel} = \dfrac{\dot{m}_{air}}{AFR} = \dfrac{0.1 \text{ m}^3/\text{s}}{15} = 6.67 \times 10^{-3} \text{ kg/s}$

The lower heating value of C_8H_{18} can be read from Table 7.5 as 44,430 kJ/kg. We can finally determine the engine's power output:

$$\dot{W} = \eta_{th}\, \dot{m}_{fuel}\, q_c = 0.3\big(6.67 \times 10^{-3} \text{ kg/s}\big)(44,430 \text{ kJ/kg}) = 89 \text{ kW}$$

b) Mean effective pressure

We know that $\dot{W} = 89 \text{ kW} = \dfrac{(mep)\, \dot{V}\, (N/60)}{2}$.

Therefore, the mean effective pressure can be calculated:

$$mep = \dfrac{2(89 \text{ kW})}{(2 \times 10^{-3}\text{m}^3)\left(\dfrac{3000 \text{ rpm}}{60 \text{ s/min}}\right)} = 1.78 \text{ MPa}$$

c) Specific fuel consumption

$$sfc = \dfrac{\dot{m}_{fuel}}{\dot{W}} = \dfrac{(6.67 \times 10^{-3} \text{ kg/s})}{89 \text{ kW}} = 74.94 \times 10^{-6} \text{ kg/kWs}$$

$$\equiv \big(74.94 \times 10^{-6} \text{ kg/kWs}\big)\left(\dfrac{10^3 \text{g}}{\text{kg}}\right)\left(\dfrac{3,600 \text{ s}}{\text{h}}\right) = 270 \text{ g/kWh}$$

Reimagine Our Future

We may not be able to see into the future, but we can influence it and we can imagine it afresh. Imagine if the world in 2050 were populated by people who were determined to minimize the harm they cause to nature and to others. Suspend your disbelief and be open-minded! What would such a future look, smell, and feel like? How would your quality of life be different from what it is today? Ask your friends to imagine this future with you and consolidate your ideas by posting a blog entry.

10.2 The Otto Cycle (for a Spark-Ignition Engine)

The Otto cycle (Otto 1877) is typically used for *spark-ignition* internal combustion engines, also known as *gasoline (or petrol) engines*. Figure 10.3 illustrates the four strokes inherent to an Otto cycle.

- *Process 1–2*: At state point 1, an intake valve opens, which draws in air and fuel (typically gasoline). The fuel-air mixture is compressed in an *adiabatic* (and isentropic) process, which

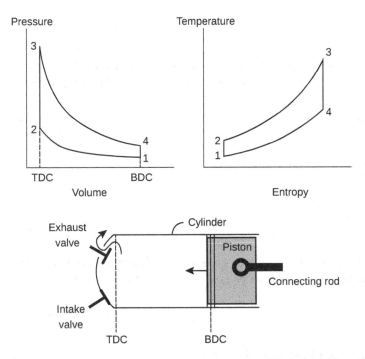

Figure 10.3 The Otto cycle represented on pressure-volume and temperature-entropy diagrams, and the piston-cylinder showing the positions of TDC and BDC. Process 1–2: Adiabatic (and isentropic) compression; Process 2–3: Constant volume heat addition; Process 3–4: Adiabatic (and isentropic) expansion stroke, or power stroke; Process 4–1: Constant volume exhaust stroke.

requires work input by the piston. At the end of the compression stroke, a spark plug produces a spark, which ignites the fuel-air mixture.

- *Process 2–3*: Heat addition from the exploding gas-air mixture happens promptly, and is approximated as a *constant volume* process.
- *Process 3–4*: The high-pressure fuel-air mixture results in the *adiabatic* (and isentropic) expansion of the gas, which causes the piston to do work. This process is also called the power stroke or expansion stroke.

  If the pressure is too high in this process, the compressed gases will auto-ignite in front of the combustion flame, resulting in damaging pressure spikes. This condition is known as *knock,* which adversely affects engine performance.
- *Process 4–1*: An exhaust valve opens at state point 4, which results in the exhaust gases being pushed out of the cylinder during a *constant volume* exhaust stroke.

The processes and salient thermodynamic relationships of a four-stroke spark-ignition engine using the Otto cycle can be summarized as follows (also see Figure 10.3):

10.2.1 Process 1–2: Adiabatic (and Isentropic) Compression

From Table 2.1, and using Eq. (10.5):

$$\frac{T_2}{T_1} = \left(\frac{v_1}{v_2}\right)^{\gamma - 1} = r^{\gamma - 1} \tag{10.21}$$

$$\frac{p_2}{p_1} = \left(\frac{v_1}{v_2}\right)^{\gamma} = r^{\gamma} \tag{10.22}$$

The engine cylinder volume can be treated as a closed system, which means that the first law of thermodynamics for a closed system (Table 2.1) can be used to calculate the work done on the fuel-air mixture:

$$w_{12} = -\Delta u_{12} \equiv -c_v \Delta T_{12} \tag{10.23}$$

10.2.2 Process 2–3: Heat Addition at Constant Volume

From Table 2.1:

$$\left(\frac{p}{T}\right)_2 = \left(\frac{p}{T}\right)_3 \tag{10.24}$$

The engine cylinder volume can be treated as a closed system, which means that the first law of thermodynamics for a closed system (Table 2.1) can be used to calculate the heat added:

$$q_{23} = \Delta u_{23} \equiv c_v \Delta T_{23} \tag{10.25}$$

10.2.3 Process 3–4: Adiabatic (and Isentropic) Expansion

From Table 2.1, and using Eq. (10.5):

$$\frac{T_4}{T_3} = \left(\frac{v_3}{v_4}\right)^{\gamma-1} = \left(\frac{1}{r}\right)^{\gamma-1} \tag{10.26}$$

$$\frac{p_4}{p_3} = \left(\frac{v_3}{v_4}\right)^{\gamma} = \left(\frac{1}{r}\right)^{\gamma} \tag{10.27}$$

The engine cylinder volume can be treated as a closed system, which means that the first law of thermodynamics for a closed system (Table 2.1) can be used to calculate the work done:

$$w_{34} = -\Delta u_{34} \equiv -c_v \Delta T_{34} \tag{10.28}$$

10.2.4 Process 4–1: Heat Rejection at Constant Volume

From Table 2.1:

$$\left(\frac{p}{T}\right)_4 = \left(\frac{p}{T}\right)_1 \tag{10.29}$$

The engine cylinder volume can be treated as a closed system, which means that the first law of thermodynamics for a closed system (Table 2.1) can be used to calculate the heat discarded:

$$q_{41} = \Delta u_{41} \equiv c_v \Delta T_{41} \tag{10.30}$$

The thermal efficiency of an Otto cycle can be calculated as follows:

$$\eta_{th} = \frac{w_{net}}{q_{in}} \tag{10.31}$$

Table 10.1 Reference values for important energy parameters of modern Otto engines for passenger cars.

Nominal engine speed (N), (rpm)	5000–8000
Compression ratio (r), (−)	9–11
Maximum pressure (MPa)	4–5
Mean effective pressure (mep), (MPa)	1.1–1.4
Power output per liter (kW/L)	40–80
Weight-to-power ratio (kg/kW)	2–0.8
Specific fuel consumption (g/kWh)	350–300
Thermal efficiency (η_{th}), (−)	0.45–0.55

The engines are naturally aspirated (i.e., not turbo-charged). Mechanical and volumetric efficiencies are assumed to be 100%.
Source: Bosch (2022) / Robert Bosch GmbH.

For a cycle, $\Delta u_{cycle} = 0$, which means that the first law for a cycle becomes $w_{net} = q_{net}$. Equation (10.31) can be simplified accordingly:

$$\eta_{th} = \frac{q_{net}}{q_{in}} = \frac{q_{in} - q_{out}}{q_{in}} \equiv 1 - \frac{|q_{41}|}{q_{23}} = 1 - \frac{|c_v(T_1 - T_4)|}{c_v(T_3 - T_2)} = 1 - \frac{1}{r^{\gamma - 1}} \tag{10.32}$$

As thermal efficiencies are defined to be positive, we work with the absolute value of heat rejected, as indicated in Eq. (10.32).

When determining specific heat capacities (c_p, c_v) and their ratios (γ), we will work with average values. For instance, with Otto engines that combust gasoline, the average cycle temperature $(T_1 + T_2 + T_3 + T_4)/4$ is close to 900 K. If the gases of combustion are assumed to act as air, which is an ideal gas, then the corresponding values of heat capacities and their ratios are (Heywood 1988):

$$\begin{aligned}
c_{p,900K} &= 1.121 \text{ kJ/kgK} \\
c_{v,900K} &= 0.834 \text{ kJ/kgK} \\
\gamma_{900K} &= \frac{c_{p,900K}}{c_{v,900K}} = 1.344 \\
R &= c_{p,900K} - c_{v,900K} = 0.287 \text{ kJ/kgK}
\end{aligned} \tag{10.33}$$

Table 10.1 summarizes the salient operating characteristics of modern Otto engines used in passenger cars.

Example 10.2 Otto Cycle (for a Spark-Ignition Engine)

Consider a gasoline engine operating on the Otto cycle. Gasoline is accurately represented by iso-octane (C_8H_{18}). The engine has a volumetric displacement of 2 L, a compression ratio of 8, and runs on a lean air–fuel ratio of 17. The exhaust valve releases combustion products in gaseous form to the surroundings. The combustion chamber rejects heat to the surroundings at 293 K. Calculate (a) the heat added (in kJ/kg fuel) during Process 2–3, (b) the temperatures and pressures, as well as heat and work transferred at the four state points, (c) the thermal efficiency of the engine, and (d) the entropy generation and irreversibility during the four processes. The average cycle temperature is typically 900 K.

Required: We must find (a) the heat added by burning C_8H_{18} in a constant volume process, and (b) the pressures and temperatures at the four state points. The heat and work transferred during the processes must also be calculated, from which we can evaluate (c) the engine's thermal efficiency. (d) Finally, we will calculate the entropy generated in each of the four processes and their resulting irreversibilities.

Solution strategy: We will use Eq. (2.14) to find the specific heat added during the combustion of C_8H_{18}. When evaluating the temperatures and pressures of the ideal gas, we will use the equations summarized in Table 2.1. We must, however, take care when calculating the maximum cycle temperature (T_3) as we need to use the correct value of specific heat capacity for C_8H_{18} (from Table 7.5), and not the value of c_v for an ideal gas. The thermal efficiency can be evaluated using Eq. (10.32). Finally, we will use Eqs. (3.58) and (3.59) to evaluate the irreversibilities in the four processes.

Assumptions: Assume that the compression and expansion processes are ideal, i.e., they are adiabatic and hence also isentropic. We will also work with average values of c_v and c_p, evaluated at an average cycle temperature of 900 K.

Analysis:

a) Heat added during Process 2–3

As the combustion products are in gaseous form, we need to work with the lower heating value of iso-octane (see Eq. (7.15)). From Table 7.5, the *LHV* of iso-octane is 44,430 kJ/kg.

From Eq. (10.19):

$$Q_{23} = mq_{23} = \underbrace{\left(\frac{p\mathcal{V}}{RT}\right)}_{m}\left(\frac{q_c}{AFR+1}\right)$$

$$\therefore q_{23} = \left(\frac{q_c}{AFR+1}\right) = \left(\frac{44{,}430 \text{ kJ/kg}}{17+1}\right)$$

$$q_{23} = 2468.3 \text{ kJ/kg fuel}$$

b) Temperatures and pressure of state points

From Eq. (10.33), for an average cycle temperature of 900 K, $c_{p,900K} = 1.121$ kJ/kgK, $c_{v,900K} = 0.834$ kJ/kgK, $\gamma_{900K} = \dfrac{c_{p,900K}}{c_{v,900K}} = 1.121$, and $R = c_{p,900K} - c_{v,900K} = 0.287$ kJ/kgK.

Process 1–2:

$$T_2 = T_1 r^{\gamma-1} = (293 \text{ K})(8)^{1.344-1} = 599.1 \text{ K}$$

$$p_2 = p_1 r^{\gamma} = (100 \text{ kPa})(8)^{1.344} = 1635.9 \text{ kPa}$$

The engine cylinder volume can be treated as a closed system, which means that the first law for a closed system (Table 2.1) can be used:

$$w_{12} = -\Delta u_{12} \equiv -c_v \Delta T_{12} = (-0.834 \text{ kJ/kgK})(599.1 - 293) \text{ K} = -255.3 \text{ kJ/kg.}$$ The work input to compress the gas is 255.3 kJ per kg of air.

Process 2–3:

$$q_{23} = 2468.3 \text{ kJ/kgfuel} \equiv c \, \Delta T_{23}$$

Note that c is the specific heat of the liquid fuel; it is not the c_v value of the combustion gas. The specific heat of iso-octane can be obtained from Table 7.5:

$$c_{C_8H_{18}} = 2.23 \text{ kJ/kg.K}$$

$$q_{23} = 2468.3 \text{ kJ/kg fuel} = c \, \Delta T_{23} = (2.23 \text{ kJ/kg fuel})(T_3 - 599.1 \text{ K})$$

$$\therefore T_3 = 1706 \text{ K}$$

$$p_3 = p_2 \left(\frac{T_3}{T_2}\right) = (1635.9 \text{ kPa}) \left(\frac{1706 \text{ K}}{599.1 \text{ K}}\right) = 4658 \text{ kPa}$$

Note: The engine cylinder volume can be treated as a closed system, which means that the first law for a closed system (Table 2.1) can be used to calculate the heat added in Process 2–3 per kg of *air*:

$$q_{23} = c_v \, \Delta T_{23} = (0.834 \text{ kJ/kgK})(1706 - 599.1) \text{ K} = 923.2 \text{ kJ/kg air}$$

Process 3–4:

$$T_4 = T_3 \left(\frac{1}{r}\right)^{\gamma - 1} = 1706 \text{ K} \left(\frac{1}{8}\right)^{1.344 - 1} = 834.3 \text{ K}$$

$$p_4 = p_3 \left(\frac{1}{r}\right)^{\gamma} = 4658 \text{ kPa} \left(\frac{1}{8}\right)^{1.344} = 284.8 \text{ kPa}$$

The engine cylinder volume can be treated as a closed system, which means that the first law for a closed system (Table 2.1) can be used to calculate the work done during the expansion stroke:

$$w_{34} = -\Delta u_{34} \equiv -c_v \, \Delta T_{34} = (-0.834 \text{ kJ/kgK})(834.3 - 1706) \text{ K} = 727 \text{ kJ/kg}$$

Process 4–1:

The engine cylinder volume can be treated as a closed system, which means that the first law of thermodynamics for a closed system (Table 2.1) can be used to calculate the heat given off to the environment:

$$q_{41} = \Delta u_{41} \equiv c_v \, \Delta T_{41} = (0.834 \text{ kJ/kgK})(293 - 834.3) \text{ K} = -451.4 \text{ kJ/kg}$$

The heat lost to the environment amounts to 451.4 kJ per kg of air.

c) Thermal efficiency

$$\eta_{th} = 1 - \frac{1}{r^{\gamma - 1}} = 1 - \frac{1}{8^{1.344 - 1}} = 0.51$$

Note: The thermal efficiency could also be calculated as follows:

$$\eta_{th} = \frac{w_{net}}{q_{in}} = \frac{w_{34} - |w_{12}|}{q_{23}} = \frac{(727 - 255) \text{ kJ/kg air}}{(923.2 \text{ kJ/kg air})} = 0.51$$

So, even though the compression and expansion processes are ideal (adiabatic, isentropic), the engine still only converts 51% of the added heat to work. The remaining 49% of heat leaves the engine in the exhaust stroke, $q_{41} = -451.4 \text{ kJ/kg}$. (The percentage of heat loss can also be calculated by taking the ratio of heat lost to heat added, which gives 49%:

$$\frac{|-451.4 \text{ kJ/kg air}|}{923.2 \text{ kJ/kg air}} = 0.49.)$$

d) Entropy generation and irreversibilities

We can calculate irreversibilities from Eq. (3.58), which reduce to the following if we disregard changes in kinetic and potential energy and assume steady-state conditions. The entropy production then becomes:

$$\sigma = \frac{\dot{\sigma}}{\dot{m}} = \sum_j \left(1 - \frac{T_0}{T_j}\right) \frac{Q_j}{T_0} - \frac{W}{T_0} + \frac{1}{T_0} [(h_1 - h_2) - T_0(s_1 - s_2)].$$

For processes 1–2 and 3–4, the processes are reversible and adiabatic, hence also isentropic. The work transfer will be identically equal to the change in enthalpies, which means that the entropy production for both these processes will be 0:

$$\sigma_{12} = \sum_j \left(1 - \frac{T_0}{T_j}\right)\frac{\cancel{Q_j}}{T_0} - \frac{\cancel{W}}{\cancel{T_0}} + \frac{1}{T_0}\left[\cancel{(h_1 - h_2)} - T_0\cancel{(s_1 - s_2)}\right] = 0$$

$$\sigma_{34} = \sum_j \left(1 - \frac{T_0}{T_j}\right)\frac{\cancel{Q_j}}{T_0} - \frac{\cancel{W}}{\cancel{T_0}} + \frac{1}{T_0}\left[\cancel{(h_3 - h_4)} - T_0\cancel{(s_3 - s_4)}\right] = 0$$

For processes 2–3 and 4–1, the entropy production can be determined by evaluating the change in entropy using Eq. (3.25), which reduces to the following, considering that these two processes happen without any change in volume ($v_e = v_i$), so $\ln\frac{v_e}{v_i} = 0$:

$$\Delta s = c_{v,avg}\ln\frac{T_e}{T_i} + R\cancel{\ln\frac{v_e}{v_i}}$$

$$\Delta s_{23} = c_{v,avg}\ln\frac{T_3}{T_2} = (0.834\,\text{kJ/kgK})\left(\ln\frac{1706\,\text{K}}{599.1\,\text{K}}\right) = 0.873\,\text{kJ/kgK}$$

$$\Delta s_{41} = c_{v,avg}\ln\frac{T_1}{T_4} = (0.834\,\text{kJ/kgK})\left(\ln\frac{293\,\text{K}}{834.3\,\text{K}}\right) = -0.873\,\text{kJ/kgK}$$

The entropy production for processes 2–3 and 4–1 can now be evaluated. Again, the work transfer will be identically equal to the change in enthalpies for both processes so that those terms cancel each other out. For Process 2–3, the high-temperature reservoir at 1706 K receives heat from the combustion chamber, which means that we will work with $-(q_{23})$ according to our sign convention for heat leaving a system:

$$\sigma_{23} = T_0\left(\frac{-q_{23}}{T_3} + s_3 - s_2\right) = (293\,\text{K})\left[\frac{-923.2\,\text{kJ/kg}}{1706\,\text{K}} + (0.873\,\text{kJ/kgK})\right] = 97.2\,\text{kJ/kg}$$

For Process 4–1, heat is lost to the environment with $q_{41} = -451.4\,\text{kJ/kg}$

$$\sigma_{41} = T_0\left(\frac{q_{41}}{T_{low}} + s_1 - s_4\right) = (293\,\text{K})\left[\frac{-(-451.4\,\text{kJ/kg})}{293\,\text{K}} + (-0.873\,\text{kJ/kgK})\right] = 195.6\,\text{kJ/kg}$$

The results may be summarized and checked as follows:

Process	q (kJ/kg)	w (kJ/kg)	$s_e - s_i$ (kJ/kgK)	σ (kJ/kg)
1–2	0	−255.3	0	0
2–3	923.2	0	0.873	97.2
3–4	0	727	0	0
4–1	−451.4	0	−0.873	195.6
Cycle	471.7	471.7	0	292.8

The two processes are reversible and adiabatic, hence also isentropic. So, processes 1–2 and 3–4 have no entropy production. The heat-addition (2–3) and heat-loss (4–1) processes are, however, irreversible as they transfer heat across finite temperatures. The irreversibility associated with process 2–3 could be reduced by decreasing T_3. The largest irreversibility is in process 4–1 and can be reduced by minimizing the heat lost to the environment, thereby recovering waste heat.

Discussion: This ideal engine does not consider heat losses during compression and expansion. It therefore has a high thermal efficiency of 51%. In a real engine, the compression and expansion steps will not be reversible or adiabatic. The work output will therefore be considerably less than the work output of this idealized cycle. Instead of working with isentropic processes in a real engine, it is appropriate to use isentropic efficiencies for the compression and expansion processes. When work is *produced*, $\eta_{isentropic,\,W_{out}} = \dfrac{W_{actual}}{W_{reversible}}$, and when the *process requires work*, $\eta_{isentropic,\,W_{In}} = \dfrac{W_{reversible}}{W_{actual}}$. If the isentropic efficiencies were 0.8 (i.e., 80%) for compression and expansion, the actual work outputs would be:

$$W_{actual,12} = \frac{W_{reversible,12}}{\eta_{isentropic,W_{12}}} = \frac{-255\,\text{kJ/kg}}{0.8} = -318.8\,\text{kJ/kg}$$

$$W_{actual,34} = \eta_{isentropic,W_{34}}\left(W_{reversible,34}\right) = 0.8\,(727\,\text{kJ/kg}) = 581.6\,\text{kJ/kg}$$

Therefore, the thermal efficiency of an actual engine with 75% isentropic efficiencies will be:

$$\eta_{th,actual} = \frac{w_{net}}{q_{in}} = \frac{w_{34} - |w_{12}|}{q_{23}} = \frac{(581.6 - 318.8)\,\text{kJ/kg air}}{(923.2\,\text{kJ/kg air})} = 0.28$$

The irreversibilities that we have assumed to exist in the actual engine by taking an isentropic efficiency of 80% for compression and expansion have reduced the thermal efficiency by a factor of nearly 2, from 51% to 28%.

Box 10.1 Alternative Spark-Ignition Cycles

The Atkinson (or "overexpansion") cycle is a common alternative cycle for spark-ignition engines (Figure 10.4), initially used in hybrid-electric vehicles like the Toyota Prius, but nowadays also in non-hybrid vehicles. Compared with a conventional Otto cycle, the Atkinson cycle can achieve larger expansion ratios, and thus a higher thermal efficiency, while maintaining a normal effective compression ratio to avoid knock (Zhao 2017; Li et al. 2021). Atkinson cycles therefore feature different compression and expansion ratios, unlike Otto engines. Atkinson cycle engines could have compression ratios of 10 and expansion ratios of 13.

Atkinson cycle engines are used in modern Toyota Camry and Honda Accord hybrid vehicles. These engines employ variable valve timing that allows operation as an Otto cycle to achieve conventional power density, or as an Atkinson cycle to achieve good fuel efficiency, but at lower power. Toyota reports thermal efficiencies of 35% with conventional Otto engines, and 40% with Atkinson cycle engines (Figure 10.5).

Figure 10.4 The Atkinson cycle, which resembles an Otto cycle, but with "overexpanded" expansion and intake sections, thus doing more work than an Otto cycle with the same amount of fuel. *Source:* Adapted from Atkinson (1886). Compression begins while the intake valve is still open, requiring complicated valve timing. (Note that the drawing is not to scale.)

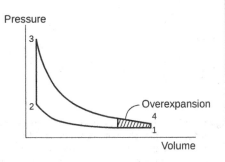

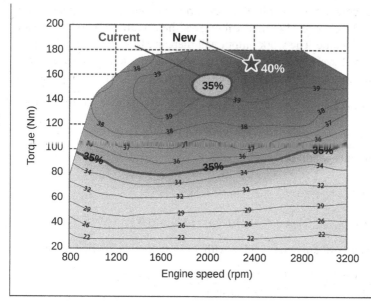

Figure 10.5 Engine performance map (thermal efficiency as a function of engine torque and speed) for Toyota's 2.5-L four-cylinder engines that operate on the Atkinson cycle. *Source:* Toyota.

10.3 The Diesel Cycle (for a Compressed-Ignition Engine)

The Diesel cycle (Diesel 1895) is typically employed in *compression-ignition* cycles. Figure 10.6 illustrates the four strokes inherent in a Diesel cycle.

- *Process 1–2*: At state point 1, an intake valve opens, which draws in only air (not air and fuel as in the case of a spark-ignition engine). The air is compressed in an *adiabatic* (and isentropic) process, which requires work input by the piston. As only air is compressed, compression-ignition engines operate at much higher pressures than spark-ignition engines, as there is no danger

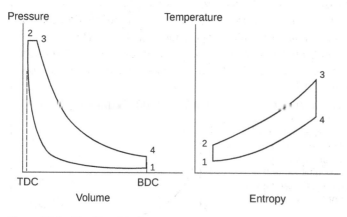

Figure 10.6 The Diesel cycle represented on pressure-volume and temperature-entropy diagrams, and piston-cylinder showing positions of TDC and BDC. Process 1–2: Adiabatic (and isentropic) compression; Process 2–3: Constant pressure heat addition; Process 3–4: Adiabatic (and isentropic) expansion stroke, or power stroke; Process 4–1: Constant volume exhaust stroke.

of auto-ignition. At the end of the compression stroke, fuel (usually diesel) is injected into the cylinder. The high-pressure results in the air temperature being higher than the fuel's ignition temperature, which then ignites the fuel (hence the term *compression ignition*).

- *Process 2–3*: The heat-addition process in a compression ignition can be approximated as a *constant pressure* process, as the fuel-air mixture begins to expand. (This is different to a spark-ignition engine, where the heat-addition process happens at a constant volume.)
- *Process 3–4*: The main expansion work is done during process 3–4, which occurs under *adiabatic* (and isentropic) conditions. This process is also called the power stroke or expansion stroke.
- *Process 4–1*: An exhaust valve opens at state point 4, which results in the exhaust gases being sucked out of the cylinder during a *constant volume* exhaust stroke.

The processes and salient thermodynamic relationships of a four-stroke compression-ignition engine using the Diesel cycle can be summarized as follows (also see Figure 10.6):

10.3.1 Process 1–2: Adiabatic (and Isentropic) Compression

From Table 2.1, and using Eq. (10.5):

$$\frac{T_2}{T_1} = \left(\frac{v_1}{v_2}\right)^{\gamma-1} = r^{\gamma-1} \tag{10.34}$$

$$\frac{p_2}{p_1} = \left(\frac{v_1}{v_2}\right)^{\gamma} = r^{\gamma} \tag{10.35}$$

The engine cylinder volume can be treated as a closed system, which means that the first law of thermodynamics for a closed system (Table 2.1) can be used to calculate the work required to compress the air:

$$w_{12} = -\Delta u_{12} \equiv -c_v \Delta T_{12} \tag{10.36}$$

10.3.2 Process 2–3: Heat Addition at Constant Pressure

From Table 2.1:

$$p_3 = p_2 \tag{10.37}$$

$$\frac{T_3}{T_2} = \left(\frac{v_3}{v_2}\right) = r_c \tag{10.38}$$

r_c is called the *cut-off ratio*, with

$$r_c = \frac{v_3}{v_2} \tag{10.39}$$

The engine cylinder volume can be treated as a closed system, which means that the first law of thermodynamics for a closed system (Table 2.1) can be used to calculate the work done and heat transfer during the heat-addition process:

$$q_{23} = \Delta h_{23} \equiv c_p \Delta T_{23} \tag{10.40}$$

$$w_{23} = q_{23} - \Delta u_{23} = q_{23} - c_v \Delta T_{23} \tag{10.41}$$

10.3.3 Process 3–4: Adiabatic (and Isentropic) Expansion

From Table 2.1, and using Eq. (10.5):

$$\frac{T_4}{T_3} = \left(\frac{v_3}{v_4}\right)^{\gamma-1} = \left(\frac{v_3 \, v_2}{v_2 \, v_4}\right)^{\gamma-1} = \left(\frac{r_c}{r}\right)^{\gamma-1} \tag{10.42}$$

$$\frac{p_4}{p_3} = \left(\frac{v_3}{v_4}\right)^{\gamma} = \left(\frac{r_c}{r}\right)^{\gamma} \tag{10.43}$$

The engine cylinder volume can be treated as a closed system, which means that the first law of thermodynamics for a closed system (Table 2.1) can be used to calculate the work done:

$$w_{34} = -\Delta u_{34} \equiv -c_v \, \Delta T_{34} \tag{10.44}$$

10.3.4 Process 4–1: Heat Rejection at Constant Volume

From Table 2.1:

$$\left(\frac{p}{T}\right)_4 = \left(\frac{p}{T}\right)_1 \tag{10.45}$$

The engine cylinder volume can be treated as a closed system, which means that the first law of thermodynamics for a closed system (Table 2.1) can be used to calculate the heat discarded:

$$q_{41} = \Delta u_{41} \equiv c_v \, \Delta T_{41} \tag{10.46}$$

The thermal efficiency of a Diesel cycle can be calculated as follows:

$$\eta_{th} = \frac{w_{net}}{q_{in}} \tag{10.47}$$

For a cycle, $\Delta u_{cycle} = 0$, which means that the first law for a cycle becomes $w_{net} = q_{net}$. Equation (10.31) can be simplified accordingly:

$$\eta_{th} = \frac{q_{net}}{q_{in}} = \frac{q_{in} - q_{out}}{q_{in}} \equiv 1 - \frac{|q_{41}|}{q_{23}} = 1 - \frac{|c_v(T_1 - T_4)|}{c_p(T_3 - T_2)} = 1 - \frac{T_1(T_4/T_1 - 1)}{\gamma T_2(T_3/T_2 - 1)}$$

$$\eta_{th} = 1 - \frac{1}{r^{\gamma-1}}\left[\frac{r_c^{\gamma} - 1}{\gamma(r_c - 1)}\right] \tag{10.48}$$

As thermal efficiencies are defined to be positive, we work with the absolute value of heat rejected, as indicated in Eq. (10.48).

When determining specific heat capacities (c_p, c_v) and their ratios (γ), we will work with average values. With actual diesel engines, the average cycle temperature ($T_1 + T_2 + T_3 + T_4$)/4 is close to 900 K, just as in the case of spark-ignition engines. The specific heat capacities stated in Eq. (10.33) can therefore also be used in calculations involving the air-diesel mixture.

No example problem will be provided for a diesel engine as it follows the same problem-solving procedure as in the case of spark-ignition engines (illustrated in Example 10.2).

Table 10.2 summarizes the salient operating characteristics of modern diesel engines used in passenger cars and Box 10.2 summarizes a range of alternative compression-ignition cycles.

Table 10.2 Reference values for important energy parameters of modern diesel engines for passenger cars.

Nominal engine speed (N), (rpm)	3500–4500
Compression ratio (r), (−)	19–24
Maximum pressure (MPa)	7–11
Mean effective pressure (mep), (MPa)	2–3
Power output per liter (kW/L)	20–35
Weight-to-power ratio (kg/kW)	4–2
Specific fuel consumption (g/kWh)	300–240
Thermal efficiency (η_{th}), (−)	0.55–0.65

The engine is turbo-charged. Mechanical and volumetric efficiencies are assumed to be 100%.
Source: Bosch (2022) / Robert Bosch GmbH.

Box 10.2   Alternative Compression-Ignition Cycles

The Miller cycle is a common alternative cycle for compression-ignition engines (Figure 10.7). The Miller cycle may, however, also be used with dual-fuel and natural gas. The Miller cycle represents a popular method to decrease NO_x emissions when burning diesel (Gonca et al. 2015; Wang et al. 2021), especially in large power plants such as those for diesel-electric locomotives such as GE's PowerHaul series of locomotives.

The Miller cycle, like the Atkinson cycle (see Box 10.1), uses variable valve timing and overexpansion that bring about a lower exhaust temperature. The lower exhaust temperature facilitates lower thermal-NO_x emissions and increases thermal efficiency but reduces power density (Miller 1954; Gonca et al. 2015). As can be seen in Figure 10.7, the Miller cycle for diesel fuel has two

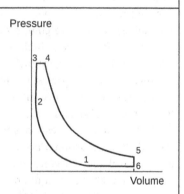

Figure 10.7   The Miller cycle used with diesel fuel, using late inlet-valve closing.

heat-addition processes: one at constant volume (Process 2–3) and one at constant pressure (Process 3–4). The Miller cycle also discards heat during two processes: one at constant volume (Process 5–6) and one at constant pressure (Process 6–1).

10.4   Alternative Fuel-Injection Technologies

Conventional spark-ignition engine combustion entails the injection of homogeneous or near-homogeneous ("stratified charge") air–fuel mixtures. Conventional compressed-ignition engine combustion entails the single or multiple injection of diesel near the TDC position. However, some air–fuel control technologies cannot be classified as belonging to either the spark-ignition (gasoline) or the compressed-ignition (diesel) fuel-injection processes. The following operating strategies comprise aspects common to both gasoline and diesel combustion. Their use results in high thermal efficiencies. Their use also results in low-temperature combustion and lower NO_x emissions.

Homogeneous charge compression ignition (HCCI) comprises the injection of an air–fuel mixture at least 40° of the crank angle before TDC. This allows for the homogenization of the fuel-air mixture and the use of a lean mixture. The fuel auto-ignites due to the high compression ratios ($r = 13 - 16$). The resulting combustion temperature is lower than in the case of conventional compression ignition. This, in turn, produces significantly lower levels of NO_x and particulate matter (PM) emissions. At low engine loads, the fuel-air mixture can be very lean. As the load increases, the fuel-air mixture changes to be closer to stoichiometric. HCCI combustion technology has evolved into PCCI, RCCI, PREMIER, and SACI technologies, discussed next.

For diesel engines, *premixed charge compression ignition* (PCCI) is a combustion technique that simultaneously reduces NO_x and PM emissions. PCCI combustion is achieved by partially premixing fuel and air, and then injecting the remaining portion of fuel close to TDC. PCCI combustion typically requires exhaust gas recirculation (EGR), advancing the fuel injection timing (which prolongs the duration of combustion and reduces the combustion temperature), increased fuel injection pressure, and multiple-staged injections of fuel during the compression stroke. These modifications lead to a more homogeneous combustion flame compared to the stratified flame that occurs along the fuel spray during conventional compression ignition (Parks et al. 2010; Reitz and Duraisamy 2015).

During *reactivity controlled compression ignition* (RCCI), well-mixed low-reactivity fuel (such as gasoline) and air are compressed, but to below the point of auto-ignition. Later, still during the compression process, high-reactivity fuel (such as diesel) is injected to form a local mixture of low- and high-reactivity fuel. The fuel charge then compression-ignites near TDC. The RCCI staged-combustion process requires two different types of fuel. The two-fuel (or dual-fuel) RCCI process can produce ultra-low NO_x emissions due to lower combustion temperatures. The RCCI process also emits less soot than a conventional diesel engine (Reitz and Duraisamy 2015). The RCCI process is similar to the *premixed mixture ignition in the end-gas region* (PREMIER) combustion process (Azimov et al. 2011). As RCCI combustion uses two fuels, the mixture can be changed according to the engine load. For instance, at low engine loads, a mixture of 20% gasoline/80% diesel could be used. At high engine loads, a mixture of 85% gasoline/15% diesel could be used (Kirkpatrick 2020).

During *spark-assisted compression ignition* (SACI), the unwanted early ignition of fuel (which results in the damaging "knocking" effect) is avoided by injecting a small amount of additional fuel near the spark plug. This generates a propagating flame and pressure front that induces the compression ignition of the surrounding homogeneous lean mixture. As a result, the combustion temperature is lower than with the traditional combustion of the fuel (diesel or gasoline). The NO_x and PM emissions are also lower (Manofsky et al. 2011).

Instead of using conventional electrical spark plugs, highly accurate ignition can be introduced with laser, microwave, or nano-second pulsed discharge ignition systems. Such ignition systems ensure optimal (low) combustion temperatures and low emissions of NO_x and PM due to their excellent controllability. These alternative ignition systems also work well with lean fuel-air mixtures, and are not prone to fouling or corrosion like their electrical spark plug counterpart.

10.5 Supercharging and Turbo-Charging

From Eqs. (10.7) to (10.20), it follows that an engine's maximum power $(\dot{W})$ is limited by the fuel flow rate $(\dot{m}_{fuel})$. The amount of fuel that can be burned is, in turn, limited by the amount of air $(\dot{m}_{air})$ available in each cycle:

$$\dot{W} = \eta_{th} \dot{Q}_{in} = \eta_{th} \dot{m}_{fuel} q_c = \eta_{th} \frac{\dot{m}_{air}}{AFR} q_c \tag{10.49}$$

If the inducted air prior to entry into the cylinder is compressed to a pressure (p_i) to provide a higher density (ρ_{air}) than at ambient conditions, the maximum power will be increased (Hiereth and Prenninger 2007):

$$\dot{m}_{air} = \rho_{air}\dot{V}_{air} = \left(\frac{p_i}{RT_i}\right)V\frac{(N/60)}{2} \tag{10.50}$$

$$\therefore \dot{W} = \left(\frac{p_i}{RT_i}\right)V\frac{N}{2}\eta_{th}\frac{q_c}{AFR} \tag{10.51}$$

The primary purpose of technologies such as supercharging and turbo-charging is to increase the pressure and density of intake air, so as to increase power output (Heywood 1988). These technologies are used with both spark-ignition and compression-ignition engines.

With *supercharging*, a blower or compressor that is mechanically driven by the engine provides the compressed air. However, a *turbo-charger* comprises a compressor and turbine in a single housing, where the compressor boosts the density of inlet air. The turbo-charger's turbine is driven by exhaust gases from the engine. A turbo-charger will have higher efficiency than a supercharger, as it does not take power from the engine (as a supercharger does), although the engine needs to absorb the load from the exhaust back-pressure.

Turbo-charging results in a 2- to 2.5-fold increase in power and mean effective pressure (hence also in torque). The highly turbo-charged engine, although it achieves a higher power output than the supercharged engine, does so in a "peaky" manner, achieving maximum output at 6000 rpm. The supercharged engine has a "flatter" power curve over the speed range.

The most advanced turbo-charger technology employs electrically assisted exhaust-gas turbo-charging. The electric exhaust-gas turbo-charger balances the conflicting objectives of a small, fast-responding turbo-charger, which achieves comparatively low peak power, and a large turbo-charger with high peak power, which responds with a significant delay (Garrett 2022). This new turbo-charger typically features a 400-V electric motor, which is integrated directly on the shaft of the exhaust-gas turbo-charger, between the turbine wheel on the exhaust gas side and the compressor wheel on the fresh air side. Electronically controlled, this electric motor directly drives the shaft of the exhaust-gas turbo-charger and thus accelerates the compressor wheel before the exhaust gas flow takes over the drive. The electrification of the turbo-charger significantly improves immediate response from idle speed and across the entire engine speed range (AMG 2021).

10.6 Outlook

A sustainable mobility future will require a diverse portfolio to ensure the right technologies for the right applications. It will presumably span internal combustion engines, fuel cells, electric vehicles, and hybrid-driven propulsion systems. Successful power plants must not only be market-competitive in terms of cost, user requirements, lifecycle emissions, and lifecycle efficiency; they must also help ensure domestic energy security and consider societal impacts related to manufacturing, and the acquisition and recycling of critical materials. In this regard, internal combustion engines and their supporting infrastructure are well established, and innovations associated with technology developments continue to improve the overall efficiency and emissions signature of combustion-based technologies (Reitz et al. 2020).

This suggests that internal combustion engines will be around for many more years, until they are replaced by electric vehicle technology. In the meantime, *low-carbon fuels* are being developed to keep the emissions footprint of internal combustion engines low. Additionally, technologies such as electric-turbo-charging and advanced variable valve timing help ensure high engine performance with the lowest emissions ever recorded.

Even as important advances are made in internal combustion technology, there is widespread agreement that we must eliminate the use of carbon-based fuels quickly. Until we do so, it is likely that fully *flexible engines* with hybridized solutions will be a large part of sought-after efficiency improvements, as well as emission reductions.

It should also be acknowledged that consumer preference is not decided by politicians or car makers. People select their choice of vehicle power train based on numerous factors, especially cost. Further, policy unilaterally favoring one technology solution may be flawed (Reitz et al. 2020). It might therefore be prudent to employ the best available technology, which might feature advanced internal combustion engines burning biofuel, rather than the immediate widespread introduction of electric vehicles. The widespread use of electric vehicles has its own sustainability challenges, including finding adequate materials for the manufacture of electric motors and batteries.

Problems

10.1 A six-cylinder gasoline engine uses iso-octane (C_8H_{18}) as fuel. The air–fuel ratio is lean (17), and the engine operates at sea level. The engine operates on the Otto cycle and has a compression ratio of 10. At the beginning of the compression process, the working fluid is at 100 kPa and at 300 K (27 °C). All products of combustion leave the engine in gaseous form. The cylinder bore is 90 mm and the piston stroke is 110 mm. The crankshaft rotates at 4000 rpm. If the engine's thermal efficiency is 36%, calculate (a) the engine's displacement in liters, (b) the power output, (c) the specific fuel consumption, and (c) the mean effective pressure.

[Answers: (a) 4.2 L; (b) 317 kW; (c) 2.3 MPa; (d) 225 g/kWh]

10.2 An ideal Otto cycle engine has a compression ratio of 8. At the beginning of the compression process, the working fluid is at 100 kPa and at 300 K (27 °C), and 800 kJ per kg of heat is supplied to the working fluid during the constant-volume heat-addition process. (Such a cycle is called an air-standard Otto cycle and the heat provided is equivalent to that which would have been obtained by the combustion of a gasoline-air mixture.) Using the specific heat values for air at a typical average cycle temperature of 900 K, determine (a) the temperature and the pressure of the air at the end of each process, (b) the net work output per cycle (kJ/kg), and (c) the thermal efficiency of the engine cycle.

[Answers: (a) $T_1 = 300$ K, $p_1 = 100$ kPa; $T_2 = 613$ K, $p_2 = 1.6$ MPa; $T_3 = 1572$ K, $p_3 = 4.2$ MPa; $T_4 = 769$ K, $p_4 = 256$ kPa; (b) $w_{net} = 391$ kJ/kg; (c) 51%]

10.3 An ideal Diesel cycle engine has a compression ratio of 18 and a cut-off ratio of 2. At the beginning of the compression process, the working fluid (which may be assumed to be air) is at 100 kPa and at 300 K (27 °C). Using the specific heat values for air at a typical average cycle temperature of 900 K, determine (a) the temperature and the pressure of the air at the end of each process, (b) the net work output per cycle (J/kg), and (c) the thermal efficiency.

[Answers: (a) $T_1 = 300$ K, $p_1 = 100$ kPa; $T_2 = 811$ K, $p_2 = 4.87$ MPa; $T_3 = 1622$ K, $p_3 = 4.87$ MPa; $T_4 = 762$ K, $p_4 = 254$ kPa; (b) $w_{net} = 524$ kJ/kg; (c) 58%]

10.4 A city bus is equipped with a compression-ignition engine with the following performance characteristics:
- Engine displacement: 12.9 L
- Six cylinders, in-line placement; four valves per cylinder; turbo-charged with charge cooling
- Fuel: Diesel ($C_{12}H_{23}$) with a density of 0.835 kg/L and a heating value of 42.5 MJ/kg
- Compression ratio: 18.5
- Cut-off ratio: 2.5
- The engine's inlet conditions are 300 K and 1 bar

Calculate (a) the heat added (in kJ/kg fuel) during the constant-pressure process, (b) the temperatures and pressures at the four state points, (c) the thermal efficiency of the engine, and (d) the entropy generation and irreversibility during the four processes. The average cycle temperature is 900 K.

[Answers: (a) 2179.5 kJ/kg fuel; (b) $T_1 = 300$ K, $p_1 = 100$ kPa; $T_2 = 818.5$ K, $p_2 = 5$ MPa; $T_3 = 2046.3$ K, $p_3 = 5$ MPa; $T_4 = 1027.9$ K, $p_4 = 339.4$ kPa; (c) 56%; (d) 456.9 kJ/kg]

10.5 The city bus from Problem 10.4 performs as follows during a test drive in the city:
- Maximum speed: 56.2 km/h
- Average speed: 25.7 km/h
- The detailed test data taken during the 1656-s test-drive is shown in Figure 10.8:

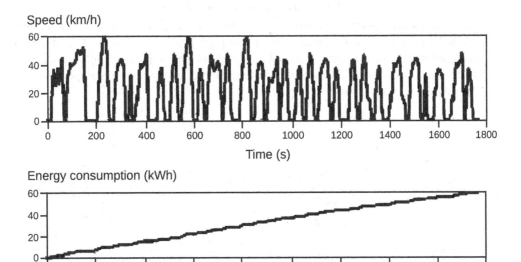

Figure 10.8 Speed and cumulative energy consumption of a city bus driving a typical urban route. *Source:* Adapted from Kivekäs et al. (2018).

The bus drives this route for 4 h a day, 5 days per week, 50 weeks a year. Using the data from Figure 10.8, (a) estimate the *average* energy consumption per distance traveled (kWh/km), (b) the fuel consumption of the bus in liters per 100 km, and (c) the annual fuel cost if the unit cost of diesel is $5 per gallon. (d) If the fuel is burned stoichiometrically in air, calculate the mass of CO_2 emitted per year.

[Answers: (a) 5.08 kWh/km; (b) 51.5 L/100 km (or 4.6 mi per gallon); (c) approximately $22,500 per year; (d) 35 tons of CO_2 per year]

10.6 It is difficult to effectively convey technical contents to a non-technical audience. It is even more challenging to convey such content to a child. Taylor Tucker (a mechanical engineer) and Nicole Dowling (an illustrator) took on the challenge and produced a children's book entitled *Jenny saves a convertible*. The book, targeting third-graders, uses colorful diagrams and a hint of magical realism to teach children about automotive mechanics. Figure 10.9 is an excerpt from their book, which explains the four-stroke cycle of an internal combustion engine in a clear, unambiguous, and fun manner.

Inspired by this children's book, you are challenged to produce a one-page infographic about modern four-stroke cycles. The infographic should, however, be suitable for reading by *high school children of grade levels 9–12,* who might *not* have solid foundations in science or mathematics. The infographic should feature simple, high-impact diagrams and text that explain how gasoline is burned in these types of engines, and how much CO_2 and other greenhouse gases are emitted when 1 L of gasoline is burned in air. The infographic should provide simple calculations that explain how many tons of CO_2 (and other greenhouse gases) a typical passenger car in the United States emits per year, and why that may be a problem. The following website may be helpful: https://www.epa.gov/greenvehicles/greenhouse-gas-emissions-typical-passenger-vehicle. The infographic should preferably be interactive (and online), perhaps by providing a brief quiz at the end of the infographic. The correct answers should appear as people complete the online quiz, along with brief explanations of the answers. The infographic must contain the proper citation of information sources.

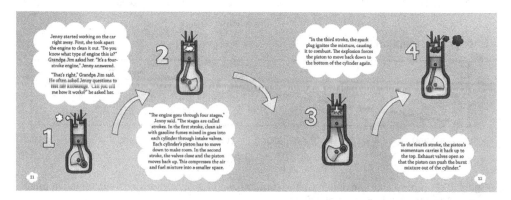

Figure 10.9 Excerpt from the children's book *Jenny saves a convertible*, which explains the rudiments of a four-stroke engine cycle. *Source:* Image reproduced from Tucker and Dowling (2021), with permission from the authors.

References

AMG (2021). Mercedes-AMG defines the future of driving performance. AMG Stuttgart. https://group-media.mercedes-benz.com/marsMediaSite/en/instance/ko.xhtml?oid=49446384&ls=L2VuL2luc3
RhbmNlL2tvLnhodG1sP29pZD05MjY3NDc1JnJlbElkPTYwODI5JmZyb21PaWQ9OTI2NzQ3NSZy
ZXN1bHRJbmZvVHlwZUlkPTQwNjI2JnZpZXdUeXBlPWxpc3Qmc29ydERlZmluaXRpb249
UFVCTElTSEVEX0FULTImdGh1bWJJTY2FsZUluZGV4PTEmcm93Q291bn
RzSW5kZXg9NSZmcm9tSW5mb1R5cGVJZD00MDYyOA!!&rs=0#prevId=49447109
Atkinson, J. (1886). Gas engine. US Patent 336,505. https://patents.google.com/patent/US336505A/en
Azimov, U., Tomita, E., Kawahara, N. and Harada, Y. (2011). Premixed mixture ignition in the end-gas region (PREMIER) combustion in a natural gas dual-fuel engine: operating range and exhaust emissions. *International Journal of Engine Research* 12 (5): 484–497.
Bosch (2022). *Bosch Automotive Handbook*. Stuttgart: Robert Bosch.
Diesel, R. (1895). Method and apparatus for converting heat into work. US Patent 542,846. https://patents.google.com/patent/US542846A/en
Garrett (2022). Whitepaper – electric turbo – a key technology for EU7. Garrett Motion Inc. https://www.garrettmotion.com/knowledge-center-category/oem/whitepaper-electric-turbo-a-key-technology-for-eu7/
Gonca, G., Sahin, B., Ust, Y., Parlak, A. et al. (2015). Comparison of steam injected diesel engine and Miller cycled diesel engine by using two zone combustion model. *Journal of the Energy Institute* 88 (1): 43–52.
Heywood, J.B. (1988). *Internal Combustion Engine Fundamentals*. New York, NY: McGraw-Hill.
Hiereth, H. and Prenninger, P. (2007). *Charging the Internal Combustion Engine*. Springer Science & Business Media.
Kirkpatrick, A.T. (2020). *Internal Combustion Engines: Applied Thermosciences*. New York, NY: Wiley.
Kivekäs, K., Lajunen, A., Vepsäläinen, J. and Tammi, K. (2018). City bus powertrain comparison: driving cycle variation and passenger load sensitivity analysis. *Energies* 11 (7): 1755.
Li, Y., Wang, S., Duan, X., Liu, S. et al. (2021). Multi-objective energy management for Atkinson cycle engine and series hybrid electric vehicle based on evolutionary NSGA-II algorithm using digital twins. *Energy Conversion and Management* 230: 113788.
Manofsky, L., Vavra, J., Assanis, D. and Babajimopoulos, A. (2011). Bridging the gap between HCCI and SI: spark-assisted compression ignition. *SAE Technical Paper* 2: 011–01.
Miller, R. (1954). High-pressure supercharging system. *US Patent* 2,670,595. https://www.google.com/patents/US2670595.
Otto, N. (1877). Improvement in gas-motor engines. *US Patent* 194,047. https://patents.google.com/patent/US194047
Parks II, J.E., Prikhodko, V., Storey, J.M., Barone, T.L. et al. (2010). Emissions from premixed charge compression ignition (PCCI) combustion and effect on emission control devices. *Catalysis Today* 151 (3–4): 278–284.
Reitz, R.D. and Duraisamy, G. (2015). Review of high efficiency and clean reactivity controlled compression ignition (RCCI) combustion in internal combustion engines. *Progress in Energy and Combustion Science* 46: 12–71.
Reitz, R.D., Ogawa, H., Payri, R., Fansler, T. et al. (2020). IJER editorial: The future of the internal combustion engine. *International Journal of Engine Research* 21 (1): 3–10.

Tucker, T. and Dowling, N. (2021). *Jenny Saves a Convertible*. Chicago: Independently Published.

Wang, P., Hu, Z., Shi, L., Tang, X. et al. (2021). Experimental investigation of the effects of Miller timing on performance, energy and exergy characteristics of two-stage turbocharged marine diesel engine. *Fuel* 292: 120252.

Zhao, J. (2017). Research and application of over-expansion cycle (Atkinson and Miller) engines–a review. *Applied Energy* 185: 300–319.

11

External Combustion Gas Engines

With external combustion gas engines, the fuel is combusted externally of the prime mover (which may be a turbine wheel or a piston). Two important types of external combustion engines are *gas turbines* (operating on the Brayton cycle) and *Stirling engines*. The working fluid in gas turbines and Stirling engines could be air or several other gases (such as helium). As the working fluid remains in the gaseous phase, these engines are often also called *gas engines*.

Gas turbines are used in ground-based (or *stationary*) power plants. They are however ubiquitous in *mobile* power plants. Gas turbines are employed as aircraft "jet engines" (i.e., turbo-jets or turbo-props), as ship propulsion units, railroad engines, and even in specialized land-based vehicles such as battle tanks.

Although Stirling engines can operate on externally combusted fuels, uniquely, they can be operated using solar concentrators to produce high-quality heat, rather than burning fuel. This makes Stirling engines good companions for renewable energy solutions, although they are challenging to manufacture.

11.1 Gas Turbines

With a *gas turbine*, air enters the compressor, which increases the pressure. Fuel is injected into the high-pressure air inside the combustor. The combustion process happens at constant pressure and raises the temperature. The resulting high-temperature gas expands in the turbine, thus performing work, while dropping in pressure and temperature. The turbine drives the compressor and the external load (e.g., an electrical generator or a propeller) through a common shaft. Although the turbine is only part of the whole engine assembly, the complete assembly is commonly referred to as a gas turbine.

11.1.1 Gas Turbine Classification

Gas turbines may be classified according to the physical arrangement of their component parts. Categories include single-shaft, two-shaft, regenerative (where a heat exchanger is used to recover heat from the exhaust gases and to pre-heat intake air to the combustor), intercooled (where heat is removed between compressors), and reheat (where heat is added between turbines). These different configurations are illustrated in Figure 11.1.

Energy Systems: A Project-Based Approach to Sustainability Thinking for Energy Conversion Systems, First Edition. Leon Liebenberg.
© 2024 John Wiley & Sons, Inc. Published 2024 by John Wiley & Sons, Inc.
Companion website: www.wiley.com/go/liebenberg/energy_systems

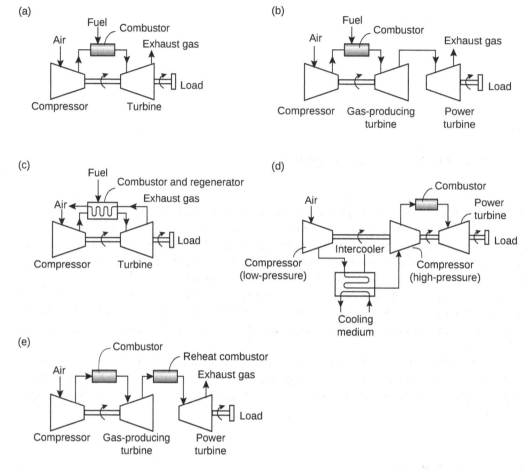

Figure 11.1 Gas turbine configurations: (a) single-shaft, (b) two-shaft, (c) regenerative (in a closed cycle), (d) intercooled, and (e) reheat.

When using a *single-shaft* gas turbine, a large overload can cause the engine to shut down. As the turbine spins the compressor, any decrease in compressor speed due to the inability to cope with an external load will result in lower pressures, and consequently also lower temperatures and lower turbine power, which will result in decreasing compressor speed, among other characteristics, until the engine shuts down. Careful matching of the load characteristics of the driven equipment (such as a generator or propeller) is essential when operating single-shaft gas turbines (Rogers et al. 2017).

The load-matching issue can be circumvented by employing a two-shaft gas turbine (Figure 11.1b and e). With a *two-shaft* gas turbine, the first turbine is a high-pressure unit (commonly called the "gas producer") that spins the compressor and produces hot gas (at 1000–1500 °C). The high-pressure, high-temperature gas is directed to the second turbine stage (commonly called a "power turbine"), which results in that stage producing work and absorbing the load. Thus, the load is effectively disconnected from the gas-producing turbine stage, which ensures that the engine will not shut down under high-load conditions. Two-shaft gas turbines also have a much wider operating speed range than their single-shaft counterparts (Rogers et al. 2017). Two-shaft gas turbines are commonly employed in some aircraft and ship engines.

Several types of compressors may be used, including positive displacement, centrifugal, and axial flow types. Most industrial gas turbines use axial flow compressors. The turbine could have impulse or reaction blading. Impulse blading operates on the principle of rate-of-change of momentum of the working fluid, while reaction blading operates on the principle of aerodynamic lift (Japikse and Baines 1994).

Gas turbines could operate in an open cycle or in a closed cycle. Most gas turbines (stationary and mobile applications) are of the *open-cycle* type, where fresh air is inducted by the compressor and the burned fuel–air products are directed onto the turbine stage, and then exhausted to the atmosphere. Such open-cycle gas turbines need to burn clean fuels, though, as any impurities or particles in the fuel may damage the turbine blades (Rogers et al. 2017).

In the event of dirty fuel being burned, a closed cycle may be used. With such a set-up, the dirty fuel is burned, and exhaust products are expelled into the atmosphere. The heat from the combustor is passed on to a working fluid (like air or helium) via a heat exchanger, which then powers the turbine stage.

11.1.2 Gas Turbine Combustion

A wide variety of fuels may be used with gas turbines, especially if they operate in closed-cycle mode. The major requirements for gas turbine fuels are that the fuel does not produce ash that could adhere to the turbine blades and reduce their effective operation; the fuel does not produce particulates that could erode the spinning turbine blades; and the fuel does not contain uninhibited vanadium that could corrode some types of older gas turbine blades (Patel et al. 2017).

A gas turbine operates with a large amount of excess air. The (usually) annular-shaped combustor is designed with an inner portion that burns only part (around 20%) of the air to achieve high combustion temperatures and good thermal efficiency (Figure 11.2). The products of combustion are effectively mixed with the remainder (around 80%) of air to minimize temperature stratification and reduce the maximum combustion temperature, which also reduces thermal nitric oxide (NO_x) emissions (Lefebvre and Ballal 2010). The air around the annulus is also used to cool down the combustion chamber, which could be as high as 1500 °C in modern gas turbines.

A gas turbine may be started by bringing it up to starting speed using an electrical motor or high-pressure air or gas. At the appropriate speed, fuel is injected into the pressurized air and the fuel–air mixture is ignited by a high-voltage spark plug. Continuous burning is maintained (without the need for further sparking) if there is adequate flow of pressurized air and fuel in the correct ratio of air to fuel (Japikse and Baines 1994). A gas turbine operates with *lean fuel–air mixtures*, with air–fuel ratios of 25–50, compared to 15 for gasoline combustion and 20–25 for diesel combustion.

11.1.3 Gas Turbine Engine Performance Parameters

The net specific work output of a stationary gas turbine power plant (e.g., Figure 11.3) is equal to the specific work output of the turbine shaft less the specific work required to drive the compressor less the specific work required to drive any auxiliary equipment (such as an electrical generator):

$$w_{net} = w_{turbine}(1 + FAR) - w_{compressor} - w_{auxiliary} \qquad [\text{kJ/kg}] \qquad (11.1)$$

where *FAR* is the fuel–air ratio, w_{net} is the net output of the power plant in kJ/kg of air, $w_{turbine}$ is the turbine shaft work in kJ/kg of gas, $w_{compressor}$ is the compressor shaft work in kJ/kg of air, and $w_{auxiliary}$ is the work required for driving auxiliary equipment (measured in kJ/kg of air).

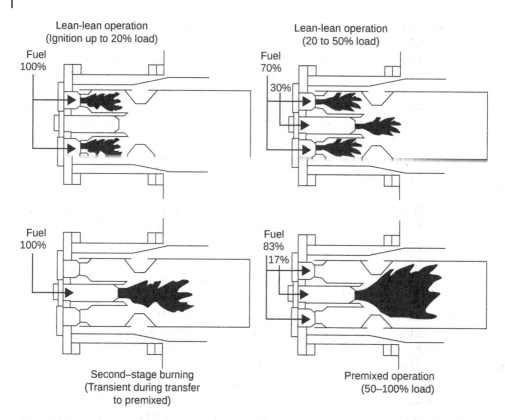

Lean-lean operation
(Ignition up to 20% load)

Fuel
100%

Lean-lean operation
(20 to 50% load)

Fuel
70%

30%

Fuel
100%

Fuel
83%

17%

Second–stage burning
(Transient during transfer
to premixed)

Premixed operation
(50–100% load)

Figure 11.2 A modern gas turbine combustor, featuring a central zone where combustion occurs and an annular section around which air is circulated to cool down the engine and reduce the combustion temperature, thus also facilitating lower NO_x emissions (Davis and Black 1995). *Source:* Reproduced with permission from GE Vernova.

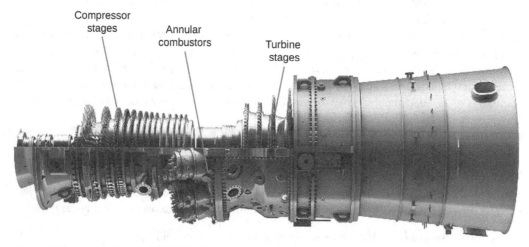

Compressor
stages

Annular
combustors

Turbine
stages

Figure 11.3 Sectional view of GE's 7HA.03 industrial gas turbine. This gas turbine produces 430 MW, has a net heat rate of 8.3 MJ/kWh, a thermal efficiency of 43%, and a start-up time of only 20 min. The gas turbine features 12 can-type, dry low-NO_x (DLN) combustors with axial fuel staging, which facilitate effective combustion and low thermal-NO_x emissions. *Source:* Reproduced with permission from GE.

The *specific fuel consumption (sfc)* of a gas turbine engine is the ratio of the fuel consumption rate $(\dot{m}_{fuel})$ to power $(\dot{W})$, and is often specified in kg of fuel per kWh produced:

$$sfc = \frac{\dot{m}_{fuel}}{\dot{W}} \qquad [\text{kg/kWh}] \qquad (11.2)$$

The thermal efficiency of a gas turbine engine is the ratio of power output to input rate of heat. The power (heat rate) input is equal to the product of the fuel flow rate and the specific heat of combustion (q_c):

$$\eta_{th} = \frac{\dot{W}}{\dot{Q}} = \frac{\dot{W}}{\dot{m}_{fuel} \, q_c} \qquad (11.3)$$

It is customary to use the lower heating value of fuels for the specific heat of combustion (q_c), as gas turbines cannot utilize the latent heat of water vapor in the products of combustion. The lower heating value (LHV) can be read from Table 7.5.

As with an internal combustion engine, the heat rate of a gas turbine's combustor can be determined using air–fuel relationships:

$$\dot{Q} = \dot{m}_{fuel} \, q_c \qquad [\text{kW}] \qquad (11.4)$$

with

$$\dot{m}_{fuel} = \dot{m} - \dot{m}_{air} \qquad (11.5)$$

The air–fuel ratio is:

$$AFR = \frac{\dot{m}_{air}}{\dot{m}_{fuel}} \equiv \frac{m_{air}}{m_{fuel}} \qquad (11.6)$$

Therefore:

$$\dot{m}_{fuel} = \frac{\dot{m}}{AFR + 1} \qquad (11.7)$$

and

$$m_{fuel} = \frac{m}{AFR + 1} \qquad (11.8)$$

From our knowledge of an ideal gas (see Section 2.4), we can determine the mass of gas in a fixed volume at a certain absolute temperature and pressure:

$$m = \frac{p \, V}{RT} \qquad (11.9)$$

where R is the gas constant of the air–fuel mixture and m is the mass of air and fuel.

Also:

$$R = c_p - c_v \qquad (11.10)$$

and

$$\gamma = \frac{c_p}{c_v} \qquad (11.11)$$

Substituting Eqs. (11.5) to (11.11) in (11.5), we obtain the following handy relationships for heat produced (kJ) or the heat rate of fuel (kW):

$$Q = \underbrace{\left(\frac{p \, V}{RT}\right)}_{\text{mass}} \left(\frac{q_c}{AFR + 1}\right) \qquad (11.12)$$

or

$$\dot{Q} = \left(\frac{p \, \dot{V}}{RT}\right) \left(\frac{q_c}{AFR + 1}\right) \qquad (11.13)$$

The air–fuel ratios are calculated as explained in Chapters 7–9. Table 11.1 summarizes the salient values of energy parameters for modern gas turbines.

Table 11.1 Reference values for important energy parameters of modern gas turbines for stationary power plants operating in single-shaft mode.

Nominal power output (MW)	1–600
Net heat rate *(LHV)*, (kJ/kWh)	8000–9500
Net efficiency based on *LHV* (%)	38–44
Start-up time (min)	15–25
Fuel	Natural gas, hydrogen (50% H_2), kerosene, diesel, gasoline, light fuel oil
Working fluid	Air, helium, CO_2, flue gas
Turbine entry temperature (°C)	1000–1700
Turbine inlet pressure (MPa)	1–4.5
Turbine exit temperature (°C)	400–850
Combustion	Dry low-NO_x, steam injection, water injection
Nominal rotational speed (rpm)	5000–12,000

Combustion and mechanical efficiencies are assumed to be 100%.

Example 11.1 Characteristics of an Industrial Gas Turbine

Consider an industrial gas turbine that produces 430 MW. The engine has a thermal efficiency of 43% and operates with a lean air–fuel ratio of 63 when using kerosene as fuel. Calculate (a) the gas turbine's specific fuel consumption (in kg/kWh), (b) the mass flow rate of fuel, (c) the mass flow rate of air, and (d) the engine's net heat rate (in kJ/kWh).

Required: We are provided rudimentary data about this engine and must find (a) the *sfc*, (b) $\dot{m}_{fuel}$, (c) $\dot{m}_{air}$, and (d) the heat rate in kJ/kWh.

Solution strategy: We will use Eq. (11.3) for all calculations.

Assumptions: Assume that the engine has no mechanical losses and that the combustion process is ideal. We will neglect the power required to drive auxiliary equipment, as we do not have information about that.

Analysis:

a) Specific fuel consumption

From Eq. (11.3): $\eta_{th} = \dfrac{\dot{W}}{\dot{m}_{fuel}\, q_c} \equiv \dfrac{1}{(sfc)q_c}$

The *LHV* of kerosene can be found from Table 7.5 as $q_c = 43$ MJ/kg or 43,000 kJ/kg.

Therefore, $\quad 0.43 = \dfrac{1}{(sfc)(43{,}000 \text{ kJ/kg})}$

$\therefore sfc = \dfrac{1}{(0.43)(43{,}000 \text{ kJ/kg})} = 54.1 \times 10^{-6}$ kg/kW s $\equiv 0.194$ kg/kW h

b) Mass flow rate of fuel

From Eq. (11.3): $\eta_{th} = \dfrac{\dot{W}}{\dot{m}_{fuel}\, q_c} \equiv \dfrac{1}{(sfc)q_c}$

Therefore, $\quad \eta_{th} = 0.43 = \dfrac{\dot{W}}{\dot{m}_{fuel}\, q_c} = \dfrac{430{,}000 \text{ kJ/s}}{\dot{m}_{fuel}(43{,}000 \text{ kJ/kg})}$

$\therefore \dot{m}_{fuel} = \dfrac{430{,}000 \text{ kJ/s}}{(0.43)(43{,}000 \text{ kJ/kg})} = 23.3 \text{ kg fuel/s}$

c) Mass flow rate of air

From Eq. (11.6): $AFR = \dfrac{\dot{m}_{air}}{\dot{m}_{fuel}}$

$\therefore \dot{m}_{air} = (AFR)(\dot{m}_{fuel}) = (63)(23.3 \text{ kg/s}) = 1468 \text{ kg air/s}$

d) Heat rate in kJ/kWh

Heat rate in kJ/kWh $= (sfc)(q_c) = (0.194 \text{ kg fuel/kWh})(43{,}000 \text{ kJ/kg fuel}) = 8342 \text{ kJ/kWh}$

11.1.4 The Brayton (or Joule) Cycle for Gas Turbines

A gas turbine operates on the Brayton (1872) cycle, sometimes loosely referred to as the constant pressure cycle or the Joule cycle. An ideal Brayton cycle comprises the following processes (see Figure 11.4):

- *Process 1–2*: Air is compressed *adiabatically* (and isentropically). Fuel is injected toward the end of the process.
- *Process 2–3*: The fuel–air mixture auto-ignites at the high pressure of state point 2. The combustion process occurs at *constant pressure* and the gas exits the combustor at a high temperature. In an open-cycle gas turbine, the products of combustion directly impinge on the turbine wheel. In a closed cycle gas turbine, the products of combustion are exhausted to the atmosphere and the heat of combustion is delivered to a heat exchanger, which heats a working fluid (such as air or helium), which is directed to the turbine wheel.
- *Process 3–4*: The high-temperature gas (or working fluid) expands *adiabatically* (and isentropically) in the turbine section, doing work in the process.
- *Process 4–1*: In an open-cycle gas turbine, the gases exhaust to the atmosphere (at *constant pressure*) and fresh air is inducted at state point 1. In a closed-cycle gas turbine, the working fluid discards heat to the environment in a heat exchanger placed between state points 4 and 1. The cool working fluid then enters the compressor intake at state point 1.

The processes and salient thermodynamic relationships of a gas turbine using the Brayton (or Joule) cycle can be summarized as follows (also see Figure 11.4), assuming that the working fluid (or gas) can be treated as an ideal gas and treating the cycle like an open cycle (Saravanamuttoo et al. 2001):

11.1.4.1 Process 1–2: Adiabatic (and Isentropic) Compression
From Table 2.1:

$$\frac{T_2}{T_1} = \left(\frac{p_2}{p_1}\right)^{\frac{\gamma-1}{\gamma}} = r_p^{\frac{\gamma-1}{\gamma}} \tag{11.14}$$

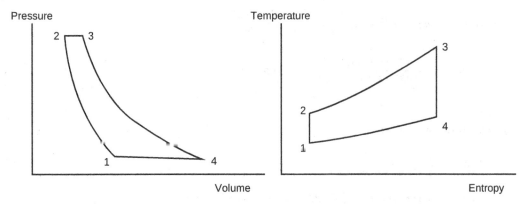

Figure 11.4 The Brayton cycle represented on pressure-volume and temperature-entropy diagrams. Process 1–2: Adiabatic (and isentropic) compression; Process 2–3: Constant pressure heat addition; Process 3–4: Adiabatic (and isentropic) expansion; Process 4–1: Constant-pressure exhaust (or cooling).

With the pressure ratio:
$$r_p = \left(\frac{p_2}{p_1}\right) \tag{11.15}$$

The work input in process 1–2 can be calculated using the first law of thermodynamics for an open system (Eq. (2.8) and Table 2.1):

$$\cancel{q_{12}} - w_{12} = \Delta h_{12} \equiv c_p \Delta T_{12} \tag{11.16}$$

11.1.4.2 Process 2–3: Heat Addition at Constant Pressure
From Table 2.1:

$$p_3 = p_2 \tag{11.17}$$

$$\frac{T_3}{T_2} = \left(\frac{v_3}{v_2}\right) = \left(\frac{v_4}{v_1}\right) = \frac{T_4}{T_1} \tag{11.18}$$

The heat transferred during process 2–3 can be calculated using the first law of thermodynamics for an open system (Eq. (2.8) and Table 2.1):

$$q_{23} - \cancel{w_{23}} = \Delta h_{23} \equiv c_p \Delta T_{23} \tag{11.19}$$

11.1.4.3 Process 3–4: Adiabatic (and isentropic) Expansion
From Table 2.1:

$$\frac{T_4}{T_3} = \left(\frac{p_4}{p_3}\right)^{\frac{\gamma-1}{\gamma}} \tag{11.20}$$

or, alternatively,
$$\frac{p_4}{p_3} = \left(\frac{T_4}{T_3}\right)^{\frac{\gamma}{\gamma-1}} \tag{11.21}$$

The work output in process 3–4 can be calculated using the first law of thermodynamics for an open system (Eq. (2.8) and Table 2.1):

$$\cancel{q_{34}} - w_{34} = \Delta h_{34} \equiv c_p \Delta T_{34} \tag{11.22}$$

11.1.4.4   Process 4–1: Heat Rejection at Constant Pressure

From Table 2.1:

$$p_4 = p_1 \tag{11.23}$$

$$\frac{T_3}{T_2} = \left(\frac{v_3}{v_2}\right) = \left(\frac{v_4}{v_1}\right) = \frac{T_4}{T_1} \tag{11.24}$$

The heat transferred during process 4–1 can be calculated using the first law of thermodynamics for an open system (Eq. (2.8) and Table 2.1):

$$q_{41} - \cancel{w_{41}} = \Delta h_{41} \equiv c_p \, \Delta T_{41} \tag{11.25}$$

The thermal efficiency of a Brayton (or Joule) cycle can be calculated as follows:

$$\eta_{th} = \frac{w_{net}}{q_{in}} \tag{11.26}$$

For an open cycle, $\Delta h_{cycle} = 0$, which means that the first law of thermodynamics for an open cycle becomes $w_{net} = q_{net}$. Equation (11.12) can be simplified accordingly:

$$\eta_{th} = \frac{q_{net}}{q_{in}} = \frac{q_{in} - q_{out}}{q_{in}} \equiv 1 - \frac{|q_{41}|}{q_{23}} = 1 - \frac{|c_p(T_1 - T_4)|}{c_p(T_3 - T_2)} = 1 - \frac{1}{r_p^{(\gamma-1)/\gamma}} \tag{11.27}$$

As thermal efficiencies are defined to be positive, we work with the absolute value of heat rejected, as indicated in Eq. (11.27).

When determining specific heat capacities (c_p, c_v) and their ratios (γ), we will work with average values. This is appropriate as the values of c_p, c_v, and γ vary throughout the cycle due to changes in temperature; c_p increases in value as temperature increases, and γ decreases with temperature; the opposing trends of c_p and γ therefore partly cancel out when calculating enthalpies, thus validating the choice of average values of c_p, c_v, and γ.

For instance, with gas turbines combusting natural gas, the average cycle temperature $(T_1 + T_2 + T_3 + T_4)/4$ is close to 800 K. If the gases of combustion are assumed to act as air, which is an ideal gas, the corresponding values of heat capacities and their ratios are:

$$c_{p,800K} = 1.099 \ \text{kJ/kgK}$$
$$c_{v,800K} = 0.812 \ \text{kJ/kgK}$$
$$\gamma_{800K} = \frac{c_{p,800K}}{c_{v,800K}} = 1.354 \tag{11.28}$$
$$R = c_{p,800K} - c_{v,800K} = 0.287 \ \text{kJ/kgK}$$

Example 11.2   GE T-700 Gas Turbine (Used in the Black Hawk Helicopter)

Consider GE's popular T-700 gas turbine (Figure 11.5) that is used in the Black Hawk helicopter, among many other applications. The engine has two shafts. The gas-producer turbine is connected to the compressor via one shaft, and the helicopter's rotor is connected to the power turbine via a separate shaft. Air enters the compressor at 298 K and 101 kPa. The mass flow rate of gas (combusted air and fuel) is 4.5 kg/s, and the pressure ratio is 15. Air exits the combustion chamber and enters the first turbine at 1200 K. The air then exits and enters a second gas turbine, which is connected to the helicopter rotor via a gearbox.

(a)

(b)

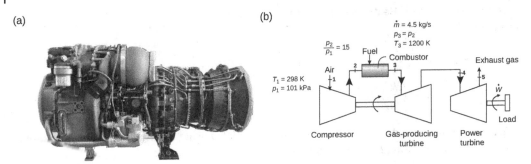

Compressor   Gas-producing   Power
turbine   turbine

Figure 11.5 (a) The GE's T700 gas turbine, which is used on the Black Hawk helicopter, and (b) its circuit diagram. *Source:* Reproduced with permission from General Electric.

Calculate (a) the power output of the engine, (b) the engine's thermal efficiency, (c) the specific fuel consumption if the engine operates with a lean air–fuel ratio of 50, and (d) the cycle's irreversibility.

Required: We must find (a) the output of the power turbine, (b) the thermal efficiency of the whole engine, (c) the specific fuel consumption, and (d) the irreversibility of the cycle.

Solution strategy: We will do a full cycle analysis and calculate the heat and work transferred in each of the processes. Equations (11.2) to (11.8) will be used to calculate the air–fuel ratio. Finally, the irreversibilities, or entropy productions, can be calculated using Equations (3.26) and (3.58).

Assumptions: Assume that the engine has no mechanical or pressure losses, and that the combustion process is complete. We will assume an average cycle temperature of 800 K, which will be used to calculate c_p, c_v, and γ. We will also assume that the compressor and both turbines have isentropic efficiencies of 100%. To calculate the entropy production, we will assume that the temperature of the hot reservoir is 1200 K and that of the cold reservoir is 293 K. The atmospheric pressure will be taken as 101 kPa.

Analysis:

a) Power output of power turbine

At the average cycle temperature of 800 K, $c_{p,800K} = 1.099$ kJ/kgK, $c_{v,800K} = 0.812$ kJ/kgK, and

$$\gamma_{800K} = \frac{c_{p,800K}}{c_{v,800K}} = 1.354.$$

The pressure–volume diagram for the cycle is shown below:

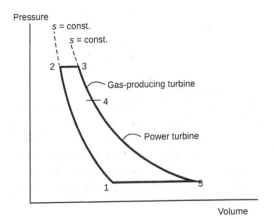

Process 1–2: Adiabatic compression

From Eq. (11.14): $\dfrac{T_2}{T_1} = \left(\dfrac{p_2}{p_1}\right)^{\frac{\gamma-1}{\gamma}} = r_p^{\frac{\gamma-1}{\gamma}}$, with $r_p = 15$

$$T_2 = T_1 r_p^{\frac{\gamma-1}{\gamma}} = (289\text{ K})(15)^{(1.354-1)/1.354} = 586.7\text{ K}$$

From Eq. (11.16): $w_{12} = \Delta h_{12} \equiv c_p\,\Delta T_{12} = (1.099\text{ kJ/kgK})(289 - 586.7)\text{ K} = -327.2\text{ kJ/kg}$

and $\left(\dfrac{p_2}{p_1}\right) = 15$, so, $p_2 = 15(101\text{ kPa}) = 1515\text{ kPa}$

Process 2–3: Combustion at constant pressure

$$p_3 = p_2 = 1515\text{ kPa}$$

$$T_3 = 1200\text{ K (given)}$$

From Eq. (11.19): $q_{23} = \Delta h_{23} \equiv c_p\,\Delta T_{23} = (1.099\text{ kJ/kgK})(1200 - 586.7)\text{ K} = 674\text{ kJ/kg gas}$

Process 3–4: Gas-producer turbine

The gas-producer turbine produces just enough work to drive the compressor: $|w_{12}| = w_{34} = 327.2\text{ kJ/kg}$.

From Eq. (11.22) and using the isentropic efficiency (Eq. (3.39)):

$$-w_{34} = \Delta h_{34} \equiv c_p\,\Delta T_{34}$$

$$\therefore w_{34} = (327.2\text{ kJ/kg}) = -\Delta h_{34} \equiv -c_p\,\Delta T_{34} = (1.099\text{ kJ/kgK})(1200 - T_4)\text{ K}$$

$$T_4 = 902.3\text{ K}$$

From Eq. (11.21): $\dfrac{p_4}{p_3} = \left(\dfrac{T_4}{T_3}\right)^{\frac{\gamma}{\gamma-1}}$

$$\therefore p_4 = p_3\left(\frac{T_4}{T_3}\right)^{\frac{\gamma}{\gamma-1}} = (1515\text{ kPa})\left(\frac{902.3\text{ K}}{1200\text{ K}}\right)^{1.354/(1.354-1)} = 509.1\text{ kPa}$$

Process 4–5: Power turbine

From Eq. (11.20): $\dfrac{T_5}{T_4} = \left(\dfrac{p_5}{p_4}\right)^{\frac{\gamma-1}{\gamma}}$

So, $T_5 = T_4\left(\dfrac{p_5}{p_4}\right)^{\frac{\gamma-1}{\gamma}} = (902.3\text{ K})\left(\dfrac{101\text{ kPa}}{509.1\text{ kPa}}\right)^{(1.354-1)/1.354} = 591.2\text{ K}$

From Eq. (11.22): $\cancel{q_{45}} - w_{45} = \Delta h_{45} \equiv c_p\,\Delta T_{45}$

$$w_{45} = -\Delta h_{45} \equiv -c_p\,\Delta T_{45}$$

$$\therefore w_{45} = (1.099\text{ kJ/kgK})(902.3 - 591.2)\text{ K} = 341.9\text{ kJ/kg gas}$$

Process 5–1: Exhaust to atmosphere

$$q_{51} = \Delta h_{51} \equiv c_p\,\Delta T_{51} = (1.099\text{ kJ/kgK})(289 - 591.2)\text{ K} = -332.1\text{ kJ/kg gas}$$

Process	q (kJ/kg)	w (kJ/kg)	Δh (kJ/kg)
1–2	0	−327.2	327.2
2–3	674	0	674
3–4	0	327.2	−327.2
4–5	0	341.9	−341.9
5–1	−332.1	0	332.1
Σ:	341.9	341.9	0

The power output of the turbine is then:

$$\dot{W}_{45} = \dot{m}_{gas} w_{45} = (4.5\,\text{kg gas/s})(341.9\,\text{kJ/kg gas}) = 1.5\,\text{MW}$$

b) Thermal efficiency

$$\eta_{th} = 1 - \frac{1}{r_p^{(\gamma-1)/\gamma}} = 1 - \frac{1}{15^{(1.354-1)/1.354}} = 0.51$$

The engine is 51% thermally efficient, meaning that 49% of the fuel energy is discarded as waste heat.

c) Specific fuel consumption

From Eqs. (11.2) to (11.8):

$$sfc = \frac{\dot{m}_{fuel}}{\dot{W}} \equiv \frac{1}{(AFR)w_{45}} = \frac{1}{(50\,\text{kg air/kg gas})(341.9\,\text{kJ/kg gas})}\frac{3600\,\text{kJ}}{\text{kWh}} = 0.211\,\text{kg fuel/kWh}$$

d) Irreversibility

We can calculate irreversibilities (or entropy production) from Eq. (3.58), which reduces to the following if we disregard changes in kinetic and potential energies and assume steady-state conditions. The entropy production then becomes:

$$\sigma = \frac{\dot{\sigma}}{\dot{m}} = \sum_j \left(1 - \frac{T_0}{T_j}\right)\frac{Q_j}{T_0} - \frac{W}{T_0} + \frac{1}{T_0}[(h_1 - h_2) - T_0(s_1 - s_2)]$$

For processes 1–2, 3–4, and 4–5, the processes are reversible and adiabatic, hence also isentropic. The work transfer will be identically equal to the change in enthalpies, which means that the entropy production for both these processes will be 0:

$$\sigma_{12} = \sum_j \left(1 - \frac{T_0}{T_j}\right)\frac{Q_j}{T_0} - \frac{W}{T_0} + \frac{1}{T_0}[(h_1 - h_2) - T_0(s_1 - s_2)] = 0$$

$$\sigma_{34} = \sum_j \left(1 - \frac{T_0}{T_j}\right)\frac{Q_j}{T_0} - \frac{W}{T_0} + \frac{1}{T_0}[(h_3 - h_4) - T_0(s_3 - s_4)] = 0$$

$$\sigma_{45} = \sum_j \left(1 - \frac{T_0}{T_j}\right)\frac{Q_j}{T_0} - \frac{W}{T_0} + \frac{1}{T_0}[(h_4 - h_5) - T_0(s_4 - s_5)] = 0$$

For processes 2–3 and 5–1, the entropy production can be determined by evaluating the change in entropy using Eq. (3.26), which reduces to the following considering that these two processes happen without any change in pressure ($p_e = p_i$), so $\ln \dfrac{p_e}{p_i} = 0$:

$$\Delta s = c_{p,avg} \ln \frac{T_e}{T_i} - \cancel{R \ln \frac{p_e}{p_i}}$$

$$\Delta s_{23} = c_{p,avg} \ln \frac{T_3}{T_2} = (1.099 \, \text{kJ/kgK}) \left(\ln \frac{1200 \, \text{K}}{586.7 \, \text{K}} \right) \approx 0.8 \, \text{kJ/kgK}$$

$$\Delta s_{51} = c_{p,avg} \ln \frac{T_1}{T_5} = (1.099 \, \text{kJ/kgK}) \left(\ln \frac{298 \, \text{K}}{591.12 \, \text{K}} \right) \approx -0.8 \, \text{kJ/kgK}$$

The entropy production for processes 2–3 and 5–1 can now be evaluated. Again, the work transfer will be identically equal to the change in enthalpies for both processes so that those terms cancel each other out. For Process 2–3, the high-temperature reservoir at 1200 K receives heat from the combustion chamber, which means that we will work with $-(q_{23})$ according to our sign convention for heat leaving a system:

$$\sigma_{23} = T_0 \left(\frac{-q_{23}}{T_3} + s_3 - s_2 \right) = (293 \, \text{K}) \left[\frac{-674 \, \text{kJ/kg}}{1200 \, \text{K}} + (0.8 \, \text{kJ/kgK}) \right] = 69.8 \, \text{kJ/kg}$$

For Process 5–1, heat is lost to the environment with $q_{51} = -332.1 \, \text{kJ/kg}$

$$\sigma_{51} = T_0 \left(\frac{q_{51}}{T_{low}} + s_1 - s_5 \right) = (293 \, \text{K}) \left[\frac{-(-332.1 \, \text{kJ/kg})}{293 \, \text{K}} + (-0.8 \, \text{kJ/kgK}) \right] = 97.7 \, \text{kJ/kg}$$

The results may be summarized and checked as follows:

Process	q (kJ/kg)	w (kJ/kg)	$s_e - s_i$ (kJ/kgK)	σ (kJ/kg)
1–2	0	−327.2	0	0
2–3	674	0	0.8	69.8
3–4	0	327.2	0	0
4–5	0	341.9	0	0
5–1	−332.1	0	−0.8	97.7
Cycle	341.9	341.9	0	167.5

The three adiabatic processes are reversible hence also isentropic. So, processes 1–2, 3–4, and 4–5 have no entropy production. The heat-addition (2–3) and heat-loss (5–1) processes are irreversible as they transfer heat across finite temperatures. The irreversibility associated with process 2–3 could be reduced by decreasing T_3. The largest irreversibility is in process 5–1 and can be reduced by minimizing the heat lost to the environment by recovering waste heat. This can be done using a regenerator (see Figure 11.1c), which uses the exhaust gas to pre-heat the air entering the combustion chamber.

Reimagine Our Future

Weatherizing a building, such as using improved insulation or sealing air leaks around doors and windows, can reduce a home's energy bill by more than 20%. The US federal government offers financial incentives for homeowners to bring about such weatherization, but the program is not successful. Homeowners surely are interested in saving on their energy bill. Why do you think more people do not take this simple step? Do you think that weatherization would become widespread if the costs of electricity were much higher?

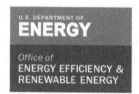

Weatherization Assistance Program

Overview

The U.S. Department of Energy's (DOE's) **Weatherization Assistance Program (WAP)**, within the **Office of Energy Efficiency and Renewable Energy**, reduces energy costs for low-income households by increasing the energy efficiency of their homes, while ensuring health and safety. WAP is part of the Weatherization and Intergovernmental Programs Office and supports DOE's objectives to lower energy bills while expanding cost-effective energy choices for all American communities. The program supports 8,500 jobs and provides weatherization services to approximately 35,000 homes every year using DOE funds. Through weatherization improvements and upgrades, these households save an average of $372 every year *(National Evaluation, expressed in 2022 dollars).*

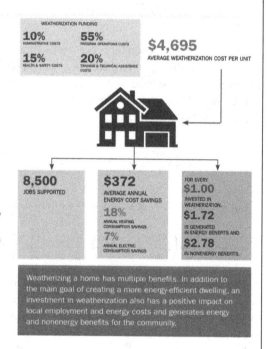

Excerpt from a fact sheet about the cost (and other) benefits of the weatherization of a home. *Source:* US Department of Energy, Office of Energy Efficiency and Renewable Energy. Image used with permission.

11.2   Stirling Engines

The Stirling engine was devised in 1816 by Robert Stirling (1816), who thought it to be a safer alternative than the steam engine. In the Stirling engine, a working fluid (like air or helium) is sealed in a cylinder and alternatively heated and cooled. Two pistons are usually employed with a regenerator (heat exchanger) placed between them. The Stirling engine obtains its heat from a source external to the cylinder, hence being termed an *external* combustion engine. The engine can be powered by concentrated sun rays, as shown in Figure 11.6, or by burning a fuel–air mixture. A Stirling engine runs quietly (as it requires no moving valves). It can be very efficient (with a maximum thermal efficiency equal to the Carnot efficiency) and has durable major components.

Figure 11.6 Two Stirling engines heated by solar concentrators. These types of energy conversion systems have become popular in sunny and low-wind regions. The shown 30 kW Stirling engines have a solar-to-grid quality electricity efficiency of 32%. *Source:* Swedish Stirling AB / CC BY-SA 2.0.

A Stirling engine features a complex set-up of cylinder(s), pistons, and a regenerator. Figure 11.6 shows a type of Stirling engine that employs a *displacer piston,* which divides the cylinder into an expansion space and a compression space. Both spaces are held at different temperatures. Movement of the displacer changes the portion of gas in the two spaces while maintaining essentially the same gas pressure on the opposing faces of the displacer. Compression and expansion of the total volume of gas are brought about by movement of the *power piston* (Sørensen 1983). The necessity for gas-tight piston-cylinder arrangements and a complex drive system, combined with the need for an effective regenerator, result in high manufacturing costs, which restrict the widespread use of these engines.

Due to difficulties in transferring heat isothermally, actual Stirling cycles do not attain isothermal compression and expansion. They thus obtain lower efficiencies than is theoretically possible.

11.2.1 The Stirling Cycle

Figure 11.7 shows the theoretical (ideal) Stirling cycle on *p-v* and *T-s* planes. The heat obtained from an external source is supplied *isothermally* in process 3–4. Heat is rejected *isothermally* in process 1–2. A *regenerator* transfers heat between process 4–1 and process 2–3, so that $Q_{23} = -Q_{41}$. The regeneration of heat ensures that the thermal efficiency of a theoretical (ideal) Stirling cycle is equal to that of the Carnot cycle, operating between the same high and low temperatures (Organ 2023).

The processes and salient thermodynamic relationships of a Stirling cycle can be summarized as follows (also see Figure 10.3):

11.2.1.1 Process 1–2: Isothermal Heat Rejection
From Table 2.1:

$$T_1 = T_2 \tag{11.29}$$

$$(pv)_1 = (pv)_2 \tag{11.30}$$

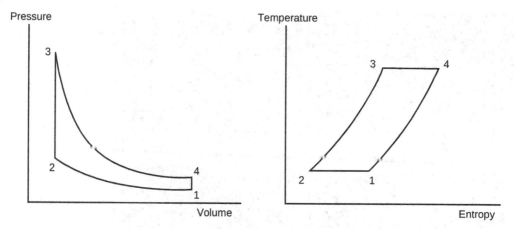

Figure 11.7 Stirling power cycle on *p-v* and *T-s* planes. Heat is transferred to the working fluid during the constant-volume process 2–3 and during the isothermal expansion process 3–4. Heat is rejected during the constant-volume process 4–1, and during the isothermal compression process 1–2.

The engine cylinder volume can be treated as a closed system, which means that the first law of thermodynamics for a closed system (Table 2.1) can be used to calculate the work done and heat transferred during the isothermal process:

$$q_{12} = w_{12} = \int_1^2 p \, dv \tag{11.31}$$

but $pv = RT$ and $T = $ constant.

Therefore:

$$\therefore q_{12} = w_{12} = \int_1^2 \left(\frac{RT}{v}\right) dv = RT \ln\left(\frac{v_2}{v_1}\right) \tag{11.32}$$

11.2.1.2 Process 2–3: Heat Addition at Constant Volume

From Table 2.1:

$$\left(\frac{p}{T}\right)_2 = \left(\frac{p}{T}\right)_3 \tag{11.33}$$

The engine cylinder volume can be treated as a closed system, which means that the first law of thermodynamics for a closed system (Table 2.1) can be used to calculate the heat added:

$$q_{23} = \Delta u_{23} \equiv c_v \, \Delta T_{23} \tag{11.34}$$

For the regeneration process: $\qquad q_{23} = -q_{41} \tag{11.35}$

11.2.1.3 Process 3–4: Isothermal Heat Addition

From Table 2.1:

$$T_3 = T_4 \tag{11.36}$$

$$(pv)_3 = (pv)_4 \tag{11.37}$$

The engine cylinder volume can be treated as a closed system, which means that the first law of thermodynamics for a closed system (Table 2.1) can be used to calculate the work done and heat transferred during the isothermal process:

$$q_{34} = w_{34} = \int_3^4 \left(\frac{RT}{v}\right) dv = RT \ln\left(\frac{v_4}{v_3}\right) \tag{11.38}$$

11.2.1.4 Process 4–1: Heat Rejection at Constant Volume
From Table 2.1:

$$\left(\frac{p}{T}\right)_4 = \left(\frac{p}{T}\right)_1 \tag{11.39}$$

The engine cylinder volume can be treated as a closed system, which means that the first law of thermodynamics for a closed system (Table 2.1) can be used to calculate the heat transferred:

$$q_{41} = \Delta u_{41} \equiv c_v \, \Delta T_{41} \tag{11.40}$$

The thermal efficiency of a Stirling cycle can be calculated as follows:

$$\eta_{th} = \frac{w_{net}}{q_{in}} \tag{11.41}$$

with $w_{net} = w_{12} + w_{34} \equiv q_{net} = T_1 R \ln\left(\frac{v_2}{v_1}\right) + T_3 R \ln\left(\frac{v_4}{v_3}\right)$, as $q = T \, \Delta s$, and from Eq. (3.25), we have that $\Delta s = c_{v,avg} \ln \frac{T_e}{T_i} + R \ln \frac{v_e}{v_i}$, because $T_1 = T_2$ and $T_3 = T_4$. Also, $v_1 = v_4$ and $v_2 = v_3$, so $\ln(v_2/v_1) = -\ln(v_4/v_3)$.

Therefore: $w_{net} = w_{12} + w_{34} \equiv q_{net} = R(T_3 - T_1) \ln\left(\frac{v_4}{v_3}\right)$

The heat added is $q_{34} = RT_3 \ln\left(\frac{v_4}{v_3}\right)$

Equation (11.41) can be simplified accordingly:

$$\eta_{th} = \frac{w_{net}}{q_{in}} = 1 - \frac{T_1}{T_3} \tag{11.42}$$

The Stirling cycle therefore achieves the thermal efficiency of a Carnot cycle operating between the same high and low temperatures (Eq. (3.32)): $\eta_{th} = 1 - \dfrac{T_{low}}{T_{high}}$.

A Stirling cycle may be analyzed like an Otto cycle (Example 10.2), except that the Stirling cycle features isothermal (i.e., not adiabatic) expansion and compression. Due to similarities, no further example will be offered for Stirling cycles.

Reimagine Our Future

Many people contend that the following attitudes support sustainability thinking: *caring* (caring about what happens to nature and to people, caring about the effects of our actions); *compassion* (feeling another's suffering); *openness* (open to learning; open to listening; open to ideas; open to trusting people or situations); *positivity* (being willing to develop the attributes that will promote consistent positive action in all situations); *trust* (trusting oneself and others); *respectfulness* (respect for and appreciation of life; respecting the rights of other people or beings to live freely in safety, respect for diversity); and a sense of *shared responsibility*. Do you agree? If people in general do not possess these attitudes to the requisite extent, how can we *reinvent ourselves*?

11.3 Outlook

Gas turbines will remain popular due to their high power-to-mass ratios, small footprints (required land area), multiple fuel capability (dependent on whether the engines are operated in an open or closed configuration), and their fast start-up times, making them ideal for *peak-lopping* electricity production, among others. Gas turbine combustion has made significant advances in the past three decades, but gas turbines need to operate on flexible fuels, and preferably on no-carbon fuel to remain competitive with renewable technologies. Until gas turbines can effectively burn *green hydrogen* (Oliveira et al. 2021) or *biofuels* (Service 2022), that goal will remain elusive.

Like gas turbines, Stirling engines require *green fuels* to be feasible future power plants. More important, though, is to optimize Stirling cycles that receive their heat from solar concentrators. If Stirling engine designs could be simplified and manufacturing costs lowered, they might become important technologies in sustainable energy solutions.

Problems

11.1 In gas turbine combustors, most of the inducted air does not take a direct part in combustion. Why is this? Discuss briefly by referring to stoichiometric mixtures and their flame temperatures, metallurgical temperature limits of gas turbine materials, primary combustion zones, and dilution zones in combustors, and the influence of excess air on thermal NO_x.

[Answer: Independent research; open-ended question.]

11.2 An LM2500 single-shaft gas turbine, manufactured by GE, burns methane gas, which is supplied at 25 °C and 1 atm. The gas turbine has a net power output of 34 MW. The methane is burned in 10% excess air. The air is pre-heated and enters the combustor at 177 °C (450 K). The combustion products exit the combustor as gases at 1500 K and 1 atm. Calculate (a) the heat produced by the combustor (in kJ/kg CH_4), (b) the fuel flow rate, (c) the engine's thermal efficiency (by disregarding any losses, and assuming adiabatic compression and expansion processes), and (d) the specific fuel consumption.

[Answers: (a) −26,389 kJ/kg CH_4; (b) 1.2 kg/s; (c) 57%; (d) 0.127 kg/kWh]

11.3 The regenerative gas turbine illustrated in the circuit diagram must be analyzed. Air enters the compressor at 1 bar and 289 K with a mass flow rate of 17.9 kg/s.

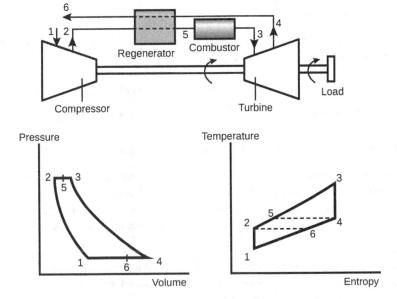

The compressor has an isentropic efficiency of 89%. The fuel flow rate is 0.37 kg/s. The turbine inlet temperature (T_3) is 1400 K, and the turbine has an isentropic efficiency of 92%. The effectiveness (ε) of the counterflow regenerator is 90%. Regenerator effectiveness is the ratio of the actual enthalpy increase of the air flowing through the compressor side of the regenerator to the maximum theoretical enthalpy increase, i.e.,:

$$\varepsilon = \frac{h_5 - h_2}{h_4 - h_2} \simeq \frac{T_5 - T_2}{T_4 - T_2}.$$

The average cycle temperature may be assumed to be 800 K when determining the average values of c_v and c_p. Ignoring any losses, calculate (a) the thermal efficiency of the engine, and (b) the net power output of the engine $\left(\dot{W}_{turbine} - \dot{W}_{compressor} \right)$.

[Answers: (a) 61%; (b) 6.4 MW]

11.4 Use the following data to estimate (a) the specific net work (kJ/kg), (b) the specific fuel consumption (kg/kWh), (c) the heat rate (kJ/kWh), (d) the thermal efficiency, and (e) the fuel flow rate of a 3.7-MW Siemens (Ruston) two-shaft TB 5000 industrial gas turbine operating at 289 K and 101.3 kPa ambient conditions.

Compressor pressure ratio: 7
Compressor isentropic efficiency: 91%
Combustion efficiency: 98%
Fuel: Kerosene with an LHV of 43.1 MJ/kg
Air–fuel ratio: 58

Gas-production turbine entry temperature: 1100 K
Gas-production turbine isentropic efficiency: 87%
Power-turbine isentropic efficiency: 85%
Average cycle temperature: 800 K

[Answers: (a) 269 kJ/kg; (b) $sfc = 0.231$ kg/kWh; (c) heat rate $= 9.96$ MJ/kWh; (d) thermal efficiency $= 36\%$; (e) fuel flow rate $= 0.238$ kg/s]

11.5 (a) Show that, for a single-shaft gas turbine engine, the thermal efficiency can be calculated with the following expression: $\eta_{th} = 1 - \dfrac{1}{r_p^{(\gamma-1)/\gamma}}$, and the net work can be expressed as fol-

lows: $W = c_p\left(r_p^{\gamma-1/\gamma} - 1\right)\left(\dfrac{T_3}{r_p^{\gamma-1/\gamma}} - T_1\right)$. Then, (b) show that, for any given value of (T_3/T_1), the optimum pressure ratio for maximum specific net work can be found by differentiating (W/c_pT) with respect to $(r_p^{\gamma-1/\gamma})$. (c) Produce a spreadsheet or write a small computer program and plot η_{th} and W as a function of r_p (between 0 and 50) to clearly depict the discrepancy that exists between maximum work and maximum efficiency of a simple gas turbine.

[Answers: When plotting the thermal efficiency and net work against the pressure ratio, the thermal efficiency will rise with the pressure ratio, but the net work will reach a maximum at a pressure ratio of 9–10, decreasing after that.]

11.6 A gas turbine power plant comprises two 50-MW units burning natural gas. The plant was built by an electricity utility for peak-lopping duty, i.e., to provide electricity over and above the base load during peak demand periods. However, changes in the mix of plant types supplying the electricity grid and in the electricity demand pattern mean that the power plant is no longer required for peak-lopping. The plant is estimated to have around 20 years of useful life remaining. It is situated close to a large industrial estate.

Discuss the options for the continued use of the power plant in its existing or a converted form, considering qualitatively the costs and benefits (which may depend on future ownership), environmental impact, and opportunities to meet the UN's Sustainable Development Goals.

[Answer: Independent research; open-ended question.]

11.7 Consider a Stirling engine cycle that is a model of a large Stirling engine powered by solar concentrators. The Stirling cycle operates as a closed system and uses 500 g of helium as the working fluid. You will have to do independent research to find the heat capacities of helium. The NIST Thermophysical Properties of Fluid Systems' website could be of assistance: https://webbook.nist.gov/chemistry/fluid.

The highest temperature in the cycle is 730 °C and the lowest is 25 °C. The highest pressure during the cycle is 3 MPa and the lowest pressure is 500 kPa.

Calculate (a) the heat transferred during isothermal compression, (b) the isochoric heat transfer, (c) the heat and work transferred during isothermal expansion, (d) the heat discarded during isochoric pressure reduction, (e) the heat transferred in the regenerator,

(f) the net work per cycle, and (g) the thermal efficiency of the cycle. (h) Comment on the extraordinarily high value of thermal efficiency.

[Answers: (a) -179.4 kJ; (b) 1098.4 kJ; (c) $Q_{34} = W_{34} = 600.7$ kJ; (d) -1098.4 kJ; (e) 1098.4 kJ; (f) 421.3 kJ; (g) 70.3%; (h) refer to text in Section 11.2]

11.8 Perform an independent study of Stirling engines. You might also want to study the following paper: Kongtragool and Wongwises (2003).

Based on your research, produce a fact sheet of no less than 500 words with clear graphics that explain how typical Stirling engines operate (especially their regenerators). The fact sheet must also explain how you used the "3-C" entrepreneurial mindset to produce a new Stirling engine concept, which uses solar concentrators to provide heat (see also Figure 11.6). Your solar Stirling engine must be used in a novel application or setting. The "3-C" mindset comprises your being curious, making connections between apparently disparate aspects, and creating value (https://orchard-prod.azurewebsites.net/media/Framework/KEEN_framework_new.pdf)

[Answer: Independent research; open-ended question.]

11.9 How does the thermal efficiency of an ideal Stirling engine compare to that of an ideal Diesel cycle and Otto cycle, and to a Carnot cycle? What are the practical ramifications of these engine cycles in terms of living more sustainably? Discuss fuel consumption, the use of available fuel resources, combustion pollutants, and the use of solar energy with a Stirling cycle.

[Answer: Study the relevant sections in this book and perform independent study.]

References

Brayton, G.B. (1872). Improvement in gas engines. *US Patent* 125,166. https://patents.google.com/patent/US125166A/en

Davis, L.B. and Black, S.H. (1995). Dry low NO_x combustion systems for GE heavy-duty gas turbines. GE Power Systems, Schenectady, NY. http://www.bright-eng.com/images/pdf/file-360-Low-NOx.pdf

Japikse, D. and Baines, N.C. (1994). *Introduction to Turbomachinery*. White River Junction, Vermont: Concepts ETI.

Kongtragool, B. and Wongwises, S. (2003). A review of solar-powered Stirling engines and low temperature differential Stirling engines. *Renewable and Sustainable Energy Reviews* 7 (2): 131–154.

Lefebvre, A.H. and Ballal, D.R. (2010). *Gas Turbine Combustion. Alternative Fuels and Emissions*. Boca Raton, FL: CRC Press.

Oliveira, A.M., Beswick, R.R. and Yan, Y. (2021). A green hydrogen economy for a renewable energy society. *Current Opinion in Chemical Engineering* 33: 100701.

Organ, A.J. (2023). Stirling and Thermal-Lag Engines. Motive Power Without the CO_2. *London, UK: World Scientific.*

Patel, N.S., Pavlík, V. and Boča, M. (2017). High-temperature corrosion behavior of superalloys in molten salts – a review. *Critical Reviews in Solid State and Materials Sciences* 42 (1): 83–97.

Rogers, G.F.C., Cohen, H., Saravanamuttoo, H., Straznicky, P. et al. (2017). *Gas Turbine Theory*. London, UK: Pearson Education.

Saravanamuttoo, H.I.H., Rogers, G.F.C. and Cohen, H. (2001). *Gas Turbine Theory*. Hoboken, NJ: Prentice Hall.

Service, R.F. (2022). Can biofuels really fly? *Science* 376 (6600): 1394–1397.

Sørensen, H. (1983). *Energy Conversion Systems*. New York, NY: Wiley.

Stirling, R. (1816). Improvements for diminishing the consumption of fuel and, in particular, an engine capable of being applied to the moving of machinery on a principle entirely new. British Patent 4081.

12

Emission Control of Internal and External Combustion Engines

Internal combustion engines that operate on fossil fuels (e.g., petroleum and natural gas) consume around 25% of the world's energy, equivalent to about 3000 out of 13,000 million ton of oil equivalent per year (Reitz et al. 2020). These engines produce about 10% of the world's greenhouse-gas emissions (IEA 2021). Internal combustion engines will still be around for many years to come. It is therefore crucial for their fuel efficiencies to improve and to employ advanced emission control technologies. Over the last four decades, research on engine combustion and exhaust after-treatment and controls has led to a demonstrably cleaner environment, depicted by a 1000-fold reduction in hazardous exhaust emissions (Reitz et al. 2020), including particulates, nitric oxide (NO_x), carbon monoxide (CO), and unburned hydrocarbons (HC).

Reimagine Our Future
Our *values,* combined with our *attitudes* and *beliefs,* influence how we think and behave. Write down three things that you most value in life, e.g., commitment, freedom, family, knowledge, well-being, health, respect, ambition, success, humility, fairness, friendship, honesty, pleasure, prestige, generosity, or trust. Are you content with these values? Should you value other things? Do you feel a need to reprioritize what you value?

12.1 Emissions from the Combustion of Gasoline and Diesel

The combustion of *diesel* results in the production of N_2, CO_2, H_2O and – depending on the air-fuel ratio – a significant amount of unburned or partially burned hydrocarbons, which give diesel fuel exhaust its characteristic odor. Diesel engines have much higher compression ratios than gasoline engines (compare the *r*-values in Tables 10.1 and 10.2). Since they do not use spark ignition, but compression ignition, diesel engines can run with very lean mixtures. Diesel engines are therefore inherently more fuel efficient than their gasoline counterparts, and their emissions of hydrocarbons and CO are comparatively lower (Taylor 2008). Diesel engines, however, emit particulates that consist primarily of soot (carbon). The combustion of diesel also results in the formation of a small amount of CO.

Energy Systems: A Project-Based Approach to Sustainability Thinking for Energy Conversion Systems,
First Edition. Leon Liebenberg.
© 2024 John Wiley & Sons, Inc. Published 2024 by John Wiley & Sons, Inc.
Companion website: www.wiley.com/go/liebenberg/energy_systems

Diesel may contain small amounts of sulfur that are oxidized to produce mainly sulfur dioxide (SO_2) and a small amount of sulfur trioxide (SO_3). Due to the high compression ratios of diesel engines, the combustion temperatures are relatively high, which means that NO_x emissions can be high (Heywood 1988). The SO_3 can transform to sulfuric acid (H_2SO_4), a main contributor to acid rain, which adversely affects plants and trees. Sulfuric acid, when inhaled, causes severe respiratory disorders. It can also corrode concrete structures like bridges and buildings.

At the concentrations currently found in air, CO interferes with the body's blood-oxygen control, which can lead to cardiovascular and neurological problems. The size of particulates ranges from 0.005 to 100 μm and they are known to cause respiratory disorders. The black smoke or "soot" emitted from diesel engines (or from smokestacks following coal and oil combustion) consists mostly of solid carbon particles. These soot particles can cause respiratory disorders, especially when organic substances are adsorbed. Diesel soot is a known carcinogen (which can cause cancer) (Masters and Ela 2014).

The combustion of *gasoline* ("petrol") results in the production of N_2, CO_2, H_2O, and – depending on the air–fuel ratio – CO and unburned or partially burned hydrocarbons (HCs). Gasoline may contain small amounts of sulfur that are oxidized to produce SO_2. In real engines, the exhaust gases will also contain oxides of nitrogen: nitric oxide (NO), and small amounts of nitrogen dioxide (NO_2). These are collectively known as NO_x. The NO_x produced by oxidation of fuel is known as *fuel NO_x*. With the high temperatures of combustion, the nitrogen also reacts to oxygen in the air to form *thermal NO_x*. The higher the burned gas temperature, the higher the rate of formation of thermal NO_x (Heywood 1988).

The emission of NO_x, combined with unburned or vaporized hydrocarbons and the action of sunlight, produces photochemical smog, which contains ozone. Such smog can cause a variety of respiratory disorders. NO_x can also react with water to form nitric acid, which is corrosive and can contribute to acid rain (Masters and Ela 2014).

The use of alcohol to reform hydrocarbon fuels substantially increases aldehyde emissions. Aldehydes in the air are known to be cytotoxic (they can damage living cells), mutagenic (they can induce DNA damage), and carcinogenic (they can cause cancer) (Masters and Ela 2014; Wang and Olsson 2019).

Figure 12.1 illustrates qualitatively how the fuel-air equivalence ratio (ϕ) of hydrocarbon

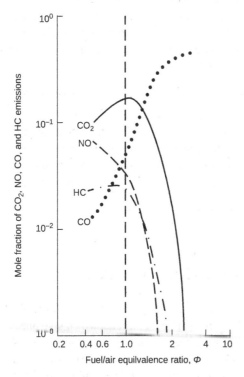

Figure 12.1 The qualitative effect of the fuel-air equivalence ratio on the variation of hydrocarbons, CO, NO, and CO_2 concentration in the exhaust of engines burning gasoline or diesel. *Source:* Adapted from Kirkpatrick (2020), Bosch (2022), Obert (1973), and Heywood (1988).

combustion may affect the variation of unburned hydrocarbons, nitrous oxides, carbon monoxide, and carbon dioxide.

With the combustion of hydrocarbon fuels such as gasoline or diesel, Figure 12.1 suggests that leaner mixtures give lower emissions, until the combustion quality becomes poor and misfire eventually occurs, after which hydrocarbon emissions sharply rise and engine operation becomes erratic. An internal combustion engine produces its highest power at slightly rich mixtures but is most fuel efficient (and produces less emissions) at slightly lean mixtures.

With the combustion of diesel, pollutant formation processes are strongly dependent on fuel distribution in the combustion chamber. The various parts of the fuel jet and the flame affect the formation of NO, unburned hydrocarbons, and soot. The NO forms in the high-temperature regions, but temperature and air–fuel ratio distributions within the burned gases are nonuniform (unlike in the case of gasoline combustion, which produces uniform air–fuel distributions). The NO formation rates are highest in the close-to-stoichiometric regions (Heywood 1988). Soot forms in the flame region of the rich unburned fuel part of diesel sprays. Hydrocarbons and aldehydes originate in the regions where the flame quenches on the walls of the combustion chamber and where excessive dilution with air prevents the combustion process from starting or during completion (Heywood 1988).

Apart from the air–fuel ratio, other factors that affect emissions include ignition timing, compression ratio, combustion chamber geometry, and whether the engine is operating at full-load or part-load (Heywood 1988).

Reimagine Our Future

Develop a simple card game for children that would explain the pros and cons of the main energy conversion systems, e.g., a large diesel generator, a fossil-fuel steam power plant, or nuclear, wind, solar, or geothermal energy systems. Reach out to a local school and see what the teachers think of your idea. You might be pleasantly surprised as your idea might be marketable!

12.2 Controlling the Emissions of Gasoline and Diesel Combustion

Oil and gas production, together with transport, is responsible for around 15% of today's greenhouse-gas emissions from the global energy sector (IEA 2021). Options to reduce these emissions include tackling methane leaks (including intentional methane venting), minimizing flaring, switching to low-carbon options for power operations, incorporating energy-efficiency improvements across the supply chain, and using carbon capture storage for centralized sources of emissions. These all play a role in reducing the emission intensity of oil and gas production.

It should be clear from the examples in Chapters 7–9 (and from Figure 12.1) that the combustion of oil and gas emits large amounts of CO_2, among many other harmful emissions. According to the International Energy Agency (IEA) (2021), the combustion of gasoline and diesel is responsible for around 10% of global greenhouse-gas emissions. Figure 12.2 illustrates the main methods of reducing emissions following the combustion of gasoline and diesel, and Box 12.1 explains the usage of catalytic converters.

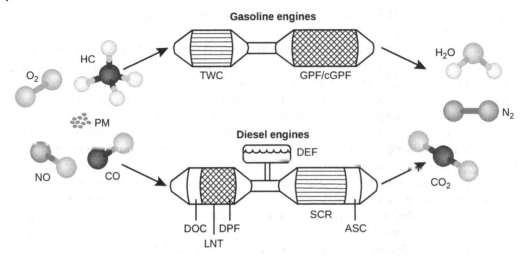

Figure 12.2 Two typical configurations of modern gasoline (top) and diesel (bottom) emission control systems. For gasoline combustion after-treatment, the unburned hydrocarbons, CO, and NO_x emissions could be converted by a three-way catalyst (TWC), which is followed by a gasoline particulate filter (GPF) or a catalyzed gasoline particulate filter (cGPF) for removing particulates. For diesel combustion after-treatment, the unburned hydrocarbons and CO emissions could be converted by a diesel oxidation catalyst (DOC) or lean NO_x trap (LNT). Particulates could be removed by a diesel particulate filter (DPF), and NO_x emissions could be converted by a selective catalytic reduction (SCR) catalyst using a diesel exhaust fluid (DEF) additive, typically comprising urea, which produces ammonia. An ammonia slip catalyst (ASC) can be used to further reduce NO_x emissions. *Source:* Adapted from Wang and Olsson (2019).

Box 12.1   Catalytic Converters

Catalytic converters are commonly used for the exhaust after-treatment of gasoline- and diesel-fueled vehicles. The three-way catalyst converter requires engines to operate with a stoichiometric air–fuel ratio. The main task of a TWC is to convert hydrocarbons, CO, and NO_x to "harmless" CO_2, H_2O, and N_2. Of course, we now know that CO_2 and H_2O are not harmless but lead to an enhanced greenhouse effect. Nitrogen can also photochemically combine with oxygen to form NO_2, a leading component of harmful smog.

The catalytic converter usually comprises a steel housing, a ceramic absorbing substrate, and a substrate coating such as aluminum oxide (Al_2O_3), which is covered with 1–10 g of a precious metal such as platinum, palladium, or rhodium, which acts as a catalyst. The catalytic material accelerates the oxidation of hydrocarbons and CO while reducing NO_x.

The typical chemical TWC reactions for a generic hydrocarbon fuel can be summarized as follows (Kirkpatrick 2020; Bosch 2022):

$$\text{Oxidation of CO}: 2\,CO + O_2 \rightarrow 2\,CO_2$$

$$\text{Oxidation of hydrocarbons}: 2\,C_2H_6 + 7\,O_2 \rightarrow 4\,CO_2 + 6\,H_2O$$

$$\text{Reduction of N}_2: \begin{array}{l} 2\,NO + 2\,CO \rightarrow N_2 + 2\,CO_2 \\ 2\,NO_2 + 2\,CO \rightarrow N_2 + 2\,CO_2 + O_2 \end{array}$$

The importance for the air–fuel ratio to be near-stoichiometric (i.e., $\lambda = 1.0$, where $\lambda = 1/\phi$, see Eq. (9.2)) can be gleaned from these equations. The reactions require adequate oxygen for the oxidation processes and to ensure the highest possible conversion ratio of the pollutants.

To ensure the optimal functioning of a TWC converter, the exhaust system is fitted with an oxygen sensor, also called a lambda (λ) sensor, which sends its signals to an automatic control system that adjusts the air–fuel ratio to optimum levels. Some vehicles are fitted with two or three sensors: one placed upstream of the catalytic converter and another one or two placed downstream of it. This arrangement helps ensure more accurate control of the air–fuel ratio to facilitate the maximal conversion of hydrocarbons, CO, and NO_x. Figure 12.3 shows the typical performance of a TWC converter.

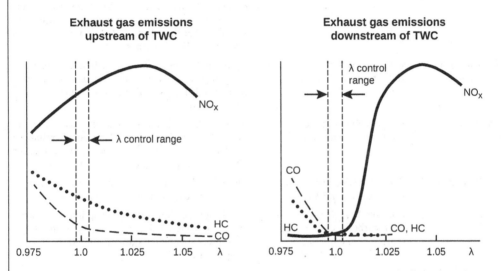

Figure 12.3 The effectiveness of a three-way catalyst (TWC) converter as a function of the λ (excess air) factor. *Source:* Adapted from Bosch (2022) and Kirkpatrick (2020).

Emissions can either be controlled by managing the combustion process (see Table 12.1), by implementing advanced fuel technologies (see Table 12.2), or by the post-combustion treatment of the exhaust gas (see Table 12.3). These technologies require advanced engine diagnostics and digital control strategies, or the application of control, diagnostics, and power train technologies (see Table 12.1).

Of course, the best way to control emissions from oil-based fuels is not to use them in the first place. In this regard, Section 9.3 briefly discusses the use of no-carbon hydrogen and ammonia as fuel, while Section 9.4 presents biofuels such as biohydrogen. Chapter 24 discusses the use of hydrogen-based fuel cells and Chapter 28 discusses the use of electric vehicles to circumvent the onboard combustion of fuel. However, each of these no-carbon or low-carbon fuel options presents its own set of technological and economic challenges.

Table 12.1 Engine design technologies for emission control.

Technology	Emissions impact	Description
Compression-ignition (diesel) engines		
Fuel injection timing	Reduces NO_x emissions	Control of timing of fuel injection, which affects combustion
Fuel injection pressure	Reduces soot (particulate) emissions	Higher injection pressures lower soot emissions
Multiple injections	Lowers NO_x, soot, hydrocarbons, and CO emissions	Multiple points of fuel injection to achieve better air–fuel mixing and more complete combustion
Exhaust gas recirculation (EGR)	Lowers NO_x emissions	Exhaust gas is recirculated, and NO_x emissions are catalytically modulated. EGR is difficult in diesel engines as they run on lean fuel-air mixtures, implying that the effect of EGR as a diluent is less effective in diesel engines
Boosting intake pressure (turbo-charging and supercharging)	Lowers soot (particulate) and NO_x emissions	Higher intake pressures are facilitated by turbo-charging the intake air. This increases the air–fuel ratio. The leaner fuel-air mixture reduces flame temperatures
Cooling of intake air (intercooling)	Reduces NO_x and soot (particulate) emissions	Air is cooled with a heat exchanger as it leaves the turbo-charger's compressor. Such a heat exchanger is referred to as a charge air cooler, intercooler, or aftercooler, depending on its placement
Variable valve timing (VVT)	Reduces hydrocarbons, NO_x, CO, and CO_2 emissions	Variable valve timing reduces fuel consumption, which also results in reduced emissions
Combustion chamber design	Controls NO_x, CO, hydrocarbons, CO_2, and soot (particulate) emissions	Combustion chamber design is optimized to bring about complete combustion over the engine's operating range while minimizing flame temperature
Spark-ignition (gasoline) engines		
Gasoline direct injection (GDI)	Controls hydrocarbons, NO_x, CO, and CO_2 emissions	Gasoline direct injection is used to improve fuel consumption and concomitantly reduces emissions
Boosting of intake air pressure (turbo-charging and supercharging)	Reduces NO_x emissions	Higher intake pressures are facilitated by turbo-charging the intake air. This increases the air–fuel ratio. The leaner fuel-air mixture reduces flame temperatures
Variable valve timing	Reduces hydrocarbons, NO_x, CO, and CO_2 emissions	Variable valve timing reduces fuel consumption, which also results in reduced emissions
Lean-burn technology	Reduces NO_x emissions	Lean-burn technology is usually used with the catalytic oxidation treatment of exhaust gas. Lean-burn requires increased compression ratios to ensure complete combustion
Exhaust gas recirculation	Lowers NO_x emissions	Exhaust gas recirculation reduces "knock," especially when the engine is operating at high loads. Exhaust gas recirculation helps to lower exhaust temperature
Oxy-fuel combustion	Reduces CO_2 emissions	Oxy-fuel combustion increases the oxygen mass fraction (e.g., from 23% to 30%), which improves fuel consumption and lowers CO_2 emissions. Oxy-fuel technology involves injecting water during combustion

Source: Adapted from Taylor (2008), Arcoumanis and Kamimoto (2008), Heywood (1988), Obert (1973), Majewski and Jääskeläinen (2015), Khair and Majewksi (2006), and Bosch (2022).

Table 12.2 Fuel and lubricant technologies for emission control.

Technology	Emissions impact	Description
Alternative fuels	Reduces CO_2 from the perspective of the fuel's lifecycle	Alternative fuels include fuel blends such as 10% ethanol and 90% gasoline (also called E10), biodiesel, ethanol, and biogasoline
Fuel additives	Reduces NO_x emissions	Blending gasoline or diesel with additives such as toluene and xylene reduces flame temperatures
Lubricating oil	Reduces hydrocarbons, CO, NO_x, CO_2, and soot (particulate) emissions	Low-viscosity lubricants reduce friction and improve fuel consumption, with a concomitant reduction of emissions

Source: Adapted from Heywood (1988), Obert (1973), DieselNet (2022), and Bosch (2022).

Table 12.3 Post-combustion technologies for emission control.

Technology	Emissions impact	Description
Compression-ignition (diesel) engines		
Diesel oxidation catalyst (DOC)	Reduces emissions of CO, hydrocarbons, soot (particulates), and sulfur CO_2 emissions are increased	With the help of a diesel oxidation catalyst, CO and hydrocarbons are converted to CO_2 and H_2O, respectively. Sulfides are oxidized to SO_2 and SO_3. DOCs also increase the fraction of NO_2 to support the performance of selective catalytic reduction systems and diesel particulate filters
Diesel particulate filter	Reduces emissions of soot (particulates)	Conventional ceramic diesel particulate filters have proven to be exceptionally effective in reducing particulate emissions with efficiencies of 90% or more. However, these filters require regular active "regenerations," as well as periodical ash removal to avoid a blockage of the exhaust line
Particle oxidation catalyst	Reduces emissions of soot (particulates)	Particle oxidation catalysts are not as effective as diesel particulate filters, but do not have the risk of blockage or the need for complex filter regeneration procedures
Urea selective catalyst reduction (SCR)	Reduces NO_x emissions	Urea-based diesel exhaust fluid ("AdBlue") and a catalytic converter effectively control NO_x emissions
Ammonia slip catalyst	Reduces NO_x emissions	Ammonia-based diesel exhaust fluid ("AdBlue") and a catalytic converter effectively control NO_x emissions, but can produce ammonia-based secondary emissions, including nitrous oxide (N_2O) and ammonium nitrate (NH_4NO_3)

(Continued)

Table 12.3 (Continued)

Technology	Emissions impact	Description
SCR-coated diesel particulate filters	Reduces NO_x emissions and soot (particulate) formation	These filters consist of a wall-flow monolith coated with materials that both enhance soot oxidation and reduce NO_x. They can be placed immediately downstream of the diesel oxidation catalyst, thus nearer the engine, to achieve a faster warm-up with respect to a more traditional layout in which the SCR unit is placed downstream of the diesel particulate filter
Passive NO_x adsorbers (PNAs)	Reduces NO_x emissions	The use of urea-based diesel exhaust fluid as a source of NH_3 to reduce NO_x currently sets a technological limit ($180\,°C$) under which SCR applications are hardly possible. A solution to this issue is PNAs: catalytic devices that contain palladium zeolites. PNAs temporarily store NO_x and release the gas when the temperature is sufficiently high for conversion
Spark-ignition (gasoline) engines		
Three-way catalyst (TWC) converter	Reduces NO_x, CO_x, and hydrocarbon emissions	A three-way catalyst that oxidizes exhaust gas pollutants – both hydrocarbons and CO – and reduces NO_x to H_2O, N_2, and CO_2. It cannot be used with lean fuel-air mixtures, i.e., TWC converters require near-stoichiometric ratios (i.e., high NO concentrations) to function properly
Lean NO_x traps (LNTs)	Reduces NO_x emissions	NO_x adsorbers, also known as lean NO_x traps, adsorb and store NO_x under lean conditions. Typically, NO is converted to NO_2 using an oxidation catalyst so that the NO_2 can be rapidly stored as nitrate on alkaline earth oxides. A brief return (1–2 s) to stoichiometric or rich operation is required to desorb (remove) the stored NO_x and provide the conditions for a conventional TWC mounted downstream to reduce the NO_x
Gasoline particulate filters (GPFs) or catalyzed gasoline particulate filters (cGPF)	Reduces particulate emissions, as well as CO and NO emissions, but has increased CO_2 emissions	Gasoline particulate filters are placed downstream of catalytic converters. The honeycomb-structured GPF traps particulates, as well as NO and CO
All internal combustion engines		
On-board CO_2 capture and storage	Captures most CO_2 emissions	CO_2 is captured in cylinders by treating exhaust gases with a mono-ethanolamine (MEA) solvent and exhaust after-heater. The large size of the technology makes it feasible for use in trucks or ships. (Also see Box 9.1.)

Source: Adapted from Heywood (1988), Obert (1973), DieselNet (2022), and Bosch (2022).

Capturing CO$_2$ at Truck Tailpipes

Remora aims to capture automotive carbon emissions (see Figure 12.4). The Detroit-based start-up company has invented a device that attaches to the exhaust system of a semi-truck and captures 80% of the carbon emissions, which is stored in an onboard tank to later be pumped out and sold to concrete producers and other users of industrial CO$_2$ or sequestered underground for permanent storage. Paul Gross started working on Remora during his senior year at Yale after reading co-founder Christina Reynolds' US Environmental Protection Agency (EPA)-funded dissertation on mobile carbon capture. He wrote a business plan for her and convinced her to leave the EPA to help start Remora. They teamed up with Eric Harding, a mechanic turned engineer, who spent years building electric and hydrogen semi-trucks for some of the world's largest automotive companies.

For further reading: Sharma and Maréchal (2019).

Figure 12.4 Concept design of a semi-truck fitted with CO$_2$ capture and storage technology. *Source:* ECHENEIDAE INC. / https://remoracarbon.com/ / last accessed 12 May, 2023.

12.3 Reducing Emissions from Natural Gas Combustion

It must be noted that around 3.5% of the gas drilled from the ground leaks into the atmosphere. These are the so-called *fugitive emissions* (Howarth and Jacobson 2021). Drilling for natural gas emits far more methane than previously known (Howarth and Jacobson 2021). Fixing the leaky pipelines would be an important step in reducing the global warming effects of the production of natural gas.

Although *fuel NO$_x$* is negligible when burning natural gas, *thermal NO$_x$* in the combustion air is a problem. The production of NO$_x$ is enhanced by greater flame temperatures (typically greater than 1600 °C) and by longer residence times of gases in the flame (Noble et al. 2021). *Low-NO$_x$ burning technology* includes injecting steam or water into the flame to reduce temperatures, but this requires a supply of water and reduces plant efficiency. Alternatively, the residence time of the gas in the flame can be reduced, but stabilizing a short flame and ensuring complete combustion in a short time are difficult to achieve. Effective and clean combustion becomes even more challenging when one considers that modern gas turbines must be able to burn a range of low-carbon or no-carbon fuels, including hydrogen (Lefebvre and Ballal 2010).

To reduce NO_x emissions and ensure a multifuel capability, gas turbine manufacturers now pre-mix fuel and air before the fuel enters the combustion chamber. Such a lean premixed combustion process is commonly called *dry low-NO$_x$* (DLN) and can attain NO_x emissions as low as 5 ppm. Future gas turbine combustors will feature pulse detonation or rotating detonation technology to further reduce emissions and facilitate combustion with hydrogen-rich fuels (NASEM 2020). Also see Chapter 15 for more on hydrogen fuels.

12.4 Outlook

Internal and external combustion engines will be around for many more years. However, their fuels will become low-carbon and eventually *no-carbon*, such as hydrogen. Depending on advances in hydrogen fuel cells (Chapter 21), internal combustion engines for vehicles might become outmoded by 2050. Unless, of course, internal combustion engines could burn green hydrogen, for instance, in an economically viable manner.

In the interim, to cope with increasingly stringent environmental demands, internal combustion engines will feature a combination of new active catalytic materials and different passive pollutant traps (to capture unburned hydrocarbons and NO_x). These catalysts and traps not only need to be highly active and selective, but also stable (Wang and Olsson 2019).

Gas turbines will continue to improve in efficiency and in their capabilities to burn low-carbon fuels of varying compositions. Eventually, gas turbines will be able to burn only hydrogen (NASEM 2020). Although such gas turbines will probably still have high thermal NO_x and H_2O vapor emissions, their carbon footprints will become minimal if *green hydrogen* is used.

Problems

12.1 As the global economy decarbonizes, people will probably still be using chemical fuels for many more years. Internal and external combustion engines that burn hydrocarbon fuels or biofuels will presumably be pivotal for future long-distance truck and ship transport.

Discuss, in a 500-word essay, the importance of catalytic converters in these truck and ship engines to help achieve the UN's ambitious Sustainable Development Goals (SDGs), especially SDG 7 and SDGs 11–15. In your research, you may find the following paper to be a good starting point: Wang and Olsson (2019).

[Answer: Open-ended question]

12.2 Climate change is happening faster than anticipated, and the role of anthropogenic greenhouse-gas emissions (especially CO_2) is now widely accepted. It is deemed crucial to urgently eliminate the use of carbon-based fuels in gas turbines, due to the widespread use of gas turbines. Scientists and engineers are therefore working fervently to develop gas turbines that can burn *green hydrogen* (see Box 9.2). This strategy also aligns with larger plans to develop a *hydrogen economy*, such as is happening in Japan (Chaube et al. 2020).

Although many gas turbines can burn mixtures of hydrogen and natural gas, no gas turbine can yet effectively and economically burn *only* hydrogen. Investigate the challenges and opportunities surrounding the combustion of hydrogen in gas turbines by producing a high-impact fact sheet of no less than 300 words with accompanying graphics or images. To initiate your research, you are encouraged to study the following paper: Noble et al. (2021).

[Answer: This question is open-ended and requires independent research.]

12.3 You are a member of a small start-up company in energy engineering. Your company wants to capture CO_2 emissions from diesel-electric locomotives. Your company will use carbon capture technology, like that shown in Figure 12.4.

Produce a business model canvas (Osterwalder et al. 2010) to summarize your innovation. Your business model canvas must be detailed and convincing enough to present to a potential funder. A template of the *Business Model Canvas* is available from the *Strategyzer* website: https://www.strategyzer.com/canvas/business-model-canvas

When producing your business model canvas, you might find the following paper useful: Dobrowolski and Sułkowski (2021).

The following White Paper, produced by GE Power/Vernova (2021), also provides important context: "Decarbonizing gas turbines through carbon capture. A pathway to lower CO_2," https://www.ge.com/content/dam/gepower-new/global/en_US/downloads/gas-new-site/future-of-energy/decarbonizing-gas-turbines-ccus-gea34966.pdf

[Answer: This question is open-ended and requires independent research.]

References

Arcoumanis, C. and Kamimoto, T. (2008). *Flow and Combustion in Automotive Engines*. Berlin, Germany: Springer.

Bosch (2022). *Bosch Automotive Handbook*. Stuttgart: Robert Bosch.

Chaube, A., Chapman, A., Shigetomi, Y., Huff, K. et al. (2020). The role of hydrogen in achieving long term Japanese energy system goals. *Energies* 13 (17): 4539.

DieselNet (2022). DieselNet internet resources. http://www.dieselnet.com

Dobrowolski, Z. and Sułkowski, Ł. (2021). Business model canvas and energy enterprises. *Energies* 14 (21): 7198.

GE Power/Vernova (2021). Decarbonizing gas turbines through carbon capture. A pathway to lower CO_2. https://www.ge.com/content/dam/gepower-new/global/en_US/downloads/gas-new-site/future-of-energy/decarbonizing-gas-turbines-ccus-gea34966.pdf

Heywood, J.B. (1988). *Internal Combustion Engine Fundamentals*. New York, NY: McGraw-Hill.

Howarth, R.W. and Jacobson, M.Z. (2021). How green is blue hydrogen? *Energy Science and Engineering* 9 (10): 1676–1687.

IEA (2021). Net Zero by 2050. A roadmap for the global energy sector. International Energy Agency. www.iea.org

Khair, W.A. and Majewksi, M.K. (2006). *Diesel Emissions and Their Control*. Pennsylvania, USA: SAE International.

Kirkpatrick, A.T. (2020). *Internal Combustion Engines: Applied Thermosciences*. New York, NY: Wiley.

Lefebvre, A.H. and Ballal, D.R. (2010). *Gas Turbine Combustion: Alternative Fuels and Emissions*. Boca Raton, FL: CRC Press.

Majewski, W.A. and Jääskeläinen, H. (2015). Engine emission control. Diesel-Net Technology Guide. https://dieselnet.com/tech/engine_emission-control.php.

Masters, G.M. and Ela, W.P. (2014). *Introduction to Environmental Engineering and Science*. London, UK: Pearson.

NASEM (2020). *Advanced Technologies for Gas Turbines*. Washington, DC: National Academies of Science, Engineering, and Medicine.

Noble, D., Wu, D., Emerson, B., Sheppard, S. et al. (2021). Assessment of current capabilities and near-term availability of hydrogen-fired gas turbines considering a low-carbon future. *Journal of Engineering for Gas Turbines and Power* 143 (4): 041002.

Obert, E.F. (1973). *Internal Combustion Engines and Air Pollution*. New York, NY: Harper & Row.

Osterwalder, A., Pigneur, Y. and Clark, T. (2010). *Business Model Generation: A Handbook for Visionaries, Game Changers, and Challengers*. Strategyzer Series. Hoboken, NJ: Wiley.

Reitz, R.D., Ogawa, H., Payri, R., Fansler, T. et al. (2020). IJER editorial: the future of the internal combustion engine. *International Journal of Engine Research* 21 (1): 3–10.

Sharma, S. and Maréchal, F. (2019). Carbon dioxide capture from internal combustion engine exhaust using temperature swing adsorption. *Frontiers in Energy Research* 7: 143.

Taylor, A.M. (2008). Science review of internal combustion engines. *Energy Policy* 36 (12): 4657–4667.

Wang, A. and Olsson, L. (2019). The impact of automotive catalysis on the United Nations sustainable development goals. *Nature Catalysis* 2 (7): 566–570.

Mini Project 4

Domestic Back-Up Generator Using Natural Gas or Propane

Consider the popular Generac *Guardian* domestic back-up generator. The generator uses an internal combustion engine, which can be modeled as an *ideal* Otto cycle, and which runs at a compression ratio of 9.5 : 1. The two cylinders have a combined displacement of 999 cm³. The engine can run on natural gas or on liquid propane. When using natural gas, the mass flow rate is 8.5 m³/h. When using liquid propane, the mass flow rate is 13.5 L/h. The engine's intake temperature and pressure are 300 K and 100 kPa, respectively. State and substantiate all your assumptions when analyzing the combustion of these two fuels.

Generac *Guardian* generator. *Source:* Generac Power Systems, Inc. / https://www.generac.com/all-products/generators/home-backup-generators#?cat=6&cat=214&cat=217&cat=249 / last accessed 12 May, 2023.

The *natural gas* used in the engine has the following composition: 82% CH_4, 2.9% C_2H_6, 0.32% C_3H_8, 0.2% C_4H_{10}, 0.9% CO_2, 0.01% O_2, and 13.67% N_2. The higher heating value (*HHV*) of this natural gas is 52.2 MJ/kg and the lower heating value (*LHV*) is 47.1 MJ/kg.

The liquid *propane* (C_3H_8) used in the engine has an *HHV* of 49.84 MJ/kg and an *LHV* of 46.28 MJ/kg.

Energy Systems: A Project-Based Approach to Sustainability Thinking for Energy Conversion Systems,
First Edition. Leon Liebenberg.
© 2024 John Wiley & Sons, Inc. Published 2024 by John Wiley & Sons, Inc.
Companion website: www.wiley.com/go/liebenberg/energy_systems

The fuels are burned in air, which may be assumed to comprise 79% nitrogen and 21% oxygen.

1) Compute the engine's power output at 3600 rpm when using natural gas and when using propane. Also calculate the engine efficiencies for these two cases.

[Answer: About 24 kW when using natural gas and about 22 kW when using propane. Both fuels result in a thermal efficiency of about 54%. This efficiency would be closer to 30% if the compression and expansion processes were not assumed to be adiabatic.]

2) If the generator works for three months per year (and 4 h per day in that period), how much CO_2 will be emitted to the atmosphere per year if the engine burns natural gas and if it burns propane? How does that compare to the emissions of the same engine burning diesel? Diesel may be assumed to be $C_{12}H_{23}$ with an *HHV* of 46.1 MJ/kg and an *LHV* of 43.2 MJ/kg. Show all combustion calculations.

[*Answer:* C_3H_8: 7.1 tons CO_2/year; natural gas: 5.4 tons CO_2/year; $C_{12}H_{23}$: 5.9 tons CO_2/year]

3) Do online research to find the cost of this generator (or a model of a similar size from another manufacturer). What is the cost per kilowatt ($/kW)? How does that compare to the $/kW of a 2-MW micro-nuclear power plant?

Your research will uncover many sites with credible information about micro-nuclear reactors. The site "Small nuclear power reactors" by the World Nuclear Association might be useful to initiate your research: https://www.world-nuclear.org/information-library/nuclear-fuel-cycle/nuclear-power-reactors/small-nuclear-power-reactors.aspx

[Answer: Open-ended questions. Stated assumptions will determine final answers.]

4) Compare the useful life of a well-maintained Generac Guardian internal combustion engine to that of a 2-MW micro-nuclear power plant.

[Answer: Open-ended question.]

5) Compare the annualized capital costs of 89 Generac Guardian generators (which will have a combined output of about 2 MW) to one 2-MW micro-nuclear plant at an annual carrying charge rate of 15%. Perform independent research to find an expression for the annualized capital cost. You will find many helpful web pages, including *HomerPro*: https://www.homerenergy.com/products/pro/docs/latest/annualized_cost.html

[Answer: Open-ended question.]

6) Compare the fuel costs of 89 Generac Guardian generators with those of the 2-MW micro-nuclear power plant over a 10-year operating period. (Clearly state the fuel costs for natural gas, liquid propane, and the micro-nuclear power plant.)

[Answer: Dependent on assumptions, fuels for the Generac Guardian might cost seven to nine times that of the micro-nuclear power plant.]

7) Based on your above answers, is it technically feasible and economically viable to run an array of small internal combustion engines to produce 2 MW rather than a micro-nuclear power plant of the same size?

[Answer: Open-ended question. Answer is dependent on assumptions and context.]

8) a) Explain how this internal combustion engine generator running on natural gas or propane helps achieve the UN's SDGs 7 and 11 (affordable and clean energy, and sustainable cities and communities).

b) Does this generator promote other SDGs, especially SDG 3 (good health and wellbeing), SDG 9 (industries, innovation, and infrastructure), SDG 10 (reduced inequalities), SDG 12 (responsible consumption and production), and SDG 13 (climate action)? What connections exist between these SDGs?

c) Briefly discuss the social, technological, environmental, economic, and political factors concerning the mass deployment of the Generac Guardian in the United States. Clearly state your references by citing the author, article title or webpage, and URL.

[Answer: Open-ended question. Answer is dependent on assumptions and context.]

Carbon sink

The iconic forests of the Rocky Mountains are disappearing. Along with that, we are losing a valuable carbon sink. One hectare (the size of 2.5 football fields) of forest in the Rocky Mountains absorbs about 40 tons of CO_2 per year.

The extraction and burning of coal – which originates as trees, among other plant material – are contributing to excessive CO_2 emissions and global warming. These higher temperatures are allowing pine beetle populations to expand and further decimate forests from New Mexico to British Columbia.

Chris Drury is an environmental artist. In his outdoor sculpture, called *Carbon Sink*, he draws attention to all these interconnected sustainability issues. *Carbon Sink* is 14 m in diameter and made from pine logs killed by beetles and arranged in a whirlpool shape around a pile of coal, representing cause and effect. The sculpture is meant to draw critical attention to the state of Wyoming's immense coal extraction and burning operations, which, in turn, help force the beetle population to overgrow and subsequently kill trees. This results in warming temperatures further, and so the cycle continues. What goes around comes around! *Source:* Courtesy of land artist Chris Drury.

Energy Systems: A Project-Based Approach to Sustainability Thinking for Energy Conversion Systems, First Edition. Leon Liebenberg.
© 2024 John Wiley & Sons, Inc. Published 2024 by John Wiley & Sons, Inc.
Companion website: www.wiley.com/go/liebenberg/energy_systems

Week 5 – Vapor Power Cycles and Vapor Refrigeration Cycles

Fossil fuel-fired steam power plants that use water as a working fluid are ubiquitous for generating electricity. Vapor-compression cycles using a refrigerant gas are predominantly used for refrigeration and air-conditioning. The working fluids in steam power plants and vapor-compression cycles are alternatively vaporized and condensed. They are therefore classified as vapor power cycles and vapor-compression cycles, respectively.

In 2022, over 60% of the world's electrical energy was generated by steam power plants that required the burning of fossil fuels as their source of energy. Large-scale fossil-fueled plants provide most of the world's base-load generating capacity. Fossil-fired power generation is also the most significant contributor to China's carbon emissions and is being increasingly used in India and other developing countries.

Using air-conditioners and refrigeration equipment accounted for 7.8% of total greenhouse-gas emissions in 2020. Rising demand for space cooling is putting enormous strain on electricity systems in many countries. It is also increasing the emission of greenhouse gases.

Due to the extensive use of steam power plants and vapor refrigeration systems, even small improvements in thermal efficiency can result in immense global energy savings, with a concomitant curtailment of harmful emissions, especially CO_2.

Chapter 13 – Steam power plants: Ideal Rankine cycle; Rankine cycle with reheating of steam, regeneration of heat, and cogeneration of heat and mechanical power; Rankine cycle combined with a Brayton cycle; outlook

Chapter 14 – Refrigeration and air-conditioning: Vapor-compression refrigeration systems; refrigeration, air-conditioning, and heat pumping; absorption refrigeration; outlook

Chapter 15 – Controlling emissions from vapor power and vapor refrigeration cycles: Controlling NO_x, SO_x, and particulate emissions from coal combustion; controlling emissions from the combustion of natural gas (or methane); outlook

Mini Project 5 – Modern fossil fuel-fired power plant

13

Steam Power Plants

A vapor power cycle, which uses water as a working fluid, is commonly called a *steam power plant* (Figure 13.1). The water in a steam power plant is alternatively vaporized and condensed, as opposed to the gas cycles covered in Chapters 10 and 11, where the working fluid (usually air or helium) remains in the gaseous phase. During the condensation and evaporation of a pure fluid such as water, the fluid pressure and temperature remain the same.

Water is boiled by heat received from the burning of fossil fuels (natural gas, coal, or oil) or from nuclear reactions. Steam power plants may also be operated in conjunction with gas turbines (in so-called *combined-cycle* power plants), in which case the exhaust heat from the gas turbine is used to boil water for the steam power plant.

Steam power plants may be modified to improve their thermodynamic performance. These modifications include additional heat exchangers, pumps, and turbines, which save large amounts of primary energy and help curtail harmful emissions such as CO_2. Modern fossil fuel-fired steam power plants also use *flue gas treatment systems* such as selective catalytic reduction (SCR) for NO_x removal, electrostatic precipitation (ESP) to remove particulate matter, and wet flue gas desulfurization (FGD) to remove SO_x. This is explained in Chapter 15, along with the rudiments of carbon capture and storage (CCS).

Electricity generation from any source affects the local and/or global environment. For example, coal-fired electricity generation is associated with *externalities* beyond the release of CO_2, including an average of 24.5 deaths, 225 serious illnesses, and 13,288 minor illness per terawatt hour of electricity generated (Markandya and Wilkinson 2007; Finkelman et al. 2021). Despite these sustainability issues, and the rise of solar and wind energy, the production of coal has never been higher. This has been brought about by droughts that have curtailed hydropower, soaring natural gas prices, and the fact that growing economies demand ever-increasing energy. Worldwide, coal remained the largest single source of electricity in 2021 (IEA 2022).

Reimagine Our Future

Twenty-five percent of CO_2 emitted today will still be affecting the climate 1000 years from now (see Section 5.1). About 10% will still be affecting climate 100,000 years from now. Does this not suggest that we should adopt a long-term perspective to our production and use of energy? Why is this not happening?

Energy Systems: A Project-Based Approach to Sustainability Thinking for Energy Conversion Systems,
First Edition. Leon Liebenberg.
© 2024 John Wiley & Sons, Inc. Published 2024 by John Wiley & Sons, Inc.
Companion website: www.wiley.com/go/liebenberg/energy_systems

Figure 13.1 The 1560 MW$_e$ steam power plant in Eemshaven, The Netherlands, uses two 800-MW$_e$ Siemens SST5-6000 steam turbines. The boiler burns hard coal and biomass (e.g., wood pellets), which provides "ultra-supercritical" steam at 600 °C and 27.5 MPa. The power plant has a thermal efficiency of 46%. *Source:* With permission from RWE.

13.1 Ideal Rankine Cycle

Steam power plants typically operate on the Rankine cycle (Rankine 1853). The *ideal* Rankine cycle comprises a feedwater pump, a boiler, a turbine, and a condenser. The ideal Rankine cycle does not involve any internal irreversibilities or entropy generation, as illustrated in Figure 13.2. The working fluid in the ideal Rankine cycle is water. However, Rankine cycles can also operate with organic fluids, as discussed in Chapter 26.

The Rankine cycle described here is also called a *superheat cycle*. This type of cycle is common with fossil-fired and biomass-fired boilers that produce high-temperature (and dry) steam. We will, however, also cover *saturated steam cycles,* which are common with lower temperature nuclear reactor steam power plants (Chapter 17) and with geothermal steam power plants (Chapter 26). The ideal Rankine cycle comprises the following four processes (see also Figure 13.2):

- *Process 1 to 2*: The working fluid (water) is adiabatically and isentropically pumped from a low to a high pressure.
- *Process 2 to 3*: The high-pressure water enters the boiler where it is heated at a constant pressure by an external heat source (e.g., the combustion of fossil fuels or a nuclear reactor).
  The boiler could be sub-divided into three sections: an "economizer" (2 to 2a), an "evaporator" (2a to 2b), and a "superheater" (2b to 3).
- *Process 3 to 4*: The dry (superheated) water vapor, or *steam*, expands adiabatically and isentropically through a turbine, generating power. This decreases the temperature and pressure of the steam, and some condensation may occur.
- *Process 4 to 1*: The steam enters a condenser, where the steam discards heat to the environment (cool air or water), where it condenses at a constant pressure to liquid water.

13.1.1 Process 1 to 2: Pump

Using the first law of thermodynamics for an open system (Eq. (2.8) and Table 2.1), disregarding changes in the working fluid's kinetic energy and potential energy, and assuming an adiabatic (and isentropic) pumping process, the steady flow energy equation becomes:

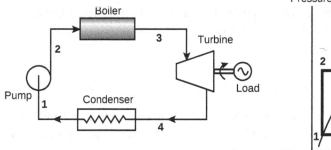

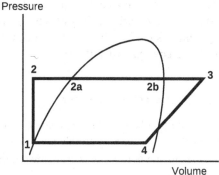

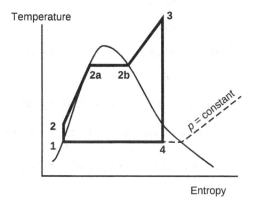

Figure 13.2 Fundamental components of an ideal Rankine cycle and accompanying *p-h* and *T-s* process diagrams. The working fluid is assumed to be water.

$$q\!\!\!/_{12} - w_{12} = h_2 - h_1$$

$$\therefore w_{12} = h_1 - h_2 \tag{13.1}$$

Due to liquid water being highly incompressible, we may assume that $\rho_1 = \rho_2 = \rho_f$ and that $v_1 = v_2 = v_f$. It is also clear from Figure 13.2 that the entropy hardly changes (quasi-isentropic) in the subcooled regime, so we can assume that $s_1 = s_2$, i.e., $ds = 0$. Equation (3.23) then becomes:

$$\int_1^2 T \, ds = \int_1^2 dh - v \int_1^2 dp$$

and, when combined with Eq. (13.1):

$$w_{12} = h_{12} = v \, (p_1 - p_2) \equiv h_1 - h_2 \tag{13.2}$$

The power required for the pumping process is obtained by multiplying the mass flow rate of the working fluid (water or steam) with the enthalpy change over the pump:

$$\dot{W}_{12} = \dot{m} \, w_{12} = \dot{m}(h_1 - h_2) \equiv \dot{m}v \, (p_1 - p_2) \tag{13.3}$$

It is customary to assume that the fluid at state 1 is a saturated liquid.

$$h_1 = h_{f,T_1} \tag{13.4}$$

and that the fluid at state 2 is a compressed liquid (or "subcooled liquid"); h_1 and h_2 can then be read off so-called "steam tables." These thermophysical property tables are available from any good thermodynamics textbook or from online sites such as those of the Engineering Equation Solver (EES) software package (https://fchartsoftware.com/ees) or from the Thermophysical Properties of Fluid Systems of the National Institute of Standards and Technology (NIST) (https://webbook.nist.gov/chemistry/fluid).

13.1.2 Process 2 to 3: Boiler

Using the first law of thermodynamics for an open system (Eq. (2.8) and Table 2.1), disregarding changes in the working fluid's kinetic energy and potential energy, and disregarding any work transfer in the boiler, the steady flow energy equation becomes:

$$q_{23} - \cancel{w_{23}} = h_3 - h_2$$
$$\therefore q_{23} = h_3 - h_2$$

(13.5)

and the rate of heat transfer in the boiler is:

$$\dot{Q}_{23} = \dot{m}\, q_{23} = \dot{m}(h_3 - h_2)$$

(13.6)

Boilers may be classified in terms of their maximum pressure and temperature levels:

- *Subcritical* boilers operate up to 22 MPa and 374 °C (the critical point of water)
- *Supercritical* boilers operate at higher than 22 MPa and 566 °C
- *Ultra-supercritical* boilers operate at higher than 22 MPa and between 566 and 760 °C

13.1.3 Process 3 to 4: Turbine

Using the first law of thermodynamics for an open system (Eq. (2.8) and Table 2.1), disregarding changes in the working fluid's kinetic energy and potential energy, and assuming the adiabatic (and isentropic) expansion of steam in the turbine, the steady-flow energy equation becomes:

$$\cancel{q_{34}} - w_{34} = h_4 - h_3$$
$$\therefore w_{34} = h_3 - h_4$$

(13.7)

and the power produced by the turbine is:

$$\dot{W}_{34} = \dot{m}\, w_{34} = \dot{m}(h_3 - h_4)$$

(13.8)

Turbines could come in configurations comprising high-pressure, intermediate-pressure, or low-pressure stages. Steam turbines can further be classified as condensing or non-condensing steam turbines. With *condensing steam turbines*, steam is condensed at below atmospheric pressure to gain the maximum amount of energy from it. In *non-condensing steam turbines*, steam leaves the turbine at above atmospheric pressure. It is then used for heating or for other required processes before being returned as water to the boiler.

13.1.4 Process 4 to 1: Condenser

Using the first law of thermodynamics for an open system (Eq. (2.8) and Table 2.1), disregarding changes in the working fluid's kinetic energy and potential energy, and disregarding work transfer in the condenser, the steady-flow energy equation becomes:

Table 13.1 Reference values for important energy parameters of modern steam power plants used for producing electricity.

Nominal power output (MW$_e$)	20–1900
Net efficiency based on *LHV*, (%)	30–48
Boiler fuel	Fossil fuel (natural gas, oil, coal), biomass, or nuclear power
Working fluid	Water or organic fluids (e.g., pentane, propane, R-245fa)
Turbine entry temperature (°C)	500–650
Maximum pressure (MPa)	35
Nominal turbine rotational speed (rpm)	1800 or 3600 (for 60-Hz electrical generators) 1500 or 3000 (for 50-Hz electrical generators)

$$q_{41} - w_{41} = h_1 - h_4$$
$$\therefore q_{41} = (h_1 - h_4)$$

(13.9)

and the condenser's heat rate is:

$$\dot{Q}_{41} = \dot{m}\, q_{41} = \dot{m}(h_1 - h_4)$$

(13.10)

Condensers can use water to cool down the process steam. These are called *wet-cooled condensers*. If power plants do not have access to cooling water, air may be used to cool down the steam. These are called *dry-cooled condensers*.

13.1.5 Thermal Efficiency of a Steam Power Plant

The thermal efficiency of the power plant is:

$$\eta_{th} = \frac{w_{net}}{q_{in}} = \frac{w_{34} - |w_{12}|}{q_{23}}$$

(13.11)

Table 13.1 summarizes some salient characteristics of modern steam power plants used to produce electricity.

Example 13.1 Steam Power Plant Operating on an Ideal Rankine Cycle
Consider a steam power plant operating on the ideal Rankine cycle (Figure 13.2). Steam enters the turbine at 7 MPa and 500 °C and is condensed at a pressure of 20 kPa. Water enters the pump in saturated liquid state. Assume that $p_1 = p_4 = 20$ kPa, $p_2 = p_3 = 7$ MPa, and that the pump and turbine are adiabatic and isentropic. Assume that heat is transferred from the boiler to the water at 1600 K. Also assume that the condensing water discards its heat to the environment at 298 K.

a) If the steam flows at 50 kg/s through the cycle, how much power does the turbine produce? Also calculate the heat rate of the condenser. (*Hint:* First analyze each of the four processes.)
b) Determine the thermal efficiency of the cycle.
c) Determine the entropy production of each of the four processes and of the total cycle.

Required: We must find (a) the power output of the turbine, (b) the thermal efficiency of the cycle, and (c) the irreversibility of the cycle.

Solution strategy: We will use Eqs. (13.1) to (13.9) to evaluate the four processes, and then Eq. (13.10) to calculate the thermal efficiency. We can calculate irreversibilities using Eq. (3.58). We will use the property tables of steam available in the EES software package to find the enthalpies.

Assumptions: Assume an ideal Rankine cycle, with adiabatic and isentropic compression (1–2) and expansion (3–4) processes. Assume state point 1 is a saturated liquid ($x = 0$).

Analysis:

a) Power output of the turbine

To calculate the power output of the turbine, we need to analyze each of the first three processes.

Process 1 to 2: Pump

$$w_{12} = h_{12} = v\,(p_1 - p_2) \equiv h_1 - h_2$$

$p_1 = 20\,\text{kPa}$, saturated liquid
$p_2 = 7\,\text{MPa}$, subcooled liquid

Using EES, we find the specific volume of saturated liquid water to be $v = v_{f,20\text{kPa}} = 0.001017\,\text{m}^3/\text{kg}$.

$$\therefore w_{12} = v\,(p_1 - p_2) = (0.001017\,\text{m}^3/\text{kg})(20 - 7000)\,\text{kPa} = -7.09\,\text{kJ/kg}$$

With a steam flow rate of 50 kg/s, the pump must provide $\dot{W}_{12} = \dot{m}\,w_{12} = (50\,\text{kg/s})(-7.09\,\text{kJ/kg}) = -354.5\,\text{kW}$

Note: You could also calculate the specific work using $w_{12} = h_1 - h_2$. From EES:

$$h_1 = h_{f,20\text{kPa}} = 251.42\,\text{kJ/kg}$$

The pumping process is isentropic, so $s_1 = s_2 = s_{f,20\text{kPa}} = 0.8320\,\text{kJ/kgK}$

$$h_2 = h_{s_1,7\,\text{MPa}} = 258.54\,\text{kJ/kg}$$

Therefore, $w_{12} = h_1 - h_2 = 251.42\,\text{kJ/kg} - 258.54\,\text{kJ/kg} = -7.12\,\text{kJ/kg}$, which translates to a required pump input of $\dot{W}_{12} = \dot{m}\,w_{12} = (50\,\text{kg/s})(-7.12\,\text{kJ/kg}) = -356\,\text{kW}$. This amount differs slightly from that calculated using specific volume and pressure rise (-354.5 kW). This small discrepancy is ascribed to our assumption that water is incompressible. Liquid water is, however, slightly compressible, which means that state points 1 and 2 will have slightly differing temperatures. We will ignore this discrepancy and assume the value of -354.5 kW for the pumping power.

Process 2 to 3: Boiler

$$q_{23} = h_3 - h_2$$

and $p_3 = 7\,\text{MPa}$ and $T_3 = 500\,°\text{C}$

Using EES, we find $h_3 = 3411.4\,\text{kJ/kg}$

Previously, we found that $h_2 = 258.54\,\text{kJ/kg}$.

Therefore, $q_{23} = h_3 - h_2 = 3411.4\,\text{kJ/kg} - 258.54\,\text{kJ/kg} = 3152.9\,\text{kJ/kg}$ and $\dot{Q}_{23} = \dot{m}\,q_{23} = \dot{m}(h_3 - h_2) = (50\,\text{kg/s})(3,152.9\,\text{kJ/kg}) = 157.6\,\text{MW}$.

Process 3 to 4: Turbine

$$w_{34} = h_3 - h_4$$

We know that $p_4 = 20$ kPa and that the steam expands isentropically through the turbine, so $s_4 = s_3 = s_{7\text{MPa, 500 °C}} = 6.8\,\text{kJ/kgK}$

and $s_4 = s_{f,20\,\text{kPa}} + x_4\,s_{fg,20\,\text{kPa}} \equiv 6.8\,\text{kJ/kgK}$.

$$s_4 = s_{f,20\,\text{kPa}} + x_4 s_{fg,20\,\text{kPa}} \equiv 6.8\,\text{kJ/kgK}$$

$$\therefore x_4 = \frac{6.8\,\text{kJ/kgK} - s_{f,20\,\text{kPa}}}{s_{fg,20\,\text{kPa}}} = \frac{(6.8\,\text{kJ/kgK}) - (0.832\,\text{kJ/kgK})}{7.0752\,\text{kJ/kgK}} = 0.844\,\text{kJ/kgK}$$

We can now calculate the enthalpy at state 4: $h_4 = h_{f,20 \text{ kPa}} + x_4 h_{fg,20 \text{ kPa}} = (251.42 \text{ kJ/kg}) + (0.844)(2357.5 \text{ kJ/kg}) = 2241.2 \text{ kJ/kg}$

Previously, we found that $h_3 = 3411.4 \text{ kJ/kg}$.

Therefore, $w_{34} = h_3 - h_4 = 3411.4 \text{ kJ/kg} - 2241.2 \text{ kJ/kg} = 1170.2 \text{ kJ/kg}$.

With a steam flow rate of 50 kg/s, the power output of the turbine is: $\dot{W}_{34} = \dot{m}\, w_{34} = (50 \text{ kg/s})(1170.2 \text{ kJ/kg}) = 58.5 \text{ MW}$

The turbine produces 58.5 MW.

Process 4 to 1: Condenser

$$q_{41} = (h_1 - h_4) = (251.42 \text{ kJ/kg}) - (2241.2 \text{ kJ/kg}) = -1989.8 \text{ kJ/kg}$$

With a steam flow rate of 50 kg/s, the condenser discards heat to the environment at a rate of $\dot{W}_{41} = \dot{m}\, w_{41} = (50 \text{ kg/s})(-1989.8 \text{ kJ/kg}) = -99.5 \text{ MW}$

The condenser discards 99.5 MW to the environment.

b) Thermal efficiency of the power plant

$$\eta_{th} = \frac{\dot{W}_{34} - |\dot{W}_{12}|}{\dot{Q}_{23}} = \frac{58,500 \text{ kW} - 356 \text{ kW}}{157,600} \approx 0.37$$

The Rankine cycle has a thermal efficiency of 37%.

c) Entropy generation and irreversibilities

We can calculate irreversibilities from Eq. (3.58), which reduce to the following if we disregard changes in kinetic and potential energies and assume steady-state conditions:

$$\sigma = \frac{\dot{\sigma}}{\dot{m}} = \sum_j \left(1 - \frac{T_0}{T_j}\right) \frac{Q_j}{T_0} - \frac{W}{T_0} + \frac{1}{T_0}[(h_1 - h_2) - T_0(s_1 - s_2)]$$

For processes 1 to 2 and 3 to 4, the processes are reversible and adiabatic, hence also isentropic. The work transfer will be identically equal to the change in enthalpies, which means that the entropy production for both these processes will be 0:

$$\sigma_{12} = \sum_j \left(1 - \frac{T_0}{T_j}\right) \frac{Q_j}{T_0} - \frac{W}{T_0} + \frac{1}{T_0}[(h_1 - h_2) - T_0(s_1 - s_2)] = 0$$

$$\sigma_{34} = \sum_j \left(1 - \frac{T_0}{T_j}\right) \frac{Q_j}{T_0} - \frac{W}{T_0} + \frac{1}{T_0}[(h_2 - h_3) - T_0(s_2 - s_3)] = 0$$

For Process 2 to 3, the entropy change can be determined by evaluating the change in entropy using values from the properties of steam.

$$s_2 = s_1 = s_{f,20\text{kPa}} = 0.832 \text{ kJ/kgK}$$

State point 3 is in the superheated regime, so, from the superheated tables of water in EES:

$$s_3 = s_{7 \text{ MPa, 500 °C}} = 6.8 \text{ kJ/kgK}$$

Therefore, $\Delta s_{23} = s_3 - s_2 = (6.8 \text{ kJ/kgK}) - (0.832 \text{ kJ/kgK}) = 5.97 \text{ kJ/kgK}$

For process 4 to 1, the entropy change in the condenser can be determined as follows:

$$s_4 = s_3 = 6.8 \text{ kJ/kgK}$$

$$s_1 = s_{f,20\text{kPa}} = 0.832 \text{ kJ/kgK}$$

Therefore, $\Delta s_{41} = s_1 - s_4 = (0.832 \text{ kJ/kgK}) - (6.8 \text{ kJ/kgK}) = -5.96 \text{ kJ/kgK}$

The entropy production for processes 2 to 3 and 4 to 1 can now be evaluated. Again, the work transfer will be identically equal to the change in enthalpies for both processes so that those terms cancel each other. For Process 2 to 3, the high-temperature reservoir is assumed to be at 1600 K and the ambient air at 298 K, which means that we will work with $-(q_{23})$ according to our sign convention for heat leaving a system:

$$\sigma_{23} = T_0\left(\frac{-q_{23}}{T_3} + s_3 - s_2\right) = (298\text{ K})\left[\frac{-3152.9\text{ kJ/kg}}{1600\text{ K}} + (5.97\text{ kJ/kgK})\right] = 1191.8\text{ kJ/kg}$$

For process 4 to 1, heat is lost to the environment with $q_{41} = -1989.8$ kJ/kg

$$\sigma_{41} = T_0\left(\frac{q_{41}}{T_{low}} + s_1 - s_4\right) = (298\text{ K})\left[\frac{-(-1989.8\text{ kJ/kg})}{298\text{ K}} + (-5.96\text{ kJ/kgK})\right] = 213.7\text{ kJ/kg}$$

The results may be summarized and checked in tabular form:

Process	q (kJ/kg)	w (kJ/kg)	Δh (kJ/kg)	$s_e - s_i$ (kJ/kgK)	σ (kJ/kg)
1 to 2	0	−7.09	$h_2 - h_1 = 7.09$	0	0
2 to 3	3152.9	0	$h_3 - h_2 = 3152.9$	5.97	1191.8
3 to 4	0	1170.2	$h_4 - h_3 = -1170.2$	0	0
4 to 1	−1989.8	0	$h_1 - h_4 = -1989.8$	−5.97	213.7
Cycle:	$q_{net} \approx 1163$	$w_{net} \approx 1163$	$\Delta h_{cycle} \approx 0$	0	1405.5

Discussion:
The efficiency of this ideal cycle will be lower if one works with the isentropic efficiencies of the pump and turbine. However, we assumed an isentropic pump and turbine. The entropy production in those two components is therefore zero. The two heat transfer processes occur over finite temperature differences (between the system and its surroundings) and are therefore accompanied by entropy production. The boiler has more than five times the irreversibility of the condenser (1191.8 versus 213.7 kJ/kgK). This is expected as the temperature difference between the system and its surroundings is large, which represents a large irreversibility. It is challenging to reduce the irreversibility of a boiler as its job is to produce heat at a high temperature.

Even though the condenser discards about 63% of the energy that the boiler produced (99.5 MW discarded; 157.6 MW received), the condenser discards the heat at a low temperature of around 60 °C (equivalent to a saturation pressure of 20 kPa) to an environment that is at 298 K. The irreversibility in the condenser is therefore comparatively low. This discarded energy is deemed to be of low thermal quality, compared to that of the boiler that is at 1600 K. Nonetheless, instead of merely discarding heat to the environment, it could be used to heat process water, for instance.

The steam exiting the turbine is at around 86% vapor quality. The "rule-of-thumb" is to operate a turbine at vapor qualities greater than 90% to avoid blade erosion caused by impinging water droplets. This Rankine cycle is therefore not optimal and requires some modifications to make it practical.

The efficiency of the basic Rankine cycle could be improved by increasing the maximum temperature at which heat is transferred to the working fluid (water) or by decreasing the temperature at which heat is rejected to the surroundings. However, due to the material limits of turbine blades (around 700 °C), it is not technically feasible to increase the turbine inlet temperature to beyond 700 °C. Unlike gas turbine blades that employ air cooling, it is not possible to cool steam turbine blades in a practical manner. Also, decreasing the condenser temperature too low will result in the steam exiting the turbine in a wet state, which will result in blade erosion.

A workaround to these predicaments is by either *reheating the steam* or by *regenerating heat*, each of which considerably raises the thermal efficiency of the Rankine cycle. Instead of the condenser discarding heat to the environment, that energy could be used to heat process water. If the Rankine cycle is used to produce useful heat and electricity, the cycle is termed a *cogeneration cycle*. Further, the Rankine cycle could receive heat from the exhaust of a gas turbine (from a Brayton cycle). This is called a *combined cycle and* achieves the highest thermal efficiencies of utility power plants. Each of these modifications of the Rankine cycle are briefly discussed.

Reimagine Our Future

How might we reimagine our future if we think comprehensively about living sustainably in terms of our energy production and use? Might we need to address the fact that corporate influence sometimes trumps citizen representation? Would we need to ban commercials aimed at children? Would we need to address the fact that the media is sometimes more concerned with advertising than with providing insight and truth?

13.2 Rankine Cycle with Reheating of Steam

With reheating, advantage is taken of the increased efficiencies at higher boiler pressures without facing the problem of excessive moisture at the final stages of the turbine. Reheating steam involves the use of two or more turbine stages (such as a high-pressure stage and a low-pressure stage). The steam is expanded through the high-pressure turbine, followed by expansion through the low-pressure turbine. The steam exiting the first turbine section is routed back to the boiler where the steam is reheated before it enters the low-pressure turbine. The steam exiting the second turbine has a high vapor quality, thus ensuring low probability of erosion damage to the turbine blades (Figure 13.3).

Analysis of the pump and condenser takes place as before, but the boiler and turbine calculations must be updated, as shown below:

$$\text{Boiler}: q_{in} = q_{primary} + q_{reheat} = (h_3 - h_2) + (h_5 - h_4) \tag{13.12}$$

$$\text{Turbine}: w_{out} = w_{turbine\,1} + w_{turbine\,2} = q_{in} = (h_3 - h_4) + (h_5 - h_6) \tag{13.13}$$

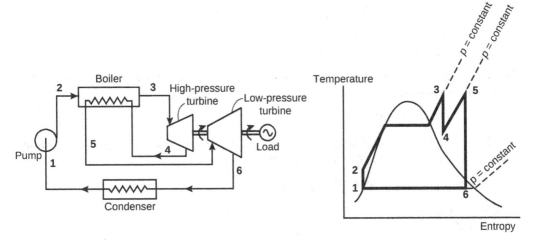

Figure 13.3 Components of the Rankine cycle with steam reheating and accompanying *T-s* process diagram. The working fluid is assumed to be water.

Example 13.2 Steam Power Plant with Reheating of Steam

Consider a modification of the Rankine cycle in Example 13.1 by reheating the steam. (Use the *T-s* process diagram shown in Figure 13.3.) The boiler pressure is 7 MPa and the condenser pressure is 20 kPa. The water enters the high-pressure turbine at 500 °C and leaves the high-pressure turbine at 2 MPa. Steam at a flow rate of 50 kg/s enters the low-pressure turbine at 500 °C. The steam enters the pump as a saturated liquid. Calculate the thermal efficiency of this modified Rankine cycle.

Required: We must find the thermal efficiency of the reheat Rankine cycle.

Solution strategy: We will modify the analysis for the simple Rankine cycle, Eqs. (13.1) to (13.11), by using the reheat relationships in Eqs. (3.12) and (3.13). We will use the property tables of steam available in the EES software package to find the enthalpies.

Assumptions: Assume an ideal Rankine cycle, with adiabatic and isentropic pump and turbine processes. Assume state point 1 is a saturated liquid ($x = 0$).

Analysis:

To calculate the efficiency of the cycle, we need to analyze each of the processes.

Process 1 to 2: Pump

$$w_{12} = h_{12} = v\,(p_1 - p_2) \equiv h_1 - h_2$$

$p_1 = 20$ kPa, saturated liquid
$p_2 = 7$ MPa, subcooled liquid

Using EES, we find the specific volume of saturated liquid water to be: $v = v_{f,20\text{kPa}} = 0.001017\ \text{m}^3/\text{kg}$.

$$\therefore w_{12} = v\,(p_1 - p_2) = \left(0.001017\ \text{m}^3/\text{kg}\right)(20 - 7000)\ \text{kPa} = -7.09\ \text{kJ/kg}$$

With a steam flow rate of 50 kg/s, the pump must provide $\dot{W}_{12} = \dot{m}\,w_{12} = (50\ \text{kg/s})(-7.09\ \text{kJ/kg}) = -354.5$ kW.

Process 2 to 3 and Process 4 to 5: Boiler and steam-reheating

$$q_{in} = q_{primary} + q_{reheat} = (h_3 - h_2) + (h_5 - h_4)$$

with $p_3 = 7$ MPa and $T_3 = 500$ °C

$$h_1 = h_{f,20\ \text{kPa}} = 251.42\ \text{kJ/kg}$$
$$h_2 = h_{s_1,7\ \text{MPa}} = 258.54\ \text{kJ/kg}$$
$$h_3 = h_{500°C,7\ \text{MPa}} = 3{,}411.4\ \text{kJ/kg}$$

To find the enthalpy at state point 4, we know that the pressure is 500 kPa. We also know that steam expands isentropically, so $s_3 = s_4 = s_{500°C,7\ \text{MPa}} = 6.8\ \text{kJ/kgK}$

$$\therefore h_4 = h_{2\ \text{MPa},\ s_4} = 3043.1\ \text{kJ/kg}$$

To find the enthalpy at state point 5, we know that the reheat process happens at a constant pressure, so $p_4 = p_5 = 2$ MPa. Further, $T_5 = 500$ °C.

$$\therefore h_5 = h_{2\ \text{MPa},500°C} = 3468.3\ \text{kJ/kg}$$

and $s_5 = s_{2\ \text{MPa},500°C} = 7.4337\ \text{kJ/kgK}$

Therefore,

$$q_{25} = (h_3 - h_2) + (h_5 - h_4) = (3411.4 - 258.54)\ \text{kJ/kg} + (3468.3 - 3043.1\ \text{kJ/kg})\text{kJ/kg}$$
$$q_{25} = 3578.1\ \text{kJ/kg}.$$

Process 3 to 4 and Process 5 to 6: High-pressure and low-pressure turbines

High-pressure turbine: $w_{34} = h_3 - h_4 = 3411.4\,\text{kJ/kg} - 3043.1\,\text{kJ/kg} = 368.3\,\text{kJ/kg}$
Low-pressure turbine: $w_{56} = h_5 - h_6$
To find the enthalpy at state point 6, we know that the steam expands isentropically, so

$$s_5 = s_6 = 7.4337\,\text{kJ/kgK} = s_f + x_6 h_{fg}$$

$$\therefore x_6 = \frac{s_6 - s_{f,20\text{kPa}}}{s_{fg,20\text{kPa}}} = \frac{7.4337\,\text{kJ/kgK} - 0.832\,\text{kJ/kgK}}{7.0752\,\text{kJ/kgK}} = 0.93$$

The steam quality at the exit of the low-pressure turbine is therefore 93% and $h_6 = h_{f,20\text{kPa}} + x_6 h_{fg,20\text{kPa}} = (251.42\,\text{kJ/kg}) + 0.93(2357.5\,\text{kJ/kg}) = 2443.9\,\text{kJ/kg}$

$$\therefore w_{56} = h_5 - h_6 = (3468.3\,\text{kJ/kg}) - (2443.9\,\text{kJ/kg}) = 1024.4\,\text{kJ/kg}$$

Process 6 to 1: Condenser

$$q_{61} = (h_1 - h_6) = (251.42\,\text{kJ/kg}) - (2443.9\,\text{kJ/kg}) = -2192.5\,\text{kJ/kg}$$

Thermal efficiency of the power plant

$$\eta_{th} = \frac{w_{net}}{q_{in}} = \frac{w_{34} + w_{56} - |w_{12}|}{q_{23}} = \frac{368.3\,\text{kJ/kg} + 1024.4\,\text{kJ/kg} - 7.09\,\text{kJ/kg}}{3578.1\,\text{kJ/kg}} = 0.39$$

The Rankine cycle with reheat has a thermal efficiency of 39%.

Discussion:
The Rankine cycle with reheat has a thermal efficiency of 39% compared to 37% of the unmodified Rankine cycle in Example 13.1. Importantly, the reheat cycle has steam exiting the low-pressure turbine at a vapor quality of 93% (which is good) compared to the paltry vapor quality of 84% of the unmodified cycle. Further, the two turbines in the reheat cycle have a combined work output of $368.3\,\text{kJ/kg} + 1024.4\,\text{kJ/kg} = 1392.7\,\text{kJ/kg}$, which translates to a combined power output of $\dot{W}_{total} = \dot{m}\,(w_{34} + w_{56}) = (50\,\text{kg/s})(1392.7\,\text{kJ/kg}) = 69.6\,\text{MW}$. The turbine of the unmodified Rankine cycle has a power output of only 58.5 MW.

Reimagine Our Future

Prolonged power outages are occurring with increasing frequency and magnitude, often due to extreme weather events but aided by bad planning and corrupt governments. Does this not reinforce the point that it might not be possible to increase the resilience of an unsustainable system? Surely, sooner or later, the careless exploitation of land, water, forests, biota, and people will lead to disaffection, overshoot of Earth's carrying capacity, and societal collapse?

13.3 Rankine Cycle with Regeneration of Heat

The thermal efficiency of a simple Rankine cycle can also be improved by increasing the average temperature of the working fluid (water) in the boiler. This can be achieved with the regeneration of heat. This entails extracting a portion of the steam exiting the turbine before it enters the condenser and using that steam to preheat the working fluid (water) before it enters the boiler. The heat regeneration occurs in a heat exchanger called a *feedwater heater* (Figure 13.4).

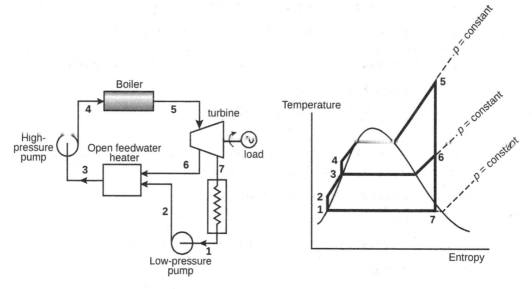

Figure 13.4 Components of the Rankine cycle with regenerative heating and accompanying *T-s* process diagram. The working fluid is assumed to be water, and the heat regenerator is an open feedwater heater, where the steam interacts directly with water. Instead of such direct-contact heat exchangers, use can also be made of indirect contact (or surface) heaters. However, at least one direct-contact (open feedwater) heater is required to act as a de-aerator, which is needed to remove dissolved oxygen from the feedwater. This minimizes corrosion of the boiler tubes.

An open (or direct-contact) feedwater heater is a kind of mixing chamber, where the steam extracted from the turbine mixes with the feedwater exiting the pump. Ideally, the mixture leaves the heater as a saturated liquid at heater pressure.

The analysis of the simple Rankine cycle must therefore be modified using the following relations:

Boiler:
$$q_{in} = h_5 - h_4 \tag{13.14}$$
$$q_{out} = (1-y)(h_7 - h_1) \tag{13.15}$$

where y is the mass fraction of extracted steam:

$$y = \frac{\dot{m}_6}{\dot{m}_5} \tag{13.16}$$

Turbine:
$$w_{out} = (h_5 - h_6) + (1-y)(h_6 - h_7) \tag{13.17}$$

Pumps:
$$w_{in,pump1} = v_1(p_2 - p_1) \tag{13.18}$$

$$w_{in,pump2} = v_3(p_4 - p_3) \tag{13.19}$$

$$w_{in,total} = (1-y)w_{in,pump\,1} + w_{in,pump\,2} \tag{13.20}$$

One feedwater heater could improve the plant's thermal efficiency by 6%. Two heaters could improve the overall efficiency by 8%. Three heaters will typically improve the overall efficiency by 9%. Although the use of one feedwater heater is almost always economically viable, the use of more than eight feedwater heaters provides diminishing improvements in thermal efficiency and is economically restrictive.

Several variations of these modifications of the Rankine cycle are possible. These include variations in the number and type of feedwater heaters, different methods of driving the feedwater pump(s), and variation of the steam and water conditions and flow rates in various parts of the cycle.

13.4 Rankine Cycle with the Cogeneration of Heat and Mechanical Power

Cogeneration or *combined heat and power* (CHP) involves the steam turbine doing useful work *and* the condenser's heat being used to produce useful *process heat* $(\dot{Q}_{process})$, instead of simply discarding that heat to the environment (Figure 13.5, Box 13.1). Cogeneration (or combined heat and power) can also be employed with gas turbines or with internal combustion engines to recover waste heat. The process heat can be used for space heating and district heating, water heating, or for industrial processes.

The thermal efficiency of a cogeneration (or CHP) plant is called an *energy utilization factor* (ε_{ut}) and may be defined as follows (Horlock 1987):

$$\varepsilon_{ut} = \frac{\text{Net power output} + \text{Process heat delivered}}{\text{Total heat input}}$$

$$\varepsilon_{ut} = \frac{\dot{W}_{net} + \dot{Q}_{process}}{\dot{Q}_{in}} \tag{13.21}$$

Energy utilization factors of greater than 80% are typical for matched cogeneration plants. When the ratio of heat to power $(\dot{Q}_{process}/\dot{W}_{net})$ of the cogeneration plant is not fully matched to the electrical and heat loads, any shortfall in electricity and heat must be made up by buying electricity or raising steam in stand-alone boilers or disposing of any surplus by selling electricity (easy) or heat (difficult). Therefore, to be economically viable, high heat and electrical load factors are required, say more than 60% for both heat loads and electrical loads. A load factor (*LF*) may be defined as follows:

$$LF = \frac{Total\ units\ of\ energy\ consumed}{(Available\ hours) \times (Maximum\ demand)} \tag{13.22}$$

If process heat is in high demand, all the steam could be routed to the process heater ($\dot{m}_4 = \dot{m}_5$ in Figure 13.5) and none to the condenser ($\dot{m}_9 = 0$), thus wasting no heat. Conversely, if there is no

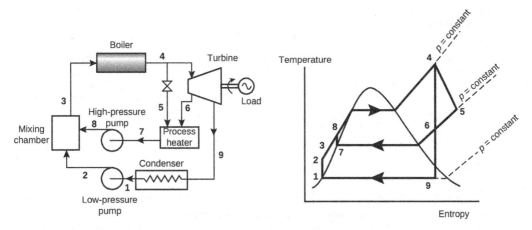

Figure 13.5 Components of the Rankine cycle featuring the cogeneration of heat and mechanical power. *Source:* Adapted from Çengel and Boles (2015).

demand for process heat, all the steam passes through the turbine and the condenser as in an unmodified Rankine cycle ($\dot{m}_5 = \dot{m}_6 = 0$). A series of valves arranges the required steam flow.

In general, the overall efficiency (or energy utilization factor) of a cogeneration system can be of the order of one-and-a-half times greater than the efficiency from separate sources. Such an increase results in large fuel savings and reduces the emission of greenhouse gases.

For analysis of the cogeneration cycle, the simple Rankine cycle must be modified using the following relations (after Çengel and Boles 2015):

$$\dot{Q}_{in} = \dot{m}_2(h_4 - h_2) \tag{13.23}$$

$$\dot{Q}_{out} = \dot{m}_9(h_9 - h_1) \tag{13.24}$$

$$\dot{Q}_{process} = \dot{m}_5 h_5 + \dot{m}_6 h_6 - \dot{m}_7 h_7 \tag{13.25}$$

$$\dot{W}_{turbine} = (\dot{m}_4 - \dot{m}_5)(h_4 - h_6) + \dot{m}_9(h_6 - h_9) \tag{13.26}$$

The power-heat rate ratio $\left(\dot{W}_{turbine}/\dot{Q}_{process}\right)$ determines the proportion of electric power to heat generated in a single cogeneration system. New technologies with higher power-heat ratios include combined cycle cogeneration plants (Section 13.5), modular thermal power stations, or fuel cells

---

Box 13.1   The Advantage of Cogeneration (CHP) Systems

This example illustrates the thermal advantage of operating a cogeneration (CHP) plant rather than individual heat and power plants. The illustration excludes electricity and heat distribution losses.

Prime mover system	Fuel input (units)	Losses (units)	Heat output (units)	Electricity supplied (units)	Fuel conversion efficiency (%)
1) Large fossil fuel-fired steam power plant producing only electricity	100	65	0	35	35
2) Small hot-water boiler at an industrial site	155	31	124	0	80
3) Combination of 1 and 2 as the standard system for the industrial site	255	96	124	35	62 (energy utilization factor)
4) Industrial-scale cogeneration (CHP) plant	185	26	124	35	86 (energy utilization factor)

The fuel energy savings of the cogeneration system may be calculated using the *fuel energy savings ratio* (FESR):

$$FESR = \frac{\dot{Q}_{conventional} - \dot{Q}_{cogeneration}}{\dot{Q}_{conventional}} \tag{13.27}$$

For this cogeneration plant, the savings in fuel energy is 255 − 185 = 70 units or 27% to provide the same service as separate heat and electricity systems.

Table 13.2 Typical heat rate and power ratios for prime movers used in cogeneration plants of various sizes.

Plant	Power-heat rate ratio (kW per kg/h steam)
Combined cycle (gas and steam) power plants	1.0 and higher
Condensing steam power plants	0.2–0.4
Gas turbines	0.2–0.7
Large diesel engines	1–2

(Chapter 21). Table 13.2 summarizes the typical heat-power ratios of prime movers used in combined cycles of various sizes.

13.5 Rankine Cycle Combined with a Brayton Cycle

A *combined cycle* is a combination of two power cycles, which could be used individually for power generation, in such a manner that the heat rejected from the first (or "top") cycle is utilized in the second (or "bottom") cycle. The most common combination is an open-cycle gas turbine (Brayton cycle) as the top cycle and a condensing-steam power plant (Rankine cycle) as the bottom cycle (Figure 13.6).

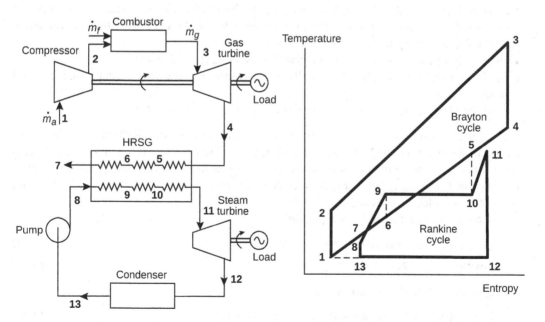

Figure 13.6 Components of the Rankine cycle (steam power plants) combined with a Brayton cycle (gas-turbine power plants) to give a *combined-cycle* power plant. In a combined cycle, the Brayton cycle is termed the *topping* cycle and the Rankine cycle is termed the *bottoming* cycle. The heat-recovery steam generator transfers heat from the hot exhaust gas to the Rankine cycle's cool water: $\dot{m}_a$ is the air mass flow rate; $\dot{m}_f$ is the fuel mass flow rate; $\dot{m}_g$ is the gas mass flow rate; $\dot{m}_s$ is the steam mass flow rate.

The efficiency of the combined cycle exceeds that of either separate cycle. Thermal efficiencies greater than 60% are common, making combined-cycle power plants the most efficient thermal conversion systems for utility power plants. The thermal efficiency of a combined cycle power plant is (Saravanamuttoo et al. 2001):

$$\eta_{combined} \approx \eta_{Brayton} + \left(1 - \eta_{Brayton}\right)\eta_{Rankine} \tag{13.28}$$

For instance, if a Brayton cycle with a thermal efficiency of 27% is combined with a Rankine cycle with 41% efficiency, the combined cycle will have a thermal efficiency of $\eta_{combined} = 0.27 + (1 - 0.27)0.41 = 0.57$ (or 57%). Compared to traditional gas-fired steam power plants, gas turbine combined-cycle power plants can emit 40% less CO_2.

In 2022, gas turbine combined-cycle power plants cost between US$650 and US$1000 per kilowatt, compared to US$3000 per kilowatt for coal and US$6000 (or more) per kilowatt for nuclear power. Combined-cycle power plants also have a larger spatial density ($1300\,W/m^2$) compared to fossil fuel-fired steam power plants ($130\,W/m^2$) (Van Zalk and Behrens 2018). These attributes are augmented by combined-cycle plants that have a rapid start-up time (20–30 min) and quick ramp rates (8–30 MW/min) compared to traditional condensing-boiler steam power plants that could take a week to reach full power output from a cold start. It is therefore unsurprising that gas turbine combined-cycle power plants are leading the field in thermal power generation.

Combined cycles are used in the following applications:

- Power plants that require the highest thermal energy conversion and the least amount of flue gas treatment.
- The repowering of old fossil-fired steam power plants. When boilers need replacing, the turbines and electricity generators usually have many years of useful life. The boilers can then be replaced by gas turbines and *heat-recovery steam generators* (HRSGs). This is usually the only economic alternative to shuttering the plant. (Also see Box 13.2.)
- Upgrading a peak-lopping gas-turbine plant to allow higher load factors.
- In an industrial CHP plant, especially if a factory requires both compressed air and process-steam.
- With the phasing out of coal-fired steam power plants.

The gas turbine exhaust gases in the Brayton cycle exchange heat with the Rankine cycle's boiler using a heat recovery steam generator. The enthalpy rise between the feedwater inlet and the steam outlet must equal the enthalpy drop of the exhaust gases in the HRSG. Using the state points indicated in Figure 13.6, the energy balance between the two cycles (assuming steady-state and steady-flow conditions) is established as follows:

Economizer section (sensible heat transfer):

$$\dot{m}_{steam}(h_9 - h_8) = \dot{m}_{gas}(h_6 - h_7) \equiv \dot{m}_{gas}c_{p,g}(T_6 - T_7) \tag{13.29}$$

$$\text{with} \qquad T_6 = T_{sat} + \Delta T_{pinch} \tag{13.30}$$

Evaporator section (latent heat transfer):

$$\dot{m}_{steam}(h_{10} - h_9) = \dot{m}_{gas}(h_5 - h_6) \equiv \dot{m}_{gas}c_{p,g}(T_5 - T_6) \tag{13.31}$$

Superheater section (sensible heat transfer):

$$\dot{m}_{steam}(h_{11} - h_{10}) = \dot{m}_{gas}(h_4 - h_5) \tag{13.32}$$

The ratio of steam to the gas mass flow rates (y) is:

Box 13.2   Repowering of Old Fossil Fuel-Fired Steam Power with a Combined Gas Turbine/ Steam-Cycle Power Plant

Danskammer Energy in New York wants to replace its 1950s vintage steam turbine power plant with a US$400 million Mitsubishi M501JAC combined cycle plant rated at around 600 MW$_e$. According to Mitsubishi Power, the new plant will incorporate quick-start, fast-ramping technology designed to provide the New York State system operator with added flexibility to handle new additions of renewable energy sources being developed in the Hudson Valley area.

The new technology will dramatically improve the environmental profile of the site and support the state's renewable energy goals. By using air-cooled condensers, the plant will cease using Hudson River water for once-through cooling, and air emissions will be greatly reduced by low-NO$_x$ gas-turbine combustion technology, and use of an SCR (for further NO$_x$ reduction) and a CO catalyst, both installed in the heat recovery steam generator.

Although the plant will initially burn natural gas, the developer says plans call for a fuel–gas mixture containing increasing amounts of green hydrogen produced from surplus renewable energy. Mitsubishi Power, like some other large power companies, produces integrated green hydrogen generation and storage technologies. The new combined-cycle power plant could therefore provide short- and long-duration energy storage infrastructure in New York State. The upgraded Danskammer plant will be able to transition to zero-emission hydrogen when the technology to transport and store hydrogen becomes available. The M501JAC combined cycle will be able to provide the flexibility to partner with the new intermittent renewable generation of the future.

Source: Adapted from Danskammer (2022).

$$y = \frac{\dot{m}_{steam}}{\dot{m}_{gas}} \tag{13.33}$$

The total specific work output from a combined cycle may then be defined:

$$w_{combined} = w_{net,\text{Brayton}} + y\, w_{net,\text{Rankine}} \tag{13.34}$$

The temperature profiles of the gas and steam/water in the HRSG are usually illustrated as shown in Figure 13.7. To aid in the heat exchanger design, a *pinch-point temperature difference* (ΔT_{pinch}) must be stated, which is the closest approach temperature of the two fluid streams. The smaller the ΔT_{pinch}, the larger the heat exchanger must be. Usually, ΔT_{pinch} cannot be less than 20 °C if the HRSG is to be of an economic size.

Example 13.3   Combined Cycle Steam Power Plant

Consider a combined-cycle steam power plant shown schematically in Figure 13.6. The plant is used in an industrial facility that requires compressed air and process-heat.

The 50-MW *gas turbine* (Brayton cycle) has a compressor isentropic efficiency of 88%, and the pressure ratio (p_2/p_1) is 10. The turbine entry temperature (T_3) is 1300 K and the turbine isentropic efficiency is 93%. The exhaust gas temperature (T_4) is 800 K. The combustor burns natural gas with a lower heating value of 47.1 MJ/kg. The fuel–air ratio is 0.018. Assume that the engine has no mechanical or pressure losses and that the combustion process is complete. Assume an average cycle temperature of 800 K and use that to calculate c_p, c_v, and γ.

After passing through the single-pressure *heat-recovery steam generator,* the gases are exhausted to the atmosphere at $T_7 = 420$ K (147 °C). The HRSG is designed for a feedwater inlet temperature of

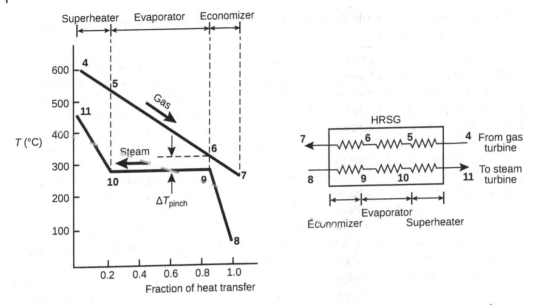

Figure 13.7 Heat-recovery steam generator of Figure 13.7 showing the economizer, evaporator, and superheater sections. The accompanying temperature profile diagram for the HRSG shows the gas temperature and steam/water temperature against the fraction of heat transferred from the gas.

$T_8 = 75\,°C$, a pinch-point temperature difference $(T_6 - T_9)$ of $21.3\,°C$, and a superheated steam outlet temperature (T_{11}) of $450\,°C$.

The *steam power plant* (Rankine cycle) features a condenser that is operating at 10 kPa absolute pressure. The steam turbine has an isentropic efficiency of 90%. The mass flow rate of the steam is $\dot{m}_s = 31.3$ kg/s.

Assume that the cogeneration plant operates at 101 kPa and 289 K. You may ignore any changes in kinetic energy or potential of the working fluid in both the gas turbine and the steam turbine. Also neglect heat losses in the combustor and any mechanical losses in the entire plant.

a) Calculate the exhaust gas mass flow rate of the gas turbine $(\dot{m}_{gas})$.
b) Sketch a temperature profile diagram for the HRSG, i.e., gas temperature and steam/water temperature against the fraction of heat transferred from the gas. Indicate on the diagram the states corresponding to those in Figures 13.7 and 13.8.
c) Perform an energy balance on the economizer section of the HRSG to verify that the steam pressure in the HRSG is around 1 MPa.
d) Calculate the power delivered by the steam turbine.

Required: We must (a) find $\dot{m}_{gas}$, (b) sketch the temperature profile of the two working fluids in the HRSG, (c) perform an energy balance of the HRSH, and (d) calculate the steam turbine's power output.

Solution strategy:

We will use the cogeneration set-up explained in Figures 13.7 and 13.8. For the gas turbine (Brayton cycle) calculations, we will use the theory developed in Chapter 11. We will use the property tables of steam available in the EES software package to find the enthalpy values of steam/water.

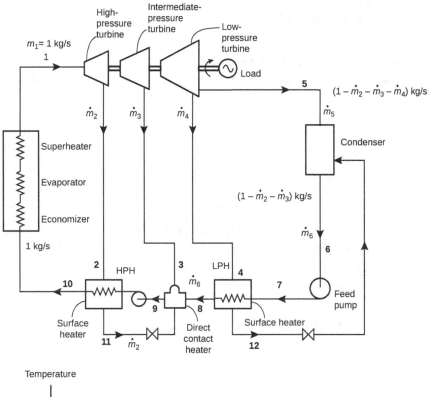

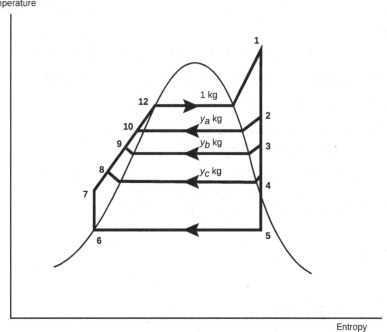

Figure 13.8 Illustration of a steam power plant with three regenerative heaters (HPH = high-pressure heater; LPH = low-pressure heater).

Assumptions: We will assume an average Brayton cycle temperature of 800 K (as explained in Chapter 11), which will be used to calculate c_p, c_v, and γ.

Analysis:

a) Exhaust gas mass flow rate of the gas turbine $(\dot{m}_g)$

At the average cycle temperature of 800 K, $c_{p,800K} = 1.099$ kJ/kgK, $c_{v,800K} = 0.812$ kJ/kgK, and $\gamma_{800K} = \dfrac{c_{p,800K}}{c_{v,800K}} = 1.354$.

Process 1–2: Compressor

From Eq. (11.14): $\dfrac{T_2}{T_1} = \left(\dfrac{p_2}{p_1}\right)^{\frac{\gamma-1}{\gamma}} = r_p^{\frac{\gamma-1}{\gamma}}$, with $r_p = 10$

$$T_2 = T_1 r_p^{\frac{\gamma-1}{\gamma}} = (289 \text{ K})(10)^{(1.354-1)/1.354} = 527.7 \text{ K}$$

From Eq. (11.16): $w_{12,s} = \Delta h_{12,s} \equiv c_p \, \Delta T_{12} = (1.099 \text{ kJ/kgK})(289 - 527.7) \text{ K} = -262.3$ kJ/kg

Considering the isentropic efficiency of the compressor, the actual power input is:

$$w_{12} = \frac{w_{12,s}}{\eta_{s,comp}} = \frac{-262.3 \text{ kJ/kg}}{0.88} = -298.1 \text{ kJ/kg}$$

and $\left(\dfrac{p_2}{p_1}\right) = 10$, so, $p_2 = 10(101 \text{ kPa}) = 1010$ kPa

Process 3 to 4: Turbine

$$T_3 = 1300 \text{ K}$$
$$T_4 = 800 \text{ K}$$
$$w_{34,s} = -\Delta h_{34,s} \equiv -c_p \, \Delta T_{34} = (1.099 \text{ kJ/kgK})(1300 - 800)\text{K} = 549.5 \text{ kJ/kgK}$$

Considering the isentropic efficiency of the turbine, the actual power output is:

$$w_{34} = w_{34,\,s}(\eta_{s,\,comp}) = (549.5 \text{ kJ/kgK})(0.93) = 511.04 \text{ kJ/kgK}$$

The net power can be calculated by rewriting an expression in terms of the air, fuel, and gas mass flow rates to eventually get an expression in terms of the fuel–air ratio (*FAR*):

$$\dot{W}_{net} = 50 \times 10^3 \text{ kW} = \dot{W}_{turbine} - \left|\dot{W}_{compressor}\right|$$

$$\dot{W}_{net} = \dot{W}_{34} - \left|\dot{W}_{12}\right| = \left(\frac{\dot{W}_{34}}{\dot{m}_{gas}}\right)\dot{m}_{gas} - \left(\frac{|\dot{W}_{12}|}{\dot{m}_{air}}\right)\left(\frac{\dot{m}_{air}}{\dot{m}_{gas}}\right)\dot{m}_{gas}$$

$$\dot{W}_{net} = \dot{m}_{gas}\left[\left(\frac{\dot{W}_{34}}{\dot{m}_{gas}}\right) - \left(\frac{|\dot{W}_{12}|}{\dot{m}_{air}}\right)\left(\frac{\dot{m}_{air}}{\dot{m}_{gas}}\right)\right]$$

with $\dot{m}_{gas} = \dot{m}_{air} + \dot{m}_{fuel}$

Therefore, $\dfrac{\dot{m}_{gas}}{\dot{m}_{air}} = 1 + \dfrac{\dot{m}_{fuel}}{\dot{m}_{gas}} = 1 + FAR$

$$\therefore \dot{W}_{net} = 50 \times 10^3 \text{ kW} = \dot{m}_{gas}\left[\left(\frac{\dot{W}_{34}}{\dot{m}_{gas}}\right) - \left(\frac{|\dot{W}_{12}|}{\dot{m}_{air}}\right)\left(\frac{1}{1 + FAR}\right)\right]$$

$$50 \times 10^3 \text{ kJ/s} = \dot{m}_{gas}\left[511.04 \text{ kJ/kgK} - (298.1 \text{ kJ/kgK})\left(\frac{1}{1 + 0.018}\right)\right]$$

$$\therefore \dot{m}_{gas} = 229.1 \text{ kg/s}$$

b) Temperature profile diagram for the HRSG

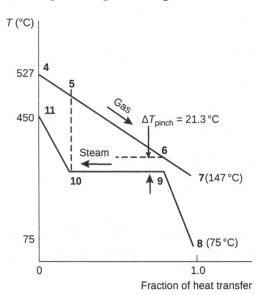

c) Steam pressure in the HRSG

Energy balance for the economizer section of the HRSG:

$$\dot{m}_{steam}(h_9 - h_8) = \dot{m}_{gas}(h_6 - h_7) \equiv \dot{m}_{gas}c_{p,gas}(T_6 - T_7) \tag{13.35}$$

$$h_9 = h_{f,\text{HRSG pressure}}$$

$$h_8 = h_{75\,°\text{C}}$$

$$T_6 = T_{sat} + \Delta T_{pinch}$$

At 1 MPa, using the thermophysical properties available from the EES:

$$h_9 = h_{f,1\,\text{MPa}} = 762.51 \text{ kJ/kg}$$

$$T_{sat,1\,\text{MPa}} = 179.9\,°\text{C}$$

$$h_8 = h_{f,75\,°\text{C}} = 314.03 \text{ kJ/kg}$$

The left-hand side of Eq. (13.35) is:

$$\dot{m}_{steam}(h_9 - h_8) = (31.3 \text{ kg/s})(762.51 \text{ kJ/kg} - 314.03 \text{ kJ/kg}) = 14{,}037 \text{ kW}$$

The right-hand side of Eq. (13.30) is:

$$\dot{m}_{gas}c_{p,gas}(T_6 - T_7) = (229.1 \text{ kg/s})(1.099 \text{ kJ/kgK})(179.9\,°\text{C} + 21.3\,°\text{C} - 147\,°\text{C}) = 13{,}647 \text{ kW}$$

These values are close to one another, confirming that the pressure inside the HRSG is close to 1 MPa.

Note: It should be borne in mind that we worked with an (assumed) average cycle temperature of 800 K to determine c_p, c_v, and γ. Using an iterative process, we could have determined the exact values of c_p, c_v, and γ, which would probably render a value of γ close to 1.15 kJ/kgK and would have given more accurate answers.

d) Steam turbine power output

$$\dot{W}_{(11-12)s} = \dot{m}_s(h_{11} - h_{12,s})$$

Using the thermophysical properties available from EES:

$$h_{11} = h_{1\,\text{MPa},\,450\,°C} = 3371.8 \text{ kJ/kg}$$

$$h_{12,s} = h_{g,\,10\,\text{kPa}} = 2583.9 \text{ kJ/kg}$$

$$\therefore \dot{W}_{(11-12)s} = \dot{m}_s(h_{11} - h_{12,s}) = (31.3 \text{ kg/s})(3371.8 \text{ kJ/kg} - 2583.9 \text{ kJ/kg}) = 24.7 \text{ MW}$$

The actual power output, considering the isentropic efficiency of the turbine, is:

$$\dot{W}_{(11-12)} = \left(\dot{W}_{(11-12)s}\right)(\eta_{s,t}) = (24.7 \text{ MW})(0.9) = 22.2 \text{ MW}$$

Discussion:

It is instructive to evaluate the ratio of $\dot{m}_{steam}$ to $\dot{m}_{gas}$ for the HRSG: $y = \frac{\dot{m}_{steam}}{\dot{m}_{gas}} = \frac{31.3 \text{ kg/s}}{229.1 \text{ kg/s}} = 0.137$. This means that 1 kg of exhaust gas can heat only 0.137 kg of steam from 75 to 450 °C as the gases are cooled from 800 K (527 °C) to 420 K (147 °C).

Reimagine Our Future

Some critics complain that renewable energy conversion systems and electric vehicles perpetuate our energy-intensive modes of living, and that a distracted mainstream environmental movement allows this to occur. Do you agree with this complaint? Discuss this with your friends.

13.6 Outlook

Changes to the global electricity industry are creating a paradigm shift in how nations' generating assets are operated. The need for considerable dispatchable generation, critical ancillary services, grid reliability and energy security concerns, combined with potentially higher future natural gas prices, creates the opportunity for advanced coal-fired generation for both domestic and international deployment (DOE 2022). The deployment of new coal plants will require a different way of thinking. What is needed are advances beyond today's utility-scale power plant concepts to also integrate carbon capture and storage technologies.

In tandem, though, steam power plants will increasingly rely on gas-turbine combined cycles, with those gas turbines eventually burning *green hydrogen*. *Combined-cycle technology* is advanced and developments in hydrogen production using renewable energy are progressing at pace. Combined cycles will also increasingly use dry-cooling systems to minimize water usage and the pollution of water vapor. Cogeneration of heat and power will become the standard mode of operating large thermal engines, thus minimizing pollution compared to independent heat and power production. These developments bode well for the future of condensing steam power plants.

Problems

13.1 Consider a well-insulated industrial steam turbine. Steam enters the turbine at 8 MPa and 500 °C with a mass flow rate of 10,800 kg/h. The steam leaves the turbine at 30 kPa. The isentropic efficiency of the turbine is 90%. Determine (a) the temperature at the turbine exit, and (b) the power output of the turbine.

[Answers: (a) 69 °C; (b) 3 MW]

13.2 Consider the ideal Rankine cycle shown in Figure 13.2, then complete the two tables below. You may assume isentropic compressor and turbine processes. You should use EES or other credible sources to find enthalpies and entropies. Show all your calculations separate from the tables.

State point	Description of state	Enthalpy (h) (kJ/kg)	Entropy (s) (kJ/kgK)	Vapor quality (x)
1	Saturated liquid water ($x_1 = 0$) at 50 °C	.... [Answer: 209.33]	.... [Answer: 0.7038]	0
2	2 MPa; $s_2 = s_1$	.... [Answer: 211.34]	.... [Answer: 0.7038]	Compressed liquid
3	Steam at 500 °C and 2 MPa	.... [Answer: 3467.7]	.... [Answer: 7.4317]	Superheated steam
4	$p_1 = p_4$; $s_4 = s_3$	.... [Answer: 2383.78]	.... [Answer: 7.4317]	.... [Answer: 91.3%]

Process	Description	Specific heat (q) (kJ/kg)	Specific work (w) (kJ/kg)
1–2	Pump (adiabatic)	0	.... [Answer: −1.98]
2–3	Boiler (isobaric)	.... [Answer: 3256.3]	.... [Answer: 0]
3–4	Turbine (adiabatic)	.... [Answer: 0]	.... [Answer: 1083.82]
4–1	Condenser (isothermal)	.... [Answer: −2174.45]	.... [Answer: 0]
Check: $\sum$		q_{net} = ... [Answer: 1081.9]	w_{net} = ... [Answer: 1081.9]

13.3 An industrial steam power plant uses regenerative heating via three feedwater heaters, as illustrated in Figure 13.8. The turbine entry conditions are 10 MPa and 540 °C. Steam leaves the turbine with a dryness fraction of 0.90. The condenser pressure is 6 kPa.

   The feedwater heaters that extract steam from the high-pressure and low-pressure regions of the turbine are surface-type (indirect) heaters, while those extracting steam at an intermediate pressure are of a direct-contact (open) type. The drains from the high-pressure and low-pressure feedwater heaters cascade to the intermediate-pressure heater and the condenser, respectively. The feedwater leaving each surface-type heater is at a temperature

10 °C below the saturation temperature of the steam condensing in the heater, but in the direct-contact heater, the feedwater reaches the bled steam saturation temperature.

a) Assume that the total water enthalpy rise is equally divided between the economizer and three feedwater heaters, i.e., each feedwater heater has a $\Delta h = (h_{10} - h_6)/4$, so that $h_9 = h_{10} - \Delta h$, $h_8 = h_9 - \Delta h$, $h_7 = h_8 - \Delta h$. Determine the steam extraction pressures p_2, p_3, p_4.

[Answers: $p_2 = 4.8$ MPa; $p_3 = 1$ MPa; $p_4 = 200$ kPa]

b) The following steam mass flow rates are extracted for each heater: $\dot{m}_2 = 0.14$ kg/s, $\dot{m}_6 = \dot{m}_7 = 7.17$ kg/s, $\dot{m}_3 = 0.084$ kg/s, $\dot{m}_4 = 0.108$ kg/s, $\dot{m}_5 = 0.668$ kg/s, $\dot{m}_5/\dot{m}_1 = 0.664$. Calculate the cycle efficiency and the cycle heat rate. What would the cycle efficiency be with no feedwater heating, but with the same turbine entry temperature and condenser conditions as before?

[Answers: 39.4% with feedwater heating; heat rate = 9137 kg/kWh; 34.6% without feedwater heating.]

c) For an output of 200 MW, (i) estimate the total savings in operating cost achieved by the three feedwater heaters (compared with no feedwater heating). Assume a boiler efficiency of 88%, and an average annual load factor of 0.6, a useful plant life of 30 years, and a nominal energy cost of US$80 per MWh. (ii) Also evaluate the net present value of these savings when the plant goes into service, assuming a discount rate of 10%.

[Answers: (i) US$1 billion; (ii) US$332 million]

13.5 Consider an ideal Rankine cycle that has been modified with a reheat process. Figure 13.3 shows the cycle set-up and state points. The boiler pressure is 6 MPa and the condenser pressure is 10 kPa. The water enters the high-pressure turbine at 600 °C and leaves the high-pressure turbine at 500 kPa. Steam enters the low-pressure turbine at 500 °C. The steam eventually leaves the condenser as a saturated liquid. Assume that heat is transferred to the working fluid in the boiler from a reservoir at 1400 K and that the fluid in the condenser rejects heat to the surroundings at 25 °C (thus, $T_0 = 298$ K). The properties of the steam are summarized below:

State	T (°C, K)	h (kJ/kg)	s (kJ/kgK)	x
1	45.81 °C 319 K	191.83	0.6493	0
2	45.81 °C 319 K	197.83	0.6493	Compressed liquid
3	600 °C 873 K	3658.4	7.1677	Superheated steam
4	225.3 °C 499 K	2909.4	7.1677	Superheated steam
5	500 °C 773 K	3483.9	8.0873	Superheated steam
6	45.81 °C 319 K	2565.5	8.0873	0.992

Calculate the specific exergy destruction (in kJ/kg) for each of the processes and complete the following table. Show all your calculations separately. Comment on the largest contributors of exergy destruction. How can the exergy destruction be minimized in those components? *Hint:* The specific exergy destruction is $e_d = \left(1 - \frac{T_0}{T_b}\right)q - w + \left(e_{fi} - e_{fe}\right)$ with $\left(e_{fi} - e_{fe}\right) = (h_i - h_e) - T_0(s_i - s_e)$.

Process	q (kJ/kg)	w (kJ/kg)	$\left(1 - \frac{T_0}{T_b}\right) q$ (kJ/kg)	$h_i - h_e$ (kJ/kg)	$s_i - s_e$ (kJ/kgK)	$e_{fi} - e_{fe}$ (kJ/kg)	Specific exergy destruction (kJ/kg)
1-2 (Pump)	0	−6	0	191.8 − 197.8 = −6	0.6493 − 0.6493 = 0	−6 − 0 = −6	0 − (−6) + (−6) = 0
2-3 (Boiler)	3460.5	0	$\left(1 - \frac{298}{1400}\right) 3460.5$ =⋯	⋯ =	⋯ =	⋯ =	⋯ =
3-4 (High-pressure turbine)	0	749	0	3658.4 − 2909.4 = 749	7.1677 − 7.1677 = 0	⋯ =	⋯ =
4-5 (Reheater)	574.5	0	⋯ =	⋯ =	⋯ =	⋯ =	⋯ =
5-6 (Low-pressure turbine)	0	918.4	⋯ =	⋯ =	⋯ =	⋯ =	⋯ =
6-1 (Condenser)	−2373.3	0	⋯ =	2565.5 − 191.83 = 2373.7	8.0873 − 0.6493 = 7.438	2373.7 − 298 (8.0873 − 0.6493) = 157.18	⋯ =
$\sum_{cycle}$	1661.3	1661.3	⋯ = [Answer: 3019]	0	0	0	= [Answer: 1356]

13.6 The existing Intermountain Power Plant is a relatively modern 1800-MW coal-fired power station in central Utah, which has been operating since 1986. Intermountain Power Agency (IPA) is the plant owner, while the Los Angeles Department of Water and Power (LADWP) operates the plant and purchases most of its output. In 2020, IPA announced that the coal plant will be retired and replaced with an 840-MW gas-fired combined cycle facility with a purported efficiency of 64%. Although the new plant will begin operation on natural gas, it will transition to a mix of 30% hydrogen (by volume) and 70% natural gas by 2025, which is the scheduled project completion date. The plan also commits the plant owner to increasing the fuel mix to 100% green hydrogen fuel by 2045, with renewable energy-to-power electrolysis facilities for hydrogen production and storage at the site (also see Box 12.2). The combined-cycle plant will be built around two single-shaft units equipped with an advanced technology power block supplied by Mitsubishi Power. Each power block will include one M501JAC gas turbine, a heat-recovery steam generator, and a steam turbine (GTW 2022).

Produce a fact sheet of no fewer than 500 words, with accompanying graphics or pictures, explaining how gas turbine combined cycles can be used to replace or retrofit coal-fired power plants. Estimate the reduction of greenhouse gases when operating a gas turbine combined-cycle power plant compared to a coal-fired power plant of the same power output and load factor. Then, briefly discuss how the use of combined-cycle power plants influences many of the UN's Sustainable Development Goals (SDGs).

[Answer: Independent research required for this open-ended question.]

13.7 Two industrial sites, A and B, have average electricity and heating requirements as follows (with little seasonal variation):

A: 5 MW electricity

   20,000 kg/h saturated steam at 1 MPa from feedwater at 20 °C

B: 5 MW electricity

   2000 kg/h saturated steam at 1 MPa from feedwater at 20 °C

   30,000 kg/h hot water at 70 °C from feedwater at 20 °C

At present, both sites purchase electricity and produce steam or hot water by on-site boilers that burn diesel. The engineering manager is now considering adopting a cogeneration (combined heat and power) scheme for both sites by using waste heat recovery.

a) Consider the suitability of using a gas turbine-based scheme for site A and a diesel engine-based scheme for site B. In each case, attempt to match both the heat and power loads. (*Hint:* Use Table 13.2 to guide you in the selection process.)

b) Estimate the energy utilization factor (EUF) for the two cases. Assume perfectly matched cogeneration schemes. The thermal efficiency of the gas turbine is 30% and that of the diesel engine is 40%. The average thermal efficiency of power stations supplying the electricity network is 35%. The existing boiler (steam and hot water) has an efficiency of 90%.

c) Briefly discuss the options for choosing the capacity of a cogeneration plant and operating it in parallel with other energy supplies in a case where the winter and summer heat and power requirements of a site differ significantly. Consider a period when the power-heat ratio demanded is greater than and less than that of the cogeneration plant. State some factors that might determine the economics of the unmatched situation.

d) Use sustainability criteria to motivate why the investment in the proposed cogeneration scheme should be approved.

[Answers: The following are possible answers: (a) Site A has a power-heat ratio of 0.33, which is a good match for a gas turbine (Table 13.2). Site B has a power-heat ratio of 1.55, which aligns with the recovery of heat from diesel engines (Table 13.2). (b) The EUF for Site A is about 0.85. The EUF for Site B is about 0.7. (c) Open-ended.]

13.8 Electric power generation consumes more than 10 trillion liters of water globally per year (Reimers 2018). Conventional steam condensers are cooled down by water in large cooling towers (Figure 13.9). In this "wet-cooling" process, copious amounts of water vapor are

(a)

(b)

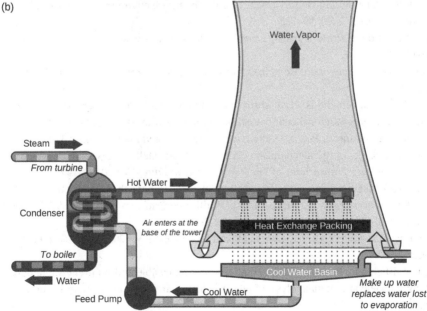

Figure 13.9 (a) Water vapor billowing out of cooling towers at the Ferrybridge Power Station (circa 2008). (b) An evaporative cooling tower to condense steam from the outlet of a steam turbine. *Source:* (a) shipley43 / Flickr / CC BY-SA 2.0, (b) Jeffrey M. Phillips Webber Energy Group, used with permission.

emitted while condensing steam. This evaporation of water represents a loss of local water, which is especially problematic in drought-ridden areas. Water vapor is also a potent greenhouse gas.

Perform independent research of the *dry-cooling* of steam. These kinds of steam-cooling systems are used in water-scarce countries, including parts of South Africa, China, and the United States. Produce a 500-word fact sheet on your research, in which you compare the salient aspects of the wet- and dry-cooling of steam in utility power plants. Your fact sheet should compare typical power plant efficiencies when using wet-cooling or dry-cooling. How much water does a dry-cooled steam power plant save per year, compared to a plant with wet-cooling? Also discuss the environmental effects of discarding heat into the atmosphere (with dry-cooling) rather than water vapor (wet-cooling).

Begin your research by reading the following article: Reimers, A. (2018). Making electricity consumes a lot of water – what's the best way to fix that? *Scientific American Blog Network* May 17. https://blogs.scientificamerican.com/plugged-in/making-electricity-consumes-a-lot-of-water-whats-the-best-way-to-fix-that

[Answer: Independent research required for this open-ended question.]

References

Çengel, Y.A. and Boles, M.A. (2015). *Thermodynamics. An Engineering Approach.* New York, NY: McGraw-Hill.

Danskammer (2022). Clean, reliable energy for New York. Danskammer Energy. http://www.danskammerenergy.com/energy-project

DOE (2022). Transformative power systems. US Department of Energy. https://www.energy.gov/fecm/science-innovation/office-clean-coal-and-carbon-management/advanced-energy-systems/transformative

Finkelman, R.B., Wolfe, A. and Hendryx, M.S. (2021). The future environmental and health impacts of coal. *Energy Geoscience* 2 (2): 99–112.

GTW (2022). Gas Turbine World. https://gasturbineworld.com/mitsubishi-power-hydrogen-gas-turbine-technology

Horlock, J.H. (1987). *Cogeneration: Combined Heat and Power, Thermodynamics and Economics.* Pergamon Press.

IEA (2022). Global coal demand is set to return to its all-time high in 2022. *International Energy Agency.* https://www.iea.org/news/global-coal-demand-is-set-to-return-to-its-all-time-high-in-2022

Markandya, A. and Wilkinson, P. (2007). Electricity generation and health. *Lancet* 370 (9591): 979–990.

Rankine, W.J.M. (1853). On the mechanical action of heat, especially in gases and vapours. *Earth and Environmental Science Transactions of The Royal Society of Edinburgh* 20 (1): 147–190.

Reimers, A. (2018). Making electricity consumes a lot of water - what's the best way to fix that? *Scientific American Blog Network* May 17. https://blogs.scientificamerican.com/plugged-in/making-electricity-consumes-a-lot-of-water-whats-the-best-way-to-fix-that

Saravanamuttoo, H.I., Rogers, G.F.C., Cohen, H. and Straznicky, P.V. (2001). *Gas Turbine Theory.* New Jersey:Pearson.

Van Zalk, J. and Behrens, P. (2018). The spatial extent of renewable and non-renewable power generation: a review and meta-analysis of power densities and their application in the US. *Energy Policy* 123: 83–91.

14

Refrigeration and Air-Conditioning

Refrigeration and air-conditioning are pervasive in modern society. This spans the refrigerated cold chain that provides us with fresh vegetables and meat, to the air-conditioning of homes and offices, and the cooling of data servers that enable the internet. The use of air-conditioners and electric fans already accounts for about a fifth of the total electricity in buildings around the world – or 10% of all global electricity consumption (IEA 2018).

In 2020, refrigeration accounted for 7.8% of total greenhouse-gas emissions, with 63% of that coming from indirect emissions stemming from the energy that drives refrigeration equipment (Coulomb 2021). The demand for cooling will continue to increase for several decades, not just because of improving economic conditions globally, but also due to a warming climate. This means that we must urgently adopt sustainable cooling practices.

Reimagine Our Future
Wendell Berry once stated that we should understand that fossil fuel energy must be replaced not just by "clean" energy, but also by *less* energy. This suggests that the unlimited use of *any* energy might be as destructive as unlimited economic growth. Discuss this with your social network.

14.1 Vapor-Compression Refrigeration Systems

Heat spontaneously flows from hot to cold. With vapor-compression refrigeration, however, the direction of heat flow is reversed either to cool something down (such as with refrigeration, air-conditioning, or process cooling) or to upgrade heat from a cold ambient source (such as when one uses heat pumps). There are many refrigeration methods and systems, but the most prevalent today is the vapor-compression system. The reason for the dominance of vapor-compression refrigeration systems compared to other methods is due to the mature technology, ease of manufacturing, and scalability from small to very large systems (McLinden et al. 2020).

Figure 14.1 illustrates that *vapor-compression refrigeration* cycles operate by boiling a refrigerant in an evaporator (process 4 to 1) due to the extraction of heat from a low-temperature source (such as a refrigerated space). The vaporized refrigerant is compressed to a higher pressure (process 1 to 2) and is then condensed back to a liquid in the condenser (process 2 to 3). The heat of condensation is

Energy Systems: A Project-Based Approach to Sustainability Thinking for Energy Conversion Systems,
First Edition. Leon Liebenberg.
© 2024 John Wiley & Sons, Inc. Published 2024 by John Wiley & Sons, Inc.
Companion website: www.wiley.com/go/liebenberg/energy_systems

(a)

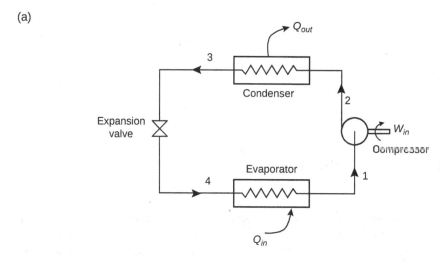

(b)

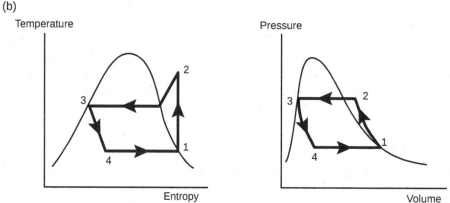

Figure 14.1 (a) Salient components, and (b) temperature-entropy diagram of an ideal vapor-compression refrigeration cycle.

transferred to a relatively high-temperature heat sink, such as ambient air or water. The high pressure of the refrigerant is reduced to the starting pressure using an expansion (or throttling) device such as a capillary tube (process 3 to 4).

Vapor-compression refrigeration systems therefore exploit two fundamental principles to transfer heat from a cold region to a warm region:

- A refrigerant's boiling temperature varies with pressure.
- Liquid refrigerants boil to a vapor and condense back to a liquid phase, with the refrigerant either absorbing heat from or releasing heat to the environment.

The refrigerant cycles indefinitely between high-pressure and low-pressure values. We will disregard pressure drops in the condenser and evaporator, and also assume the use of pure refrigerants so that condensation and evaporation processes will be isothermal and isobaric. An ideal compression process will be isentropic. The expansion (or throttling) process device is irreversible – therefore not isentropic – because it relies on fluid friction for its operation. Despite this irreversibility, the cycle depicted in Figure 14.1 is still referred to as an ideal cycle.

The evaporator and condenser are also major sources of irreversibility as they transfer heat across finite temperature differences between the refrigerant and its surroundings. To reduce irreversibilities associated with heat transfer, engineers design evaporators and condensers with large surface areas that produce smaller temperature differences.

The processes and salient thermodynamic relationships of an ideal vapor-compression cycle can be summarized as follows (also see Figure 14.1):

14.1.1 Process 1 to 2: Adiabatic (and Isentropic) Compression

From Eq. (2.8) for an open system and an adiabatic process:

$$\cancel{q}_{12} - w_{12} = \Delta h_{12} \tag{14.1}$$

$$s_1 = s_2 \tag{14.2}$$

14.1.2 Process 2 to 3: Constant-Pressure Heat Rejection in a Condenser

$$q_{23} - \cancel{w}_{23} = \Delta h_{23} \tag{14.3}$$

$$p_2 = p_3 \tag{14.4}$$

14.1.3 Process 3 to 4: Expansion (Throttling)

$$\cancel{q}_{34} - \cancel{w}_{34} = \Delta h_{34} \tag{14.5}$$

$$\text{Therefore,} \quad h_3 = h_4 \tag{14.6}$$

14.1.4 Process 4 to 1: Constant-Pressure Heat Absorption in Evaporator

$$q_{41} - \cancel{w}_{41} = \Delta h_{41} \tag{14.7}$$

The thermal efficiency of a refrigeration cycle is called the *coefficient of performance* (*COP*) and is defined as follows:

$$\text{Refrigerator, air-conditioner:} \quad COP_{cooling} = \frac{\text{Cooling capacity}}{\text{Work input}} = \frac{Q_L}{W_{in}} \tag{14.8}$$

$$\text{Heat pump:} \quad COP_{heating} = \frac{\text{Heating effect}}{\text{Work input}} = \frac{Q_H}{W_{in}} \tag{14.9}$$

The *COP* of most refrigeration and air-conditioning systems falls between 2 and 5. If the *COP* is 3, it means that only one unit of electrical energy is required to produce three units of heat.

Example 14.1 Ideal Vapor-Compression Cycle
Consider an industrial water chiller that operates on an ideal vapor-compression cycle, such as the example shown in the accompanying figure. The system uses R-134a as a refrigerant. The refrigerant enters the compressor as a saturated vapor at 5 °C and leaves the condenser as a saturated liquid at 55 °C. The chilled water temperature is 15 °C, while the ambient air is 30 °C. The mass flow rate of the refrigerant is 4 kg/s. Calculate (a) the enthalpies and entropies at each state point,

(b) the compressor power, (c) the heat transfer rate in the condenser and the evaporator, (d) the coefficient of performance, and (e) the irreversibility of each component. Comment on how the irreversibility may be reduced.

Required: We must find (a) the enthalpies and entropies at the four state points. Using those, we must calculate (b) the compressor power, (c) heat transfer rates in the condenser and the evaporator, (d) the coefficient of performance, and (e) the irreversibility of the four components.

Solution strategy: We will use Eqs. (14.1) to (14.9).

Assumptions: Assume that the compression process is adiabatic and reversible, i.e., it is isentropic. Assume that the condenser exit is at a saturated liquid condition. We will also assume that the pressure drops in the condenser and evaporator are negligible.

Series R® chiller, manufactured by Trane. This modular chiller system has cooling capacities of 500–1800 kW and features a helical rotary compressor. *Source:* Trane / CC BY 4.0

Analysis:

Using the process numbering of Figure 14.1, the cycle may be represented as follows:

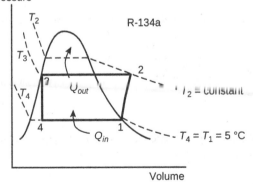

a) Enthalpies and entropies

The thermodynamic properties of R-134a can be read off tabulated values found in textbooks, or we could use Engineering Equation Solver (EES) or the widely used Refprop software, among others. A free version of Mini-Refprop is available from the National Institute of Standards and Technology (NIST).

Using EES, the following values are found:

State point 1:

The refrigerant enters the compressor at $5\,°C$ as a saturated vapor ($x_1 = 1.0$). Therefore:

$$h_1 = h_{g,5°C} = 250.1 \text{ kJ/kg}$$

$$s_1 = s_{g,5°C} = 0.9164 \text{ kJ/kgK}$$

State point 2:

The refrigerant is superheated at state point 2, with $T_2 = 55\,°C$ and $p_2 = p_3 = p_{sat,55°C}$.
Also, Process 1 to 2 is isentropic, so $s_2 = s_1 = 0.9164 \text{ kJ/kgK}$, from which we find that $p_2 = p_3 = p_{sat,55°C} = 1.49 \text{ MPa}$. Therefore, $h_2 = 279.93 \text{ kJ/kg}$.

State point 3:

The expansion process is isenthalpic:

$$h_3 = h_4 = h_{f,55°C} = 129.34 \text{ kJ/kg}$$

$$s_3 = s_{f,55°C} = 0.4575 \text{ kJ/kgK}$$

State point 4:

From previous calculations, we have that $h_4 = 129.34 \text{ kJ/kg}$.
We can determine the entropy at state point 4 using $s_4 = s_{f,5°C} + x_4(s_{fg})_{5°C}$. So, we must first calculate the vapor quality, x_4.

Using EES, we find that:

$$s_{f,5°C} = 0.221 \text{ kJ/kgK}$$

$$s_{g,5°C} = 0.9164 \text{ kJ/kgK}$$

$$(s_{fg})_{5°C} = s_{g,5°C} - s_{f,5°C} = (0.9164 - 0.221) \text{ kJ/kgK} = 0.6954 \text{ kJ/kgK}$$

$$\text{But, } x_4 = \frac{h_4 - h_f}{h_{fg}} = \frac{129.34 \text{ kJ/kg} - h_{f,5°C}}{h_{fg,5°C}}$$

Using EES, we find the following:

$$h_{f,5°C} = 56.7 \text{ kJ/kg}$$

$$h_{fg,5°C} = 193.41 \text{ kJ/kg}$$

$$\therefore x_4 = \frac{h_4 - h_f}{h_{fg}} = \left(\frac{(129.34 - 56.7) \text{ kJ/kg}}{193.41 \text{ kJ/kg}}\right) = 0.38$$

Therefore, $s_4 = s_{f,5°C} + x_4(s_{fg})_{5°C} = 0.221 \text{ kJ/kg} + 0.38(0.6954) \text{ kJ/kg} = 0.483 \text{ kJ/kg}$

b) Compressor power

From Eq. (14.1): $\cancel{q_{12}} - w_{12} = \Delta h_{12} = h_2 - h_1$

So, $w_{12} = h_1 - h_2 = (250.1 - 279.92) \text{ kJ/kg} = -29.8 \text{ kJ/kg}$

Therefore, $\dot{W}_{12} = \dot{m}w_{12} = (4 \text{ kg/s})(-29.8 \text{ kJ/kg}) \cong 120 \text{ kW}$

The compressor therefore requires 20.9 kW of electrical power.

c) Heat transfer rates of the condenser and the evaporator

Condenser

From Eq. (14.3): $q_{23} - w_{23} = \Delta h_{23} = h_3 - h_2 = (129.34 - 279.92) \text{ kJ/kg} = -150.6 \text{ kJ/kg}$

The condenser heat transfer rate is:

$$\dot{Q}_{23} = \dot{m}q_{23} = (4 \text{ kg/s})(-150.6 \text{ kJ/kg}) = -602 \text{ kW}$$

The condenser therefore rejects 602 kW to the surrounding air.

Evaporator

From Eq. (14.3): $q_{41} - w_{41} = \Delta h_{41} = h_1 - h_4 = (250.1 - 129.3) \text{ kJ/kg} = 120.8 \text{ kJ/kg}$

The evaporator heat transfer rate is:

$$\dot{Q}_{41} = \dot{m}q_{41} = (4 \text{ kg/s})(120.8 \text{ kJ/kg}) = 483 \text{ kW}$$

The evaporator absorbs 483 kW from the process water. The *cooling capacity* of this water chiller is therefore 483 kW.

d) Coefficient of performance

From Eq. (14.8): $COP_{cooling} = \dfrac{\text{Cooling capacity}}{\text{Work input}} = \dfrac{\dot{Q}_L}{|\dot{W}_{in}|} = \dfrac{483 \text{ kW}}{120 \text{ kW}} \approx 4$

e) *Entropy generation and irreversibilities*

We can calculate irreversibilities from Eqs. (3.58) and (3.59), which reduce to the following if we disregard changes in kinetic and potential energies and assume steady-state conditions:

$$\sigma = \frac{\dot{\sigma}}{\dot{m}} = \sum_j \left(1 - \frac{T_0}{T_j}\right) \frac{Q_j}{T_0} - \frac{W}{T_0} + \frac{1}{T_0}\left[(h_i - h_e) - T_0(s_i - s_e)\right]$$

The irreversibility (*i*) can then be calculated using the following:

$$i = T_0 \left[(s_e - s_i) + \frac{q_k}{T_k}\right]$$

where the subscript *k* refers to the heat reservoir (either a source or a sink). Note that T_0 is not the temperature of the low-temperature reservoir, but rather the temperature of the lowest naturally occurring temperature, the ambient air; $T_0 = T_H = 30\,°C = 303 \text{ K}$. Also, $T_L = 15\,°C = 288$ K.

Process 1 to 2:

$$i_{12} = T_0\left[(s_2 - s_1) + \frac{q_k}{T_k}\right] = 303 \text{ K}\left[(0.9164 - 0.9164) \text{ kJ/kgK} + \frac{0}{303 \text{ K}}\right] = 0 \text{ kJ/kg}$$

Process 2 to 3:

$$i_{23} = T_0\left[(s_3 - s_2) + \frac{q_k}{T_k}\right] = 303 \text{ K}\left[(0.4575 - 0.9164) \text{ kJ/kgK} + \frac{150.6 \text{ kJ/kg}}{303 \text{ K}}\right] = 11.55 \text{ kJ/kg}$$

Process 3 to 4:

$$i_{34} = T_0\left[(s_4 - s_3) + \frac{q_k}{T_k}\right] = 303 \text{ K}\left[(0.483 - 0.4575) \text{ kJ/kgK} + \frac{0 \text{ kJ/kg}}{303 \text{ K}}\right] = 7.72 \text{ kJ/kg}$$

Process 4 to 1:

$$i_{41} = T_0\left[(s_1 - s_4) + \frac{q_k}{T_k}\right] = 303 \text{ K}\left[(0.9164 - 0.483) \text{ kJ/kgK} + \frac{(-120.8 \text{ kJ/kg})}{288 \text{ K}}\right] = 4.22 \text{ kJ/kg}$$

The results may be summarized and checked as follows:

Process	q (kJ/kg)	w (kJ/kg)	$s_e - s_i$ (kJ/kgK)	i (kJ/kg)
1 to 2	0	−29.83	0	0
2 to 3	−150.58	0	−0.458	11.55
3 to 4	0	0	0.025	7.72
4 to 1	120.76	0	0.433	4.22
Cycle:	−29.83	−29.83	0	23.49

The condenser has the highest irreversibility due to it transferring heat across a large temperature difference. The evaporator's irreversibility is lower than that of the condenser, as the difference between the refrigerant temperature and the chilled water is small. The irreversibility in the expansion (throttling) process is ascribed to frictional losses that result in the pressure decrease and subsequent entropy increase of the refrigerant.

Discussion:

To reduce the irreversibility in the condenser and the evaporator, the average temperature differences between the refrigerant and the surrounding fluid (air or water) must be reduced. This means that the average temperature of the refrigerant in the condenser should be decreased, while the average temperature of the refrigerant in the evaporator should be increased. This would require increasing the surface areas of both the condenser and the evaporator, as heat transfer is proportional to the heat transfer surface area and the temperature difference between the fluids: $q = UA \ \Delta T$.

14.2 Refrigeration, Air-Conditioning, and Heat Pumping

Refrigeration, air-conditioning, and heat pumping use the vapor-compression cycle, but in different ways. In *refrigeration* and *air-conditioning*, the purpose is the removal of heat from a cold region to keep it cold. In *heat pumping*, the purpose is to move heat to a warm region and to keep it warm. Figure 14.2 shows how refrigeration fluid could be directed in a vapor-compression system to either cool or heat a space by positioning a rotary valve. The valve interchanges the condenser and evaporator coils so that the system can operate either as a heat pump in cold weather or as an air-conditioner in warm weather.

The evaporator coils of heat pumps can obtain heat from the ground (Sarbu and Sebarchievici 2014), air (Bertsch and Groll 2008), or water (Hepbasli and Kalinci 2009). A limiting factor of air-source heat pumps is their operation at low ambient temperatures. In colder climates, heat pump applications have historically been limited to more expensive ground-source heat pumps to ensure they would be able to operate during the extreme winter cold (Liebenberg and Meyer 1998; Brodowicz et al. 1993). Chapter 26 expands on geothermal (or "ground-source") heat pumps. However, heat pump technology is rapidly advancing, and heat pumps will soon be able to operate at ambient temperatures of around −15 °C.

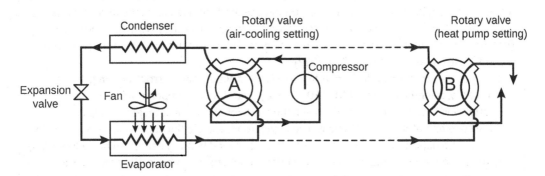

Figure 14.2 A vapor-compression system capable of operating in air-conditioning (cooling) or heat pumping mode. Rotating the refrigerant valve through 90° interchanges the condenser and evaporator coils to facilitate either cooling (*A* position) or heating (*B* position).

Box 14.1 summarizes some salient characteristics of heat pumps. Box 14.2 shows a highly specialized application of a gas refrigeration cycle. Box 14.3 briefly outlines the importance of passive cooling, such as increasing vegetation in urban settings.

With split air-conditioning systems, the evaporator (or refrigeration coil) is placed inside a room or building with a fan blowing interior air over the cold coil. The compressor and the condenser coil

---

Box 14.1   Heat Pumps around the World

(Also see Section 26.1 for details about ground-source [or "geothermal"] heat-pumps and Section 28.3 for information on industrial heat pumps.)

Heat pumps, powered by low-emissions electricity, are the central technology in the global transition to secure and sustainable heating. Heat pumps currently available on the market are three to five times more energy efficient than natural gas boilers. They reduce households' exposure to fossil-fuel price spikes, which has been made more urgent by the ongoing global energy crisis. Over one-sixth of global natural gas demand is for heating in buildings – in the European Union, this number is one-third. Many heat pumps can provide cooling too, which eliminates the need for a separate air conditioner for the 2.6 billion people who will live in regions requiring heating and cooling by 2050.

Heating in buildings is responsible for 4 Gt of CO_2 emissions annually – 10% of global emissions (IEA 2022).

Installing heat pumps instead of fossil fuel-based boilers significantly reduces greenhouse-gas emissions in all major heating markets, even with the current electricity generation mix – an advantage that will increase further as electricity systems decarbonize.

Around 10% of space heating needs globally were met by heat pumps in 2021, but the pace of installation is growing rapidly. The share of heat pumps is comparable to that of fuel oil for heating and of other forms of electric heating, but lower than the over 40% of heating that is reliant on gas heating and the 15% that is reliant on district heating. In some countries, heat pumps are already the largest source of heating. In Norway, 60% of buildings are equipped with heat pumps, with Sweden and Finland at over 40%, undercutting the argument that heat pumps are unsuitable for cold climates.

Global sales grew by nearly 15% in 2021, double the average of the last decade. Growth in the European Union was around 35% and is slated to accelerate further considering the energy crisis, with sales in the first half of 2022 roughly double that over the same period in the previous year in Austria, Italy, Poland, and The Netherlands. China continues to be the largest market for new sales, while North America has the largest number of homes with heat pumps today. Together these regions, along with Japan and Korea, are also major manufacturing hubs, home to the industry's largest players (IEA 2022).

Heat pumps can also address heating needs in industry and district heating. Large heat pumps can provide heat up to 140–160 °C today, with higher temperatures possible through innovation and improved designs. However, the most common industrial heat pumps today provide lower-temperature (<100 °C) heat. The paper, food, and chemicals industries have the largest near-term opportunities. Nearly 30% of their combined heating needs can be addressed by heat pumps. In Europe alone, 15 GW of heat pumps could be installed in 3000 facilities in these three sectors, which have been hit hard by recent rises in natural gas prices (IEA 2022). Refer to Section 28.3 for more information on industrial heat pumps.

are placed outside the room or building. Heat collected from inside the room or building is transferred to the exterior air. The configuration of air-conditioning systems depend on the space to be cooled, type of application (domestic, commercial, or industrial), cooling capacity, aesthetic preferences, and required safety level, among others.

Reimagine Our Future

Urban *heat islands* require increased energy for air-conditioning and refrigeration in cities in hot climates or during heatwaves. The Heat Island Group estimates that the heat-island effect costs Los Angeles about US$100 million per year in energy. How might cities best counteract the urban heat-island effect? Reflective surfaces on buildings, planting vegetation on the roofs of buildings, and planting more trees around the city have been mentioned in this context. Share your ideas with your network on social media.

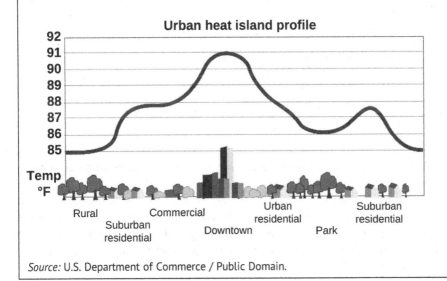

Source: U.S. Department of Commerce / Public Domain.

14.2.1 Wet-Bulb Temperature, Dry-Bulb Temperature, and Dewpoint Temperature

The dry bulb, wet bulb, and dewpoint temperatures are important to determine the state of humid air. The knowledge of two of these values is enough to determine the state of the moist air, including the content of water vapor, and the sensible and latent energy (enthalpy) in the air.

The *dry-bulb temperature* (or simply "air temperature") (T_{db}) can be measured using a thermometer freely exposed to air but shielded from radiation and moisture. The dry-bulb temperature is an indicator of the heat content of air.

The *wet-bulb temperature* (T_{wb}) is the adiabatic saturation temperature of air. Wet-bulb temperature can be measured using a thermometer with its bulb wrapped in a wet cotton wick. The adiabatic evaporation of water from the wet wick and the cooling effect of air are depicted by the wet-bulb temperature being lower than the dry-bulb temperature. The rate of evaporation from the wet cotton on the bulb and the temperature difference between the dry bulb and the wet bulb depend on the humidity of the air. The evaporation from the wet cotton is lower when air contains more water

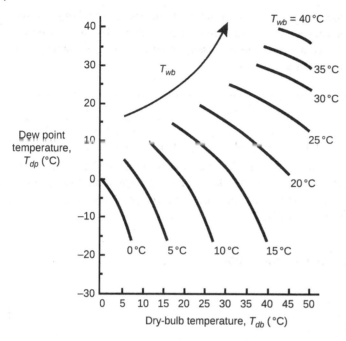

Figure 14.3   A typical dewpoint chart. *Source:* Stoecker and Jones (1982) / McGraw-Hill.

vapor. The wet bulb temperature is therefore the air temperature that can be achieved through evaporative cooling.

The *dewpoint temperature* (T_{dp}) is the temperature at which water vapor starts to condense from the air when cooled at a constant vapor pressure. Above this temperature, the moisture remains in the air. Therefore, if (T_{dp}) is close to (T_{db}), the relative humidity is high. Conversely, if T_{dp} is well below T_{db}, the relative humidity is low. The dewpoint can be read off psychrometric charts or much simpler *dewpoint charts*, such as the one in Figure 14.3.

Psychrometric charts and dewpoint charts make use of the values of relative humidity and the humidity ratio. The relative humidity (ϕ) is the ratio of the mass of water vapor (m_v) in air relative to the maximum mass of water vapor (m_{sat}) that it can contain:

$$\phi = \frac{m_v}{m_{sat}} \equiv \frac{p_v}{p_{sat}} \tag{14.10}$$

Air at a dry-bulb temperature of $T_{db} = 20\,^\circ\text{C}$ and with $\phi = 1$ (i.e., 100% relative humidity) will have a wet-bulb temperature of $T_{wb} = 20\,^\circ\text{C}$ as the saturated air cannot hold any additional moisture. If the relative humidity is $\phi < 1$, evaporation from the wet-bulb thermometer to the air surrounding the air–water vapor mixture will occur so that T_{wb} will decrease below 20 °C.

The humidity ratio (ω) is the ratio of the mass of water vapor in air relative to the mass of dry air:

$$\omega = \frac{m_v}{m_a} \tag{14.11}$$

For space cooling applications, $\omega \cong 0.01$ to $0.02 \ \dfrac{\text{kg water vapor}}{\text{kg dry air}}$.

14.3 Absorption Refrigeration

The mechanical work required to raise refrigerant pressure in a vapor-compression cycle can be vastly reduced by taking advantage of the ability of certain liquids to absorb large quantities of refrigerant vapor. The substances most often used in *absorption refrigeration* cycles are based on water/lithium-bromide (Li–Br) and ammonia (NH_3)/water cycles (Herold et al. 2016). The equilibrium water vapor pressure of these aqueous salt solutions is considerably lower than that of pure water and is strongly temperature dependent.

Figure 14.4 illustrates the operating principle of an absorption refrigeration system. The portion of the diagram to the left of the vertical dashed line corresponds to the vapor-compression cycle shown in Figure 14.1. The low-pressure ammonia vapor from the evaporator is transferred to the high-pressure side of the cycle by dissolving it in water and pumping the water to the high-pressure regenerator where the ammonia vapor is driven out of solution by heating. The work required to raise the pressure of the water solution is very small due to water having a low specific volume (high density). The pump work is usually negligibly small compared to Q_L and Q_R. In some absorption refrigeration systems, Q_R is supplied by solar concentrators or by the combustion of natural gas.

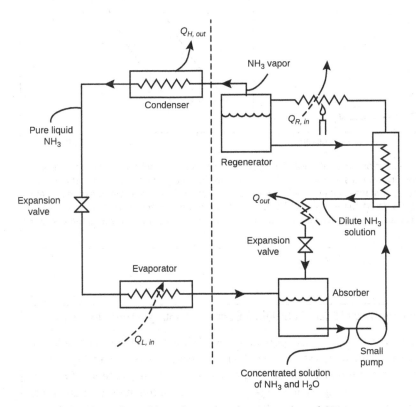

Figure 14.4 Absorption refrigeration cycle using ammonia and water.

Box 14.2 Gas Refrigeration Cycles for Ultra-Cold Refrigeration

Although not common, gas power cycles such as the Brayton and Stirling cycles (see Chapter 11) can be run in reverse to operate as *gas refrigeration cycles*. For instance, the Stirling power cycle can be reversed, as shown in Figure 14.5.

   With the Pfizer/BioNTech COVID-19 vaccine requiring ultra-low temperature storage from −60 to −80 °C, and the Moderna vaccine requiring −20 °C, Walgreens asked the energy entrepreneurs at *Stirling Ultracold* to develop a Stirling refrigeration system. The required ultra-cold refrigeration temperatures made traditional vapor compression refrigeration unfeasible. However, instead of using a traditional refrigerant that might boil at only around −30 °C, a Stirling freezer or Stirling cryocooler might use ethane (C_2H_6) that boils at −89 °C (at 100 kPa) or helium that boils at −269 °C (at 100 kPa). A Stirling cryocooler does not use a traditional compressor either, but rather two pistons and a regenerator, circumventing many difficulties associated with ultra-low temperature vapor-compression refrigeration.

   Further reading: Stirling Ultracold, https://www.stirlingultracold.com/stirling-engine

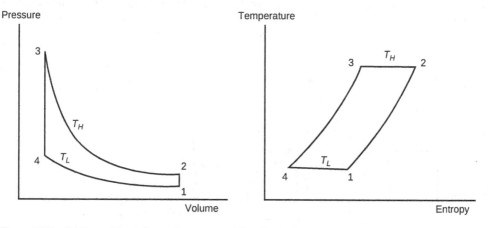

Figure 14.5 Stirling refrigeration cycle on *p-v* and *T-s* planes. Process 2 to 3: Isothermal gas compression and heat rejection to the surroundings. Process 3 to 4: Constant-volume regenerative cooling as the hot gas enters the regenerator at T_H and leaves it at T_L at lower pressure. Process 4 to 1: Isothermal expansion and heat absorption from refrigerated space. Process 1 to 2: Constant-volume regenerative heating as the regenerator transfers heat to the gas, with a concomitant pressure increase and the increase of temperature from T_L to T_H.

Box 14.3 Embedding Passive and Energy-Efficient Sustainable Cooling in Urban Infrastructure

Towns and cities will hold 66% of the global population by 2050, making them an epicenter of cooling demand. Urban infrastructures with energy-efficient passive cooling, such as increasing vegetation through street trees, green façades, and "green roofs," hold much potential for reducing cooling demand. For example, in Xiamen Island in China, the integration of green roofs reduced average land surface temperature by 0.91 °C.

   For further information, see Dong et al. (2020).

14.4 Outlook

Using air conditioners and electric fans to stay cool accounts for nearly 20% of the total electricity used in buildings around the world today. Rising demand for space cooling is also putting enormous strain on electricity systems in many countries, as well as driving up emissions. Without firm policy interventions, there is no doubt that global demand for space cooling and the energy needed to provide it will continue to grow for decades to come. However, there is an enormous opportunity to quickly influence the growth of cooling-related energy demand through policies to improve equipment efficiency. A special IEA report aims to raise awareness globally about one of the most critical energy issues of our time, outlining a sustainable path to the future of cooling that will allow people to reap the benefits of cooling without straining the energy system or the environment (IEA 2018).

Growing demand for air conditioners is one of the most critical blind spots in today's energy debate. Setting higher efficiency standards for cooling is one of the easiest steps that governments can take to reduce the need for new power plants, cut emissions, and reduce costs at the same time.

Problems

14.1 A refrigeration system uses R-134a as refrigerant and operates on the ideal vapor-compression cycle. The evaporation pressure is 0.14 MPa and the condensing pressure is 0.8 MPa. If the mass flow rate of the refrigerant is 0.5 kg/s, calculate (a) the cooling capacity of the system, (b) the compressor power, (c) the rate of heat rejection to the environment, and (d) the coefficient of performance.

[Answers: (a) 71.3 kW; (b) 18 kW; (c) 89.3 kW; (d) 3.96]

14.2 R-32 is currently the most common refrigerant to replace R-410A in split air-conditioning units (see Figure 14.3 for details of this device). The global warming potential (GWP) of R-32 is about a third that of R-410A but is still considerably higher than that of a growing number of nonfluorinated alternatives like propane (R-290) with a GWP < 1.

    a) Compare the *COP*s of these three refrigerants, each operating in an ideal vapor-compression refrigeration system with an evaporating temperature of 0 °C and a condensing temperature of 40 °C. The air in the cooled room may be assumed to be dry.

    b) Briefly discuss why the *COP* values are so close to one another despite widely varying compressor work and cooling capacities.

Hint: For good background on these refrigerants, read the following paper: Higashi et al. (2020).

[Answers: (a) R-410A: 4.6, R-32: 4.5, R-290: 4.3; (b) R-290 has nearly double the compressor work of R32 or R-410A, but R-290 also has nearly double the cooling capacity due to it having large latent heat.]

14.3 A vapor-compression heat pump is operating in heating mode to keep an office space at a comfortable 21 °C. Heat is transferred from the ambient air, which is at a frigid 4 °C. The rate of heat loss from the room to the environment (through the windows and walls) is 15 kW. The compressor requires a power input of 4 kW. Calculate (a) the rate of heat absorption from the cold external air and also the capacity of the heat pump, (b) the coefficient of performance, and (c) the rate of exergy destruction of the cycle. How can the exergy destruction be minimized?

[Answers: (a) 11 kW; (b) 3.75; (c) 3.13 kW]

14.4 Consider an advanced domestic air-conditioning system/heat pump. The system uses R-134a as a refrigerant and is designed to remove 2 kW of heat from a room when operating as an air-conditioner. (This problem is based on correspondence with Newell (2020).)

a) Model this vapor-compression cycle using EES or computer code of your choice.
b) Determine the compressor power and condenser heat transfer.
c) Calculate the system's *COP* when operating in air-conditioner mode in summer and in heat pump mode in winter. In winter, the room being heated is at 20 °C and the external ambient air is at 8 °C. In summer, the inside of the room is at 27 °C (dry-bulb) and 20 °C (wet-bulb), and the exterior air is at 35 °C (dry-bulb) and 24 °C (wet-bulb).

Hint: Using Figure 14.3, the indoor air temperature of 27 °C (dry-bulb) and 20 °C (wet-bulb) gives a dewpoint temperature of about 16.5 °C. So, to avoid dew from forming on the evaporator (which could hamper its operation), the R-134a evaporating temperature must be lower than 27 °C and higher than 16.5 °C.

Using Figure 14.3, the outdoor air temperature of 35 °C (dry-bulb) and 24 °C (wet-bulb) gives a dewpoint temperature of 19.5 °C. So, to avoid dew from forming on the condenser (which could hamper its operation), the R-134a condensing temperature must be higher than 35 °C (and also higher than 19.5 °C).

d) This heat pump costs US$5000 and typically saves US$300 a year on energy costs (compared to a traditional air-conditioner). For the 10-year lifetime of the heat pump, calculate the internal rate of return (IRR) of this US$5000.
e) If a homeowner purchases this system by taking out a 10-year loan of US$5000 to be repaid at a 7% interest rate, what will the monthly payments be? Considering that the homeowner saves US$300 per year on energy costs, what will be the net annual savings be and what is the benefit-to-cost ratio? Interpret your answers.

[Answers: (a) –; (b) Compressor: 140 W in summer and 160 W in winter; Condenser: 2.1 kW in summer and 1.9 kW in winter; (c) *COP* when working as a refrigerator in summer = 14.7 and in winter = 11.3; *COP* when working as a heat pump in summer = 15.6 and in winter = 12.2; (d) IRR = −8.35%; (e) Net annual savings = US$396.72, yearly savings-cost ratio = 0.43]

14.5 Data center energy usage accounts for about 1% of global electricity demand, or 200 TWh (IEA 2021). Data centers have consistently invested in technologies that are designed to reduce energy demand and CO_2 emissions. One such technology is *thermosyphons,* which do not use compressors such as in vapor-compression refrigeration systems, but rather feature gravity-fed liquid–vapor systems.

Do independent research about thermosyphons and produce a two-page fact sheet to clearly depict their operating principles and utility in economically and effectively removing heat from data centers. The total energy demand of data centers includes both the energy demand of the IT equipment itself and the energy demand of the supporting infrastructure, the vast majority of which is for cooling systems. Your fact sheet must also explain how thermosyphons might affect the *power usage effectiveness* (PUE) of data centers, where *PUE* represents the ratio of the total data center energy usage to the energy usage of the IT equipment.

To initiate your research, you should read the following articles:

Fleischer (2020).
Amalfi et al. (2021).

[Answer: Open-ended question that requires independent research.]

14.6 Air-conditioning accounts for about 10% of all electricity use worldwide. Some developing countries could see electricity demand for air-conditioning grow up to five times by 2050. With sustainable solutions, it is possible to reduce energy consumption for air-conditioning by 25% in 2050, compared to 2022. But that cannot be done without delivering affordable solutions in the developing world. *Cooling as a Service* (CaaS) aims to solve this problem.

CaaS is an innovative business model that enables end-users to access clean and efficient cooling solutions without the need of an upfront investment. This "pay-as-you-go" model addresses the key market barriers that hinder the adoption of sustainable cooling by allowing customers to pay for the service consumed (i.e., cooling) on a fixed-fee-per-unit basis. The ownership of the system remains with the technology provider, who is responsible for all operation costs. CaaS is a nascent business model in most low-income countries.

Do independent research about CaaS and explain, in no more than 1000 words, how you would use the "3-C entrepreneurial mindset" to popularize CaaS in rural Africa. The "3-C" mindset comprises your being curious, making connections between apparently disparate aspects, and creating value. (https://orchard-prod.azurewebsites.net/media/Framework/ KEEN_framework_new.pdf)

[Answer: Independent research; open-ended question.]

References

Amalfi, R.L., Faraldo, F.P., Salamon, T., Enright, R. et al. (2021). Hybrid two-phase cooling technology for next-generation servers: Thermal performance analysis. 2021 20th IEEE Intersociety Conference on Thermal and Thermomechanical Phenomena in Electronic Systems (iTherm).

Bertsch, S.S. and Groll, E.A. (2008). Two-stage air-source heat pump for residential heating and cooling applications in northern US climates. *International Journal of Refrigeration* 31 (7): 1282–1292.

Brodowicz, K., Dyakowski, T. and Wyszyński, M.L. (1993). *Heat Pumps*. Oxford: Butterworth-Heinemann.

Coulomb, D. (2021). Environmental issues related to refrigeration technologies. *International Journal of Air-Conditioning and Refrigeration* 29 (2): 2130002.

Dong, J., Lin, M., Zuo, J., Lin, T. et al. (2020). Quantitative study on the cooling effect of green roofs in a high-density urban area – a case study of Xiamen, China, *Journal of Cleaner Production* 255: 120152.

Fleischer, A.S. (2020). Cooling our insatiable demand for data. *Science* 370 (6518): 783–784.

Hepbasli, A. and Kalinci, Y. (2009). A review of heat pump water heating systems. *Renewable and Sustainable Energy Reviews* 13 (6–7): 1211–1229.

Herold, K.E., Radermacher, R. and Klein, S.A. (2016). *Absorption Chillers and Heat Pumps*. Boca Raton, FL: CRC Press.

Higashi, K., Kondou, C. and Koyama, S. (2020). Feasibility analysis for intermediated fluid type LNG vaporizers using R32 and R410A considering fluid properties. *International Journal of Refrigeration* 118: 325–335.

IEA (2018). The future of cooling opportunities for energy-efficient air conditioning. Paris: International Energy Agency. https://www.iea.org/reports/the-future-of-cooling.

IEA (2021). Data centres and data transmission networks. Paris: International Energy Agency. https://www.iea.org/reports/data-centres-and-data-transmission-networks

IEA (2022). The future of heat pumps. Paris: International Energy Agency. https://www.iea.org/reports/the-future-of-heat-pumps

Liebenberg, L. and Meyer, J.P. (1998). Potential of the zeotropic mixture R-22/R-142b in high-temperature heat pump water heaters with capacity modulation. *ASHRAE Transactions* 104: 418.

McLinden, M.O., Seeton, C.J. and Pearson, A. (2020). New refrigerants and system configurations for vapor-compression refrigeration. *Science* 370 (6518): 791–796.

Newell, T. (2020). Private communication with Emeritus Professor Ty Newell, *Build Equinox*. https://buildequinox.com

Sarbu, I. and Sebarchievici, C. (2014). General review of ground-source heat pump systems for heating and cooling of buildings. *Energy and Buildings* 70: 441–454.

Stoecker, W.F. and Jones, J.W. (1982). *Refrigeration and Air Conditioning*. New York: McGraw-Hill.

15

Controlling Emissions from Vapor Power Cycles and Vapor Refrigeration Cycles

Although an estimated 103 Gt of carbon can be removed in the next century using nature-based solutions (planting trees), such an initiative cannot be scaled up quick enough to noticeably reduce peak atmospheric temperatures and counteract global warming. Nature restoration is crucial but cannot offset fossil-fuel emissions to achieve net-zero by 2050 (Dooley et al. 2022). A portfolio of *carbon-capture and storage* (CCS) technologies will be required to remove an extra 220 Gt of CO_2 by 2050 (ETC 2021). Doing that will help us limit our cumulative CO_2 emissions to below 500 Gt by 2050 and might give us a good chance of limiting global warming to below 2 °C (Pörtner et al. 2022).

These carbon-capture technologies might include *pre-combustion* treatment, such as converting pulverized coal into a mixture of CO_2-rich gas and H_2 prior to combustion, which facilitates the simpler removal of CO_2 in flue gases. *Post-combustion* technologies can be used to treat flue gases by chemically treating the gas with an "amine" solution (*monoethanolamine*, MEA) to remove CO_2, after which the captured CO_2 could be stored in underground caverns or injected into oil wells to help remove oil deposits (called *enhanced oil recovery* [EOR]). Carbon dioxide can also be captured from ambient air using *direct air capture* (DAC) technologies.

Further, there are several pre- and post-combustion technologies that simultaneously reduce NO_x, SO_x, and particulate emissions from the burning of fossil fuels (especially coal). This is crucial technology as several countries (including China and India) will exploit coal-burning steam power plants for many years to come.

However, it is not only power-generation plants that produce harmful emissions. Refrigeration and air-conditioning account for around 8% of total greenhouse-gas emissions. This stems from the electrical energy that drives the refrigeration equipment (5%) and from the leakage of harmful refrigerants from these plants (3%).

Refrigeration and air-conditioning will continue to increase in the next decades, as explained in Chapter 14, which implies that much must be done to curb the leakage of global-warming refrigerants into the atmosphere. One way to do this is to switch to refrigerants with substantially lower *global warming potential* (GWP). Doing that could avert around 3% of global greenhouse gas emissions (McLinden et al. 2017).

Energy Systems: A Project-Based Approach to Sustainability Thinking for Energy Conversion Systems,
First Edition. Leon Liebenberg.
© 2024 John Wiley & Sons, Inc. Published 2024 by John Wiley & Sons, Inc.
Companion website: www.wiley.com/go/liebenberg/energy_systems

---

Reimagine Our Future

Climate scientists inform us that sea levels will likely rise by more than 1 m by 2050 compared to pre-industrial times. They also predict that heat waves will kill millions and disturb global agriculture. They say that these changes are unavoidable, given the greenhouse gases that have already been emitted. Why should we even bother with decarbonizing strategies if the world will be ravaged by climate change in any case?

15.1 Controlling NO_x, SO_x, and Particulate Emissions from Coal Combustion

The choice of coal and the design and operation of flue gas units are aimed at assuring that emissions are below the permitted levels. However, reducing emissions from power plants is complex and involves high investment and operational costs. The effective removal of pollutants is facilitated by the power plants being large-scale and stationary sources of emissions.

Steam power plants that operate with fossil fuels require extensive and expensive measures to control emissions from combustion. Due to their high combustion temperatures, coal-fired plants require the catalytic treatment of flue gas to remove thermal NO_x emissions. This is called *selective catalytic reduction* (SCR). Depending on the level of particulate emissions, flue gas usually also requires the filtration, called *electrostatic precipitation* (ESP) of particulates. Depending on the sulfur content of fossil fuels, the flue gas must also be treated with *flue gas desulfurization* (FGD) units. These technologies are illustrated in Figure 15.1.

Steam power plants that receive their heat from nuclear reactors operate at a much lower temperature than their fossil-fired counterparts. Nuclear steam power plants therefore have less of a problem with thermal NO_x emissions. Nuclear power plants do not have to contend with local emission of CO_2, SO_x, or particulates either. However, due to their lower maximum operating temperatures, nuclear steam power plants have a lower thermal efficiency than their fossil-fired counterparts.

Table 15.1 gives the estimated incremental impact of the above flue gas treatment technologies on the levelized cost of electricity (*LCOE*). The impact of achieving low emission levels is about 1 ¢/kWh$_e$ or about 20% of the total cost of electricity from a highly controlled pulverized coal unit (Deutsch and Lester 2007).

It should be noted that coal could also contain heavy metals such as mercury. Today, around 30% of the mercury in burned coal is removed by existing flue gas treatment technologies, primarily with fly-ash via electrostatic precipitators or fabric filters. Wet FGD achieves around 50% mercury removal. When combined with SCR, mercury removal could approach 95% for bituminous coals (Deutsch and Lester 2007).

15.2 Controlling Emissions from the Combustion of Natural Gas (or Methane)

There are two ways to systematically approach the task of turning high-efficiency gas generation into a zero or near-zero carbon resource using pre- and post-combustion technologies (Figure 15.2).

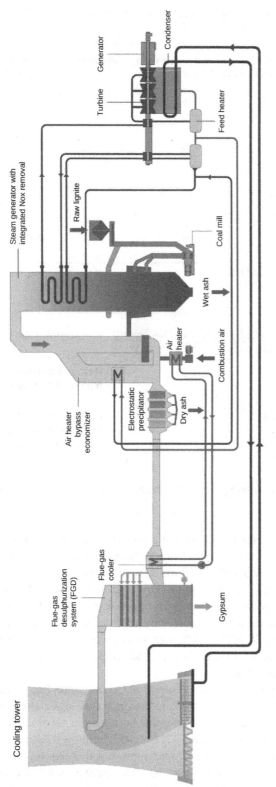

Figure 15.1 Schematic of an advanced, low-emission, pulverized coal-fired power plant. The flue gas is controlled by selective catalytic reduction for NO_x removal, followed by electrostatic precipitation to remove particulate matter, and wet flue gas desulfurization to remove SO_x. *Source:* Image courtesy of The Niederaussem Coal Innovation Center, RWB.

Labels within figure: Cooling tower; Flue-gas desulphurization system (FGD); Flue-gas cooler; Gypsum; Air heater bypass economizer; Electrostatic precipitator; Dry ash; Combustion air; Air heater; Steam generator with integrated Nox removal; Raw lignite; Coal mill; Wet ash; Turbine; Generator; Condenser; Feed heater

Table 15.1 Estimated incremental costs for a pulverized coal unit to meet 2022's best demonstrated emissions control performance versus no control.

	Capital cost ($/kW$_e$)	Operation and maintenance cost (¢/kWh$_e$)	Levelized cost of electricity (¢/kWh$_e$)
Electrostatic precipitation to remove particulates	40	0.18	0.26
Selective catalytic reduction for NO$_x$ removal	25	0.10	0.15
Wet flue gas desulfurization to remove SO$_x$	150	0.22	0.52
Average incremental control cost:	215	0.50	0.93

Source: Adapted from Deutsche and Lester (2007).

Pre-combustion

Use a low-carbon or carbon-neutral fuel

- Hydrogen (blue, green, pink)
- Synthetic (renewable CH$_4$)
- Biofuels
- NH$_3$

Post-combustion

Remove carbon from the exhaust

- Carbon capture (with liquid solvents)
- Carbon capture (solid sorbents)
- Oxyfuel cycles

Figure 15.2 Means of decarbonizing a gas turbine. *Source:* With permission from General Electric / CC BY 4.0.

Pre-combustion technologies refer to the systems and processes upstream of the gas turbine. The most common pre-combustion decarbonization technique is to change the fuel. Most gas turbines burn natural gas or methane (CH$_4$). Burning methane or natural gas still produces copious amounts of CO$_2$, albeit nearly half than when burning the same mass of coal. Gas turbine manufacturers are therefore developing combustion technology that will facilitate the burning of low-carbon fuels of varying compositions. Eventually, gas turbines will be able to burn only hydrogen (NASEM 2020; GTW 2021b). Although such gas turbines will probably still have high thermal NO$_x$ and H$_2$O vapor emissions, their carbon footprints will become minimal or zero if *green hydrogen* is used. (See Box 15.1.)

Post-combustion technologies involve the removal of CO$_2$ from the flue gases in a process commonly referred to as carbon capture. Post-combustion technologies also control emissions of NO$_x$, and SO$_x$ (refer to Chapter 12). Gas turbines that burn natural gas or methane need to contend with high levels of thermal NO$_x$ emissions. "Low-NO$_x$"-burning technology includes injecting steam or water into the flame to reduce temperatures, but this requires a supply of water and reduces plant efficiency. To reduce NO$_x$ emissions and ensure a multifuel capability, gas turbine manufacturers today premix fuel and air before the fuel enters the combustion chamber. Such a lean premixed combustion process is commonly called dry low-NO$_x$ (DLN) and can attain NO$_x$ emissions as low as 5 ppm (see Figure 12.5).

Box 15.1	**Gas Turbines Burning Green Hydrogen**

Mitsubishi Power offers a green hydrogen solution known as the Hydaptive Standard Flexibility Package. The Hydaptive package provides renewable energy flexibility by acting as a near-instantaneous power balancing resource that greatly enhances the ability of a simple cycle or combined cycle power plant to ramp output up and down to provide grid balancing services (GTW 2021a).

The Hydaptive system integrates a mixed hydrogen and natural gas-fueled gas turbine power plant with electrolysis to produce green hydrogen using 100% renewable power and onsite hydrogen storage. The Hydaptive storage package adds expanded or extended storage capability and is available for new gas turbine power plants or as a retrofit to existing plants to improve flexibility and extend asset life (GTW 2021a, 2021b).

15.3 Carbon Capture and Storage

The various technologies described in Section 15.1 do well to control the emission of particulates, NO_x, and SO_x. However, different types of technology are required to capture carbon from the CO_2 that is emitted when combusting fossil fuels.

Carbon capture and storage has been widely recognized as a key technology for mitigating global climate change (BP 2022), but the relatively high cost of current CCS systems remains a major barrier to its widespread deployment at power plants and other industrial facilities (Pörtner et al. 2022). There are a variety of technologies to capture CO_2, including post-combustion capture, pre-combustion capture, and oxyfuel capture (Bui et al. 2018) (see Figures 15.2 and 15.5). These CCS technologies add significantly to overall plant costs, as summarized in Table 15.2.

Table 15.2 CO_2 emissions and costs for a new 600-MW_e coal-fired steam power plant with and without CO_2 capture, operating at a 75% capacity factor with a fixed charge rate of 14.8%, and a coal price of 1.2 $/GJ, CO_2 transport cots of 3.2$/t CO_2, a geologic storage cost of 5 $/t, and tax credit for carbon storage of $10/t CO_2. All costs are in US dollars.

Cost and performance measures	Pulverized coal power plant	IGCC power plant
Emission rate without carbon capture (kg CO_2/MWh)	811	817
Emission rate with carbon capture (kg CO_2/MWh)	107	88
Percentage CO_2 reduction per MWh	87	85
Capital cost without carbon capture ($/kW)	1205	1311
Capital cost with carbon capture ($/kW)	1936	1748
LCOE without carbon capture ($/MWh)	46.1	48.3
LCOE with carbon capture ($/MWh)	82.1	69.6
Cost of CO_2 avoided ($/t CO_2)	51.2	29.5
CCS energy penalty ($/MWh)	28	15
With tax credit for carbon storage		
LCOE with carbon capture ($/MWh)	67.6	56.7
Cost of CO_2 avoided ($/t CO_2)	30.5	11.6

A pulverized coal system uses an amine-based CO_2 scrubber, and an IGCC plant uses hydrolyzer and Selectox™ CO_2-capture technology.
Source: Adapted from Rubin et al. (2015).

15.3.1 Post-combustion CO$_2$ Capture

Post-combustion capture involves, for instance, *scrubbing* CO$_2$ from unpressurized flue gases using a solution of monoethanolamine (MEA) and water. The amine solution is literally rained down through the exhaust stream, absorbing CO$_2$ in the process. The captured CO$_2$ is moved to a second vessel, where it is heated to drive off pure CO$_2$. That CO$_2$ can then be compressed and transported to a sequestration site or an industrial site.

In a gas turbine combined-cycle power plant, extracted steam from the low-pressure turbine section is used to regenerate the MEA solution. This, and other power requirements – namely CO$_2$ compression and liquefaction – reduce the net efficiency of the overall energy conversion process by about 10–12% (Deutsch and Lester 2007; DoE 2020).

15.3.2 Pre-combustion CO$_2$ Capture

As discussed in Section 15.2, pre-combustion technologies involve the use of different fuels. Pre-combustion capture, for instance, may also involve the conversion of pulverized coal into a mixture of CO$_2$-rich gas and H$_2$ prior to combustion. An example of this technology is the integrated gasification combined cycle (IGCC) power plant. With IGCC plants, a pulverized coal-water slurry is oxidized in liquid O$_2$, producing synthetic gas ("syngas") that comprises about 63% CO and 34% H$_2$. The syngas is treated with steam and a catalyst that produces CO$_2$ and H$_2$ in a so-called shift reaction (Bui et al. 2018):

$$\text{Syngas production}: 3C + O_2 + H_2O\,(g) \rightarrow 3CO + H_2$$
$$\text{Shift reaction}: CO + H_2O\,(g) \rightarrow CO_2 + H_2 \tag{15.1}$$

CO$_2$ can be removed from the pressurized flue gas by physical scrubbing with methanol, for instance, that could be used as the reclaimable scrubbing solution.

The H$_2$-rich syngas is sent to a gas turbine power plant to produce electricity. The exhaust gas from the gas turbine is used to boil water for use in a steam turbine, which also produces electricity, thus producing high fuel utilization. Such energy conversion is called a combined cycle power plant and was covered in Section 13.5. The overall energy conversion efficiency of the IGCC plant with CO$_2$ separation is 8–10% lower than an IGCC plant without CO$_2$ separation (Deutsch and Lester 2007; DoE 2020).

15.3.3 Oxyfuel CO$_2$ Capture

Oxyfuel capture involves burning pulverized coal in O$_2$ provided by an air-separation unit (ASU). The flue gas is recirculated, which helps ensure that the oxyfuel process produces mainly CO$_2$ and H$_2$O vapor. The unpressurized flue gas leaving the boiler is dehumidified and has a CO$_2$ content of around 90%, which allows for effective CO$_2$ capture. The power requirements of the ASU and ancillary equipment result in the power plant's net thermal efficiency reducing by 9 to 11% (Deutsch and Lester 2007; DoE 2020).

Figure 15.3 summarizes the pre- and post-combustion technologies used for CO$_2$ capture.

Table 15.2 reveals that a pulverized coal-fired power plant and an IGCC power plant have similar CO$_2$ emission rates. With carbon capture, both plants remove about 85% of CO$_2$ from the flue gas. Without CO$_2$ capture, the IGCC plant has the highest levelized cost of electricity. With CCS, the IGCC plant has a lower LCOE than the pulverized coal unit.

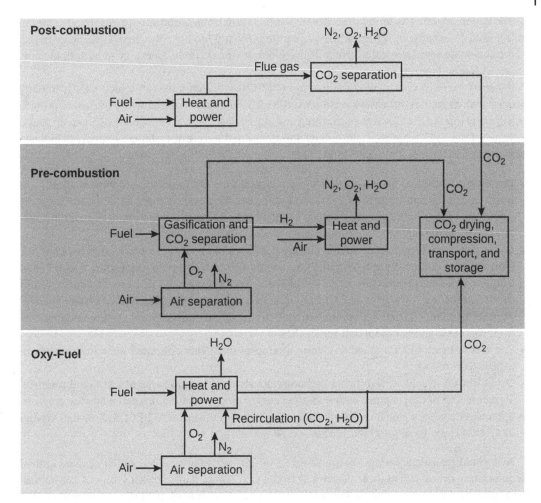

Figure 15.3 Broad classification of CO_2-capture technologies: post-combustion CO_2 capture, pre-combustion CO_2 capture, and oxyfuel CO_2 capture.

Reimagine Our Future

There is lack of a widespread sense of climate emergency and of the need to urgently decarbonize society. The "party" goes on with the belief that our actions will be counteracted by breakthrough and future technologies. Is that a rational and ethical way of approaching things?

15.3.4 Natural Climate Solutions Versus Direct Air Capture

The most tried and tested method for capturing CO_2 from the atmosphere is the one that the Earth has been utilizing for millions of years: photosynthesis. *Natural climate solutions* (NCS) apply this biogeochemical process and, in some cases, leverage technology to further support sequestration and long-term or permanent storage. Examples include afforestation and reforestation (including commercial forestry), improved natural forest management and agroforestry, improved agricultural practices to enhance soil carbon sequestration, and marine ecosystem restoration (ETC 2021, 2022).

The largest opportunity for natural climate solutions is in tropical and sub-tropical regions, where substantial co-benefits in terms of positive community and biodiversity impact can be expected. Natural climate solutions store carbon in live biomass (for example, trees), in soils, and, in some cases, in buried biomass.

Bio-energy with carbon capture and storage (BECCS) is a technology in which CO_2 is initially sequestered via photosynthesis (a version of NCS). The biomass is subsequently burned to provide energy, and most of the CO_2 is then captured and placed in geological storage. BECCS can undoubtedly play a role in CO_2 removal: the crucial questions are the scale of sustainable supply of biomass and the optimal use of land (ETC 2021, 2022).

- Devoting all residual waste materials from agriculture and forestry production to BECCS might enable carbon capture of 2–5 Gt CO_2 per year. However, competing demands for these resources – for instance for bio-plastics or biofuels to be used in aviation – will significantly reduce this potential.
- Given that a small portion of land is already dedicated to energy crop production today, if this biomass were exclusively used for BECCS, it could theoretically yield sequestration of 0.5–1 Gt CO_2 per annum. However, by 2050, the biomass produced on land dedicated to energy crops could – in theory – deliver up to 5 Gt CO_2 of sequestration per year if ambitious system changes were achieved in food and agricultural sectors, freeing up existing crop and pastureland for other uses, such as biomass production for BECCS.
- The cost of using BECCS to achieve carbon removal in 2050 is estimated to be in the range of \$100–\$200/t of CO_2.
- Using BECCS to achieve significant carbon removals would require a significant land resource. To secure 1 Gt of CO_2 capture with dedicated energy crops could require about 50 Mha per year, approximately 3% of global crop land today. Conservative estimates of BECCS show that around 22 Gt CO_2 could be sequestered in the next 30 years.

NCS could realistically remove around 6.5 Gt of CO_2 per year by 2050, which equates to total sequestration over the next three decades of 164 Gt CO_2. These nature-based solutions are impressive and must be implemented, but an additional cumulative amount of 220 Gt of CO_2 will need to be captured by 2050 if we are to have a 50% chance of keeping global warming below 2 °C (ETC 2021; Pörtner et al. 2022). This is where the direct air capture of carbon comes into play.

Direct air capture is a chemical process that can capture CO_2 from ambient air, with the CO_2 then stored in products or geological formations. This is called direct air capture and carbon storage (DACCS). DAC technologies are at an early stage of development and are only demonstrated at small scales (e.g., 50 tCO_2 per year).

In principle, there appears to be no obvious upper limit to the technical potential of DAC, but it requires vast electricity inputs due to the low concentrations of CO_2 (420 ppm) in ambient air, necessitating large quantities of air to be processed (ETC 2022). Current estimates suggest that capturing 1 t of CO_2 would require about 2.8 MWh, and 1 Gt would therefore require 2800 TWh (ETC 2022; BP 2022). If all the electricity used came from photovoltaics, the land requirement would be about 4 Mha (40,000 km^2) per Gt of captured carbon dioxide.

Cost estimates for DAC at current production scales have been around \$600/t of CO_2 captured, but it is believed that further research and development, plus large-scale deployment, could drive costs down to \$140 to \$300/t of CO_2 captured (ETC 2021, 2022).

15.3.5 Other Carbon Capture Technologies

Hybrid solutions include those based on mineral absorption or biogeochemical processes. In these cases, technology is leveraged to aid and accelerate known natural biogeochemical processes that sequester CO_2 (Rosen 2018; Service 2020).

Mineral absorption solutions explore geochemical inorganic reactions, while other solutions aim to enhance biological uptake through means beyond photosynthesis. These are nascent technologies that have not been proven on a large scale and have not yet demonstrated that there are no adverse effects on the environment.

Mineral absorption solutions include the following:

- *Enhanced weathering:* Adding crushed carbonate and silicate rocks to accelerate geochemical processes on land, which sequesters CO_2 from the atmosphere. The process would involve milling silicate rocks or alkaline solid wastes (Rosen 2018; Service 2020) and spreading the dust over large areas of managed cropland, speeding up the weathering reaction from proximity to plant roots and increased surface area. This technology could technically be applied today, but its impact is uncertain and further research is needed. When using the annual waste from silicate mining and industrial processes, an estimated sequestration of 0.7–1.2 Gt CO_2/year might be possible. Cost estimates range from \$50 to \$200 per tCO_2 (ETC 2021, 2022).
- *Ocean alkanization or sea water mineralization:* An increasing concentration of positive ions such as calcium in the ocean to enhance the natural ability to remove CO_2 and reverse acidification (Rosen 2018). This could be achieved by adding lime directly to sea water or reacting CO_2 gas and limestone with water and injecting it into the ocean. The chemical processes involved are well understood. However, application at scale has never been tested and the ecosystem impacts are not well understood. The full costs have been estimated at \$14 to \$500/t of CO_2, but these are highly uncertain (ETC 2021, 2022).

Other solutions applying biogeochemical processes include the following:

- *Ocean fertilization:* Enhancing open-ocean photosynthesis productivity by adding nutrients to increase CO_2 drawdown by phytoplankton, moving carbon into the deep ocean. The science of this carbon transfer is unproven and fertilization nutrients (nitrates and phosphorus) are expensive, energy-intensive, and (in the case of phosphorus) scarce (Rosen 2018).

Figure 15.4 illustrates the scale of potential carbon removals of these technologies.

15.3.6 Storage Options and Permanence

CO_2 removed from the atmosphere could be stored in numerous ways: in land-based nature, in geological storage, in the oceans, or in products and buildings ("storage in use"). Of these, the first two are likely to be the most important. Each entails different resource demands and management challenges, and for each it is important to assess the permanence and duration of storage (Rosen 2018; Service 2020; ETC 2021, 2022).

- Storage in land or the biosphere involves the direct sequestration of carbon into plant biomass and soils. This is clearly possible on a large scale. More than a billion hectares of land could be restored to natural ecosystems while also maintaining global food security. However, we need to carefully manage carbon stored in biomass (e.g., reforestation) to protect it from future deforestation driven by the same factors that drove it in the past. Secure finance to ensure continued active management is therefore critical.

  In addition, there is some concern about increasing instances of wildfire, pests, and disease due to the impacts of climate change, but these risks are highly localized. Tropical forests, in particular, because of their natural humidity, have little risk for wildfire if well managed for restoration; temperate forests may be more vulnerable. The duration of storage in land or the biosphere could therefore range from anywhere from 10 years (in the case of exogenous events such as extreme weather causing falling trees) to more than 1000 years (in the case of ancient peatlands). The addition of biochar and/or other technologies to convert the end-of-life biomass to a permanent storable good could therefore increasingly be required (ETC 2021, 2022).

Resource considerations and other co-benefits

	Description	Estimated carbon removal potential in 2050	Estimated costs in 2050		Key CO-benefits	Resource constraints	Land req. (Mha/ GtCO₂/y)	Permanence (years)
Natural climate solutions (NCS)	TRL at Scale = HIGH CO₂ is sequestered via photosynthesis and stored in biomass and soils via natural processes	Afforestation/Reforestation: 1–10 Gt CO₂/y; Soil Carbon Sequestration: 1–9 Gt CO₂/y; All other NCS: 0–10 Gt CO₂/y	Afforestation/reforestation: $5–50/tCO₂; Soil Carbon sequestration: $0–100/tCO₂	Forest and wetland restoration	Ecosystem restoration	Land water fertilizer	~70	10–1000+
				Commercial forestry materials	Displaces steel and concrete	Land water	–130	50–200
Biochar	TRL at Scale = MEDIUM Biomass burned in pyrolysis & used to stabilise organic matter	Biochar: 0.5–2 Gt CO₂/y	Biochar: $30–$120/tCO₂	Biochar	Soil health	Biomass	N/A	1000+
BECCS	TRL at Scale = HIGH CO₂ is sequestered via photosynthesis, the biomass used for bioenergy, & most of the CO₂ then captured and geologically stored (CCS)	BECCS from residues: 2–5 Gt CO₂/y; BECCS from energy crops: 0.5–5 Gt CO₂/yt	$100–$200/tCO₂	BECCS from residues	Energy	Land water fertilizer	280	100–1000+
				BECCS from energy crops	Energy	Land water fertilizer	~50	100–1000+
DACCS	TRL at Scale = MEDIUM CO₂ is captured from ambient air via technology and stored via CCS	0.5–5 Gt CO₂/y	$100–$300/tCO₂	DACCS from renewables	Habitat creation	Power water	~5	100–1000+
Mineral absorption	TRL at Scale = LOW Adding mineral materials to accelerate biogeochemical processes on land in oceans which sequester CO₂ through rock weathering and ocean geochemical process	Ocean alkalinisation: ~1 GT CO₂/y; Enhanced Weathering: 2–4 Gt CO₂/y	Ocean alkalinisation: $14–$500/tCO₂ enhanced weathering: $50–$200/tCO₂	Ocean alkalinisation		Power minerals	N/A	1000+
				Enhanced weathering	Soil health	Power minerals	N/A	1000+

Figure 15.4 An overview of salient carbon dioxide removal options. "TRL" refers to technology readiness level. "BECCS" refers to bioenergy with carbon capture and storage. All costs are in US dollars. *Source:* Adapted from ETC (2021).

- Geological storage makes CCS possible and is also at a relatively high level of technological readiness. It is already used within the oil and gas sector. It secures carbon in sedimentary formations, basalt, and peridotite, and has the theoretical potential to store vast quantities of carbon, although the availability of storage capacity varies greatly by country or region. It is also relatively secure in terms of permanence. Depending on the integrity of geological formations chosen, leakage rates are likely to be less than 1% over 100 years. The effective duration of storage could range from 100 years (e.g., failure of storage) to over 1000 years (effective storage with leakage rates of approximately 1% per 1000 years) (Service 2020; ETC 2021, 2022).
- There could be significant potential for storing carbon in oceans. However, the technologies to achieve this are the most unproven and the possible feedback effects on the ocean are the least clear (Rosen 2018).
- Storage-in-use entails the storage of carbon in products that can sequester carbon over a relatively long period of time, such as timber or concrete (Service 2020). Storage-in-use has a relatively small capacity compared with the volume of CO_2 that must be sequestered for a 1.5 °C pathway. It also has a relatively low permanence, with typical storage durations estimated between 50 and 200 years. Even if the storage is not permanent, it can still play a valuable role since the use of biomaterials typically substitutes for high-carbon alternatives (such as steel or conventional concrete in construction) (ETC 2021, 2022).

15.4 Controlling Emissions from Refrigeration and Air-Conditioning

An important category of greenhouse gases (GHGs) is fugitive emissions, which result from the direct release to the atmosphere of GHG compounds from various types of equipment and processes. Fugitive emissions are commonplace in the refrigeration and air-conditioning sector.

Historically, air conditioning and refrigeration equipment utilized various ozone-depleting substances (ODSs) as refrigerants, primarily chlorofluorocarbons (CFCs) and hydrochlorofluorocarbons (HCFCs). However, in accordance with the Clean Air Act Amendments of 1990 and the Montreal Protocol, these ODSs are being phased out of manufacture and use in the United States (EPA 2020).

Hydrofluorocarbons (HFCs) and, to a lesser extent, perfluorocarbons (PFCs) are used as substitute refrigerants for the regulated ODSs. Some countries may, however, still use these substances directly or in blends within refrigeration and air-conditioning systems (Kim et al. 2018). Many contemporary air-conditioning and refrigeration systems in the Western world use refrigerants with a lower environmental impact: non-halogenated refrigerants such as ammonia, CO_2, propane, isobutane, or hydrofluoroolefin-1234yf (McLinden et al. 2017; EPA 2020).

As mentioned in Chapter 6, *global warming potential* relates to the warming effectiveness of a gas compared with the reference gas, CO_2 (Shine et al. 1990). CO_2 therefore has a GWP of 1.0. The concept of *ozone-depletion potential* (ODP) is a metric that was originally developed as a single-value index to measure the potential impact of a chemical on stratospheric ozone, compared to that of CFC-11 (trichlorofluoromethane). CFC-11 therefore has an OPD of 1.0 (Wuebbles 1981).

Emissions from the refrigeration and air-conditioning sector result from the manufacturing process, from leakage and service over the operational life of the equipment, and from disposal at the end of the useful life of the equipment. These gases have a 100-year GWP (GWP_{100}) that is typically greater than 1000 times that of CO_2, so their potential impact on climate change can be significant (see examples in Table 15.2). By the same token, any reductions of these gases can have a large potential benefit. A global phasedown of HFCs is expected to avoid up to 0.5 °C of global warming by 2100 (McLinden et al. 2017; Pörtner et al. 2022).

Table 15.3 Global warming potential, ozone depletion potential, and atmospheric lifetime of some common refrigerants and refrigerant blends.

Refrigerant R-number (ASHRAE) and common name	Composition	Atmospheric lifetime (years)	Global warming potential in 100 years (GWP$_{100}$)	Ozone depletion potential (ODP)
R-744 (carbon dioxide)	CO_2	29,300–36,100	1	0
R-717 (ammonia)	NH_3	0.02	0	0
R-718 (water)	H_2O	0.03	0.2	0
R-134a (tertrafluoro-ethane)	CH_2FCF_3	14	1300	0
R-22 (chlorodifluoro-methane)	$CHClF_2$	12	1760	0.055
R-32 (difluoromethane)	CH_2F_2	4.9	675	0
R-1234yf (2,3,3,3-tetrafluoropropene)	$C_3H_2F_4$	0	4	0
R-407C (mixture of three HFCs)	Mixture of R-32/ R-125/R-134a (R-32: CH_2F_2 R-125: CHF_2CF_3 R-134a: CH_2FCF_3)	15.7	1774	0
R-410A (mixture of difluoromethane [called R-32] and pentafluoroethane)	R-32/R-125 (R-32: CH_2F_2 R-125: CHF_2CF_3)	16.95	2088	0
Propane	C_3H_8	13 d	3.3	0
Isobutane	C_4H_{10}	12	3	0

Source: ASHRAE (2020), EA/DEFRA (2019), Kim et al. (2018), and Pörtner et al. (2022).

Table 15.3 reveals that some fluorinated gases have a GWP thousands of times larger than CO_2. The Intergovernmental Panel on Climate Change (IPCC) estimates that these potent fluorinated gases make a major contribution to the greenhouse effect caused by human actions, and account for about 13% of global warming (Pörtner et al. 2022). The climate change impact caused by HFCs alone, via radiative forcing, already approximately doubled between 2005 and 2011 (IGSD 2013).

Research suggests that almost 40% of the emission of these potent gases by 2050 will fall outside the scope of international agreements such as the Paris Accord, Montreal Protocol, and Kigali Amendment. Without comprehensive and sustained interventions, uncontrolled growth in fluorinated-gas emissions could offset the gains made by the Clean Development Mechanism of the Kyoto Protocol, or the cornerstone of existing international climate governance: the Nationally Determined Contributions of the 2015 Paris Accord (Sovacool et al. 2021).

Disturbingly, it is projected that emissions of fluorinated gases will increase from one billion tons in 2005 to nearly four billion tons in 2050; 80% of fluorinated gas emissions will be from stationary and mobile refrigeration and air-conditioning applications (Gschrey et al. 2011; Purohit and Höglund-Isaksson 2017; Flerlage et al. 2021).

15.4.1 There Is No Perfect Refrigerant

Replacement refrigerants with a low GWP and low ODP are required but are difficult to come by.

In 1993, the global automotive industry replaced CFC-12 (dichlorodifluoromethane) with HFC-134a (1,1,1,2-tetrafluoroethane) in a two-year period. Although HFC-134a does not contain chlorine, which damages stratospheric ozone, HFC-134a has a GWP_{100} of 1300, which is above the limits set by the 1992 United Nations Framework Convention on Climate Change (McLinden et al. 2017; Kim et al. 2019). Work began in 1995 to move away from HFC-134a to CO_2, which has a GWP_{100} of 1. CO_2 is now used in refrigeration systems in supermarkets, industrial freezers, heat pump water heaters, data center cooling, and automotive air-conditioners. However, CO_2 refrigeration systems are complex to manufacture as they operate at high pressures. This also makes them very expensive (McLinden et al. 2017).

Hydrofluoroolefins (HFOs) were identified as the most promising replacements. The GWP_{100} of HFOs is typically less than 1. These organic chemicals are, however, flammable unless their molecules feature five or six fluorine atoms. For instance, HFO-1234yf (2,3,3,3-tetrafluoropropene) is difficult to ignite and has a GWP_{100} of 4, but it is toxic. Some olefins are also unstable molecules. To stabilize them, those olefins are blended with other refrigerants that contain chorine and fluorine, which of course gives the blend a non-zero ODP. Although useful for automotive air-conditioning, none of the HFOs are a direct replacement for HFC-410A, which is widely used in small air-conditioning systems (McLinden et al. 2017).

Hydrocarbons such as isobutane and propane are now often used as a refrigerant in small cooling systems such as domestic refrigerators. These refrigerants have a GWP of around 3. Although these refrigerants are flammable, the flammability hazard is deemed to be manageable due to the small charges (no more than 150 g) that are used (McLinden et al. 2017).

Ammonia has an ODP and a GWP of 0 and is therefore an excellent refrigerant, except for the fact that it is toxic and flammable. Ammonia is therefore typically only used in large or industrial refrigeration systems.

Water (R-718) is also considered as a refrigerant, albeit one with a relatively high critical temperature, thus it is used with low-pressure refrigeration systems. Water and other low-pressure refrigerants are intrinsically more energy efficient than their fluorinated or other counterparts. Cold water can be drawn from the bottom of deep lakes and used to cool data centers, for instance. However, these systems are currently much more expensive than traditional refrigeration systems that use fluorinated refrigerants.

It should be apparent that no perfect refrigerant exists, and trade-offs must be made between GWP, ODP, toxicity, flammability, stability, energy efficiency, system complexity, cost, and the prospect of long-term availability.

15.4.2 Challenges for Reducing the Emission of Fluorinated Gases

The emission of fluorinated gases must be reduced. There are, however, several barriers that must be overcome:

- *Financial and economic barriers:* Only around 1% of fluorinated gases are collected at the end of their life in Europe, despite strong environmental regulations. The situation is worse in the rest of the world where most fluorinated refrigerants eventually end up in the atmosphere (Purohit and Höglund-Isaksson 2017). Extracting fluorinated gases from disposal banks is expensive, ranging from $15 to $35 per ton of CO_2-equivalent, and only about a third of the captured gases are economically viable for recovery. Further, the decomposition of fluorinated gases (e.g., incineration, catalysis, or plasma destruction) is expensive and can range between $100 and $170 per ton of CO_2-equivalent (Castro et al. 2021; Sovacool et al. 2021).

- *Occupational hazards:* The destruction of fluorinated gases may create many occupational, health, and environmental hazards. The byproducts following the destruction of some modern refrigerants are usually highly reactive and may be carcinogenic or cause cardiovascular diseases, as well as being toxic, corrosive, and flammable (Han et al. 2021).
- *Unreported emissions*: Significant uncertainties and discrepancies exist regarding fluorinated gas emissions data, globally. In the United States alone, the CO_2-equivalent emissions from fluorinated gases range between 150 million and 500 million tons per year. Several industries have been accused of underestimating or not reporting their emissions of fluorinated gases, with little effect (Sovacool et al. 2021).
- *Illegal trade:* A vast black market exists for illegal and banned fluorinated refrigerants. The amount of black-market HFCs originating in China is equivalent to the annual CO_2 emissions of 3.5 million cars. This alone results in lost profits of around $500 million from the legal trade of refrigerants (Sovacool et al. 2021).
- *Gaming the system:* Due to fluorinated gases being effective global warmers, it has become more valuable for companies to earn carbon credits for abating these refrigerants than producing them in the first place. For instance, the cost of the destruction of HFC-23 is only around $1 per ton of CO_2-equivalent, but the value of the associated carbon credits for destroying this gas is 45–75 times the cost of abatement. This gaming of the system is banned in the Kyoto Protocol, but market manipulations are persistent (Sovacool et al. 2021).
- *Ineffective standards and regulations:* Developed countries like the United States sign up to pollution reduction agreements, just to be negated when the following government comes into power. Developing countries like China and India take advantage of this chaotic situation by further expanding their use and emission of fluorinated gases. To exacerbate this problem, the issuing of standards to mitigate the emission of fluorinated gases is complex and takes many years. Current standards to regulate low-GWP substitute refrigerants are more than 20 years old (Sovacool et al. 2021).

15.5 Outlook

To have a fair chance of limiting global warming to less than 2 °C, countries will need to employ negative emission technologies to pull as much as 10 Gt of CO_2 out of the atmosphere every year toward 2050 (Figure 15.5). Such negative emission technologies will no doubt include planting vast forests, which absorb CO_2 from the air as they grow; chemically absorbing CO_2 from the air or power plant exhaust and pumping it underground; and growing grasses or shrubs, burning them for energy, and capturing and storing the CO_2.

Regarding the use of refrigeration and air-conditioning systems, which will increase more than sixfold between now and 2050, urgent financing, governance, and policy instruments are required to abate emissions of fluorinated refrigerant gases. Widespread education is also required to help demystify the use of fluorinated gases in the manufacture of "low-carbon" technologies such as energy-efficient windows, thin-film photovoltaic cells, lightweight cars made with magnesium or aluminum casting, high-voltage transmission systems, and semiconductors. The calamitous radiative forcing of fluorinated refrigerant gases must be addressed now.

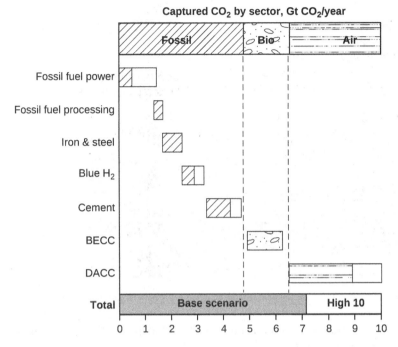

Figure 15.5 Capturing 7–10 Gt of CO_2 per year by 2050, based on carbon captured from the air, as well as after the combustion of fossil fuels and biogenic sources. *Source:* Adapted from ETC (2022).

Problems

15.1 Long discussed, but rarely used, carbon capture and storage projects, which bury waste CO_2 underground, are on the rise globally. Some scientists and engineers see the technology as a necessary tool in reducing carbon emissions, but others say it simply perpetuates the burning of fossil fuels.

The International Energy Agency (IEA) projects that around 1.2 billion tons of CO_2 must be locked away by 2030 if global warming is to stay under 1.5 °C. Among many other companies, the Norwegian company *Northern Lights* is drilling a second well line into an aquifer some 2500 m below the sea floor, preparing it to receive liquefied waste CO_2. The project will be the first in Europe designed to collect greenhouse garbage from a swath of industrial customers, from cement manufacturers to waste-to-energy plants, and ship it out for burial at sea (Figure 15.6).

The Centre for International Environmental Law condemns carbon capture and storage, stating that it is delaying decarbonization. It states that most CCS projects merely give fossil-fuel companies the incentive to produce more oil or coal, leading to an *increase* in net carbon emissions. With the plummeting price of wind and solar energy, it seems a step backward to incentivize waste disposal technology that prolongs the use of coal or oil.

Who is correct? Is CCS technology a boon or a bane? Produce a high-impact two-page infographic (no fewer than 400 words, with accompanying graphics or drawings) that explains the pros and cons of CCS technology. To guide you, search the internet for infographics on CCS, like the following:

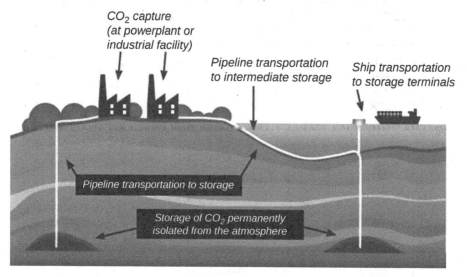

Figure 15.6 Carbon capture and storage at sea. *Source:* Image courtesy of Northern Lights Equinor.

ExxonMobil: https://corporate.exxonmobil.com/-/media/Global/Files/carbon-capture-and-storage/CCS-Infographic.pdf

Imperial College London: www.imperial.ac.uk/be-inspired/social-and-multimedia/info-graphics/carbon-capture-and-storage

Global Forest Coalition: https://globalforestcoalition.org/phantom-cop21-opera-bioenergy-carbon-capture-storage

[Answer: This question is open-ended and requires independent research.]

15.2 Download and study the following 2022 report by the Energy Transitions Committee (ETC): *Carbon capture, utilization and storage in the energy transition: Vital but limited.* Energy Transitions Commission, https://www.energy-transitions.org/publications/carbon-capture-use-storage-vital-but-limited

Then, produce a one-page fact sheet that compares the various forms of carbon capture and storage technologies. The fact sheet must present societal, technological, economic, environmental, and political challenges for implementing CCS in the iron and steel industry, electrical power CCS, bio-energy with CCS, fossil-fuel processing, "blue hydrogen" production, cement production, and direct air carbon capture.

[Answer: This question requires independent research. The provided ETC source contains helpful information that helps you to answer most of the question.]

15.3 Download and study the following 2022 report by the IEA: *Net zero by 2050. A roadmap for the global energy sector*, https://www.iea.org/reports/net-zero-by-2050

Then, produce a three-page fact sheet (around 1000 words, with diagrams and drawings) that discusses the behavioral changes that energy consumers will have to undergo for effectively cutting CO_2 emissions to achieve net-zero carbon emissions by 2050. Your fact sheet must outline some behavioral changes (with respect to, for instance, car and air travel) and illustrate the concomitant CO_2 reduction when adopting such behaviors. Your fact sheet must also outline how such behavioral changes can be brought about by appropriate regulations (like upper car speed limits), market-based instruments (like heavily taxing certain vehicle types), and information and awareness measures (such as personalized and real-time travel planning information). Summarize the various policy options to help bring about the required behavioral changes and indicate their cost-effectiveness, social acceptability, and impact on CO_2 emissions.

[Answer: This question requires independent research. The provided IEA source contains helpful information that helps you to answer most of the question.]

15.4 Produce a one-page fact sheet that outlines how each of the UN's 17 Sustainable Development Goals might relate to refrigeration and air-conditioning. You might want to read the following paper before you do so: Khosla et al. (2021).

[Answer: This question requires independent research. The cited journal article provides excellent context and examples.]

15.5 Scientists and engineers are investigating passive radiative cooling as a possible technology to reduce global warming. This involves a sky-facing surface on the Earth that spontaneously cools by radiating heat to the ultracold outer space (which is at around 3 K) through the atmosphere's longwave infrared (LWIR) transparency window ($\lambda \sim 8$–$13\,\mu m$). Write a short technical memo of around 500 words about the technical feasibility and economic viability of photonic materials to bring about daytime sub-ambient radiative cooling, especially in large cities.

The following paper provides a great starting point for your research: Wang et al. (2021).

[Answer: This question requires independent research. The cited journal article provides excellent context and examples.]

References

ASHRAE (2020). Update on new refrigerants designations and safety classifications. *American Society for Heating, Refrigeration, and Air-Conditioning Engineers*. https://www.ashrae.org/file%20library/technical%20resources/refrigeration/factsheet_ashrae_english_20200424.pdf

BP (2022). BP energy outlook. https://www.bp.com/content/dam/bp/business-sites/en/global/corporate/pdfs/energy-economics/energy-outlook/bp-energy-outlook-2022.pdf.

Bui, M., Adjiman, C.S., Bardow, A., Anthony, E.J. et al. (2018). Carbon capture and storage (CCS): the way forward. *Energy and Environmental Science* 11 (5): 1062–1176.

Castro, P.J., Aráujo, J.M., Martinho, G. and Pereiro, A.B. (2021). Waste management strategies to mitigate the effects of fluorinated greenhouse gases on climate change. *Applied Sciences* 11 (10): 4367.

Deutsch, J.M. and Lester, R.K. (eds.) (2007). *Making Technology Work: Applications in Energy and the Environment*. Cambridge, UK: Cambridge University Press.

DoE (2020). 2020 Carbon Capture R&D. Compendium of carbon-capture technology. US Department of Energy. https://netl.doe.gov/sites/default/files/2020-07/Carbon-Capture-Technology-Compendium-2020.pdf.

Dooley, K., Nicholls, Z. and Meinshausen, M. (2022). Carbon removals from nature restoration are no substitute for steep emission reductions. *One Earth* 5: 812–824.

EA/DEFRA (2019). Fluorinated gases (F gases). *UK Environment Agency; UK Department for Environment*, Food and Rural Affairs. https://www.gov.uk/guidance/fluorinated-gases-f-gases.

EPA (2020). Greenhouse gas inventory guidance. *Direct fugitive emissions from refrigeration, air conditioning, fire suppression, and industrial gases*. Environmental Protection Agency (EPA). https://www.epa.gov/sites/default/files/2020-12/documents/fugitiveemissions.pdf.

ETC (2021). Reaching climate objectives – the role of carbon dioxide removals. Energy Transitions Commission. https://www.energy-transitions.org/wp-content/uploads/2021/05/ETC-NegEmiss-White-paper-v4-Final.pdf.

ETC (2022). Carbon capture, utilization and storage in the energy transition: Vital but limited. Energy Transitions Commission. https://www.energy-transitions.org/publications/carbon-capture-use-storage-vital-but-limited.

Flerlage, H., Velders, G.J. and De Boer, J. (2021). A review of bottom-up and top-down emission estimates of hydrofluorocarbons (HFCs) in different parts of the world. *Chemosphere* 283 (November): 131208.

Gschrey, B., Schwarz, W., Elsner, C. and Engelhardt, R. (2011). High increase of global F-gas emissions until 2050. *Greenhouse Gas Measurement and Management* 1 (2): 85–92.

GTW (2021a). Gas Turbine World. https://gasturbineworld.com/mitsubishi-power-hydrogen-gas-turbine-technology.

GTW (2021b). Integrating hydrogen with renewables. *Gas Turbine World*. https://gasturbineworld.com/integrating-hydrogen-with-renewables.

Han, J., Kiss, L., Mei, H., Remete, A.M. et al. (2021). Chemical aspects of human and environmental overload with fluorine. *Chemical Reviews* 121 (8): 4678–4742.

IGSD (2013). Primer on short-lived climate pollutants. Institute for Governance and Sustainable Development. IGSD Working Paper: November. http://igsd.org/documents/PrimeronShort-LivedClimatePollutantsFeb192013.pdf.

Khosla, R., Miranda, N.D., Trotter, P.A., Mazzone, A. et al. (2021). Cooling for sustainable development. *Nature Sustainability* 4 (3): 201–208.

Kim, H.J., Liebenberg, L. and Jacobi, A.M. (2018). Convective boiling of R-134a near the micro-macroscale transition inside a vertical brazed plate heat exchanger. *Journal of Heat Transfer* 140 (9): 091501.

Kim, H.J., Liebenberg, L. and Jacobi, A.M. (2019). Flow visualization of two-phase R-245fa at low mass flux in a plate heat exchanger near the micro-macroscale transition. *Science and Technology for the Built Environment* 25 (10): 1292–1301.

McLinden, M.O., Brown, J.S., Brignoli, R., Kazakov, A.F. et al. (2017). Limited options for low-global-warming-potential refrigerants. *Nature Communications* 8 (1): 1–9.

NASEM (2020). Advanced Technologies for Gas Turbines. Washington, DC: National Academies of Science, Engineering and Medicine.

Pörtner, H.O., Roberts, D.C., Adams, H., and Adler, C. (2022). IPCC 6th Assessment Report. Climate Change 2022: Impacts, Adaptation and Vulnerability. Intergovernmental Panel on Climate Change.

Purohit, P. and Höglund-Isaksson, L. (2017). Global emissions of fluorinated greenhouse gases 2005–2050 with abatement potentials and costs. *Atmospheric Chemistry and Physics* 17 (4): 2795–2816.

Rosen, J. (2018). The carbon harvest. *Science* 369 (6508): 733–737.

Rubin, E.S., Davison, J.E. and Herzog, H.J. (2015). The cost of CO_2 capture and storage. *International Journal of Greenhouse Gas Control* 40: 378–400.

Service, R.F. (2020). The carbon vault. *Science* 359 (6377): 1156–1159.

Shine, K.P., Derwent, R.G., Wuebbles, D.J. and Morcrette, J.J. (1990). Radiative forcing of climate. *Climate Change: The IPCC Scientific Assessment*, 41–68.

Sovacool, B.K., Griffiths, S., Kim, J. and Bazilian, M. (2021). Climate change and industrial F-gases: a critical and systematic review of developments, sociotechnical systems and policy options for reducing synthetic greenhouse gas emissions. *Renewable and Sustainable Energy Reviews* 141: 110759.

Wang, T., Wu, Y., Shi, L., Hu, X. et al. (2021). A structural polymer for highly efficient all-day passive radiative cooling. *Nature Communications* 12 (1): 1–11.

Wuebbles, D.J. (1981). The relative efficiency of a number of halocarbons for destroying ozone. Lawrence Livermore National Laboratory Report UCID-18924. Livermore, CA.

Mini Project 5

Modern Fossil Fuel-Fired Power Plant

As the world decarbonizes its energy production, natural gas-fired steam power plants will offer important transitional power. Nuclear power reactors will presumably take over that role to supplement the intermittent power from energy sources such as wind and solar. Importantly, though, the Rankine cycle will still be used in nuclear power plants.

Consider a reheat steam turbine cycle operating on the ideal Rankine cycle. Use the process diagram and state points from Figure 13.3. The boiler is heated by burning fossil fuel such as natural gas (or coal or heavy fuel oil). The upper pressure is 8 MPa and the condenser temperature is 40 °C. The turbine inlet temperature at state point 3 is 455 °C. The feedwater pump and turbines are irreversible with isentropic efficiencies $\eta_P = 0.85$ and $\eta_T = 0.88$, respectively. The steam quality at the exit of the second turbine is 90%. The steam flow rate is 12 t/h.

1) Show the entire cycle on a detailed p-h diagram, generated with code of your choice, such as EES.
2) Determine the power required to drive the feedwater pump.
3) Determine the rate of heat transferred in the boiler.
4) Determine the power output of the two turbines.
5) Determine the heat rejected to the cooling water in the condenser.
6) Determine the thermal efficiency of the power plant.
7) Estimate the total savings in natural gas cost achieved by the reheating system compared to a conventional system without reheating. Assume that the efficiency of the steam generator (boiler) is 90%, the annual load factor is 70%, the useful plant life is 30 years, and the nominal cost of natural gas to power the furnace is $60 per MWh.
8) a) What is the annualized fixed cost per kW generated?
   b) Calculate the initial year annual variable costs per kW.
   c) What are the total annual costs ($/kWh) for an output period of 7000 h?

You may assume the following values for the above calculations:

Capital costs: $810/kW
Heat rate: 8840 Btu/kWh (or around 2.6 kW/kWh)
Fuel cost: $2/million Btu
Variable costs: 0.005 $/kWh
Annual fixed charge rate: 0.15

Energy Systems: A Project-Based Approach to Sustainability Thinking for Energy Conversion Systems,
First Edition. Leon Liebenberg.
© 2024 John Wiley & Sons, Inc. Published 2024 by John Wiley & Sons, Inc.
Companion website: www.wiley.com/go/liebenberg/energy_systems

Landscape of thorns

Consider a wanderer 100,000 years in the future discovering a strange construction of granite thorns in the New Mexico Desert, their points weathered by centuries, their shadows stretching at sinister angles. It is unclear for what this site was intended, or who created its menacing forms. On a wall, in several languages, maybe one of which the wanderer can decipher, are these words:

> This place is not a place of honor.
> No highly esteemed deed is commemorated here.
> Nothing valued is here.
> This place is a message and part of a system of messages.
> Pay attention to it!
> Sending this message was important to us.
> We considered ourselves to be a powerful culture.

"Landscape of thorns," a random field or forest of random concrete thorns, 15-m high, whose shapes suggest danger and bodily harm, is from a 1992 report by Sandia National Laboratories for the US Department of Energy for the Waste Isolation Pilot Plant (WIPP). The concept is part of ongoing semiotic studies to communicate across thousands of years that we, in the 21st century, have interred deep in the Earth the hazardous byproducts of our nuclear power. Those nuclear byproducts will be radioactive for 100,000 years or more. Do not enter this area! *Source:* Michael Brill and Safdar Abidi, Sandia National Laboratories / Public Domain.

Energy Systems: A Project-Based Approach to Sustainability Thinking for Energy Conversion Systems,
First Edition. Leon Liebenberg.
© 2024 John Wiley & Sons, Inc. Published 2024 by John Wiley & Sons, Inc.
Companion website: www.wiley.com/go/liebenberg/energy_systems

Week 6 – Nuclear Power

Nuclear power plants generate around 20% of electricity in the United States by using controlled nuclear *fission* chain reactions to heat a fluid (like water or helium), which is used to power turbines. Nuclear power plants generate few carbon emissions over the period of their construction, operation, and decommissioning. As the United States and other nations search for low-emission energy sources, the benefits of nuclear power must, however, be weighed against their operational risks, the challenges of safely storing spent nuclear fuel and radioactive waste, and dealing with security threats such as terrorist or military attacks. People make mistakes, but nuclear power demands "fail-safe" operation. This makes nuclear power a contentious issue. And controlled nuclear *fusion*, if ever attained over a sustained period and at a large scale, will probably not be economical. Even if nuclear fusion is proven to be a viable technology in the next 50 years, it will be too little too late. Nuclear energy (fusion or fission) will not be able to replace other forms of power generation quickly enough to avert the imminent effects of climate change.

Chapter 16 – Nuclear physics: Nuclear properties; nuclear binding energy; semi-empirical mass formula

Chapter 17 – Nuclear fission and fusion power plants: The fission process; cross-sections for particle interaction; neutron cross-section; neutron slowing; enrichment of nuclear fuel; self-sustaining chain reactions; reactor power; fission reactors; outlook for nuclear fission; fusion reactors; outlook for nuclear fusion; outlook

Chapter 18 – Controlling waste and emissions from nuclear power plants: Reactor safety; nuclear waste and its disposal; radioactivity of nuclear waste; radiation doses; safety and public policy; outlook

Mini Project 6 – Micro-nuclear power plant for a large university campus

16

Nuclear Physics

In the study of nuclear power, we must take account of the macroscopic and microscopic properties of individual particles of matter. *Microscopic properties* include masses of atoms or their constituent components (protons, neutrons, electrons), their sizes, or the number of particles in a volume. *Macroscopic properties* include mass density, charge density, and thermal conductivity. Models are generated to effectively capture the microscopic and macroscopic aspects of nuclear reactions, which enable the accurate prediction of available energy from such nuclear reactions.

Reimagine Our Future

A resilient electrical power system would be distributed among many renewable energy sources and organized around interlinked smart micro-grids. How would you encourage this to occur rapidly and on a large scale?

16.1   Nuclear Properties

A nucleus is located at the heart of an atom. It contains almost all the mass in a fraction (about 10^{-15}-th) of the volume. The nucleus is made up of positively charged *protons* and neutral *neutrons*. The force of electrostatic repulsion between like charges, which varies inversely as the square of their separation would be expected to be so large that nuclei could not be formed. The fact that they exist is evidence that there is an even larger force of attraction. This *strong force* (or *nuclear force*) overcomes the Coulomb repulsion of the charged protons and holds quarks together to form protons and neutrons. Negatively charged *electrons* spin around in shells around the nucleus.

The strong force operates over a short range, only a few femtometers (1 fm = 10^{-15} m). The force is repulsive over scales <0.5 fm due to the *Pauli Exclusion Principle* (two identical particles cannot occupy the same quantum state simultaneously); see Figure 16.1.

The charge on a proton is $+e$, where $e = 1.60217662 \times 10^{-19}$ coulombs (C) is the magnitude of the charge on an electron (see also Appendix A). A neutral atom with Z protons must have Z electrons. Therefore, the number of protons determines the chemical properties of an atom, even though chemistry is mainly about the interaction of electrons shared between atoms.

Energy Systems: A Project-Based Approach to Sustainability Thinking for Energy Conversion Systems,
First Edition. Leon Liebenberg.

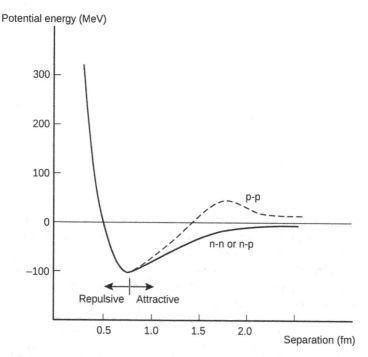

Figure 16.1 The strong force between nucleons, which is the electric potential experienced by one nucleon (a neutron or proton) as it is moved in the potential field of another. The interactions between pairs of neutrons (n–n), neutrons and protons (n–p), and protons and protons (p–p) are complicated and are therefore simplified in this diagram. Electrons have a very small mass compared to protons and neutrons and have a negligible effect in nucleon interactions.

Consider a nuclide X with Z protons that has N neutrons. A is the mass number or the total number of particles (or *nucleons*) in the nuclide:

$$A = N + Z \tag{16.1}$$

Z is often called the *atomic number* as it is used to arrange elements in the periodic table of elements. We will use the following shorthand notation when depicting nuclides:

$$^A_Z X_N \tag{16.2}$$

Example 16.1 Notation

For uranium-235, we will write $^{235}_{92}U_{143}$ and often abbreviate it as either $^{235}_{92}U$ or ^{235}U.

A crucial concept in nuclear physics is that of *isotopes*. Isotopes are elements with the same number of protons (Z), but a different number of neutrons (N). Although isotopes have nearly the same chemistry, they often have vastly different nuclear properties.

Example 16.2 Isotopes of Hydrogen

Ordinary hydrogen: ^1_1H – This isotope is stable, abundant, and can absorb neutrons.
Heavy hydrogen, or deuterium: ^2_1H – This isotope is stable, rare, and does not absorb neutrons.
Heaviest hydrogen, or tritium: ^3_1H – This isotope is unstable, radioactive, and is used as a fuel for nuclear fusion.

16.1.1 Radioactive Decays

The family of all nuclides may be visualized on a table of nuclides, which is a plot with N and Z making up the axes. Each nuclide is a point on this plot (Figure 16.2). A detailed table of nuclides is available from the US National Nuclear Data Center (NNDC): https://www.nndc.bnl.gov/

Figure 16.2 reveals that only some of the nuclides are stable. This forms a *region of stability*. Other nuclides tend to be unstable and will decay to the state of minimum energy consistent with conservation laws. This *radioactive decay* will take place in a way to conserve the mass number (A) and the electric charge. At fixed A, a nucleus can "adjust" N/Z by β-decay (these *beta particles* are nothing but electrons) and related processes like electron capture. Nuclei can also change A by emitting α particles (these *alpha particles* are nothing but $^4_2$He atoms) if energetically favorable. Alpha and beta decays may be accompanied by gamma (γ) decays.

Radioactive reactions may therefore comprise three components (Kaplan 1955; Gamow and Cleveland 1960):

- *Alpha (α) rays:* These are fast-moving and heavy helium-4 nuclei. $^4_2$He nuclei carry a double-positive charge and have a mass of four atomic units ("*amu*" or "*u*"). So, an element that emits an α particle is transformed into another element that has an atomic number smaller by two and an atomic weight smaller by four *amu*.
- *Beta (β) rays:* These are fast-moving (negative) electrons. Because a loss of a negative charge is equivalent to a gain in a positive charge, the atomic number of the resulting element increases by 1. The atomic weight does not change because of the negligibly small mass of the electron.
- *Gamma (γ) rays:* These are associated with both alpha and beta transformations, and are short electromagnetic waves emitted by atomic nuclei in the process of alpha or beta particle ejection. The γ rays are like X-rays, except that X-rays are emitted by atomic electrons, i.e., from processes outside the nucleus, while γ-rays originate inside the nucleus. X-rays are also generally lower in energy and, therefore, less penetrating than γ-rays.

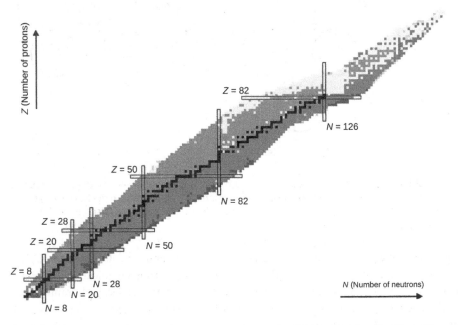

Figure 16.2 Representation of the chart of nuclides. *Source:* National Nuclear Database Center, CC BY-SA 4.0.

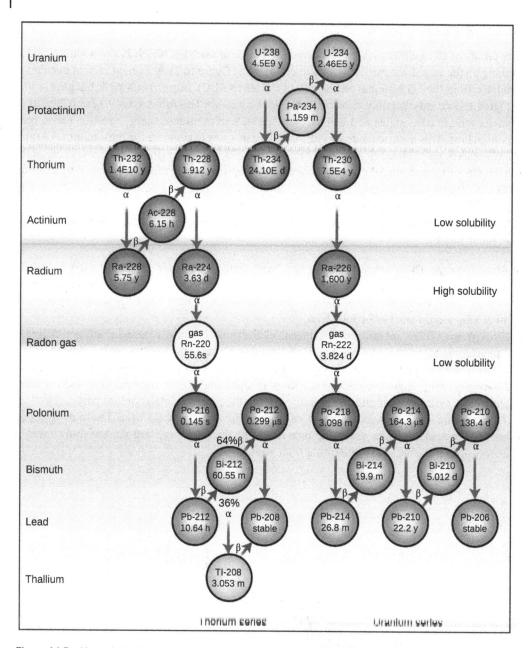

Figure 16.3 Natural thorium and uranium decay chains. Abbreviations for half-lives: d = days; h = hours; m = minutes; s = seconds; y = years. Half-lives and decay information were obtained from the NuDat 2 Database of the National Nuclear Database Center. *Source:* Environmental Health Perspectives/Nelson et al. (2015), CC BY-SA 4.0.

Alpha, beta, and gamma particles have very high kinetic energies, and typically contains millions of times more energy than solar radiation or than energy emitted during chemical combustion.

There are several "families" of radioactive elements: one of them for the decay of uranium-238. Figure 16.3 shows how uranium-238 decays via the emission of α or β particles. For instance, $^{238}_{92}\text{U}$ decays via the emission of α (or $^{4}_{2}\text{He}$) particles: $^{238}_{92}\text{U} \rightarrow ^{234}_{90}\text{Th} + ^{4}_{2}\text{He}$, and thorium-234, in turn, decays by the emission of β particles into protactinium: $^{234}_{90}\text{Th} \rightarrow ^{234}_{91}\text{Pa} + ^{0}_{-1}\beta$. After several alpha and beta emissions, atoms eventually transform into a stable isotope of lead, $^{206}_{82}$Pb, and no further radioactive transformations take place (Gamow and Cleveland 1960).

Box 16.1 explains the rudiments of radioactive decay and half-lifetimes.

16.1.2 Energy Scales and Nuclear Mass

In radioactive transformations, we refer to energies in terms of electron volts: 1 eV is the energy gained by a particle carrying one elementary electric charge (irrespective of whether it is an electron or a positive-charged particle) when it is accelerated through an electric field with a potential difference of 1 V.

The elementary charge of a proton or electron is 1.6×10^{-19} C.
So, 1 eV $= 1.6 \times 10^{-19}$ J because 1 V $= 1$ J/C.
We can also calculate the inverse: 1 J $= 6.24 \times 10^{18}$ eV.

In *chemical reactions* such as the combustion of coal or gasoline, the chemical reactions will be in the eV range. For instance, ionizing hydrogen, $^1_1$H (to remove its electron), requires 13.6 eV.

In *nuclear reactions*, however, it is more typical to use the mega electron volt (MeV) scale, where 1 MeV $= 10^6$ eV. For instance, to split $^2_1$H into a proton and a neutron requires 2.2 MeV of energy. Nuclear reactions are millions of times more energetic than chemical reactions!

The mass of a proton is approximately 1.67×10^{-27} kg. To make for more manageable calculations, we define an *atomic mass unit* (amu, or u). One atom of $^{12}_6$C has a mass of 12 u. Remember that this includes the mass of the electrons in the atom, not just the protons and neutrons.

$$1 \text{ amu} = 1 \text{ u} = \frac{M\left(^{12}_6\text{C}\right)}{12} \approx 1.66 \times 10^{-27} \text{ kg} \tag{16.3}$$

A neutron has a mass of 1.008665 u and a proton has a mass of 1.007276 u. An electron has a mass of only 0.000546 u.

When comparing masses of atomic nuclei with those of their constituent protons and neutrons, there will always be a slight discrepancy, called the *mass defect* (Δ) with

$$\Delta = m\left(^A_Z X\right) - (A \times 1 \text{ u}) \tag{16.4}$$

This defect can be expressed in terms of u or MeV. The mass defects of all elements can be found on the NNDC website. https://www.nndc.bnl.gov/nudat3/indx_sigma.jsp

---

Reimagine Our Future

How can universities and colleges redirect their buying and investments to promote local, regional, and global renewal? Can universities and colleges help make agriculture, food systems, energy conversion systems, and businesses more sustainable? Imagine if universities and colleges were themselves to become net-zero carbon emitters before 2035 and led the way for the rest of society. Share your thoughts on social media.

16.2 Nuclear Binding Energy

When comparing the masses of atomic nuclei with those of their constituent protons and neutrons, we can also identify a mass deficit, ΔM. This can be explained by considering Einstein's equation, $E = mc^2$, where c is the speed of light (3×10^8 m/s):

$$E = \Delta M\, c^2 = \left(1.66 \times 10^{-27}\, \text{kg}\right)\left(3 \times 10^8 \text{m/s}\right)^2 = 1.49 \times 10^{-10}\, \text{J}$$

But, $1\ \text{eV} = 1.6 \times 10^{-19}\, \text{J}$

$$\therefore 1\ \text{amu} = 1\, \text{u} = 1.49 \times 10^{-10}\, \text{J}\left(\frac{1\ \text{eV}}{1.6 \times 10^{-19}\, \text{J}}\right) = 931.5\, \text{MeV}$$

(16.5)

This is called the *rest mass energy* of a nucleon.

For example, for oxygen $\left(^{16}_{8}\text{O}\right)$: $\;$ 8 neutrons $= 8 \times 1.008665\, \text{u} \;= 8.07092\, \text{u}$

$$8\ \text{protons} = 8 \times 1.007276\,\text{u} \;= 8.05820\,\text{u}$$

$$8\ \text{electrons} = 8 \times 0.000546\, \text{u} = 0.00440\, \text{u}$$

$$\overline{\hspace{6cm}}$$

$$16.13353\, \text{u}$$

But the atomic weight of oxygen is 16.00000. So, the oxygen nucleus is lighter than its constituent components by 0.13353 u.

The explanation for this so-called *mass deficit* is based on Einstein's law of the equivalence of mass and energy, $E = mc^2$. Accordingly, all matter possesses energy, and all energy is associated with matter. Energy is required to break up a nucleus into neutrons and protons. This energy is called nuclear *binding energy*. For the $^{16}_{8}\text{O}$ atom, the binding energy (B) is:

$$B = (0.13353\, \text{u})\left(\frac{931.5\, \text{MeV}}{1\, \text{u}}\right) = 124.4\, \text{MeV}$$

Dividing the total binding energy (B) of the composite nucleus by the total number of protons and neutrons forming it (A), we obtain the average binding energy per nucleon (b). For $^{16}_{8}\text{O}$ with its 16 nucleons, the average binding energy per nucleon is:

$$b = \frac{B}{A} = \left(\frac{124.4\ \text{MeV/nucleus}}{16\ \text{nucleons/nucleus}}\right) = 7.78\ \text{MeV/nucleon}$$

The average binding energy for each element can be calculated in a similar manner. When plotting b against A for all elements, we find a plot such as that illustrated in Figure 16.4.

Figure 16.4 reveals important nuclear phenomena, specifically the competition between the attractive but short-range nuclear force, and the long range, but repulsive Coulomb force (Kaplan 1955; Lamarsh 1966):

- The average binding energy per nucleon (b) is virtually constant, except for a few light nuclei; $b = 8$ MeV/nucleon $\pm 10\%$.
- A higher value of b implies a more stable atom.
- Curve increases for light nuclei. Adding nucleons strongly attracts nearby nucleons, making the whole nucleus more tightly bound.
- Curve peaks near $A \approx 60$. The wide stable region from Mg to Xe is because the nucleus is larger than the extent of the nuclear force, so the force saturates, implying that adding more nucleons does not increase the value of b.
- Gradual decay for heavy atoms, $A > 100$. This is because the nuclei are so large that the Coulomb forces across the nucleus are stronger than the attractive nuclear forces, decreasing the strength of the binding.

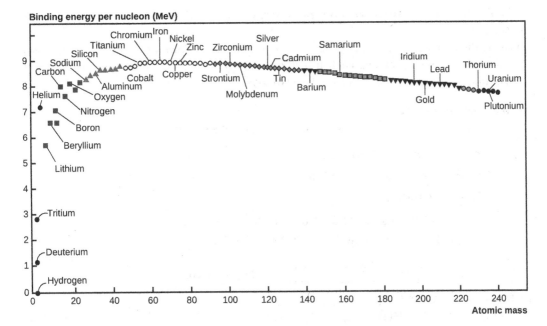

Figure 16.4 Average binding energy per nucleon (*b*) versus atomic mass (*A*).

- Above $A \approx 120$, the value of b decreases, which implies that splitting nuclei will release energy, a process called *fission*.
- Combining light elements ($A < 40$) also releases energy, which we call *fusion*.

The atomic mass of an atom $^A_Z X_N$ is the sum of the masses of the constituent particles, minus the nuclear binding energy, $B(Z, A)$:

$$m\left(^A_Z X\right) = M(Z,A) = Zm_{\mathrm{p}} + Nm_{\mathrm{n}} + Zm_e - \frac{B}{c^2}$$

$$m\left(^A_Z X\right) = M(Z,A) \equiv Zm_{\mathrm{H}} + Nm_{\mathrm{n}} - \underbrace{\frac{B}{c^2}}_{\text{mass deficit, } \Delta M} \tag{16.6}$$

In Eq. (16.6), m_{H} is the mass of $^1_1\mathrm{H}$, which comprises only a proton and an electron.
We can calculate B in terms of *mass deficit* by rewriting Eq. (16.6) with the help of Eq. (16.5):

$$B = \left[Zm_{\mathrm{H}} + Nm_n - m\left(^A_Z X\right)\right] c^2$$
$$B = \left[Zm_{\mathrm{H}} + Nm_n - m\left(^A_Z X\right)\right] 931.5 \text{ MeV/u} \tag{16.7}$$

We can also calculate B using the *mass defect* (Lamarsh 1966):

$$B\left(^A_Z X\right) = Z\Delta_{\mathrm{H}} + N\Delta_n - \Delta\left(^A_Z X\right) \tag{16.8}$$

For fission to occur, the binding energy of the products of a nuclear reaction must exceed that of the initial nucleus. The difference between these two binding energies gives the heat released (Q) from fission:

$$Q = B_{products} - B_{initial}$$
$$Q = \Delta_{initial} - \Delta_{products} \tag{16.9}$$

Example 16.3 Calculation of Binding Energy Using Either Mass Deficit or Mass Defect (Δ)

Calculate the average binding energy per nucleon of $^4_2$He (also known as an α particle). Perform the calculation using (a) the mass deficit method, and (b) the mass defect method. The mass of an $^4_2$He atom is accurately determined to be 4.001506 u.

a) Mass deficit method

From Eq. (16.5): $B\left(^4_2\text{He}\right) = \left[2m_\text{H} + 2m_n - m\left(^4_2\text{He}\right)\right] 931.5 \text{ MeV/u}$

with $2(m_\text{H}) = 2(1.007276 \text{ u}) = 2.014552 \text{ u}$

$2(m_n) = 2(1.008665 \text{ u}) = 2.017330 \text{ u}$

$m\left(^4_2\text{He}\right) = 4.001506 \text{ u}$

$\therefore B = (2.014552 \text{ u} + 2.017330 \text{ u} - 4.001506 \text{ u})(931.5 \text{ MeV/u})$

$\therefore B = (0.030376 \text{ u})(931.5 \text{ MeV/u}) = 28.3 \text{ MeV/nucleus}$

Therefore, $b = \dfrac{B}{A} = \dfrac{28.3 \text{ MeV/nucleus}}{4 \text{ nucleons/nucleus}} = 7.1 \text{ MeV/nucleon}$

b) Mass defect method

From Eq. (16.7): $B\left(^4_2\text{He}\right) = 2\Delta_\text{H} + 2\Delta_n - \Delta_\text{He}$

We can calculate the mass defects by hand, or we can read the values from the NNDC's webpage, https://www.nndc.bnl.gov/nudat3/indx_sigma.jsp:

The mass defect of $\left(^4_2\text{He}\right)$ is $\Delta_\text{He} = 2.4249 \text{ MeV}$

The mass defect of $\left(^1_1\text{H}\right)$ is $\Delta_\text{H} = 7.2889 \text{ MeV}$

The mass defect of a neutron $\left(^1_1 n\right)$ is $\Delta_n = 8.071 \text{ MeV}$

$\therefore B\left(^4_2\text{He}\right) = 2(7.2889 \text{ MeV}) + 2(8.071 \text{ MeV}) - 2.4249 \text{ MeV} = 28.3 \text{ MeV/nucleus}$

Therefore, $b = \dfrac{B}{A} = \dfrac{28.3 \text{ MeV/nucleus}}{4 \text{ nucleons/nucleus}} = 7.1 \text{MeV/nucleon}$

Example 16.4 Energy Scales – Chemical Reactions Versus Nuclear Reactions

Let us compare the energy per nucleus when (a) combusting iso-octane (C_8H_{18}), and (b) when fissioning uranium-235.

a) Iso-octane

The higher heat value of iso-octane (C_8H_{18}) is 5.47 MJ/mol or 47.9 MJ/kg. This can be converted to an equivalent amount of eV per nucleus.

The atomic mass of iso-octane (C_8H_{18}) = 8 mol C(12 g/mol C) + 18 mol H(1 g/mol H) = 114 g.

We also know that there are 6.02×10^{23} nucleons (Avagadro's number) in 1 mol. Further, $1 \text{ J} = 6.24 \times 10^{18} \text{ eV}$. We can now convert the heat released in 47.9 MJ/kg to eV/nucleus:

$$\left(\frac{47.9 \text{ MJ}}{\text{kg}}\right)\left(\frac{10^6 \text{ J}}{\text{MJ}}\right)\left(\frac{1 \text{ kg}}{1000 \text{ g}}\right)\left(\frac{114 \text{ g}}{\text{mol}}\right)\left(\frac{1 \text{ mol}}{6.02 \times 10^{23} \text{ nucleons}}\right)\left(\frac{6.24 \times 10^{18} \text{ eV}}{1 \text{ J}}\right)$$
$$\cong 57 \text{ eV/nucleus}$$

A nucleus of iso-octane releases 57 eV during complete combustion.

b) Uranium-235

For the decay of uranium-235, assume the following reaction: $^{235}_{92}\text{U} + ^1_0 n \rightarrow ^{144}_{56}\text{Ba} + ^{90}_{36}\text{Kr} + 2^1_0 n$

We can use the mass deficit method to determine b:

From Eq. (16.9), we can calculate the energy released in fissioning:

$$Q\left(^{235}_{92}U\right) = \left[(m_U + m_n) - (m_{Ba} + m_{Kr} + 2(m_n))\right] \times (931.5 \text{ MeV/u})$$

with $\quad m_U = 235.043928$ u

$^1_0 n = 1.008665$ u

$m_{^{144}Ba} = 143.922955$ u

$m_{^{90}Kr} = 89.919528$ u

$$\therefore Q\left(^{235}_{92}U\right) = (235.043928 \text{ u} + 1.008665 \text{ u} - 143.922955 \text{ u} - 89.919528 \text{ u} - 2.017330 \text{ u})(931.5 \text{ MeV/u})$$

$$\therefore B = (0.19278)(931.5 \text{ MeV/u}) = 179.6 \text{ MeV/nucleus}$$

A nucleus of uranium-235 therefore releases 179.6 MeV during transformation to barium and krypton, which is more than three million times the heat released from a nucleus of combusting iso-octane.

Reimagine Our Future

To paraphrase Albert Einstein, we must change the way we think to change the way we act. How should we change our thinking to pursue a net-zero carbon society?

Box 16.1   Half-Lifetimes and Radioactive Decay

Uranium atoms may decay over a period of billions of years; other radioactive elements may last only a fraction of a second. In nuclear physics, the term *half-life* is used to indicate the time during which the initial number of atoms is reduced to one half. At the end of twice that period, only a quarter of the original number of atoms will be left; at the end of three half-life periods, only one eighth will be left, etc. (see Figure 16.5). The amount of a decaying element that is left (N) after several half-lives (n) is:

$$N = N_0 \left(\frac{1}{2}\right)^n \tag{16.10}$$

For example, the half-life of radon is 3.82 days. If we start with 10 mg of radon, after 30 days, the following will be left:

$30/3.82 = 8 \text{ half-lives}$

$$\text{So}, N = N_0 \left(\frac{1}{2}\right)^n = 10\left(\frac{1}{2}\right)^8 = 0.039 \text{ mg}$$

The 200 mg of radium that Marie Curie separated in 1898 now contains only around 190 mg of radium.

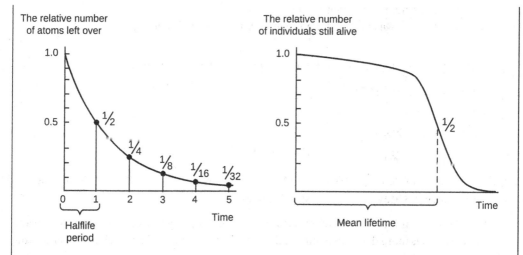

Figure 16.5 Representation of half-lifetimes of radioactive elements. *Source:* Adapted from Gamow and Cleveland (1960).

16.3 Semi-Empirical Mass Formula

The properties of nuclear force are used to create a model that can be used to predict masses and binding energies during nuclear transformations. The *semi-empirical mass formula* (SEMF) is widely used as it provides an accurate and easy interpretable model for both fission and fusion.

The shape of a nucleus may be assumed to be like that of a liquid droplet. The *droplet model* of an atomic nucleus considers the different nuclei as minute droplets of a universal nuclear "fluid" with constant "density." According to nuclear droplet theory, the volumes of different nuclei are proportional to their masses. If the nuclear "fluid" is charged, then the binding energy (B) will comprise a volume-energy term, a surface-energy term, a Coulomb-energy term, a symmetry-energy term, and a pairing-energy term.

16.3.1 Volume and Volume Energy

The radii (R) of atomic nuclei vary as the cube roots of their masses (A):

$$R \propto A^{1/3} \tag{16.11}$$

The bulk binding energy in this volume (B_V) will be equal to: (Number of bonds/Nucleon) $\times$ (Energy/Bond) $\times$ (Number of nucleons), which can be written as follows:

$$B_V = \varepsilon_V A \tag{16.12}$$

Experimentation (Jaffe and Taylor 2018) reveals that $\varepsilon_V = 15.56$ MeV.

16.3.2 Surface and Surface Energy

The volume term (Eq. (16.12)) overestimates the binding energy as the nucleons on the surface of the droplet have a deficit of neighbors. This is accounted for by a surface-energy term:

$$B_S = -\varepsilon_S A^{2/3} \tag{16.13}$$

Experimentation (Jaffe and Taylor 2018) reveals that $\varepsilon_S = 17.23$ MeV.

16.3.3 Coulomb Energy

Protons will tend to repel each other, as illustrated in Figure 16.1. This Coulomb repulsion is accurately captured by the following term:

$$B_C = -\varepsilon_C \frac{Z^2}{A^{1/3}} \tag{16.14}$$

Experimentation (Jaffe and Taylor 2018) reveals that $\varepsilon_C = 0.7\,\text{MeV}$.

16.3.4 Symmetry Energy

Experimentally, it has been found that nuclei have lower binding energy if they comprise a differing number of protons and neutrons. This asymmetry is represented by the following term:

$$B_{symm} = -\varepsilon_{sym} \frac{(N-Z)^2}{A} \tag{16.15}$$

Experimentation (Jaffe and Taylor 2018) reveals that $\varepsilon_{sym} = 23.28\,\text{MeV}$.

16.3.5 Pairing Energy

Experiments have also revealed that account must be taken of whether there is an odd or even number of protons and neutrons. A pairing-energy term may be defined to capture this effect:

$$B_{pairing} = -\eta(Z,N) \frac{\Delta}{A^{1/2}} \tag{16.16}$$

where Δ is the mass defect as defined in Eq. (16.4), and assumes an approximate value of 12 MeV for ^{235}U, and

$$\eta(Z,N) = \begin{cases} +1 & \text{if } Z \text{ and } N \text{ are even} \\ 0 & \text{if } Z \text{ is odd and } N \text{ is even, or } Z \text{ is even and } N \text{ is odd} \\ -1 & \text{if if } Z \text{ and } N \text{ are odd} \end{cases} \tag{16.17}$$

Combining Eqs. (16.12) to (16.17), the SEMF becomes:

$$B = \varepsilon_V A^{1/3} - \varepsilon_S A^{2/3} - \varepsilon_C \frac{Z^2}{A^{1/3}} - \varepsilon_{sym} \frac{(N-Z)^2}{A} - \eta(Z,N) \frac{\Delta}{A^{1/2}}$$

$$\text{with} \quad \varepsilon_V = 15.56\,\text{MeV}, \varepsilon_S = 17.23\,\text{MeV},$$

$$\varepsilon_C = 0.7\,\text{MeV}, \varepsilon_{sym} = 23.28\,\text{MeV}, \Delta = 12\,\text{MeV} \tag{16.18}$$

Example 16.5   Binding Energy of a Nucleus of $^{235}_{92}\text{U}$

The binding energy of a nucleus of $^{235}_{92}\text{U}$ may be estimated with the use of the SEMF stated in Eq. (16.18):

$$B\left(^{235}_{92}\text{U}\right) = \varepsilon_V A - \varepsilon_S A^{2/3} - \varepsilon_C \frac{Z^2}{A^{1/3}} - \varepsilon_{sym} \frac{(N-Z)^2}{A} - \eta(Z,N) \frac{\Delta}{A^{1/2}}$$

$$B\left(^{235}_{92}\text{U}\right) = (15.56\ \text{MeV})(235) - (17.23\ \text{MeV})(235)^{2/3} - (0.7\ \text{MeV}) \frac{(92)^2}{(235)^{1/3}}$$

$$- (23.28\ \text{MeV}) \frac{(143-92)^2}{235} - (0) \frac{(12\ \text{MeV})}{(235)^{0.5}}$$

$$B\left(^{235}_{92}\text{U}\right) = 3600\ \text{MeV} - 656\ \text{MeV} - 960\ \text{MeV} - 258\ \text{MeV} - 0 = 1726\ \text{MeV}$$

Every nucleus of $^{235}_{92}$U will therefore have binding energy (B) of 1726 MeV and an average binding energy per nucleon of $b = \dfrac{B}{A} = \dfrac{1726\text{ MeV}}{235\text{ nucleons}} = 7.35$ MeV/nucleon. This predicted value of b corresponds well with the measured value (7.59 MeV/nucleon) plotted in Figure 16.4.

Problems

16.1 How much energy (in joules) is released in the fission of a uranium-235 nucleus that yields 190 MeV? *Hint:* $1\text{ eV} = 1.6 \times 10^{-19}$ J.

[Answer: 3.04×10^{-11} J]

16.2 Using Einstein's equation that relates the equivalence of mass and energy, $E = mc^2$, how many kilograms of matter are converted in Problem 16.1?

[Answer: 337×10^{-30} kg]

16.3 How much energy is contained in an atom of uranium-235? *Hint:* $1\text{ u} = 1.66 \times 10^{-27}$ kg, so a uranium-235 atom has a mass of $(235\text{ u})(1.66 \times 10^{-27}\text{ kg/u})$.

[Answer: 35×10^{-9} J]

16.4 A typical fission reaction involving uranium-235 is: $^{235}_{92}$U $+ ^{1}_{0}$n $\rightarrow$ $^{144}_{56}$Ba $+ ^{90}_{36}$Kr $+ 2^{1}_{0}$n.
  a) Using mass defects (Δ), calculate the energy released in the fission. The mass defects must be read off the online tables ("Nuclear Data Wallet Cards") of the NNDC, https://www.nndc.bnl.gov/.
  b) Calculate how much energy one could obtain in one year by the complete fissioning of 1000 kg of $^{235}_{92}$U.

  Hint: In 1 mol of $^{235}_{92}$U, there are 6.02×10^{23} nuclei. Also, 1 mol of $^{235}_{92}$U has a mass of 235 g.

  [Answers: (a) 179.6 MeV; (b) 73.6×10^{15} J or 2.3×10^{15} Wh]

16.5 Estimate the energy released in the neutron-induced fission of $^{235}_{92}$U into two nuclei with mass numbers 114 and 118, together with the emission of three neutrons.

  [Answer: About 209 MeV]

References

Gamow, G. and Cleveland, J.M. (1960). *Physics. Foundations and Frontiers*. New Jersey, NJ: Prentice-Hall.

Jaffe, R.L. and Taylor, W.M. (2018). *The Physics of Energy*. Cambridge: Cambridge University Press.

Kaplan, I. (1955). *Nuclear Physics*. Boston, MA: Addison-Wesley.

Lamarsh, J.R. (1966). *Introduction to Nuclear Reactor Theory*. Reading, MA: Addison-Wesley.

Nelson, A.W., Eitrheim, E.S., Knight, A.W., May, D. et al. (2015). Understanding the radioactive ingrowth and decay of naturally occurring radioactive materials in the environment: an analysis of produced fluids from the Marcellus Shale. *Environmental Health Perspectives* 123 (7): 689–696.

17

Nuclear Fission and Fusion Power Plants

Energy can be derived from nuclei by either the *fission* (splitting) of heavy nuclei or the *fusion* of light nuclei. All contemporary nuclear plants make use of the fission process, whereby an unstable heavy atom (such as uranium-235) splits into smaller and more stable atoms. We will, however, also survey the difficulties of achieving nuclear fusion, where two light nuclei (such as helium and hydrogen) are fused to form a lighter and more stable atom.

Around 700 nuclear reactors have been built worldwide since the first reactor was built in the Soviet Union (or Russia, as it is known today) in 1954. Currently, there are only 439 nuclear reactors in operation; 93 of which are in the United States. As of September 2022, 55 reactors were under construction, including 18 in China and four in the United States (IEA 2022).

Nuclear energy provides about 20% of electricity in the United States, and US nuclear power plants have a capacity factor of around 93% (IEA 2022). In 2018, the United States generated nearly a third of the world's nuclear electricity (EIA 2021). Countries generating the next largest amounts of electricity using nuclear energy are France, China, and Russia.

Reimagine Our Future
Much of the focus of contemporary technology is to respond to the *supply side* of energy. Imagine if we could also respond to the *demand side*. This would involve reducing our unrelenting, unsustainable consumption of planetary resources.

17.1 The Fission Process

The absorption of neutrons by isotopes involves radiative capture, with the excitation energy appearing as a gamma (γ) ray. In certain heavy elements, specifically uranium and plutonium, an alternate consequence is observed: the splitting of the nucleus into two fragments, a process called *fission*. Figure 17.1 illustrates the sequence of events.

In the first stage, a neutron ($^1_0$n) approaches the $^{235}_{92}$U nucleus. In the second stage, the nucleus captures the neutron and transforms it into an energized $^{236}_{92}$U atom. The excess energy may be released as the emission of γ rays, usually resulting in the nucleus changing its shape to that of a dumbbell, as shown in the third stage. The unstable $^{236}_{92}$U nucleus is heavier than its ground state

Energy Systems: A Project-Based Approach to Sustainability Thinking for Energy Conversion Systems,
First Edition. Leon Liebenberg.
© 2024 John Wiley & Sons, Inc. Published 2024 by John Wiley & Sons, Inc.
Companion website: www.wiley.com/go/liebenberg/energy_systems

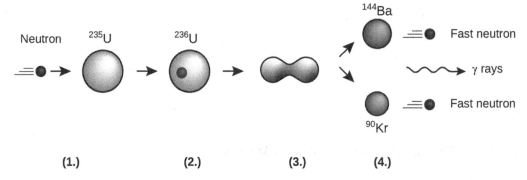

Figure 17.1 A typical fission process.

and has an excess energy of around 6.5 MeV, which is enough to split the nucleus into two fragments (for instance, $^{144}_{56}$Ba and $^{90}_{36}$Kr). These are called *fission fragments*. This fission process is accompanied by the release of a large amount of mass energy and the emission of more neutrons (Martin and Shaw 2019).

The fission fragments fly off at great speed, carrying around 169 MeV of kinetic energy. The fission reaction also emits more neutrons with energy of around 4.8 MeV, and gamma rays with energy of around 7 MeV. There is also a delayed radioactive process, whereby the fission fragments later emit a further 15.3 MeV in beta (β) decays and 6.3 MeV in more gamma (γ) rays. In total, a uranium-235 nucleus releases around 202.5 MeV in the fission process (Martin and Shaw 2019).

The fission reaction of uranium-235 may be written as follows:

$$^{235}_{92}\text{U} + \underbrace{^{1}_{0}\text{n}}_{\substack{\text{slow} \\ \text{neutron}}} \rightarrow {}^{236}_{92}\text{U} + \gamma \rightarrow {}^{144}_{56}\text{Ba} + {}^{90}_{36}\text{Kr} + \underbrace{2^{1}_{0}\text{n}}_{\substack{\text{fast} \\ \text{neutrons}}} + 202.5 \text{ MeV} \qquad (17.1)$$

Importantly, $^{235}_{92}$U only fissions when it is hit by neutrons at a low speed, so-called slow neutrons. These slow neutrons have energies <10 eV. If their energies are around 0.02 eV, they are called "thermal neutrons" as they can be "thermalized" at around 300 °C.

Only one *natural* isotope, $^{235}_{92}$U, undergoes fission when hit by a slow neutron (and not by a fast neutron). Two other *synthetic* isotopes, however, also undergo fission in this manner, plutonium $\left(^{239}_{94}\text{Pu}\right)$ and uranium-233 $\left(^{233}_{92}\text{U}\right)$ (Lederer and Shirley 1978).

Most other heavy isotopes require significantly larger excitation energy to fission. That extra energy must be provided by *fast neutrons*. Those fast neutrons have energies >1 MeV, which can cause the fission of other heavy elements, such as californium $\left(^{252}_{98}\text{U}\right)$. Fast reactors (also called "breeding reactors") permit the "breeding" of nuclear fuel. These reactors are mostly experimental. The focus of this chapter is on the fission of $^{235}_{92}$U, which uses slow neutrons to initiate the fission process.

17.2 Cross-Sections for Particle Interaction

Cross-section is a measure of the chance of collision between two nuclear particles. Consider the hypothetical experiment illustrated in Figure 17.2. In Figure 17.2a, a cylinder has an end area of 1 cm^2 and the cylinder contains one target particle. Another particle is injected into the cylinder

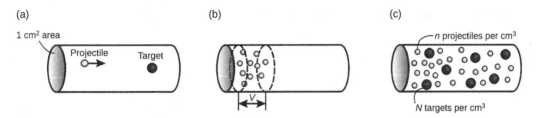

Figure 17.2 Particle collisions in a hypothetical tube.

parallel to the tube axis, but its exact position is not specified. The chance of a collision between the two particles is labeled as σ and is called the *microscopic cross-section*. The σ is the ratio of the target area to the area of the tube (Martin and Shaw 2019).

Now (Figure 17.2b), inject a stream of particles into the cylinder at a speed of V (with units of cm/s). In 1 s, each particle would have moved a distance of $V \times \Delta t = (V \text{cm/s}) \times (1 \text{ s}) = V \text{cm}$. A column of these particles will therefore sweep through a volume of $(1 \text{ cm}^2) \times (V \text{cm}) = V \text{cm}^3$ in a second. If there are n particles per cm^3, the *current density* is the number of particles that cross a unit area perpendicular to the stream in 1 s.

If we now fill the cylinder with N target particles, each with an area of σ, as experienced by incoming particles, the total target area will be $N\sigma$. Now (Figure 17.2c), again inject a stream of projectile particles. In 1 s, nV projectile particles will pass through the target volume. Since the chance of the collision of each projectile particle with a target particle is σ, the number of collisions is $nVN\sigma$. The reaction rate per unit volume (R) may now be defined as (Murray 1980):

$$R = \underbrace{nV}_{j} \underbrace{N\sigma}_{\Sigma} \text{ for parallel motion} \tag{17.2}$$

with the product $N\sigma$ being labeled Σ, the *macroscopic cross-section*. The current density (nV) is often labeled j. So:

$$R = j\Sigma \tag{17.3}$$

If the particles now move in a random manner (i.e., not parallel), nV is called the flux (ϕ):

$$\phi = nV \text{ for random motion} \tag{17.4}$$

In this case of random collisions, the current densities will be lower than nV. Statistically, the total current densities for random particle movement will be $nV/2$. The rate of reactions can be calculated in a similar manner to the parallel motion of particles. The reaction rate for random motion will therefore be (Murray 1980):

$$R = \phi\Sigma \tag{17.5}$$

When a neutron collides with a target nucleus, there is a chance of one of several reactions:

- The neutron can simply bounce off the target nucleus, which is termed *elastic scattering*.
- When a fast neutron collides with a target nucleus, *inelastic scattering* occurs, whereby the neutron becomes part of the nucleus and energizes the nucleus, which releases more neutrons. The *scattering cross-section*, σ_s, is the chance of a collision that results in neutron scattering.
- The neutron may be absorbed by the nucleus, giving an *absorption cross-section*, σ_a.

The total cross-section is (Martin and Shaw 2019):

$$\sigma = \sigma_s + \sigma_a \tag{17.6}$$

The unit of neutron cross-section is the *barn* (b), where:

$$1\,\text{b} = 10^{-28}\,\text{m}^2 \tag{17.7}$$

Example 17.1   Reaction Rate of Uranium-235
Consider an experimental nuclear reactor "burning" uranium-235. The reactor contains neutrons moving at 2200 m/s. The neutron flux (ϕ) is 2×10^{12} cm^{-2}/s; the density of uranium-235 atoms is $N = 4.8 \times 10^{22}$ cm^{-3}; and the absorption cross-section of a neutron is $\sigma_a = 585$ b. Calculate the reaction rate.

Required: We must calculate the rate (R) at which uranium-235 atoms are being consumed in a nuclear reactor.

Solution strategy: We will use Eq. (17.2) to calculate the macroscopic absorption cross-section and Eq. (17.5) to calculate the reaction rate.

Analysis:
We must calculate $R = \phi\,\Sigma_a$

with $\Sigma_a = N\sigma_a = \left(4.8 \times 10^{22}\,\text{cm}^{-3}\right)\left(585 \times 10^{-24}\,\text{cm}^2\right)$

   $\Sigma_a = 28.08\,\text{cm}^{-1}$

So $R = \phi\,\Sigma_a = \left(2 \times 10^{12}\,\text{cm}^{-2}\,\text{s}^{-1}\right)\left(28.08\,\text{cm}^{-1}\right) = 56.16 \times 10^{12}\,\text{cm}^{-2}$/s

Discussion: The uranium-235 atoms are consumed at a rate of 56.16×10^{12} cm^{-2}/s.

17.3   Neutron Cross-Section

The cross-section (σ) for neutron absorption in materials depends on the isotope bombarded and on the neutron energy. Table 17.1 shows the values of the absorption cross-section (σ_a) of the thermal (or slow) neutrons of a few isotopes regularly encountered in nuclear reactors. There are similarly cross-sections for non-fission capture (σ_c), for fission (σ_f), and for scattering (σ_s).

The naturally occurring $^{235}_{92}$U has a large absorption cross section and thus easily absorbs a slow neutron, making $^{235}_{92}$U an ideal nucleus for nuclear reactions (Jaffe and Taylor 2018). This peculiar characteristic of uranium-235 is also represented in Figure 17.3, which shows that $^{235}_{92}$U has a large absorption cross-section for slow neutrons, but a negligibly small cross-section for fast neutrons. Uranium-238 $\left(^{238}_{92}\text{U}\right)$, however, only fissions with fast neutrons.

Equation (17.1) shows that fast neutrons are released in fission reactions. If those reactions are to continue in a chain reaction, the neutrons must be slowed down. This is achieved by using moderating materials.

Table 17.2 shows a few typical fission cross-sections (σ_f), neutrons per fission (ν), capture-to-fission ratio (σ_c/σ_f), neutrons per absorption (η), and the delayed neutron fraction (β).

Table 17.1 Thermal (slow) neutron cross-sections of a few isotopes, arranged from small to large cross-sections, as well as logarithmic energy decrement.

Isotope or molecule	Absorption cross-section of slow neutrons, σ_a (barns)	Logarithmic energy decrement, ξ
$^{4}_{2}\text{He}$	~0	0.425
$^{238}_{92}\text{U}$	2.66×10^{-6}	0.0084
$^{16}_{8}\text{O}$	1.78×10^{-4}	0.12
$^{2}_{1}\text{H}$	5.2×10^{-4}	0.725
$^{1}_{1}\text{H}$	0.2	1
H_2O	0.33	0.92
$^{12}_{6}\text{C}$	0.0034	0.158
$^{92}_{40}\text{Zr}$	0.183	—
$^{233}_{92}\text{U}$	531	—
$^{235}_{92}\text{U}$	585	—
$^{239}_{94}\text{Pu}$	0	—
$^{10}_{5}\text{B}$	3838	0.187
$^{135}_{54}\text{Xe}$	2.6×10^{6}	—

Source: Lederer and Shirley (1978), Von Dardel and Sjöstrand (1954), Rinard (1991), and Jaffe and Taylor (2018).

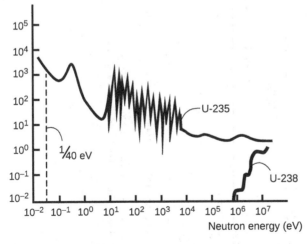

Fission cross-section (barns)

Figure 17.3 Fission cross-section of neutrons hitting uranium-235 and uranium-238 nuclei. Uranium-235 will fission when hit by thermal neutrons (1/40 eV), while uranium-238 will not, but will rather fission when hit by fast neutrons (1 MeV).

Table 17.2 Some important nuclear parameters for thermal and fast nuclear reactors.

| | Fissile fuel | | | | | | Fertile material | | | |
| | U-235 | | U-233 | | Pu-239 | | U-238 | | Th-232 | |
	Fast	Thermal	Fast	Thermal	Fast	Thermal	Fast	Thermal	Fast	Thermal
Fission cross-section (σ_f)	1.4	577	2.2	527	1.78	790	0.112	0	0.025	0
(σ_c/σ_f)	0.27	0.15	0.11	0.09	0.025	0.4	0.23	0	0	0
Neutrons per fission (ν)	2.5	2.4	2.59	2.51	3.0	2.90	2.60	—	2.4	0
Neutrons per absorption (η)	2.2	2.06	2.42	2.28	2.6	1.93	2.27	—	2.0	—
Delayed neutron fraction (β)	0.0065	—	0.0027	—	0.002	—	0.0147	—	0.0204	—

"Fast" fission at 100 keV and "thermal" fission at 0.025 eV.
Source: Lamarsh (1966), Murray (1980), and Shultis and Faw (2016).

- The fission cross-sections (σ_f) give the probability that a neutron moving in the vicinity of a nucleus will cause it to fission.
- For a self-sustaining nuclear chain reaction, the neutrons per fission (ν) must be greater than 1.
- The capture-fission ratio shows that when plutonium is bombarded by thermal neutrons, (σ_c/σ_f) = 0.5, half as many will be captured unproductively as will cause fission. But, when uranium-235 is bombarded by thermal neutrons, (σ_c/σ_f) = 0.17, careful design is required to obtain a chain reaction and even greater economy of neutrons to increase the breeding ratio. With plutonium, a thermal breeder reactor would be virtually impossible, so use is made of fast neutrons to provide fast (breeding) reactors. A nucleus may capture a neutron productively, causing fission, or unproductively, by simply capturing it.
- The neutrons produced per absorption (η) are the average amount of neutrons produced per fission multiplied by the fissions per absorption. For instance, in Table 17.2, for $^{235}_{92}U$, (σ_c/σ_f) = 0.15, the fissions per absorption are $1/1.15 = 0.87$ and the neutrons per absorption are $(0.87)(2.5) \approx 2.2$.
- As discussed in Box 17.1, some neutrons are produced by the decay of fission products (like radioactive iodine) many seconds after the initial fission and production of prompt neutrons. The delayed neutron fraction (β) is small, 0.65% for uranium-235, but without delayed neutrons it would be virtually impossible to produce controlled fission.

Box 17.1   Typical Nuclear Fission Reactions

Typical fission reactions of uranium-235, uranium-233, and plutonium-239

$$^{235}_{92}U + ^{1}_{0}n \rightarrow ^{144}_{56}Ba + ^{90}_{36}Kr + 2^{1}_{0}n + 202.5 \text{ MeV}$$

$$^{233}_{92}U + ^{1}_{0}n \rightarrow ^{136}_{56}Ba + ^{96}_{36}Kr + 2^{1}_{0}n + 200 \text{ MeV}$$

$$^{239}_{94}Pu + ^{1}_{0}n \rightarrow ^{137}_{56}Ba + ^{100}_{38}Sr + 3^{1}_{0}n + 210 \text{ MeV}$$

Typical breeding reactions of plutonium-239 and uranium-233

$$^{238}_{92}U + ^{1}_{0}n \rightarrow ^{239}_{92}U + \gamma \xrightarrow{24 \text{ min}} e^- + ^{239}_{93}Np \xrightarrow{2.3 \text{ days}} e^- + ^{239}_{94}Pu$$

$$^{232}_{90}Th + ^{1}_{0}n \rightarrow ^{233}_{90}Th + \gamma \xrightarrow{22 \text{ min}} e^- + ^{233}_{91}Pa \xrightarrow{27 \text{ days}} e^- + ^{233}_{92}U$$

In these reactions, $^{239}_{94}Pu$ and $^{233}_{92}U$ are *fissile* materials, while $^{238}_{92}U$ and $^{232}_{90}Th$ are *fertile* materials. The electrons produced in these reactions are assumed to fly off.

Typical delayed neutrons from radioactive iodine

$$^{137}_{53}I \xrightarrow{23 \text{ s}} e^- + ^{137}_{54}Xe \xrightarrow{0 \text{ s}} ^{137}_{54}Xe + ^{1}_{0}n$$

The ability to effectively control a fission reaction depends considerably on *delayed neutrons*. If all neutrons were prompt, the slightest withdrawal of the control rods beyond criticality would result in an extremely rapid build-up of power, which is unwanted. But, with a small portion of the neutron production delayed, there is reasonable time to move the control rods to facilitate smooth control. A typical source of delayed neutrons is radioactive iodine, which is produced as a fission product. It decays to xenon with a half-life of 23 s and then emits a neutron.

17.4 Neutron Slowing

The neutrons produced by fission reactions are fast (Eq. (17.1)), while the cross-section for the fission reaction of $^{235}_{92}$U is high for slow neutrons. This means that a reactor *moderator* is required to slow down neutrons, which will help create a fission chain reaction.

Inspection of the absorption cross-sections of the various materials listed in Table 17.1 indicates that boron-10, $^{10}_{5}$B, makes for an excellent moderating material as it has a large neutron cross-section of 3838 b. The $^{10}_{5}$B is therefore often used in the *control rods* of nuclear reactors; inserting $^{10}_{5}$B rods into the reactor will slow the reaction down; removing $^{10}_{5}$B rods from the reactor will speed up the reaction (Rinard 1991). Materials such as xenon $\left(^{135}_{54}\text{Xe}\right)$ have massive absorption cross-sections (2.6×10^6 b), implying that they will literally kill a nuclear reaction. Such materials are called *fission poisons*.

Water (H_2O) is often used to slow neutrons down so that they can help sustain a nuclear chain reaction. Table 17.1 shows that water will not absorb neutrons. But water will slow neutrons down, as indicated by its logarithmic decrement of energy, $\xi = 0.92$ (see Table 17.1), with 1.0 (hydrogen) being the maximum. Water also slows neutrons down much better than graphite does $\left(^{12}_{6}\text{C}\right)$ with its logarithmic energy decrement of $\xi = 0.158$ (Lamarsh 1966).

ξ may be approximated as follows for atoms with atomic mass A (Lamarsh 1966):

$$\xi = \ln \frac{E_{initial}}{E_{after}} \cong \frac{2}{A + 2/3} \tag{17.8}$$

with $E_{initial}$ being the energy before a collision and E_{after} being the energy after a collision. Neutrons slow down by deflection and scattering, which results in their energy loss.

17.5 Enrichment of Nuclear Fuel

Naturally occurring uranium (or U_3O_8, uranium oxide) contains only 0.72% ^{235}U, with the remainder (99.28%) comprising ^{238}U (Martin and Shaw 2019). Enrichment to at least 3–5% increases the probability that a slow neutron will be absorbed by a ^{235}U nucleus and perpetuate nuclear fission. Enriching uranium increases the proportion of ^{235}U nuclei to ^{238}U by removing unwanted ^{238}U. The enrichment factor, x, is defined in terms of the number densities of these two isotopes (Murray 1980):

$$x = \frac{n\left(^{235}\text{U}\right)}{n\left(^{235}\text{U}\right) + n\left(^{238}\text{U}\right)} \tag{17.9}$$

When uranium is mined, it contains about 140 atoms of unwanted ^{238}U for every one atom of required ^{235}U. This gives $x = 1/141 = 0.7\%$. To enrich that uranium to, say, 3.7%, 114 of the ^{238}U atoms must be removed for every ^{235}U atom: $x = 1/27 = 3.7\%$. "Weapons grade" uranium is considered to have an enrichment of 90% or higher. Uranium can be enriched by gaseous diffusion, gas centrifuge, or laser operations.

17.6 Self-Sustaining Chain Reactions

To sustain a ^{235}U fission reaction, at least one thermal neutron is required. The number of thermal neutrons in the $(n + 1)$ generation can be compared to those of the n-th generation (Lamarsh 1966):

$$N = N_0 e^{(k-1)t/t_0} \tag{17.10}$$

with t being the time elapsed, and t_0 the nuclear time constant. The *k-factor* is also called the *neutron multiplication factor*. If $k < 1$, the fission reaction is: *sub-critical* and will quickly die away. If $k = 1$, the fission reaction is: *critical* and will remain stable. If $k > 1$, the fission reaction is: *super-critical* and will increase exponentially.

The *infinite-neutron multiplication factor* (k_∞ *factor*) may be defined as follows (Jaffe and Taylor 2018):

$$k_\infty = \eta epf \tag{17.11}$$

with k_∞ being the k factor of an infinitely large nuclear reactor (where losses can be ignored), η being the fast-neutron multiplication number (number of fast neutrons produced by thermal neutrons), ε being the fast-fission multiplication factor (probability of fast neutrons producing more fast neutrons), p being the resonance escape probability (probability that a fast neutron will avoid capture and slow to thermal velocity), and f being the thermal utilization factor (survival probability of a thermal neutron).

The η is a combination of the average number of neutrons produced per fission (ν), the absorption cross-sections (σ_a) of ^{235}U and ^{238}U, and the fission cross-section of ^{235}U: σ_f. If a nuclear reaction is: critical, $k_\infty = \eta epf = 1.0$. Experimental work with ^{235}U shows $\eta = 2.2$. This means that $(epf) = 1/2.2 = 0.45$. Therefore, 45% of neutrons must remain in the reactor, while no more than 55% may escape if the reactor is to remain critical.

The effective multiplication factor (k_{eff}) may be expressed mathematically in terms of the infinite multiplication factor (k_∞) and two additional factors that account for neutron leakage during neutron thermalization (fast non-leakage probability, P_f) and neutron leakage during neutron diffusion (thermal non-leakage probability, P_t) by the following equation, usually known as the six-factor formula:

$$k_{eff} = k_\infty P_f P_t \tag{17.12}$$

The ability to convert significant amounts of fertile material into fissile materials depends on the magnitude of the reproduction factor (η_{repr}), which is the number of neutrons (ν) produced per neutron absorbed by the fuel (Murray 1980):

$$\eta_{repr} = \frac{\sigma_f}{\sigma_a}\nu \tag{17.13}$$

A *breeding reaction* is: more likely if the reproduction factor (η_{repr}) is above 2.

The ability to convert fertile isotopes into fissile isotopes can also be measured by a conversion ratio (CR):

$$CR = \frac{\text{fissile atoms produced}}{\text{fissile atoms consumed}} \tag{17.14}$$

Fissile atoms are produced by absorption in fertile atoms, while consumption occurs through fission and capture. The CR is also dependent on the fast fission factor (ε) and on the amount of neutron loss by leakage and by the absorption in non-fuel material (L):

$$CR = \eta_f \varepsilon - 1 - L \tag{17.15}$$

In the example of a conventional nuclear reactor burning ^{235}U, $\eta_f = 2.02$, $\varepsilon = 1.03$, and $L = \left(\frac{51\,\text{neutrons}}{1030\,\text{neutrons}}\right)\left(\frac{245\,\text{neutrons}}{979\,\text{neutrons}}\right)\left(\frac{29\,\text{neutrons}}{734\,\text{neutrons}}\right)\left(\frac{210\,\text{neutrons}}{705\,\text{neutrons}}\right) = 0.64$, which would give a $CR = 0.44$. With such a set-up, fewer fissile atoms are produced for each atom consumed. By increasing η_f and reducing L, conversion ratios of better than 1 can be achieved, which means that more fuel will be produced than is used. Such reactors are called breeder reactors (Murray 1980) and their fission is initiated by fast neutrons, unlike uranium-235, which uses slow neutrons (see Figure 17.3). Figure 17.4 shows the main categories of fission reactors.

(a) PWR

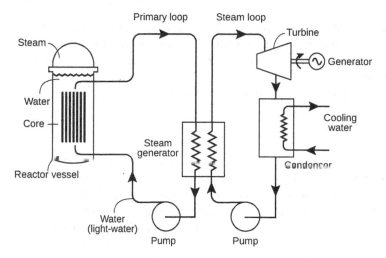

(b) BWR

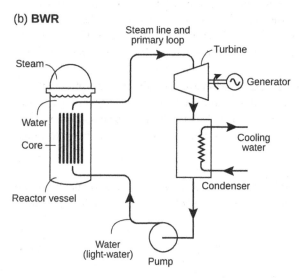

(c) CANDU

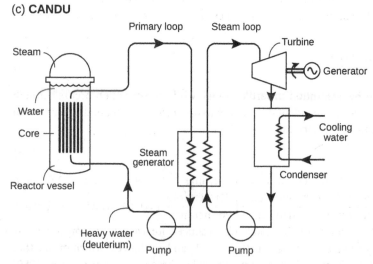

Figure 17.4 (a) Pressurized water reactor (PWR), (b) boiling water reactor (BWR), (c) heavy water reactor (CANDU), (d) high-temperature gas reactor (HTGR), and (e) liquid metal-cooled fast-breeder reactor (LMFBR).

(d) HTGR

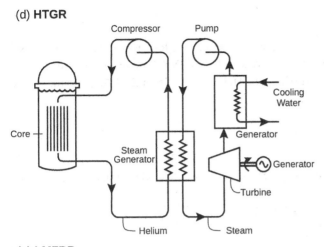

(e) LMFBR

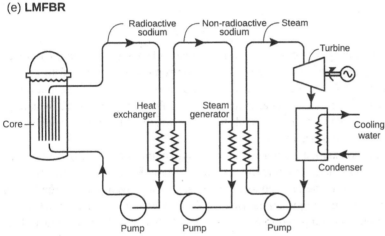

Figure 17.4 (Continued)

17.6.1 Conversion Ratio

Example 17.2 Conversion Ratios and Unburned Fuel

Consider a nuclear reactor in which ^{235}U and ^{239}Pu are equally effective in multiplication ("breeding"), i.e., they have similar reproduction factors (η_{repr}). Assume that ^{238}U is burned to produce fissile ^{239}Pu that is consumed alongside ^{235}U. The conversion ratio is then

$$CR = \frac{^{238}U}{^{235}U + {}^{238}U}.$$ This could also be expressed as $\frac{^{238}U}{^{235}U} = \frac{CR}{1-CR}$.

So, if $CR = 0.65$, $\frac{^{238}U}{^{235}U} = \frac{0.65}{1-0.65} = 1.857$, which is far from a complete conversion of ^{238}U. If all 0.72% ^{235}U in the natural uranium (U_3O_8) were burned, only $(1.857\%)(0.72\%) = 1.34\%$ ^{238}U would be converted, leaving 98.66% unused. The percentage of the original natural uranium used is then only $1.34\% + 0.72\% = 2.06\%$.

To achieve full conversion, the $100\% - 0.72\% = 99.28\%$ ^{238}U must be transformed, which will give $\frac{^{238}U}{^{235}U} = \frac{CR}{1-CR} = \frac{99.28}{1-99.28}$, or $CR = 0.9928$.

17.7 Reactor Power

Consider N fuel nuclei occupying 1 cm^3 of the reactor. Each of the fuel nuclei has a fission cross-section (σ_f). The total number of fuel nuclei in the reactor is $N_t = N\mathcal{V}$. If there are n neutrons, each moving at a speed V, their average flux (from Eq. (17.4)) is $\bar{\phi} = nV$. From Eq. (17.2), the rate of the fission reaction (R_f) is (Jaffe and Taylor 2018):

$$R_f - nVN\sigma_f = \frac{P}{E}$$

(17.16)

where P is the power output per unit volume and E is the energy delivered per nucleus (i.e., approximately 202.5 MeV).

The total reactor power obtained from fissioning (P_f) is:

$$P = \underbrace{\bar{\phi}N_t\sigma_f E}_{R_f}$$

(17.17)

Equation (17.17) shows that the power output of a nuclear reactor is proportional to the number of neutrons, and the number and type of fuel atoms. A high neutron flux is required if the reactor contains a small amount of fuel. Conversely, a low neutron flux is required for a large amount of fuel. Power can be increased or decreased by changing the neutron flux. This is achieved by manipulating the reactor control rods: Raising the rods leads to an increase in neutron flux; lowering the rods into the reactor will decrease the neutron flux.

With all other terms being constant, a reactor with a high fission cross-section (σ_f) can produce the required power with less fuel than a reactor with a small σ_f. From Figure 17.3, we note that the fission cross-section decreases as the neutron energy increases. So, for a given power output (P), a fast reactor (operating with neutron energies in the vicinity of 1 MeV) requires either a much larger flux or a larger mass of fissile fuel compared to a thermal reactor that activates with slow ("thermal") neutrons.

The consumption of $^{235}_{92}$U is due to both fission and capture processes. The rate of fuel consumption may be approximated as follows (Lamarsh 1966; Jaffe and Taylor 2018):

$$\dot{N} = \bar{\phi}N_t\sigma_a$$

(17.18)

$$\sigma_a = \sigma_f + \sigma_c$$

(17.19)

Example 17.3 Reactor Power Output and Fuel Consumption

Consider a nuclear power plant that produces heat by fissioning $^{235}_{92}$U. Each $^{235}_{92}$U fission releases on average 200 MeV of energy. There are 2.6×10^{24} uranium $- 235$ nuclei/kg. The heat from fission is used to boil water, from which the steam is used to drive steam turbines, which, in turn, drive electrical generators that produce 1 GW$_e$. The efficiency of converting heat into electricity is 33%. (a) Calculate the energy released by 1 kg of $^{235}_{92}$U. (b) How much natural uranium will be required per year? (c) If the fuel is enriched to 3.7%, how much natural uranium would be required?

Given: We know that each $^{235}_{92}U$ fission releases 200 MeV and we are told that there are 2.6×10^{24} $^{235}_{92}U$ nuclei/kg which are being consumed in a $1\,GW_e$ nuclear reactor with a thermal efficiency of 33%.

Required: We must calculate (a) the energy released per kilogram of $^{235}_{92}U$, (b) the amount of natural uranium required per year, and (c) the amount of natural uranium required if the $^{235}_{92}U$ is enriched to 3.7%.

Solution strategy: We will start off by converting the electrical energy output of the reactor to thermal energy output, using the provided thermal efficiency of the plant. This will be followed by a few simple conversions to eventually arrive at the amount of required natural uranium.

Analysis:

a) **Fuel consumption rate**

At 33% energy conversion efficiency, the thermal capacity of the nuclear power plant is:

$$P_{th} = \frac{P_e}{0.33} = \frac{1000\,MW_e}{0.33} = 3030\,MW_{th} \text{ or } 3030 \times 10^6\,J/s$$

The energy released (J/kg) by $^{235}_{92}U$ is:

$$= \left(200 \times 10^6\,eV/nucleus\right)\left(2.6 \times 10^{24}\,nuclei/kg\right)\left(1.6 \times 10^{-19}\,J/eV\right)$$
$$= 83.2 \times 10^{12}\,J/kg$$

This amount of energy can be converted to the more useful measure of megawatt-years (or MWy):

$$1\,kWh = 3.6\,MJ$$
$$\therefore 1\,MWy = \left(\frac{3.6\,MJ}{kWh}\right)\left(\frac{1000\,kWh}{MWh}\right)\left(\frac{24\,h}{day}\right)\left(\frac{365\,days}{year}\right) = 3.15 \times 10^{13}\,J/year$$

Therefore, the energy released (MWy) by $1\,kg$ $^{235}_{92}U$ is:

$$\frac{83.2 \times 10^{12}\,J/kg}{3.15 \times 10^{13}\,J/year} = 2.6\,MWy$$

b) **Amount of natural uranium required per year**

From (a), $1\,kg$ of $^{235}_{92}U$ can provide 2.6 MW every hour of the day for 365 days a year.

With a thermal output of 3030 MW, this reactor will require $\dfrac{3030\,MW}{2.6\,MWy/kgU} = 1.17\,tU/y$.

Therefore, 1.17 tons of $^{235}_{92}U$ will provide $3\,GW_{th}$ or $1\,GW_e$ continuously for a year.

As natural uranium contains only 0.72% $^{235}_{92}U$, the amount of natural uranium required per year would be:

$$\left(\frac{1.17\,tU}{0.72\%\,U/nat.U}\right) = 162.5\,t/nat.U$$

This processed uranium is called "yellowcake," with a chemical composition of U_3O_8.

c) **Uranium requirement when using 3.7%-enriched fuel**

A $^{235}_{92}U$ fuel enrichment of 3.7% would be equivalent to an annual fuel requirement of

$$\frac{1.17\,tU/y}{0.037} = 39\ t\ of\ fuel\ per\ year.$$

Discussion:

As a "rule of thumb," 1 g of $^{235}_{92}$U provides 1 MW-day of thermal energy. For a 3 GW$_{th}$ nuclear power plant (or 1 GW$_e$ with 33% energy conversion efficiency), around 3 kg $^{235}_{92}$U is burned per day. A nuclear power plant with a 90% capacity factor will require around 40 t of fuel per year. In comparison, a 1-GW$_e$ coal-fired power plant requires about 10,000 t of coal *per day*!

Example 17.4   Reactor Power Output and Fuel Consumption

A nuclear reactor is used to produce heat, which is, in turn, used to produce steam. The reactor contains 1000 kg of $^{235}_{92}$U. The energy released per nucleus fission is 200 MeV. There are 3×10^{27} nuclei in 1000 kg of $^{235}_{92}$U. The neutron flux is 10^{13} cm^{-2}/s. For $^{235}_{92}$U, the fission cross-section is 579 b and the capture cross-section is 101 b. (a) Determine the reactor's power output (in MW), and (b) the reactor's fuel consumption rate.

Solution strategy: We must remember that $1 \text{ b} = 10^{-24}$ cm^2, so $\sigma_f = 579 \text{ b} = 579 \times 10^{-24}$ cm^2. As we need to calculate power (in watts), we must convert eV to J and then use Eq. (17.17) to evaluate the power output: $P = \underbrace{\bar{\phi} N_t \sigma_f}_{R_f} E$.

Analysis:

a) **Power output**

$$1 \text{ eV} = 1.60218 \times 10^{-19} \text{ J, so } E = \left(\frac{200 \text{ MeV}}{\text{nucleus}}\right)\left(\frac{1.60218 \times 10^{-19} \text{ J}}{\text{eV}}\right) = 32.04 \times 10^{-12} \text{ J/nucleus}$$

$$\therefore P = \bar{\phi} N_t \sigma_f E = \left(10^{13} \text{ cm}^{-2} \text{ s}^{-1}\right)\left(3 \times 10^{27} \text{nuclei}\right)\left(579 \times 10^{-24} \text{ cm}^2\right)\left(32.04 \times 10^{-12} \text{ J/nucleus}\right)$$
$$\therefore P = 556 \text{ MW}$$

b) **Fuel consumption rate**

Using Eq. (17.19), the absorption cross-section of $^{235}_{92}$U is

$$\sigma_a = \sigma_f + \sigma_c = 579 \text{ b} + 101 \text{ b} = 680 \text{ b} = 680 \times 10^{-24} \text{ cm}^2.$$

Using Eq. (17.18), the number of $^{235}_{92}$U nuclei consumed per second is:

$$\dot{N} = \bar{\phi} N_t \sigma_a = \left(10^{13} \text{ cm}^{-2} \text{ s}^{-1}\right)\left(3 \times 10^{27} \text{ nuclei}\right)\left(680 \times 10^{-24} \text{ cm}^2\right) = 20.4 \times 10^{18} \text{ nuclei/s}$$

Discussion:

The reactor consumes $\left(\dfrac{20.4 \times 10^{18} \text{ nuclei}}{\text{s}}\right)\left(\dfrac{31.5 \times 10^6 \text{ s}}{\text{year}}\right) = 642.6 \times 10^{24}$ nuclei/year. The reactor is initially loaded with 3×10^{27} nuclei, which means that the reactor will require refueling in $\dfrac{3 \times 10^{27} \text{ nuclei}}{642.6 \times 10^{24} \text{ nuclei/year}} = 4.67$ years.

17.8   Fission Reactors

The only requirement for a neutron chain reaction is: enough fissionable elements. Nonetheless, several possible combinations of materials and arrangements can be used to construct an operable nuclear reactor. Experience with individual reactor concepts has led to the selection of a few that are most suitable in terms of economy, reliability, and ability to meet performance and safety demands.

Reactors can be classified according to several features (Murray 1980; Jaffe and Taylor 2018; Foss et al. 2021):

- *Purpose*, e.g., electric power generation (on Earth or for spacecraft), propulsion (nuclear submarines, aircraft carriers, icebreakers).
- *Size*, e.g., micro-reactors (5–50 MW_e), small reactors (50–300 MW_e), large reactors (>300 MW_e) (IAEA 2021, 2022; INL 2022).
- *Neutron energy*, e.g., slow reactors (using slow or thermal neutrons, around 0.1 eV of energy) or fast reactors (using high-energy or fast neutrons, around 0.1–1 MeV of energy).
- *Moderator and coolant*, e.g., light water ($_1^1H_2O$), heavy water ($_1^2H_2O$), or graphite (C) as moderators; light water ($_1^1H_2O$), carbon dioxide, or helium as coolant.
- *Fuel*, e.g., natural uranium-235 (0.7% $_{92}^{235}U$), 3%-enriched $_{92}^{235}U$, or 19% $_{92}^{235}U$; some reactors may also fission plutonium ($_{94}^{239}Pu$), uranium-233 ($_{92}^{233}U$), or thorium-232 ($_{90}^{232}Th$); reactors might use nuclear fuel encased in 3.7-m fuel rods filled with fuel pebbles. The fuel pebbles could also be encased in nuclear fuel spheres 70 mm in diameter, each with layers of graphite – called tri-structural-isotropic (TRISO) fuel.
- *Reactor materials*, e.g., fuel rods, may be clad with zircalloy, graphite, or stainless steel; the control rods could comprise boron-graphite (B_4C), tantalum (Ta), cadmium (Cd), or indium (Id); the reactor's containment structure could be made of concrete or steel.

More than 90% of nuclear reactors today are *light water reactors* and are either pressurized or boiling water reactors. Pressurized water reactors (PWRs) were first used in nuclear submarines and feature water pressurized to around 15 MPa (and 315 °C), which keeps the water in liquid phase in the primary loop, as illustrated in Figure 17.4a. Boiling water reactors (BWRs) operate at lower pressures and do not have secondary water loops; rather, radioactive water is circulated through the turbine, which makes maintenance difficult and expensive, see Figure 17.4b.

Non-light water reactors (NLWRs) are not cooled by water, but by other substances, such as helium gas, liquid sodium, or molten salts. NLWRs are sometimes referred to as "advanced reactors," which is a misnomer as many of these technologies descend from those proposed many decades ago (WNA 2003; GA 2021). NLWR designers often claim that such reactors have innovative features that could disrupt the nuclear power industry and solve its many problems. They might also state that their designs could lower energy costs, be built quickly, reduce the accumulation of nuclear waste, use uranium more effectively, improve safety, and reduce the risk of nuclear proliferation (Zohuri 2020). NLWR designers usually cite the advantages of features such as passive shutdown and cooling, the ability to consume or recycle nuclear waste, and the provision of high-temperature process heat for industrial applications such as hydrogen production. These companies often claim that their design can be demonstrated in a decade or two (UCS 2021).

Unfortunately, the field of nuclear energy is highly politicized, and it is difficult to find unbiased assessments. Certain organizations, such as the Union of Concerned Scientists (UCS), have found that most new NLWR designers' claims are false or misleading with respect to safety and security, sustainability, and the risks of nuclear proliferation and nuclear terrorism. Recent research has also found that small "advanced" reactors will produce more voluminous and chemically or physically reactive waste than large light water reactors, which will impact on options for the management and disposal of radioactive waste (Testoni et al. 2021). The intrinsically higher neutron leakage associated with small reactors suggests that most designs are inferior to light water reactors with respect to the generation, management, and final disposal of key radionuclides in nuclear waste (Krall et al. 2022). The UCS recommends that policymakers, private investors, and regulators fully vet the risks and benefits of these technologies before committing the vast time and resources needed to commercialize them (UCS 2021).

Heavy water reactors are usually of the Canada deuterium uranium (CANDU) type (see Figure 17.4c). Figure 17.4d shows the layout of a *high-temperature gas-cooled reactor* (HTGR), and Figure 17.4e shows how *breeder reactors* use liquid metal as a coolant.

It should be noted that there are currently only two operational breeder reactors in the world (with a net output of 1.4 GW$_e$), around 295 pressurized water reactors (with a net output of 275 GW$_e$), around 80 boiling water reactors (with a net output of 75 GW$_e$), 40 heavy water reactors (with a net output of 25 GW$_e$), and 18 gas-cooled reactors (with a net output of 10 GW$_e$). The main types of reactors are summarized in Table 17.3.

Nuclear power plants, unlike dams and other infrastructure, are not underwritten by the World Bank or most other international lending organizations. The large investments required for nuclear power therefore compete with the pressing needs for health, education, and poverty reduction. In 2022, the average cost of constructing a nuclear plant in the United States amounted to $6000/kW$_e$. This translates to $6 billion for a 1 GW$_e$ nuclear power plant (IEA 2020). However, construction costs have skyrocketed. The Vogtle (Georgia, USA) power plant's two new reactors (totaling 2.4 GW$_e$) were initially priced at $14 billion and expected to begin operation in 2017. The two reactors will probably only commence operation in 2023/2024 at a final cost of $29 billion. Mega projects such as these contend with challenges in program management and quality control, as well as regulatory issues that cause lengthy (and costly) delays. Table 17.4 shows projected levelized costs of electricity (LCOE) for nuclear power and other power sources in 2040.

Table 17.3 Broad classification of nuclear reactors.

	Light water reactor		Heavy water reactor (CANDU)	High-temperature gas-cooled reactor	Liquid metal-cooled fast breeder reactor
	Pressurized water reactor	Boiling water reactor			
Average core power density (MW_{th}/m^3)	100	50	10	2–7	300–550
Fissile isotope (and enrichment) (%)	$^{235}_{92}U$ ($\sim 3\%$)	$^{235}_{92}U$ ($\sim 3\%$)	Natural uranium with $\sim 0.7\%$ ^{235}U	$^{235}_{92}U$ ($>4\%$)	$^{239}_{94}U$ ($>10\%$)
Coolant	$^1_1H_2O$	$^1_1H_2O$	$^2_1H_2O$	He or CO_2	Liquid-Na
Coolant temperature and pressure	330 °C 16 MPa	280 °C 7 MPa	420 °C 25 MPa	780 °C 5 MPa	620 °C 1 MPa
Moderator	$^1_1H_2O$	$^1_1H_2O$	$^2_1H_2O$	Graphite (C)	None
Conversion ratio	~ 0.55	~ 0.55	~ 0.7	~ 0.7	~ 1.2
Thermal efficiency	33	34	30	42	42

Source: Adapted from IAEA (2022).

Table 17.4 Levelized costs of electricity for new power plants entering service in 2040.

Plant type	Plant capacity factor (%)	*LCOE* ($/MWh)
Ultra-supercritical coal	85	79.46
Combined cycle	87	44.05
Advanced nuclear	90	85.28
Geothermal	90	41.91
Biomass	83	86.53
Wind, onshore	40	40.08
Wind, offshore	43	98.01
Solar, standalone	29	33.42
Solar with 4-h battery storage	28	48.63
Hydroelectric	56	63.83
Gas turbine	10	121.87
Battery storage	10	120.47

Source: Adapted from EIA (2022b).

17.9 Outlook for Nuclear Fission

Only around 135 years of uranium from terrestrial resources remain available. (See also Box 17.2.) If nuclear power is to have a long-term future, alternative nuclear fuels should be sought, such as *thorium*. Thorium is three to four times more abundant than uranium, and there are large resources in Brazil, China, India, Madagascar, and the USA. One can breed fissile ^{233}U by capturing neutrons in ^{232}Th; one neutron capture is followed by two successive beta decays:

$$^{232}\text{Th} + {}^{1}\text{n} \rightarrow {}^{233}\text{Th} \xrightarrow{\beta^-} {}^{233}\text{Pa} \xrightarrow{\beta^-} {}^{233}\text{U} \tag{17.20}$$

The difficulty with operating a thorium cycle is that a thorium nuclear reaction must be initiated by either highly enriched ^{235}U or ^{239}Pu. *Highly enriched uranium* (HEU) has, however, been banned from civilian nuclear power plants to reduce the risk of nuclear proliferation. India has been banned from international nuclear trade due to its continuing work with HEU to power its thorium reactors.

In 2022, India's *fast breeder test reactor* (FBTR) reached its full 40 MW$_{\text{th}}$ design power level for the first time, more than 35 years after it started operating. This paves the way for India to commission large-scale thorium-based closed fuel cycles. Regardless, it is difficult to see how the limited research successes with the thorium cycle over the past 60 years will lead to the industrialization of a thorium reactor anytime soon.

Instead of thorium, a future option might be to extract uranium from ocean water. Uranium exists in sea water at a concentration of 3.3 µg/L. Though diluted, this amounts to an estimated 4.5 billion tons of uranium, which is approximately one thousand times more than is available from conventional sources such as terrestrial ores. Although it would be exceedingly expensive to extract uranium from ocean water, it might be a viable and necessary option in a century's time when terrestrial uranium supplies will be depleted.

The nuclear accidents at Fukushima Daiichi (2011) and Chernobyl (1986) have severely damaged nuclear power's "green credentials" and delayed the "nuclear renaissance" by decades. If nuclear power is to become a bigger role-player in energy production, it will have to allay people's fears of a nuclear accident and possible nuclear proliferation. Solutions must be found for the *long-term*

Box 17.2   Remaining Uranium Resources

The USA Nuclear Energy Agency biennially publishes *Uranium Resources, Production and Demand*, also known as the "Red Book" (NEA 2020). This book classifies uranium resources by a scheme (based on geological certainty and costs of production) developed to combine resource estimates from several different countries into consolidated global figures. *Identified resources* (which include reasonably assured resources, or RAR, and inferred resources) refer to uranium deposits delineated by sufficient direct measurement to conduct pre-feasibility and sometimes feasibility studies.

Total identified resources recoverable amounted to 6.2 million of uranium metal in the < $130/kgU category. In the highest cost category (<$260/kgU), total identified resources amounted to 8.1 million tonnes of uranium metal. Inferred resources in the <$260/kgU cost category increased to 3.2 million tonnes of uranium metal. We may therefore assume that around 11 million tonnes of uranium metal remain to be exploited under economically feasible conditions (NEA 2020).

As of January 2019, 450 commercial nuclear reactors were connected to the grid globally, with a net generating capacity of 396 GW$_e$, requiring about 59,200 t of uranium metal annually. Considering changes in policies announced in several countries and revised nuclear programs, world nuclear capacity is projected to grow to between 354 GW$_e$ net in the low-demand case and about 626 GW$_e$ net in the high-demand case by 2040. Accordingly, world annual reactor-related uranium requirements are projected to rise from 56,640 t of uranium metal in 2019 to 100,225 t of uranium metal by 2040 (NEA 2020).

Sufficient uranium resources exist to support the continued use of nuclear power and significant growth in nuclear capacity for low-carbon electricity generation and other uses (e.g., heat, hydrogen production) in the long term. Identified recoverable resources, including reasonably assured resources and inferred resources (at a cost (<$260/kgU)) are sufficient for over 135 years, considering the uranium requirements as of 1 January 2019. However, considerable exploration, innovative techniques, and timely investment will be required to turn these resources into refined uranium ready for nuclear fuel production and to facilitate the deployment of promising nuclear technologies (NEA 2020).

---

storage of radioactive waste. The problem of a reliable long-term fuel supply must also be solved. All of this must be done economically.

It appears that the efficient use of uranium will require us to abandon prevailing safety principles to achieve a breeding ratio greater than 1. Some proponents are therefore discarding the use of water as a coolant in favor of *liquid sodium*, for instance. Modern breeder reactors use liquid sodium, which conducts heat well and does not slow down neutrons. But liquid sodium explodes when it meets air. Most experimental breeder reactors have reported problems with sodium leaks, which makes this technology extremely dangerous.

The *Generation IV International Forum* (GIF) group of countries is working on solving issues with sodium-cooled fast reactors. The GIF group of countries has identified several other advanced reactor concepts, including the gas-cooled fast reactor, lead-cooled fast reactor, molten salt reactor, supercritical water-cooled reactor, and the very-high temperature reactor (Behar 2014; Bragg-Sitton et al. 2020). Most of these experimental reactors might, however, only become a reality by 2035. If breeding reactions can be performed safely and economically, there is potential to produce nuclear fuel for the next 20,000 years (Nuttall 2022).

In the meantime, many organizations are attempting to develop *small-scale nuclear reactors* (20–300 MW$_{th}$). Most of these are of the high-temperature gas-cooled kind, which uses "high-assay low-enriched uranium," i.e., uranium enriched to 5–20% (Testoni et al. 2021), although there is a large market for small modular reactors, especially in countries such as the United States, where electricity is provided by many (*not* well-connected) small utilities. Nuclear power over the past several decades has taught us that large plants (in the order of $3\,GW_{th}$) are more economical and safer to operate than smaller plants. Considering the immense capital costs of a nuclear power plant, the extraordinarily long construction times, and the complex and very long licensing application process, it seems simpler, cheaper, less time consuming, and safer to construct one $3\text{-}GW_{th}$ plant rather than fifteen $200\text{-}MW_{th}$ plants.

Globally, nuclear power output has been declining since 2011 (IEA 2021; EIA 2022a). The global nuclear order has been challenged in recent years by individual proliferators, the moribund US-Russian arms control process, and the resultant frustration over stalled progress toward disarmament.

In 2022, Russia launched its full-scale invasion of Ukraine under cover of nuclear threats against the North Atlantic Treaty Organization (NATO). A major consequence of Russian President Putin's war was renewed public awareness of the often-unpalatable role nuclear weapons play in international politics. Nuclear targeting, deterrent threats, and associated risk-reduction efforts are hardly new phenomena (Bollfrass and Herzog 2022).

The war in Ukraine also triggered historic shifts in the world of energy conversion systems. Many countries had long depended on oil and natural gas from Russia, which is one of the world's top exporters of fossil fuels. But when those exports became politically toxic, these countries were scrambling to find other ways to meet their energy needs (Vaughan 2022).

The world's energy systems have been going through a slow transition to lower-carbon fuels for decades. But the war in Ukraine changed everything, and *energy security* became the most pressing priority. A crucial question is how this will play out for the environment. Will it mean a renewed race toward renewables, or a rush to exploit domestic fossil fuels and new suppliers of oil and gas? How will this affect nuclear power?

For proponents, investment in nuclear energy will enable countries to simultaneously become energy self-sufficient and reach their climate change targets. However, sustained political and industry support for nuclear energy is far from guaranteed. The continued disruption of oil and gas markets, possibly triggering a global recession, may strengthen support for nuclear investment. Nuclear power's inclusion in the European Union's sustainable finance taxonomy could also undermine investment in renewables, which will have dubious environmental consequences (Bordoff and O'Sullivan 2022).

It is too early to tell what the impacts of the global energy crisis of 2022 will be on nuclear power, among others. But, like many crises, it provides an opportunity for improvement. Energy, security, and climate change are forcing countries to reevaluate their energy supply, geopolitical dependencies, and capacity to cope with climate-exacerbated hazards. One thing is certain. That is that climate change will have a longer-lasting and deeper impact on energy security in Europe and the world than the war in Ukraine.

Action-Oriented Problem Solvers Promote Novel Nuclear Power

With a few exceptions, today's commercial nuclear reactors utilize ceramic-covered uranium fuel pellets, embedded in fuel rods. Ceramics have many advantageous properties, but are also limited to a maximum reactor temperature of around 750 °C. Most of today's nuclear reactors also use water – whether light or heavy – as a coolant for the reactor system. But this current technology has several drawbacks.

Kevin O'Sullivan (24) and **Mason Rodriguez Rand** (23), who founded Alpha Nur in 2022, are aiming to address this challenge. Their company aims to organize commercial and political action to bring a new generation of nuclear technologies to market: nuclear pellets clad in metal-alloy, not ceramics; and sodium-cooled reactors, not reactors cooled by water. For their innovative ideas, Kevin and Mason raised $35,000 in 2022, including $27,000 from the US Department of Energy.

Kevin and Mason state that "to combat the challenges of ceramic-clad fuel pellets, metal-alloy fuel pellets use a mixture of uranium, plutonium, and zirconium to raise the melting point of the metal beyond what was previously possible." These so-called metal fuels are also significantly easier to recycle and re-cast than their ceramic counterparts, allowing the re-utilization of 95% or more of the original spent fuel.

Another advantage to metal fuels is an increased efficiency to devolve heat into the coolant due to the higher thermal conductivity of metals. The fuel material is compatible with sodium coolant, which allows the system to be operated at high temperatures while remaining at atmospheric pressure. "The Experimental Breeder Reactor-II, one of Argonne National Laboratory's demonstration sodium reactors, experimentally demonstrated an intrinsic ability to safely shut itself down in the event of the most severe types of nuclear accidents, including coolant or heat sink loss, which were the greatest contributing causes of the Chernobyl, Three-Mile Island, and Fukushima nuclear accidents," they enthuse.

Alpha Nur is pursuing a reactor design in a thermal power range in the upper hundreds of megawatts with the purpose of not only generating electricity, but also generating hydrogen, ammonia, and synthetic hydrocarbon. Their reactor will have around one-third the power output of a utility-scale nuclear reactor and is therefore called a small modular reactor. This modularity is important for Kevin and Mason, as it will enable economies of scale for engineering, design, and procurement efforts. It will also drastically reduce costs, compared to traditional nuclear reactors.

Both Eagle Scouts from the Chicago suburbs, **Kevin O'Sullivan** (left) and **Mason Rodriguez Rand** (right) bonded in their freshman year over a shared value of sustainability and a common vision to build a nuclear reactor. They set off to do just that in their senior year through the University of Chicago's College New Venture Challenge. Kevin is a 2018 Fermilab Science Award recipient with a BS in Biology from the University of Chicago. Mason graduated from the University of Chicago with a BS in Molecular Engineering and has worked on sustainability projects throughout his life. *Source:* Kevin O'Sullivan and Mason Rodriguez Rand.

17.10   Fusion Reactors

When two light nuclear particles (such as $_1^2$H and $_2^3$He) combine or fuse, energy is released because the product nuclei (such as $_2^4$He and $_1^1$H) are lighter and more stable than the original particles (see Figure 16.4). Such *fusion* reactions can be initiated by bombarding target particles with charged particles using a particle accelerator or by raising the temperature of a gas to a high enough level for fusion to take place. When a fusion reaction can be sustained, it is termed *ignition*.

The most promising fusion reactions employ the isotopes deuterium $\left(_1^2\text{H}\right)$ and tritium $\left(_1^3\text{H}\right)$. Although deuterium $\left(_1^2\text{H}\right)$ is present as hydrogen in water, with an abundance of only 0.015%, the vast amount of water available on Earth suggests an inexhaustible supply of deuterium. Tritium $\left(_1^3\text{H}\right)$ is artificially produced, usually as a byproduct of certain nuclear reactors, such as those of the CANDU type that use heavy water, where deuterium $\left(_1^2\text{H}\right)$ takes the place of hydrogen $\left(_1^1\text{H}\right)$. There is, however, a shortage of tritium fuel, which will be exacerbated when the world's 19 CANDU reactors are decommissioned toward 2030 (Clery 2022b).

Four fusion reactions are of great importance (Murray 1980):

$$_1^2\text{H} + _1^2\text{H} \rightarrow _1^3\text{H} + _1^1\text{H} + 4.03\,\text{MeV} \tag{17.21}$$

$$_1^2\text{H} + _1^2\text{H} \rightarrow _2^3\text{He} + _0^1\text{n} + 3.27\,\text{MeV} \tag{17.22}$$

$$_1^2\text{H} + _1^3\text{H} \rightarrow _2^4\text{He} + _0^1\text{n} + 17.6\,\text{MeV} \tag{17.23}$$

$$_1^2\text{H} + _2^3\text{He} \rightarrow _2^4\text{He} + _1^1\text{H} + 18.63\,\text{MeV} \tag{17.24}$$

The fusion of two deuterons (deuterium nuclei) is designated a D–D reaction; the fusion of a deuteron and triton (tritium nucleus) is designated a D–T reaction, illustrated in Figure 17.5. The products of Eqs. (17.21) and (17.22) appear as reactants in Eqs. (17.23) and (17.24), which suggests the possibility of a composite process. Adding all the equations, the *net* effect would be to convert deuterium into helium (Murray 1980; Jaffe and Taylor 2018):

$$4\,_1^2\text{H} \rightarrow 2\,_2^4\text{He} + 47.7\,\text{MeV} \tag{17.25}$$

It must however be noted that the reaction rates for D–D reactions are so low that in a D–T mix those D–D reactions will probably never occur. The concomitant deuterium-$_2^3$He reaction rate is

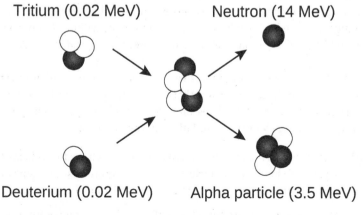

Figure 17.5   A possible fusion reaction between deuterium $\left(_1^2\text{H}\right)$ and tritium $\left(_1^3\text{H}\right)$.

even lower than that of D–D and will probably never happen either. It is therefore preferred to use D–T reactions as they have many orders of magnitude higher probability of happening than other fusion reactions (Andruczyk 2023).

Therefore, the energy yield per atomic mass of deuterium would be $b\left(^2_1\text{H}\right) = \dfrac{B}{A} = \dfrac{47.7\,\text{MeV}}{4} \approx 6\,\text{MeV}$. In comparison, a fission of $^{235}_{92}\text{U}$ yields "only" about $b\left(^{235}_{92}\text{U}\right) = \dfrac{B}{A} = \dfrac{200\,\text{MeV}}{235} \approx 0.9\,\text{MeV}$ per atomic mass unit. In addition, controlled fusion produces no greenhouse gases, does not require hazardous nuclear fuel, produces shorter-lived and less hazardous waste than nuclear (fission) reactors, and poses no danger of runaway reactions. Fusion reactions quickly stop after running out of fuel, but fusion reactions are difficult to start and sustain (Sovacool 2008).

Fusion *ignition* refers to the moment the energy from a controlled fusion reaction outstrips the rate at which X-ray radiation losses and electron-conduction cool the implosion. A fusion reactor that produces more energy than it requires is said to have positive net energy. *Plasma energy gain* (Q) is the ratio of thermal energy released by the plasma to the thermal energy input to the plasma. Researchers are therefore seeking to achieve Q values larger than 1.0.

The Tokamak Fusion Test Reactor (TFTR) was the first reactor to achieve about 10 MW of fusion power from a 50–50 mix of D–T. These *supershots* were later replicated in the Joint European Torus (JET) reactor, where researchers achieved 16 MW of fusion energy from a 50–50 mix of D–T. The plasma energy gain was about 0.4 and 0.7 for these two reactors, respectively. The Japan Torus-60 (or JT-60) Tokomak achieved a D–D equivalent of a D–T supershot achieving $Q = 1.25$ (McGuire et al. 1995; Shirai et al. 2017; Gibney 2022; Andruczyk 2023).

In December 2022, researchers at the Lawrence Livermore National Laboratory (LLNL) managed to fuse deuterium and tritium into helium using high-powered lasers, in the process releasing more energy than the lasers put in (LLNL 2022). The LLNL reactor achieved $Q = 1.5$ as the reactor produced around 3 MJ of fusion energy from 2 MJ of laser energy (Zylstra et al. 2022). *However*, it required an estimated 400 MJ of electrical energy to produce the 2 MJ of laser energy, giving an actual (net) $Q = 0.005$. So, even though fusion research is advancing at a rapid pace, much work still needs to be done to achieve *net* plasma reaction gains of larger than 1.

The challenge with fusion is that the process does not produce enough neutrons, unlike fission where the chain reaction releases an exponentially growing number of neutrons. With fusion, each D–T reaction only produces a single neutron. Researchers are producing novel schemes to address this issue. For instance, a fusion reactor could be covered with a *breeding wall* (or *breeding blanket*) comprising lithium, to breed its own tritium. Figure 17.6 illustrates such a *fusion-fission hybrid reactor*. The breeding wall (or breeding blanket) could comprise a mix of PbLi or BeLi, where the lead or beryllium act as neutron multiplier and the lithium absorbs a neutron to create tritium. Although a tritium-breeding ratio of 2 would be ideal, that figure is currently closer to 1.1 (Andruczyk 2023; Ruegsegger et al. 2023).

Fusing two light elements is difficult because of the massive electrostatic repulsion that exists between positively charged nuclei. Our sun forces nuclei together with the strength of its large gravitational pull, but this is not an option on Earth. Instead, researchers attempt to fuse light elements using several techniques, among which are (Murray 1980; Jaffe and Taylor 2018):

- *Magnetic confinement* in a doughnut-shaped reactor called a *Tokomak*, in which the fuel is heated and kept afloat with strong magnetic fields. Thermonuclear reactions require vast amounts of energy, which can be evaluated by considering the kinetic energy relation for ideal gases: $\overline{E} = \dfrac{3}{2}kT$, with k being the Boltzmann constant (1.38×10^{-23} J/K). A temperature of

$$T = \frac{2}{3}\frac{(10 \times 10^3 \text{ eV})(1.6 \times 10^{-19} \text{ J/eV})}{1.38 \times 10^{-23} \text{ J/K}} = 77 \times 10^6 \text{ K}$$ would be required to achieve a deuteron

energy level of only 10 keV. This temperature is larger than that of the surface of the sun. No material can be used at such temperatures, which suggests that the containment of such a high-temperature plasma could be confined in a strong magnetic field. Considering the radiation losses from such a machine, only temperatures greater than 400×10^6 K in the case of D–D reactions would have a net energy yield (Goldston 2020). Figure 17.6 shows an experimental Tokomak reactor.

- *Inertial confinement,* in which very small pellets of deuterium and tritium compound are compressed (to a million times their normal densities) and heated (to temperatures more than that of the surface of the sun) by laser beams or high-speed particles (Craxton et al. 2015). The resulting high densities of the fuel pellets are necessary to ensure that the fusion reaction products hit the target particles that might result in fusion. In December, the test chamber of the US National Ignition Facility triggered fusion using 192 lasers. The fusion process produced 1.5 MJ for a very short period, with a plasma energy gain of 1.5 (Clery 2022a; NIF 2022).

Once a fusion reaction ignites, it runs to completion. If the heating fails or magnets switch off, the ions will drift to the walls. The fuel density is so low that the ions will cool down to the wall

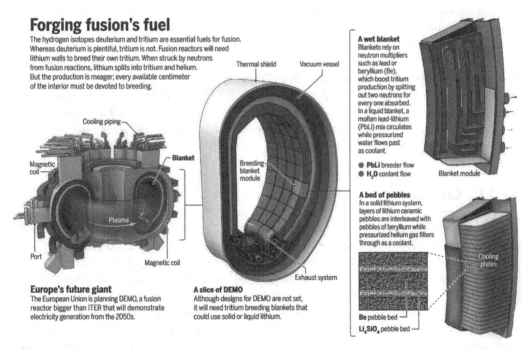

Figure 17.6 Magnetic confinement of plasma: Experimental EuroFusion Tokomak reactor, called DEMO. Commercial fusion plants producing 3 GW of electricity will burn about 167 kg of tritium a year. Considering the global shortage of tritium, it will be imperative for commercial fusion reactors to produce their own tritium (Clery 2022b). The shown DEMO fusion reactor uses lithium walls to breed its own tritium, according to the following reactions: $^6_3\text{Li} + ^1_0\text{n} \rightarrow ^3_1\text{H} + ^4_2\text{He} + 4.8$ MeV and $^7_3\text{Li} + ^1_0\text{n} \rightarrow ^3_1\text{H} + ^4_2\text{He} + ^1_0\text{n} - 2.5$ MeV. *Source:* C. Bickel/Science. From Science, 23 June 2022 (doi:10.1126/science.add5489). Reprinted with permission from AAAS.

temperature with the plasma soon switching off. This inherent safety feature is one of the main attractions of nuclear fusion. Unlike nuclear fission, fusion therefore does not suffer from the threat of calamitous explosions (Andruczyk 2023). But the neutrons emitted from a fusion reaction can react with materials they hit to produce low-level residual radioactivity. Fusion is not a completely clean technology and there is some amount of radioactive waste produced.

The heat released by controlled fusion is intended to produce electricity. Heat exchangers can remove heat from appropriate sections of the fusion reactor, and water, for instance, could be heated to produce steam which is used in a Rankine cycle (Chapter 13).

17.11 Outlook for Nuclear Fusion

Fusion with a positive *plasma energy gain* has been proven to be possible at the Lawrence Livermore National Laboratory (in December 2022), albeit at a very small scale. An experiment like the one at Livermore, in which one tiny deuterium-tritium pellet was vaporized by 192 laser beams, is one thing, but a utility-scale power plant that rapidly vaporizes thousands upon thousands of pellets and safely draws off the released energy for conversion into electricity is quite another. That is not to say that it cannot be done, but rather that it will take lots of time to achieve utility-scale controlled fusion in an economic manner.

This suggests that fusion might not replace fission, at least not anytime soon. One must remember that about 10 times more energy can be obtained by using the fast neutron emitted by a fusion reaction to initiate fast-neutron fission to breed fissionable fuel (Freidberg and Kadak 2009). This has been suggested for use in so-called *fusion-fission hybrid reactors*. It is, however, uncertain whether governments or companies would discard 90% of their energy "merely" to avoid producing nuclear waste from fission. Fission will clearly be around for a long time to come, even if scientists and engineers manage to create successful fusion reactors. But nuclear fusion will eventually be successfully developed, at which stage it will probably become the dominant energy source.

17.12 Outlook

Nuclear fusion has immense potential to permanently solve our energy woes. Fusion has, however, established a reputation for always being a technology that is three decades away from reality. The same joke, including the 30-year time frame, used to be said of electric rocket propulsion, and there are now hundreds of spacecraft that use electric propulsion (e.g., gridded ion engines, Hall effect thrusters, plasma thrusters, magneto-plasma thrusters, and field emission electric propulsion systems).

The issues surrounding nuclear fusion could perhaps be contextualized using the *S-curve for technology development* (covered in Section 4.2). Until a technology rises above the early, exploratory stage, predictions about it becoming mature enough for practical use can be driven more by optimism and enthusiasm than by hard facts. Although fusion has just entered the S-curve, it must be observed that technology nowadays develops at a much quicker rate than, say, 50 years ago. Also, there has been a recent surge in fusion start-up companies and growing funding opportunities for researchers, albeit mainly from public and private investors. Together with the important fusion breakthroughs at the LLNL (in December 2022), these developments suggest that fusion could become a viable energy conversion technology sooner than most people think. But, sources of scarce tritium must be found, such as by surrounding a fusion reactor with lithium walls, which essentially breeds its own tritium (see Figure 17.6). The expected acceleration along the technology S-curve will also require global cooperation and effective regulation of fusion projects.

Regarding *nuclear fission*, only 135 years of uranium resources remain for use in fission reactors. Although breeder reactors show much promise, their use of plutonium fuel does not paint a picture of safety. It appears that we will be better off with once-through fuel cycles. Further, instead of focusing on the use of mainly uranium-235, alternative sources of nuclear fuel should be investigated in earnest. Advanced reactors, such as those being investigated by the Generation IV consortium, must be vigorously pursued. The feasibility of small- and micro-reactors is questioned as they will not have the economies of scale of their 3-GW$_{th}$ counterparts. These small ($<$100 MW$_{th}$) nuclear reactors do not seem to offer obvious improvements over their (much larger) light water reactor counterparts to justify their many risks.

The short-term future of fission nuclear power indeed depends on capital and operating costs, safety records, coupling to nuclear militarization, and the overall sense of competence and responsibility that the industry projects. Nuclear energy does not present a near- or even medium-term solution to sustainable energy. Considering the many social, technological, economic, environmental, and political challenges that stand in the way of building safer, more efficient, and cost-competitive fission reactors, nuclear fission will not be able to replace other forms of energy conversion quick enough to mitigate the worst effects of climate change.

Despite their limitations, fission and fusion still have the potential to reduce carbon emissions. But, instead of placing unfounded faith in the ability of nuclear power to save the planet, we might be better off focusing on the real threat of climate change. We need to acknowledge that even if fusion power plants become a reality, it would likely not happen in time to help stave off the near-term worsening effects of climate change. It is far better, many climate scientists, engineers, and policymakers say, to focus on currently available renewable energy technologies like solar and wind power to help reach net-zero emission targets.

Problems

17.1 Uranium that is fed in a nuclear reactor has a $^{235}_{92}$U concentration of 3%, and the spent fuel has a $^{235}_{92}$U concentration of 0.8%. What mass of U_3O_8 fuel is required for a 1000 MW$_e$ plant that runs for one year at a 95% capacity factor and a thermal efficiency of 33%?

[Answer: About 56 t per year.]

17.2 Nuclear fuel that is fed in a micro-nuclear reactor has a $^{235}_{92}$U concentration of 19% and the spent fuel has a $^{235}_{92}$U concentration of 0.8%. What mass of U_3O_8 fuel is required for a 30 MW$_e$ plant that runs for ten years at a 95% capacity factor and a thermal efficiency of 33%?

[Answer: About 2033 kg (or 2 t) of U_3O_8.]

17.3 Consider a nuclear reactor operating with 3% enriched $^{235}_{92}$U. The reactor produces 3 GW of heat. (a) Estimate the mass of $^{235}_{92}$U consumed per year. If the cost of 3% enriched uranium-235 is \$35,000/kg, (b) calculate the cost of the uranium consumed per year, as well as (c) the cost of electricity per kilowatt-hour if the heat-to-electricity conversion efficiency is 33% and the plant capacity factor is 95%.

[Answers: (a) 1400 kg; (b) \$49 million; (c) 7 ¢/kWh]

17.4 A nuclear reactor core initially contains 5000 kg of $^{235}_{92}$U enriched to 4.7%. The fuel is replaced when the level of enrichment reaches 1.35%. The following reactor parameters are provided: $\eta_f = 2.07$, $\varepsilon = 1.05$, $L = 0.68$. Calculate (a) the reactor's initial power output, (b) the reactor's conversion ratio, and (c) the fuel "burn-up" in gigawatt-day per tonne.

[Answers: (a) 4.8 GW; (b) 0.49; (c) around 50 GW/t]

17.5 Around 10% of the world's electricity is generated by about 440 nuclear power reactors. In 2021, nuclear plants supplied 2653 TWh of electricity. Assuming an average nuclear power plant efficiency of 33%, (a) how much uranium was consumed in 2021? (b) With uranium reserves of 4.9 MT, how many years of uranium reserves remain?

[Answers: (a) 51,000 t; (b) 99 years.]

17.6 A nuclear power plant generates 750 MW of electrical power. The power plant has an efficiency of 33% and a capacity factor of 95%. The reactor core contains 100 t of $^{235}_{92}$U. The neutron flux is 10^{13} cm^{-2}/s. Estimate the enrichment (%) of the fuel.

[Answer: 2.7%]

17.7 Calculate the volume (m^3) of spent fuel generated annually by a 3 GW$_{th}$ nuclear power plant with a 90% capacity factor. The uranium fuel is enriched to 3% $^{235}_{92}$U. The fuel density is 19 g/cm^3.

[Answer: 5.7 m^3]

17.8 Estimate the cost of 1 kg of uranium-235 enriched to 3.7%. Clearly state all assumptions and cite all your resources. For instance, the website of the Idaho National Laboratory features a useful online cost calculator: https://fuelcycleoptions.inl.gov/costbasis/SitePages/CostModules.aspx

The mean costs in the online calculator were obtained from the following report, which you should also read: *Advanced fuel cycle*. Nuclear Technology Research and Development (NTRD), NTRD-FCO-2017-000265, 29 September 2017. https://fuelcycleoptions.inl.gov/costbasis/Shared%20Documents/2017_Advanced_Fuel_Cycle_Cost_Basis.pdf

[Answer: Front-end costs are $1400–$4800 per kg depending on assumptions and whether you include transportation costs, interest rates, etc.]

17.9 A micro-nuclear power plant has a capacity factor of 85% and a lifetime of 14 years. Assume a discount rate of 8% and annual operation and maintenance costs of $10/MWh. Calculate the levelized costs of electricity (*LCOE*). Comment on this cost versus that of fossil fuel-fired power plants and renewable plants (photovoltaic, wind) of similar 15 MW$_{th}$ (or approximately 5 MW$_e$) size.

[Answer: About $80/MWh.]

17.10 Figure 17.7 shows the cover of a booklet about breeder reactors. In an essay of no more than 300 words, explain how breeder reactors provide a means of increasing the world's supply of fissionable fuel, while it generates power. Discuss the salient nuclear conversions in a

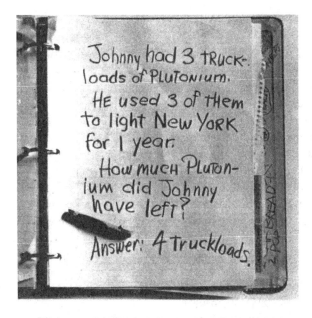

Breeder reactors

by Walter Mitchell, III and
Stanley E. Turner

Figure 17.7 Why breeder reactors? *Source:* Mitchell and Tuner (1971), United States Atomic Energy Commission / Public Domain.

breeder reactor and how such a reactor makes more efficient use of fission neutrons. Why are there so few operational breeder reactors? Should we invest in this technology rather than light-water reactors, for instance?

[Answer: Open-ended question; independent research required.]

17.11 The USA has invested an estimated $7 billion trying to create a controlled form of nuclear fusion that could be the energy source for an endless supply of electricity. One laboratory has succeeded in delivering fusion that produces more energy than it requires to generate the reaction. It will likely take many decades before this laboratory-scale fusion can be scaled up.

Do you therefore think that it is a good idea to spend so much on nuclear fusion now, when other clean energies can make a more tangible impact on our energy mix? Present your arguments in a well-researched 500-word essay that addresses the interconnected social, technological, economic, environmental, and political (or "STEEP") considerations surrounding commercial nuclear fusion.

For context, read "Reaching the threshold of ignition. An in-depth look at NIF's 1.35 megajoule milestone," US National Ignition Facility, https://lasers.llnl.gov/content/assets/docs/news/reaching-the-threshold-of-ignition-magazine.pdf

[Answer: Open-ended question, but STEEP aspects must be discussed as they pertain to commercial nuclear fusion.]

References

Andruczyk, D. (2023). Personal communication with Prof. Daniel Andruczyk, UIUC Nuclear Plasma and Engineering, 14 June.

Behar, C. (2014). Technology roadmap update for generation IV nuclear energy systems. *OECD Nuclear Energy Agency for the Generation IV International Forum* 17: 2014–03.

Bollfrass, A.K. and Herzog, S. (2022). The war in Ukraine and global nuclear order. *Survival* 64 (4): 7–32.

Bordoff, J. and O'Sullivan, M.L. (2022). The new energy order: how governments will transform energy markets. *Foreign Affairs* 101 (July/August): 131.

Bragg-Sitton, S.M., Gorman, J., Burton, G., Moore, M. et al. (2020). Flexible nuclear energy for clean energy systems (No. NREL/TP-6A50-77088). Golden, CO: National Renewable Energy Lab.

Clery, D. (2022a). European fusion reactor sets record for sustained energy. World's largest tokomak paves the way for ITER with a capstone run of pulses using power-producing tritium. *Science* 375 (6581): 600.

Clery, D. (2022b). Out of gas. A shortage of tritium fuel may leave fusion energy with an empty tank. *Science* 376 (6600): 1372–1376.

Craxton, R.S., Anderson, K.S., Boehly, T.R., Goncharov, V.N. et al. (2015). Direct-drive inertial confinement fusion: a review. *Physics of Plasmas* 22 (11): 110501.

EIA (2022a). Energy Outlook 2022. Washington, DC: US Energy Information Administration. https://www.eia.gov/outlooks/aeo/

EIA (2022b). Levelized costs of new generation resources in the Annual Energy Outlook 2022. Washington, DC: US Energy Information Administration. https://www.eia.gov/outlooks/aeo/pdf/electricity_generation.pdf

EIA (2021). Levelized cost and levelized avoided cost of new generation resources in the Annual Energy Outlook 2021. Washington, DC: US Energy Information Administration. https://www.eia.gov/outlooks/aeo/pdf/electricity_generation.pdf

Foss, A.W., Smart, J.G., Bryan, H.C., Dieckmann, C. et al. (2021). NRIC integrated energy systems demonstration pre-conceptual designs (No. INL/EXT-21-61413-Rev. 01). Idaho Falls, ID: Idaho National Lab.

Freidberg, J.P. and Kadak, A.C. (2009). Fusion–fission hybrids revisited. *Nature Physics* 5 (6): 370–372.

GA (2021). Advanced reactors. *General Atomics*. https://www.ga.com/nuclear-fission/advanced-reactors

Gibney, E. (2022). Reactor smashes energy record. *Nature* 602: 371 (Feb. 9).

Goldston, R.J. (2020). *Introduction to Plasma Physics*. Boca Raton, FL: CRC Press.

IAEA (2021). What are small modular reactors? International Atomic Energy Agency, Paris. https://www.iaea.org/newscenter/news/what-are-small-modular-reactors-smrs

IAEA (2022). Advanced reactors information system (ARIS). International Atomic Energy Agency, Paris. https://aris.iaea.org/sites/core.html

IEA (2022). Nuclear power. International Energy Agency. https://www.iea.org/fuels-and-technologies/nuclear

IEA (2021). Global Energy Review 2021. International Energy Agency. https://www.iea.org/reports/global-energy-review-2021

IEA (2020). Projected costs of generating electricity, 2020 edition. International Energy Agency. https://www.oecd-nea.org/upload/docs/application/pdf/2020-12/egc-2020_2020-12-09_18-26-46_781.pdf

INL (2022). Microreactors. Idaho Falls, ID: Idaho National Laboratory. https://inl.gov/trending-topic/microreactors/

Jaffe, R.L. and Taylor, W.M. (2018). *The Physics of Energy*. Cambridge: Cambridge University Press.

Krall, L.M., Macfarlane, A.M. and Ewing, R.C. (2022). Nuclear waste from small modular reactors. *Proceedings of the National Academy of Sciences of the United States of America* 119 (23): e2111833119.

Lamarsh, J.R. (1966). *Introduction to Nuclear Reactor Theory*. Boston, MA: Addison-Wesley.

Lederer, C.M. and Shirley, V.S. (1978). *Table of Isotopes*. New York, NY: Wiley.

LLNL (2022). *National Ignition Facility achieves fusion ignition*. Lawrence Livermore National Laboratory. https://www.llnl.gov/news/national-ignition-facility-achieves-fusion-ignition

Martin, B.R. and Shaw, G. (2019). *Nuclear and Particle Physics: An Introduction*. New York: NY: Wiley.

McGuire, K.M., Adler, H., Alling, P., et al. (1995). Review of deuterium–tritium results from the Tokamak Fusion Test Reactor. *Physics of Plasmas* 2 (6): 2176–2188.

Mitchell III, W. and Turner, S.E. (1971). Breeder reactors. *US Atomic Energy Commission (AEC)*,

Murray, R.L. (1980). *Nuclear Energy. An Introduction to the Concepts, Systems, and Applications of Nuclear Processes*. Pergamon Press.

NEA (2020). Uranium resources, production and demand (Red Book). *US Nuclear Energy Agency*. https://www.oecd-nea.org/jcms/pl_28569/uranium-resources-production-and-demand-red-book

NIF (2022). National Ignition Facility Achieves Long-Sought Fusion Goal. Lawrence Livermore National Laboratory's National Ignition Facility. December 6. https://www.aip.org/fyi/2022/national-ignition-facility-achieves-long-sought-fusion-goal#:~:text=NIF%20is%20the%20world's%20highest,nuclear%20warheads%20without%20explosive%20testing

Nuttall, W.J. (2022). *Nuclear Renaissance: Technologies and Policies for the Future of Nuclear Power*. CRC Press.

Rinard, P. (1991). Neutron interactions with matter (375–377). In *Passive Nondestructive Assay of Nuclear Materials* (No. NUREG/CR-5550; LA-UR-90-732) (eds. D. Reilly, N. Enselin, H. Smith and S. Kreiner), US Nuclear Regulatory Commission, Washington, DC (United States). Office of Nuclear Regulatory Research; Los Alamos National Lab., New Mexico (United States).

Ruegsegger, J., Moreno, C., Nyberg, M., et al. (2023). Scoping studies for a lead-lithium-cooled, minor-actinide-burning, fission-fusion hybrid reactor design. *Nuclear Science and Engineering* (January 26), 1–17.

Shirai, H., Barabaschi, P. and Kamada, Y. (2017). Recent progress of the JT-60SA project. *Nuclear Fusion* 57 (10): 102002.

Shultis, J.K. and Faw, R.E., 2016. *Fundamentals of Nuclear Science and Engineering*. CRC Press.

Sovacool, B.K. (2008). Valuing the greenhouse gas emissions from nuclear power: a critical survey. *Energy Policy* 36 (8): 2950–2963.

Testoni, R., Bersano, A. and Segantin, S. (2021). Review of nuclear microreactors: Status, potentialities and challenges. *Progress in Nuclear Energy* 138: 103822.

UCS (2021). "Advanced" isn't always better. Assessing the safety, security, and environmental impacts of non-light-water nuclear reactors. Union of Concerned Scientists. https://www.ucsusa.org/sites/default/files/2021-05/ucs-rpt-AR-3.21-web_Mayrev.pdf

Vaughan, A. (2022). The first global energy crisis. *New Scientist* 253 (3379): 18–21.

Von Dardel, G. and Sjöstrand, N.G. (1954). Diffusion parameters of thermal neutrons in water. *Physical Review* 96 (5): 1245.

WNA (2003). Small nuclear power reactors. World Nuclear Association (WNA), *ATW - Internationale Zeitschrift fuer Kernenergie* 48 (2): 102–104.

Zohuri, B. (2020). *Nuclear Micro Reactors*. Springer International Publishing.

Zylstra, A.B., Hurricane, O.A., Callahan, D.A., et al. (2022). Burning plasma achieved in inertial fusion. *Nature* 601 (7894): 542–548.

18

Controlling Waste and Emissions from Nuclear Power Plants

Uranium is mostly extracted through open-pit mining (16.1%), underground mining (20%), and *in situ* leaching (ISL) (57.4%). ISL is the underground injection of acidic or alkaline solutions to dissolve and pump uranium to the surface. It eliminates ore tailings associated with other mining but raises aquifer protection concerns.

Nuclear electricity is falsely claimed to produce no greenhouse-gas (GHG) emissions. However, activities related to the nuclear fuel cycle produce GHG emissions. The *embodied emissions* during the life of a nuclear plant include those in the extraction of uranium, nuclear plant construction and operation, conditioning the spent fuel and storing it, deconstructing (or decommissioning) the plant and reclaiming the land. All these operations result in about 65 g CO_2-equivalent being emitted for every kWh_e of nuclear electricity. This is way below the approximately 500 g of CO_2-equivalent/kWh_e for a gas-fired steam power plant or the 1 kg of CO_2-equivalent/kWh_e for coal-fired steam power plants. But a nuclear power plant still contributes to significant amounts of greenhouse gases over its 30-year lifetime.

The management and disposal of *radioactive waste* from the nuclear fuel cycle is one of the most intractable problems facing the nuclear power industry (Deutch and Lester 2004). The nuclear fuel cycle is the entire process of producing, using, and disposing of uranium fuel. Powering a 1-GW nuclear plant for a year can require mining 20,000–400,000 t of uranium ore, processing it into 28 t of uranium fuel, and disposing of 28 t of radioactive spent fuel, of which 90% (by volume) is low-level waste, 7% is intermediate-level waste, and 3% is high-level waste. Globally, more than 300,000 t of radioactive waste are produced annually (Feiveson et al. 2011).

To date, no country has yet succeeded in disposing of high-level nuclear waste (which remains radioactive for tens of thousands of years). Finland will become the first country to implement a long-term underground storage facility, in 2023/2024. Countries such as Canada, Finland, France, Germany, Japan, Sweden, Switzerland, United Kingdom, and the United States have stated their intentions to also dispose of such waste in underground repositories, but all have encountered social, technological, economic, or political difficulties in enacting their goals.

Apart from secure long-term fuel disposal, nuclear power must contend with numerous issues surrounding safety and security risk, nuclear proliferation, and the risk of nuclear terrorism. All these issues are complex, interdependent on other issues, and require vast resources to address.

Energy Systems: A Project-Based Approach to Sustainability Thinking for Energy Conversion Systems,
First Edition. Leon Liebenberg.
© 2024 John Wiley & Sons, Inc. Published 2024 by John Wiley & Sons, Inc.
Companion website: www.wiley.com/go/liebenberg/energy_systems

---

Reimagine Our Future

Imagine if your university were to build a micro-nuclear plant on campus, mainly to generate heat during winter. Compare the costs (social and environmental) of doing so with plugging energy leaks by, for instance, improving building insulation.

18.1 Reactor Safety

Nuclear safety is closely linked to nuclear security and nuclear safeguards.

Security focuses on the intentional misuse of nuclear or other radioactive materials by non-state elements to cause harm. It relates mainly to external threats to materials or facilities. Security at nuclear facilities is the responsibility of national governments. To date, no cyber-attack on a nuclear reactor's information and control system has compromised safety. The first time an operating civil nuclear power plant was attacked by an armed group was during Russia's military action in Ukraine in 2022. Although safety and security are treated separately, if a facility or a radioactive source is not secure, it could pose a potential hazard and, thus, is not safe. Since the early 2000s, there has been a shift in attention from ensuring that nuclear materials are not diverted from peaceful uses, toward protecting plants from armed assault and cyberattacks (WNA 2022).

Safeguarding focuses on restraining activities by states that could lead to the acquisition or development of nuclear weapons. It concerns mainly materials and equipment in relation to rogue governments (WNA 2022). Most countries participate in international initiatives designed to limit the proliferation of nuclear weapons. The international safeguard system has successfully prevented the diversion of fissile materials into weapons since 1970. Its scope has been widened to address undeclared nuclear activities. The International Atomic Energy Agency (IAEA) undertakes regular inspections of civil nuclear facilities and audits the movement of nuclear materials through them. These safeguards are backed by diplomatic and economic measures but are complex and not fail-safe. The apparent nonadherence to safeguarding protocols by India, Iran, North Korea, and Pakistan, among other countries, is troublesome.

Safety focuses on unintended conditions or events leading to radioactive releases from authorized activities. It relates mainly to intrinsic problems or hazards. Effective *safety* measures of a fission reaction begin with the control of neutron multiplication (see Sections 17.1–17.6), cooling the fuel (Section 17.8), and containing radioactive emissions and waste (see Section 18.2).

Most physical barriers are considered *passive* and include (but are not limited to) the nuclear fuel itself (i.e., fission product holdup in fuel), fuel cladding, reactor vessel and reactor coolant system piping, and containment structures. Whereas passive structures do not rely on component response or operator actions, *active* systems require operator actions and/or a multitude of systems, structures, and components to launch on demand in response to some initiating event. Passive and automatic system responses are designed in many light-water reactor technologies, but in many severe accident scenarios, operator actions and active system responses are essential to prevent damage to the nuclear core (Clark and Rowland 2021).

Modern nuclear power plants operate using a *defense-in-depth* (DID) approach, with multiple safety systems supplementing the natural features of the reactor core. Key aspects of the approach are (Clark and Rowland 2021; WNA 2022):

- *High-quality design and construction:* For instance, uranium in light-water reactors is encased in ceramic pellets. The radioactive fission products remain largely bound inside these pellets as the fuel is burned. The pellets are clad in zirconium alloy ("zircalloy") tubes to form fuel rods. These are confined inside a large steel pressure vessel with walls up to 300 mm thick – the associated primary water-cooling pipework is also substantial. All this is enclosed in a reinforced concrete containment structure with walls at least 1 m thick. This amounts to three significant barriers around the fuel, which itself is stable up to very high temperatures.

- *Equipment that prevents operational disturbances or human failure and error developing into problems:* For instance, the main safety features of most reactors are inherent – *negative temperature coefficient* and *negative void coefficient*. Negative temperature coefficient means that, beyond an optimal level, as the temperature increases, the efficiency of the reaction decreases (Murray 1980). Negative void coefficient means that, if any steam has formed in the cooling water, there is a decrease in moderating effect so that fewer neutrons can cause fission and the reaction slows down automatically (Murray 1980). Around 2.5 neutrons are released from fission within milliseconds, which makes effective control nearly impossible. Fortuitously, around 0.65% of these neutrons appear only 10–12 s later. These *delayed neutrons*, although few, extend the fuel cycle time significantly and slow the rate of neutron multiplication (Murray 1980).
- *Comprehensive monitoring and regular testing:* These detect equipment or operator failure.
- *Redundant and diverse systems to control damage* to the fuel and prevent significant radioactive releases: For instance, in pressurized water reactors (PWRs), safety is provided by several groups of movable rods with neutron-absorbing material (such as boron, $^{10}_{5}B$, see Section 17.4). In a PWR, these rods are supported by electromagnets that release the rods in the event of an interruption of the electrical current (Murray 1980). The nuclear reaction should therefore promptly shut down in the event of a catastrophic nuclear event.
- *Provision to confine the effects of severe fuel damage (or any other problem) to the plant itself* (Xing et al. 2016): In the event of failure of the primary accident prevention mechanisms, the fuel could overheat and melt the zircalloy rods, which would result in the significant emission of fission products. To help prevent "core meltdown," nuclear reactors feature emergency core cooling systems that could comprise auxiliary pumps circulating cooling water in the core (Chang et al. 2013). Those auxiliary pumps are usually driven by back-up diesel generators. The dome-shaped reactor building must be built to withstand extreme temperatures and pressures from within the core. Nuclear power plants operate within a safety zone called an "exclusion area" to ensure that the plant is sited several kilometers from any populous area. An *exclusion area,* with a radius of 16 km from a utility-scale nuclear power plant, is commonly specified (NRC 2020). However, it is still uncertain what the exclusion area will be for small and micro-nuclear power plants. Exclusion areas of 300 m to 2 km are being suggested.

Westinghouse Electric produces the highly acclaimed "Advanced Passive" AP1000 reactor. In the event of a design-basis accident, such as a main coolant pipe break, the plant is designed to achieve and maintain safe shutdown condition without operator action, and without the need for electric power or pumps. Rather than relying on active components, such as diesel generators and pumps, the AP1000 plant relies on passive safety, using gravity, natural circulation, and compressed gases to keep the core and the containment from overheating. The AP1000 PWR provides multiple levels of defense for accident mitigation (defense-in-depth), resulting in extremely low core damage probabilities while minimizing the occurrences of containment flooding, pressurization and heat-up (Westinghouse 2022).

The heat that must be removed by emergency core cooling systems can be estimated by considering the heat of decay after shutdown. After shutdown, the power of fission reduces exponentially with time (Murray 1980):

$$P(t) = P_0 A t^{-\alpha} \tag{18.1}$$

where P_0 is the original thermal power output of the reactor, $A \simeq 0.066$ and $\alpha \simeq 0.2$, both being experimentally derived factors.

Example 18.1 Heat of Decay after the Shutdown of a Nuclear Reactor

At 10 s after the shutdown of a 3 GW_{th} nuclear reactor, the fission power is: $P(10 \text{ s}) = P_0At^{-\alpha} =$ (3000 MW)$(0.066)(10)^{-0.2} \simeq 125$ MW, or about 4.2% of the reactor's initial power.

At 1 h after the shutdown of a 3 GW_{th} nuclear reactor, the fission power is: $P(3600 \text{ s}) = P_0At^{-\alpha} =$ (3000 MW)$(0.066)(3600)^{-0.2} \simeq 38$ MW, or about 1.3% of the reactor's initial power.

At 1 day after the shutdown of a 3 GW_{th} nuclear reactor, the fission power is: $P(86,400 \text{ s}) =$ $P_0At^{-\alpha} = (3000 \text{ MW})(0.066)(86,400)^{-0.2} \simeq 20$ MW, or about 0.7% of the reactor's initial power.

Discussion: An emergency core cooling system must be able to remove this decay heat after the shutdown of a reactor to avoid melting of the fuel rod cladding (at around 1200 °C), which would result in the catastrophic emission of high-level radioactivity.

18.2 Nuclear Waste and Its Disposal

As calculated in Example 17.3, a typical 1 GW_e light-water nuclear power plant requires about 40 t of enriched uranium a year. On average, the plant discharges 40 t of spent fuel (which fills a volume of around 10 m^3) every year. Spent fuel typically comprises 95% non-fissile uranium-238, 3% fission products, 1% fissile uranium-235, and 1% plutonium. Most of the spent fuel is temporarily stored (for five years to several decades) in *water-filled basins* at the nuclear power plants. The storage pool of circulating cooled water absorbs heat and blocks the high radioactivity of fission products.

The 93 operating nuclear power reactors in the United States have a combined spent fuel inventory of around 100,000 t per year (EIA 2021; IEA 2022). This mass of spent fuel would occupy a volume of around 25,000 m^3 if stacked closely together. In practice, however, the spent fuel is not packed closely together as the spent fuel rods require adequate spacing between them to allow for convective cooling.

After spending around five years in cooling basins, the radioactive fuel assemblies are transferred to *dry cask storage* (Figure 18.1). Dry casks are large concrete and stainless-steel containers that are designed to passively cool radioactive waste (via natural convection) and withstand natural disasters or large impacts. Around one-third of spent fuel in the United States is now placed in dry casks at nuclear power plants. The dry casks are century-scale solutions, but the search for storage solutions that isolate nuclear waste for millennia continues.

In the United States, *high-level radioactive waste* from electricity generation and from US nuclear energy defense activities have been "permanently" stored at the Hanford site in Seattle, among other places. More than a third of the 177 storage tanks at Hanford have however leaked radioactive waste into the groundwater. Today, high-level waste is deposited at the 800-km^2 Savannah River site in South Carolina. Regrettably, several leaks have also occurred at this site (Deutch and Lester 2004). The waste will need to be removed at these sites, processed, and immobilized in an acceptable form for permanent disposal. Completing the cleanup of the Hanford nuclear reservation alone will cost an estimated $300 billion to $640 billion, according to a new Department of Energy report (GAO 2022). The United States is now also storing high-level waste at its Idaho National Laboratory, where 15 underground tanks can store around 13 million liters of high-level waste (Jacobson 2021). Nevada's Yucca Mountain site was to hold 70,000 t of radioactive waste, but is no longer under consideration, mostly due to political pressure and opposition by Nevadans.

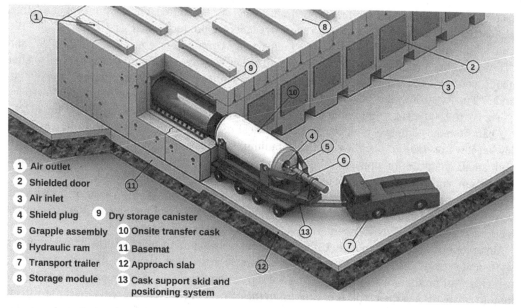

Figure 18.1 After spending around five years cooling down in water basins, high-level radioactive fuel assemblies are transferred to dry casks located at nuclear power plants. Areva TN/Orano is supplying the illustrated NUHOMS® dry-shielded canisters and horizontal storage modules for the Limerick Generating Station in Pennsylvania. Areva also has a long-term contract with FirstEnergy Nuclear Operating Co. to provide dry fuel-storage equipment and related services. Areva TN/Orano will also supply its NUHOMS dry fuel-storage systems, in use at more than 30 US plant sites, to the Davis Besse and Beaver Valley nuclear power plants. *Source:* Areva TN/Orano, CC BY.

According to the World Nuclear Association, in 2020, there was around 35 million m³ of radioactive waste globally. Of this amount, around 25,000 m³ (or 390,000 tons) was long-lived, solid radioactive waste. The amount of disposed (not only stored) waste roughly decreases with the increase in the waste level. While 85% of low-level radioactive waste was disposed of in 2020, only 20% of intermediate-level waste had been disposed of at that stage. Of the 390,000 tons of solid high-level radioactive waste, the number dropped to zero as there had been no functioning repository for long-term waste (WNA 2022), at least not until 2023 with the advent of the Onkalo site in Finland.

The Onkalo site in Finland is the world's first permanent (>100,000 years) underground repository for high-level nuclear waste. High-level radioactive waste will be buried in about 100 tunnels 430 m below ground. Onkalo relies on multiple barriers to prevent water from reaching the rods and carrying radionuclides to the surface. Onkalo is carved out of gneiss and granite, two hard, crystalline rocks that are nearly impervious to water. Once filled with nuclear waste (in about 100 years), the disposal tunnels will be backfilled with bentonite clay and sealed.

18.3 Radioactivity of Nuclear Waste

Ten years after use, the surface of a spent fuel assembly releases 10,000 Sv/h of radiation; in comparison, a dose of 5 Sv is lethal to humans if received all at once (Murray 1980). Box 18.1 provides further information about the biological effects of radiation. Figure 18.2 shows the radioactive decay of radioactive species. After between 1 and 1000 years, the total radioactivity declines by a factor of about 1000 and continues to decline sharply thereafter. At the time of discharge, radioactivity is largely contributed by short-lived fission products such as ^{95}Nb, ^{95}Zr, ^{103m}Rh, and ^{103}Ru. After 10 years, the important radioactivity contributors are ^{137}Cs, ^{137m}Ba, ^{90}Sr, and ^{90}Y. These same radionuclides remain important as key activity contributors after 100 years. After about 1000 years, the activity becomes dominated by α-emitting radionuclides such as ^{241}Am, ^{239}Pu, and ^{240}Pu. At 10,000 years, ^{239}Pu remains the important radioactivity contributor, followed by ^{240}Pu. At 100,000 years, ^{239}Pu remains the top activity contributor, followed by ^{99}Tc (Yim 2022).

Apart from the high-level radioactive waste, nuclear reactors also produce transuranic waste and low-level waste. *Transuranic* (heavier than uranium) *waste* is non-high-level waste contaminated with small quantities of alpha-emitting "transuranic" isotopes (such as plutonium, neptunium, americium, and curium) with half-lives of tens of thousands to millions of years.

Low-level waste is all radioactive waste that is not considered to be high-level or transuranic waste. Although high-level waste poses the greatest immediate and long-term threat, countries have been unable to effectively contain low-level and transuranic waste.

Many countries, although not the United States, reprocess used nuclear fuel. The process reduces waste and extracts 25–30% more energy than non-reprocessed fuel. However, the principal arguments against plutonium recycling are that the separation, stockpiling, transport, and use of plutonium create risks of theft. This is of concern in the context of efforts to prevent nuclear terrorism or proliferation. International safeguards and costly security measures for reprocessing facilities cannot fully compensate for the increased vulnerability resulting from separating weapons-grade materials.

Box 18.1 Biological Effects of Radiation

The adverse effects of nuclear radiation are not from chemistry (as when poisoned by arsenic, for example), but rather from emitted sub-atomic particles (alpha, beta, and γ rays, X-rays, and neutrons). Biological damage following radiation usually manifests itself as ionization, i.e., as radiation passes through matter, it produces ions by knocking electrons out of their orbits.

In Chapters 7–9, we saw that around 32 eV of energy is required to ionize a molecule. So, an alpha particle with a typical energy of 4 MeV would release around 10,000 ion pairs before stopping. Water in human cells could be converted to free radicals such as OH^-, O_2^-, H_2O_2, O_3 or H^+. Biological free radicals are highly unstable molecules that have electrons available to react with various organic substrates such as nucleic acids, proteins, lipids, and carbohydrates, which leads to various kinds of damage: DNA strand breaks, point mutations, chromosomal aberrations, and ultimately cell death.

The somatic effects of radiation may range from skin reddening when the surface of the body is irradiated, to shortening the life of an exposed person due to the general impairment of body functions, and the initiation of cancer in the form of solid tumors or as a blood disease such as leukemia. The genetic effects of radiation consist of DNA mutations, in which progeny are significantly different from their parents, usually in life-shortening ways (Murray 1980).

Alpha particles have low-penetrating power and usually only affect the skin. However, γ rays, X-rays, and neutrons can penetrate the whole body, including bone marrow, organs, and eye lenses. The thyroid gland is usually also affected by irradiation because of its affinity for the fission product iodine. The gastro-intestinal tract and lungs are also sensitive to radiation through eating or breathing.

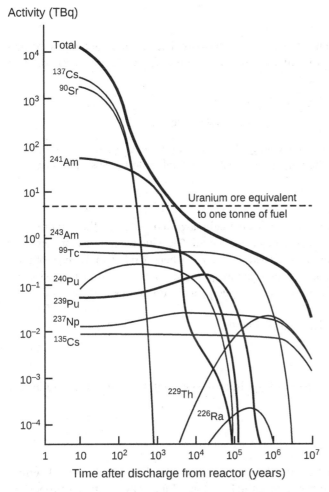

Activity (TBq)

Figure 18.2   The radioactive decay behavior of spent nuclear fuel initially enriched to 4.5%. *Source:* After Yim (2022), Bruno and Ewing (2006), Jaffe and Taylor (2018), and Deutch and Lester (2004).

Reimagine Our Future

What are the social, ecological, and spiritual implications of choosing nuclear power? How might our use of nuclear power affect nature and future generations?

Reimagine Our Future

Do you think that nuclear power will ever play a larger role in providing for our energy needs? When thinking about this, you should consider the immense cost of building nuclear reactors, as well as their associated safety and security issues, including the effective long-term disposal of radioactive waste.

18.4 Radiation Doses

The rate of radioactive emissions from a source is called *activity*. The SI unit of activity is the *becquerel* (Bq), which is defined as one radioactive decay per second (Jaffe and Taylor 2018). An older measure of activity is often used, the *curie*:

1 becquerel (Bq) = 1 count per second (cps) = $1\,s^{-1}$
1 curie (Ci) = 37,000,000,000 becquerel = 37 Gigabecquerels (GBq)

Absorbed dose (D) is a measure of radiation energy absorbed per unit mass of the absorbing material. The SI unit of absorbed dose is the *gray* (Gy), which is equivalent to the absorption of 1 J of radiation energy per kg of matter:

$$1\,Gy = 1\,J/kg$$

To account for the fact that some particles can do greater biological harm than others, radioactive emissions are also categorized in terms of their *equivalent dose*. The equivalent dose to an organ or human tissue is defined by weighting the absorbed dose, $D_{R,T}$ (where R refers to the type of radiation and T refers to the type of tissue), by a *radiation weighting factor*, w_R. Table 18.1 shows some common values of w_R. Table 18.1 also shows the average penetration depths of these radioactive particles in human tissue.

The *equivalent dose*, H_T, may be calculated as follows:

$$H_T = \sum_R w_R D_{R,T} \tag{18.2}$$

The SI unit of equivalent dose is the *sievert* (Sv), while an older unit, the *rem*, is also still in use:

1 sievert = 100 rem
1 gray (Gy) = 100 rad
1 rad = 10 milligray (mGy)

Finally, an *effective dose* (E) may be defined by discriminating the differential effect of certain tissues to radiation. For instance, the gastro-intestinal tract and lungs have a greater propensity to absorb radiation than do the thyroid gland or skin. A *tissue weighting factor*, w_T, is used to distinguish these differences, with some typical values shown in Table 18.2.

The effective dose (E) may be expressed as:

$$E = \sum_T w_T H_T \tag{18.3}$$

Table 18.1 Some radiation weighting factors.

Type of radiation	Radiation weighting factor (w_R)	Tissue penetration (mm)
Gamma particles and X-rays	1	40,500
Beta particles	1	180
Protons (>2 MeV)	2	3
Alpha particles, neutrons (1 MeV) and heavy nuclei	20	1

Source: Jaffe and Taylor (2018) / Cambridge University Press, Section 20.3.

Table 18.2 Some tissue weighting factors.

Organ or tissue	Tissue weighting factor (w_T)
Lung, stomach, colon, bone marrow	0.72
Gonads	0.08
Thyroid, esophagus, bladder, liver	0.16
Bone surface, skin, brain, salivary glands	0.04

Source: Jaffe and Taylor (2018) / Cambridge University Press, Table 20.2, page 383.

Example 18.2 Effective Dose of Radiation

An emergency worker at a nuclear power plant is exposed to 1-MeV-neutron emissions for 10 min. The neutron activity is estimated to be 40 GBq. (a) Calculate the equivalent dose that this person receives if they have a mass of 60 kg and if the radiation is absorbed by their whole body. (b) What is the effective dose on this person's lungs, which have a combined mass of 0.9 kg?

Solution strategy: (a) We will use Eq. (18.2) to calculate the whole-body (equivalent) dose that the colon receives. We can read off w_R from Table 18.1. (b) We can determine w_T from Table 18.2, which will enable us to calculate the effective dose with Eq. (18.3).

Analysis:

a) The total absorbed energy is $E = (1 \times 10^6 \text{ eV})(10 \text{ min})\left(\dfrac{60 \text{ s}}{\text{min}}\right)(1.6 \times 10^{-19} \text{ J/eV})(40 \times 10^9 \text{ s}^{-1}) =$ 3.84 J.

This amount of energy corresponds to the following absorbed dose: $D_R = (3.84 \text{ J})$ $\left(\dfrac{1 \text{ Gy}}{1 \text{ J/kg}}\right)\left(\dfrac{1}{60 \text{ kg}}\right) = 0.064 \text{ Gy}$

The equivalent dose is $H_T = \sum\limits_R w_R D_{R,T}$

From Table 18.1, for 1-MeV-neutron emissions: $w_R = 20$

$\therefore H_T = \sum\limits_R w_R D_{R,T} = (20)(0.064 \text{ Gy}) = 1.28 \text{ Sv}$

b) Effective dose in a person's lungs:

From Table 18.2, for lungs: $w_T = 0.72$

$D_{R,lungs} = \dfrac{1.28 \text{ Sv}}{0.9 \text{ kg}} = 1.42 \text{ Gy}$

The effective dose that the lungs will receive is $E = \sum\limits_T w_T H_T = w_T w_R D_{R,lungs} = (0.72)(20)$ $(1.42 \text{ Gy}) = 20.5 \text{ Sv}$.

Discussion:

A dose of 4–5 Sv received over a very short duration has a 50% probability of killing a human being in 30 days. Doses of more than 5 Sv are fatal, which means that the emergency worker sketched here has some serious problems.

Reimagine Our Future

Imagine discovering that a micro-nuclear power plant on a university campus was unnecessary, uninteresting, and unprofitable because improving the energy efficiency of buildings was a cheaper, easier, safer, and faster way to reach affordable energy security while protecting public health preserving nature.

18.5 Safety and Public Policy

Due to the interconnected social, technological, economic, environmental, and political aspects surrounding contemporary issues, engineers do not practice their profession in isolation. This is nowhere clearer than when contemplating nuclear power or when dealing with the aftermath of nuclear disasters. Politicians, environmentalists, engineers, and lay citizens might all have widely differing opinions about the best course of action regarding safety issues and public policy.

In 1986, a series of explosions occurred at the *Chernobyl power plant* in Ukraine. Pieces of the reactor were ejected high into the atmosphere. The loss of cooling water in the reactor allowed the fuel to heat to the point of core meltdown. A total of 134 workers and emergency responders were diagnosed with acute radiation syndrome and 31 died within weeks, while thousands of people suffered immediate injuries. It is estimated that the radioactive fallout led to an increase of 100,000 cancer deaths. Around 140,000 people had to be evacuated, which has resulted in ongoing psychosocial effects. An estimated 10 t of nuclear debris was emitted into the atmosphere, reaching countries in Europe, and the United Kingdom, and even the Arctic Circle. Apart from the adverse effects on agriculture, the long-term effects of this radiation on nature and humans are still being evaluated (Bennett et al. 2006; Cardis et al. 2006; Weiss 2018).

On 11 March 2011, a magnitude 9.0 earthquake occurred near *Fukushima*, Japan. The resulting tsunami damaged the reactor cooling system, leading to three meltdowns and hydrogen explosions. No deaths or radiation sickness have been directly linked to the accident (Yamashita et al. 2018). But thousands of people have suffered adverse health effects, including excess cancer risk and long-term psychosocial effects. The effects of the radiation fallout on the environment are still being assessed.

No private insurer has yet chosen to insure a nuclear plant against third-party off-site damages (Trebilcock and Winter 1997). Nor has any private bank chosen to fully finance a new nuclear reactor build. Nuclear power only advances where governments heavily subsidize nuclear operations.

The *US Price-Anderson Nuclear Industries Indemnity Act* limits the liability of nuclear plant owners if a radioactive release occurs to $450 million for individual plants and $13.5 billion across all plants. Total liability for reactors under 100 MW_e is limited to $560 million. For small reactors, the US federal government would pay for any damages to the public above the required liability insurance coverage, up to the $560 million limit (McIntosh 2022). The Act apparently understates the risks inherent in nuclear power, does not require nuclear reactors to carry adequate insurance, and would therefore result in taxpayers footing most of the bill for a catastrophic accident. Price-Anderson has been criticized by many groups due to the portion of the Act that indemnifies the US Department of Energy and private contractors from nuclear incidents, even in cases of gross negligence and willful misconduct (although criminal penalties would still apply).

Public skepticism about nuclear power is driven by fears about *safety* and *radioactive waste disposal*, as well as *nuclear proliferation* and the threat of a *nuclear war*. Although large modern nuclear power plants (in the order of 1 GW_e) have impressive safety features, the potential of a catastrophic event is ever-present. Besides the immediate devastation that would be caused by a severe nuclear event, such events are also accompanied by the entire nuclear enterprise being called into question.

In the United States, the Environmental Protection Agency (EPA) is responsible for determining the general standards for the protection of human health and the environment that a repository of radioactive waste must meet. The Nuclear Regulatory Commission (NRC) is responsible for establishing technical criteria for such a repository. The NRC also has licensing responsibility for the storage facility and is required to issue construction and operating licenses to the US Department of Energy. The jurisdictional boundaries of the NRC and EPA overlap, and these two agencies have frequently been in conflict (Anastas and Zimmerman 2021). Much of the controversy has focused not on what should be done, but rather on who should decide what to do: a political question (Deutch and Lester 2004).

Public skepticism about the immense *cost of nuclear power plants* is also well founded. A nuclear power plant replacing a 1 GW_e coal-fired steam power plant could save about 1.5 million tons of carbon being emitted annually. A natural gas-fired steam power plant replacing the same coal power plant would save about 700 million tons of carbon being emitted. The nuclear power plants would therefore double the savings (Feiveson et al. 2011). However, a modern gas-fired power plant has a capital cost about a quarter that of a nuclear plant of the same capacity. Therefore, for the same capital cost, natural gas could save more than double the amount of carbon emissions as nuclear power. It could also achieve this much quicker than is possible with nuclear expansion.

Considering the above, it is apparent that engineers cannot separate technological aspects surrounding nuclear power from the societal, economic, environmental, and political ones. These aspects are all interconnected. Nuclear power becomes even more difficult to talk about when considering that we are expecting *a thousand future generations* to clean up leakage from our nuclear waste, while we cannot now effectively contain radioactive waste for even 20 years.

However complex, it is crucial to take a long-term view of the challenge of effectively disposing of nuclear waste. We will be wise to perform data-driven analyses and projections, using the power of patterns to construct detailed scenarios and quantitative models of the far future (Ialenti 2020). We will need to believe the science (and our gut instinct) if we are to survive the so-called Anthropocene.

18.6 Outlook

Because nuclear fission is a low-carbon way of generating electricity or heat, there is considerable interest in expanding its role to help mitigate the dangers of climate change. However, fission technology has fundamental safety and security disadvantages compared to other low-carbon energy sources. Nuclear reactors and their associated facilities for fuel production and waste handling are vulnerable to catastrophic accidents and sabotage and can be misused to produce materials for nuclear weapons.

Nuclear waste disposal is a huge problem and will continue to be one several centuries hence. The adoption of breeding reactors will not eliminate the nuclear waste problem but will lessen it considerably. With breeding reactors, the bulk of a spent nuclear fuel rod, which comprises unused uranium, will not be wasted as in the case of light-water reactors, but will rather become a valuable feedstock. Further, transuranic elements other than plutonium, the chief source of radioactivity on scales greater than a thousand years, will fission in the presence of neutrons and could possibly be burned up altogether by reprocessing. The question, however, remains how to reprocess nuclear waste *safely and economically* without adding to the danger of nuclear proliferation. Fast (breeder) reactors will always be more dangerous than light-water reactors, so they will always be undesired in most countries.

The nuclear industry, policymakers, and regulators must address these shortcomings and challenges fully if the global use of nuclear power is to increase without posing unacceptable risks to public health, the environment, and international peace and security. We must also contend with the fact that only around 135 years' worth of terrestrial uranium remains. Although nuclear power will be an important role-player in the future energy mix, it is apparent that the much safer, technically feasible and economically viable solar energy systems will dominate our search for sustainable energy.

Problems

18.1 A 1-GW$_e$ nuclear reactor operates with a thermal efficiency of 33%. The plant suffers an accident that results in the loss of all cooling water. Calculate and plot the rate of heat decay from the nuclear core at the following times after the accident occurs: 1 h, 6 h, 24 h, 3 months, 1 year.

[Answers: 1 h: 38.8 MW; 6 h: 27.2 MW; 24 h: 20.6 MW; 3 months: 8.4 MW; 1 year: 6.3 MW.]

18.2 An emergency worker at a nuclear power plant inhales radioactive gas containing 1-MeV alpha particles for 2 min. The activity of the alpha emissions is estimated to be 29 GBq. (a) Calculate the *equivalent dose* that this person receives if they have a mass of 80 kg and if the radiation is absorbed by their whole body. (b) What is the effective dose on this person's lungs, which have a combined mass of 1.1 kg?

[Answers: (a) 0.14 Sv; (b) 1.8 Sv.]

18.3 "The international nuclear order is being shaken by developments including:

- a rise in international tensions – essentially a new Cold War – largely due to increasingly nationalistic policies and greater competition among the major powers, and increasing disregard for international rules and institutions;
- inaction on nuclear arms control, nuclear risk reduction and disarmament, exacerbated by nuclear "modernization" programs and new strategic doctrines that increase the salience of nuclear weapons and lower the threshold for their use;
- the prospect of greater spread of proliferation-sensitive nuclear technologies, and increasing interest in the Middle East and elsewhere in developing latent nuclear weapon capabilities;
- political discussion within a few key non-nuclear-weapon states about whether they should acquire their own nuclear deterrents, thereby abandoning the Non-Proliferation Treaty." (Carlson 2018)

It appears that society and politics have not kept pace with the evolution in the nuclear safeguards system of the International Atomic Energy Agency. Most states still think in adversarial terms, seeing safeguards as a challenge to national sovereignty. Reducing and eliminating the danger of nuclear weapons will clearly require a different outlook, especially with respect to the changing balance between sovereign rights and the wider international interest. Discuss how attitudes can be changed. Your 500-word essay must reflect rigorous research on nuclear safeguards and how to implement constructive dialogue and positive safeguard action.

[Answer: Open-ended question. Independent research required.]

18.4 It is proposed to use micro-nuclear reactors in a large steel-manufacturing plant to reduce the plant's greenhouse-gas emissions. (In the United States, the steel industry produces more than 3% of greenhouse gases in the industrial sector.) Nuclear micro-reactors are purported to be zero-emitting, failsafe, economically viable, and due to their small size and modularity, could be ideal sources of high-temperature heat (800 °C) to power an electrolysis plant. The hydrogen that will be produced in this manner will be used as fuel in steel furnaces, which will produce around 1.5 million tonnes of steel per year.

Produce a six-page narrative memo of no less than 800 words and with at least five diagrams to summarize the characteristics of micro-nuclear power plants, steel manufacturing, and high-temperature electrolysis. Your fact sheet should provide a summary of a techno-economic analysis to show the viability of nuclear-powered electrolysis. You should also report on the social (focusing on safety and security), environmental and political ramifications of producing hydrogen using a micro-nuclear reactor.

[Answer: Open-ended question. Independent research required.]

18.5 The world has never experienced war that threatens the active nuclear power infrastructure, and world leaders may be underestimating the peril conventional warfare presents to these powerful and perilous assets. Do research pertaining to the Russian bombardment in 2022 of Ukraine's Zaporizhzhia nuclear power plant (5.7 GW$_e$), the largest in Europe. Do independent research about war threats to nuclear power plants and the social, economic, environmental, and political ramifications that resultant radioactive emissions could have. How can such a threat be minimized?

[Answer: Open-ended question. Independent research required.]

18.6 Conduct an online survey among at least 100 people regarding nuclear safety and their willingness to have a 30 MW$_e$ micro-reactor placed in their city. The local nuclear regulating authority is proposing an exclusion area with a radius of 300 m around the plant. Report your survey questions, number of respondents and their details (gender, age, race), and summarize the salient findings of your research in a 300-word essay.

[Answer: Open-ended question. Independent research required.]

18.7 Examine several methods of nuclear waste disposal and list the advantages and disadvantages of each. Be sure to report on the disposal of radioactive waste containing low, medium, and high levels of radioactivity.

[Answer: Independent research required. Also see Section 18.2.]

18.8 High-level radioactive waste remains hazardous for upward of 100,000 years, or 3000 generations. Time capsules, key information files, sapphire disks, and monuments of terror are proposed to communicate our nuclear heritage to future generations. These techniques typically converge around ideas of either passing on as much technical information as possible or instigating fear and repulsion to keep future human beings from digging up the waste we have buried.

Write a 500-word essay on how we can inform future generations about the nuclear waste they inherited and the risks it poses so that they would be equipped with adequate knowledge to make decisions. You must, however, take a different approach to the fear-instigating and techno-driven approaches that have been proposed to date. Rather, your essay must investigate how we can *care* for our nuclear waste. How can we make a nuclear site worth remembering? Propose your own marker of nuclear waste.

You should refer to the views of a prominent scholar in the field of science and technology studies, Donna Haraway, who has suggested that the key to a responsible and responsive relation to future generations lies not in finding ways to make the future "safe," but rather in learning to be "truly present," which is to realize that the present is "thick" with memories of the past and beginnings of what might still be, and to build relations that inhabit and respond to this "thickness." A practice of care is suggested, which addresses the future, while engaging with complexities of the present (Haraway 2016).

Your essay should also refer to the views of Bruno Latour, who has provocatively demanded that "we must care for our technologies as we do for our children" (Latour 2011).

[Answer; Open-ended question. Independent research required.]

References

Anastas, P.T. and Zimmerman, J.B. (2021). Moving from protection to prosperity: evolving the US Environmental Protection Agency for the next 50 years. *Environmental Science and Technology* 55(5): 2779–2789.

Bennett, B., Repacholi, M. and Carr, Z. (2006). Health effects of the Chernobyl accident and special health care programmes. In: *Report of the UN Chernobyl Forum Expert Group "Health"*. Geneva: World Health Organization.

Bruno, J. and Ewing, R.C. (2006). Spent nuclear fuel. *Elements* 2 (6): 343–349.

Cardis, E., Howe, G., Ron, E., Bebeshko, V. et al. (2006). Cancer consequences of the Chernobyl accident: 20 years on. *Journal of Radiological Protection* 26 (2): 127.

Carlson, J. (2018). Future directions in IAEA safeguards. Discussion Paper: Managing the Atom Project, Belfer Center. https://www.belfercenter.org/publication/future-directions-iaea-safeguards

Chang, S.H., Kim, S.H. and Choi, J.Y. (2013). Design of integrated passive safety system (IPSS) for ultimate passive safety of nuclear power plants. *Nuclear Engineering and Design* 260: 104–120.

Clark, A. and Rowland, M. (2021). Safety and security defense-in-depth for nuclear power plants (No. SAND2021-14591). Albuquerque, NM: Sandia National Laboratories.

Deutch, J.M. and Lester, R.K. (2004). *Making Technology Work. Applications in Energy and the Environment*. Cambridge: Cambridge University Press.

EIA (2021). Spent nuclear fuel. US Energy Information Agency. https://www.eia.gov/nuclear/spent_fuel

Feiveson, H., Mian, Z., Ramana, M.V. and Von Hippel, F. (2011). Managing spent fuel from nuclear power reactors. Experience and lessons from around the world. International Panel on Fissile Materials, Princeton.

GAO (2022). Nuclear waste cleanup: Hanford site cleanup costs continue to rise, but opportunities exist to save tens of billions of dollars. US Government Accountability Office. https://www.gao.gov/assets/gao-22-105809.pdf

Haraway, D.J. (2016). *Staying with the Trouble: Making Kin in the Chthulucene*. Durham, North Carolina: Duke University Press.

Ialenti, V. (2020). *Deep Time Reckoning: How Future Thinking Can Help Earth Now*. Cambridge, MA: MIT Press.

IEA (2022). Nuclear power. International Energy Agency. https://www.iea.org/fuels-and-technologies/nuclear

Jacobson, V.L. (2021). *Idaho Nuclear Technology and Engineering Center (INTEC) Sodium Bearing Waste-Waste Incidental to Reprocessing Determination* (No. DOE/ID-10780). Idaho Falls, ID (United States): Idaho National Lab.(INL).

Jaffe, R.L. and Taylor, W.M. (2018). *The Physics of Energy*. Cambridge: Cambridge University Press.

Latour, B. (2011). Love your monsters. *Breakthrough Journal* 2 (11): 21–28.

McIntosh, S. (2022). Nuclear liability and post-Fukushima developments. In *Nuclear Law*, 249–269, International Atomic Energy Agency (ed.). The Hague: TMC Asser Press.

Murray, R.L. (1980). *Nuclear Energy. An Introduction to the Concepts, Systems, and Applications of Nuclear Processes*. Oxford, UK: Pergamon Press.

NRC (2020). *Reactor site criteria*. US Nuclear Regulatory Commission. https://www.nrc.gov/reading-rm/doc-collections/cfr/part100/full-text.html

Trebilcock, M. and Winter, R.A. (1997). The economics of nuclear accident law. *International Review of Law and Economics* 17 (2): 215–243.

Weiss, W. (2018). Chernobyl thyroid cancer: 30 years of follow-up overview. *Radiation Protection Dosimetry* 182 (1): 58–61.

Westinghouse (2022). AP1000® Plant. Passive safety systems and timeline for station blackout. Westinghouse Electric Company. https://www.westinghousenuclear.com/Portals/0/New%20Plants/AP1000/AP1000%20Station%20Blackout.pdf?timestamp=1404842353431

WNA (2022). Safety of nuclear power reactors. World Nuclear Association. https://world-nuclear.org/information-library/safety-and-security/safety-of-plants/safety-of-nuclear-power-reactors.aspx

Xing, J., Song, D. and Wu, Y. (2016). HPR1000: advanced pressurized water reactor with active and passive safety. *Engineering* 2 (1): 79–87.

Yamashita, S., Suzuki, S., Suzuki, S., Shimura, H. et al. (2018). Lessons from Fukushima: latest findings of thyroid cancer after the Fukushima nuclear power plant accident. *Thyroid* 28 (1): 11–22.

Yim, M.S. (2022). Characteristics of spent fuel and its storage and transportation. In: *Nuclear Waste Management*, 257–339, Man-Sung Yim (ed.). Dordrecht: Springer.

Mini Project 6

Micro-Nuclear Power Plant for a Large University Campus

A handful of companies and governments are developing small-scale (around 300 MW_{th}) modular nuclear reactors that proponents say are safer, cheaper, and more economically compatible with renewables than traditional large nuclear power plants (Serfontein 2014; Shropshire et al. 2021; Testoni et al. 2021). But critics contend that these small nuclear power plants do not address concerns about safety, security, and radioactive waste. Even smaller reactors (around 15–30 MW_{th}), called micro-nuclear reactors, are being promoted as a promising way to introduce reliable, safe, clean power to remote regions and to urban areas that currently rely on coal or other fossil fuels to generate power. These micro-reactors are small-scale, easily transportable, factory-fabricated nuclear reactors, similar in size to research and test reactors.

One such micro-reactor is being developed by the Ultra Safe Nuclear Corporation (USNC) in the United States. The reactor uses graphite-encapsulated nuclear fuel pebbles. The 1-mm diameter nuclear fuel pebbles contain uranium enriched to 19.75%. Thousands of these pebbles or kernels are contained in hundreds of graphite cartridges placed in the reactor core. This core is cooled by an inert gas, such as helium. The heated helium could be used to run a gas turbine to produce electricity. The hot helium could also be stored in a molten salt bath for use as process heat. The reactor is fueled once in its lifetime and the fuel core is rated at 20 years of full power, which is 15 MW_{th}.

It is proposed to place such a micro-reactor in the middle of a university campus with 55,000 students, which is in a city of 200,000 people.

Energy Systems: A Project-Based Approach to Sustainability Thinking for Energy Conversion Systems, First Edition. Leon Liebenberg.
© 2024 John Wiley & Sons, Inc. Published 2024 by John Wiley & Sons, Inc.
Companion website: www.wiley.com/go/liebenberg/energy_systems

Depiction of a micro-reactor placed underground. *Source:* ULTRA SAFE NUCLEAR / https://usnc.com/ / last accessed 9 May, 2023.

1) What rate of fission is required to produce 15 MW_{th}? Assume that a single-fission event releases 200 MeV of thermal energy, and remember that $1\,eV = 1.6 \times 10^{-19}\,J$.

2) How many tonnes of U_3O_8 are required to provide a thermal capacity of 15 MW? (1 t = 1000 kg)

3) How many years could the reactor operate at a capacity factor of 100% until all ^{235}U has been fissioned? What practical factors would not allow such complete depletion? What is the fuel burn-up in GW-day per tonne of the fuel?

4) How many tonnes of fission products would be produced assuming full depletion of ^{235}U? Estimate the mass of high-level waste that would be produced. What is the half-life of these wastes?

5) Reprocessing nuclear waste reduces the volume of waste by a factor of 4. Discuss reprocessing the nuclear waste of a micro-nuclear reactor in terms of technical feasibility and economic viability.

6) Discuss safety, security, and nuclear waste issues pertaining to the planned placement of such a 15 MW_{th} micro-nuclear reactor on a university campus. When discussing safety issues, also address the "passive safety measures" promoted by proponents of micro-nuclear reactors. Be sure to also address insurance issues. Who will insure such a campus-based micro-nuclear power plant and what will the typical insurance costs be?

7) Considering only the micro-nuclear reactor, what materials and how much of each material would be required for its operation over a 14-year period? In which cities or countries are those materials typically found? From your research, state the social and environmental concerns of extracting the materials from those cities. Is this type of materials extraction sustainable? Also explain any sustainability (and cost) issues surrounding the safe discarding of the spent fuel.

8) The micro-nuclear reactor is cooled by helium. Provide a detailed analysis (and show the full cycle on a computer-generated *p-h* diagram, for instance, by using EES software) and calculate the thermal efficiency of this Brayton cycle. State your assumptions and show all your calculations.

References

Serfontein, D.E. (2014). Nuclear power more profitable than coal if funded with low cost capital: A South-African case study (Paper HTR 2014-1-11183). TR-2014 Conference, Weihai, *China*, 28 October.

Shropshire, D.E., Black, G. and Araújo, K. (2021). Global market analysis of microreactors (No. INL/EXT-21-63214-Rev000). Idaho Falls, ID: Idaho National Lab.

Testoni, R., Bersano, A. and Segantin, S. (2021). Review of nuclear microreactors: status, potentialities and challenges. *Progress in Nuclear Energy* 138: 103822.

Turning freeways into power plants

In the eyes of Swedish architect and urban strategist, Måns Tham, the stretch of the I-10 known as the Santa Monica freeway is an achievement to be celebrated and revered. It is covered in so many solar modules that it looks like a massive snake. "The freeways themselves are majestic structures. Their size is both inspiring and intimidating. It is very, very hard to make a 10-lane freeway aboveground disappear," he says. "Better then to make it engaging and beautiful, and to add new positive functions."

In Tham's world, one of Los Angeles' famed highways (among many others) can be turned into a monumental force of public good, but only if it is covered in photovoltaic cells that power the very city the freeway bisects. The modules would shield cars from the weather, and algae ponds along the roadside would act as a giant carbon sink. Tham philosophizes that "by letting infrastructure be a visually powerful part of the city, inside and out, its citizens are allowed to understand and cherish the complexity of their daily urban life. The whole set-up could bring green-tech jobs for farming, harvesting and processing to the very neighborhoods that today are the most disadvantaged by their proximity to the freeway." *Source:* Courtesy of Måns Tham Architects.

Energy Systems: A Project-Based Approach to Sustainability Thinking for Energy Conversion Systems, First Edition. Leon Liebenberg.
© 2024 John Wiley & Sons, Inc. Published 2024 by John Wiley & Sons, Inc.
Companion website: www.wiley.com/go/liebenberg/energy_systems

Week 7 – Direct Energy Conversion

Instead of converting energy first to heat (e.g., a boiler) and then to work (e.g., a steam turbine connected to an electrical generator), it is possible to *directly* convert chemical, nuclear, solar, and thermal energies. Such direct energy conversion systems thus eliminate mechanical elements such as rotating machinery (e.g., turbines) or reciprocating machinery (e.g., piston pumps). Direct energy conversion systems may employ semiconductors (e.g., photovoltaic panels), thermionic devices (e.g. thermionic generators), or ionized gases (e.g., magnetohydrodynamic generators). Direct energy conversion systems can also directly convert chemical reactions to electricity (e.g., fuel cells and batteries). All these systems use an ionized working medium of low charge density, either as a solid (e.g., thermoelectrics), liquid (e.g., fuel cell), or gas (e.g., thermionics and magnetohydrodynamics).

Chapter 19 – Concepts of direct energy conversion: Preliminaries; general representation of direct energy conversion systems; quantum energy levels in atoms, molecules, and solids; conductors, insulators, and semiconductors; thermoelectricity and semiconductors; outlook

Chapter 20 – Solar electricity: Solar electricity (photovoltaics); the *p-n* junction diode; the *p-n* junction with an applied voltage; the *p-n* junction with impinging light; photovoltaic cells, modules, and arrays; types and costs of photovoltaic cells and systems; the sustainability of solar electricity; outlook

Chapter 21 – Fuel cells and electrolyzers: Background to fuel cells; fuel cell performance; potential difference (voltage) and current; hydrogen and other fuels; types of fuel cells; electrolyzers; summary of direct energy conversion systems; outlook for fuel cells and electrolyzers

Mini Project 6 – Photovoltaic car canopy

19

Concepts of Direct Energy Conversion

Direct energy conversion refers to the conversion of chemical, nuclear, solar, and thermal energy into electrical energy without the use of mechanical elements such as rotating or reciprocating machinery. In Figure 19.1, the direct energy conversion systems are colored black, while other energy conversion systems are light gray. (This figure has been extracted from Figure 1.1.)

Direct energy conversion devices use various types of primary energy sources, depending on the type of device and the properties of the source. Those energy sources (chemical, nuclear, solar, and thermal) provide energy to direct energy conversion devices such as photovoltaic cells, thermophotovoltaic cells, nuclear batteries, thermionic generators, thermoelectric generators, thermoelectric coolers, magnetohydrodynamic generators, electrolyzers and fuel cells, among many others. Electrochemical batteries will not be covered in this section of the book, but in Chapter 27 (Energy Storage).

Direct energy conversion systems usually have a low specific power (kW/kg) and high spatial power footprint (m^2/kW), and are mainly used for specialized applications such as powering the International Space Station (110 kW, continuous load), powering a remote weather station in Antarctica (5 W, continuous load), cooling computer servers using their waste heat, or in certain transportation applications (80–500 kW), like fuel cells in city buses, trains, and ferries. However, with the recent growth of photovoltaic systems and battery storage, direct energy conversion systems have finally gone mainstream.

Table 19.1 depicts the typical power ranges and power densities for direct energy conversion systems compared to conventional heat engines.

Reimagine Our Future

An economy is meant to benefit a country's people. What would a sustainable, fair, and resilient economy look like? What energy sources can dependably and benignly power it?

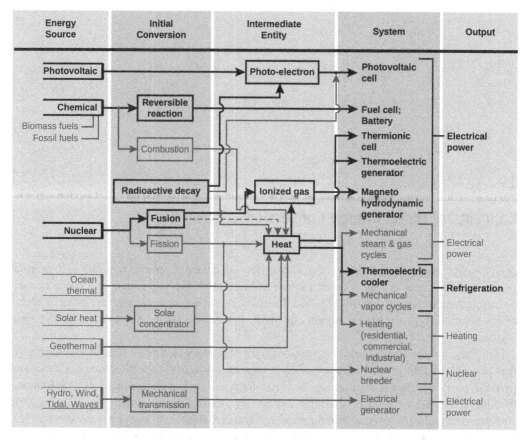

Figure 19.1 Some typical direct energy conversion systems.

Table 19.1 Performance characteristics of typical direct energy conversion devices compared with a few fossil-fuel energy conversion systems.

System	Energy source	Specific power (kW/kg)	Spatial power footprint (m^2/MW)	Practical best efficiency to date (%)
Fuel cell	Hydrogen	0.5–1.5	0.006	60
Photovoltaic power plant	Sun	0.1–0.3	65,000	<30
Nuclear battery	Uranium-235	0.08	—	29
Thermionic cell	Waste heat, concentrated solar	0.4	0.01	>20
Thermoelectric generator	Waste heat, concentrated solar	0.5	0.05	<20
Thermoelectric cooler	Waste heat, concentrated solar	0.02	20	<20
Magnetohydrodynamic generator	Heat	0.2–0.5	0.02	<60
Coal-fired steam power plant	Coal	0.5	1800	48
Integrated gas turbine combined cycle	Natural gas	10	770	60
Diesel engine	Diesel	1.0	3	65

The spatial power footprint depicts the land area (m^2) required to accommodate 1 MW of the power plant and includes the separation distance between modular units.

19.1 Preliminaries

The operation of direct energy conversion systems involves the transport of electrons over potential fields due to excitation by the source of energy. Devices for direct energy conversion may be categorized as shown in Table 19.2, which also compares the performance characteristics of a few direct energy conversion systems with conventional energy conversion systems. There are innumerable schemes and combinations of schemes that affect the transport of electrons. Only a few salient ones are discussed in this book.

It should be emphasized that direct energy conversion systems have almost no practical limit in their unit capacity. The efficiency of most direct energy conversion systems is also unaffected by their size. Direct energy conversion systems have no moving mechanical components, which means that they are silent, and mechanical wear is minimal. On the other hand, photovoltaic cells do not last forever as they suffer from radiation damage. Also, the electrodes of fuel cells become contaminated with prolonged use and the semiconductors in nuclear cells also suffer from radiation damage.

Table 19.2 Classification of major forms of direct energy conversion systems.

Category	Description	Typical applications
Direct electron transport	Energy is converted from a source by the movement of charge carriers, i.e., into electrical energy. Heat generation is incidental but is not needed for this operation. These devices are not heat engines, so they do not require low-temperature reservoirs	Fuel cells, nuclear batteries, photovoltaic cells, thermophotovoltaic cells
Energy conversion via heat	Heat is converted into electrical energy by thermoelectric and thermionic elements. To comply with the second law of thermodynamics, these cycles must include a low-temperature reservoir for the continuous conversion of heat from a high-temperature reservoir. These devices can produce power or operate in reverse as refrigeration devices	Thermionic generators, thermoelectric generators, thermoelectric coolers
Energy converted through the kinetic energy of a fluid	Electrical energy is produced from the interaction of the random motion of charged particles (e.g., an ionized gas) with a magnetic field	Magnetohydrodynamic generators
Fusion and plasma generators	When two light nuclear particles fuse, energy is released because the product nuclei are more stable than the original particles. Such *fusion* reactions can be initiated by bombarding target particles with charged particles using a particle accelerator, or by raising the temperature of a gas to a high enough level for fusion to take place	Fusion generators (see Section 17.10)

19.2 General Representation of Direct Energy Conversion Systems

The common feature of all direct energy conversion systems is excitation and the transport of charge carriers, such as electrons. The outputs of direct energy conversion systems occur at electrodes over a finite potential difference.

When electrons vacate their space, they essentially leave "holes" (i.e., deficiencies of electrons). The *holes* are considered positive charge carriers, while the *electrons* are negative charge carriers. Electrons and holes move in opposite directions. To move an electron over a potential difference of 1 V requires energy to the amount of $1 \text{ eV} = 1.6 \times 10^{-19} \text{ J}$.

Figure 19.2 shows the two categories of direct energy conversion systems: *cells* and *heat systems*. Figure 19.2a illustrates that the excitation (chemical, nuclear, solar, or kinetic energy) of charge carriers in cells gives rise to holes that migrate to the cathode, and electrons that migrate to the anode, thus generating a potential difference and flow of current. Energy conversion devices that operate on this principle include fuel cells, nuclear cells, photovoltaic ("solar") cells, thermophotovoltaic cells, and magnetohydrodynamic cells. As per international convention, *current (I)* flows

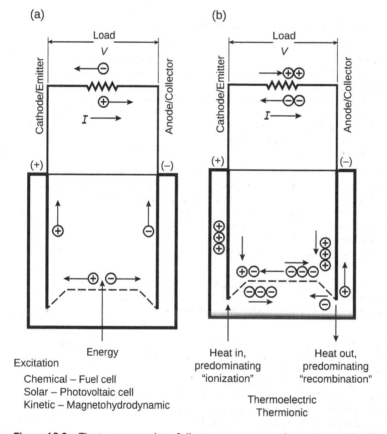

Figure 19.2 The two categories of direct energy conversion systems, (a) cells and (b) heat systems. All these systems function by using an ionized medium of low charge density. It may be a liquid (e.g., with fuel cells), gas (e.g., with thermionic and magnetohydrodynamic devices), or a solid (e.g., photovoltaic and thermophotovoltaic devices). Sørensen (2017) / with permission of Elsevier.

in the same direction as positive charge carriers (holes), from the cathode to the anode, with electrons flowing in the opposite direction.

Figure 19.2b illustrates the excitation of charge carriers in heat systems. Heat input results in ionization (or splitting into electron-hole pairs), which occurs at the hot junction and excites electrons to a higher potential. Net current is obtained in the presence of a cold junction where heat is discarded (due to the recombination of holes and electrons). Energy conversion devices that operate on this principle include thermoelectric and thermionic devices. These devices can produce electric power when they absorb heat, or they can operate as refrigeration devices if electric power is provided.

Here, the focus is on energy exchange due to the displacement of charge carriers over an electric potential. The states and processes can therefore be represented in *potential diagrams*. We will employ solid-state physics to determine how the large-scale properties of solid materials (such as photovoltaic modules) result from their atomic-scale properties. To do that, we will need to use *quantum mechanics*.

19.3 Quantum Energy Levels in Atoms, Molecules, and Solids

The fundamentals of solid-state physics pertain to regions measured in nanometers. Here, the laws of classical physics break down and we need to use quantum mechanics. Quantum mechanics is based on a few fundamental principles (Gamow 1960; Soo 1968; Murray 1980; Jaffe and Taylor 2018):

- *Principle 1: Energy is not indefinitely indivisible*: The smallest amount of energy that can be transferred is called a *quantum,* or a quantized packet of energy. The quantum of the electromagnetic field is the *photon.*
- *Principle 2: An electron has wave-like and particle-like qualities,* just like a photon. An electron (or a photon) wave may be regarded as an undulation of energy with its frequency (ν) proportional to the electron's (or the photon's) energy (E):

$$E = h\nu = E_f - E_i \tag{19.1}$$

where h is Planck's constant (6.63×10^{-34} Js) and with the frequency (ν) measured in hertz (Hz). The discrete packets of energy $h\nu$ are called quanta; E_i is the electron's energy in its initial state and E_f is its energy in its final state.

Consider the energy levels associated with an atom comprising a single proton and a single neutron (like hydrogen). The system can be described using quantum mechanics (called the *Bohr model* of an atom) to yield wave functions that allow only discrete energy levels E_n (Jaffe and Taylor 2018):

$$E_n = \frac{-m_e e^4}{8\varepsilon_0^2 h^2 n^2} = \frac{-2.18 \times 10^{-18} \text{ J}}{n^2} = \frac{-13.6 \text{ eV}}{n^2} \tag{19.2}$$

where m_e is the mass of an electron, e is the charge of an electron, ε_0 is the permittivity of free space, h is Planck's constant, and n is the principal quantum number describing the energy level of the orbiting electron ($n = 1, 2, 3...$) 1 eV = 1.6×10^{-19} J. According to convention, an electron's energy is taken to be negative when attracted to a nucleus (see Figure 16.1), thus the negative sign in Eq. (19.2).

From Eq. (19.2), the energy gaps between the levels are proportional to their separation:

$$E_g \cong 13.6 \left(\frac{1}{n_i^2} - \frac{1}{n_f^2} \right) \qquad [\text{eV}] \qquad (19.3)$$

where E_g is the gap energy required for an electron to move from an initial energy state at quantum number n_i to a final energy state at quantum number n_f. Therefore, a photon with energy E_g can cause an electron to jump from an energy state (n_i) to another (n_f) after absorption.

Frequency and wavelength are inversely correlated:

$$\nu = \frac{c}{\lambda} \qquad (19.4)$$

with c being the speed of light ($\sim$300,000 km/s or 3×10^8 m/s) and λ being the wavelength measured in meters. Figure 19.3 illustrates the relationship between energy, frequency, and wavelength.

- *Principle 3:* If an electron is given an amount of energy $h\nu$, then in order to escape from the material containing it, the electron must use up an amount $e\phi$ of this energy, where ϕ is the *work function* of the material in volts (V) and e is the electronic charge, measured in coulombs (C). The maximum energy that an electron can have after it is emitted from a surface is:

$$h\nu - e\phi \qquad (19.5)$$

- *Principle 4:* When radiation penetrates material, the wavelength of the radiation particle changes. For instance, when a photon collides with an electron, the energy of the photon is reduced by just the same amount of energy that the electron gains. And as energy levels change, so do the

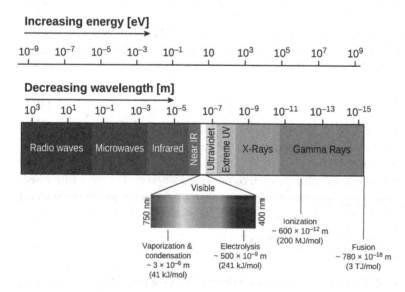

Figure 19.3 The figure illustrates the relationship of shorter wavelengths of electromagnetic radiation that give higher energy, according to $\nu = c/\lambda$ and $E = h\nu$, where h is Planck's constant, E is energy, ν is frequency, λ is wavelength, and c is the speed of light. For example, 1 MJ/mol corresponds to 10.4 eV/atom, which corresponds to a wavelength of 119 nm or a frequency of 2.5 THz. The figure shows the approximate energetic values of water that vaporizes or condenses, electrolyzes, ionizes, and fuses. *Source:* Jaffe and Taylor (2018) / Cambridge University Press, Figure 9.2, page 163.

wavelengths, in accordance with Eqn. (19.1). Although photons do not have mass, they may contain *momentum (p)*, which is different to momentum in classical (or Newtonian) mechanics:

$$p = \frac{h\nu}{c}$$ (19.6)

Since $c = \lambda\nu$, the wavelength (λ) associated with a photon is:

$$\lambda = \frac{h}{p}$$ (19.7)

Therefore, the energy of an electromagnetic wave of wavelength λ is carried by massless particles (photons) of momentum h/λ.

- *Principle 5:* According to *Heisenberg's uncertainly principle*, the precise motion and exact position of small particles like electrons cannot be simultaneously determined. If the position of an electron were measured accurately, then there must be doubt about its velocity, and vice versa. This happens because, for instance, if one attempts to use a photon to observe the position of an electron, as the photon strikes the electron, some of the photon's energy will go into the electron, causing it to recoil. Unless the photon strikes the electron, no evidence is provided by the observation instrument of the position of the object. If the photon is scattered by the electron, then the electron's position and momentum will change.
- *Principle 6:* According to *Pauli's exclusion principle*, two electrons cannot occupy the same quantum state. Electrons can therefore only occupy certain states in discrete energy levels. If a quantum state is occupied, the next electron must go to an empty higher energy state, filling up the empty states from the lowest energy to the higher energy. The quantum levels referred to in Pauli's principle correspond roughly to electron orbits around the nucleus.

Example 19.1   Quantum Leap of an Electron
Calculate how much energy is emitted if an electron falls from the −3.4 eV quantum level to the −13.6 eV quantum level.

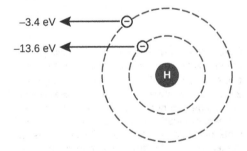

Analysis: Using Eq. (19.1), $E = h\nu = E_f - E_i = -13.6$ eV $- (-3.4$ eV$) = -10.2$ eV

Discussion: 10.2 eV is emitted when an electron falls from the −3.4 eV quantum level to the −13.6 eV quantum level. Using Eq. (19.1), this emission corresponds to light with a frequency of

$$\nu = \frac{E}{h} = \frac{10.2\,\text{eV}}{4.13 \times 10^{-15}\text{eV s}} = 2.4 \times 10^{15}\,\text{Hz}$$

Since $c = \lambda \nu$, the wavelength (λ) associated with this emission is:

$$\lambda = \frac{c}{\nu} = \frac{3 \times 10^8 \text{ m/s}}{2.4 \times 10^{15} \text{ Hz}} = 125 \text{ nm}$$

An electron in a hydrogen atom which falls from the -3.4 eV quantum level to the -13.6 eV quantum emits electromagnetic waves in the ultraviolet range. Conversely, if an electron is to move from the tightly-bound -13.6 eV quantum level to the -3.4 eV quantum level, ultraviolet light with wavelength of 125 nm is required.

19.3.1 Fermi Level

At 0 K (or the *ground state*), the energy states are packed so that not all electrons can occupy the lowest energy level, in accordance with Pauli's exclusion principle. All energy levels are filled to the maximum, known as the Fermi energy $\left(E_F^0\right)$, illustrated in Figure 19.4. The *Fermi energy* (or *Fermi level*) may be defined as the energy at which the probability of a state being filled is 0.5 (or 50%). In layman's terms, the Fermi level is the energy of the most energetic electrons in the material.

At 0 K, there will be a sharp distinction between filled and empty levels, which occur at the energy of the Fermi level. As the temperature increases from 0 K, some of the electrons that occupy states near the Fermi level will have absorbed sufficient thermal energy to move to higher, unoccupied energy states (Angrist 1982).

In *atoms*, electrons will fill discrete energy levels. In *molecules*, due to the motion of nuclei, vibrations, and rotations, electron excitations are of the order of 1 eV. In *solids*, due to motion of nuclei spread across a crystalline lattice, energy levels of individual atoms spread out into bands, and the gaps between energy levels remain as gaps between the bands. Electrons cannot then be thought of as being attached to individual atoms. Rather, they wander throughout the whole crystal structure. Figure 19.5 illustrates that the allowed energy levels for electrons in a crystal lattice consist of a series of bands with forbidden gaps (or energy gaps) in between.

The band that contains electrons in their highest energy state is called the *valence band*. The band which contains electrons in their lowest energy state is called the *conduction band*. The energy that must be absorbed for an electron to jump from the valence band to the conduction band is termed the *gap energy* (E_g).

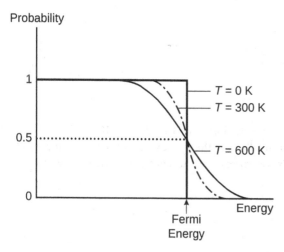

Figure 19.4 Fermi–Dirac probability distribution of energy for various temperatures, where the energy is given by $E = \frac{3}{2}k_B T$ with k_B being the Boltzmann constant. F is known as the Maxwell–Boltzmann distribution of kinetic energies used in elementary kinetic theory to describe the state of molecules in a gas. The plot shows that it is no longer an absolute certainty that an electron will be found at states below the Fermi level. *Source:* After Jaffe and Taylor (2018), and Angrist (1982).

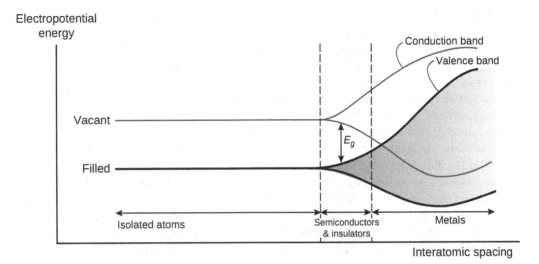

Figure 19.5 Quantum energy levels in atoms, molecules, and solids. Atoms are in *discrete* energy levels. Metals feature vibrating molecules where the energy levels of individual atoms spread out in bands, and the gaps between energy levels remain as gaps between the bands. Conductors have energy gaps in the order of 1 eV, while insulators have gaps closer to 10 eV, and semiconductors feature between those energy levels. *Source:* Sørensen (2017) / with permission of Elsevier.

19.4 Conductors, Insulators, and Semiconductors

Thermal and electrical conductivity depend on the location of the Fermi level (E_F) relative to the band gap and on the size of the band gap compared with $k_B T$. Figure 19.6 shows the relation between the energy level of electrons in *conductors* (or metals), *insulators*, and *semi-conductors*:

- *Conductor:* The uppermost occupied levels (valence band) are very close to some unoccupied energy levels in the conduction band. The conduction band is partly filled, which means that an applied voltage (order of 1 eV) can deliver energy to an electron in the conduction band. Such

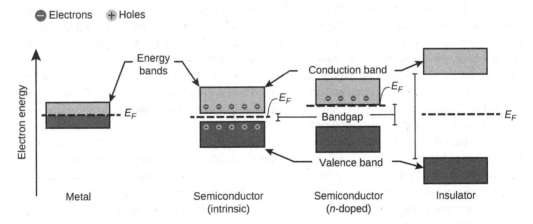

Figure 19.6 Fermi energy level (E_F) in conductors, insulators, and semi-conductors. *Source:* Reproduced from Jean et al. (2015) with permission from the Royal Society of Chemistry.

an electron will move to an unoccupied state at a slightly higher energy level within the same band; it will therefore conduct electric current.

- *Insulators and intrinsic semiconductors:* Electrons fill up the lowest (or valence) bands exactly, leaving the next allowed bands empty. The highest occupied band corresponds to the ground state of the outermost or *valence electrons*. The energy gap between the last filled and first empty or conduction band is large (much greater than $k_B T$), with the Fermi level situated approximately halfway between the two bands. Under ordinary circumstances, a valence electron cannot accept energy from an applied field as there are no empty allowed states accessible to it. An insulator such as quartz (or an intrinsic semiconductor) requires a large amount of energy (on the order of 10 eV, in the ultraviolet range of electromagnetic waves) to raise an electron into a band where excited states lie. An intrinsic semiconductor such as germanium, however, requires only 0.76 eV to excite its electrons into the conduction band.
- *(Extrinsic) Semiconductor:* The presence of impurities in a pure intrinsic semiconductor changes the periodic nature of the crystal structure. The impurities give rise to a few energy levels in the forbidden band (also called the "energy gap"). If the energy gap between these and the first empty (conduction) band is small, meaning that the Fermi level is close to both bands, excitation of electrons can give rise to conductivity, leaving behind holes in the valence band in the process. Visible light of high frequency (with $E = h\nu$) or matter at high temperature (with $E = k_B T$), for instance, may excite electrons across the energy gap, i.e., when $E > E_g$. These kinds of semiconductors, which contain impurities, are also called extrinsic semiconductors.
- The energy gap (E_g) is about 1.12 eV at room temperature (300 K) for silicon. This gap energy corresponds to the energy of photons with wavelengths shorter than 1130 nm.

The bandgap energy (E_g), measured in eV, sets up a potential or voltage (V_{out}), measured in V, as a function of an electron's charge (e), measured in C:

$$V_{out} = \frac{E_g}{e} \tag{19.8}$$

The specific power output, (P_{out}), measured in W/m^2, may be calculated by multiplying the voltage (V_{out}) with the current density (J_{out}), measured in A/m^2:

$$P_{out} = V_{out} J_{out} \qquad [\text{W/m}^2] \tag{19.9}$$

Figure 19.6 shows that a metal (or conductor) has a band gap energy of essentially zero. From Eqs. (19.8) and (19.9) that means that conductors could have large current densities but have no potential and thus produce negligible power. Conductors will therefore not make good photovoltaic energy converters. Also, insulators have very large band gap energy and do not conduct current, although they accommodate large potentials. So, insulators will also not make good photovoltaic energy converters. We need semiconductors like silicon, which has a gap energy around 1.2 eV, which just happens to correspond with the energy in the sun's electromagnetic spectrum with wavelengths between 0.5 and 1 μm (see Figures 5.5 and 19.3).

The conductivity of the semiconductor at room temperature is low. At 300 K, silicon has about 10^{10} intrinsic charge carriers per cm^3, while a metal has about 10^{22} charge carriers per cm^3. To improve the conductivity, one can introduce small amounts of impurities into the host material.

The impurity band may serve as either a donor (n-type semiconductor) or an acceptor (p-type semiconductor) for electrons. Consider a silicon (Si) atom that has four valence electrons shared with neighboring atoms (Figure 19.7a). By substituting (or "doping") the silicon, which is in column-IV of the periodic table of elements, with either column-III materials (boron, aluminum,

(a)

(b)

(c)

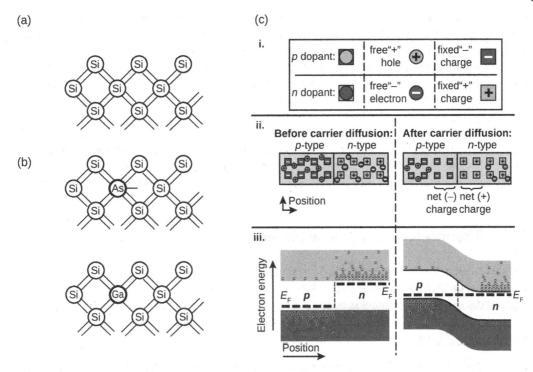

Figure 19.7 (a) Pure silicon and (b) silicon doped with arsenic and gallium to provide a *p-n* junction, showing (c) the physical structure and electric properties of a *p-n* junction diode before and after diffusion of charge carriers across the junction interface (also called the "depletion zone"). *Source:* Reproduced from Jean et al. (2015) with permission from the Royal Society of Chemistry.

gallium, or indium) or column-V materials (phosphorus, arsenic, or antimony), one can increase and precisely control the number of conduction-band electrons or valence-band holes (Figure 19.7b).

A column-V dopant, such as arsenic (As), completes the covalent bond and leaves an additional loosely bound electron that can be transferred to the conduction band by an energy of only about 40–50 meV. This is the *ionization energy* or *donor activation energy* and is much less than that required to break a valence bond. Figure 19.7b shows that this donor activation energy in the forbidden band is located very close to the conduction band. Because conduction occurs by means of electrons that have a negative charge, silicon containing this type of impurity is called *n-type*. In general, for a *donor* impurity D, we have (Soo 1968):

$$D \rightleftharpoons D^+ + e^- \tag{19.10}$$

where D^+ is the ionized donor (with a positive charge) and e^- is an electron. It must be noted that donors provide conduction electrons without the simultaneous creation of a hole in the valence band. There is no vacancy in the bond structure into which other bond electrons can slide. The positive ion is therefore a fixed charge, which cannot help to carry current; it is not a hole (Angrist 1982).

Column-III dopants such as gallium (Ga) leave the covalent bond deficient of one electron (i.e., with a hole). An electron from the valence band can transfer to the empty site and satisfy the bond requirement (Figure 19.7c). The holes effectively move as the transferred electrons leave behind a hole. The amount of energy required to place the hole in the valence band requires only 45–160 meV and is called the *acceptor activation energy*. The hole moves in a direction opposite to that of the electron. The energy level of these vacancies is close to the top of the valence band. Elements substituted for parent materials that are deficient in electrons are called *acceptor* impurities, A. The resulting material is called *p-type* because conduction is by holes. These acceptors (A^-) therefore create holes ($\oplus$) in the valence band without freeing electrons in the conduction band (Soo 1968):

$$A \rightleftharpoons A^- + \oplus \tag{19.11}$$

Box 19.1 outlines a few direct energy conversion technologies based on the aforementioned fundamentals. Chapter 17 briefly discusses fusion, while Chapter 20 shows how fuel cells employ direct energy conversion principles. Chapter 21 does the same for photovoltaics ("solar cells").

19.5 Thermoelectricity and Semiconductors

The thermoelectric effect is the direct conversion of temperature difference to electric voltage and vice versa via a thermocouple. Seebeck, Peltier, and Thomson (Lord Kelvin) made the three foundational discoveries of thermoelectricity.

19.5.1 Seebeck Effect

In 1821, Seebeck reported his observation that a magnetic needle is deflected when held near an electric circuit made of two conductors of dissimilar materials. The electromotive force set up in the junction was also a function of temperature difference over the junction. The electrons at the hot end of the metal or semiconductor will have higher kinetic energy than those at the cold end. This will result in thermal conductance with electrons diffusing to the cold end.

Figure 19.8a shows a thermocouple comprising metals A and B connected to each other, and with voltage measured across their open ends. The potential difference measured between the metals is the Seebeck voltage, $V_{Seebeck}$:

$$V_{Seebeck} = \int_{T_C}^{T_H} (\alpha_A - \alpha_B)\, dT \tag{19.12}$$

with α being the Seebeck coefficient (V/°C) for the type of junction material.

19.5.2 Peltier Effect

In 1834, Peltier discovered that the passage of electrical current through a junction formed by two different conductors (called a thermocouple) caused the absorption or liberation of heat. For instance, water could be frozen when placed on a bismuth-antimony junction by the passage of an electric current; and by reversing the current, the ice could be melted.

Figure 19.8b shows a junction of two dissimilar materials A and B, through which an electric current (I) flows. The junction is held at a uniform temperature ($T_0 + \Delta T$). The electric current sets up heat exchange with the environment, but the heat exchange is greater or less than the Joule heating ($I^2 R$) experienced by an ordinary conductor, with R being the electrical resistance. The difference between the heat generated or absorbed in this set-up and the Joule heating, which would be expected in a homogenous conductor, depends on the magnitude and direction of current, on the temperature, and on materials A and B. This phenomenon is called the Peltier effect. If $\dot{Q}$ is the rate

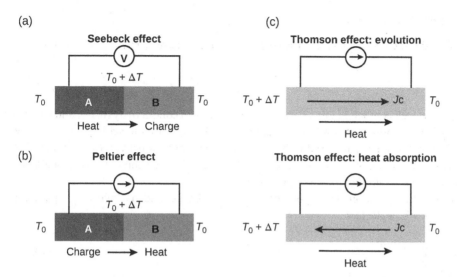

Figure 19.8 Illustration of the (a) Seebeck effect, (b) Peltier effect, and (c) Thomson effect. *Source:* Sears et al. (1982) / Addison-Wesley Publishing Company.

at which energy must be removed from the junction to maintain its constant temperature, the first law of thermodynamics gives:

$$\dot{Q} = I^2R + I(\Pi_A - \Pi_B) \tag{19.13}$$

where Π_A and Π_B are the Peltier coefficients (W/A) for the two materials. Using Seebeck's relation in Eq. (19.12) and remembering that $P = VI$, Eq. (19.13) becomes (Angrist 1982):

$$\dot{Q} = \underbrace{I^2R}_{\text{Joule heat}} + \underbrace{IT(\alpha_A - \alpha_B)}_{\text{Peltier heat}} \tag{19.14}$$

The power $\dot{Q}$ represents rate of heat absorbed or heat liberated, depending on the relative magnitude of the two terms on the right-hand side of Eq. (19.14). Since Peltier heat is reversible, the second term on the right can be added to or subtracted from the first, simply by reversing the direction of the current. It is therefore possible to devise a Peltier cooler or a Peltier electrical generator.

19.5.3 Thomson (or Kelvin) effect

Lord Kelvin (William Thomson) realized that a relation should exist between the Seebeck and Peltier effects. He discovered a third thermoelectric effect, namely that a conductor carrying an electric current along a temperature gradient should experience a heating or cooling in addition to, and independent of, the irreversible Joule heating, depending on the direction of the current. Therefore, the heat developed will be greater or less than I^2R, the difference depending on the magnitude and direction of current, the temperature, and the material (Figure 19.8c).

The rate of heat transfer through a conductor or semiconductor with the Thomson effect is (Angrist 1982):

$$\dot{Q} = \underbrace{I^2R}_{\text{Joule heat}} + \underbrace{I\left(T\frac{d\alpha}{dt}\Delta T\right)}_{\text{Thomson heat}} \tag{19.15}$$

19.5.4 Figure of Merit, Z

When comparing thermoelectric devices, it is common to use a *figure of merit*, Z, defined as:

$$Z = \frac{\alpha^2 \sigma}{k} \tag{19.16}$$

with α being the Seebeck coefficient (V/ ° C), σ being the electrical conductivity (S/m), and k being the material's thermal conductivity (W/mK). Equation (19.16) shows that to achieve a satisfactory value of Z, it is essential for the material to have a high electrical conductivity (to minimize I^2R losses) and a low thermal conductivity (to reduce heat transfer through these devices). Semiconductors have the best combination of these properties.

For a semiconductor junction, the figure of merit may be defined as follows if operating temperature effects are neglected:

$$Z = \frac{(\alpha_1 + \alpha_2)^2}{\left[\left(\frac{k_1}{\sigma_1}\right)^{1/2} + \left(\frac{k_2}{\sigma_2}\right)^{1/2}\right]^2} \tag{19.17}$$

where the subscripts 1 and 2 designate the properties of the two different materials that form the junction. With semiconductors, the p-type and n-type elements usually feature the same basic material (like silicon) so that $k_1 \cong k_2$. When doping the basic material (e.g., silicon) with impurities (such as gallium and arsenide), $\alpha_1 \cong \alpha_2$ and $\sigma_1 \cong \sigma_2$. The figure of merit for a semiconductor junction then simplifies to Eq. (19.16).

If temperature effects are considered, then the figure of merit becomes:

$$M = Z\left(\frac{T_h + T_c}{2}\right) = Z\overline{T} = \left(\frac{\alpha^2 \sigma}{k}\right)\overline{T} \tag{19.18}$$

where T_h and T_c are, respectively, the hot-side and cold-side temperatures of the thermoelectric device and $\overline{T}$ is the average of these two temperatures. The maximum energy conversion efficiency of a thermoelectric generator is (Snyder and Toberer 2008):

$$\eta_{te} = \eta_{Carnot} \frac{\left(1 + Z\overline{T}\right)^{1/2} - 1}{\left(1 + Z\overline{T}\right)^{1/2} + \frac{T_c}{T_h}} \tag{19.19}$$

where η_{Carnot} is the Carnot efficiency (also see Eq. (3.32)): $\eta_{Carnot} = 1 - (T_c/T_h)$, and $Z = (Z_{hot} + Z_{cold})/2$.

The maximum coefficient of performance of a thermoelectric cooler is approximated by:

$$COP_{max} = \frac{T_c}{T_h - T_c} \frac{\left(1 + Z\overline{T}\right)^{1/2} - \frac{T_h}{T_c}}{\left(1 + Z\overline{T}\right)^{1/2} + 1} \tag{19.20}$$

It follows from Eqs. (19.19) and (19.20) that maximum energy conversion efficiencies (or maximum coefficients of performance) are obtained when $Z\overline{T}$ is a maximum.

Some typical values of $Z\overline{T}$ at their maximal operating temperatures are: 1.3 for $Bi_2Se_{0.5}Te_{2.5}$, 1.4 for $(BiSb)_2Te_3$, 0.9 for BiCuSeO, and 2.6 for a single crystal of SnSe. (Bi is bismuth; Sb is antimony; Te is telluride; Se is selenium; Cu is copper; and O is oxide.)

19.5.5 Output Power of a Thermoelectric Generator

A thermoelectric generator has internal electrical resistance (R), and when connected to a load, also experiences an external resistance (R_e). If properly balanced, $R = R_e$ and the maximum output power of the thermoelectric can be expressed by:

$$\dot{W} = I^2 R_e \tag{19.21}$$

with $I = \dfrac{\alpha\,\Delta T}{(R + R_e)} = \dfrac{\alpha\,\Delta T}{2R}$, so that

$$\dot{W} = \frac{\alpha^2\,(\Delta T)^2}{(2R)^2}R = \frac{\alpha^2(\Delta T)^2}{4R} \tag{19.22}$$

with $R = \dfrac{\rho L}{A}$ and where ρ is the electrical resistivity (measured in $\Omega^{-1}m^{-1}$), with $\rho = 1/\sigma$ and with σ being the electrical conductivity ($\Omega\ m$), A is the cross-sectional area of the conducting element (m^2), L is the length of the element (m), and ΔT is the temperature gradient across the element (K). Equation (19.22) may then be rewritten as follows:

$$\dot{W} = \frac{\alpha^2(\Delta T)^2 A}{4\rho L} \tag{19.23}$$

19.5.6 Semiconductors and Thermoelectricity

The Seebeck effect (Figure 19.8a) relies on temperature difference between the ends of a conductor or a semiconductor to bring about a flow of electrons. With semiconductors, the Seebeck effect occurs because energy is absorbed or liberated at semiconductor junctions as electrons or holes are forced to change their energy levels because of the continuity of the Fermi level (Angrist 1982).

Figure 19.9a shows how the Seebeck effect can be exploited with a semiconductor for generating electrical power, and Figure 19.9b shows a Seebeck circuit operating as a cooler.

Figure 19.10 shows a schematic of a Peltier element. The thermoelectric legs are thermally in parallel and electrically in series.

Figure 19.11 illustrates a p-type semiconductor connected between two metal contacts, with a DC battery placed in the circuit to provide a potential across the ends. The Peltier coefficient (Π) gives the magnitude of the heating or cooling that occurs at a junction of two different materials over and above Joule heating.

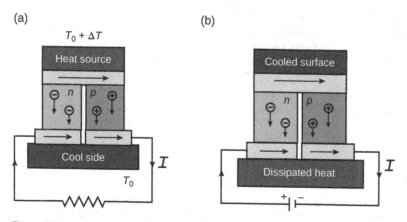

Figure 19.9 (a) A Seebeck thermoelectric generator, and (b) a Seebeck thermoelectric cooler.

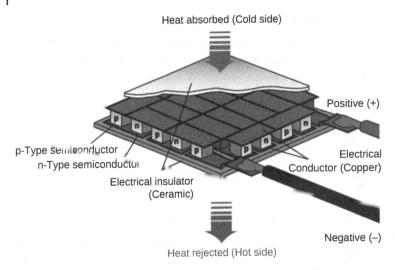

Figure 19.10 Schematic of a Peltier element comprising interconnected *p*-type and *n*-type semiconductors. *Source:* Reproduced from Pourkiaei et al. (2019) with permission from Elsevier.

The Fermi level in the metal is continuous with the Fermi level in the semiconductor at equilibrium conditions. The Fermi level in the metal lies in the center of the electron energy distribution of the partly filled band. A hole current flows from left to right across the junction at the Fermi energy level of the metal. For it to cross to the left-hand junction, though, it must have at least the Fermi energy of the semiconductor (E_F) more than its average energy in the metal. Only a fraction of the total number of holes in the metal have enough energy to cross the contact. Holes that can cross the contact carry an average energy of $2kT$ more than the Fermi energy (Angrist 1982). The energy required by the holes to enter the semiconductor is absorbed in the neighborhood of the contact, thus cooling it in the process. To satisfy the conservation of energy, an equal amount of heating should occur at the other metal contact.

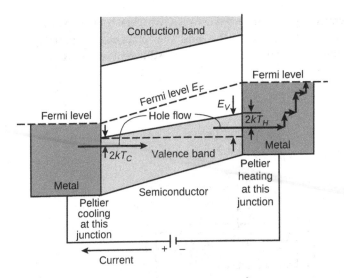

Figure 19.11 Potential energy diagram for a circuit containing a *p*-type semiconductor and two metal contacts. The energy of the charge carriers in the metal is the Fermi energy, but in the semiconductor, they are at Fermi energy plus $2kT$. *Source:* After Sørensen (2017), Sears et al. (1982), and Soo (1968).

Example 19.2   Figure of Merit, Power Output, and Energy Conversion Efficiency for a Semiconductor Junction

Hybrid silicon-germanium (SiGe) material is used for both the n-side and p-side materials of a thermoelectric generator. The cold side of the generator is at 20 °C and the hot side is at 900 °C. The material has the following properties: $\sigma = 484$ S/cm, $\alpha = 250$ μV/K, $k = 2.4$ W m^{-1}/K, $\rho = 2 \times 10^{-5}$ Ω m. Calculate (a) the figure of merit for this semiconductor, (b) its maximum energy conversion efficiency, and (c) the maximum power output of an element with a length of 15 mm and cross-sectional area of 180 mm^2.

Required: (a) $Z\overline{T}$, and (b) η_{te}

Solution strategy: We will use (a) Eq. (19.18) to determine $Z\overline{T}$, (b) Eq. (19.19) to determine η_{te}, and (c) Eq. (19.23) to determine $\dot{W}$.

Analysis:

a) $Z\overline{T}$:

$$Z\overline{T} = \left(\frac{\alpha^2 \sigma}{k}\right)\overline{T}$$

with $\overline{T} = \dfrac{(20 + 273)\,\text{K} + (900 + 273)\,\text{K}}{2} = 586.5\,\text{K}$, since K = °C + 273

$$\sigma = \left(\frac{484\,\text{S}}{\text{cm}}\right)\left(\frac{100\,\text{cm}}{\text{m}}\right) = 48.4 \times 10^3\,\text{S/m}$$

Therefore, $Z\overline{T} = \left(\dfrac{\alpha^2 \sigma}{k}\right)\overline{T} = \left(\dfrac{(250 \times 10^{-6}\,\text{V/K})^2 (48.4 \times 10^3\,\text{S/m})}{2.4\,\text{W m}^{-1}/\text{K}}\right)(586.5\,\text{K}) = 0.74.$

b) η_{te}:

$$\eta_{te} = (1 - (T_c/T_h))\frac{(1 + Z\overline{T})^{1/2} - 1}{(1 + Z\overline{T})^{1/2} + \dfrac{T_c}{T_h}} = \left(1 - \frac{293\,\text{K}}{1173\,\text{K}}\right)\left(\frac{(1 + 0.74)^{1/2} - 1}{(1 + 0.74)^{1/2} + \left(\dfrac{293\,\text{K}}{1173\,\text{K}}\right)}\right)$$

$\eta_{te} = 0.065$ (6.5%)

c) $\dot{W}$:

$$\dot{W} = \frac{\alpha^2 (\Delta T)^2 A}{4\rho L} = \frac{(250 \times 10^{-6}\,\text{V/K})^2 [(900 + 273)\,\text{K} - (20 + 273)\,\text{K}]^2 (180\,\text{mm}^2)\left(\dfrac{\text{m}^2}{10^6\,\text{mm}^2}\right)}{4(2 \times 10^{-5}\,\Omega\,\text{m})(0.015\,\text{m})}$$

$\dot{W} = 7.3\,\text{W}$

Discussion: Successful, high-temperature (>900 K) thermoelectric generators typically use silicon-germanium alloys for both n- and p-type legs. The $Z\overline{T}$ of these materials is low because of the relatively high lattice thermal conductivity of the semiconductor's diamond structure. Note that the power density for this example semiconductor is high: $\dfrac{\dot{W}}{A} = \dfrac{7.3\,\text{W}}{(180 \times 10^{-6}\,\text{m}^2)} \cong 41\,\text{kW/m}^2$ and the energy conversion efficiency is about 6.5%. The state-of-the-art in semiconductor technology in 2023 provided conversion efficiencies of up to 8% with power densities in the range of $30 - 80$ kW/m^2.

Box 19.1   Thermionics and Magnetohydrodynamics

Thermionic energy converters (TECs) are heat engines that convert high-temperature heat directly into electricity by driving electrons across a vacuum gap, allowing for high efficiency. Operating at high temperatures allows TECs to accept heat directly from a variety of sources such as hydrocarbon combustion, concentrated sunlight, or nuclear generation processes. In many cases, TECs reject unused heat to the environment without large heat exchangers. Further, the lack of moving parts can give TECs inherently long lifetimes with little associated mainte-nance due to lack of energy dissipation mechanism such as friction. TECs have relatively small sizes that can provide high specific power outputs of, say, 100 W/cm². This technology is, how-ever, still developing, and specific power outputs in the milliwatt per square centimeter range are more common for commercial systems.

Figure 19.12 illustrates the operating principle of a thermionic direct energy conversion sys-tem. The *emitter* (or cathode) is heated by a heat source and then emits electrons, which have sufficient energy to cross the space to the *collector* (or anode) and to complete the circuit by passing through the load back to the emitter.

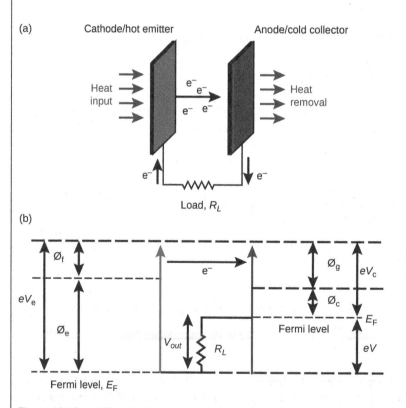

Figure 19.12   (a) Thermionic converter and its (b) potential diagram of electrons.

The emitter can emit electrons if they have a potential ϕ_e more than the Fermi level; ϕ_e is called the *work function*, defined in Eq. (19.5). But the electrons leaving the emitter produce a space charge that opposes the flow of further electrons. Therefore, for the emitter to emit more electrons, an additional potential ϕ_f is required to overcome the space charge (Walsh 1967). Figure 19.12 shows that electrons require a potential of at least $(\phi_e + \phi_f)$ to rise to the emitter potential of eV_e.

As electrons arrive at the collector, they give up their space charge ϕ_g and the collector work function ϕ_c, which sets up a Fermi level of potential eV above that of the emitter. This voltage (V) is available to force electrons through the load impedance back to the emitter.

The heat source must supply the energy (eV_e) per electron at the emitter's high temperature (T_e). The heat sink must be able to absorb energy amounting to eV_c per electron and maintain the low temperature (T_c).

The emission current density (J) can be calculated with the following simplified expression (Angrist 1982):

$$J = AT^2 \exp\left(\frac{-e\,\phi_e}{k_B T}\right) \tag{19.24}$$

where J is the saturation current density (A/m^2), A is the Richardson-Dushman constant (1.2×10^6 A/m^{2}K^2), ϕ_e is the potential barrier (or work function) of the surface (eV), e is the value of a fundamental electronic charge (1.6×10^{-19} C), k_B is the Boltzmann constant (1.38×10^{-23} m^{2}kg/s/K), and T is the absolute temperature of the surface (K).

The heat rate $(\dot{Q})$ required to maintain the emitter at the high temperature (T) is equal to the energy carried away by the emitted electrons (Angrist 1982):

$$\dot{Q} = \frac{de}{dt}\left(f_e + \delta + \frac{2k_B T}{e}\right) \tag{19.25}$$

where e is the charge (measured in coulombs), ϕ_e represents the surface barrier of the emitter, δ is the barrier that an electron effectively experiences due to the space charge in the vacuum, and $\frac{2k_B T}{e}$ represents the original kinetic energy of the electrons at temperature T.

The efficiency of this direct energy conversion system cannot exceed V/V_c, but is usually much less, typically on the order of 5%. For high efficiency, the work function of the anode should be much less than that of the cathode. And, to minimize the space-charge effect, the distance between the anode and cathode should be as small, typically less than $25 - 100$ μm. The space containing the anode and cathode is often filled with a gas such as cesium, which helps reduce the space charge effect. Emitter temperatures are on the order of 1300 K.

Currently, researchers are investigating *semiconductor thermionics* for next-generation solar cells. Instead of using metal emitters, such a system might use emitters made of a *p*-type semiconductor, and if used with a photovoltaic cell (Chapter 20), the photogeneration-induced quasi-Fermi level splitting can reduce the effective barrier for electron emission – a mechanism used by a *photon-enhanced thermionic emission* (or PETE) device. The PETE process thus uses the light and heat of the sun to generate electricity. This enhances the energy conversion efficiency by more than double (>40%) compared with using only photovoltaic cells. The semiconductors consist of gallium nitride or gallium arsenide (Go et al. 2017; Datas and Vaillon 2019; Rahman and Nojeh 2021).

In a *magnetohydrodynamic (MHD) generator*, high-temperature heat (usually obtained by the combustion of a fossil fuel) is used to ionize a gas, and this conducting gas (or "plasma") with charge q is moved through an expanding duct at a high velocity $\overline{V}$. A powerful magnetic field, $\overline{B}$, is set up across the duct. The gas motion is sustained by a pressure drop between the chamber where heat is added and the open end of the expanding duct. In accordance with Faraday's Law of Induction, an electromagnetic force $(\overline{F})$ is generated that acts in a direction perpendicular to both the gas flow and the magnetic field. The electromagnetic force can therefore be calculated by taking the vector cross-product between $\overline{V}$ and $\overline{B}$ (Rosa 1987):

$$\overline{F} = e(\overline{V} \times \overline{B}) \tag{19.26}$$

The walls of the channel parallel to the magnetic field serve as electrodes and enable the generator to provide an electric current to an external circuit. The induced current (I) may be calculated as:

$$I = \sigma(\overline{V} \times \overline{B}) \tag{19.27}$$

where σ is the electrical conductivity of the gas, measured in siemens per meter (S/m).

The power output of the generator $(\dot{W})$ may be approximated by substituting Eq. (19.27) in $\dot{W} = I^2 R$, with R being the electrical resistance of the ionized gas, which is the reciprocal of the gas conductivity (σ), i.e., $R = 1/\sigma$. The power output then becomes:

$$\dot{W} = \frac{1}{\sigma}\left(\frac{F}{L^2 B}\right)^2 \tag{19.28}$$

where L is the length of the channel along which the ionized gas is accelerated.

For instance, with an electromagnetic force of 6.7 MN being generated by an ionized gas with $\sigma = 65\ \Omega^{-1}/m$ in an electric field of 20 T and with a channel length of 3 m, the power output will be:

$$\dot{W} = \frac{1}{\sigma}\left(\frac{F}{L^2 B}\right)^2 = \frac{1}{(65\ \Omega^{-1}\mathrm{m}^{-1})}\left(\frac{6.7 \times 10^6\ \mathrm{N}}{(3\ \mathrm{m})^2 (20\ \mathrm{T})}\right)^2 = 21\ \mathrm{MW}.$$

The power output of the generator $(\dot{W})$ is proportional to the volume $(\mathcal{V})$ of the generator (measured in m³), square of the gas velocity (V), and the magnetic field $(\overline{B})$, measured in tesla, T:

$$\dot{W} \propto (\mathcal{V}\sigma V^2) \tag{19.29}$$

19.6 Outlook

Thermoelectric devices are far from representing mature technology, especially due to the difficulty of controlling heat flux. Advances in microfabrication will help address this challenge. Microfabrication techniques are increasingly able to produce thermoelectric devices with exceedingly small dimensional tolerances, and gap sizes and structures, including novel electrode materials such as graphenes, new sealants and welding techniques, and new solid or flexible substrates. With thermionic devices, the use of novel space charges and electrodes will continue to increase the electric field in the gap region.

Thermoelectric and thermionic devices will become increasingly practical with more useful power outputs as scientists and engineers develop new thermoelectric materials, especially those that are able to operate at room temperature. More effective heat sinks will be developed for thermoelectric coolers using materials such as flexible inorganic/organic foams. Heat transfer fluids might also include nanofluids instead of water or air.

One method for producing more sustainable electricity is by scavenging waste heat with thermoelectric generators. Home heating, exhausts from internal combustion engines, and heat-intensive industrial processes all generate an enormous amount of unused waste heat, which could be converted to electricity by thermoelectric generators. For instance, alternators in cars with internal combustion engines could be replaced by thermoelectric generators placed in the exhaust stream, therefore improving fuel efficiency. Similarly, advances in thermoelectric cooling could enable the replacement of vapor-compression refrigeration systems with solid-state Peltier coolers.

Advances in materials science will no doubt lead to the discovery of more high-efficiency thermoelectric materials, with their conflicting requirements of high electrical conductivity and low thermal conductivity. Importantly, a heterostructure emitter can improve the solar conversion efficiency in photon-enhanced thermionic emission. Photovoltaic cells can be incorporated into electron collectors in combined thermionic-photovoltaic converters. These are potentially important developments in the field of sustainable energy, which bodes well for thermoelectricity. In 2017, the global thermoelectric market generated $400 million in revenue. In 2025, that number is expected to increase to more than $1 billion.

Problems

19.1 Study Section 19.5 pertaining to power generation and system cooling using semiconductors. Then read the following paper about thermoelectric cooling: Huang et al. (2000).

Based on your research, estimate the coefficient of performance (*COP*) of a 15-W thermoelectric cooler that operates on 12 V and draws 4 A. The thermoelectric cooler has a heat sink resistance of 0.25 °C/W, cold-side temperature of 5 °C and hot-side temperature of 20 °C. Clearly state all your assumptions.

[Answer: $COP \cong 0.25$]

19.2 A thermionic energy converter with a surface area of 5 cm^2 is used to convert waste heat to electricity. The tungsten emitter temperature operates at 2500 °C. The work function for tungsten is given as 4.5 eV.
a) What is the maximum current that this small thermionic cell can deliver?
b) If the emitter voltage is 2.2 V and the collector voltage is 1.4 V, what is the power output of this thermionic device?

[Answers: (a) 11 mA; (b) 8.8 mW]

19.3 Do independent research about a magnetohydrodynamic (MHD) generator. In an MHD generator, high-temperature heat (usually obtained by the combustion of a fossil fuel) is used to ionize a gas, and this conducting gas (or "plasma") is moved through a duct at a high velocity. A powerful magnetic field is set up across the duct. The gas motion is sustained by a pressure drop between the chamber where heat is added and the open end of the expanding duct. In accordance with Faraday's Law of Induction, an electromagnetic force is generated that acts in a direction perpendicular to both the gas flow and the magnetic field. Some proponents believe that old steam power plants can be converted to be run by an MHD generator, rather than using steam turbines. Discuss this idea in a well-argued essay of no less than 500 words.

[Answer: Open-ended.]

19.4 Radioisotope batteries, often called "nuclear microbatteries," convert part of the energy emitted by a radioactive material and absorbed in the converter material directly into electrical energy. They consist of the radioactive material (e.g., a radioisotope X-ray, γ-ray, α particle, or β^- particle source), and the semiconductor conversion device. Alphavoltaic cells exhibit energy densities that are orders of magnitude greater than betavoltaic technology. Do independent research about this technology and summarize the operating principles and typical applications of alphavoltaic and betavoltaic cells, using the format of a three-page fact sheet.

[Answer: Open-ended.]

References

Angrist, S.W. (1982) *Direct Energy Conversion*. Boston, MA: Allyn and Bacon.

Datas, A. and Vaillon, R. (2019). Thermionic-enhanced near-field thermophotovoltaics. *Nano Energy* 61: 10–17.

Gamow, G. (1960). *Physics. Foundations and Frontiers*. New Jersey: Prentice-Hall.

Go, D.B., Haase, J.R., George, J., Mannhart, J. et al. (2017). Thermionic energy conversion in the twenty-first century: advances and opportunities for space and terrestrial applications. *Frontiers in Mechanical Engineering* 3: 13.

Huang, B.J., Chin, C.J. and Duang, C.L. (2000). A design method of thermoelectric cooler. *International Journal of Refrigeration* 23 (3): 208–218.

Jaffe, R.L. and Taylor, W. (2018). *The Physics of Energy*. Cambridge, UK: Cambridge University Press.

Jean, J., Brown, P.R., Jaffe, R.L., Buonassisi, T. et al. (2015). Pathways for solar photovoltaics. *Energy and Environmental Science* 8 (4): 1200–1219.

Murray, R.L. (1980). *Nuclear Energy*. Oxford, UK: Pergamon Press.

Pourkiaei, S.M., Ahmadi, M.H., Sadeghzadeh, M., Moosavi, S. et al. (2019). Thermoelectric cooler and thermoelectric generator devices: a review of present and potential applications, modeling and materials. *Energy* 186: 115849.

Rahman, E. and Nojeh, A. (2021). Semiconductor thermionics for next generation solar cells: Photon enhanced or pure thermionic? *Nature Communications* 12 (1): 1–9.

Rosa, R.J. (1987). *Magnetohydrodynamic Energy Conversion*. Washington, DC: Hemisphere Publishing.

Sears, F.W., Zemansky, M.W., Young, H.D. (1982). *University Physics*. Boston, MA: Addison-Wesley Publishing Company.

Snyder, G.J. and Toberer, E.S. (2008). Complex thermoelectric materials. *Nature Materials* 7 (2): 105–114.

Soo, S.L. (1968). *Direct Energy Conversion*. Englewood Cliffs, NJ: Prentice-Hall.

Sørensen, B. (2017). *Renewable Energy. Physics, Engineering, Environmental Impacts, Economics and Planning*. Cambridge, MA: Academic Press/Elsevier.

Walsh, E.M. (1967). *Energy Conversion. Electromechanical, Direct, Nuclear*. New York, NY: The Ronald Press Company.

20

Solar Electricity

The Earth's ultimate recoverable resource of oil, estimated at three trillion barrels, contains about 1.7×10^{22} J of energy. This is also the amount of energy the sun supplies to the Earth in one-and-a-half days. Our human global annual energy usage is about 5.8×10^{20} J. This is what the sun delivers to the earth in just over an hour and a half. The sun continually delivers about 1.2×10^{5} TW of power to the Earth, compared to our annual power consumption of "only" 15 TW. This awe-inspiring solar energy can be used to produce electricity ("solar electricity" or "photovoltaics"), fuel ("solar fuel"), or heat ("solar thermal" or "solar heat"). However, despite its abundance and versatility, we are using far too little of it to power human activities.

In 2021, the world produced just over 1000 GW (or 1 TW) of solar electricity, which is about 3% of the world's electricity demand (IEA 2022). Every year, globally, around 200 GW_e of photovoltaic power comes online, enough to power 150 million homes. Photovoltaic power (or "solar electricity") is now the cheapest option for new electricity generation in most of the world (2–4 ¢/kWh_e). This is expected to result in a rapid increase in solar investment in the coming years.

In 2021, about half the energy demand for buildings (which use 36% of the world's energy) was for space and water heating; only 0.8% of that space and water heating was provided by solar heating. The combustion of biomass, which is a solar-derived fuel (via the process of photosynthesis), accounted for about 12% of human energy needs in 2021. However, more than two-thirds of that biomass was gathered unsustainably. Trees were chopped down without replanting enough new ones. The wood was burned in inefficient stoves where combustion is incomplete, and the resulting pollutants were, and still are, uncontrolled. This biomass can, however, be artificially synthesized using semiconductor solar cells to imitate photosynthesis occurring in plants. Although in its infancy, this "solar fuel" technology shows great promise to help us to further reduce our carbon footprint.

Clearly, solar energy has the potential to play a central role in the future global energy system because of the scale of the solar resource, its predictability and versatility, and its ubiquitous nature (Figure 20.1).

Reimagine Our Future

Cities house half the world's population, but consume 80% of the world's energy and produce three-quarters of the world's pollution. Might we learn something from traditional low-energy technologies such as using natural ventilation (wind) and orienting a building correctly to avoid excessive sunlight (in so-called passive buildings), rather than energy-intensive air-conditioning and ventilation practices?

Energy Systems: A Project-Based Approach to Sustainability Thinking for Energy Conversion Systems,
First Edition. Leon Liebenberg.
© 2024 John Wiley & Sons, Inc. Published 2024 by John Wiley & Sons, Inc.
Companion website: www.wiley.com/go/liebenberg/energy_systems

An award-winning "environmental building" that features natural ventilation, not air-conditioning, and uses 30% less energy than a regular building. *Source:* Dennis Gilbert / View Pictures.

Figure 20.1 The solar garden at the Margaret A. Cargill Lodge at Wolf Ridge Environmental Learning Center in Finland, Minn. *Source:* Chad Holder, Wolf Ridge.

20.1   Solar Electricity ("Photovoltaics")

Solar electricity (or "photovoltaics") employs semiconductor-based technology to convert sunlight directly to electricity. Photovoltaic systems are modular, so their electrical power output can be engineered for myriads of applications, from small scale to large scale. Incremental photovoltaic power additions are easily accommodated, which is not the case with gigawatt-scale fossil-fuel or nuclear power plants that need to be large to be economically viable, and in which incremental additions are much more difficult.

Chapter 19 provided basic information on semiconductors pertaining to their use in some direct energy conversion systems. The widely used semiconductor silicon (Si) has a lattice structure similar to that found in diamonds, with each atom being covalently bonded to each of its four nearest neighbors. As discussed in Section 19.4, *p*-type and *n*-type semiconductors may be formed by doping the silicon with elements from the third or the fifth columns of the periodic table; *p*-type semiconductors accept electrons, and *n*-type semiconductors donate electrons (see Figure 20.2). Without *built-in electric fields*, though, electrons would de-excite very quickly and produce no current. What is required is a semiconductor device with a built-in electric field: a *p-n* junction diode.

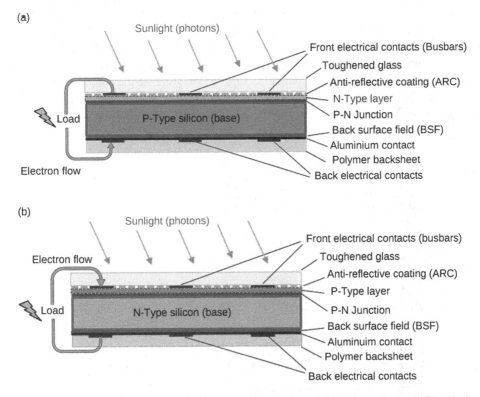

Figure 20.2   Basic photovoltaic cell. (a) *p*-type silicon cell, and (b) *n*-type silicon cell. Electrical current is taken from the photovoltaic cell through a grid contact structure on the top (which does not impede sunlight to hit the top of the semiconductor surface), a contact on the bottom surface that completes the electrical circuit, and an anti-reflective coating that maximizes sunlight absorption in the photovoltaic cell. *Source:* Clean Energy Reviews (2022). Most efficient solar panels 2022. https://www.cleanenergyreviews.info.

20.2 The *p-n* Junction Diode

From Chapter 19, a *p-n junction* is formed when a uniform *p*-type sample is metallurgically joined to a uniform *n*-type sample. When this happens, the positive and negative electrical charges redistribute, establishing an internal electric field. On the *n*-side (extending to the junction), a uniform concentration n_{n0} of free electrons is found, as well as a concentration p_{n0} of free holes. On the *p*-side (extending to the junction), a uniform concentration p_{p0} of holes is found, as well as a concentration n_{p0} of free electrons. The following relationship is satisfied (Sørensen 2017);

$$n_{n0}\,p_{n0} = n_{p0}\,p_{p0} = n_i^2 \tag{20.1}$$

where n_i is the intrinsic-carrier concentration at the given temperature of the semiconductor.

When the *p-n* junction is formed, the electrons have a much larger concentration on the *n*-side than on the *p*-side. Concomitantly, the hole concentrations are much larger on the *p*-side than on the *n*-side. This large difference in carrier concentrations sets up an initial *diffusion current*. Electrons flow from the *n*-region to the *p*-region, and holes flow from the *p*-region to the *n*-region. This flow of charge carriers results in a region near the *p-n* junction that is depleted of *majority carriers* – i.e., of electrons on the *n*-side and of holes on the *p*-side, as illustrated in Figure 20.3.

The charged stationary atoms produce an *internal electric field* that acts as a barrier to holes moving from the *p*-type material, because of the positive charged atoms in the *n*-type materials, specifically in the *depletion region* of the *p-n* junction. This electric field builds up in a direction that opposes the further flow of electrons from the *n*-region and holes from the *p*-region. The magnitude of the field exactly balances the further flow of majority carriers by diffusion. This electric field corresponds to a built-in voltage between the *n*-type and *p*-type sides of the junction.

We can use an energy level diagram (or "potential" diagram) to visualize the potential around a *p-n* junction. Figure 20.4b shows a schematic of a *p-n* junction under equilibrium conditions, its energy bands, the Fermi level, and the donor and acceptor charges. The small number of thermally

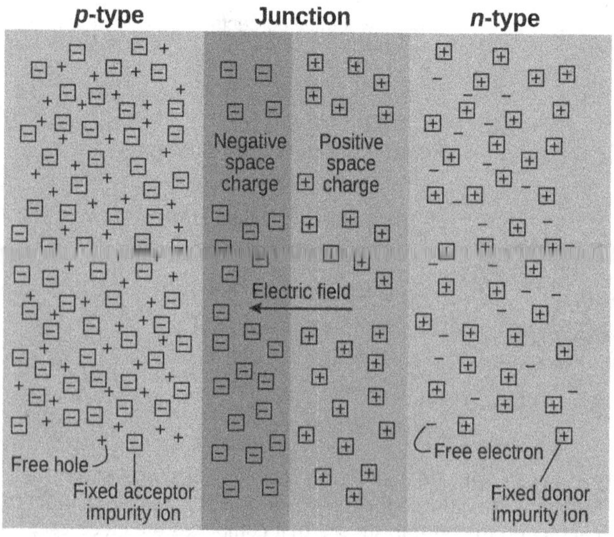

Figure 20.3 The *p-n* junction.

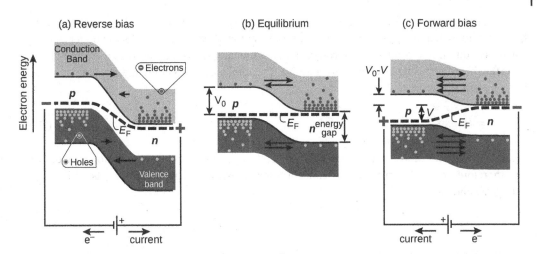

Figure 20.4 An unilluminated ("dark") *p-n* junction with (a) reverse bias, (b) equilibrium, and (c) forward bias. E_F is the Fermi energy level; I_{diff} is the diffusion current of either electrons or holes; and I_{field} is the field current (sometimes called the drift current) of either electrons or holes. *Source:* Reproduced from Jean et al. (2015) with permission from the Royal Society of Chemistry.

excited electrons on the *p*-side of the conduction band can easily "flow down" the concentration gradient to the *n*-type region. This diffusion current (I_{diff}) is exactly counteracted by the field current (I_{field}), which is produced by the internal electric field. Therefore, in the absence of externally applied potentials (voltage), no current will flow.

20.3 The *p-n* Junction with an Applied Voltage

If a positive voltage is now applied to the *p*-side relative to the *n*-side, the Fermi energies are split and the energy barrier in the *n*-type region of the conduction band is reduced. Current will flow across the junction. This condition is called *forward bias* (Figure 20.4c). Conversely, if a negative voltage is applied to the *p*-side relative to the *n*-side, current will be discouraged from flowing. This condition is called *reverse bias* (Figure 20.4a). These two conditions may be described by Shockley's *ideal-diode equation*:

$$I_{field} = I_{diff} \exp^{(eV/k_B T)} \tag{20.2}$$

where I_{field} is the field current (A) induced by the internal electric field, I_{diff} is the diffusion current (A) brought about by the flow of holes (and electrons), e is the value of the fundamental electronic charge (1.6×10^{-19} C), V is the applied voltage, k_B is the Boltzmann constant (1.38×10^{-23} m^{2}kg/s^2/K), and T is the absolute temperature (K).

The net electron current is then (Angrist 1982):

$$I = I_{field} - I_{diff} = I_{diff} \exp^{(eV/k_B T)} - I_{diff} = I_{diff} \left(\exp^{(eV/k_B T)} - 1 \right) \tag{20.3}$$

With reverse bias, the exponential term in Eq. (20.3) becomes negligible compared to 1.0 and $I \cong -I_0$. If $V = 0$, then $I = 0$. The current increases with positive values of eV. The current decreases when eV is negative.

The hole current behaves in a similar manner, with an applied voltage that also lowers the barrier height for holes. Large numbers of holes will flow from the *p*-region to the *n*-region under the same voltage condition that produces electron currents in the opposite direction. Electron and hole currents flowing in opposing directions are additive. The total current of holes and electrons may then be expressed as (Angrist 1982):

$$I = I_0 \left(\exp^{(eV/k_B T)} - 1 \right) \tag{20.4}$$

where I_0 is the saturation (or "dark") current and is strongly correlated to the junction temperature, and therefore on the intrinsic-carrier concentration. I_0 is thus larger for materials with smaller bandgaps, and vice versa.

20.4 The *p-n* Junction with Impinging Light

An electron in the valence band will absorb electromagnetic radiation and will jump to the conduction band, leaving behind a hole. As noted in Chapter 19, a hole depicts where an electron used to be, and therefore takes on a positive charge.

If light is now allowed to impinge on the *p-n* junction, the equilibrium conditions are disturbed and minority carriers are created. *Minority carriers* are electrons in the *p* material and holes in the *n* material. These minority carriers lower the potential energy barrier at the *p-n* junction, allowing current to flow. Under forward bias, the holes generated in the *n*-doped region are swept over the *p*-doped region to produce this current.

The magnitude of light-induced current I_L is dependent on the generated electron-hole pairs from the absorption of the impinging light and on the collection efficiency of these charge carriers. The impinging light therefore modifies Eq. (20.2) to the following (Jaffe and Taylor 2018):

$$I = I_0 \left(\exp^{(eV/k_B T)} - 1.0 \right) - I_L \tag{20.5}$$

The *thermal voltage* (V) for silicon is about 26 mV at room temperature of $T = 300$ K.

Figure 20.5 plots Eq. (20.5) and shows that the diode equation, Eq. (20.2), shifts down by an amount equal to I_L. Power is the product of voltage and current, which implies that negative work is produced (because the current is negative and the voltage is positive). This, in accordance with the sign convention adopted for electricity, suggests that work is done on the environment, which is exactly what is required.

Equation (20.5) can be rewritten to express the open-circuit voltage (when the current is 0):

$$V_{oc} = \frac{k_B T}{e} \ln \left(\frac{I_L}{I_0} + 1 \right) \tag{20.6}$$

If the cell voltage is set to 0, the cell current reaches a maximum limiting value called the short-circuit current, I_{sc}. In an ideal case, the short-circuit current will equal the light-induced current, i.e., $I_{sc} = I_L$.

This photovoltaic device may be modeled as an ideal diode in parallel with a light-induced current generator I_L as illustrated in Figure 20.6.

Figure 20.6 shows a typical current voltage diagram for a photovoltaic cell, with and without impinging light. The *maximum power* (P_{max}) that the cell will deliver is the product of the current

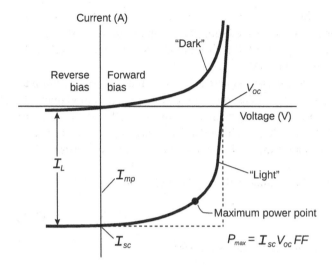

Figure 20.5 Current-voltage diagram of photovoltaic cell (*p-n* junction), in the absence of light and when illuminated. *Source:* After Sørensen (2017).

(a)   (b)

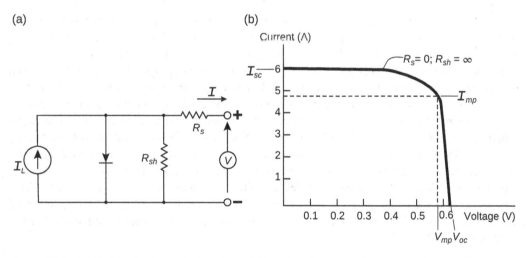

Figure 20.6 (a) Model of a photovoltaic cell and (b) its related current-voltage curve. *Source:* After Sørensen (2017).

and voltage at the point of maximum power, respectively, I_{mp} and V_{mp}, but modified by the fill factor:

$$P_{max} = I_{mp}V_{mp} = FF\, I_{sc}\, V_{oc} \tag{20.7}$$

where I_{sc} is the short-circuit current, V_{oc} is the open-circuit voltage, and where the *fill factor* (*FF*) is defined as follows:

$$FF = \frac{V_{mp}\, I_{mp}}{V_{oc}\, I_{sc}} \tag{20.8}$$

The *FF* characterizes how "square" the current–voltage curve (Figure 20.6) is and represents how "difficult" or "easy" it is to extract the photogenerated charge carriers from a photovoltaic device.

The ideal value for *FF* is unity (100%), when the *I–V* curve is a rectangle. The *FF* typically ranges between 0.5 and 0.85. When exposed to non-concentrated sunlight, the short-circuit current density (J_{sc}) of a single-junction silicon-based photovoltaic cell is around 45 mA/cm^2, with $(J_{sc})(A) = I_{sc}$, while the open-circuit voltage (V_{oc}) is around 0.55 V.

The *open-circuit voltage* (V_{oc}) is the voltage measured across the terminals of a solar cell under illumination when no load is applied. In photovoltaics, the open-circuit voltage is fundamentally related to the balance between light current and recombination current, and is thus a primary measure of the quality of a solar cell. Photovoltaic technologies with high open-circuit voltages (i.e., close to the material-dependent bandgap) typically exhibit low internal losses (Jean et al. 2015).

The energy conversion efficiency of a photovoltaic cells may be defined as:

$$\eta = \frac{V_{mp} I_{mp}}{P_{in}} \tag{20.9}$$

where the *solar irradiant power* (P_{in}) may be calculated as follows:

$$P_{in} = A \int_0^\infty F(\lambda)(hc/\lambda) \, d\lambda \tag{20.10}$$

where A is the device area (m^2), $F(\lambda)$ is the number of photons per m^2 per second per unit bandwidth incident on the device at wavelength λ, and hc/λ is the energy of each photon. The solar irradiance (P_{in}) is often approximated as 1 kW/m^2 for sunlight impinging on the Earth.

Figure 20.7 shows the energy potential diagram of an *illuminated p-n junction*. The light-activated cell generates free electrons and holes. These free charge carriers are separated under the built-in electric field of the diode, generating photocurrent. As the generation of photocurrent is roughly independent of the voltage across the solar cell, the illuminated curve in Figure 20.5 is vertically offset by a constant amount from the "dark" curve. (Note: The applied forward bias potential forces current to flow in the opposite direction as I_L. The large negative light-induced current I_L however offsets the "dark" current to produce a net current. A photovoltaic cell therefore needs to operate under forward bias.) The current is correlated with the number of carriers generated, which depends on the absorption properties of the semiconductor and its efficiency in turning absorbed photons into extractable charge carriers. The voltage is correlated with the strength of the built-in electric field of the diode.

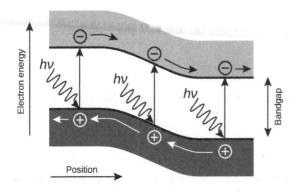

Figure 20.7 Operation of a solar cell under illumination, showing excitation of electrons and holes by light, followed by charge carrier separation under the built-in electric field. *Source:* Reproduced from Jean et al. (2015) with permission from the Royal Society of Chemistry.

20.5 Photovoltaic Cells, Modules, and Arrays

The electricity output from a single photovoltic *cell* is of the order of a few watts. To produce a photovoltaic *module* with adequate power output, cells are connected in series and parallel. Those modules could, in turn, be connected in series or parallel to form an *array*. This is illustrated in Figure 20.8a.

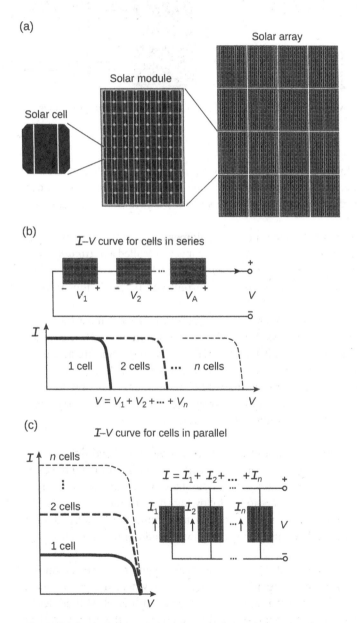

Figure 20.8 (a) Photovoltaic cells, modules, and arrays, and connecting photovoltaic cells (or modules, and arrays) in (b) series and in (c) parallel.

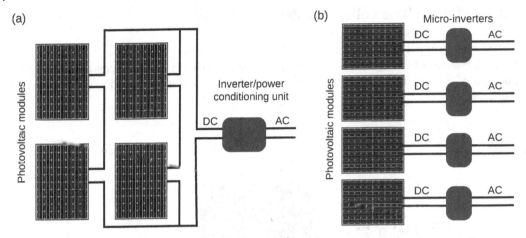

Figure 20.9 (a) Photovoltaic modules with a central inverter, and (b) those that each feature their own micro inverters.

When two identical cells (or modules, or arrays) are connected in series, their voltage is doubled, while the current remains the same; and when two identical cells (or modules, or arrays) are connected in parallel, their current is doubled and the voltage stays the same. This is illustrated in Figure 20.8b.

Photovoltaic cells have a direct current (DC) output. The agglomerate current of a photovoltaic module may be kept DC and then later modified by a DC-to-DC converter (called a *maximum power point tracker* (MPPT), which is part of a *power conditioning unit* (PCU)). This is then fed to a "string" or "central" *inverter* that transforms the DC to alternating current (AC). This is illustrated in Figure 20.9a.

Alternatively, some manufacturers produce photovoltatic modules that each have their own inverter, called a *micro inverter* (Figure 20.9b). Modules with micro inverters therefore produce AC. Their wiring is simpler and cheaper than their DC alternatives. These systems are also more reliable than those that require extensive DC modification, but the modules with micro inverters are more expensive. However, micro inverters are simple to add one at a time. Each module and micro inverter pair can easily be added to an existing solar array without needing to worry about purchasing or siting, and without installing additional string inverters.

Example 20.1   Scaling a Photovoltaic Cell to a Number of Photovoltaic Modules
The solar *cells* that make up a solar *module* (which, in turn, forms part of a solar *array*) have the following characteristics: $FF = 0.82$, $V_{oc} = 0.52$ V, $A = 331$ cm^2, and a short-circuit current density $J_{sc} = 42$ mA/cm^2. The array of modules must adhere to the following specifications: $V_{mp} = 54.7$ V, $I_{mp} = 5.98$ A, $V_{oc} = 64.9$ V. The recommended voltage for the maximum power point tracker ($V_{MPPT,max}$) is 1000 V, and the recommended current for the power conditioning unit ($I_{PCU,max}$) is 36 A.

Calculate (a) the maximum power output of a 331-cm^2 cell, and (b) the number of cells required for a solar module that produces 390 W. (c) Calculations for the facility where the photovoltaic modules will be used reveal that 24 modules are required (for an array output of 24 modules × 390 W/module = 9.4 kW). Determine how the modules should be connected in the array to satisfy the operating parameters.

Solution strategy: We can use Eq. (20.7) to calculate the power output. For parts (b) and (c), we must remember the relationship between a PV cell, module, and array, as illustrated in Figure 20.8a. Also, the typical set-up of a PV array and associated electronics is illustrated in Figure 20.9. This knowledge will help us to calculate the number of cells in a module and the required number of modules in the array.

Analysis:

a) P_{max}

$$P_{max} = I_{mp}V_{mp} = FF\, I_{sc}\, V_{oc} = (0.82)\left(\frac{42 \times 10^{-3}\, \text{A}}{\text{cm}^2}\right)(331\, \text{cm}^2)(0.52\, \text{V}) = 5.9\, \text{W}$$

b) Number of cells per module

$\dfrac{P_{panel}}{P_{max,cell}} = \dfrac{390\, \text{W/panel}}{5.9\, \text{W/cell}} = 66$ cells/module. The cells could be stacked in columns of 11 and rows of 6.

c) Number of modules and their circuitry

The set-up requires 24 modules of 390 W each to deliver a power output of 9.4 kW.

Also, it is given that the maximum circuit voltage is $V_{MPPT,max} = 1000$ V. So, the number of modules that are connected in series may not exceed $\dfrac{V_{MPPT,max}}{V_{MP}} = \dfrac{1000\, \text{V}}{54.7\, \text{V}} = 18.3$. Therefore, *no more than* 18 modules should be connected in series.

To determine the number of modules that should be connected in parallel, we need to work with the maximum current that the electronic PCU can handle, $I_{PCU,max}$, and divide that by the current at the point of maximum power (I_{MP}):

$$\frac{I_{PCU,max}}{I_{MP}} = \frac{36\, \text{A}}{5.98\, \text{A}} = 6.02$$

Therefore, *no more than* six modules should be connected in parallel.

Discussion: We can select several array configurations in which to place the required 24 modules, one option being to connect four modules in series and to connect six such rows in parallel.

Reimagine Our Future

If many more people installed rooftop photovoltaic modules, would that help stop glaciers from melting?

20.6 Types and Costs of Photovoltaic Cells and Systems

Silicon is the dominant semiconductor material for the production of photovoltaic cells. About 90% of all photovoltaic cells consist of monocrystalline or polycrystalline silicon. Other types of technology include amorphous (or "thin-film") photovoltaic cells and organic (or "polymer") photovoltaic cells. The salient characteristics of these cells are summarized in Table 20.1,

Table 20.1 Some common photovoltaic materials.

Type		Example material	Laboratory cell efficiency (%)	Commercial cell efficiency (%)	Cost per watt (US $/W)	Advantages	Disadvantages
Wafer	Crystalline	Monocrystalline silicon (mono-Si)	28	22	1.20–3.40	0.2-mm thick wafers are cut from a silicon melt making these cells highly efficient. Optimized for commercial use. Stable over a long period. Lifetime: 25 yr	Expensive. Sensitive to high temperature. A lot of material is lost during the wafer-cutting manufacturing process
		Polycrystalline silicon (poly-Si)	24	19	0.90–1.50	Cheaper than monocrystalline silicon. Lifetime: 15 yr	Wafers are cut from a cast silicon block, with a lot of material loss. Irregular crystal facets make these cells more inefficient than their monocrystalline counterparts. Sensitive to high temperatures; shorter lifespan (15 yr); less space efficient
Thin-film	Amorphous	Amorphous silicon (a-Si)	12	8	1.00–2.90	Films are 1 – 30 μm thick and can be vapor-deposited on a variety of substrates, like glass. Relatively lightweight and cheap. Substrate may be flexible, making the film simple to integrate in building structures. If amorphous silicon is used, less silicon is used than in mono- or multicrystalline silicon cells	Low efficiency. Unstable materials with short lifespan. Expensive. Take up a lot of space, making them unsuitable for residential use. Some of the materials, such as cadmium, are toxic if ingested or inhaled. Sensitive to high temperatures
		Copper indium selenide (CIS); copper indium gallium selenide, CIGS	23	19			
		Cadmium telluride (CdTe)	22	20			
		Gallium arsenide (GaAs)	47	39		Cells are less prone to overheating. To improve efficiency, different combinations of semiconductors can be used (e.g., amorphous silicon, a-Si; cadmium telluride, Cd/Te; copper indium diselenide, CIS; copper indium gallium	

		18.5	12	1.50–2.40	selenide, CIGS), several layers or junctions can be used, as well as different substrates than glass. Films are produced by high-output machines Lifetime: 12–20 yr	
Organic	Polymer (fullerenes, C$_{60}$)	18.5	12	1.50–2.40	Produced in high-output processes. Relatively simple to control their shapes and thicknesses. Up to 10 times lighter than silicon modules. May also be transparent, great for architectural material such as building façades Lifetime: 5 yr	Low efficiencies. Cells degrade quickly, giving a lifetime of only around five years
Perovskite	Calcium titanium oxide (CaTiO$_2$); tin dioxide (SnO$_2$)	26	18	Emergent	Relatively inexpensive. Cheap and simple in construction, lightweight, flexible, high efficiency	Unstable, degrades quickly
Dye-sensitized solar cells (DSSCs)	Titanium oxide (TiO$_2$) as semiconductor with ruthenium-polypyridine as dye	12	—	Emergent	Can be manufactured with simple technology (e.g., screen-printing). Cells do not require extremely pure semiconductor materials. Wide variety of dyes and ion-conducting materials. Relatively low cost. Cells can operate in low light and at low temperatures. Robust Lifetime: 25 yr	Unstable at high temperatures. Dyes are toxic and volatile
Quantum dot solar cells (QDSC)	Lead sulfide (PbS)	19	—	Emergent	Less bulky than other photovoltaic materials. Quantum dots are atoms that have been created artificially. Quantum dots have band gaps that can be adjusted by changing the size of the dots over a wide range of energy levels. Low estimated production cost	Unstable, degrades quickly Toxic

Comparison based on single-junction cells.

Source: Jean et al. (2015), Yoo et al. (2021), CER (2022), and NREL (2022).

Box 20.1 Top-Performing Solar Cell

In 2021, revolutionary perovskite solar technology set a new world record for the amount of the sun's energy that can be converted into electricity by a single solar cell. The groundbreaking cell produced by Oxford PV has been independently proven to convert 29.52% of solar energy into electricity. In contrast, standard silicon cells, which are used on millions of homes globally, have an average conversion rate of just 15–20% and a practical maximum conversion rate of around 26%.

The technology used in Oxford PV's record-breaking cell involves coating ordinary silicon solar cells with a thin film of the material perovskite to better utilize photons across the solar spectrum. The synthetic perovskite material is affordable, sustainable and could eventually replace silicon entirely – 35 kg of perovskite generates the same amount of power as seven tons of silicon.

In 2022, Oxford PV became the first company to sell these next-generation solar cells to the public. Initial products, designed for residential roofs, generate 20% more power from the same number of cells. With further development, Oxford PV believes future solar cells could be improved significantly.

---

including cost per watt of photovoltaic power. In 2021, the cost for utility-scale photovoltaic projects was US $30-60/MWh in Europe and the United States, and just US $20-40/MWh in China and India (IEA 2022).

Table 20.1 summarizes the state of the art with *single-junction cells*. However, solar cells can also be produced with more than one charge-collecting junction. Such *multi-junction cells* are stacked in order of decreasing bandgap so that the multiple junctions allow light of particular wavelength ranges to be absorbed and photovoltaic energy conversion to occur in the sub-cell that incurs minimal thermal losses for that wavelength range (Jean et al. 2015). Box 20.1 provides details of a high-effeiciecy photovoltaic cell.

20.7 Sustainability of Solar Electricity

20.7.1 Emissions

The operation of photovoltaic systems might not generate any toxic or greenhouse gases, but such gases may be emitted during the production of the semiconductor materials required in the solar cells. However, the lifecycle emissions from an organic photovoltaics module amount to only about 15 kg CO_2-eq/MWh$_e$ and 49 kg CO_2-eq/MWh$_e$ for silicon modules, in comparison to 1000 kg CO_2/MWh$_e$ from coal-fired steam power plants (which require the mining and burning of coal, and ash disposal).

Production facilities of amorphous silicon (thin-film) also release SF_6 and NF_3, which are both potent greenhouse gases. The lifecycle emission of solar cells further includes the release of SO_2, NO_x, and particulates, albeit at merely 2–4% the level of fossil-fuel plants. Solar cells are certainly not zero-emission technologies, but their *carbon footprints* are a fraction of those of their fossil-fuel counterparts.

20.7.2 Energy Payback

The *energy payback time* (EPBT) is the time it takes for an energy system to generate the amount of energy equivalent to the amount that it took to produce the system. For photovoltaic systems, EPBT is usually calculated by using a solar insolation of 1700 kWh/m^2 per year average performance ratio to account for all losses (including temperature losses) of 75%, and a lifetime of 25 years (Leccisi et al. 2016). The EPBTs of solar technologies are astounding, and range from two to four years for crystalline silicon to only one or two years for thin-film technologies, and only a few months for organic and hybrid emerging technologies (Urbina 2022).

20.7.3 End-of-Life Waste

Photovoltaic cells have lifetimes ranging from 5 to 25 years. By 2050, an estimated 80 million tons of photovoltaic modules will reach the end of their lifetimes. This massive amount of electronic waste will require proper treatment, which may include the following options:

- The reuse of modules that still deliver enough power
- Recycling modules and the recovery of parts or materials
- Landfilling
- A combination of the above

Figure 20.10 shows the typical construction of a photovoltaic module that uses crystalline semi-conductors. These solar modules use many kinds of recyclable material, including aluminum, glass, semiconductor materials, plastics (e.g., ethylene-vinyl-acetate, EVA), and metals such as silver, lead, copper and cadmium (see Table 20.2). Although more than 90% of these materials can be recycled, this is not yet happening.

The use of silver is especially problematic. To be sustainable at multi-terawatt scale photovoltaics, the use of silver needs to reduce to below 5 mg/W from today's 15 mg/W. This is because today's silver consumption for photovolatics is already 10% of global silver production (Haegel et al. 2023).

The recycling of photovoltaic modules is complex and expensive. The recycling industry is in its infancy in terms of commercialization. While some of the components such as glass and aluminum

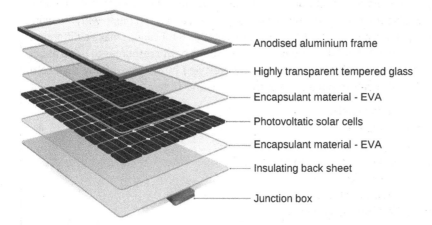

Anodised aluminium frame

Highly transparent tempered glass

Encapsulant material - EVA

Photovoltatic solar cells

Encapsulant material - EVA

Insulating back sheet

Junction box

Figure 20.10 The typical structure of a photovoltaic module comprising crystalline cells. "EVA" refers to ethylene-vinyl-acetate, which is a key encapsulation material used for traditional solar module lamination. *Source:* Global Sustainable Energy Solutions (Pty) Ltd (GSES), CC BY-SA 4.0.

Table 20.2 Typical material composition of a photovoltaic module that uses crystalline semiconductors.

Material	Mass (kg)	Mass per total mass (%)
Glass, containing antimony (0.01–1%/kg of glass)	700	70
Aluminum	185.3	18.53
Copper	11.14	1.11
Polymer-based adhesive (ethylene-vinyl acetate) (EVA)	51	5.1
Polyvinyl fluoride (PVA) sheeting	15	1.5
Silicon metal semiconductor	36.5	3.65
Silver	0.53	0.053
Other metals (e.g., tin, lead)	0.53	0.053
Total	1000	100

Analysis based on the production of 1000 kg of end-of-life photovoltaic waste.
Source: Tan et al. (2022) /MDPI / CCBY 4.0 / Public domain.

are relatively easy and cheap to reclaim, the costs are still too high for separating the encapsulant material, semiconductor material, and other metals. Silicon and thin-film modules are already being successfully recycled, but at great expense. The economic recovery of secondary raw materials like tellurium, indium, gallium, and selenium from the recycling of thin-film modules is still out of reach. These recycling costs will eventually come down as has occurred with other technologies that enter the innovation "S-curve" when processes become simpler and more effective (Figure 4.8), and as value chains are established for recycled materials.

The responsible management of photovoltaic modules at the end of their lifetimes will reduce pollution, reduce (finite) resource extraction, and generate sufficient economic return and value to finance the production of another two billion photovoltaic modules by 2050 (Tan et al. 2022).

Reimagine Our Future

Self-limiting beliefs, or beliefs that limit our options and constrain our sense of what we might accomplish, often involve the use of negative words or phrases, and often begin with the word "but." Do you think you have any such self-limiting beliefs? (e.g., But I cannot make a difference to what happens in the world. But I do not have the capabilities or the resources to bring about change. But I do not have enough time to change my lifestyle.) In the case of any such self-limiting belief you find yourself to have, ask yourself these questions:

- How long have I held this belief?
- Where did this belief come from?
- What evidence do I have that this belief is true or false?
- What is the pay-off for me holding on to this belief?
- Does this belief hold me back?
- Is this belief redundant, unhelpful, or out of date?
- Would I be better off without this belief?

20.8 Outlook

Solar energy has the potential to play a central role in the future global energy system because of the scale of the solar resource, its predictability, and its versatility. The global photovoltaics industry is changing at an immense pace in terms of growth toward *terawatt-scale photovoltaics*, continued dramatic cost decreases, and increases in manufacturing scale. The global installed capacity of photovoltaics has already reached 1 TW; the next terawatt will be added between 2022 and 2025. The challenge is however to develop low-cost operational strategies and complementary technologies (including energy storage technologies) to accommodate the growing fraction of renewable generation.

Today, the use of photovoltaics (or solar electricity) is the cheapest option for new electricity generation in most of the world (2–4 ¢/kWh$_e$). This is expected to propel investment in the coming years. However, an average annual increase in photovoltaic power generation of 25% between 2022 and 2030 is needed to obtain global net-zero carbon emissions by 2050 (IEA 2022). This corresponds to a more-than-threefold increase in annual capacity deployment of photovoltaics until 2030, requiring much greater policy ambition and more effort from both public and private stakeholders, especially in the areas of grid integration and the mitigation of policy, regulation, and financing challenges. This is especially the case in emerging and developing countries (IEA 2022).

It is apparent that the world is now seeking a solar energy future. Tremendous advances in the field of nanoscience are helping ensure that photonic direct energy conversion will remain cost competitive. Scientists and engineers must further improve energy-conversion efficiencies, reduce materials usage, and reduce the complexity of manufacturing and cost.

We are not yet producing photovoltaic modules in a sustainable manner. Photovoltaic modules that reach the end of their lifetimes *must* be recycled to effectively participate in the much-needed circular economy. Despite several pilot recycling plants that reclaim some photovoltaic material, the processes are still ineffective and expensive, making industry adoption difficult. Rectifying this situation will require urgent intervention from business and political leaders. Policies that demonstrate long-term commitments to achieve climate-change mitigation can provide market security to support the major capital investments that are required (see also Chapter 28).

Problems

20.1 The band gap of a silicon solar cell is 1.1 eV. If the incident light has an intensity (or luminosity) of 2 mW, (a) determine the wavelength and frequency of the incident light of which the photons have just enough energy to overcome the band gap. (b) Calculate the maximum power that this cell can generate if the light-capturing efficiency of the cell is 57%.

[Answers: (a) 1.12 μm and 3.7 fHz; (b) about 2 mW]

20.2 A silicon solar cell operating at 300 K has the following characteristics: $I_0 = 15$ μA; $I_{sc} = 5$ A; $V_{oc} = 0.48$ V; $V_{mp} = 0.4$ V; $I_{mp} = 4.7$ A. Calculate (a) the maximum power output, (b) the fill factor, and (c) the number of cells required to provide a module with a 395 W output. Describe how the cells should be connected.

[Answers: (a) 1.88 W; (b) $FF = 0.78$; (c) 210 cells, 30 cells in series and seven strings in parallel.]

20.3 A home in Chicago receives 5.5 h of sunshine per day throughout the year. The household has a daily electricity usage of 18 kWh. An array of photovoltaic modules is required to provide this load. The array comprises 395-W modules. The array of modules must adhere to the following specifications: $V_{mp} = 55$ V, $I_{mp} = 6$ A, $V_{oc} = 65$ V, $I_{sc} = 6.5$ A; the recommended voltage for the maximum power point tracker ($V_{MPPT,max}$) is 1000 V, and the recommended current for the power conditioning unit ($I_{PCU,max}$) is 36 A. One battery operates at 100 V and 25 Ah. The battery stack must provide three days of power delivery (in the event of the sun not shining), and the batteries may not discharge by more than 70%. The power conditioning unit has an efficiency of 97%. (a) Calculate the number of photovoltaic modules required and specify how they should be connected. (b) Specify the energy requirement (kWh) and number of batteries.

[Answers: (a) 10 modules; (b) 2.5 kWh, thus 31 batteries, with no more than 10 batteries connected in series and three batteries in parallel.]

20.4 A solar array for a house in Chicago has a power output of 9.4 kW. The system will produce an estimated 16,920 kWh/year. The household will operate the solar modules using net metering, and no batteries are required. The solar array is guaranteed for a period of 25 years. The installed cost of the solar array is US $2.50 per watt. Calculate (a) the total installed cost, and (b) the levelized cost of electricity for a 5% rate of return on the investment.

[Answers: (a) US $23,500; (b) Annualized cost is US $1,667.40 and the LCOE is 9 ¢/kWh.]

20.5 The population of the cities of Urbana and Champaign is around 120,000. Each resident of Urbana-Champaign typically uses energy at a rate of 1 GJ per day. If the average irradiation in the area is 150 W/m², calculate the area of land needed to supply all the energy for the residents of Urbana-Champaign, assuming that the conversion efficiency of sunlight is 2%, 4%, 8%, and 16%. Compare this to the land area of Urbana-Champaign, which is 90 km².

[Answers: For 2%, 417 km² is required; for 4%, 209 km² is required; for 8%, 104 km² is required; for 16%, 52 km² is required. Only the 16% option is viable as 52 km²<90 km².]

20.6 Assuming a silicon photovoltaic array attains 20% efficiency over a lifetime of 25 years of use, where it is exposed to an average insolation of 250 W/m², what is the total energy output of 1 m² of photovoltaic cells (in joules). What quantity of coal would it replace (at 33% net energy conversion efficiency)?

[Answers. 39.5 GJ of photovoltaic system; 1 tonne of coal-equivalent delivers 29.3 GJ. So, the 1 m² solar module displaced 39.5/(29.3 × 0.33) = 4 t of coal equivalent.]

20.7 *Net metering* allows residential and commercial customers who generate their own electricity from solar power (or other renewable electricity) to sell the electricity they are not using back into the grid (Figure 20.11). For example, if a residential customer has a photovoltaic system, it may generate more electricity than the home uses during daylight hours. If the home is net-metered, the electricity meter will effectively run backward to provide a credit against what electricity is consumed at night or other periods when the home's electricity use exceeds the system's output. Customers are only billed for their "net" energy use. On average, only 20–40% of a solar energy system's output goes into the grid. This exported solar electricity serves nearby customers' loads.

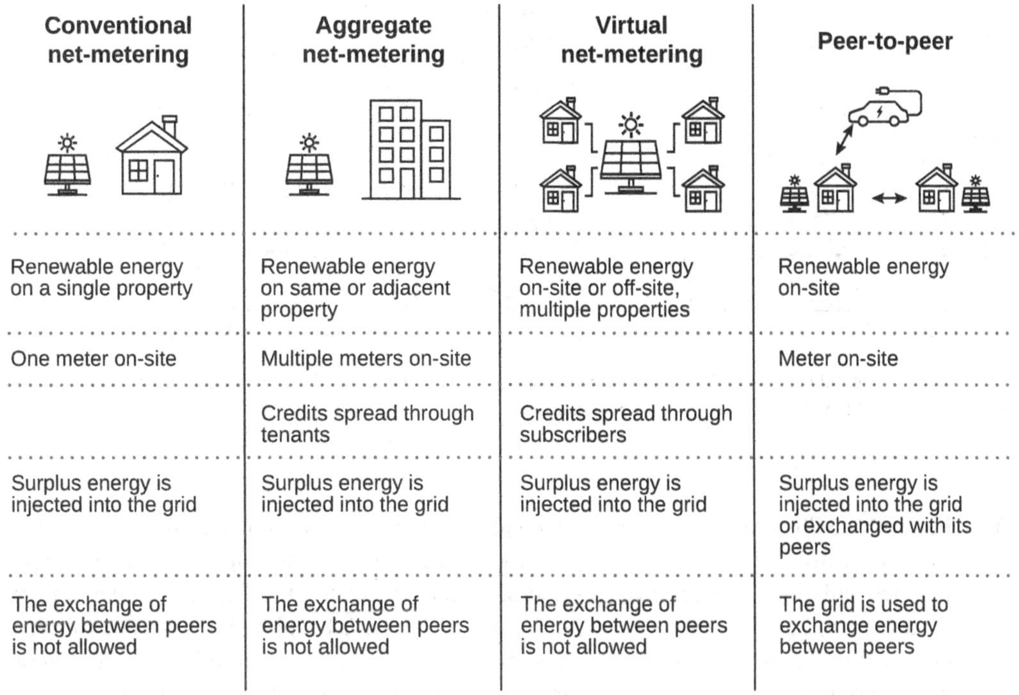

Conventional net-metering	Aggregate net-metering	Virtual net-metering	Peer-to-peer
Renewable energy on a single property	Renewable energy on same or adjacent property	Renewable energy on-site or off-site, multiple properties	Renewable energy on-site
One meter on-site	Multiple meters on-site		Meter on-site
	Credits spread through tenants	Credits spread through subscribers	
Surplus energy is injected into the grid	Surplus energy is injected into the grid	Surplus energy is injected into the grid	Surplus energy is injected into the grid or exchanged with its peers
The exchange of energy between peers is not allowed	The exchange of energy between peers is not allowed	The exchange of energy between peers is not allowed	The grid is used to exchange energy between peers

Figure 20.11 Schematic representation of net metering: conventional, aggregate, virtual, and peer-to-peer. Apart from the conventional, aggregate, and virtual net-metering techniques, *peer-to-peer net metering* (P2P) is a novel alternative. For instance, the P2P model could use the power grid for electricity exchange, while another option is to use electric vehicles. Simulations of such net-metering techniques show that the P2P model that uses the power grid has the best performance, followed by the P2P model that uses electric vehicles. *Source:* Adapted from Soto et al. (2022).

Many states in the United States have passed net metering laws. In some states, utilities may offer net metering programs voluntarily or as a result of regulatory decisions. Differences between state legislation, regulatory decisions, and implementation policies mean that the mechanism for compensating solar customers varies widely across the United States.

Produce a high-impact fact sheet of no fewer than 500 words that explains how net metering works and stating its advantages and disadvantages. Your fact sheet must also outline the various types of net-metering illustrated in Figure 20.14. You should initiate your research by studying the following paper: Soto et al. (2022).

[Answer: Open-ended question, with much of the answer locked up in the paper that should be studied.]

20.8 Figure 20.12 shows the typical characteristics of commercial photovoltaic modules. Write a 1000-word essay about the state of the art in photovoltaics, showcasing the top-performing photovoltaic cells and modules. Your essay should compare technical characteristics (band gap voltage, energy conversion efficiency, cell lifetime) and economic characteristics (US $/W and US $/MWh).

[Answer: Open-ended question.]

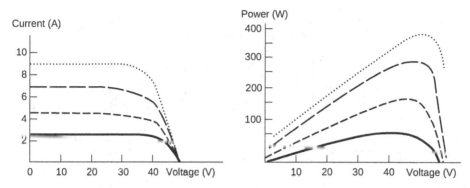

Figure 20.12 Typical current-voltage and power-voltage curves of modern 395-W photovoltaic modules that use crystalline cells, with various levels of irradiation.

20.9 Figure 20.13 illustrates the angles at which photovoltaic modules should be tilted toward the sun (between 25° and 40° latitudes) to provide maximum power output. Do independent research and then produce a two-page fact sheet that elucidates the science behind optimum tilt angles in the four seasons of a year. Your fact sheet must explain how engineers design photovoltaic arrays for homes with fixed-pitch roofs. Provide an example that shows how many photovoltaic modules are required on a house's roof with a 37.5° pitch if the roof faces north and if it faces south. The house is located in Chicago and the homeowners use 1380 kWh$_e$ per month. The photovoltaic modules are used in a net-metering scheme, and the system does not use any batteries.

[Answer: Open ended question that requires independent research. A south-facing roof will require about 18 modules and a north-facing roof will require about 24 modules.]

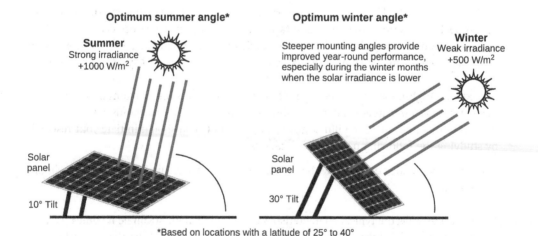

Figure 20.13 Optimum angles at which photovoltaic modules should be tilted in summer and winter locations with a latitude of 25–40°.

20.10 Study the following article about the material requirements for photovoltaic modules: Jean et al. (2015).

In this article, pay attention to the material requirements if solar electricity were to provide 5%, 50% or 100% of the global electricity demand in 2050. Based on the data presented in the associated graphs, discuss the implications of adopting 5%, 50%, and 100% solar electricity scenarios. For each of these scenarios, interpret the sustainability of solar electricity in terms of material requirements and the possibility of recycling solar cell material. Based on your analysis, do you think that it is viable (technologically, economically, and environmentally) for photovoltaic cells to provide 100% of global electricity requirements by 2050?

[Answer: Open ended question, with much of the answer tied up in the suggested reading.]

20.11 The expansion of large-scale photovoltaic power generation is essential to global efforts to mitigate climate change. A constraint to such photovoltaic plants is their extensive space requirements, complicated by increasing competition for land driven by population growth and rising food demand. Agrivoltaic systems (Figure 20.14), which integrate crop production and solar electricity generation, offer a potential solution to the land economy problem (Marrou 2019).

Services provided by agrivoltaics include renewable electricity generation, decreased greenhouse-gas emissions, increased crop yield, plant protection from excess solar energy, plant protection from inclement weather such as hail, water conservation, agricultural employment, local food, and increased revenue. Agrivoltaics also increase crop yield for

Figure 20.14 Jack's Solar Farm. *Source:* Werner Slocum.

a variety of crops. Along with solar electricity generation, this substantially increases land-use efficiency. Many crops grow better with some photovoltaic module-related shading because the photovoltaic array creates a microclimate beneath the modules that alters air temperature, relative humidity, wind speed, wind direction, and soil moisture content.

Perform an investigation into the energy production of an agrivoltaic system for different locations in the northern hemisphere. You must assess and compare the energy production performance of a fixed-tilt photovoltaic array with a variable-tilt system, such as the one proposed in the following paper: Vandewetering et al. (2022).

[Answer: Open-ended question that requires independent research.]

20.12 How sustainable is solar electricity? Re-read Section 20.7 and study the following paper in your attempt to answer this question: Nugent and Sovacool (2014).

Produce an infographic summarizing your findings and be sure to address the lifecycle emissions of CO_2 in the production of photovoltaic cells and the related social, technological, economic, environmental, and political issues.

[Answer: Refer to Section 20.7 and the cited paper.]

References

Angrist, S.W. (1982) *Direct Energy Conversion*. Boston, MA: Allyn and Bacon.

CER (2022). Most efficient solar panels 2022. *Clean Energy Reviews*. https://www.cleanenergyreviews.info/blog/most-efficient-solar-panels

Haegel, N.M. Verlinden, P., Victoria, M., Altermatt, P. et al. (2023). Photovoltaics at multi-terawatt scale: Waiting is not an option. *Science* 380 (6640): 39–42.

IEA (2022). Solar PV. International Energy Association (IEA). https://www.iea.org/reports/solar-pv

Jean, J., Brown, P.R., Jaffe, R.L., Buonassisi, T. et al. (2015). Pathways for solar photovoltaics. *Energy and Environmental Science*, 8 (4): 1200–1219.

Leccisi, E., Raugei, M. and Fthenakis, V. (2016). The energy and environmental performance of ground-mounted photovoltaic systems – a timely update. *Energies* 9 (8): 622.

Marrou, H. (2019). Co-locating food and energy. *Nature Sustainability* 2 (9): 793-794.

NREL (2022). Best research-cell efficiency chart. *National Renewable Energy Laboratory (NREL)*. https://www.nrel.gov/pv/cell-efficiency.html

Sørensen, B. (2017). *Renewable Energy. Physics, Engineering, Environmental Impacts, Economics and Planning*. Cambridge, MA: Academic Press/Elsevier.

Soto, E.A., Bosman, L.B., Wollega, E. and Leon Salas, W.D. (2022). Comparison of net-metering with peer-to-peer models using the grid and electric vehicles for the electricity exchange. *Applied Energy* 310: 118562.

Urbina, A. (2022). Sustainability of photovoltaic technologies in future net-zero emissions scenarios. *Progress in Photovoltaics: Research and Applications*, 1–15.

Yoo, J.J., Seo, G., Chua, M.R., Park, T.G. et al. (2021). Efficient perovskite solar cells via improved carrier management. *Nature* 590 (7847): 587–593.

21

Fuel Cells and Electrolyzers

Fuel cells are an important energy technology for a low-carbon future. They convert the chemical energy stored in oxygen and hydrogen *directly* to electricity and heat through an electrochemical reaction. Fuel cells are often more efficient than internal combustion engines because they do not require an intermediate energy transformation to heat (see Figure 1.1). Importantly, fuel cells produce electricity with zero local emissions, have high energy conversion rates, and produce only water when hydrogen is used as fuel. Furthermore, fuel cells have no moving components. They are therefore silent, and their maintenance is simpler and cheaper than their internal combustion engine counterparts.

In contrast to battery cells, fuel cells can operate continuously if there is a supply of oxygen and fuel (such as hydrogen). Because no carbon-containing fuels are required, fuel cells offer the promise of *zero-carbon electricity*. The hydrogen and oxygen required for fuel cells can be obtained in an environmentally friendly manner, such as from water splitting (electrolysis) and air separation, respectively. However, present fuel cell designs often use hydrocarbon fuels. Fuel cells also presently suffer from poor utilization of the catalyst that speeds up the reaction. The catalysts (such as platinum) are also very expensive.

Fuel cells are already widely used in vehicles, especially large vehicles such as trucks and buses (see Figure 21.1). The fuel cells give an increased range compared to battery electric vehicles and have a similar range to diesel-powered internal combustion engines. Fuel cells also refuel quickly. It takes only 10 min to refuel the citybus shown in Figure 21.1. Fuel cells can provide both heat and electricity at an overall efficiency of around 80%. They are vibration-free, quiet, and reliable. But they are expensive. The hydrogen citybus in Figure 21.1 costs around US $1.5 million, where a conventional diesel citybus of the same size would cost around US $400,000, and a diesel-electric citybus would set one back US $650,000.

21.1 Background to Fuel Cells

Fuel cells generate electricity by combining hydrogen and oxygen to make water. The fuel cell process is therefore the reverse of electrolysis. Fuel cells use catalysts to speed up the reactions at the electrodes.

Fuel cells are quickly becoming cost-competitive, but usually require expensive *low-carbon hydrogen generation* and *hydrogen storage*. Instead of using hydrocarbon fuels, hydrogen can be

Energy Systems: A Project-Based Approach to Sustainability Thinking for Energy Conversion Systems,
First Edition. Leon Liebenberg.
© 2024 John Wiley & Sons, Inc. Published 2024 by John Wiley & Sons, Inc.
Companion website: www.wiley.com/go/liebenberg/energy_systems

Figure 21.1 Two hydrogen fuel-cell buses parked beside the 1-MW electrolysis plant that produces their hydrogen fuel. The buses use an 85-kW fuel cell produced by Ballard, together with 150 kWh of lithium-ion batteries for onboard energy storage. The buses store approximately 68 kg of compressed hydrogen on their roofs, which provides a driving range of 400 km under urban driving conditions. *Source:* The Champaign-Urbana Mass Transit District (MTD).

produced by electrolyzing water with electricity obtained from wind, photovoltaics, or nuclear power. Although these energy sources are more renewable and less polluting, the combined energy conversion technologies are expensive.

In Chapter 7, we observed that entropy generation in combustion, and the subsequent heat transfer, lead to considerable irreversible work losses. The combustion loss can be attributed to the uncontrolled movement of electrons as new molecules are formed in the reaction. As a fuel cell can theoretically perform energy conversion *isothermally*, the Carnot limit to efficiency does not apply and fuel cells can attain energy conversion efficiencies greater than 50%. Fuel cells offer a process in which the electron movement is controlled; thus no combustion losses occur. The performance of fuel cells is, however, diminished by irreversible losses due to resistance, reaction activation, and mass flow restrictions.

In fuel cells, an electrochemical process directly converts the chemical energy stored in a fuel into electrical energy. Different types of fuel cells consume different forms of fuel, e.g., direct methanol fuel cells, direct carbon fuel cells, and hydrogen fuel cells. Let us first focus on hydrogen fuel cells, which have the following net chemical reaction:

$$H_2 + \frac{1}{2}O_2 \rightleftharpoons H_2O \, (\ell) \qquad \Delta H = -285.8 \, \text{kJ/mol} \tag{21.1}$$

The net chemical reaction is the same as if the fuel were burned (see Chapter 7), but the fuel cell intercepts the stream of electrons that flows from the fuel (hydrogen) to the oxidant (oxygen or air) and diverts it for use in an external circuit. This is facilitated by spatially separating the reactants (Figure 21.2 and Box 21.1).

Figure 21.2 shows that hydrogen and oxygen (pure or with air) are supplied to the two sides of the fuel cell in transport channels, and enter *gas diffusion layers* (GDLs), through which they travel toward the catalyst layer. The GDL shields the gas from the electrolyte and contributes to the management of water flows within the cell. The GDL acts as a transport channel for reactants and products that take part in the chemical reactions. It also acts as an electric current collector and conductor, as well as a transporter of electrons to the current collector. The GDL is porous and typically comprises carbon fiber-based materials. A fuel cell is therefore an electrochemical device that keeps the fuel molecules from mixing with the oxidizer molecules, but permitting the transfer of electrons via a metallic path that may contain a load.

The electrochemical reactions take place in the catalyst layers at the anode and the cathode. These are separated by an electrolyte through which electrons cannot pass. The electrons move from the anode to the cathode through the electrical device (with resistance R_d) and provide the power $\dot{W}$ to run the device. At the same time, ions move through the electrolyte (Figure 21.3).

Fuel oxidation takes place at the anode, and oxygen reduction takes place at the cathode:

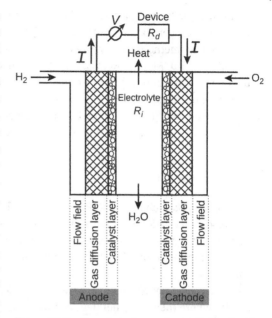

Figure 21.2 Generic hydrogen fuel cell. The fuel is supplied to the anode, and an oxidant is supplied to the cathode. The anode and cathode are separated by an electrolyte, which permits the flow of ions (but not of electrons or reactants) between the anode and the cathode.

- *Fuel oxidation (at the anode):* At the anode of a hydrogen-consuming fuel cell, the H_2 molecules dissociate into H^+ ions, which are free to migrate through the electrolyte to the cathode. Electrons can pass through the external load. The spontaneous dissociation is

$$H_2 \rightarrow 2H^+ + 2e^- \tag{21.2}$$

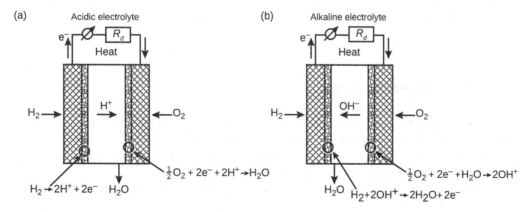

Figure 21.3 (a) A proton (acidic) exchange membrane (PEM) fuel cell, and (b) an alkaline anion exchange membrane fuel cell (AAEMFC).

and is sped up by using a catalyst that coats the anode. Platinum-based catalysts are most widely used for hydrogen oxidation. However, platinum (Pt) catalysts are easily "poisoned" by the CO produced by methanol/ethanol electro-oxidation at low temperatures. Alloying Pt with other metals (e.g., Ru, Mo, Sn, Re, Os, Rh, Pb, and Bi) can improve its performance. Due to platinum being expensive and a finite resource, researchers are working on eliminating such "critical" materials from fuel cells.

- *Oxygen reduction (at the cathode):* At the cathode, O_2 molecules dissociate into atoms and combine with electrons and the migrating H^+ ions to form water:

$$\frac{1}{2} O_2 + 2H^+ + 2e^- \rightarrow H_2O \tag{21.3}$$

Noble metals such as Pt and Pd are commonly used as primary catalysts for oxygen reduction. The oxygen reduction reaction at the cathode is kinetically sluggish compared to that of fuel oxidation at the anode. Scientists and engineers are therefore actively seeking better-performing catalysts, including combining Pt and Pd with other transition metals, such as Ru, Sn, Pb, Au, Ni, Ag, Cu, or Co.

An electric potential appears between the electrodes because of the excess of electrons at the anode (where they are generated) compared with the cathode (where they are consumed). This potential difference drives current through the external load, thus producing electrical power.

The reactions at the catalytic anode and the catalytic cathode, and the transport processes through the electrolyte depend on the type of electrolyte, which may be either acidic or alkaline. Proton exchange membranes feature acidic electrolytes (Figure 21.3a), and anion exchange membranes feature alkaline electrolytes (Figure 21.3b):

- With *proton exchange membrane* (PEM) fuel cells, cation species crossing the membrane are H^+, generated at the anode. The acidic electrolyte could comprise phosphoric acid (H_3PO_4) and a proton exchange membrane, which transport the H^+ species from the cathode to the anode.
- With *alkaline anion exchange membrane fuel cells* (AAEMFCs), typical anion species crossing the membrane are OH^-, generated at the cathode. The alkaline electrolyte (e.g., potassium hydroxide [KOH]) transports the OH^- species from the cathode to the anode. At the anode, the fuel (e.g., hydrogen, methanol, ethanol) in contact with the electrocatalysts undergoes an oxidation reaction to generate a molecule of water per single electron. In AAEMFCs, water is generated at the anode. At the same time, water is a reactant at the cathode.

21.2 Fuel Cell Performance

For the thermodynamic analysis of fuel cells, it is not necessary to distinguish between acidic and alkaline fuel cells. Even though the fuel is not combusted in a fuel cell as in a furnace, we still work with the value of the *enthalpy of combustion* as a fuel delivers the same amount of chemical energy in a fuel cell as when it is burned in a furnace.

For a chemical cell, the first and second laws of thermodynamics (Eqs. (3.12), (3.17) and (3.21)) combine to give:

$$T \, dS = dU + p \, d\cancel{V} - V \, dQ \tag{21.4}$$

where V is the potential to drive the charge Q, the enthalpy is $H = U + p\mathcal{V}$, and the Gibbs free energy is $G = H - TS$. We can then write Eq. (21.4) as follows:

$$dG = -S\,dT + \mathcal{V}\,dP - V\,dQ \tag{21.5}$$

For constant T and p, we obtain:

$$\Delta G = V\Delta Q \tag{21.6}$$

where ΔG *is the amount of electrical work* that the cell delivers.

From Eq. (21.1), the enthalpy of reaction for the combustion of hydrogen is -285.8 kJ/mol, while the standard Gibbs free energy is $\Delta G^0 = -237.1$ kJ/mol (see Table 7.5), so the second-law efficiency bound is about 83% at 298 K and 1 atm:

$$\eta = \frac{\Delta G^0}{\Delta H^0} \left(\equiv 1 - \frac{T\Delta S}{\Delta H} \right) = \frac{-237.1\text{ kJ/mol}}{-285.8\text{ kJ/mol}} \cong 0.83 \tag{21.7}$$

where $\Delta G^0 = eE_{th}$, e is the charge of an electron, and E_{th} is the theoretical cell voltage. The efficiency will always be less than unity as long as heat ($T\Delta S$) is discarded.

The amount of energy available as electricity is (ΔG), where ΔG (change in Gibbs free energy) is:

$$\begin{aligned} \Delta G &= \Delta H - T\Delta S \\ &= (-285.8 + 48.7)\text{ kJ/mol} = -237.1\text{ kJ/mol} \end{aligned} \tag{21.8}$$

where ΔH is the heat of combustion (285.8 kJ/mol) and the heat generated is $T\Delta S$ (48.7 kJ/mol). The Gibbs free energy therefore measures the energy in a system that is available to do work. A system seeks to minimize its free energy by both reducing its enthalpy and increasing the entropy.

We may write Eq. (21.1) as follows:

$$H_2 + \frac{1}{2}O_2 \rightleftharpoons H_2O\ (\ell) + 237.1\text{ kJ/mol (electricity)} + 48.7\text{ kJ/mol (heat)} \tag{21.9}$$

The efficiency of fuel cells is high (typically >90%), because the heat dissipated ($T\Delta S$) to the environment is small compared to the change in enthalpy (ΔH). Box 21.2 explains how to calculate the second-law efficiency bound of a fuel cell.

Box 21.1 Energy Contents of Hydrogen

Using the value stated in Eq. (21.3), the energy contents of hydrogen available for conversion to electricity (in MJ/kg) is:

$$\left(\frac{237.1\text{ kJ}}{\text{mol}}\right)\left(\frac{1\text{ mol H}_2}{2.0158\text{ g}}\right)\left(\frac{1000\text{ g}}{1\text{ kg}}\right) = 117.6\text{ MJ/kg}$$

or, in terms of kWh/kg:

$$\left(\frac{117.6\text{ MJ}}{\text{kg}}\right)\left(\frac{1\text{ Ws}}{\text{J}}\right)\left(\frac{1\text{ kW}}{1000\text{ W}}\right)\left(\frac{1\text{ h}}{3600\text{ s}}\right) = 32.7\text{ kWh/kg}$$

Note: The higher heating value (*HHV*) of hydrogen gas is 141.7 MJ/kg (Table 7.5). Multiplying the *HHV* by the Second-Law efficiency bound (Eq. (21.4)) of 83% gives 117.6 MJ/kg, in accordance with our calculations above.

Box 21.2 The Second-Law Efficiency Bound of a Methanol Fuel Cell

Instead of hydrogen (H_2), a fuel cell could also use methanol (CH_3OH) as fuel. We can estimate the second-law efficiency of such a fuel cell using Eqs. (21.4) and (21.5).

The net reaction of methanol is:

$$CH_3OH\ (\ell) + \frac{3}{2}O_2 \rightarrow CO_2\ (g) + 2H_2O\ (\ell)$$

The enthalpies and free energies of formation can be obtained from sources such as the NIST-JANAF tables, https://janaf.nist.gov

Compound	ΔH_f^0 (kJ/mol)	ΔG_f^0 (kJ/mol)
$CH_3OH\ (\ell)$	−239	−166
$H_2O\ (\ell)$	−286	−237
$CO_2\ (g)$	−394	−394

The standard enthalpy and Gibbs free energy of the net methanol reaction are:
From Eqs. (7.13) and (7.26), we may write the following:

$$\Delta H_{reaction} = \sum_{products} \Delta H_f - \sum_{reactants} \Delta H_f$$

$$\Delta G_{reaction} = \sum_{products} \Delta G_f - \sum_{reactants} \Delta G_f$$

Therefore:
$$\Delta H^0 = 2(-286\ kJ/mol) + (-394\ kJ/mol) - (-239\ kJ/mol) = -727\ kJ/mol$$
$$\Delta G^0 = 2(-237\ kJ/mol) + (-394\ kJ/mol) - (-166\ kJ/mol) = -702\ kJ/mol$$

The second-law efficiency bound is:

$$\eta = \frac{\Delta G^0}{\Delta H^0} = \frac{-702\ kJ/mol}{-727\ kJ/mol} \cong 0.97$$

Note: Methanol therefore has a second-law efficiency bound of 97% compared to 83% for hydrogen (Eq. (21.4)). However, unlike a hydrogen fuel cell, a methanol fuel cell produces CO_2, which is highly undesirable.

Importantly, if the entropy of the products is greater than the entropy of the reactants, then ΔS will be positive and the reaction will absorb heat ($T\Delta S$) from the environment, so that η will be larger than 1 (as illustrated in Table 21.1 for formic acid). This highlights the main virtue of fuel cells: that they are successful in liberating electrical energy, not thermal energy.

Reimagine Our Future

Every energy conversion system has side-effects and limitations. A global shift to new forms of power production (e.g., more solar or nuclear energy) will have new sustainability ramifications. Do you think we can handle both the rapid phasing out of fossil fuels *and* the sustainability ramifications of solar and nuclear energy?

21.3 Potential Difference (Voltage) and Current

From Eq. (21.6), the voltage generated is given by equating ΔG to the total charge ΔQ that flows:

$$\Delta G = \Delta Q \ V = nN_A eV \tag{21.10}$$

where n is the number of electrons released per mole ($n = 2$ for H_2, Eq. (21.2)), N_A is Avogadro's constant (number of molecules in 1 mol), e is the charge of an electron, and V is the voltage generated per cell. The product ($N_A e$) is known as *Faraday's constant* (F), with

$$F = 9.65 \times 10^4 \ \text{C/mol} \quad \text{or} \quad 9.65 \times 10^4 \ \text{J/V} \tag{21.11}$$

The lower value of the cell potential is due to the minimum resistance to transport of charge carriers, which gives rise to ΔV. In general, for zero-current potential V_0, we have:

$$V = V_0 - \Delta V \tag{21.12}$$

and ΔV increases as greater current is drawn from the cell.

The electrical current produced by a fuel cell may be approximated as follows:

$$I = n \dot{n}_{H_2O} e \tag{21.13}$$

where n is the number of electrons released per mole ($n = 2$ for H_2, Eq. (21.2)), $\dot{n}_{H_2O}$ is the flowrate of the product (water) in moles per second, and e is the electronic charge.

The electrical power obtained from a fuel cell may be calculated by:

$$\dot{W} = VI \tag{21.14}$$

A fuel cell's overall performance is often characterized by its *j-V* (current density versus voltage) curve and by its *power density*. Power density is the product of j and V and is usually given in mW/cm^2, j is given in mA/cm^2, and V is given in volts (V).

Box 21.3 explains how to estimate the consumption of air and hydrogen, as well as the production of water, in a fuel cell.

Figure 21.4 shows the typical *j-V* and power density characteristic curves of a proton exchange membrane (PEM) fuel cell. The efficiency of the fuel cell relative to its ideal efficiency – given by the ratio of the operating voltage at a given current density to the ideal theoretical voltage – can be inferred from its current density versus voltage curve, as depicted in Figure 21.4. A major objective of research is to reduce the rate at which a fuel cell's potential drops as current increases and thereby to increase its efficiency and the current density it can sustain.

The deviation of the potential from the theoretical value is called the "overpotentials" and may be ascribed to several sources:

- *Electrode losses:* When power is drawn from a fuel cell, the terminal voltage decreases from $V = \dfrac{\Delta H}{nN_A e}$ to lower than $\dfrac{\Delta G}{nN_A e}$. This reflects not only the $I^2 R$ (ohmic, or Joule) drop in the cell but also the potential losses associated with the irreversibility of various processes at the electrode–electrolyte interface. These losses are dependent on the catalytic properties of the electrode surface. Catalysts are used to increase the rate of the reactions at the electrodes, resulting in an increase in entropy due to irreversibility and a corresponding decrease in voltage (Soo 1968).
- *Activation losses*, caused by the slowness (or "sluggishness") of the reaction taking place on the surface of the electrodes. A proportion of the voltage generated is lost in driving the chemical reaction that transfers the electrons.

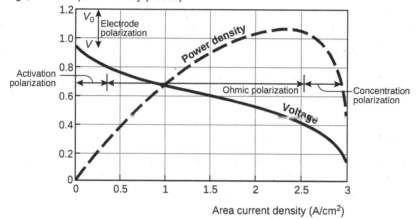

Figure 21.4 Typical *j-V* and power density characteristic curves of a PEM fuel cell. The contributions of the main losses in a fuel cell are also illustrated: electrode losses, activation losses, ohmic losses, and concentration losses.

- *Ohmic losses*, which is the voltage drop due to the resistance to the flow of electrons through the electrodes. This loss varies linearly with current density.
- *Concentration losses*, which result from the change in concentration of the reactants at the surface of the electrodes as the fuel is used.
- *Fuel crossover losses*, which result from the waste of fuel passing through the electrolyte, and electron conduction through the electrolyte. This loss is typically small but can be important in low-temperature cells.

Example 21.1 Potential Difference Across a Hydrogen Fuel Cell
Find the potential difference generated across a fuel cell powered by hydrogen.
Required: We must find V for a fuel cell powered by hydrogen.
Solution strategy: From Eqs. (21.5) and (21.7),

$$\Delta G = \Delta Q V = n N_A e V = 237.1 \times 10^3 \text{ J/mol}$$

with $n = 2$ electrons (two electrons are released per molecule of hydrogen), and $(N_A e) = 9.65 \times 10^4$ C/mol. We can then use Eq. (21.10) to calculate the voltage generated by the hydrogen reaction.
Analysis:
The voltage generated is:

$$V = \frac{\Delta G}{n(N_A e)} = \frac{237.1 \times 10^3 \text{ J/mol}}{2(9.65 \times 10^4 \text{ C/mol})} = 1.23 \text{ V}$$

Each cell thus ideally produces 1.23 V between the anode and the cathode.
Discussion:
To obtain a voltage of practical usage, fuel cells are stacked in series to increase their voltage. For instance, to obtain a potential difference of 200 V, a stack of $\dfrac{200 \text{ V}}{1.23 \text{ V/cell}} = 162$ cells connected in series is required.

Box 21.3   Consumption of Air and Hydrogen, and Production of Water in a Fuel Cell

The air usage in a hydrogen-powered fuel cell may be estimated using the stoichiometric equations in Chapter 7:

From Chapter 7, there is 0.21 mol O_2 in 1 kg air, and dry air has a molar mass of about 29 g/mol.

The *air usage* of a hydrogen-powered fuel cell is therefore

$$= \frac{(29 \times 10^{-3} \text{ kg/mol})\dot{W}}{(0.21 \text{ mol } O_2/\text{kg air})(4 \text{ electrons})V(eN_A)} \quad [\text{kg/s}]$$

$$\therefore \text{Air usage} = 3.57 \times 10^{-7}\left(\frac{\dot{W}}{V}\right) \quad [\text{kg/s}] \tag{21.15}$$

The *hydrogen usage* of a hydrogen-powered fuel cell is

$$= \frac{\dot{W}}{(2 \text{ electrons})V(eN_A)} \quad [\text{mol/s}]$$

But the molar mass of hydrogen is 2.02×10^{-3} kg/mol. So, the hydrogen usage (in kg/s) is:

$$\therefore \text{Hydrogen usage} = \frac{(2.02 \times 10^{-3} \text{ kg/mol})\dot{W}}{(2 \text{ electrons})V(eN_A)} = 1.51 \times 10^{-8}\left(\frac{\dot{W}}{V}\right) \quad [\text{kg/s}] \tag{21.16}$$

The *water production* in a hydrogen-powered fuel cell

$$= \frac{\dot{W}}{(2 \text{ electrons})V(eN_A)} \quad [\text{mol/s}]$$

But the molar mass of H_2O is 18×10^{-3} kg/mol. So, the water production (in kg/s) is:

$$\therefore \text{Water production} = 9.34 \times 10^{-8}\left(\frac{\dot{W}}{V}\right) \quad [\text{kg/s}] \tag{21.17}$$

For instance, an 85-kW hydrogen-powered fuel cell operating for 1 h at a cell voltage of 0.7 V, produces $9.34 \times 10^{-8}\left(\dfrac{\dot{W}}{V}\right) = (9.34 \times 10^{-8})\dfrac{(85 \times 10^3 \text{ J/s})}{0.7 \text{ V}} = 0.01$ kg/s. In 1 h, it will produce $\left(\dfrac{0.01 \text{ kg}}{\text{s}}\right)\left(\dfrac{3600 \text{ s}}{\text{h}}\right) = 36$ kg of water.

For a hydrogen fuel cell, therefore, the generation of 1 kWh$_e$ produces about 500 mL of water.

21.4   Hydrogen and Other Fuels

Most hydrogen is currently produced by the steam reforming of hydrocarbons. It can also be produced from biomass, by heating coal to about 1000 °C in the absence of oxygen, by the partial burning of coal in the presence of steam (see Chapter 8), or by electrolysis. With the production from coal, a mixture of hydrogen with CO, CO_2, and other gases is obtained, which needs to be separated to obtain pure hydrogen. The generation of CO_2 can be avoided if the electricity is generated from nuclear, hydroelectric, wind, solar, or geothermal energy. The cost of hydrogen fuel is estimated to settle at around US $8 to US $10 per kg between now and 2050.

Table 21.1 summarizes the salient thermodynamic and other properties of hydrogen and other fuels used in fuel cells. Clearly, hydrogen has a low volumetric energy density. Even when

Table 21.1 Hydrogen and other fuels used in fuel cells, compared to diesel and gasoline.

Fuel	Chemical reaction or chemical formula	Change in enthalpy per atom of fuel, ΔH (kJ/mol)	Change in Gibbs free energy per atom of fuel, ΔG (kJ/mol)	Charge transferred per atom of fuel, e (C)	Theoretical voltage, E_{th} (V)	Theoretical efficiency, η (%)	Power density, (Wh/kg)
Hydrogen	$H_2 + \frac{1}{2}O_2 \rightarrow H_2O$	−285.8	−237.1	3.2×10^{-19}	1.23	83	See below
Methane	$CH_4 + 2O_2 \rightarrow CO_2 + 2H_2O$	−892.5	−820.2	1.3×10^{-18}	1.06	91.9	14,889
Methanol	$CH_3OH + \frac{3}{2}O_2 \rightarrow CO_2 + 2H_2O$	−728.5	−703.4	9.6×10^{-19}	1.21	97.7	5472
Formic acid	$HCOOH + \frac{1}{2}O_2 \rightarrow CO_2 + H_2O$	−271.1	−285.6	3.2×10^{-19}	1.48	105.6	1450
Hydrazine	$N_2H_4 + O_2 \rightarrow N_2 + 2H_2O$	−623.3	−603.1	6.4×10^{-19}	1.56	96.8	5416
Carbon monoxide	$CO + \frac{1}{2}O_2 \rightarrow CO_2$	−283.7	−257.6	3.2×10^{-19}	1.34	90.9	6300
Diesel	$C_nH_{1.8n}$ (ℓ)						11,890
Gasoline	$C_nH_{1.8n}$ (ℓ)						12,070
Liquid H_2	H_2 (ℓ)						1900
H_2 at 34.5 MPa	H_2 (g)						1800
Hydride	H^-						400

Source: Carrette et al. (2000), Hren et al. (2021), and NIST-JANAF (1998).

compressed to 34.5 MPa, hydrogen gas still only has an energy density of 1800 Wh/kg compared to 12,070 Wh/kg of gasoline at atmospheric pressure. Liquifying the hydrogen increases the volumetric energy density to 1400 Wh/kg, but it is still more than six times less than that of gasoline, and one must contend with the complexities and energy requirements of cryogenic liquefaction.

21.5 Types of Fuel Cells

A variety of fuel cells are in different stages of development. They differ from one another regarding their operating parameters (e.g., temperature and relative humidity) and technical characteristics, such as power density and efficiency. The general design of fuel cells is similar, except for the *electrolyte*, which determines the operating temperature. The operating temperature, in turn, determines the physicochemical and thermo-mechanical properties of materials used in the cell components (e.g., electrodes, electrolyte, interconnect, current collector). Together with the classification by electrolyte, some fuel cells may be classified by the type of fuel used such as direct methanol fuel cells (DMFCs) and direct carbon fuel cells (DCFCs).

Table 21.2 summarizes the key characteristics of the five main classes of fuel cells: *polymer electrolyte fuel cells* (PEFCs), which may be proton exchange membrane fuel cells (PEMFCs) or anion

Table 21.2 Main features of salient classes of fuel cells.

	PEFC		PAFC	MCFC	SOFC
	PEMFC	**AEMFC**			
Electrolyte	Phosphoric acid, hydrated polymeric ion exchange membranes	Potassium hydroxide, hydrated polymeric ion exchange membranes	Hydrated polymeric ion exchange membranes	Immobilized liquid molten carbonate in $LiAlO_2$	Perovskites (ceramics)
Electrodes	Carbon; transition metals, e.g., Ru, Sn, Pb, Au, Ni, Ag, Cu, or Co; covalent organic frameworks (COFs)		Carbon	Nickel and nickel oxide	Perovskite and Perovskite / metal cermet
Catalyst	Platinum, palladium, carbon, and other transition metals, e.g., Ru, Sn, Pb, Au, Ni, Ag, Cu, or Co; also alloys with Mo, Sn, Re, Os, Rh, Pb, and Bi		Platinum	Nickel and nickel oxide	Perovskite and Perovskite / metal cermet
Interconnect	Carbon or metal		Graphite	Stainless steel or nickel	Nickel, ceramic, or steel
Operating temperature (°C)	40–100	50–220	150–250	600–700	500–1000
Charge carrier	H^+	OH^-	H^+	CO_3^{2-}	O^{2-}

(Continued)

Table 21.2 (Continued)

	PEFC		PAFC	MCFC	SOFC
	PEMFC	AEMFC			
Power density (mW/cm^2)	300–1000	150–400	150–300	100–300	250–350
Typical stack size (kW)	1–100	10–100	100–400	0.3–3000	1–2000
CO tolerance	Poison (> 50 ppm)		Poison (> 1%)	Fuel	Fuel
Efficiency	45–60	40–60	55	60–65	55–65

PEFC = Polymer electrolyte fuel cell; PEMFC = Proton exchange membrane fuel cell; AEMFC = Anion exchange membrane fuel cell; PAFC = Phosphoric acid fuel cell; MCFC = Molten carbonate fuel cell; SOFC = Solid oxide fuel cell.
Source: Ma and Lutkenhaus (2022), Zhang et al. (2022), and Hren et al. (2021).

exchange membrane fuel cells (AEMFCs), *alkaline fuel cells* (AFCs), *phosphoric acid fuel cells* (PAFC), *molten carbonate fuel cells* (MCFC), and *solid oxide fuel cells* (SOFC).

Aqueous electrolytes are limited to temperatures of about 200 °C or lower because of their high vapor pressure and rapid degradation at higher temperatures. Hence, the AFC, PAFC, and PEMFC are considered low-temperature fuel cells, whereas the MCFC and SOFC are considered high-temperature fuel cells. Operating temperature also plays an important role in determining the degree of fuel processing required. In low-temperature fuel cells, all the fuel must be converted to hydrogen prior to entering the fuel cell. The anode catalyst in low-temperature fuel cells (mainly platinum, Pt) is also strongly poisoned by carbon monoxide (CO). In high-temperature fuel cells, CO and even CH_4 can be internally converted to hydrogen or even directly oxidized electrochemically.

The cost of fuel cells ranges between US $20 and US $30 per kilowatt (Zhang et al. 2022).

Reimagine Our Future

To reduce our carbon footprint, we must seize combusting fossil fuels. Also, it is technologically feasible, economically viable, and is the intelligent choice to rapidly shift to a global society powered by renewable energy and employing energy efficiency ("doing more with less"). We are however not doing this at the required speed and scale. Has the magnitude of the risk perhaps exceeded our capacity to comprehend and respond?

21.6 Electrolyzers

Fuel cells need hydrogen as fuel. As discussed in Section 21.4, hydrogen could be obtained from *water electrolysis* instead of using greenhouse gas-emitting fossil fuels such as methanol. If water electrolysis uses electricity from renewable sources (such as photovoltaics or wind power), the result is *green hydrogen* with nearly no emissions over its lifecycle.

In 2021, water electrolysis accounted for only around 0.1% of global hydrogen production. The installed capacity of electrolyzers is expanding quickly and reached 510 MW by the end of 2021, an increase of 210 MW, or 70% relative to 2020. The completion of the Ningxia Solar Hydrogen Project in China with an electrolyzer capacity of 150 MW accounted for almost three-quarters of the increase. Today it is the world's largest electrolyzer in operation (IEA 2022).

The three major types of water electrolyzer cells are alkaline, proton exchange membrane, and solid oxide electrolysis. In 2021, almost 70% of the installed capacity was alkaline electrolysis, followed by PEM electrolyzers, which accounted for one-quarter of the installed capacity. Other emerging technologies are solid oxide electrolysis cells and anion exchange membrane electrolysis, but they are less mature than alkaline and PEM electrolyzers, and represent only a minimum share of the installed capacity today. Table 21.3 summarizes the salient characteristics of these three types of electrolyzers. Electrolyzer costs in 2022 were around US \$850/kW$_e$. Costs are expected to drop to US \$200/kW$_e$ by 2030 and US \$100/kW$_e$ by 2050 (Bloomberg 2021).

Table 21.3 Main features of alkaline, proton exchange membrane, and solid oxide cell electrolyzers.

	Alkaline electrolyzer	Proton exchange membrane electrolyzer	Solid oxide electrolyzer
Commercial status	Mature	Commercial, fast growth	Demonstration plants
Energy conversion efficiency (kWh/kg hydrogen)	Today: 53 2030: 47 2050: 42	Today: 59 2030: 49 2050: 45	Today: 45 2030: 40 2050: 37
Electrolyte	Potassium hydroxide	Perfluorosulfonic acid	Y_2-O_3-ZrO_2; Sc_2O_3-ZrO_2; MgO-Zr-O_2; CaO-ZrO_2O^{2-}
Charge carrier	OH^-	H^+	
Overall reaction	$H_2O \rightarrow H_2 + \frac{1}{2}O_2$		
Anode reaction	$2OH^- \rightarrow \frac{1}{2}O_2 + H_2O + 2e^-$	$H_2O \rightarrow \frac{1}{2}O_2 + 2H^+ + 2e^-$	$O_2^- - 2e^- \rightarrow \frac{1}{2}O_2$
Cathode reaction	$2H_2O + 2e^- \rightarrow H_2 + 2OH^-$	$2H^+ + 2e^- \rightarrow H_2$	$H_2O + 2e^- \rightarrow H_2 + O_2^-$
Anode catalyst	Ni_2CoO_4; La-Sr-CoO_3; Co_3O_4	Ir/Ru oxide	$(La,Sr)MnO_3$, $(La,Sr)(Co,Fe)O_3$
Cathode catalyst	Nickel foam/Ni-stainless steel Ni-Mo/ZrO_2–TiO_2	platinum	Ni–YSZ or Ni-GDC cermet
Interconnect	Asbestos, polysulfone-bonded polyantimonic acid, ZrO_2 on polyphenylsulfone, NiO, polysulfone impregnated with Sb_2O_5 polyoxide	Polymer membrane	Ceramic
Operating temperature (°C)	60–80	50–80	650–1000

(Continued)

Table 21.3 (Continued)

	Alkaline electrolyzer	Proton exchange membrane electrolyzer	Solid oxide electrolyzer
Operating pressure (bar)	1–200	1–350	1–30
Plant footprint (m^7/kW)	0.095	0.048	Unknown
Efficiency (%)	60–80	80	100
Characteristics	Slower ramp-up and ramp-down times compared to proton exchange membrane cells	Fast dynamic response	Highest efficiency Must continuously run at high temperatures and can therefore (usually) not be ramped-up and down
Implications	Not well suited to intermittent electricity supply (e.g., photovoltaics, wind), likely to be overcome by innovation for faster ramping and batteries to smoothen short-term variations	Well-suited to a variable electricity supply (e.g., photovoltaics, wind). Suitable for voltage regulation services	Potentially well suited for constant base-load hydrogen production Companies such as DynElectro are developing DC:AC conversion systems for solid oxide electrolyzers that will negotiate fluctuating electricity supplies from wind and photovoltaic systems Solid oxide electrolyzers can operate in reverse as fuel cell electricity generators
Stack lifetime (hours)	90,000–100,000	10,000–60,000	500–2000
Major producers	Suzhou Jingli, thyssenkrupp, Nel Hydrogen	Siemens, ITM Power, Cummins	Haldor Topsøe, Ceres, Sunfire

Source: Bloomberg (2021), Siemens (2022), and IEA (2022).

Cracking the "Green Hydrogen" Code

Hydrogen has enormous potential for energy storage, as a transportation fuel, as a source of clean dispatchable power 24/7, and to decarbonize heavy industry. As hydrogen that is produced using renewable energy such as wind or solar, green hydrogen is especially sought after. Although this technology is being exploited globally, green hydrogen is typically three times as expensive as fossil fuels.

The work of **Anne Lyck Smitshuysen** (28) could help make the large-scale generation of hydrogen from renewable energy sources economically viable within this decade.

Anne is passionate about green energy solutions. "The idea of being a part of something that makes the world a better place motivates me a lot," says the Danish physicist. In 2021, Anne won the 2021 Flemming Bligaard Award and a cash prize of US $70,000 for demonstrating how to increase the size of the solid oxide electrolysis cells used to produce green hydrogen, thereby cutting total costs by as much as 15%.

Anne Lyck Smitshuysen at her lab in Denmark. *Source:* Nina Houkjær Klausen / Rambøll.

Because the solid oxide cells are thin and delicate, they often curl at the edges during production when they are heated to 1300 °C. This renders them unusable. This is why her solution is so innovative. "I realized that by using a 3D-printed cast to mold the cells in a way that corresponds to the shrinkage caused by the heating process, the size of the cells could be increased by 500%," she explains.

Anne worked with scientists and engineers at the start-up DynElectro and the Technical University of Denmark to increase the size of electrolysis cells from 15 to 1000 cm^2 without breaking the cells during construction. "The results so far show that the method works better the larger the cells get, but from an economic point of view, there is a limit to how much we can save, so as of now, we have settled on 1000 cm^2." Anne and her colleagues at DynElectro are now working hard at converting her lab findings to industry-ready applications.

"It still feels unreal that I am a part of 'cracking the green hydrogen code'," enthuses Anne. "You get a little humbled when journalists and people from abroad suddenly take an interest in you. It is crazy to be part of making a green future. But the best part is that others can suddenly see the cool in what I – and many others – have been geeking out with for years," she quips.

Reimagine Our Future

Producing hydrogen by electrolyzing water is energy intensive. More energy is required to create, compress (or liquefy), and transport hydrogen than is available at the point of consumption. However, hydrogen fuel cells are still being used in applications such as city buses and ferries. Why are we then doing this?

21.7 Outlook for Fuel Cells and Electrolyzers

Hydrogen and fuel cells are now widely regarded as one of the key energy solutions for the twenty-first century. These technologies will contribute significantly to a reduction in environmental impact, enhanced energy security (and diversity), and the creation of new-energy industries. Hydrogen and fuel cells can be utilized in transportation, distributed heat and power generation, and energy storage systems (as is further discussed in Chapter 27). However, the transition from a carbon-based (fossil fuel) energy system to an energy system in which hydrogen-based energy plays a major role involves significant scientific, technological, and socioeconomic barriers.

As with other direct energy conversion systems, advances in fuel-cell technology will be driven by advances in materials science and quantum chemistry. For instance, *metal–organic framework* (MOF)-based materials and their derivatives are emerging as unique electrocatalysts for oxygen reduction reactions. The use of MOFs leads to improved energy conversion efficiencies due to their tunable compositions and diverse structures, which speed up the sluggish oxygen reduction reactions at the cathode in fuel cells. The high cost of producing MOFs and the limited number of synthesis methods are hampering the further development of MOFs, but this is bound to change as more engineers and scientists start to work in this nascent field.

Green hydrogen (hydrogen produced with renewable electricity) is the Holy Grail of fuels for fuel cells. Green hydrogen is produced by splitting water into oxygen and hydrogen using electrolyzers. The rapid scale-up in electrolyzer capacity is expected to continue and accelerate in coming years. Globally, more than 460 electrolyzer projects were under development in 2022. The global electrolyzer capacity reached around 1.4 GW in 2022, almost tripling the 2021 level. About 40% of this capacity is developed in China and around a third is developed in Europe. Based on the current project pipeline, global electrolyzer capacity could reach 134 GW in 2030. While the average plant size of new electrolyzers starting operation in 2021 was 5 MW, the average size of new plants could be around 260 MW in 2025 and in the GW-scale by 2030. Of the projects under construction or under development, 22 (or 5%) are above 1 GW (IEA 2022).

The next few years will no doubt see dramatic improvements in the efficiency and stability of electrocatalysts. Breakthroughs in materials science and quantum chemistry will also lead to fuel cells with improved tolerance against reaction poisons (e.g., CO_x, halide ions, SO_x, and NO_x). With these advances and the pressing need for sustainable energy, the fuel cell market is predicted to grow from US $3 billion in 2022 to more than US $9 billion by 2027.

Problems

21.1 A city bus is equipped with a hydrogen fuel cell propulsion system. The 80-kW (approximately 260 V, 300 A) proton exchange membrane fuel cell stack is connected to a lithium-ion battery pack with a capacity of 150 kWh (i.e., approximately 415 V and 360 Ah). The fuel cell stack charges the batteries, and electricity is drawn from the batteries to a traction motor (Figure 21.5).

   a) If every fuel cell in the stack has a conversion efficiency ($\Delta G/\Delta H$) of 54.5%, calculate the voltage that each cell in the fuel cell stack will produce.

   b) How many fuel cells will need to be stacked in series to provide the required voltage [i.e., 260 V]?

(Answers: (a) 0.81 V; (b) 323 stacked fuel cells)

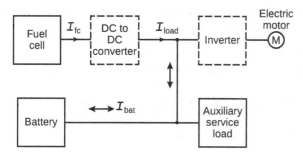

Figure 21.5 Fuel cell system for a citybus.

21.2 The city-bus in Problem 21.1 performed as follows during a test drive in the city:

- *Maximum speed*: 56.2 km/h
- *Average speed*: 25.7 km/h

The detailed test data taken during the 1656-second test drive is shown in Figure 21.6. Based on the above experimental data, what will the bus's driving range be (in kilometers) if it stores 65 kg of hydrogen onboard the bus?

(Answer: Around 275 km)

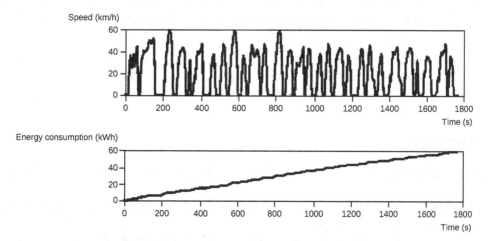

Figure 21.6 Speed and cumulative energy consumption of a city-bus driving a typical urban route. *Source:* Kivekäs et al. (2018) / MDPI / CC BY 4.0 / Public domain.

21.3 Compare the range of the fuel cell bus with that of a diesel bus of a similar size that operates on the same driving cycle as the one in Problem 21.2. Provide the energy densities of diesel and hydrogen in MJ/L and kWh/kg – show all your calculations.

(Answers: (a) The hydrogen fuel cell bus has a range of 275 km, and the diesel-powered bus has a range of 550 km. (b) For the hydrogen fuel cell bus, 0.3 MJ/L and 21.5 kWh/kg, and for the diesel-powered bus, 37 MJ/L and 10.2 kWh/kg.)

21.4 An energy entrepreneur wants to convert the diesel-powered internal combustion engine drivetrain of a 22-m ferry to an electric drivetrain powered by hydrogen fuel cells. The vessel will carry around 70 passengers.

a) Calculate the total power required to move this vessel at a maximum speed of 20 knots (1 knot is about 0.5 m/s).

The total power can be estimated using the following relation:

$$P = \left(\frac{1}{2} C_F \rho A_{transverse} v^3\right)_{friction} + \left(\frac{1}{2} C_D \rho A_{front} v^3\right)_{wave}$$

The wave-drag coefficient of the ferry is $C_D = 0.4$ and the skin-friction coefficient is $C_F = 0.002$. The ferry's hull has an underwater frontal area (A_{front}) of 20 m^2 and an underwater transverse area ($A_{transverse}$) of 60 m^2.

b) How large must the fuel cell be (in kW)?

c) If the ferry must travel at 15 knots for 3 h at a time, what is the required storage capacity (in m^3) of the hydrogen tanks? Assume 55% energy conversion efficiency and that the hydrogen is stored at 680 kPa and −240 °C. Explain the challenges related to the required volume to store hydrogen fuel.

d) If the developers of this hydrogen fuel-cell ferry were to borrow all the money required to implement the fuel cell (with hydrogen and battery storage), what would the annual loan repayments be for a 20-year, 6% loan on the total cost of US $1,491,220. Also analyze the payback period of this fuel cell system. You may assume that the fuel cell system has a capacity factor of 25%. You should base these economic calculations on your researched costs to implement this fuel cell and associated hydrogen and battery storage.

(Answers: (a) around 4500 kW; (b) 4500 kW; (c) 100 m^3; (d) US $130,011 per year; payback period is about seven years.)

References

Bloomberg (2021). New Energy Outlook, 20210. Bloomberg NEF. https://about.bnef.com/new-energy-outlook

Carrette, L., Friedrich, K.A. and Stimming, U. (2000). Fuel cells: principles, types, fuels, and applications. *ChemPhysChem* 1 (4): 162–193.

Hren, M., Božič, M., Fakin, D., Kleinschek, K.S. et al. (2021). Alkaline membrane fuel cells: anion exchange membranes and fuels. *Sustainable Energy and Fuels* 5 (3): 604–637.

IEA (2022). Global Hydrogen Review 2022. International Energy Agency (IEA). https://www.iea.org/reports/global-hydrogen-review-2022

Kivekäs, K., Lajunen, A., Vepsäläinen, J. and Tammi, K. (2018). City bus powertrain comparison: driving cycle variation and passenger load sensitivity analysis. *Energies* 11 (7): 1755.

Ma, T. and Lutkenhaus, J.L. (2022). Hydrogen power gets a boost. *Science* 378 (6616): 138–139.

NIST-JANAF (1998). NIST-JANAF thermochemical tables. http://www.janaf.nist.gov

Siemens (2022). Hydrogen Solutions. Siemens. https://www.siemens-energy.com/global/en/offerings/renewable-energy

Soo, S.L. (1968). *Direct Energy Conversion*. Englewood Cliffs, NJ: Prentice-Hall.

Zhang, Q., Dong, S., Shao, P., Zhu, Y. et al. (2022). Covalent organic framework–based porous ionomers for high-performance fuel cells. *Science* 378 (6616): 181–186.

Mini Project 7

Photovoltaic Car Canopy

Solar (photovoltaic) farms are proliferating on undeveloped land, often harming ecosystems. However, placing photovoltaic canopies on large parking lots offers a host of advantages: making use of land that is already cleared, producing electricity close at hand to those who need it, and even shading cars.

Source: With permission from Iqony Solar Energy Solutions.

Energy Systems: A Project-Based Approach to Sustainability Thinking for Energy Conversion Systems, First Edition. Leon Liebenberg.
© 2024 John Wiley & Sons, Inc. Published 2024 by John Wiley & Sons, Inc.
Companion website: www.wiley.com/go/liebenberg/energy_systems

1) An energy entrepreneur wants to place an array of photovoltaic canopies on the parking lots of 3500 Walmart superstores. A typical Walmart parking lot has an area of five acres (1 acre is about 4050 m^2). How much power can such an array of photovoltaic modules generate? Clearly state your assumptions and provide technical specifications of the solar modules you will use. Your solution must consider the structural issues of such photovoltaic canopies and their need to contend with severe weather, including strong winds. For context, read the paper by Vandewetering et al. (2022).

2) Contact the manager of a Walmart superstore and enquire how much electrical power (kW) and energy (kWh per year) such a store typically consumes. How does that compare with the power and energy that the proposed photovoltaic canopy can provide?

3) Considering your calculations above, how much electrical storage capacity would be needed to satisfy a Walmart superstore's electrical needs? If lithium-ion batteries were to be used, what mass of batteries would be required? (Clearly state your assumptions and references.)

4) How long will these photovoltaic modules last? (Clearly state your peer-reviewed journal references.)

5) How much would these photovoltaic modules and batteries cost? Also clearly state the supplier's details and the technical specifications of the photovoltaic modules and storage batteries.

6) What value would such photovoltaic canopies add in terms of social, technological, economic, environmental, and political factors for the Walmart superstores and their surrounding communities?

7) Considering only the photovoltaic modules and lithium-ion storage batteries required for the 3500 Walmart parking lots, what materials and how much of each material would be required? In which cities or countries are those materials typically found? From your research, state the social and environmental concerns of extracting the materials from those cities. Is this type of material extraction sustainable? Explain.

8) Assume that the cost of placing the photovoltaic canopies and storage batteries amounts to $12 million for a Walmart superstore. Also assume that the annual insolation in the location where the solar array will be installed is 2500 kWh/m^2. Calculate the lifecycle cost of electricity per kWh if the lifetime of the solar modules is 20 years. Assume a discount rate of 6%. Ignore maintenance or other fees.

9) What will the cost of electricity per kWh be if there is an annual maintenance charge of $400 and if there is a battery storage cost of $2000 per kWh, with half of the electricity generated used at night, and if the system lasts for 15,000 charge/recharge cycles.

10) The solar modules used in the photovoltaic array are often made of crystalline silicon (c-Si) and are bifacial devices so that they absorb irradiance from the front and rear sides, which achieves higher annual energy yield for the same module area compared to their monofacial counterparts. Reference: Liang et al. (2019).

What is the conversion efficiency of a 24.6 cm^2 bifacial c-Si cell if the saturation current is 4×10^{-15} mA and the short-circuit current is 40 mA/cm^2. Assume AM1.5 solar radiation. Reference: Yu et al. (2021).

References

Liang, T.S., Pravettoni, M., Deline, C., Stein, J.S. et al. (2019). A review of crystalline silicon bifacial photovoltaic performance characterisation and simulation. *Energy and Environmental Science* 12 (1): 116–148.

Vandewetering, N., Hayibo, K.S. and Pearce, J.M. (2022). Open-source photovoltaic – electrical vehicle carport designs. *Technologies* 10 (6): 114.

Yu, B., Shi, J., Li, F., Wang, H. et al. (2021). Selective tunnel oxide passivated contact on the emitter of large-size n-type TOPCon bifacial solar cells. *Journal of Alloys and Compounds* 870 (July): 159679.

Solar reserve

A 110-MW solar-thermal tower in Nevada was the subject of an installation on the main plaza at Lincoln Center in 2022.

The piece is by Irish artist John Gerrard, and it is a virtual world made of 3-D images developed from photographs of the actual solar-thermal power plant located in central Nevada. The image on the 8.5 m × 7 m LED wall changes slowly, offering views of the landscape, and an overhead view of the solar tower.

From above, the plant's 10,347 mirrors (or *heliostats*) form a perfect disc, mimicking the layout of a sunflower. From the front, it looks like a lighthouse with its illuminated tower.

Gerrard says the hustle and bustle of New York City are fitting for the piece. "I thought it was interesting to bring these two things together, a site where energy is consumed, you know, historic energy as such, alongside a site where ... there's an attempt to produce energy in a new way, and to allow those two things to intersect." *Source:* Solar Reserve (Tonopah, Nevada), courtesy of the artist, John Gerrard, http://www.johngerrard.net/solar-reserve.html

Energy Systems: A Project-Based Approach to Sustainability Thinking for Energy Conversion Systems,
First Edition. Leon Liebenberg.
© 2024 John Wiley & Sons, Inc. Published 2024 by John Wiley & Sons, Inc.
Companion website: www.wiley.com/go/liebenberg/energy_systems

Week 8 – Solar-Thermal and Wind Energy Systems

Solar radiation can be converted into different forms of usable energy:

- *Chemical energy:* Photocatalytic processes in chemical reactors or reactions similar to natural photosynthesis convert direct and diffuse solar radiation into chemical energy. When irradiated with sunlight, carbon dioxide and water are converted into biomass. This was briefly surveyed in Chapter 20 (Box 20.2).
- *Heat:* With the help of solar collectors, the sun can make a major contribution toward producing low-temperature heat in buildings: for domestic hot water and space heating, as well as for cooling purposes via thermally driven chillers. That is the focus of Chapter 22.
- *Electricity:* Photovoltaic generators convert solar radiation directly into electricity. The energy output of these systems is proportional to their surface area and depends on the global radiation value. Photovoltaic systems were covered in Chapter 20.

  If direct sunlight is concentrated, very high temperatures can be produced. This allows process heat to be generated by large, central heating systems for industrial purposes or for generating electricity in thermal power stations, which is the focus of Chapter 23. (Chapters 22 and 23 feature high-impact solar-thermal systems, and therefore disregard solar cookers, solar dryers, solar ponds, solar air-conditioners, solar stills, and solar chimneys.)
- *Wind:* The unequal solar heating of the Earth results in the poles being very cold and the tropics very warm. The temperature differentials cause air pressure gradients that give rise to winds. Wind energy is therefore an indirect form of solar energy, which is covered in Chapter 24.

  Chapter 22 – Solar heating: Availability of solar energy; collection of solar-thermal energy; solar collectors located horizontally on the Earth; solar collectors tilted; selective solar surfaces; flat-plate collectors; evacuated tube (or vacuum tube) collectors; outlook

  Chapter 23 – Solar-thermal electrical power plants: Solar-thermal power plants; solar resource potential for concentrated solar power; focusing ("concentrating") collectors; parabolic reflectors; central-receiver ("power tower") collectors; solar heating for industrial processes and gas-cycle power plants; outlook

  Chapter 24 – Wind energy: Overview of wind power systems; availability of wind energy; wind turbine characterization; power available from the wind; one-dimensional wind turbine model; estimating the wind turbine's average power and energy production; sustainability of wind power; outlook

  Mini Project 8 – Solar Stirling power plant

22

Solar Heating

Worldwide, 250 million dwellings used solar-thermal technologies for water heating in 2020. However, to reach carbon neutrality by 2050, as proposed by the International Energy Agency (IEA) in its net-zero scenario, around 170 million new conventional solar-thermal technologies (e.g., thermosyphons and pumped circulation systems) and 120 million emerging solar systems (e.g., photovoltaic-driven systems) are needed (SHC 2022).

Small-scale *solar water-heating systems* and, to a certain extent, solar "combi" systems for combined hot water preparation and space heating in single-family houses (Figure 22.1), apartment buildings, multi-family houses, hotels, and public buildings represent about 60% of the world's annual solar-thermal installations.

Industrial scale solar-thermal plants are fledgling technologies, but are growing rapidly. Projects range from 250 kW$_{th}$ to 100 MW$_{th}$ systems to meet the heating needs of many industrial processes (e.g., Figure 22.2). In 2021, at least 78 solar-heating industrial plants were commissioned across the world with a total collector area of 51,539 m^2 (36 MW$_{th}$), bringing the total installed solar-thermal industrial installed collector area to around 2 million m^2 (1615 MW$_{th}$).

The global solar-thermal capacity of unglazed and glazed solar collectors in operation grew from 62 GW$_{th}$ (89 million m^2) in 2000 to 522 GW$_{th}$ (746 million m^2) in 2021. The corresponding annual solar-thermal energy yields amounted to 51 TWh in 2000 and 425 TWh in 2021. In 2021, a total capacity of 21 GW$_{th}$ or 31 million m^2 of collector area was installed. In terms of energy, in 2021, solar-thermal systems supplied 425 TWh of heat, whereas wind turbines supplied 1980 TWh, and photovoltaic systems 1138 TWh of electricity (IEA 2022; SHC 2022).

In 2021, about half the energy demand for buildings (which use 36% of the world's energy) was for *space and water heating*; only 0.8% of that space and water heating was provided by solar heating (IEA 2022). So, although solar heating is already making an impact, it has significant growth potential. However, countries like China are catching up quickly. In 2021, China reported 20 system installations, totaling a collector area of about 151,000 m^2 for district heating and large-scale building applications (SHC 2022).

Table 22.1 summarizes the salient characteristics of *flat-plate* solar collectors. With *evacuated tube* (or vacuum tube) plate collectors, the internal pressure is reduced, which reduces the convective heat losses by more than 50% and therefore enhances the overall collector efficiency. *Photovoltaic-thermal* (PVT) collectors combine the production of both types of solar energy – solar heat and solar electricity (photovolatics) – simultaneously in one collector, thus reaching higher yields per area. This is particularly important if the available roof area is limited, but integrated solar energy concepts are needed to achieve a climate-neutral energy supply for consumers, such as is required in residential and commercial buildings.

Energy Systems: A Project-Based Approach to Sustainability Thinking for Energy Conversion Systems,
First Edition. Leon Liebenberg.
© 2024 John Wiley & Sons, Inc. Published 2024 by John Wiley & Sons, Inc.
Companion website: www.wiley.com/go/liebenberg/energy_systems

Figure 22.1 Advertisement for a domestic thermosyphon-type solar water heater. *Source:* Solahart.

Figure 22.2 A solar process heat system for Martini & Rossi (Turin, Italy) with a capacity of 400 kW$_{th}$ and equipped with high-vacuum (10 μPa) flat-plate collectors that produce pressurized hot water at 170 °C to drive a 300-kPa indirect steam generator. *Source:* TVP Solar.

Reimagine Our Future

Renewable energy conversion systems do not clean the air or the water, nor will they save rhinos from extinction, support human rights, improve neighborhoods, or strengthen democracies. They do not reduce energy consumption either. Rather, they "only" produce power. How should we think about this issue?

Table 22.1 Salient characteristics of the main types of flat-plate thermal collectors.

Type of collector		Annual solar efficiency (%)	Average thermal efficiency (%)	Working temperature (°C)	Thermal power density (kW$_{th}$/m^2)
Non-focusing	Flat-plate collectors	12	15–45	< 80	0.25–0.4
	Evacuated tube ("vacuum tube") collectors	12	50–80	80–200	0.8–1.3
	Photo-voltaic-thermal collectors	12	30–45	50–120	0.5–0.9

Flat-plate solar heaters ("thermosyphons") in Moçambique. *Source:* Soltrain.

Vacuum tube collectors. *Source:* Naked Energy.

A 200 m^2 PVT collector field for a new building in Offenbach an der Quaich, Germany. *Source:* IEA.

Source: Müller-Steinhagen and Trieb (2004), IEA (2019), Duffie et al. (2020), DLR (2021), Zhu et al. (2022), and Sun et al. (2020).

22.1 Availability of Solar Energy

Consider the Earth's geometrical relation with the sun (Figure 22.3). The Earth's axis of rotation is not perpendicular to its orbit around the sun and displays a tilt, or *declination*, of $\delta = 23.45°$.

The *irradiance* (P) and the *spectral irradiance* (L) are related as follows:

$$P = \int_0^\infty L(\lambda) \, d\lambda \qquad \left[\text{W/m}^2 \right] \tag{22.1}$$

Irradiance can be evaluated as follows (Mackay 2015):

$$P = \frac{2\pi^4}{15} g_s \frac{k_B^4}{h^3 c^2} T_{sun}^4 \equiv \sin\left(\theta_{sub}\right)^2 \sigma T_s^4 \qquad \left[\text{W/m}^2 \right] \tag{22.2}$$

where T_{sun} is the surface temperature of the sun (5780 K), k_B is the Stefan-Boltzmann constant (5.67×10^{-8} W/(m²K⁴)), h is Planck's constant (6.62×10^{-34} Js), c is the speed of light (3×10^8 m/s), θ_{sub} is the angle subtended between the sun and the earth (5.67×10^{-8} W/(m²K⁴)), and g_s is a geometric factor:

$$g_s = \pi \sin \theta_{sub} = 6.72 \times 10^{-5} \tag{22.3}$$

The *air mass factor* (AM) is a measure of the path length of radiation through the atmosphere and is a function of the angle between the zenith (see Figure 22.10) and the direction of the sun. The radiation spectrum at the surface of Earth's atmosphere is denoted as AM0, meaning air mass zero. The spectral irradiance on the Earth's surface is denoted as AM1.5G. The number 1.5 refers to an air mass 1.5 times directly above 1, or sunlight that must travel 1.5 times the atmospheric thickness. The letter G refers to "global," meaning that the direct and diffuse components of sunlight are included.

Figure 22.4 shows the orders-of-magnitude difference in the spectral irradiance of the sun (at 5780 K) at terrestrial level and a black body at 300 K.

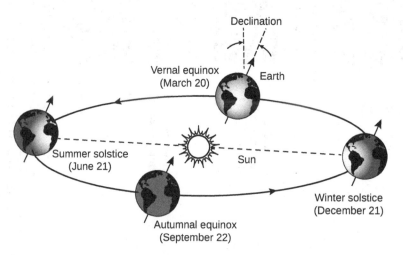

Figure 22.3 The earth–sun geometry showing an angular declination of $\delta = 23.45°$.

Spectral irradiance (W/m² nm)

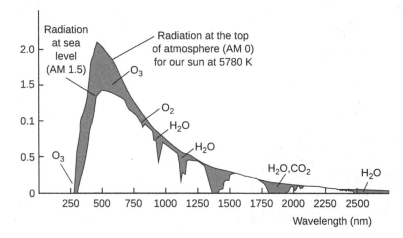

Figure 22.4 Spectral irradiance $L(\lambda)$ at terrestrial level, where $L(\lambda) = \dfrac{2g_s hc^2}{\lambda^5} \dfrac{1}{\exp(hc/\lambda k_B T_s) - 1}$ with units of (W/m²)/nm. When $L(\lambda)$ is integrated with respect to wavelength in nm, one obtains the irradiance $P(\lambda)$ in W/m².

The value of irradiance at the edge of the Earth's atmosphere is about 1366 W/m². This value is also called the *solar constant*. The solar radiation (or irradiance) at a given location on the Earth's surface varies according to the time of day and year, and prevailing weather conditions.

Peak values for solar radiation (or "irradiation") on clear sunny days are around $P_{\text{AM 1.5G}} = \int_0^\infty L(\lambda)\, d\lambda \cong 1000\ \text{W/m}^2$. In temperate climatic regions, the annual total for solar energy (or "insolation") is 1000–1200 kWh/m². It can reach 2500 kWh/m² in arid desert zones.

The solar radiation that arrives at the edge of the Earth's atmosphere must pass through the atmosphere, which is about 150 km thick. The atmosphere comprises various gases, water vapor, and dust, which absorb, scatter, and reflect the radiation. The amount of solar energy that reaches the Earth's surface is dependent on factors such as the day of year, cloud and haze, ozone concentration, diffuse radiation, and air mass.

The radiation available for use by a collector on the Earth's surface will have a direct (or "beam") component, a diffuse component, and an albedo (or "reflected") component (Figure 22.5):

$$P_{avail} = P_{dir} + P_{diff} + P_{alb} \qquad \left[\text{W/m}^2\right] \tag{22.4}$$

The direct component of radiation (P_{dir}) is a function of the transmission coefficient (τ), with τ depending on the absorbed radiation, scattering from molecules, dust, and water droplets (with $\tau_{scat} \approx 0.95$), and reflection from the atmosphere ($\tau_{refl} \approx 0.8$):

$$P_{dir} = \tau P_{\text{AM1.5G}} \qquad \left[\text{W/m}^2\right] \tag{22.5}$$

$$\text{with} \qquad \tau = \tau_{abs} + \tau_{scat} + \tau_{refl} \tag{22.6}$$

There are several semi-empirical correlations that could be used to estimate the diffuse radiation. The following correlation provides accurate estimates (Duffie et al. 2020):

$$\frac{P_{diff}}{P_{avail}} = \frac{0.85}{1 + \left(0.5\, K_T^6\right)} + 0.15 \tag{22.7}$$

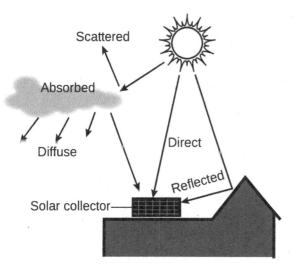

Figure 22.5 The radiation reaching a collector on the Earth's surface will have a direct (or "beam") component, a diffuse component, and a reflected (or "albedo") component. Some radiation is absorbed by the atmosphere and does not reach the collector.

where K_T is the clearness index, experimentally determined to correlate as (Duffie et al. 2020):

$$K_T = \frac{P_{avail}(0)}{P_{AM1.5G}(0) \; \cos\theta_z} \qquad (22.8)$$

22.2 Collection of Solar-Thermal Energy

Solar radiation consists of two components:

- *Direct (or beam)* solar radiation, which is dependent on the position of the sun and can be concentrated with the aid of lenses or reflectors. It is primarily available when the total radiation value is high (clear sky).
- *Diffuse* solar radiation, which comes from the whole sky as scattered light and cannot be concentrated by optical devices. At medium latitudes, global or total radiation consists of 60% diffuse and 40% direct radiation. When the total radiation value is low (on cloudy and foggy days), diffuse radiation predominates. For this reason, attempts are made to use both types of radiation.

Solar-thermal collectors absorb all solar radiation, both direct and diffuse, over the spectral range from ultraviolet through visible to infrared radiation. Solar collectors then convert solar radiation into heat (Figure 22.6a).

The radiation directly heats a metal or polymer absorber, through which a heat transfer medium flows. This conveys the generated heat to the user or a storage unit. Optically selective absorbing materials, highly transparent covers, and good thermal insulation in the collectors ensures that as

(a)

(b)

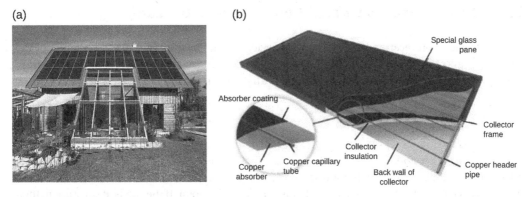

Figure 22.6 (a) Solar-thermal collectors on a "solar active house," producing its own hot water throughout the year. The house also features photovoltaic panels to produce its own electricity. *Source:* Fraunhofer ISE. (b) A typical solar-thermal collector. Solar radiation is transmitted through the transparent upper glass pane and is then absorbed by the dark "selective absorber" coating, heating the underlying copper sheet. The radiative selectivity of the absorber and the presence of the cover prevent heat from being radiated from the absorber (the "greenhouse effect"). The heat transfer fluid circulating in the pipes, usually water (mixed with anti-freeze fluid), transports the heat out of the collector. *Source:* Fraunhofer ISE.

much solar energy as possible is converted into useful heat (Figure 22.6b). Temperatures of 50–150 °C can be attained.

In evacuated tubular collectors, the space between the outer glass tube and the absorber is evacuated. This reduces thermal losses caused by air conduction or convection (the thermos principle) so that higher operating temperatures can be achieved. This technology is applied in domestic solar-thermal collectors and in solar-thermal power stations.

Solar-thermal collectors are most commonly installed on individual houses to produce domestic hot water. However, as the number of houses with new windows and thermal insulation increases, and the houses' heating energy consumption decreases, the use of solar energy for space heating becomes a growing trend. This has resulted in the installation of collector systems with a large surface area in combination with large thermal storage tanks (Duffie et al. 2020).

Solar-thermal systems range from *passive building* designs to *active solar concentrator systems* that may produce steam, which is used to generate electricity. The focus of this chapter is on active solar collector and concentrator systems.

Reimagine Our Future

Modern buildings are sealed (featuring excellent insulation, no open windows, etc.) and are energy efficient only as long as HVAC systems and their control systems are working well. This means that they require uninterruptable electricity, dependable back-up systems, and proficient building managers. To support endless technological innovations, would engineers and architects not do well to also contemplate buildings that are more attuned to using the environment for heating, cooling, and ventilation? Might such *passive buildings* not also make for happier and healthier building occupants?

22.3 Solar Collectors Located Horizontally on the Earth

The direct radiation received by a horizontal surface on the Earth is equal to the product of the irradiation and the zenith angle (θ_z) (Figure 22.7). The sun makes an angle β with the zenith and θ_s with the device's normal angle. The hour angle (ω) is defined as:

$$\omega = \frac{360°}{24\,\text{h}}(t_{LST} - 12\,\text{h}) \qquad [°] \tag{22.9}$$

where t_{LST} is the local solar time in hours (e.g., 2.30 p.m. is equivalent to 14 h and 30 min, or 14:30). The hour angle therefore converts the local solar time to the sun's position in degrees; ω is indicative of the apparent rotation of the sun about the Earth's axis. One hour of time is equivalent to 15° of rotation. The time is expressed in hours before or after solar noon. If the sun is at solar noon, then $\omega = 0°$ and $\beta = \theta_s + \theta_z$.

Using Figure 22.7, the zenith angle (θ_z) can be written as a function of time (Duffie et al. 2020):

$$\text{Horizontal collector:} \qquad \cos\theta_z = \sin\delta\sin|L| + \cos\delta\cos|L|\cos\omega \tag{22.10}$$

where δ is the angular declination ($\delta \approx 23.45°$), L is the latitude (in degrees), and ω is the hour angle (in degrees).

The angular altitude of the sun at solar noon, α_s, may be calculated as follows for the northern hemisphere:

$$\alpha_{s,N} = 90° - |L| \pm \delta \qquad [°] \tag{22.11}$$

In general, $L = 90° - \theta_s$, with θ_s ranging between 0° and 180°.

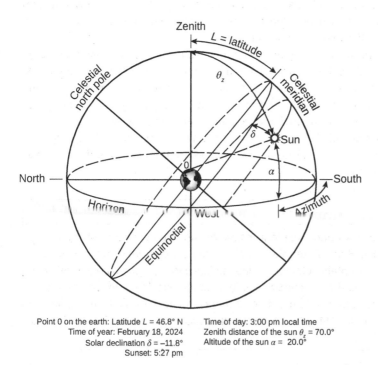

Point 0 on the earth: Latitude L = 46.8° N
Time of year: February 18, 2024
Solar declination δ = −11.8°
Sunset: 5:27 pm

Time of day: 3:00 pm local time
Zenith distance of the sun θ_z = 70.0°
Altitude of the sun α = 20.0°

Figure 22.7 The sun's *apparent* path depicted on the Earth's celestial sphere for a specific time of year.

For the collector lying flat on the Earth's surface, $\theta_s = \theta_z$. Furthermore, $P_{albedo} = 0$ because reflected radiation from the surrounding terrain cannot reach the collector. The direct irradiation received by the horizontal collector ($P_{avail,0}$) is then:

$$P_{avail,0} = P_{dir,0} + P_{diff,0}$$

with $P_{dir,0} = P_{AM1.5G} \cos \theta_z$ and with $P_{diff,0}$ evaluated using Eqs. (22.7) and (22.8).

22.4 Tilted Solar Collectors

Equation (22.10) applies to solar collectors placed horizontally on the Earth's surface. This, however, hardly happens in practice, as solar collectors are rather tilted to maximize the amount of collected radiation.

Figure 22.8 shows a solar device that is placed at angle β to the horizontal plane. After manipulation of Eq. (22.6), the following relationship is obtained (Duffie et al. 2020):

$$\textit{Tilted collector: } \cos\,\theta_s = \cos\,\theta_z \cos\beta + \sin\,\theta_z \sin\beta\,\cos\omega \qquad (22.12)$$

$$\text{or} \qquad \cos\,\theta_z = \cos L\,\cos\delta\cos\omega + \sin L\,\sin\delta \qquad (22.13)$$

As a rule of thumb, the collectors are tilted at $\beta = |L| + 10°$ (where L is the latitude, in degrees) and are pointed toward the equator. So, in the northern hemisphere, a collector will be aligned in an east-west direction and tilted at β degrees toward the south; the opposite will be done in the southern hemisphere.

22.4.1 Tracking the Sun

Collectors that automatically track the sun must adjust the tilt angle β to ensure that the collector maintains a 90° angle with the sun (Figure 22.8). From Figure 22.8, it also follows that $\beta = \theta_s + \theta_z$ and $\omega = 0°$ at solar noon (i.e., normal direction to the sun). The collector can therefore be made to track the sun continuously by rotating and tilting the collector's surface normal to the sun's direction to ensure a situation where $\beta = \theta_s + \theta_z$ and $\omega = 0°$, using Eq. (22.12) or (22.13).

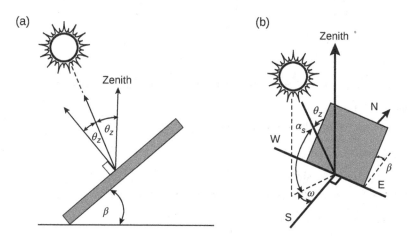

Figure 22.8 (a) A solar device is placed at an angle of β to the horizontal plane. (b) Angles used to describe the position of the sun for a southward-facing solar device inclined at an angle β.

Therefore, for a plane rotated about a horizontal east–west axis with continuous adjustment to minimize the angle of incidence, we get:

$$\cos\,\theta_s = \left(1 - \cos^2\delta\,\sin^2\omega\right)^{1/2} \tag{22.14}$$

$$\text{and}\qquad \tan\,\beta = \tan\,\theta_z|\cos\omega| \tag{22.15}$$

Similarly, for a plane rotated about a horizontal north–south axis with continuous adjustment to minimize the angle of incidence, we get:

$$\cos\,\theta_s = \left(\cos^2\theta_z + \cos^2\delta\,\sin^2\omega\right)^{1/2} \tag{22.16}$$

$$\text{and}\qquad \tan\,\beta = \tan\,\theta_z|\cos\omega| \tag{22.17}$$

Example 22.1 Tracking the Sun

Calculate (a) the sun's hour angle, (b) the angle of incidence, and (c) the slope of a collector's surface for a solar collector in Chicago at 14:00 on 15 May 2024, if the collector is continuously rotated about an east–west axis to minimize the angle of incidence. Use the solar calculator of the National Oceanic and Atmospheric Administration (NOAA) (or any other reputable nautical almanac), https://gml.noaa.gov/grad/solcalc/sunrise.html, to determine the exact value of solar declination and Chicago's latitude.

Given: Date and time of year (14:00 or 2:00 p.m. on 15 May 2024), and location (Chicago)

Required: (a) ω, (b) θ_s, (c) β

Solution strategy: We will use Eqs. (22.14) and (22.15) to determine the angle of incidence and the tilt angle, Eq. (22.9) to determine the hour angle, and Eq. (22.13) to determine the zenith angle. We will first calculate the solar declination angle and Chicago's latitude using the NOAA's nautical almanac. For 14:00 (or 14 h) on 15 May 2024, we find the solar declination angle, δ, to be 18.99°. Chicago's latitude is 41.88°.

Analysis:

a) We use Eq. (22.9) to determine the hour angle:

$$\omega = \frac{360°}{24\,\text{h}}(t_{LST} - 12\,\text{h}) = \frac{360°}{24\,\text{h}}(14\,\text{h} - 12\,\text{h}) = 30°$$

We can now use Eqs. (22.13) and (22.14).

b) $\cos\theta_s = (1 - \cos^2\delta\,\sin^2\omega)^{1/2}$:

$$\theta_s = \cos^{-1}\left(1 - \cos^2\delta\,\sin^2\omega\right)^{1/2} = \cos^{-1}\left(1 - \cos^2(18.99°)\,\sin^2(30°)\right)^{1/2} = 28.2°$$

c) $\tan\beta = \tan\theta_z|\cos\omega|$:

We will use Eq. (22.13) to determine the zenith angle:

$$\cos\,\theta_z = \cos\,L\,\cos\delta\cos\omega + \sin\,\varphi\,\sin\delta$$

$$\theta_z = \cos^{-1}(\cos\,L\,\cos\delta\cos\omega + \sin\,L\,\sin\delta)$$

$$\theta_z = \cos^{-1}(\cos\,(41.88°)\,\cos(18.99°)\,\cos(30°) + \sin\,(41.88°)\,\sin(18.99°)) = 34.2°$$

$$\therefore \beta = \tan^{-1}(\tan\theta_z|\cos\omega|) = \tan^{-1}(\tan 34.2°|\cos 30°|) = 30.48°$$

Discussion: The same procedure as above can be used to determine the hour angles and the required tilt and azimuth angles during various times of the day to minimize the angle of incidence.

Modern collector tracking systems feature several position sensors and use drive motors to adjust the collector surface to minimize the angle of incidence between the incoming sunlight and the collector's surface.

22.5 Selective Solar Surfaces

Section 1.6 covered the fundamentals of radiation, and Section 5.2 showed the importance of the reflectance (ρ), absorptance (α), and transmittance (τ) of electromagnetic radiation. The sum of these three processes must equal the incident radiation:

$$\rho + \alpha + \tau = 1 \tag{22.18}$$

If there is no reflection, then $\rho = 0$ and $\alpha = 1 - \tau$. For a thermal black body, $\alpha = 1$ for all wavelengths. For a thermal gray body, $\alpha < 1$ for all wavelengths. Not all systems are perfect emitters like the sun, so for gray bodies at temperature T_g, the emissivity (ε) is related to the irradiance of the gray body (P_{gb}) and the irradiance of a black body (P_{bb}) at the same temperature as the gray body:

$$\varepsilon = \frac{P_{gb}}{P_{bb}} = \frac{P_{gb}}{\sigma T_g^4} < 1 \tag{22.19}$$

where Eq. (1.18) is used in the above expression and σ is the Boltzmann constant.

The ratio of irradiance gives the absorptance:

$$\alpha = \frac{P_{gb}}{P_{bb}} \tag{22.20}$$

which is the same as Eq. (22.19), so that:

$$\alpha = \varepsilon \tag{22.21}$$

According to Eq. (22.21), called *Kirchhoff's law*, whatever a body absorbs is emitted at a steady state.

The spectral irradiance of a real body (L_{rb}) is related to that of a black body (L_{bb}) by the emissivity:

$$L_{rb} = \varepsilon L_{bb} \qquad [\text{W}] \tag{22.22}$$

Consider an opaque material that has high absorptance and/or is very thick, so that it does not transmit radiation, ($\tau = 0$). For this material, Eq. (22.18) becomes:

$$\rho + \alpha = 1 \qquad \text{or} \qquad \rho + \varepsilon = 1 \tag{22.23}$$

$$\text{or} \qquad \varepsilon = 1 - \rho \tag{22.24}$$

Equation (22.24) allows the calculation of emissivity by measuring material reflectance.

For an absorber in a solar-thermal application, one wants a low reflectivity ($\rho \rightarrow 0$), a high absorptance ($\alpha \rightarrow 1$), and a low emissivity ($\varepsilon \rightarrow 0$). However, device performance and analysis are complicated by the fact that $\alpha = \varepsilon$ and $\varepsilon = 1 - \rho$. Engineers and scientists of solar collectors must therefore contend with the problem of maximizing absorption while minimizing emission and reflectance. The answer to this dilemma lies in two concepts:

- *Surface texturing*, the size and spacing of which is related to the radiation wavelength to help ensure that emitted radiation is re-absorbed by the material. This reduces the reflected radiation, which is also why kitchen stove plates are matt black and not glossy.

- *Selective surfaces*, which absorb most of the radiation from the sun, but do not re-emit it at that range of wavelengths. Rather, selective surfaces emit radiation at a much lower temperature than the sun (e.g., 300 K versus 6000 K) and at much longer wavelengths (therefore also at lower energies), in accordance with Planck's Law, Eq. (19.1): $E = h\nu = h\frac{c}{\lambda}$. Materials that fulfill these demanding requirements have band-gap energies in the order of 1 eV. Such materials will therefore absorb radiation for wavelengths below 1200 nm, in accordance with Eq. (19.1). However, most such materials have a large reflectance ($\rho \approx 0.35$) at short wavelengths, which limits their applicability. Such solar-thermal materials are coated (typically 1 μm thick) with materials such as black nickel, chrome, cadmium, or cobalt, which have short-wavelength absorptances better than 0.95 and long wavelength emissivities less than 0.1. Such selective surfaces therefore facilitate high short-wavelength ($\lambda < 500$ nm) absorptance with high long-wavelength ($\lambda > 2500$ nm) transmittance, thus reducing their emissivity.

Two general types of collectors are used: *flat-plate* and *focusing* collectors. As many variants of each type have been developed, only the main types will be covered here.

22.6 Flat-Plate Collectors

Flat-plate collectors generally have fewer complex structures than focusing collectors. Flat-plate collectors are usually installed in a fixed position, such as on the angled roof of a house. In the northern hemisphere, an *ideal tilt angle* is equal to the latitude where they are positioned plus 10°, and the solar collector should be tilted toward the south; in the southern hemisphere, the collectors should be tilted toward the north (Walker 2013).

Figure 22.9 shows the configuration of a typical flat-plate (or non-focusing) collector. Flat-plate collectors can absorb all sunlight: *diffuse* and *direct* ("beam") radiation. However, as they do not concentrate sunlight, they can only achieve minimal increases (in the order of 30–50 °C) in the temperature of the circulating fluid. The back surface of a flat-plate collector is coated with a selective absorber material that is matt and dark in color, as previously explained.

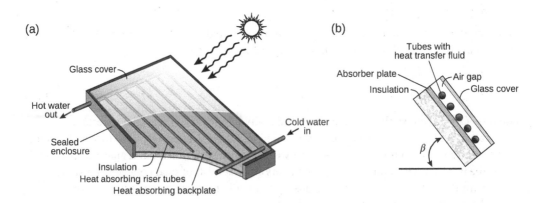

Figure 22.9 Flat-plate solar energy collector with a glass cover. The air trapped between the glass cover and the heat absorber backplate recirculates due to natural convection. Solar energy heats the water (or heat transfer fluid) inside the tubes.

The convective and radiative heat losses from a flat-plate collector are substantial, especially when the collectors operate at high temperatures. Convective heat losses may, however, be minimized by insulating the collector. The collectors are also usually covered by a special glass (called "solar glass"). This glass is opaque to long-wavelength radiation and has a high transmissivity for short-wavelength solar radiation. The glass cover therefore effectively traps some of the incoming solar radiation that reaches the collector's surface. However, the rate of useful heat might only increase by around 5–8% (Walker 2013).

The rate of useful heat $\left(\dot{Q}_u\right)$ that can be collected from the incoming radiation can be calculated using the first law of thermodynamics for an open system:

$$\dot{Q}_u = \dot{m}(h_{out} - h_{in}) \equiv \dot{m}c_p(T_o - T_i) \qquad \text{[W]} \qquad (22.25)$$

where c_p is the specific heat of the heat transfer fluid (which is often water or a solution of water and an anti-freeze liquid), $\dot{m}$ is the mass flow rate of the heat transfer fluid, and T_o and T_i are the respective inlet and outlet temperatures of the heat transfer fluid. An energy balance on the solar collector yields the following:

$$\dot{Q}_u = \dot{m}c_p(T_o - T_i) \equiv (\alpha_{abs}\tau P_{avail,0} - \dot{q}_{loss})A \qquad (22.26)$$

where α_{abs} and τ are the absorptance and transmittance of the absorber surface, respectively, $P_{avail,0}$ is the direct irradiation (W/m^2) received by the aperture, which is assumed to be normal to the solar radiation, $\dot{q}$ is the specific rate of heat loss from the system (W/m^2), and A is the external surface area (m^2) of the absorber pipe ($A = \pi D_{pipe}L_{pipe}$), with D_{pipe} being the pipe's outside-diameter and L_{pipe} being its length.

The thermal efficiency of a flat-plate collector is:

$$\eta_{coll} = \frac{\dot{Q}_u}{P_{avail,0}A} \qquad (22.27)$$

As illustrated in Example 22.2, the temperature increase that can be expected in a flat-plate solar collector with a once-through heat exchanger is small, in the order of 20 °C. This is inadequate for most commercial or domestic water-heating systems, which require water at around 50 °C. This inadequacy can be solved by recirculating water with the use of a reservoir (or tank). Using the first law of thermodynamics, we can calculate the rate of heat transfer:

$$\dot{Q}_u = \dot{m}(h_{out} - h_{in}) = \frac{m}{t}c_p\left(T_{final} - T_0\right) \qquad (22.28)$$

or $$T_{final} = T_0 + \frac{\dot{Q}_u}{mc_p}t \qquad (22.29)$$

where T_0 is the heat transfer fluid's initial temperature, T is its final temperature, t is the time taken to bring about the temperature change, c_p is the fluid's specific heat, and $\dot{Q}$ is the rate of heat transfer.

22.7 Evacuated Tube (or Vacuum Tube) Collectors

Evacuated tube (or "vacuum tube") collectors consist of two concentric glass tubes that are sealed at the ends (Figure 22.10). The inner tube is coated with a solar selective coating. A high vacuum (e.g., 10 μPa) is produced between the two tubes, thus significantly reducing convection loss while

(a)

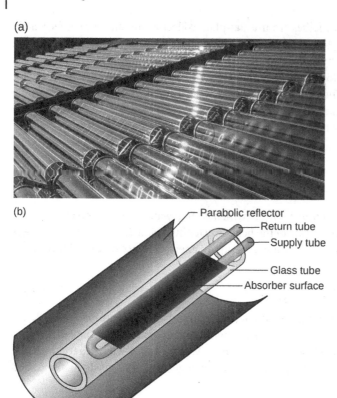

(b)

Figure 22.10 (a) Evacuated tube (or "vacuum tube") collectors; *Source:* Naked Energy / Elektrek.
(b) Evacuated tube collectors typically comprise two glass tubes welded together or an evacuated glass
tube enclosing the absorber plate and a heat pipe assembly to remove the heat.

providing excellent thermal insulation. When the tube is exposed to the sun, the solar selective coating absorbs the solar energy and transfers it to a heat pipe that is located inside the inner tube. The heat pipe contains a heat transfer fluid that transfers the heat to a system manifold. The liquid in the heat pipe has a low boiling point. When the heat transfer fluid is heated, the liquid inside the pipe begins to expand (vaporize) and rises to the top of the heat pipe. As the manifold removes heat, the vapor condenses and liquid returns to the bottom of the heat pipe.

Figure 22.11 illustrates the typical application ranges for flat-plate and evacuated tube collectors. The selection of either collector depends on the operating temperature, where evacuated tube collectors excel in heating applications of more than 50 °C above the ambient temperature.

Example 22.2 Rudimentary Flat-Plate Collector

Water $\left(c_{p_{avg}} = 4.18 \text{ kJ/kg K}\right)$ enters a flat-plate collector at 22 °C and leaves it at 35 °C. The mass flow rate of the water is 0.03 kg/s and the collector's surface area is 4 m². The solar irradiation is 925 W/m². The absorber surface has a transmittance of 0.98 and an absorptance of 0.88. Disregarding the heat trapping effect of the glass cover, (a) find the rate of useful energy reaching the collector, (b) the heat lost through the collector, and (c) its thermal efficiency.

Efficiency = % of solar energy captured by collector

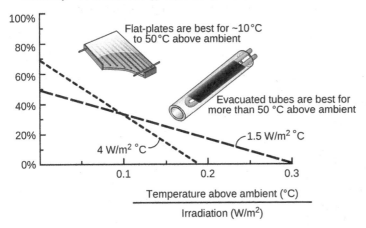

Figure 22.11 Typical application ranges for flat-plate and evacuated tube collectors. Efficiency is the percentage of solar energy captured by the collector. *Source:* Walker (2013) / John Wiley & Sons.

Given:

Water: $T_i = 22\,°C$, $T_o = 35\,°C$, $\dot{m} = 0.03$ kg/s, $c_{p_{avg}} = 4.18$ kJ/kg K
Collector: $\tau = 0.98$, $\alpha = 0.88$, $A = 4\,m^2$, $P_{avail} = 925\,W/m^2$

Required: (a) $\dot{Q}_u$, (b) $\dot{Q}_{loss}$, (c)
Solution strategy: We are given adequate information to use Eqs. (22.25), (22.26), and (22.27).
Analysis:

$\dot{Q}_u$:

From Eq. (22.25): $\dot{Q}_u = \dot{m}c_p(T_o - T_i) = (0.03\text{ kg/s})(4.18\text{ kJ/kg K})(35-22)\text{ K} = 1.63\text{ kW}$

$\dot{Q}_{loss}$:

From Eq. (22.26): $\dot{Q}_u = \dot{m}c_p(T_o - T_i) = 1.63\text{ kW} \equiv (\alpha_{abs}\tau P_{avail,0} - \dot{q}_{loss})A$

$\therefore 1630\text{ W} \equiv (0.88)(0.98)(925\text{ W/m}^2)(4\text{ m}^2) - \dot{Q}_{loss}$

$\therefore \dot{Q}_{loss} = 1561\text{ W}$

From Eq. (22.27), the collector efficiency is $\eta_{coll} = \dfrac{\dot{Q}_u}{P_{avail,0}A} = \dfrac{1630\text{ W}}{(925\text{ W/m}^2)(4\text{ m}^2)} = 0.44$.

The collector's thermal efficiency is 44%.

Example 22.3 Recirculating Water with a Flat-Plate Solar Collector
Figure 22.12 shows the set-up of a flat-plate solar collector recirculating water with the use of a tank. Consider the same starting conditions used in Example 22.2. How long will it take to heat 300 kg of water from 22 to 50 °C? The water is stored in a well-insulated recirculation tank and any thermal losses may be disregarded.
Given: $m = 300$ kg, $T_i = 22\,°C$, $T_{final} = 50\,°C$, $c_{p_{avg}} = 4.18$ kJ/kg K

Required: We must calculate the heating period, t.

Solution strategy: We will use Eq. (22.29) to calculate the time required for the water to heat to 50 °C.

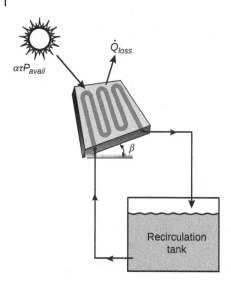

Figure 22.12 Flat-plate solar collector featuring water recirculation to achieve a high fluid temperature.

Analysis: From Eq. (22.29), the heating period is:

$$t = \frac{mc_p}{\dot{Q}_u}\left(T_{final} - T_0\right) = \frac{(300\,\text{kg})(4.18\,\text{kJ/kgK})}{1.63\,\text{kJ/s}}(50-22)\,\text{K} = 21,541\,\text{s} \cong 6\,\text{h}$$

Encouraging Student Curiosity to Find Solutions to the Global Sustainability Challenge

Finding solutions for the global sustainability challenge is a noble goal. However, it can only be achieved by changing mindsets. At the same time, the crucial role of education in the global quest for sustainability cannot be denied.

Shreyas Venkatarathinam (23), a mechanical engineer who has served as a student mentor, has several questions about the quest for sustainability on a global scale: Why do we bother only with incremental sustainability solutions? Why are people still so skeptical about utilizing alternative forms of energy to power their homes, cars, and communities? "Although there are compelling points about the cost of changing to sustainable forms of energy, the market will only increase its price tag on renewable energy products and services as other nonrenewable forms of energy become increasingly depleted, particularly gasoline," he says.

He is adamant that technology will not solve all our sustainability problems. "For instance, electric vehicles are a great advance, but their manufacture is not sustainable. We cannot even effectively recycle their batteries. We need to do more. We need to change our ways, collectively, on a massive scale."

Shreyas believes that students must be given more opportunities to talk about sustainability and to generate ideas that could become sustainability solutions. "Instructors should also engage more with their students about the interrelated social, technological, economic, environmental, and political aspects of energy innovations."

"The solution is straightforward, albeit not simple," he explains. "There is no 'one size fits all' solution for our global sustainability challenge. Each country, state or small community should find customized sustainability solutions. Investigating what makes a small community more sustainable based on the infrastructure that is already in place is the quickest path to making it more sustainable."

Born and raised in Silicon Valley, Shreyas experienced first-hand the incredible growth rate of technologies and their effect on people and the planet. As a student, he helped run an annual student sustainability competition, Reimagine Our Future. This was where he experienced the transformational role of design thinking and collaboration to generate possible sustainability solutions. "I noticed that student curiosity is piqued when participants start to question why certain technologies are successful and others are not, and how they can be improved while minimizing their environmental footprint." He believes that, with a bit of nudging, students can engage in independent, self-directed learning. "The results are impressive," he concludes.

Shreyas Venkatarathinam. *Source:* Shreyas Venkatarathinam, https://reimagine.web.illinois.edu

22.8 Outlook

Based on specialist projections, global solar heating systems should produce around 16.5 EJ (4583 TWh$_{th}$) by 2050 (IEA 2022; SHC 2022). If this happens, solar-thermal energy will meet more than 16% of total final energy use for low-temperature heat. In this scenario, solar collectors for hot water and space heating could reach an installed capacity of nearly 3500 GW$_{th}$, thus satisfying 14% of space and water heating energy use by 2050. Solar collectors for low-temperature (<120 °C) process heat could also reach an installed capacity of 3200 GW$_{th}$, meeting around 20% of global requirements for low-temperature heat by 2050 (IEA 2022; SHC 2022). If these deployment levels are achieved by 2050, solar heating can avoid 700 Mt of CO_2 emissions per year by 2050 (IEA 2022).

Problems

22.1 A flat-plate solar collector with an aperture area of 3.5 m^2 receives irradiation of 750 W/m^2. The average transmittance-absorptance product is 0.88. The average heat loss from conduction, convection, and radiation is 400 W/m^2. Calculate (a) the average rate of useful energy, and (b) the average efficiency of the collector.

[Answers: (a) 910 W; (b) 39.4%]

22.2 A flat-plate solar collector with an area of 2.5 m^2 receives irradiation of 750 W/m^2. The average transmittance-absorptance product is 0.88. Water ($c_p = 4.18$ kJ/kgK) flows through the collector at a rate of 0.03 kg/s. Calculate (a) the average rate of useful energy, (b) the average rate of heat loss, and (c) the average efficiency of the collector.

[Answers: (a) 752 W; (b) 898 W; (c) 40%]

22.3 Calculate (a) the 1 sun's hour angle, (b) the angle of incidence, and (c) the slope of a collector's surface for a solar collector in Chicago at 11:00 (or 11 a.m.) on 15 September 2024, if the collector is continuously rotated about an east–west axis to minimize the angle of incidence. Use NOAA's solar calculator (or any other reputable nautical almanac), https://gml.noaa.gov/grad/solcalc/sunrise.html, to determine the exact value of solar declination and Chicago's latitude.

[Answers: (a) −15°; (b) 14.98°; (c) 41.3°]

22.4 Analyze a solar collector for the day, date and location specified in Problem 22.3. The solar collector has a surface area of 3 m^2, and a heat removal factor (F_R) of 0.82. The collector faces 30° east and is tilted at 35° from the horizontal plane. The irradiation (diffuse and beam) is 780 W/m^2. The average transmittance-absorptance product is 0.88. The ambient temperature is 18 °C. Water enters the collector at 37 °C and 0.04 kg/s. The rate of heat loss from the collector is measured to be 1060 W. Calculate (a) the rate of useful energy, (b) the exit temperature of the water, and (c) the collector's thermal efficiency.

[Answers: (a) 999 W; (b) 43 °C; (c) 42.7%]

22.5 A Vitosol 300-T solar water-heating evacuated-tube solar collector (manufactured by Viessmann) is mounted at 45° on a house roof in Chicago, where it faces south. The annual average ambient temperature is 10 °C. The collector, manufactured by Viessmann, is connected to an insulated 300-L water storage tank, which is maintained at 50 °C, while the average (annual) ambient air temperature is 15 °C. The solar collector is guaranteed for a period of 25 years. The total installed cost of the solar collector is $7700. Do independent research on the manufacturer's website (search for Viessmann's "System Design Guidelines" and for the "Vitosol 300-T") and (a) estimate how much energy the system will deliver per year to heat water for showers and sinks, and (b) the levelized cost of heat for a 5% discount rate.

[Answers: (a) 3.61 MWh$_{th}$/year; (b) 15 ¢/kW$_{th}$]

References

DLR (2021). Solar thermal power plants. Heat, electricity and fuels from concentrated solar power. German Aerospace Center (DLR), Institute of Solar Research. http://www.dlr.de/sf/en

Duffie, J.A., Beckman, W.A. and Blair, N. (2020). *Solar Engineering of Thermal Processes, Photovoltaics and Wind*. New York: Wiley.

IEA (2019). *Solar Energy: Mapping the Road Ahead*. Paris: International Energy Agency.

IEA (2022). *Technology and Innovation Pathways for Zero-Carbon-Ready Buildings by 2030*. Paris: International Energy Agency.

Mackay, M.E. (2015). *Solar Energy*. Oxford, UK: Oxford University Press.

Müller-Steinhagen, H. and Trieb, F. (2004). Concentrating solar power. A review of the technology. *Ingenia Inform QR Acad Eng* 18: 43–50.

SHC (2022). *Solar Heat Worldwide*. Solar Heating and Cooling Programme (SHC), International Energy Agency.

Sun, J., Zhang, Z., Wang, L., Zhang, Z. et al. (2020). Comprehensive review of line-focus concentrating solar thermal technologies: parabolic trough collector (PTC) vs linear Fresnel reflector (LFR). *Journal of Thermal Science* 29 (5): 1097–1124.

Walker, A. (2013). *Solar Energy: Technologies and Project Delivery for Buildings*. New York: Wiley.

Zhu, G., Augustine, C., Mitchell, R., Muller, M. et al. (2022). Roadmap to Advance Heliostat Technologies for Concentrating Solar-Thermal Power (No. NREL/TP-5700-83041). Golden, CO: National Renewable Energy Laboratory.

23

Solar-Thermal Electrical Power Plants

Solar-thermal power or electric generation systems collect and concentrate sunlight to produce the high-temperature heat needed to generate electricity. All solar-thermal power systems have solar-energy collectors that comprise reflectors that capture and focus sunlight onto a receiver. In most types of systems, a heat-transfer fluid is heated and circulated in the receiver and used to produce steam. The steam is converted into mechanical energy in a turbine, which powers a generator to produce electricity.

In 2020, the global installed capacity of concentrated solar power approached 7 GW, a fivefold increase between 2010 and 2020. In this period, the global weighted average *levelized cost of electricity (LCOE)* of *concentrated solar power* (CSP) plants fell by 68%, from $0.358/kWh to $0.114/kWh. Between 2010 and 2020, CSP's global average total installed costs declined by half, to $4746/kW. During 2021, however, these total installed costs increased to $9090/kW – just 4% lower than in 2010. There was, however, only one CSP project (the Cerro Dominador project in Chile) that came online in 2021; that CSP plant boasts 17.5 hours of energy storage, which is – in part – responsible for the high plant costs (IRENA 2022).

23.1 Solar-Thermal Power Plants

Different types of optical systems – tracking lenses or reflectors (mirrors or "heliostats") – concentrate direct solar radiation onto an absorber. The energy heats a medium, which most commonly generates steam in a secondary circuit. The steam then drives turbines to generate electricity.

The three basic types of power station apply different principles to concentrate solar radiation onto a focal point or line. These power stations have different concentration factors, average operating temperatures, and output power values. Parabolic trough power plants and solar power towers have already proven their reliability.

In a *parabolic trough* power station, reflecting troughs concentrate solar beams onto an absorber pipe, which is installed along the focal line (Figure 23.1). Thermal oil is heated, which drives a turbine via a secondary steam circuit. At present, further developments are being made in direct evaporation technology. The efficiency increases if the thermal oil is replaced by a water–steam mixture that can be heated to higher temperatures. In the future, improved, selective solar absorber coatings should produce temperatures in excess of 500 °C.

Linear Fresnel collectors are similar to parabolic trough collectors in that they are both line-focus systems that use a linear receiver tube. Linear Fresnel reflectors are made of many thin, flat mirror

Energy Systems: A Project-Based Approach to Sustainability Thinking for Energy Conversion Systems,
First Edition. Leon Liebenberg.
© 2024 John Wiley & Sons, Inc. Published 2024 by John Wiley & Sons, Inc.
Companion website: www.wiley.com/go/liebenberg/energy_systems

Figure 23.1 Parabolic trough reflectors concentrate sunlight onto a tube containing a heat transfer fluid. *Source:* Sky Fuel/NREL.

strips that run parallel to the receiver, each oriented to reflect the sun onto the fixed receiver (Figure 23.2). The mirrors rotate on a single axis to keep the sun focused on the stationary receiver throughout the day. Due to their geometry, linear Fresnel collectors have a lower optical efficiency than parabolic trough collectors. However, their simple design and ability to use low-cost, flat glass mirrors hold significant cost-reduction potential (Augustine et al. 2022).

In a *parabolic dish* system, a paraboloidal reflector with a diameter of several meters focuses direct solar radiation onto a hot-air (or Stirling) engine, which drives an electric generator (Figure 23.3). The reflector continually tracks the sun. If the sun is obscured, the systems can be operated with fossil fuels (usually a natural gas-fueled gas turbine). Current developments are therefore focusing on solar or chemical hybrid systems – some of which use biomass.

Figure 23.2 Fresnel concentrating solar power collectors comprise many thin, flat mirror strips that run parallel to the receiver. Each collector rotates on a single axis to keep the sun focused on the stationary receiver throughout the day. *Source:* Orano.

Figure 23.3 Parabolic dish concentrator (Diameter: 8.5 m, aperture, 56.7 m^2, C_R = 2500) concentrating solar energy onto a Stirling engine. *Source:* Schlaich Bergermann via Wikimedia Commons.

Figure 23.4 The Ivanpah solar tower in Nevada. The facility deploys 173,500 heliostats, each with two mirrors focusing solar energy onto water boilers located on three 140-m tall "solar power towers." The produced steam drives three Siemens SST-900 steam turbines with a total power output of 392 MW$_e$. *Source:* Siemens Energy.

In *solar power towers*, solar radiation is concentrated onto a receiver at the top of a tower by many reflectors (or heliostats) on the ground (Figure 23.4). The system consists of an absorber (black wire matting, porous ceramics, and pipe bundles) and a heat transfer medium, which can be air, steam, liquid sodium or molten salt. The efficiency value of solar towers would be dramatically improved if the air that is heated to very high temperatures can be used to drive a gas turbine in a primary circuit. The surplus energy can be used to generate steam, which drives a steam turbine in a secondary circuit.

Table 23.1 compares the salient characteristics of these three types of solar power plants and several types of solar collectors.

Table 23.1 Salient characteristics of the main types of solar thermal power plants.

Type of collector		Annual solar-to-electric efficiency (%)	Peak thermal efficiency (%)	Concentration ratio, C_R (–)	Working temperature (°C)	Thermal power output (MW$_{th}$)	Land use (m²/ MWh/y)	Levelized cost of electricity in 2021 ($/kWh)
Line-focus	Parabolic trough collector Source: Sky Fuel/NREL	13–15	25	40–85	<600	10–400	7	0.3–0.4
	Linear Fresnel reflector Source: Areva/Areva Solar	15	20	30–80	150–600	≤50	5	0.15–0.35
Point-focus	Parabolic dish Source: Schlaich Bergermann via Wikimedia Commons	21	32	1000–3000	<1500	0.1–50	11	0.3–0.4
	Central receiving system ("power tower") Source: NREL/Augustine et al. (2022)	15–19	30	200–1000	<1300	10–400	11	0.15–0.35

Source: IEA (2019), Sun et al. (2020), DLR (2021), SHC (2022), Augustine et al. (2022), SolarPaces (2022), Zhu et al. (2022), and Müller-Steinhagen and Trieb (2004).

23.2 Solar Resource Potential for Concentrated Solar Power

For medium- and high-temperature applications, sun-tracking focusing solar technologies are needed, although their use is restricted to areas with good *direct normal irradiance* (DNI), e.g., Australia, Chile, the Middle East, North Africa, northern Mexico, north-western India, Peru, southern Africa, the south-western United States, and western China. In system comparisons, a baseline DNI of 7.7 kWh/m^2/day is commonly used (Augustine et al. 2022; Zhu et al. 2022). Apart from the latitude where the solar system is located, DNI is also significantly higher at higher elevations, where the absorption and scattering of sunlight due to aerosols are normally much lower.

Trieb et al. (2009) assumed a cut-off of 2000 kWh/m^2/year DNI (5.5 kWh/m^2/day) as being sufficient for electricity production using CSP. They found more than 3 million TWh$_e$/y of total CSP resource potential worldwide, with Africa, Australia, and the Middle East enjoying the greatest share. This resource potential represents more than 100 times the world demand for electricity, demonstrating that CSP technologies are certainly not resource limited.

23.3 Focusing ("Concentrating") Collectors

Focusing (or "concentrating") collectors use curved or flat reflecting surfaces to concentrate the solar radiation onto a small target area. These collectors can either focus their energy on a point or a line. Line-focusing technologies make up around 80% of present solar concentrating technologies (DLR 2021).

Line-focusing plants use single-axis tracking to concentrate solar radiation onto a receiver along a line, whereas point-focusing plants use dual-axis tracking to concentrate solar radiation onto a single point (Table 23.1). Point-focusing technologies attain higher temperatures, but the plants are difficult to scale and are more complex and expensive than their line-focusing counterparts. The two main line-focusing technologies commercially available are parabolic trough collectors (PTCs) and linear Fresnel collectors (LFCs), while the two main point-focus technologies are central receiver systems (CRSs) and parabolic dish reflectors (Table 23.1).

The concentration of solar energy using point- or line-focusing collectors can yield very high temperatures (in the order of 1000 °C) in the energy-absorbing medium, dependent on surface reflectivity and the optical precision of the collector's curvature. A focusing collector can, however, only concentrate direct (or "beam") radiation and not diffuse radiation. Focusing collectors therefore perform poorly in cloudy or hazy skies that are filled with aerosols such as smoke particles.

Solar concentrators effectively increase the collector's surface. To determine the amount of radiation that reaches the absorber surface, we use the concentration ratio (C_R) of the collector's surface (Duffie et al. 2020):

$$C_R = \frac{A_{aper}}{A_{abs}} \tag{23.1}$$

where A_{aper} is the aperture area and A_{abs} is the absorber area, as illustrated in Figure 23.5.

We can use Eq. (22.2) to estimate the irradiance and heat transfer rate from solar radiation:

$$\dot{Q}_{solar} = \sin\left(\theta_{sub}\right)^2 \sigma T_{sun}^4 A_{aper} \qquad \text{[W]} \tag{23.2}$$

where T_{sun} is the absolute temperature of the sun (5780 K).

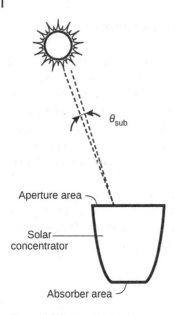

Figure 23.5 An ideal solar concentrator with an aperture area (A_{aper}) and absorber area (A_{abs}), which gives a concentration ratio, $C_R = A_{aper}/A_{abs}$.

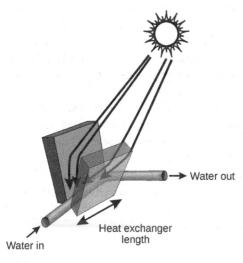

Figure 23.6 A rudimentary solar concentrator, comprising two tilted surfaces.

The maximum radiation that the absorber can emit is:

$$\dot{Q}_{abs} = F_{a-s}\sigma T^4_{abs}A_{abs} \tag{23.3}$$

where T_{abs} is the temperature of the absorber's surface and F_{a-s} is the specular exchange factor (or "view factor"), which indicates the amount of radiation that the absorber can emit back to space by specular means or reflected by a mirror (Duffie et al. 2020). $F_{a-s} = 1$ if all radiation is emitted back to space. If this ideal condition is assumed, we may substitute Eqs. (23.2) and (23.3) in Eq. (23.1) to arrive at the following for circular concentrators (Mackay 2015):

$$C_R = \frac{A_{aper}}{A_{abs}} = \frac{1}{\sin(\theta_{sub})^2}\frac{\dot{Q}_{solar}}{\dot{Q}_{abs}}\frac{T^4_{abs}}{T^4_{sun}} \tag{23.4}$$

The maximum concentration ratio, disregarding any losses and assuming a linear concentrator, is (Winston and Hinterberger 1975):

$$C_{R,max} = \frac{1}{\sin\theta_{sub}} \tag{23.5}$$

Example 23.1   Rudimentary Focusing Collector

Figure 23.6 shows a rudimentary solar concentrator, which comprises two tilted surfaces that reflect solar radiation onto a 30 mm-diameter pipe filled with water. The collector's concentration ratio is 45. The pipe has an absorptance of 0.9 and the solar irradiation is 925 W/m². Calculate the required length of the pipe if the mass flow rate of water is 0.06 kg/s and the water's temperature must be raised by 30 °C.

Given: $D = 0.03$ m, $C_R = 45$, $\alpha = 0.9$, $P_{bb} = 925$ W/m², $\dot{m} = 0.06$ kg/s, $\Delta T = 30$ °C

Required: We must calculate the pipe length, L, required to bring about the necessary heat transfer.

Solution strategy: We need to calculate the heat transfer surface area, A, and can then use Eqs. (22.20) and (22.25): $\dot{Q}_u = \dot{m}c_p(T_o - T_i) \equiv C_R\alpha P_{bb}A$ to calculate L.

Analysis:

With the surface area of the pipe being $A = \pi DL$

Combining the above, we can find the required length of the heat exchanger pipe:

$$L = \frac{\dot{m}c_p(T_o - T_i)}{C_R\alpha P_{bb}\pi D} = \frac{(0.06\text{ kg/s})(4180\text{ J/kgK})(30\text{ K})}{(45)(0.9)(925\text{ W/m}^2)\pi(0.03\text{ m})} = 2.1\text{ m}$$

Discussion: We disregarded heat losses, and the heat-trapping effect of glass covers. Those must, of course, be included in detailed calculations.

Example 23.1 reveals that simple concentrating surfaces can raise the water temperature by 30 °C, which might be adequate for use in domestic applications (e.g., for shower water). However, to produce steam for use in a Rankine cycle, the solar radiation must be concentrated onto a pipe (the absorber) filled with a heat transfer fluid (such as water). This can be done using parabolic reflectors, parabolic dishes, or linear Fresnel lenses. Only parabolic reflectors will be investigated here, due to their similar underlying theories.

23.4 Parabolic Reflectors

Equation (22.27) gives the maximum possible concentration ratio assuming ideal, one-dimensional conditions. For parabolic reflectors, such as those illustrated in Figure 23.7, the typical value of the concentration ratio ranges between 20 and 30. The maximum absorber temperature can be found with Eq. (22.26). As summarized in Table 23.1, the absorber temperatures that can be attained are in the order 600 °C.

From the first law of thermodynamics for an open system, we can write the following energy balance for a parabolic reflector (Duffie et al. 2020):

$$\dot{m}(h_{out} - h_{in}) = (C_R \alpha_{abs} \tau P_{avail,0} - \dot{q}_{loss})A \tag{23.6}$$

where $\dot{m}$ is the mass flow rate of the heat transfer fluid, h_{in} and h_{out} are, respectively, the enthalpies at the inlet and outlet, C_R is the concentration ratio, α_{abs} and τ are, respectively, the absorptance and

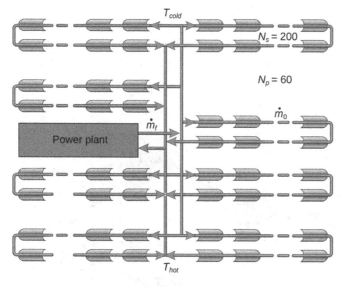

Figure 23.7 A solar field comprising 200 parabolic reflectors or heat exchangers connected in series and 60 loops operating in parallel so that there are $N_p N_s = 60 \times 200 = 12,000$ reflectors or heat exchangers. If each reflector has a length of 5 m and an aperture (width) of 6 m, then each reflector will have an aperture area of 5 m × 6 m = 30 m². The solar field will then have an effective solar capture area of (12,000 reflectors) × (30 m²/reflector) = 360,000 m² of reflectors. Such a system could produce around 65 MW$_e$ of power with irradiation of around 1000 W/m², using Eq. (23.1) and assuming a 35% conversion efficiency of steam to electricity.

transmittance of the absorber surface, $P_{avail,0}$ is the direct irradiation (W/m²) received by the aperture, which is assumed to be normal to the solar radiation, $\dot{q}_{loss}$ is the specific rate of heat loss from the system (W/m²), and A is the external surface area (m²) of the absorber pipe ($A = \pi D_{pipe}L_{pipe}$), with D_{pipe} being the pipe's outside diameter and L_{pipe} being its length.

For an entire parabolic solar field, the heat transfer rates can be calculated as follows:

$$\dot{Q}_{HX} = \dot{m}c_p\Delta T \equiv N_pN_s(C_R\alpha_{abs}\tau P_{avail,0} - \dot{q}_{loss})A \tag{23.7}$$

where $\dot{Q}_{HX}$ is the combined heat transfer rate of N_s number of collectors (and heat exchangers, HX) connected in series to form a loop, and with N_p number of loops that work in parallel with one another. Figure 23.7 explains this type of configuration.

23.5 Central-Receiver ("Power Tower") Collectors

Solar power towers employ large fields of heliostats to focus the sun's rays onto the top of a tall tower. Figure 23.8 illustrates the external receiver exchanging energy with molten salt, which is stored and circulated to a boiler where steam is generated that powers a steam turbine. The optical properties of power tower heliostats are complex. The energy transferred from a heliostat field to an external receiver may, however, be approximated as follows:

$$\dot{Q} \approx P_{avail,0}A\rho\,\alpha_{abs}\tau \tag{23.8}$$

where ρ is the specular reflectance of the heliostat and $P_{avail,0}$ the average irradiation.

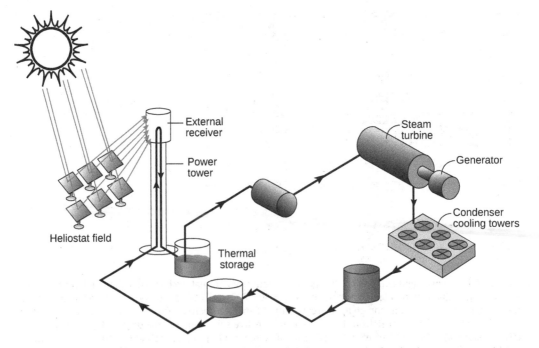

Figure 23.8 Schematic of a solar power tower. (Figure 23.4 shows an operational solar power tower.)

Example 23.2   Solar Power Tower with 173,500 Tracking Mirrors (Heliostats)

Consider a solar "power tower," which employs a field of 173,500 heliostats to concentrate solar energy onto a receiver located at the top of the tall tower (see Figure 23.8 for a typical layout). The 173,500 heliostats, each with a 14-m^2 reflecting surface, actively track the sun using an automated control system. Each heliostat has a reflectance of 0.88. The product of $\alpha_{abs}\tau$ may be taken as 0.85. The heat transfer fluid is molten salt, which is pumped up to the tower. It enters the external receiver at 400 °C, is heated to 900 °C by the focused solar radiation, stored in tanks, and pumped to heat exchangers when needed, generating steam at 580 °C to operate a Rankine cycle with a thermal efficiency of 29%. The average ambient temperature is 22 °C. The heliostats have an energy concentration ratio of 900:1 (Wendelin et al. 2013). The emissivity of the external receiver for long-wavelength radiation is 0.98. If the irradiance is 800 W/m^2, estimate the electric power output of the plant.

Given: $n = 173,500$ heliostats; $A_{helio} = 14$ m^2; $\rho = 0.88$; $\alpha_{abs}\tau = 0.85$; $P_{avail,\,0} = 800$ W/m^2; $\varepsilon = 0.98$; $C_R = 900$; $\eta_{th} = 0.29$.

Required: We must estimate the electrical power output of a solar power tower, $\dot{W}_{el}$.

Solution strategy: The electrical power produced can be calculated with $\dot{W}_{el} = \eta\dot{Q}_{th}$. We can use Eq. (23.8) to calculate the energy transferred to the receiver, $\dot{Q}_{th}$.

Analysis:

From Eq. (23.8), the energy transferred to the receiver is:

$$\dot{Q} \approx P_{avail,0}A\rho\alpha_{abs}\tau = \left(800 \text{ W/m}^2\right)\left(173,500 \text{ heliostats}\right)\left(\frac{14 \text{ m}^2}{\text{heliostat}}\right)(0.88)(0.85) = 1.45 \times 10^9 \text{ W}.$$

Due to the high temperature of the receiver, it experiences significant heat loss due to radiation, which can be approximated by $\dot{Q}_{loss,r} = \varepsilon A_{rec}\sigma\left(T_s^4 - T_{amb}^4\right)$. The receiver area can be estimated from the concentration ratio, using Eq. (23.1):

$$C_R = \frac{A_{aper}}{A_{rec}} = 900 = \frac{\left(173,500 \text{ heliostats}\right)\left(\dfrac{14 \text{ m}^2}{\text{heliostat}}\right)}{A_{rec}}$$

$$\therefore A_{rec} = \frac{2.43 \times 10^6 \text{ m}^2}{900} = 2.69 \times 10^3 \text{ m}^2$$

$$\therefore \dot{Q}_{loss,r} = \varepsilon A_{rec}\sigma\left(T_s^4 - T_{amb}^4\right) = 0.98\left(2.69 \times 10^3 \text{ m}^2\right)\left(5.67 \times 10^{-8} \text{ W/}(\text{m}^2\text{K}^4)\right)$$
$$\left[\left((900 + 273) \text{ K}\right)^4 - \left((22 + 273) \text{ K}\right)^4\right]$$
$$\therefore \dot{Q}_{loss,r} = 281.8 \times 10^6 \text{ W}$$

The energy transferred to the liquid sodium is $\dot{Q}_{th} = \dot{Q} - \dot{Q}_{loss,r} = (1.45 \times 10^9 \text{ W})$ − $(281.84 \times 10^6 \text{ W}) = 1.22 \times 10^9 \text{ W}$. Assuming no losses in the heat exchanger, where the liquid sodium transfers heat to water, the electrical power output of the steam plant is $\dot{W}_{el} = \eta\dot{Q}_{th} = 0.29(1.22 \times 10^9 \text{ W}) \cong 355 \text{ MW}_e$.

Discussion: The Invanpah solar power tower in California uses 173,500 mirrors of a similar size to those in this problem. The Ivanpah power plant has a net electrical power output of 377 MW. It therefore appears as though the approximations made in this problem are appropriate.

Reimagine Our Future

Will increased energy efficiency ("doing more with less") and expanding renewable energy conversion systems address the sources of unsustainability or merely the symptoms? We might replace energy systems that emit greenhouse gases with renewable energy (which would be a profoundly important achievement), but otherwise continue to live in much the same way. Would we have solved the problem of living unsustainably?

23.6 Solar Heating for Industrial Processes and Gas-Cycle Power Plants

As discussed in the preceding sections, solar-thermal systems can be used to produce hot water (50 °C) for domestic or commercial processes or to produce power via a steam (600 °C) generation cycle (such as the Rankine Cycle discussed in Chapter 13). Solar concentrators can, however, also be used to heat a gas for use in a Stirling engine (via the Stirling cycle discussed in Chapter 11) or a gas turbine (via a Brayton cycle, as discussed in Chapter 11). There are numerous possible combinations of heat engine technologies to utilize solar heat at kilowatt and megawatt scale.

In industrial processes, solar heating can meet heating needs at temperatures below 120 °C, such as for washing, leaching, cooking, drying, curing, chemical extraction processes, preheating boiler feed water, and space heating in industrial buildings. These processes are common in such industry subsectors as minerals processing, chemicals, transport equipment, machinery, mining and quarrying, food and tobacco, and textiles and leather.

Like buildings, the solar-thermal technologies used in industry in this temperature range are typically flat-plate or evacuated collectors, although non-tracking concentrated solar collectors are sometimes used to obtain temperatures of up to 200 °C. For solar crop-drying – important in developing countries – much simpler collectors can be used, such as perforated roof panels through which air is drawn via small openings and then ducted to feed the dryers (IEA 2019).

For medium- and high-temperature applications, sun-tracking concentrating solar technologies are needed, although their use is restricted to areas with good direct normal irradiance.

Several large-scale concentrating solar thermal plants are being built in Oman, Kuwait, and in California, which are entirely enclosed in standard agricultural greenhouses that protect them from dust and wind. Light, concave mirrors rotate around simple, fixed tubes, directly generating steam for enhanced oil recovery (EOR) operations. This technology could also be used to generate medium- and high-temperature heat in refineries and industries. Emerging technologies include solar ovens and solar-power towers with particle receivers to process non-metallic minerals such as cement materials (IEA 2019).

Box 23.1 compares the electricity cost of solar-thermal plants with those of photovoltaic plants, while Box 23.2 explains how solar energy may be used to drive refrigeration cycles.

Box 23.1   Are Solar-Thermal Power Plants Cost-Competitive?

In 2020, the electricity generation costs for photovoltaic electricity with a lithium-ion battery system that has a four-hour discharge time were approximately $0.2 for 1 kWh of electricity (IRENA 2022). For electricity from solar-thermal power plants, the International Renewable Energy Agency (IRENA) calculated average electricity generation costs of $0.182 per kilowatt-hour for the same year. The average auction and power purchase prices for new solar-power plants that went into operation in 2020 and 2021 were between $0.073 and $0.094 per kilowatt-hour – i.e., 48–59% below the calculated costs for 2019. Although the prices for 2020 and 2021 are not directly comparable with the costs for 2019, they clearly show that the competitiveness of CSP plants is continuing to increase.

Box 23.2   Solar Cooling

Solar energy may also be used to drive refrigeration cycles, using absorption cycles (Chapter 14), desiccant cycles, or solar-mechanical processes.

- In an *absorption refrigeration* system, the refrigerant vapor is drawn from the evaporator by absorption into the absorbent (see Section 14.3). The addition of thermal energy to the generator liberates refrigerant vapor from the strong solution. The refrigerant is condensed in the condenser by rejecting the heat. The liquid refrigerant is then expanded to the evaporator and the cycle is completed. Solar-based absorption chillers are thermally activated refrigeration systems that draw heat from a storage tank connected to a solar collector to operate the absorption chiller (Figure 23.9).
- *Solar-based desiccant cooling* typically takes warm air (either from outside or inside a building), dehumidifies it with a solid or liquid desiccant, cools it through the exchange of sensible heat, and evaporatively cools it to the desired state (Duffie et al. 2020). Numerous desiccant materials are currently being employed, including silica gel, metal–organic frameworks, hydrogels, zeolites, hygroscopic salts, and composite desiccant materials (Gado et al. 2022).
- *Solar-mechanical cooling* systems typically comprise a solar-powered Rankine cycle, which drives a conventional vapor-compression refrigeration system. Solar collectors produce hot working fluid, which is stored in tanks, circulated to a boiler where steam is produced, and piped to a turbine that drives the refrigeration cycle's compressor. These systems are difficult to balance, in that the boiler temperature changes throughout the day. This could result in overly wet steam entering the turbine. Auxiliary energy must often be added to facilitate dry steam entering the turbine to avoid damage to the turbine blades and ensure high energy conversion efficiency.

An advantage of using solar energy to cool a building, for instance, is that the cooling demand is highest when the solar intensity is at its highest (Figure 23.10). The problem with solar energy, however, is its intermittent nature, which necessitates thermal energy storage. This is further investigated in Chapter 26.

Although seasonal and daily solar irradiance largely coincides well with the cooling demand in buildings, the thermal inertia of buildings causes the cooling demand to remain high long after sunset. While individual solar thermal cooling systems must have some thermal storage (mostly as cold storage) to cover this period, electric chillers do not (SHC 2022). At the system level, however, cold storage in air-conditioning systems – a feature that too few companies are marketing – could prompt a higher grid penetration of solar PV electricity.

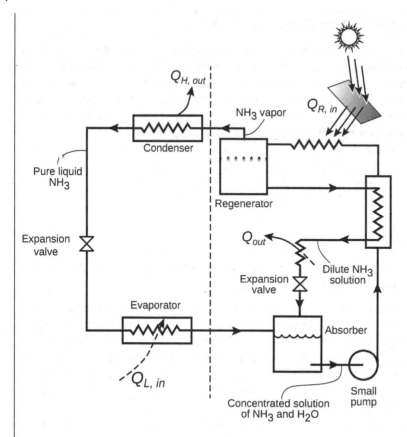

Figure 23.9 Solar-based absorption cooling system.

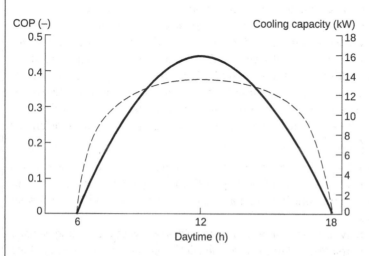

Figure 23.10 Typical cyclic cooling capacity and coefficient of performance (*COP*) for a 15-kW$_{cooling}$ solar absorption-type refrigeration system on a typical summer's day in Chicago. The *COP* continuously increases in the morning hours as irradiation increases, achieving a maximum at 14 : 00 to 15 : 00. The *COP* then decreases with time as the irradiation is insufficient to heat up the circulating heat transfer fluid. This results in less fluid circulating in the generator and the desorber, thus reducing the cooling effect.

23.7   Outlook

Worldwide, dwellings that use solar-thermal technologies for water heating reached 250 million in 2020. To help attain net-zero emissions by 2050, at least another 290 million solar-thermal systems will need to be installed between 2021 and 2030 (IEA 2022). The contribution from emerging solar-thermal technologies will be critical to meet this goal.

Solar-thermal electricity currently costs more than *photovoltaic electricity*, but also offers a lot. CSP plants can integrate thermal storage to deliver electricity on demand. They contribute to power system stability and flexibility by making it possible to integrate more photovoltaic and wind power. Different combinations of solar field size, storage tank size, and electricity capacities provide great flexibility in CSP plant design. Solar-thermal electricity is currently most valuable when generation is shifted to after sunset to complement photovoltaic electricity. In the not-too-distant future, all-night generation will be required to further increase the solar share in total electricity generation and reduce the use of fossil fuels (IEA 2019).

Heliostat-based CSP plants could be competitive in dispatchable electricity generation, solar fuel, and heavy industry process heat applications once their economic performance can be further improved. For electricity, it is projected that, if a baseload heliostat-based electricity generation system with a storage of 12 hours or more can achieve a cost target of 5 ¢/kWh, its commercial deployment may reach a total commercial scale of 35–200 GW_e by 2050 and may account for 3.5–20% of the United States' electricity generation for a given scenario (Zhu et al. 2022).

Based on other specialist projections, global *solar cooling systems* should produce around 1000 GW_{th} of solar cooling by 2050, nearly 17% of total energy use in 2050 (IEA 2022; SHC 2022). As solar-thermal cooling is still in the early stages of market development, costs need to be reduced through further development and deployment. A standardized, effective, and simplified range of technologies needs to be developed – particularly for single- and multi-family dwellings – to make solar cooling competitive with conventional and supported renewable technologies, enabling widespread deployment. Training, qualification, quality assurance, and system certification procedures are also needed to stimulate the market by inspiring consumer confidence.

Targeted innovation (technology, regulatory, and market innovation) directed toward bringing these emerging technologies into their growth phase will be necessary in the next few years. Government support, globally, will be required for large-scale pilot projects for emerging smart solar-powered heat storage systems (also see Chapters 26 and 27).

Problems

23.1   A field of parabolic trough solar collectors has an average thermal efficiency of 33%. The trough collectors are used to exchange heat with water to produce steam for a Rankine cycle with a thermal efficiency of 37%. The steam turbine drives an electrical generator with 90% energy conversion efficiency; the generator has an output of 7 MW_e. The average (beam and diffuse) irradiation is 750 W/m². Calculate (a) the total aperture of the field of parabolic troughs, (b) the rate of useful heat delivered to the Rankine cycle, and (c) the overall efficiency of the system.

[Answers: (a) 84,933 m²; (b) 21 MW; (c) 11%.]

23.2 A parabolic dish solar collector supplies heat to a Stirling engine at an average surface temperature of 600 °C when the irradiation is 850 W/m². The collector has a concentration ratio of 21. The average efficiency of the 12-m diameter solar collector is 25%. If the Stirling engine cycle has a thermal efficiency of 30%, how much power can it generate?

[Answer: 151 kW.]

23.3 Consider a solar thermal "power tower" with a central tower 60 m in height, surrounded by an array of 675 flat mirrors (heliostats) extending to a radius of 100 m around the tower. A cylindrical absorber is placed on top of the tower. The width of the absorber is the same as that of a mirror, namely 900 mm. Determine the optimal layout of these 675 mirrors around the power tower using the SolTrace simulation tool: https://www.nrel.gov/csp/soltrace.html.

   Hint: The following paper provides contextualizing information: Wendelin, T., Dobos, A. and Lewandowski, A. (2013). *SolTrace: A Ray-Tracing Code for Complex Solar Optical Systems* (No. NREL/TP-5500-5559 163). Golden, CO: National Renewable Energy Laboratory.

[Answer: Open-ended question; assumptions will determine answer. But, a circular staggered arrangement should be optimal.]

23.4 Produce a three-page narrative memo of no fewer than 500 words to summarize the state-of-the-art and future prospects of solar fuels (excluding bio-hybrids). Your survey must present the salient characteristics, and advantages and disadvantages of the following technologies: (a) solar thermochemistry (direct water dissociation or thermochemical redox cycles), (b) photovoltaics coupled with high-temperature electrochemistry, (c) high-temperature photovoltaics (e.g., based on thermionic emission) coupled to high-temperature electrochemistry, (d) photovoltaics coupled with room-temperature electrochemistry, (e) photo-thermal catalysis, (f) photoelectrochemistry, and (g) photocatalysis. Figure 23.11 illustrates these pathways.

   When doing your research, the following papers might be good starting points: Haussener (2022), Zoller et al. (2022), Schäppi et al. (2022).

[Answer: The immense potential of solar fuels is summarized in the above three papers.]

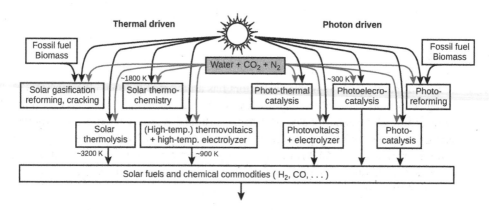

Figure 23.11 Pathways of the various technical solar fuel and chemical commodities (with a focus on hydrogen). The left-hand side highlights the thermally driven pathways, and the right-hand side highlights photon-driven pathways. The center of the diagram illustrates pathways that make use of both thermal and photon energy.

References

Augustine, C., Turchi, C. and Mehos, M. (2022). *The Role of Concentrating Solar-Thermal Technologies in a Decarbonized US Grid* (No. NREL/TP-5700-80574). Golden, CO: National Renewable Energy Laboratory.

DLR (2021). Solar thermal power plants. Heat, electricity and fuels from concentrated solar power. German Aerospace Center (DLR), Institute of Solar Research. http://www.dlr.de/sf/en

Duffie, J.A., Beckman, W.A. and Blair, N. (2020). *Solar Engineering of Thermal Processes, Photovoltaics and Wind*, Wiley.

Gado, M.G., Nasser, M., Hassan, A.A. and Hassan, H. (2022). Adsorption-based atmospheric water harvesting powered by solar energy: comprehensive review on desiccant materials and systems. *Process Safety and Environmental Protection* 160: 166–183.

Haussener, S. (2022). Solar fuel processing: comparative mini-review on research, technology development, and scaling. *Solar Energy* 246: 294–300.

IEA (2019). *Solar Energy: Mapping the Road Ahead*. Paris: International Energy Agency.

IEA (2022). *Technology and Innovation Pathways for Zero-Carbon-Ready Buildings by 2030*. Paris: International Energy Agency.

IRENA (2022). *Renewable Power Generation Costs in 2021*. Abu Dhabi: International Renewable Energy Agency.

Mackay, M.E. (2015). *Solar Energy*. Oxford, UK: Oxford University Press.

Müller-Steinhagen, H. and Trieb, F. (2004). Concentrating solar power. A review of the technology. *Ingenia Information QR Academic Engineering* 18: 43–50.

Schäppi, R., Rutz, D., Dähler, F., Muroyama, A. et al. (2022). Drop-in fuels from sunlight and air. *Nature* 601 (7891): 63–68.

SHC (2022). Solar Heat Worldwide. Solar Heating and Cooling Programme, International Energy Agency.

SolarPaces (2022). CSP Projects Around the World. https://www.solarpaces.org/csp-technologies/csp-projects-around-the-world

Trieb, F., Schillings, C., O'Sullivan, M., Pregger, T. et al. (2009). Global Potential of Concentrating Solar Power. SolarPACES 2009, Berlin, Germany.

Wendelin, T., Dobos, A. and Lewandowski, A. (2013). *SolTrace: A Ray-Tracing Code for Complex Solar Optical Systems* (No. NREL/TP-5500-59163). Golden, CO: National Renewable Energy Laboratory.

Winston, R. and Hinterberger, H. (1975). Principles of cylindrical concentrators for solar energy. *Solar Energy* 17 (4): 255–258.

Zhu, G., Augustine, C., Mitchell, R., Muller, M. et al. (2022). *Roadmap to Advance Heliostat Technologies for Concentrating Solar-Thermal Power* (No. NREL/TP-5700-83041). Golden, CO: National Renewable Energy Laboratory.

Zoller, S., Koepf, E., Nizamian, D., Stephan, M. et al. (2022). A solar tower fuel plant for the thermochemical production of kerosene from H_2O and CO_2. *Joule* 6 (7): 1606–1616.

24

Wind Energy

The planetary wind patterns are caused by a combination of the Earth's rotation and the unequal solar heating of the Earth's surface, resulting in the poles being very cold and the tropics being very warm. The temperature differentials result in air pressure gradients that give rise to winds. Wind energy is therefore an indirect form of solar energy and a renewable source of energy. Once constructed, a wind turbine produces no (or negligible) carbon emissions.

Generally, *wind turbines* comprise a tower on a foundation (or anchored or floating in the ocean) with a nacelle placed at the top (Figure 24.1). The nacelle comprises the spinning rotor, gearbox (if required), electrical generator, and control equipment. The rotor comprises two or more blades joined at a hub. The blades' angles of attack are actively adjusted in accordance with the wind speed and electrical load to provide the most efficient energy conversion.

Wind energy is a *variable and intermittent resource*; hence, it is not dispatchable. It is important to note that global wind speeds are declining at more than 2% per decade due to climate change (see Section 24.2), which suggests that countries should perhaps not rely too much on wind power to meet their long-term energy needs.

Regardless, wind and solar power hit new milestones in 2021, for the first time jointly supplying more than 10% of all electricity-generated globally. Wind and solar power also accounted for three-quarters of all new electricity-generating capacity installed that year (Bloomberg NEF 2022). Wind power is now among the lowest-cost means of electricity supply and energy-sector decarbonization in many regions (Lazard 2021; Pryor et al. 2020).

Reimagine Our Future

Does clean energy mean less energy? Explain your thoughts on social media.

24.1 Overview of Wind Power Systems

Wind is the largest source of renewable electricity generation in the United States, supplying about 21% of the country's electricity in 2021, and growing (EIA 2022). Wind energy, however, supplied only 7% of the world's electricity in 2021 (see Table 24.1). According to countries' present energy policies, wind energy could provide 13% of global electricity in 2030, and 21% in 2050 (IEA 2022).

Energy Systems: A Project-Based Approach to Sustainability Thinking for Energy Conversion Systems, First Edition. Leon Liebenberg.
© 2024 John Wiley & Sons, Inc. Published 2024 by John Wiley & Sons, Inc.
Companion website: www.wiley.com/go/liebenberg/energy_systems

Figure 24.1 A wind park (or "wind farm") comprising horizontal-axis wind turbines. Each turbine can produce around 2 MW$_e$ at a wind velocity of 15 m/s. *Source:* GE.

Table 24.1 Wind energy and world electricity production, based on countries' policy statements.

Year	Energy produced (TWh$_e$), actual and predicted			Shares (%)		
	2021	2030	2050	2021	2030	2050
Total generation	**28,344**	**34,834**	**49,845**	**100**	**100**	**100**
Renewables	**8060**	**15,073**	**32,452**	**28**	**43**	**65**
Photovoltaic	1003	4011	12,118	4	12	24
Wind	1870	4604	10,691	7	13	21
Hydro	4327	5078	6809	15	15	14
Bioenergy	746	1145	1951	3	3	4
Concentrating solar power (CSP)	15	45	329	0	0	1
Geothermal	97	183	458	0	1	1
Marine	1	8	96	0	0	0
Nuclear	**2776**	**3351**	**4260**	**10**	**10**	**9**
Hydrogen and ammonia	**—**	**9**	**44**	**—**	**0**	**0**
Fossil fuels with carbon capture	**1**	**5**	**133**	**0**	**0**	**0**
Coal	1	5	61	0	0	0
Natural gas	—	—	72	—	—	0
Unabated fossil fuels	**17,436**	**16,324**	**12,862**	**62**	**47**	**26**
Coal	10,201	9044	5892	36	26	12
Natural gas	6552	6848	6658	23	20	13
Oil	682	432	312	2	1	1

Source: Data taken from (IEA 2022).

The top five countries in wind electricity generation and their percentage shares of the total world wind electricity generation in 2021 were China (30%), the United States (21%), Germany (8%), the United Kingdom (5%), and India (4%). Although wind energy production has been growing steadily in the past 15 years, to achieve net-zero emissions by 2050 (according to scenarios produced by professional bodies such as the International Energy Agency [IEA]), around 7900 TWh$_e$ of wind electricity generation will be required by 2030. However, as can be seen from Table 24.1, country policies are only projecting a cumulative total of 4604 TWh$_e$ of wind electricity generation by 2030.

The cumulative total of installed wind turbine capacity in 2021 was around 840 GW$_e$ globally. Figure 24.2 shows that, while the number of wind turbines on land (onshore systems) has increased significantly in the past decade, the global cost of producing electricity from wind has dropped from around \$0.30/kWh$_e$ in 2012 to less than \$0.04/kWh$_e$ in 2021 (IEA 2022; Lazard 2021). The costs are expected to continue to decline as the technology matures and as turbine sizes increase. The cost for producing electricity from offshore wind platforms has also decreased from \$0.50/kWh$_e$ in 2012 to around \$0.08/kWh$_e$ in 2021 (IEA 2022).

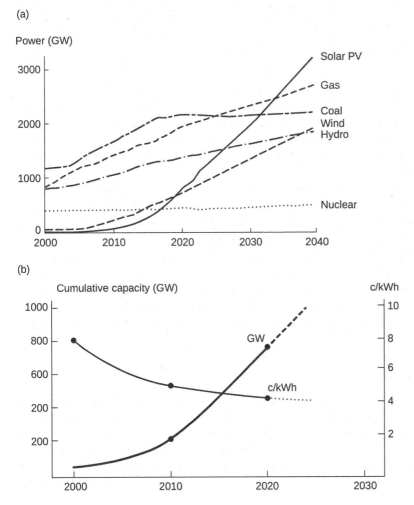

Figure 24.2 (a) Renewable, coal, and nuclear energy usage for electricity generation (2000–2040), and (b) global cumulative wind power capacity. *Source:* IEA (2022), Lazard (2021), and Veers et al. (2019).

24.2 Availability of Wind Energy

24.2.1 Variability of Wind

Kinetic energy in the atmospheric boundary layer exceeds present world electricity and energy demand. Estimates of the present technical potential for wind energy span an order of magnitude owing to the range of assumptions used (17–320 TW_e). The global extractable resource may be larger than 28 TW_e, which greatly exceeds the present *total primary energy supply* (TPES) of 18 TW_e (Barthelmie and Pryor 2014).

Generally, the relative *variability of wind* decreases as the generation of more wind power plants is combined. A group of 200 wind turbines spread over a large area will have a much smaller normalized variability in power output than 10 wind turbines sited together.

The fact that "the wind does not always blow" is often used to suggest the need for large energy storage facilities to cope with fluctuations in the generation of wind power. Such a suggestion ignores the realities of both grid operation and the performance of a utility-scale, spatially diverse wind-generation plant (Sovacool 2009). Modern electricity grids utilize aggregated electricity produced by several resources; the grids therefore deal with changes in grid demand and associated grid reliability issues in an integrated manner. Energy storage is almost never coupled with any single energy source; it would be uneconomical to do so, and it would decrease the capability of a region or country to deal with energy stressors or shocks if it were to overly rely only on wind energy, for instance.

The *conventional energy generation mix* (e.g., nuclear + gas + solar electricity + wind electricity) is therefore designed with much flexibility to manage the daily *energy demand fluctuations*. Intermediate units (e.g., wind parks, solar parks) and peaking units (e.g., gas turbines) must be designed to cycle and help meet the load demand. Only base-load units (e.g., nuclear or natural gas-fired steam power plants) operate continuously. Wind power helps maintain an electrical grid with much maneuvring capability.

Further, according to latest scientific evidence, *the Earth's wind speeds are declining* at more than 2% per decade due to climate change (Pryor and Barthelmie 2010). Since 1979, the Arctic has also been warming four times faster than the rest of the world (Rantanen et al. 2022). This warming could presage an even steeper decline in wind than anticipated.

Only a relatively small fraction of wind energy is typically delivered during peak and high-risk electricity demand periods. Wind generators therefore have *limited capacity* value. This leads to concerns about the impacts of wind power on maintaining reliability and the balance between electrical load and electricity generation. The capture of wind energy and its effective utilization is therefore challenging, and wind energy can only be expected to meet a small fraction of the global energy demand.

24.2.2 Availability of Wind

The planetary wind patterns (on a scale of around 10,000 km) shown in Figure 24.3 are caused by a combination of the unequal solar heating of the Earth's surface and the Earth's rotation (Jaffe and Taylor 2018). The Earth receives greater solar heating at the equator than at the poles, resulting in warm air rising at the equator and flowing through the upper atmosphere toward the poles. This upper atmospheric wind falls back to the Earth's surface at the horse latitudes (about 25° North and South). Simultaneously, cold air leaves the poles and flows near the Earth's surface toward the

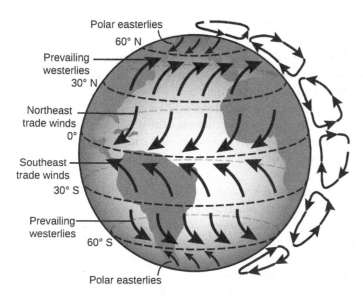

Figure 24.3 Primary atmospheric circulation, also called "planetary winds."

equator, rising at the polar front at about 50°. This cold air is called the *polar easterlies*. The *prevailing westerlies*, between about 25° and 50°, flow near the ground between the horse latitudes and the polar front. The *northeast and southeast trade winds* flow near the Earth's surface from the horse latitudes to the equator where the air is heated and rises to the upper atmosphere. Since this upward flow does not create any significant surface winds, the region is called the doldrums. The temperature difference of air therefore results in pressure differences, which give rise to winds.

Because the Earth rotates in an eccentric orbit about a tipped axis relative to the sun-earth plane, seasonal variations occur in these general flow patterns. Coriolis, inertial, and frictional forces also play a large role in determining the planetary wind pattern.

Local winds (on a scale of 10–100 km) are the result of two different phenomena, the first being the differential heating of the air above land and the air above water. The sun warms the land surface, which warms the air above it, thereby reducing the average air density. Solar energy incident on water is partially absorbed in depth and partially used to evaporate water. The result is that the air above the water is not heated as much as the air above the land. This results in the air over the water being denser. The resulting pressure gradient causes cool denser air to flow toward the warm lighter air. On clear nights, the land cools faster than the water, and the air flow reverses, resulting in sea breezes.

The second mechanism is the *heating and cooling on the sides of hills and mountains*. During the day, the air on the slopes of mountains is heated and rises, drawing in the cool air from the valley below. The air flow direction reverses at night when the cool air drains down the hillside and settles in the valley floor.

In the United States, the National Renewable Energy Laboratory (NREL), along with Pacific Northwest Laboratory, publishes the *Wind Energy Resource Atlas of the United States,* https://www.nrel.gov/gis/wind.html. Information about other wind atlases is available in the *Global Wind Atlas*, https://globalwindatlas.info/en/area/Denmark.

24.3 Wind Turbine Characterization

Onshore wind parks (or "wind farms") are more cost-effective than their offshore counterparts. However, they are usually met with stiff opposition from activists who object to their marring the landscape, their proximity to their homes, and their impact on nature.

Offshore wind turbines have become popular in the past decade. Although they have great potential to capture continuously strong winds, their proliferation in near-shore wind arrays is not feasible due to growing opposition from fishing fleets, environmental conservation groups, and coastal residents. Far-offshore (also called "floating offshore") wind turbines are therefore set to dominate the growth in offshore wind turbines.

Today the most common new installations are *three-bladed horizontal-axis wind turbines*. Most of these turbines are of the upwind variety, where the blades are in front of the nacelle, which houses the generating components, including the generator, gearbox, drive train, and brake assembly. The downwind variety has the advantage that the turbines automatically face into the wind. However, the disadvantage is that their towers block some of the incoming wind and create an increased cyclic stressing of the blades as they pass behind the tower.

Table 24.2 summarizes the salient performance characteristics of grid-scale (>1 MW$_e$) horizontal-axis wind turbines (HAWTs) and vertical-axis wind turbines (VAWTs), while Table 24.3 compares the advantages and disadvantages of each.

HAWTs typically have *cut-in wind speeds* of 3–4 m/s, *cut-out wind speeds* of 24–27 m/s, power coefficients between 44% and 50%, and produce power at 690 V three-phase alternating current (AC). At their rated wind speeds, 4.5-MW$_e$ (offshore) horizontal-axis wind-turbine blades rotate at about 14 rpm, and 2-MW$_e$ (onshore) horizontal wind-turbine blades rotate at about 19 rpm.

Although some utility-scale VAWTs have been built (mostly in the <1.5 MW$_e$ range), they have fallen out of favor due to their blades not simultaneously creating torque, which requires complex blade-pitch control mechanisms and/or massive support structures to contain oscillations, or "flutter" (Ahsan et al. 2022). Some companies or partnerships are trying to resurge VAWTs for use especially in far-offshore wind parks. VAWTs have higher cut-in wind speeds, but similar cut-out and rated wind speeds than HAWTs. However, VAWTs have smaller power coefficients, typically 35–40%.

This chapter focuses on the dominant utility-scale wind turbines, which are horizontal-axis turbines.

24.4 Power Available from the Wind

The power available from the wind may be calculated by considering the mass flow rate of air ($\dot{m}$) passing through an area (A) with specific kinetic energy ($V^2/2$):

$$P_{wind} = \frac{\dot{m}V^2}{2} = \frac{\rho A V^3}{2} \tag{24.1}$$

For example, for an area of 80 m^2 (i.e., a diameter of about 10 m), wind speed of 7 m/s, and air density of 1.2 kg/m^3, the available power from the wind is $P_{wind} = \dfrac{1}{2}\left(1.2 \text{ kg/m}^3\right)\left(80 \text{ m}^2\right)$

Table 24.2 Salient characteristics of contemporary utility-scale (>1 MW_e) wind turbines.

Type of wind turbine		Rated power (MW_e)	Rated wind velocity (m/s)	Rotor diameter (m)	Tower height: HAWT (m), or rotor height: VAWT (m)	Power output per used surface area (MW_e/km^2)	Carbon footprint (g CO_2-eq./kWh_e)	Major manufacturers
Horizontal-axis wind turbine	Onshore *Source:* GE.	1.5–4.8	10–15	115–160	100–170, site-specific	2.3–19, dependent on average wind velocity	4.2–9.6	Vestas, Goldwind, Siemens Gamesa, Envision, GE, Windey, Mingyang, Nordex, Shanghai Electric, Dongfang Electric, RWE, Enercon, Mitsubishi Heavy Industries
	Offshore *Source:* Vestas.	8–15	6–11	160–220	<260, site-specific	>7.2, dependent on average wind velocity	6.4–12.3	
Vertical-axis wind turbine	Onshore: Numerous configurations of VAWTs exist in the range of 10–250 kW. There are no onshore VAWTs in the utility scale (>1 MW).							
	Offshore *Source:* SeaTwirl.	1–5	Not public information	50	40	>6.5, dependent on average wind velocity	Not public information	SeaTwirl, DeepWind/Technical University of Denmark/European Commission, Technip/Nenuphar, Asah LM

Source: Borg et al. (2014), Cheng and Zhu (2014), Enevoldsen and Jacobson (2021), Ennis and Bacelli (2018), Hand and Cashman (2020), Hansen et al. (2021), Jain (2016), Li and Chen (2008), Manwell et al. (2010), Möllerström et al. (2019), Pryor et al. (2021), Rohrig et al. (2019), Tong (2010), and Yaramasu et al. (2015).

Table 24.3 Salient characteristics of horizontal-axis wind turbines and vertical-axis wind turbines.

	Horizontal-axis wind turbines		Vertical-axis wind turbines	
	Advantages	*Disadvantages*	*Advantages*	*Disadvantages*
Blades	High aerodynamic efficiency	Complicated profile	Simple shape, symmetrical airfoils	Lower aerodynamic efficiency
	High power coefficient	Expensive	Less expensive	Lower power coefficient
		Noisy	Low-noise	
Yaw system	Constant aerodynamic loads	Expensive control system	Not required as the rotor is omnidirectional	Oscillating loads
	Self-start capability	Frequent maintenance	Fewer components	Resists self-starting
			Less maintenance	
Shaft and drivetrain	Short shaft	Drive train is located high (100 m) off the ground	Ground (or surface)-based drive train	Long shaft
	Low mass	Difficult to access drive train for maintenance, expensive	Easier to access drive train for maintenance	High mass
	Lower torsional stress and torsional inertia			Large torsional stress and torsional inertia

$(7 \text{ m/s})^3 = 16.5 \text{ kW}$. As outlined in Section 24.2, global wind speeds are declining. The Intergovernmental Panel on Climate Change (IPCC) forecasts that average annual wind speeds will drop by at least 10% by 2100. From Eq. (24.1), a 10% drop in wind speed will result in a 30% drop in the generated wind power. This could be disastrous for countries that are heavily reliant on wind power. For instance, in 2021, the United Kingdom generated 24% of its energy from more than 11,000 onshore and offshore wind turbines, and in the same year, the European Union got 15% of its electrical energy from wind parks.

Wind speed data is usually collected from weather stations at a height of 10 m above the ground. However, the hubs of most large wind turbines are typically at a height of 150 m. This is problematic as wind speed increases with height through the surface layer of the atmosphere. It is therefore necessary to correct the measured wind speed data using *Hellman's Power Law*:

$$\frac{V_{z_1}}{V_{z_2}} = \left(\frac{z_1}{z_2}\right)^{\alpha} \tag{24.2}$$

where V_{z_1} is the wind velocity at height z_1; and α is experimentally determined and depends on terrain, surface roughness, time of day, season, and temperature. If no such data is available, α is often assumed to be $\frac{1}{7}$, which is suitable for calculating the wind velocity variation (due to "wind shear") above a terrain characterized by tall grass on level ground, which represents a typical wind park site.

The average power available from the wind $(\bar{P}_{wind})$ may be computed directly by measuring wind speed or by using statistical approaches.

24.4.1 Direct Use of Wind Speed Data

When working with a collected time series of wind speed data, the average power available from the wind is (Manwell et al. 2010):

$$\overline{P}_{wind} = \frac{A}{2N} \sum_{i=1}^{N} \rho_i V_i^3 \tag{24.3}$$

where N is the number of wind speed observations.

Assuming average air density $(\overline{\rho})$, the average power density of the wind $\left(\overline{P}_{wind}/A\right)$ is:

$$\frac{\overline{P}_{wind}}{A} = \frac{\overline{\rho}}{2N} \sum_{i=1}^{N} V_i^3 \tag{24.4}$$

The standard deviation of wind speed is:

$$\sigma = \sqrt{\frac{1}{N-1} \sum_{i=1}^{N} \left(V_i - \overline{V}\right)^2} \tag{24.5}$$

where the average wind speed is (Zhang 2015)

$$\overline{V} = \frac{1}{N} \sum_{i=1}^{N} V_i \tag{24.6}$$

where V_i is the i-th mean wind speed record in a data series usually covering one or more years.

24.4.2 Statistical Expression of Wind Speed Data

Instead of working with large sets of wind speed measurements, statistical expressions may be used to provide quick but accurate approximations or to facilitate the easy comparison of wind turbine locations. The most common type of statistical distribution function is the *Weibull distribution* and one of its subtypes, the *Rayleigh distribution*.

The frequency of occurrence of wind speeds may be described by the probability density function of wind speeds, $p(V)$, where the probability of a wind speed occurring between V_a and V_b is:

$$p(V_a \leq V \leq V_b) = \int_{V_a}^{V_b} p(V)\, dV \tag{24.7}$$

The total area under the probability density curve is 1:

$$\int_{0}^{\infty} p(V)\, dV = 1 \tag{24.8}$$

The mean wind speed may then be expressed in terms of $p(V)$:

$$\overline{V} = \int_{0}^{\infty} V\, p(V)\, dV \tag{24.9}$$

and the standard deviation of the wind speed is (Manwell et al. 2010):

$$\sigma = \sqrt{\int\limits_0^\infty (V_i - \overline{V})^2 p(V)\, dV} \tag{24.10}$$

The average wind power density is:

$$\frac{\overline{P}_{wind}}{A} = \frac{1}{2}\overline{\rho} \int\limits_0^\infty V^3 p(V)\, dV = \frac{1}{2}\overline{\rho}\,\overline{V^3} \tag{24.11}$$

A cumulative distribution function may also be defined to describe the time fraction (or probability) that the wind speed is smaller than or equal to a given wind speed, V. The cumulative distribution function is (Manwell et al. 2010):

$$F(V) = \int\limits_0^V p(V')\, dV' \tag{24.12}$$

where V' is a dummy variable.

24.4.3 Weibull Probability Distribution

The probability density function for the Weibull distribution is given by (Manwell et al. 2010):

$$p(V) = \frac{k}{c}\left(\frac{V}{c}\right)^{k-1} \exp\left[-\left(\frac{V}{c}\right)^k\right] \tag{24.13}$$

and the cumulative distribution function is given by:

$$F(V) = \int\limits_0^V p(V)\, dV \equiv 1 - \exp\left[-\left(\frac{V}{c}\right)^k\right] \tag{24.14}$$

with k being the dimensionless Weibull shape factor and c the Weibull scale factor (in m/s). Both k and c are functions of $\overline{V}$ and σ. The average velocities can then be determined:

$$\begin{aligned}
\overline{V} &= c\Gamma\left(1 + \frac{1}{k}\right) \\
\overline{V^2} &= c^2\Gamma\left(1 + \frac{2}{k}\right) \\
\overline{V^3} &= c^3\Gamma\left(1 + \frac{3}{k}\right)
\end{aligned} \tag{24.15}$$

where Γ is the statistical *gamma function*:

$$\Gamma(x) = \int\limits_0^\infty e^{-t} t^{x-1}\, dt \cong (2\pi x)(x^{x-1})(e^{-x})\left(1 + \frac{1}{12x} + \frac{1}{288x^2} - \frac{139}{51,840x^3} + \ldots\right) \tag{24.16}$$

with x being any real number, which should not be a negative integer (0, −1, −2, ...).

Instead of using the above series calculation, accurate gamma functions can be determined using software such as Excel or Engineering Equation Solver (EES).

The square of the standard deviation is (Manwell et al. 2010):

$$\sigma^2 = \overline{V}^2 \left[\frac{\Gamma(1 + 2/k)}{\Gamma^2(1 + 1/k)} - 1 \right] \tag{24.17}$$

For $1 \leq k < 10$, the following approximations hold (Manwell et al. 2010):

$$k \cong \left(\frac{\sigma}{\overline{V}} \right)^{-1.086} \tag{24.18}$$

$$c \cong \frac{\overline{V}}{\Gamma(1 + 1/k)}$$

Figure 24.4 shows the Weibull probability density functions for various values of σ and k when keeping $\overline{V}$ constant, and for various $\overline{V}$ when $k = 2$. The distributions are skewed to the left, which implies that wind speeds are more likely to be lower than the mean.

24.4.4 Rayleigh Probability Distribution

When $k = 2$, the Weibull distribution reduces to the Rayleigh distribution. The probability density function for the Rayleigh distribution is given by (Manwell et al. 2010):

$$p(V) = \frac{\pi}{2} \left(\frac{V}{\overline{V}^2} \right) \exp\left[-\frac{\pi}{4} \left(\frac{V}{\overline{V}} \right)^2 \right] \tag{24.19}$$

The Rayleigh cumulative distribution function is given by:

$$F(V) = \int_0^V p(V) \, dV \equiv 1 - \exp\left[-\frac{\pi}{4} \left(\frac{V}{\overline{V}} \right)^2 \right] \tag{24.20}$$

The probability of the wind speed being in a range between V_1 and V_2 is:

$$F(V) = \int_{V_1}^{V_2} p(V) \, dV = \exp\left[-\frac{\pi}{4} \left(\frac{V_1}{\overline{V}} \right)^2 \right] - \exp\left[-\frac{\pi}{4} \left(\frac{V_2}{\overline{V}} \right)^2 \right] \tag{24.21}$$

This Rayleigh probability density function is illustrated in Figure 24.5c. The Rayleigh distribution is usually used for wind turbine analyses, not just because of its relative simplicity, but also because of it modeling experimental wind data very well. The International Electrotechnical Commission (IEC) has also adopted the Rayleigh distribution in its IEC 61400 international standard for wind turbine design.

Using the relationships in Eq. (24.14), the average power from the wind can be determined.

$$\overline{P}_{wind} = \overline{\rho} A \frac{\overline{V^3}}{2} = \frac{\overline{\rho}}{2} Ac^3 \, \Gamma\left(1 + \frac{3}{k} \right) \text{ with } k = 2 \text{ (Rayleigh)} \tag{24.22}$$

It follows from Eqs. (24.14) to (24.22) that the following gamma functions are required for a Rayleigh distribution (which has $k = 2$): $\Gamma(1 + 1/2)$, $\Gamma(1 + 2/2)$, $\Gamma(1 + 3/2)$. Although these gamma functions can be accurately determined using software such as Excel or EES, they may also be approximated (Touré 2019):

$$\Gamma(1 + 1/2) \cong 0.869 \tag{24.23}$$

$$\Gamma(1 + 2/2) \cong 1.000 \tag{24.24}$$

$$\Gamma(1 + 3/2) \cong 0.674 \tag{24.25}$$

(a)

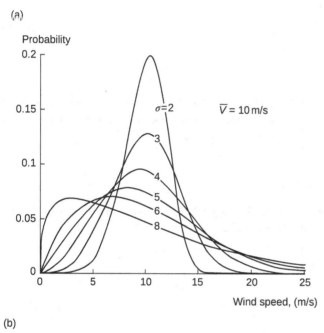

(b)

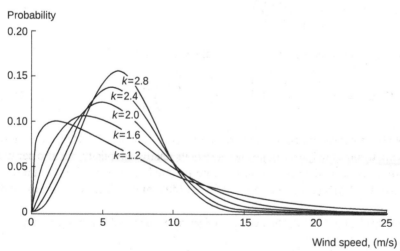

Figure 24.4 (a) Weibull probability density function for an average wind velocity $(\overline{V} = 10 \, \text{m/s})$ and various values of σ, (b) Weibull probability density function for various values of k and for $\overline{V} = 6 \, \text{m/s}$ and $c = 7 \, \text{m/s}$, and (c) Weibull probability density function for various average wind velocities $k = 2$ (after Manwell et al. 2010). (d) Example of a plot of Weibull frequency and its constituent "bins" of measured wind speed data. (See Problem 24.1 for application of the "method of bins.")

(c)

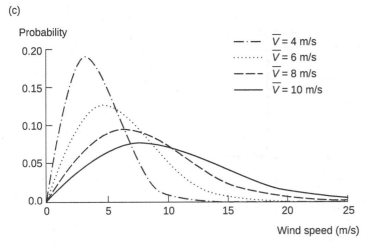

(d)

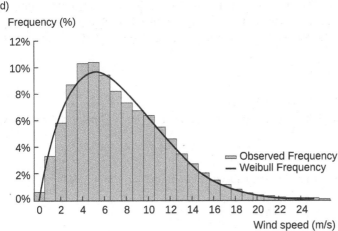

Figure 24.4 (Continued)

Example 24.1 Rayleigh Distribution of Wind Speed for a Large Wind Turbine

An onshore wind turbine has a rotor diameter of 73 m. If the wind speed is too low, the turbine cannot operate; if the wind speed is too high, the mechanical loads may be extreme and cause damage to the system. A large wind turbine therefore features a "cut-in" speed where it begins to produce power, and a "cut-out" speed where the automatic control system alters the pitch of the blades and the yaw of the turbine rotor so that the blades do not rotate. This rotor begins to rotate at a cut-in wind speed of 4 m/s and cuts out at 25 m/s (see Figure 24.5). At the rated speed of 14 m/s, the turbine produces its rated power of 1.4 MW$_e$. The turbine is in a location with an average annual wind speed of 8 m/s; the annual wind characteristic is well represented by a Rayleigh distribution. Calculate (a) the number of hours per year that the wind speed is below the cut-in speed, (b) the number of hours per year that the speed exceeds the cut-out speed, and (c) the yearly energy production when the turbine is running at rated power.

Given: $D = 73$ m, $V_{cut\text{-}in} = 4$ m/s, $V_{cut\text{-}out} = 25$ m/s, $V_{rated} = 14$ m/s, $V_{wind,avg} = 8$ m/s, $P = 1.4$ MW$_e$

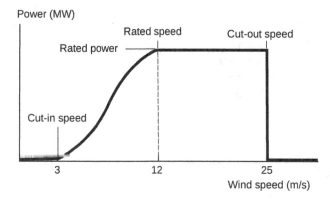

Power (MW)

Figure 24.5 Cut-in and cut-out speeds of a 1.4 MW$_e$ wind turbine.

Required: We must find (a) the number of hours per year that $V_{wind} < V_{cut\text{-}in}$, (b) the number of hours per year that $V_{wind} > V_{cut\text{-}out}$, and (c) the annual energy produced at rated power (1.4 MW$_e$).

Solution strategy: We will use the Rayleigh distribution, Eqs. (24.19) to (24.25), to analyze the power available in the wind.

Analysis:

a) **Number of hours per year that $V_{wind} < V_{cut\text{-}in}$**

From Eq. (24.20), the Rayleigh cumulative distribution is $F(V) = \int_0^V p(V)\, dV \equiv 1 - \exp\left[-\dfrac{\pi}{4}\left(\dfrac{V}{\overline{V}}\right)^2\right]$.

Therefore, the fraction of hours that the wind speed is less than $V_{cut\text{-}in} = 4$ m/s is:

$$F(V) = \int_0^4 p(V)\, dV \equiv 1 - \exp\left[-\frac{\pi}{4}\left(\frac{4\text{ m/s}}{8\text{ m/s}}\right)^2\right] = 0.178$$

There are 8760 h per year, so the number of hours per year that the wind speed is less than $V_{cut\text{-}in} = 4$ m/s is (8760 h/year)(0.178) = 1559 h.

b) **Number of hours per year that $V_{wind} > V_{cut\text{-}out}$**

The fraction of hours that the wind speed is more than $V_{cut\text{-}out} = 25$ m/s is:

$$F(V) = \int_{25}^\infty p(V)\, dV \equiv 1 - \left\{1 - \exp\left[-\frac{\pi}{4}\left(\frac{25\text{ m/s}}{8\text{ m/s}}\right)^2\right]\right\} = 467 \times 10^{-6}$$

There are 8760 h per year, so the number of hours per year that the wind speed is greater than $V_{cut\text{-}out} = 25$ m/s is (8760 h/year)(467 × 10^{-6}) = 4 h.

c) **Annual energy produced at rated power (1.4 MW$_e$)**

The fraction of hours that the turbine will be producing 1.4 MW$_e$ at between 14 m/s and the cut-out speed of 25 m/s can be calculated with Eq. (24.21):

$$F(V) = \int_{14}^{25} p(V)\, dV \equiv \exp\left[-\frac{\pi}{4}\left(\frac{14\text{ m/s}}{8\text{ m/s}}\right)^2\right] - \exp\left[-\frac{\pi}{4}\left(\frac{25\text{ m/s}}{8\text{ m/s}}\right)^2\right] = 0.09 - (466 \times 10^{-6}) = 0.089$$

There are 8760 h per year, so the number of hours per year that the turbine will be operating at rated power is (8760 h/year)(0.089) = 780 h.

The energy produced in this period is $E = P \Delta t = (1.4\,\text{MW}_\text{e})(780\,\text{h}) = 1092\,\text{MWh}_\text{e} \approx 1\,\text{GWh}_\text{e}$

Discussion: This wind turbine will not produce power for only four hours a year. A more detailed analysis will reveal the amount of energy (MWh_e) the turbine will produce between the cut-in speed of 4 m/s and the rated speed of 14 m/s. However, the turbine will produce around 1 GWh_e per year when operating at the rated wind speed and rated power.

24.5 One-Dimensional Wind Turbine Model

The power that a wind turbine can extract from the wind may be approximated by a one-dimensional model. According to the Betz limit, a wind turbine rotor cannot extract more than 59% of the power available in the wind (Betz 1926).

The flow through a wind turbine's rotor may be analyzed upstream and downstream of the turbine rotor (or propeller), as illustrated in Figure 24.6. The rotor (or propeller) is represented by an actuator disk, which creates a pressure discontinuity of area A and velocity V (Glauert 1926). The wind flowing through the actuator disk is represented by a *streamtube*. The wind approaches the disk with velocity V_1 and reaches V_2 far downstream of the disk. The freestream pressure is p_∞. It rises to p_b before the disk and is p_a just after the disk. An external force, F, holds the disk in position and operates in a direction opposite to that of the approaching wind.

Using the momentum equation for a fluid (Section 3.8), we may analyze a control volume between sections 1 and 2:

$$\sum F_x = -F = \dot{m}(V_2 - V_1) \tag{24.26}$$

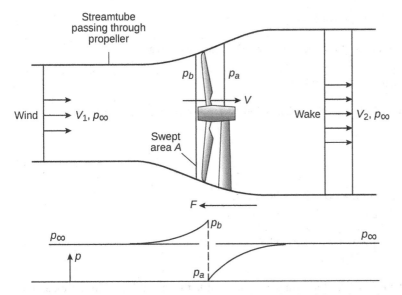

Figure 24.6 Control volumes for a one-dimensional analysis of flow through a turbine rotor.

Similarly, a momentum analysis for the control volumes before and after the disk gives:

$$\sum F_x = -F + (p_b - p_a)V = \dot{m}(V_a - V_b) = 0 \tag{24.27}$$

Setting Eq. (24.26) equal to (24.27) yields the disk (or rotor) force:

$$F = (p_b - p_a)V = \dot{m}(V_1 - V_2) \tag{24.28}$$

Assuming ideal (inviscid, incompressible) flow, we can use Bernoulli's equation to find the pressures:

From 1 to b : $\qquad p_\infty + \dfrac{1}{2}\rho V_1^2 = p_b + \dfrac{1}{2}\rho V^2 \tag{24.29}$

From a to 2 : $\qquad p_a + \dfrac{1}{2}\rho V^2 = p_\infty + \dfrac{1}{2}\rho V_2^2 \tag{24.30}$

Subtracting Eqs. (24.29) and (24.30), and noting from the conservation of mass that $\dot{m} = \rho A V$, we can rewrite Eq. (24.28) as follows:

$$p_b - p_a = \frac{1}{2}\rho\left(V_1^2 - V_2^2\right) = \rho V(V_1 - V_2)$$

or $\qquad V = \dfrac{1}{2}(V_1 + V_2) \tag{24.31}$

Equation (24.31) is also known as *Froude's theorem*.

Combining Eqs. (24.28) and (24.31) yields an expression of the power that the disk (or rotor) can extract from the wind:

$$P = FV = \rho A V^2(V_1 - V_2) = \frac{1}{4}\rho A\left(V_1^2 - V_2^2\right)(V_1 + V_2) \tag{24.32}$$

The maximum possible power can be found by differentiating Eq. (24.32) with respect to V_2 and keeping V_1 fixed, which gives:

$$P_{max} = \frac{8}{27}\rho A V_1^3 \text{ at } V_2 = \frac{1}{3}V_1 \text{ or } V = \frac{2}{3}V_1 \tag{24.33}$$

The maximum available power to the disk (or rotor) is, however, also:

$$P_{max} = \frac{1}{2}\dot{m}V_1^2 = \frac{1}{2}(\rho A V_1)V_1^2 = \frac{1}{2}\rho A V_1^3 \tag{24.34}$$

The maximum possible efficiency of an ideal wind turbine rotor can be defined in terms of a power coefficient (C_p):

$$C_p = \frac{P}{\frac{1}{2}\rho A V_1^3} \tag{24.35}$$

Using Eq. (24.33), the maximum power coefficient is:

$$C_{p,max} = \frac{P_{max}}{\frac{1}{2}\rho A V_1^3} = \frac{\frac{8}{27}\rho A V_1^3}{\frac{1}{2}\rho A V_1^3} = \frac{16}{27} \text{ (or 59.3\%)} \tag{24.36}$$

Equation (24.36) gives the *Betz limit* for a wind turbine, stating that a wind turbine cannot extract more than 59.3% of the power available in the wind. As may be gleaned from Table 24.1, modern wind turbines have already attained a power coefficient of 50%, leaving little room for improvement in turbine design.

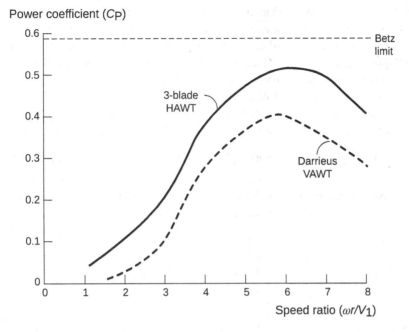

Figure 24.7 Typical power coefficients of horizontal axis wind turbines (HAWTs) and vertical-axis wind turbines. The values of C_p are plotted as a function of the rotor's blade-tip speed ratio, where ω is the rotational speed of the blade (in rad/s), r is the rotor diameter, and V_1 is the wind speed. Also, $\omega = 2\pi N/60$, where N is the rotational speed in revolutions per minute (rpm). Modern wind turbines attain maximum power coefficients at tip-speed ratios of 4–6, which means that the tips of these turbine blades move four to six times faster than the wind speed.

Figure 24.7 shows the range of power coefficients for several types of wind turbines.

24.6 Estimating the Wind Turbine's Average Power and Energy Production

At a given wind speed, the power output (P) of a wind turbine of area A is equal to the wind power (P_{wind}) multiplied by the power coefficient (C_p) multiplied by the combined efficiency of the drive train and electrical generator ($\eta = \eta_m \eta_g$). Using Eq. (24.10), we may then write:

$$\overline{P} = \eta C_p \overline{P}_{wind} = \eta C_p \frac{\rho A \overline{V^3}}{2} \tag{24.37}$$

Example 24.2 Power Output of a Large Wind Turbine
Consider a wind turbine manufactured by General Electric (GE). The rotor has a diameter of 117 m and the hub of the wind turbine is situated 100 m above the ground. According to the manufacturer's catalog, this turbine has an average power coefficient of 0.4. The combined efficiency of the drive train and electrical generator is estimated to be 94%. The wind turbine is placed in a location where the average wind speed is 6.62 m/s and the average air density is 1.1 kg/m³. The wind speeds were measured throughout several years by a weather station located 10 m above the

ground. (a) Use Hellman's Power Law to calculate the average wind velocity at a height of 100 m, which is the height of a GE 3-MW wind turbine's nacelle above the ground. (b) Estimate the wind power density using the provided (average) value of average wind speed. (c) Estimate the wind power density using a Rayleigh approximation of the wind speed distribution. (d) What is the turbine's electrical power output using power densities from (b) and (c)?

Given: $D = 117$ m, $h_{hub} = 100$ m, $C_{p,avg} = 0.4$, $V_{wind,avg} = 6.62$ m/s

Required: We must find (a) $V_{100\,m}$, (b) $(P/A)_{avg}$, (c) $(P/A)_{Rayleigh}$, and (d) $\overline{P}$ and $P_{Rayleigh}$.

Solution strategy: For part (a), we can use Hellman's Power Law, Eq. (24.2). For part (b), we can use Eq. (24.22) to calculate the average power density of the wind. $\dfrac{\overline{P}_{wind}}{A} = \dfrac{1}{2}\overline{\rho}\,\overline{V^3}$. For part (c) we must use a Rayleigh distribution of wind speed, so we can use Eqs. (24.22) and (24.18) and the approximations of the gamma function given in Eqs. (24.23) to (24.25). For part (d) we can use Eq. (24.37) to calculate the average electrical power output of the turbine.

Analysis:

a) $V_{100\,m}$:

Hellman's Power Law, Eq. (24.2), with $\alpha = \dfrac{1}{7}$:

$$\frac{V_{100\,m}}{V_{10\,m}} = \left(\frac{100\ m}{10\ m}\right)^{\frac{1}{7}}$$

$$\therefore V_{100\,m} = (6.62\ m/s)\left(\frac{100\ m}{10\ m}\right)^{\frac{1}{7}} = 9.2\ m/s$$

b) $(P/A)_{avg}$:

From Eq. (24.22), the average power density of the wind is $\dfrac{\overline{P}_{wind}}{A} = \dfrac{1}{2}\overline{\rho}\,\overline{V^3}$. Therefore:

$$\left(\frac{\overline{P}_{wind}}{A}\right)_{avg} = \frac{1}{2}\overline{\rho}\,\overline{V^3} = \frac{1}{2}(1.1\ kg/m^3)(9.2\ m/s)^3 = 428\ W/m^2$$

c) $(\overline{P}/A)_{Rayleigh}$:

For a Rayleigh distribution of wind speed, we use Eqs. (24.22) and (24.18) and the approximations of the gamma function given in Eqs. (24.23) to (24.25). The average Rayleigh power density of the wind can then be calculated:

$$\text{Equation (24.18): } c \cong \frac{\overline{V}}{\Gamma(1 + 1/k)} = \frac{\overline{V}}{\Gamma(1 + 1/2)} = \frac{(9.2\ m/s)}{0.869} = 10.59\ m/s$$

$$\therefore \frac{\overline{P}_{wind}}{A} = \frac{\overline{\rho}}{2}\overline{V^3} = \frac{\overline{\rho}}{2}c^3\,\Gamma\left(1 + \frac{3}{k}\right) = \frac{(1.1\ kg/m^3)}{2}(10.59\ m/s)^3(0.674) = 440\ W/m^2$$

d) $\overline{P}$ and $P_{Rayleigh}$:

For (b), using average wind speed, the average electrical power output of the turbine can be estimated using Eq. (24.37):

$$\overline{P} = \eta C_p \frac{\rho A \overline{V^3}}{2} = (0.94)(0.4)\frac{(1.1\ kg/m^3)}{2}\left(\frac{\pi(117\ m)^2}{4}\right)(9.2\ m/s)^3 = 1.73\ MW$$

with $C_{p,avg} = 0.4$ and the area swept by the rotor being $A = \dfrac{\pi D^2}{4}$, with D being the rotor diameter (117 m).

For (c), using the Rayleigh distribution of wind speed, the electrical power output of the turbine is:

$$P_{Rayleigh} = \eta C_p \left(\frac{\overline{P_{wind}}}{A} \right) A = (0.94)(0.4)\left(440 \ \text{W/m}^2\right)\left(\frac{\pi(117 \ \text{m})^2}{4} \right) = 1.79 \ \text{MW}$$

Discussion:
As the average wind velocity will change from month to month and year to year, it is more accurate to describe the wind profile using a probability density function. As we do not have access to the hourly variation of wind data at this site (we only have the average wind speed), one may use the Rayleigh probability density function to analyze the wind profile. However, there is only a 3% difference in the power output of the turbine when using an average wind speed compared to using a Rayleigh distribution of wind speed. It is therefore appropriate to use average values of wind speed to make rough estimates of wind speed.

24.7 Sustainability of Wind Power

Lifecycle analyses of utility-scale wind turbines reveal that they produce a mean value of 34.11 g of CO_2-equivalent per kilowatt-hour of electrical energy (Nugent and Sovacool 2014). Box 6.1 illustrates that this is equivalent to the production of around 1844 t of CO_2-equivalent per megawatt of the rated output of a large wind turbine over its lifetime (Crawford 2009). Thus, wind energy systems are in no way "carbon free" or "emissions free," although they could be called "low-carbon" energy conversion systems.

The large increase in both the raw materials and rare metals that are required for the large-scale expansion of wind power appears to be manageable, and "no insurmountable long-term constraints to materials supply, labour availability, installation infrastructure or manufacturing capacity appear likely if policy frameworks for wind energy are sufficiently economically attractive and predictable" (Wiser et al. 2021). For example, rare earth oxides used in 20% of wind turbines with permanent magnet generators have known reserves of ~1000 years' supply at present consumption levels (Barthelmie and Pryor 2014).

Electricity generation from any source, including the wind, affects the local and/or global environment. Large wind parks, like major cities and forests, extract momentum from the air and add turbulence, thus altering the meteorology downwind. However, detailed in-situ and remote-sensing measurements at operating wind parks show non-negligible impacts on, for example, the near-surface temperature beyond a few kilometers. The modeling of wind deployment, considering various wind energy scenarios, predicts the meteorological impacts of ~1% in zonal mean precipitation and a ±0.3 °C change in surface temperature (Marvel et al. 2013; Miller and Keith 2018; Pryor et al. 2020). These *externalities* from large-scale wind-energy deployment seem modest, but are not negligible.

The cheapest renewable power projects in the first half of 2022 were able to achieve a levelized cost of electricity (*LCOE*) of US \$19/MWh, similar to best-in-class onshore wind parks in Brazil, US \$21/MWh for tracking photovoltaic parks in Chile, and US \$57/MWh for offshore wind in Denmark. If the offshore transmission costs are excluded, the latter estimate falls to US \$43/MWh (Bloomberg NEF 2022).

24.8 Outlook

Wind energy cost reductions have accelerated over the past five years – at much greater rates than predicted by most experts, learning-curve extrapolation or engineering assessments. Depending on the type of wind turbine system (onshore or offshore), experts anticipate substantial future reductions, as illustrated in Figure 24.8. Experts predict 20–40% cost reductions between 2020 and 2030, and 35–50% reductions by 2050.

Onshore wind is set to dominate the global picture, with China driving growth. Experts forecast that China will account for more than 50% of onshore wind additions between 2022 and 2030, reflecting declining turbine prices, a push for gigawatt-scale project clusters, and a concerted effort to meet that country's climate goals (Bloomberg NEF 2022; Lazard 2021).

Thanks to aggressive decarbonization targets in China and a fast-growing offshore wind sector, more than 100 GW of wind capacity is expected to be added each year for the rest of the decade. Experts forecast that cumulative offshore wind capacity will surge 10-fold between 2021 and 2035 to 504 GW (Bloomberg NEF 2022; Lazard 2021).

Experts also predict plant sizes of 1100 MW for fixed-bottom and 600 MW for floating offshore wind turbines by 2050. Despite reducing the levelized cost of electricity, these massive wind parks can also enhance wind energy's grid service, for example, via project hybridization with batteries and hydrogen production (Bloomberg NEF 2022; Lazard 2021).

Much greater efforts are needed to achieve this anticipated level of sustained capacity growth, with the most important areas for improvement being streamlined permitting for onshore wind, and cost reductions for offshore wind (Bloomberg NEF 2022; IEA 2021).

Some countries are currently investigating floating offshore platforms that are placed much further (>25 km) from the shore than their traditional offshore counterparts. Not only are winds in deeper waters more powerful than those closer to the shore, but the physics of flexible, suspended rigs (using taut mooring cables) enables them to carry larger rotors. However, floating offshore

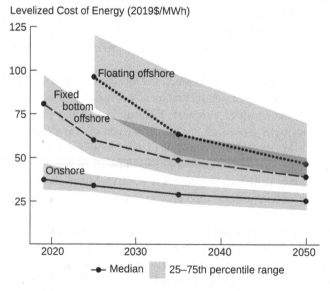

Figure 24.8 Estimates of future levelized costs for wind energy. Shaded areas show 25–75th percentile ranges. Costs are in US dollars. *Source:* After Bloomberg NEF (2022), IEA (2021), Lazard (2021), Wiser et al. (2016).

wind turbines are twice as expensive as near-shore wind turbines, and three to four times more expensive than land-based (onshore) wind turbines. Offshore wind turbines (floating or with fixed foundations) involve maritime engineering, making those systems expensive to build, deploy, and maintain. The life spans of offshore systems are short because of the corrosive nature of the maritime environment.

It is therefore expected that offshore wind turbines anchored to the ocean floor with fixed foundations will remain the mainstay of the offshore wind market up to 2035. However, floating wind turbines are set for a growth spurt from 2027, driven by demand from areas where the sea is too deep for fixed-bottom installations. Floating wind capacity is forecast to reach over 25 GW by 2035, from just under 60 MW in 2021. The United Kingdom is projected to be the market leader in 2035 with around 8 GW of total capacity, followed by the United States, with close to 5 GW (Bloomberg NEF 2022).

Wind speeds appear to be declining due to climate change. Decreasing wind speeds could exacerbate the volatility of energy markets. If natural gas prices spike – say due to war in a country such as Ukraine or if an especially cold winter is experienced – at the same time as a regional wind-stilling event, energy prices could rise beyond the means of millions of people.

There is, also, large uncertainty in wind speed data. We need to better understand the physics of atmospheric flow to enable better predictions of wind speeds. Regardless, wind remains a broadly available (although variable and intermittent) resource and zero-cost fuel, and wind turbine systems have exceptionally low lifecycle pollutant emissions. Wind energy is therefore poised to continue to play an important role in meeting the world's growing clean energy needs.

Problems

24.1 If adequate wind data is available, the "method of bins" is often used to estimate the power output of a wind turbine. Figure 24.9 shows a wind duration curve, which is divided into $N = 10$ time-duration bins of equal size. The average power produced by a turbine operating in this wind condition can be calculated using the bin data:

$$\overline{P} = \eta C_p \frac{\rho A}{2} \sum_{i=1}^{N} f_i V_i^3 \tag{24.38}$$

where f_i is the fraction in bin i, and N is the number of bins.

Consider a 100-m high wind turbine with a rotor diameter of 120 m operating under the wind conditions illustrated in Figure 24.9. The wind speed data was calculated with a weather station located 10 m above the ground. The turbine has a power coefficient of 37% and a combined mechanical and electrical efficiency of 93%. The air density remains constant at 1.2 kg/m^3. Calculate (a) the electrical power, and (b) the electrical energy produced in a year if the turbine has a capacity factor of 0.45.

[Answers: (a) 0.9 MW$_e$; (b) 3.9 GWh$_e$.]

24.2 Repeat Problem 24.1, but now use a Rayleigh distribution of the wind speed. Assume that the average wind speed in this location is 5.8 m/s, measured by a weather station located 10 m above the ground.

[Answers: (a) 0.56 MW$_e$; (b) 2.2 GWh$_e$.]

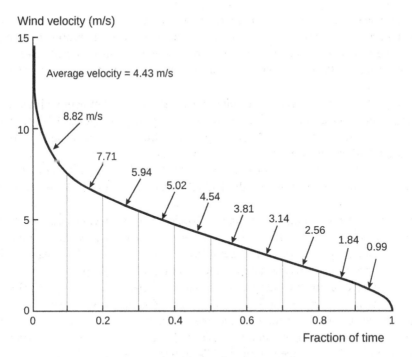

Figure 24.9 Wind duration curve with 10 equal fraction bins.

24.3 Consider a 1.5-MW onshore wind turbine with a 33% capacity factor and a capital cost of $2100 per kilowatt, an annual operating cost of $30 per kilowatt, and a lifetime cost of $3200 per kilowatt. If the rate of return for the project is 6% and the turbine has a lifetime of 25 years, calculate the levelized cost of energy for this wind turbine.

[Answer: 0.11 $/kWh$_e$.]

24.4 You must perform the preliminary design of a wind park (or "wind farm") in the Mohave Desert in California. The wind park requires a "nameplate capacity" of 180 MW$_e$. The site that has been identified for the wind park has an average wind speed of 5.9 m/s. This wind speed was calculated from measurements taken every hour for one year with an anemometer located 10 m above the ground. The air density is 1.2 kg/m^3.

For this wind park, you must use turbines from GE's family of 3-MW onshore turbines. The chosen turbine has a rotor diameter of 130 m. The turbine has a power coefficient of 37% and a combined mechanical and electrical efficiency of 93%.

The electricity generated by the wind park will be exported to a nearby large-scale (180 MW$_e$) electrolyzer facility that will produce 40,000 kg of hydrogen per day for 40 hydrogen-refueling stations in a 200 km radius of the electrolyzer plant (see Figure 24.10).

(a) Use Hellman's Power Law to calculate the average wind velocity at a height of 100 m, which is the height of a GE 3-MW wind turbine hub above the ground. (b) Estimate the wind power density using a Rayleigh approximation of the wind speed distribution. (c) What is the electrical power output of one turbine? (d) How many turbines are required to provide the required power for the electrolyzer plant?

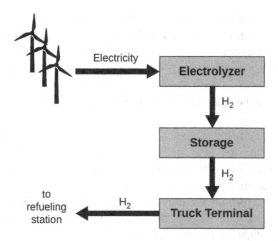

Figure 24.10 Illustration of an electrolyzer facility using electrical energy produced by a nearby wind park.

Meteorologists and engineers report that the annual capacity factor of a site close to the proposed wind park is 43.2%. The capacity factor is the ratio of energy produced annually to the maximum potential of the wind park. Using this information, (e) estimate the annual energy production of this proposed wind park. (f) Estimate how much electrical energy the electrolyzer plant will need to store in lithium-ion batteries to cope with those times that it does not get electricity from the wind park.

For a typical wind park, the turbines are usually spaced $8D$ downwind from one another and $5D$ crosswind from one another, where D is the rotor diameter (Meyers and Meneveau 2012). This arrangement is used to avoid turbine rotors not getting "clean" wind due to their being too close to the wake of other turbine rotors. (g) How large an area (in km^2) is required to accommodate the turbines in the wind park using the preferred spacing of turbines? (h) What is the average output power (MW_e) per land area (m^2) and how does that compare to the same metric for a nuclear power plant?

[Answers: (a) 8.2 m/s; (b) 340.2 W/m^2; (c) 1.56 W/m^2; (d) 116 turbines; (e) 681.2 GWh_e/year; (f) 894.8 MW_eh/year; (g) 78.4 km^2 when placing the 116 turbines in four rows of 29 turbines each; (h) 2.3 MW_e/km^2.]

24.5 Briefly outline (preferably in the shown table form) the interconnected social, technological, economic, environmental, and political (or "STEEP") factors surrounding the implementation of the wind park analyzed in Problem 24.4. Mention at least seven related issues for each of these STEEP factors.

Social	Technological	Economic	Environmental	Political

[Answer: Open-ended question.]

24.6 For most companies, the easiest and cheapest way to deal with spent wind turbine parts is to bury them in landfills. In the USA, ambitious projections for renewable energy development envision 3000 GW of cumulative wind power capacity by 2050, more than 20 times the amount installed in 2022. Globally, the number of wind turbine blades reaching their end of life could reach 12 billion tonnes in 2050. Those retired wind turbines will pose a serious problem for communities and the environment if they continue to wind up in dump sites.

As an energy entrepreneur, you want to help change this situation and keep wind turbine blades out of landfills. You must employ the "3-C" approach of creating value, making connections among seemingly disparate ideas, while using a creative framework to help you generate ideas to solve this problem while making money in the process. Do independent research for recycling or reusing wind turbine blades and present your ideas in a three-page narrative memo that also summarizes your ideas in the categories of creativity, creating value, and making connections, as illustrated in the table below.

Hint: You should first review the 3-C framework for engineering entrepreneurship, see https://engineeringunleashed.com. Then, do research using Google Scholar and other platforms to find journal papers and technical reports related to reusing or recycling wind-turbine blades. You will also be inspired by the work that certain organizations do in this regard, such as the Re-Wind Network (https://www.re-wind.info), Veolia North America (https://www.veolianorthamerica.com), Siemens Gamesa (https://www.siemensgamesa.com), and DecomBlades (https://decomblades.dk).

Creativity	Connections	Creating value

[Answer: Open-ended question.]

24.7 You must investigate the feasibility of wind energy-on-demand, i.e., you should investigate if and how wind energy can be transformed from an intermittent resource into a demand-driven dispatchable resource that is capable of improving grid resilience, cutting the grid's carbon footprint economically, and supporting the well-being of nearby communities by promoting sustainable development. You should conduct a STEEP analysis, considering the social, technological, economic, environmental, and political contexts in which the wind energy-on-demand might be deployed. Be sure to indicate the possible interconnections between each of the five categories. Figure 24.11 summarizes a simple STEEP framework that could guide your research. Summarize your research, findings, and suggestions in a two-page fact sheet.

[Answer: Open-ended question.]

24.8 Airborne wind energy systems convert wind energy into electricity using tethered flying devices, typically kites (or "parafoils") or aircraft. Replacing the tower and foundation of conventional wind turbines can substantially reduce material use and, consequently, the cost of energy while providing access to higher-velocity wind at higher altitudes. Because the flight operation of tethered devices can be adjusted to a varying wind resource, the energy availability increases in comparison to conventional wind turbines.

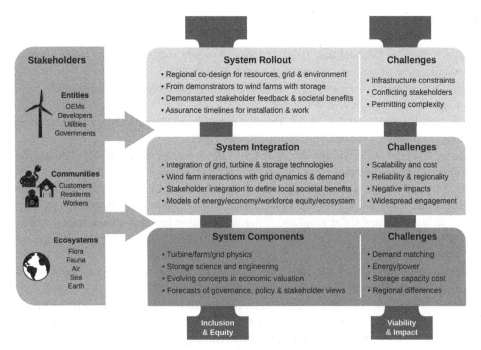

Figure 24.11 STEEP framework for investigating the feasibility of wind energy-on-demand.
Source: Adapted from Aziz et al. (2022).

Perform independent research about power kites and produce a two-page infographic that explains how such a system works, how it is controlled, and its advantages and disadvantages. Do you think that these systems will ever be mass produced? Explain.

Hint: The following paper might help initiate your research: Vermillion et al. (2021)

[Answer: Open-ended question.]

24.9 Theo Jansen is a Dutch kinetic sculptor who builds large works called *strandbeests*, which is Dutch for "beach animals." Jansen uses PVC tubes, wood, and fragile-looking wind sails. The whimsical animal skeletons walk when propelled by the wind (Strandbeest.com).

Jansen's kinetic sculptures fuse art and engineering. Watching his sculptures gracefully move in the wind resonates with people at an emotional level. When we see these fantastic wind-propelled structures, we inadvertently think that understanding is within our grasp; we are inspired by the power of the wind, https://www.exploratorium.edu/strandbeest.

You are asked to design and make your own small *strandbeest*. The challenge is to use only found or available material, such as discarded cardboard, plastic drinking straws, pins, wooden sticks, or copper wire. You may also use Lego parts. If you really want to, you can 3-D print some components, although this is discouraged. See if you can produce a *strandbeest* no longer than 300 mm that will be propelled by wind on a sidewalk. Make a short video clip of your beast in action, and post it on social media. Your message should be that there is a lot of power in the wind, and that – with a bit of imagination and willpower – we can successfully harness this renewable resource. Ask your social media contacts to forward your post to all their contacts and ask for their comments or suggestions on how we can better utilize wind power.

[Answer: Open-ended question.]

References

Ahsan, F., Griffith, D.T. and Gao, J. (2022). Modal dynamics and flutter analysis of floating offshore vertical axis wind turbines. *Renewable Energy* 185: 1284–1300.

Aziz, M.J., Gayme, D.F., Johnson, K., Knox-Hayes, J. et al. (2022). A co-design framework for wind energy integrated with storage. *Joule* 6 (9): 1995–2015.

Barthelmie, R.J. and Pryor, S.C. (2014). Potential contribution of wind energy to climate change mitigation. *Nature Climate Change* 4 (8): 684–688.

Betz, A. (1926). *Wind-Energie und Ihre Ausnutzung Durch Windmühlen* (Vol. 2). Göttingen, Germany: Vandenhoeck & Ruprecht.

Bloomberg NEF (2022). *Wind and Solar Top 10% of Global Power Generation For First Time*. Bloomberg NEF. https://about.bnef.com/blog/wind-and-solar-top-10-of-global-power-generation-for-first-time

Borg, M., Shires, A. and Collu, M. (2014). Offshore floating vertical axis wind turbines, dynamics modeling state of the art. Part I: aerodynamics. *Renewable and Sustainable Energy Reviews* 39 (November): 1214–1225.

Cheng, M. and Zhu, Y. (2014). The state of the art of wind energy conversion systems and technologies: a review. *Energy Conversion and Management* 88 (December): 332–347.

Crawford, R.H. (2009). Life cycle energy and greenhouse emissions analysis of wind turbines and the effect of size on energy yield. *Renewable and Sustainable Energy Reviews* 13 (9): 2653–2660.

EIA (2022). Wind explained. US Energy Information Administration.

Enevoldsen, P. and Jacobson, M.Z., 2021. Data investigation of installed and output power densities of onshore and offshore wind turbines worldwide. *Energy for Sustainable Development*, 60 (Feb.): 40–51.

Ennis, B.L. and Bacelli, G. (2018). *Floating Offshore Vertical-Axis Wind Turbine System Design Studies and Opportunities* (No. SAND2018-8138C). Albuquerque, NM: Sandia National Laboratory.

Glauert, H. (1926). *The Elements of Aerofoil and Airscrew Theory*. Cambridge, UK: The University Press.

Hand, B. and Cashman, A. (2020). A review on the historical development of the lift-type vertical axis wind turbine: from onshore to offshore floating application. *Sustainable Energy Technologies and Assessments* 38: 100646.

Hansen, J.T., Mahak, M. and Tzanakis, I. (2021). Numerical modeling and optimization of vertical axis wind turbine pairs: a scale up approach. *Renewable Energy* 171: 1371–1381.

IEA (2021). *Wind*. International Energy Agency. https://www.iea.org/fuels-and-technologies/wind.

IEA (2022). *World Energy Outlook 2022*. International Energy Agency. https://www.iea.org/reports/world-energy-outlook-2022.

Jaffe, R.L. and Taylor, W. (2018). *The Physics of Energy*. Cambridge, UK: Cambridge University Press.

Jain, P. (2016). *Wind Energy Engineering*. New York: McGraw-Hill Education.

Lazard (2021). Levelized cost of electricity. www.lazard.com.

Li, H. and Chen, Z. (2008). Overview of different wind generator systems and their comparisons. *IET Renewable Power Generation* 2 (2): 123–138.

Manwell, J.F., McGowan, J.G. and Rogers, A.L. (2010). *Wind Energy Explained: Theory, Design and Application*. New York: Wiley.

Marvel, K., Kravitz, B. and Caldeira, K. (2013). Geophysical limits to global wind power. *Nature Climate Change* 3 (2): 118–121.

Meyers, J. and Meneveau, C. (2012). Optimal turbine spacing in fully developed wind farm boundary layers. *Wind Energy* 15 (2): 305–317.

Miller, L.M. and Keith, D.W., 2018. Climatic impacts of wind power. *Joule*, 2(12): 2618–2632.

Möllerström, E., Gipe, P., Beurskens, J. and Ottermo, F. (2019). A historical review of vertical axis wind turbines rated 100 kW and above. *Renewable and Sustainable Energy Reviews* 105 (May): 1–13.

Nugent, D. and Sovacool, B.K. (2014). Assessing the lifecycle greenhouse gas emissions from solar PV and wind energy: a critical meta-survey. *Energy Policy* 65 (February): 229–244.

Pryor, S.C., Barthelmie, R.J. and Shepherd, T.J. (2021). Wind power production from very large offshore wind farms. *Joule* 5 (10): 2663–2686.

Pryor, S.C., Barthelmie, R.J., Bukovsky, M.S., Leung, L.R. et al. (2020). Climate change impacts on wind power generation. *Nature Reviews Earth and Environment* 1 (12): 627–643.

Pryor, S.C. and Barthelmie, R.J. (2010). Climate change impacts on wind energy: a review. *Renewable and Sustainable Energy Reviews* 14 (1): 430–437.

Rantanen, M., Karpechko, A.Y., Lipponen, A., Nordling, K. et al. (2022). The Arctic has warmed nearly four times faster than the globe since 1979. *Communications Earth and Environment* 3 (1): 1–10.

Rohrig, K., Berkhout, V., Callies, D., Durstewitz, M. et al. (2019). Powering the twenty-first century by wind energy – options, facts, figures. *Applied Physics Reviews* 6 (3): 031303.

Sovacool, B.K. (2009). The intermittency of wind, solar, and renewable electricity generators: technical barrier or rhetorical excuse? *Utilities Policy* 17 (3–4): 288–296.

Tong, W. (2010). *Wind Power Generation and Wind Turbine Design*. Southampton, UK: WIT Press.

Touré, S. (2019). Investigations into some simple expressions of the gamma function in wind power theoretical estimate by the Weibull distribution. *Journal of Applied Mathematics and Physics* 7 (12): 2990–3002.

Veers, P., Dykes, K., Lantz, E., Barth, S. et al. (2019). Grand challenges in the science of wind energy. *Science* 366 (6464): 2027.

Vermillion, C., Cobb, M., Fagiano, L., Leuthold, R. et al. (2021). Electricity in the air: insights from two decades of advanced control research and experimental flight testing of airborne wind energy systems. *Annual Reviews in Control* 52 (January): 330–357.

Wiser, R., Rand, J., Seel, J., Beiter, P. et al. (2021). Expert elicitation survey predicts 37% to 49% declines in wind energy costs by 2050. *Nature Energy* 6 (5): 555–565.

Wiser, R., Jenni, K., Seel, J., Baker, E. et al. (2016). Expert elicitation survey on future wind energy costs. *Nature Energy* 1 (10): 1–8.

Yaramasu, V., Wu, B., Sen, P.C., Kouro, S. et al. (2015). High-power wind energy conversion systems: state-of-the-art and emerging technologies. *Proceedings of the IEEE* 103 (5): 740–788.

Zhang, M.H. (2015). *Wind Resource Assessment and Micro-Siting: Science and Engineering*. Wiley.

Mini Project 8

Solar Stirling Power Plant

The Stirling engine was devised in 1816 by Robert Stirling, who thought it to be a safer alternative than the steam engine. In the Stirling engine, a gas (like air or helium) is sealed in a cylinder and alternatively heated and cooled. In the process, a piston is driven, which can be used to power an electrical generator. Importantly, the Stirling engine obtains its heat from a source *external* to the cylinder, i.e., it is an *external* combustion engine. The engine can therefore be powered by concentrated sunrays, as shown in Figure MP8.1.

1) Explain (with the use of clear drawings) the operating principle of the Stirling engine. Be sure to highlight each of the four processes in this cycle and to explain the functioning of the regenerator. Also draw a simplified *p-v* diagram of the Stirling engine cycle showing where heat and work are transferred. Show the salient thermodynamic equations to evaluate each process, including thermal efficiency and net work.
2) How does the thermal efficiency of a Stirling engine compare with that of an air-standard Diesel cycle and Otto cycle, and to an ideal Carnot cycle? What are the practical ramifications in terms of living more sustainably? Discuss fuel consumption, the use of available fuel resources, combustion pollutants, and the use of solar energy with a Stirling cycle.
3) The key to implementing a successful Stirling cycle is that the heat discarded during isochoric cooling is stored in a device called a regenerator, which then returns heat to the system during the isochoric heating process. Discuss what materials these regenerators are typically made of, and why. Also explain how practical regenerators could be practically implemented at low cost.

Consider a Stirling engine cycle, which is a model of a large Stirling engine powered by solar concentrators. The Stirling cycle operates as a closed system and uses 500 g of helium as the working fluid. You must do independent research to find the required heat capacities of helium. The website of the NIST Thermophysical Properties of Fluid Systems could be of assistance: https://webbook.nist.gov/chemistry/fluid/

A parabolic dish solar collector supplies heat to a Stirling engine (Figure MP8.1) at an average surface temperature of 600 °C when the irradiation is 850 W/m². The collector has a concentration ratio of 20. The average efficiency of the 12-m-diameter solar collector is 25%.

Energy Systems: A Project-Based Approach to Sustainability Thinking for Energy Conversion Systems,
First Edition. Leon Liebenberg.
© 2024 John Wiley & Sons, Inc. Published 2024 by John Wiley & Sons, Inc.
Companion website: www.wiley.com/go/liebenberg/energy_systems

Figure MP8.1 Stirling engines heated by solar concentrators have become popular in sunny and low-wind regions. The shown 30 kW Stirling engines have a solar-to-grid electricity efficiency of 32%. *Source:* Ripasso Energy AB.

Calculate the following:

1) The maximum power (in kW).
2) The heat transfer (in kJ) for the two energy reservoirs per cycle.
3) The heat transfer (in kJ) to the regenerator per cycle.
4) The net work (in kJ) per cycle.

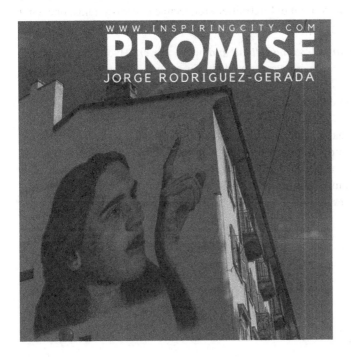

Promise
Created for the initiative "Toward 2030 – What Are You Doing?" in collaboration with Lavazza, this mural depicts Goal number 7 of the United Nations' Sustainable Development Goals – "affordable and clean energy."

The artist, **Jorge Rodríguez-Gerada**, says that "I created this mural to bring awareness to the need of ensuring access to affordable, reliable, and sustainable energy for all. The girl in the mural touches the icon button for the Toward 2030 goal number 7 and lets loose a flow of clean energy. The piece alludes to the importance of acting now to assure a positive outcome. We must think about our world's future and the environmental conditions that our children will inherit. Three billion people (41% of the world's population) are still cooking with fuel and stoves that pollute. A billion people still do not have access to electricity. Electricity in the First World is still mainly obtained from polluting fuels. Our future energy sources must be clean and renewable. To reach this goal we must start now." *Source:* Picture courtesy of Jorge Rodríguez-Gerada.

Week 9 – Energy Storage (Including Water and Geothermal Power)

Meeting net-zero goals will require the prompt roll-out of renewable energy. The problem associated with storing massive amounts of electricity becomes increasingly urgent as more and more intermittent renewable electricity is delivered to the power grid. Energy needs to be stored for

Energy Systems: A Project-Based Approach to Sustainability Thinking for Energy Conversion Systems,
First Edition. Leon Liebenberg.
© 2024 John Wiley & Sons, Inc. Published 2024 by John Wiley & Sons, Inc.
Companion website: www.wiley.com/go/liebenberg/energy_systems

multiple hours, days, or even weeks. Energy storage can be achieved through different approaches. The main form of energy storage at present is through pumped hydro storage schemes, also called "water batteries." The vast energy stored in the Earth (geothermal energy) can also effectively be used to produce electricity. There are many other technologies that can be used to store energy for long periods of time, including the popular lithium-ion battery. These energy-storing technologies are no panacea, though. They each feature important disadvantages, most notably environmental effects, and high costs. These need to be addressed urgently, as effective and cheap energy storage schemes will help ensure the effective roll-out of intermittent renewable energy systems, like wind and solar.

Chapter 25 – Waterpower and pumped storage, tidal and wave power: Hydroelectric power ("hydropower"); pumped hydroelectric storage ("pumped storage"); tidal power; wave power; outlook

Chapter 26 – Geothermal power and energy storage: Geothermal (or "ground-source") heat-pump resources; hydrothermal resources; enhanced (or "engineered") geothermal systems; generating electricity and heat with geothermal systems; sustainability of geothermal power plants; outlook

Chapter 27 – Long-duration storage of electricity and heat: Mechanical energy storage; thermal energy storage; electrochemical energy storage; electrostatic energy storage; comparison of long-duration energy storage technologies; outlook for long-duration storage of energy

Mini Project 9 – Grid-storage batteries

25

Waterpower and Pumped Storage, Tidal and Wave Power

Water has about 800 times higher mass density (ρ) than air, so moving water has a much higher power density ($P/A = \rho V^3/2$) than moving air for comparable velocities (V). Such a highly concentrated form of energy can be tapped more easily and economically than the more diffuse forms of wind and solar energy. Hydropower (or "waterpower") systems are also known for their long plant lives (100 years or more) and low carbon footprints when viewed from a lifecycle perspective. These attributes have made hydropower a popular renewable energy resource for many years, using dams, ocean currents, surface waves, tidal flow, and the flow of water in rivers.

Hydroelectric power systems (using dams) involve a large capital cost, the relocation of populations, ecosystem damage, and the dangers of dam collapse. Climate change further weakens the case for new hydroelectric dams due to intensifying droughts and elevated temperatures that boost evaporation from dams (Keovilignavong et al. 2023). Regardless, globally, hydroelectric power produces about 15.8 EJ (4400 TWh$_e$) annually from over 1.2 TW$_e$ of installed capacity (IRENA 2020a; IHA 2021).

Wave and tidal energy are also significant forms of renewable energy. All ocean-based power sources, including tidal, wave, and marine current, generate about 1 TWh$_e$ of electric power from 0.5 GW$_e$ of installed capacity. The estimated global technical potential of this resource is 15,000 TWh$_e$ (1700 GW$_e$ continuous) at 45% capacity (IHA 2021).

Reimagine Our Future

Some people believe that technological development *alone* will bring about a sustainable energy future. Do you agree? If not, what else might be needed? Perhaps changes in social conditions or consumption patterns or personal expectations? Are cultural changes necessary? What sort of future are you imagining when you think in a comprehensive way about a sustainable energy future?

25.1 Hydroelectric Power ("Hydropower")

Large-scale hydroelectric power facilities extract energy from water falling over a distance from an elevated dam, through a hydraulic turbine, to a lower surface. We may use the *energy equation* stated in Eq. (3.41) to analyze the available energy in a hydroelectric plant:

$$\frac{p_2}{\rho g} + \frac{V_2^2}{2g} + z_2 = \frac{p_1}{\rho g} + \frac{V_1^2}{2g} + z_1 + h_{turbine} - h_{loss} \qquad [m] \qquad (25.1)$$

Energy Systems: A Project-Based Approach to Sustainability Thinking for Energy Conversion Systems,
First Edition. Leon Liebenberg.
© 2024 John Wiley & Sons, Inc. Published 2024 by John Wiley & Sons, Inc.
Companion website: www.wiley.com/go/liebenberg/energy_systems

Key:

1. Reservoir
2. Control gate
3. Trash rack
4. Intake
5. Penstock
6. Transformer
7. Powerhouse
8. Generator
9. Turbine
10. Draft tube
11. Outflow
12. Spillway
13. Fish ladder
14. Transmission

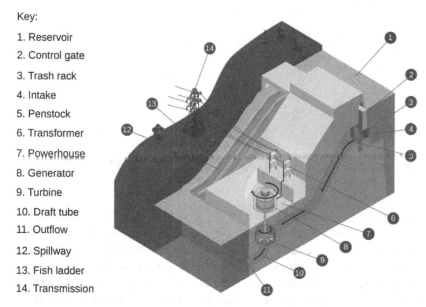

Figure 25.1 Typical layout of a hydroelectric power plant. *Source:* International Hydropower Association.

where $h_{turbine} = \dfrac{w_{turbine}}{\eta_{turbine}g} = \dfrac{\dot{W}_{turbine}}{\eta_{turbine}\dot{m}g}$ is the head removed from the fluid by the turbine, $\eta_{turbine}$ is the mechanical efficiency of the turbine, $w_{turbine}$ is the net work developed by the system per unit of mass flow rate $\dot{m}$, and h_{loss} is the *head loss* (also known as friction work or energy dissipation); "1" denotes the free surface of the reservoir or dam and "2" depicts the average position at the exit of the turbine (see Figure 25.1). All terms in this form of the energy equation have dimensions of meters.

The power transferred to the turbine is:

$$P = \dot{W} = \dot{m}gh_{turbine} \tag{25.2}$$

with $$\dot{m} = \rho\dot{V} \quad \text{and} \quad \gamma = \rho g$$

so that $$P = \rho g \dot{V} h_{turbine} \tag{25.3}$$

The power lost because of friction is:

$$P_{loss} = \rho g \dot{V} h_{loss} \tag{25.4}$$

The mechanical efficiency (η) or power coefficient (C_p) of the turbine is the ratio of the actual power (P_{actual}) produced by the turbine to that which the pressure head can theoretically provide, using Eq. (25.3):

$$C_p = \eta_{mech} = \frac{P_{actual}}{P} = \frac{P - P_{loss}}{P} = 1 - \frac{h_{loss}}{h_{turbine}} \tag{25.5}$$

Sometimes the mechanical efficiency of the turbine is combined with the electrical conversion efficiency of the electrical generator to give a total (or overall) efficiency:

$$\eta_{total} = \eta_{mech}\eta_e \tag{25.6}$$

Example 25.1   The World's Largest Hydroelectric Plant

Consider the world's largest hydroelectric plant, the Three Gorges Dam hydroelectric facility in China, which features 32 turbines operating under an average turbine head of 100 m. The volumetric flow rate of the water through each turbine is a massive 825 m^3/s, and the turbines have mechanical efficiencies of 91%. The electrical generators have conversion efficiencies of 95%. The water density may be assumed to be 1000 kg/m^3. (a) Calculate the power output of this facility, and (b) the annual energy produced if the plant has a capacity factor of 45%.

Given: $\dot{V} = 825$ m^3/s, $h_{turbine} = 100$ m, $\rho = 1000$ kg/m^3, $\eta_{mech} = 0.91$, $\eta_e = 0.95$, $CF = 0.45$

Required: We must find (a) $P_{total,e}$, and (b) E.

Solution strategy: We can use Eqs. (25.3), (25.5), and (25.6) to find the power output of one turbine, which we will multiply by 32 to get the total power output. The energy output will be determined by calculating the number of hours that the plant operates in a year.

Analysis:

a) $P_{total,e}$:

Using Eqs. (25.3), (25.5), and (25.6), we can calculate the power output of one turbine:

$$P_e = \eta_{total}\rho g\,\dot{V}\,h_{turbine} = (0.91 \times 0.95)\left(1000\text{ kg/m}^3\right)\left(9.81\text{m/s}^2\right)\left(825\text{ m}^3\text{/s}\right)(100\text{ m}) \approx 700\text{ MW}_e$$

Since there are 32 turbines, the total electrical power output of this hydroelectric plant is:

$$P_{total,e} = (32\text{ turbines})(700\text{ MW}_e/\text{turbine}) = 22{,}400\text{ MW}_e = 22.4\text{ GW}_e$$

b) E:

Using a capacity factor of 45%, the annual energy production is:

$$E = CF(P_{total,e})(\Delta t) = (0.45)(22.4\text{ GW}_e)(8760\text{ h/year}) = 88.3 \times 10^3\text{ GWh}_e = 88.3\text{ TWh}_e$$

25.1.1   Types of Hydraulic Turbines

Hydraulic turbines may be broadly classified as impulse turbines or as reaction turbines. This classification is based on how flowing water interacts with the turbine blades.

In an *impulse turbine*, the force on the blades is only produced by changing the direction of the water, i.e., by changing the water's momentum. This happens without any large change of pressure across the blade passage, which also means that parts of the rotor are not covered by water.

- The main type of impulse turbine is the *Pelton turbine* (or Pelton wheel), which is typically shaped like a double spoon. Figure 25.2 illustrates the typical operating regime of the Pelton wheel, which is usually at high heads (80–1000 m), but low flow rates.
- At low to medium heads (1–150 m) and small power outputs (5–100 kW), such as produced in waterways or rivers, *micro hydroelectric plants* use rotors that could be classified as *cross-flow turbines*. These turbines include "eggbeater" (or Darrieus)-type turbines, Bánki-Michell turbines, Ossberger turbines, and more (Figure 25.2 and Table 25.1).

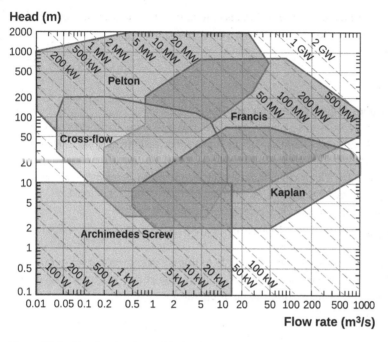

Figure 25.2 Rough selection guide for the main types of hydraulic turbines, based on current technology.

In a *reaction turbine*, though, some of the force on the blades is produced by changing the water's momentum, and some by the acceleration of the water relative to the blade.

- At medium to high heads (60–500 m) and low flow rates, the most popular choice of reaction turbine is a *Francis turbine* (also called a mixed-flow turbine, see Table 25.1). *Turgo turbines* are also sometimes used for low-cost applications that involve heads of around 70 m and very low flow rates (Table 25.1). The entire rotor of these turbines is immersed in fluid.
- At low heads (10–50 m) and medium to high flow rates, the best choice of reaction turbine is a *Kaplan turbine*, sometimes also called a propeller-type turbine (Figure 25.2 and Table 25.1).
- *Bulb turbines* operate at very low heads (10–15 m) and very high flow rates (Table 25.1). The term "bulb" describes the shape of the upstream watertight casing, which contains a generator located on a horizontal axis.
- *Pump turbines*, as the name suggests, can operate as turbines or as pumps over a range of heads up to 1000 m (Figure 25.2). Pumped hydro storage plants store energy using a system of two inter-connected reservoirs with one at a higher elevation than the other. Water is pumped to the upper reservoir in times of surplus energy (when electricity tariffs are low) and, in times of excess demand (when electricity tariffs are high), water from the upper reservoir is released, generating electricity as the water passes through reversible Francis turbines on its way to the lower reservoir. The process is repeated with an overall cycle efficiency of about 80%.

Table 25.1 summarizes the operating characteristics of these types of hydraulic turbines.

Reimagine Our Future
What do you think needs to be done, now, to help secure a sustainable-energy future? Do you think that your own actions can contribute significantly to what needs to be done?

Table 25.1 Characteristics of the main types of hydraulic turbines.

Type of hydraulic turbine		Rated power (MW)	Rated head (m)	Rated flow rate (m³/s)	Rotor diameter (m)	Efficiency (%)	Major manufacturers
Impulse turbines	Pelton turbine *Source:* Andritz.	<350	80–1900	1–50	<4	80–90	Andritz, GE Renewable Energy, KBL, Voith, Gilkes, Siapro
	Cross-flow turbine (e.g., Darrieus, Bánki-Michell, Ossberger) *Source:* Emrgy.	0.01–0.2	0.5–2	0.5–7	1–1.5	70–80	GE Renewable Energy, KBL, NHE, Mitsubishi Heavy Industries, Emrgy, ABB
Reaction turbines	Francis turbine *Source:* Voith.	100–800	60–500	1–1000	<10	85–95	GE Renewable Energy, Dongfang, KBL, Voith, Gilkes, Andritz, Siapro, Mitsubishi Heavy Industries, Toshiba Energy
	Turgo turbine *Source:* Gilkes, used with permission.		60–500	0.5–5	<6.5	80–90	Gilkes, NHE

(Continued)

Table 25.1 (Continued)

Type of hydraulic turbine		Rated power (MW)	Rated head (m)	Rated flow rate (m³/s)	Rotor diameter (m)	Efficiency (%)	Major manufacturers
Kaplan turbine	*Source*: Voith.	17–120	10–60	1–1000	10–15	88–92	GE Renewable Energy, KBL, Voith, Andritz, Siapro, CKD Blansko
Deriaz turbine	*Source*: GE.	<50	100		<3	85–90	Voith, CKD Blansko
Bulb turbine	*Source*: GE.	10–80	4–15		<8	85–92	GE Renewable Energy, Voith
Pump turbine	*Source*: Voith.	50–500	5–1200			85 (pump mode)–95 (turbine mode)	GE Renewable Energy, KBL, Voith

25.2 Pumped Hydroelectric Storage ("Pumped Storage")

Electric utilities have employed pumped hydroelectric storage for more than a century. Today it still accounts for about 93% of all utility-scale energy storage in the United States (DOE 2022) and more than 10% of total hydropower capacity worldwide (IRENA 2020b). Pumped hydro storage provides a fast response to power demand and back-up to variable sources like solar and wind power.

Figure 25.3 shows that pumped hydro storage schemes pump water from one reservoir (or basin or dam) to another at a higher elevation when electricity demand is low, i.e., during off-peak hours. When electricity demand is high, water is released through turbine generators to the lower reservoir. As the system operation is analogous to that of a battery, pumped hydro is colloquially referred to as a "water battery."

Pumped storage systems can be operated in an *open loop,* where there is an ongoing hydraulic connection to a natural body of water, or a *closed loop*, where the reservoirs are not connected to an outside body of water. Each of these modes of operation impact multiple ecosystems in their own ways (Schmitt et al. 2018; DOE 2020) and must form part of the environmental impact analyses and strategic planning that precede the construction of pumped hydro storage schemes.

Pumped hydroelectric storage systems use a *reversible pump turbine* (see Table 25.1) and a generator that can operate in reverse as an electrical motor. Although efficient and suitable for long-term storage, pumped hydro is severely restricted in its further deployment by the availability of suitable locations, multibillion-dollar capital costs, and environmental concerns. Globally, there are about 150 GW$_e$ of pumped hydroelectric storage plants.

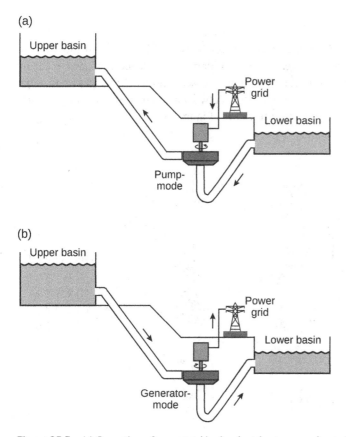

(a)

(b)

Figure 25.3   (a) Operation of a pumped hydroelectric storage scheme in pump mode and (b) generator mode.

Example 25.2 Small Pumped Hydro System

A small pumped-storage facility with two reservoirs roughly the size of two Olympic swimming pools and a 500-m height difference between them could provide a volumetric flow rate of 0.7 m³/s. With a 90% energy conversion efficiency, this small pumped hydro scheme has a capacity of $P = \eta \rho g \, \dot{V} \, h_{turbine} = (0.9)(1000 \text{ kg/m}^3)(9.81 \text{ m/s}^2)(0.7 \text{ m}^3/\text{s})(500 \text{ m}) = 3.1 \text{ MW}_e$, using Eqs. (25.3) and (25.6). If an Olympic-size pool holds about 2500 m³ of water, the system will produce power for around $\Delta t = (2500 \text{ m}^3)/(0.7 \text{ m}^3/\text{s}) = 3571 \text{ s} \approx 1 \text{ h}$. The systems can therefore store (3.1 MW_e) $(1 \text{ h}) = 3.1 \text{ MWh}_e$ of electricity. It is thus easy to see why pumped hydro systems are often called "water batteries."

25.3 Tidal Power

The gravitational pull of the moon and the sun, along with the rotation of the earth, creates tides in the oceans. In some places, tides cause water levels near the shore to rise and fall up to 12 m. Although tidal power is intermittent, it is predictable, based on the motion of the moon and the sun. Tidal power is a large global resource (around 2.5 TW), but only about 3% (75 GW) is economically feasible, with tidal ranges of 3 m or more. The mean daily tidal range is illustrated in Figure 25.4, with the global average at 0.75 m.

Tidal power can be harnessed in two main ways: as *tidal barrage* systems and as *tidal stream* systems.

25.3.1 Tidal Barrages

Tidal barrage power systems take advantage of differences between high tide and low tide (in an ocean bay or lagoon) by using a "barrage," or type of dam, to block receding water during ebb periods. At low tide, the water behind the barrage is released, and the water passes through turbines (situated in the barrage) that generate electricity. A two-way tidal power system generates electricity from both the incoming and outgoing tides (See Figure 25.5).

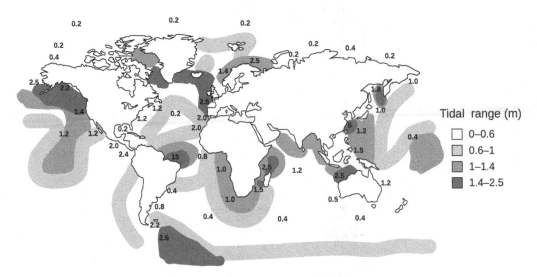

Figure 25.4 Mean daily tidal range.

(a)

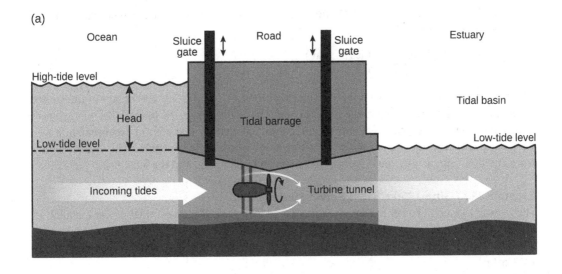

(b)

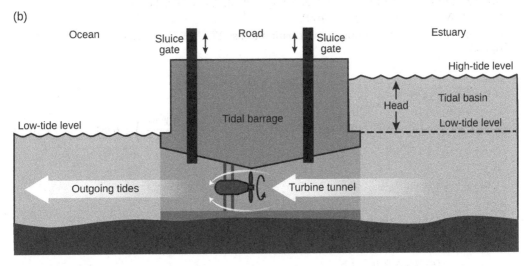

Figure 25.5 A tidal power station can produce electricity at both (a) low and (b) high tides, using a bi-directional turbine situated in a barrage.

A potential disadvantage of tidal power is the effect a tidal station can have on plants and animals in estuaries of the tidal basin. Tidal barrages can change the tidal level in the basin and increase turbidity (the amount of matter in suspension in the water). They can also affect navigation and recreation (Waters and Aggidis 2016).

The potential energy from a tidal barrage is proportional to the mass of water that will flow through the turbines, where mass is density times volume, and volume equals the surface area (A) of the water system times the tidal height (h). If z is the vertical distance through which the tide moves, then the maximum energy that the tidal barrage can extract is (Prandle 1984):

$$E_{max} = mgh = \rho g A \int\limits_0^h z \, dz = \frac{1}{2}\rho g A h^2 \qquad (25.7)$$

The amount of power that can be extracted is dependent on the height and duration of the tides. There are, on average, two high tides and two low tides around the earth at any instant. The interval between tides is 12.42 h (or 12 h and 25 min) so that the maximum power that can be extracted with every high tide and low tide is:

$$P_{max} = \frac{(2 \text{ high and low tides per day})}{(12.42 \text{ h/tide})} E_{max} \qquad (25.8)$$

The average global tidal range is 0.5–1 m (see Figure 25.4), but the Severn Estuary in the United Kingdom has a maximum tidal range of about 8 m. This could supply 7% of the UK's energy needs using about 250 turbines, each with an 8-m diameter. Tidal power systems are expensive, however They also disrupt the tides, which adversely affects marine ecosystems, including changes in salinity, sedimentation, aquatic life, the dispersal of sewage and pollutants, and the local fishing industry.

Example 25.3 Sihwa Lake Tidal Power Station

The Sihwa Lake Tidal Power Station in South Korea is the world's largest tidal power installation. How much power and energy can this power station produce if it has 10 bulb turbines, the operating tidal range is 5.6 m, and the working basin area is about 30 km^2? Take the density of seawater to be 1025 kg/m^3.

Given: 10 turbines, tidal range = $h = 5.6$ m, $A = 30$ km^2, $\rho = 1025$ kg/m^3

Required: We must find (a) E and (b) P.

Solution strategy: We have enough information to use Eqs. (25.7) and (25.8).

Analysis:

a) E:

From Eq. (25.7):

$$E_{max} \approx \frac{1}{2}\rho g A h^2 = (0.5)(1025 \text{ kg/m}^3)(9.81 \text{ m/s}^2)(30 \text{ km}^2)\left(\frac{10^6 \text{ m}^2}{\text{km}^2}\right)(5.6 \text{ m})^2 = 4.73 \times 10^{12} \text{ J}$$

b) P:

From Eq. (25.8):

$$P_{max} = \frac{(2 \text{ high and low tides per day})}{(12.42 \text{ h/tide})}E_{max} = \frac{2(4.73 \times 10^{12} \text{ J})}{12.42 \text{ h/tide}}$$

$$= \frac{761.7 \times 10^9 \text{ J}}{\text{h}}\left(\frac{1 \text{ h}}{3600 \text{ s}}\right) = 211.6 \text{ MW}$$

This power is extracted with the use of 10 turbines, each generating at least 21 MW.

25.3.2 Tidal Stream Systems

Tidal stream power systems take advantage of ocean (or lagoon) currents to drive turbines, particularly in areas around islands or coasts where these currents are fast. They can be installed as tidal fences – where turbines are stretched across a channel – or as tidal turbines, which resemble underwater "wind" turbines.

Because water is about 800 times denser than air, tidal turbines must be much sturdier and heavier than wind turbines. However, they have much smaller rotor diameters due to the high energy density of water. Tidal turbines are more expensive to build than wind turbines but can capture more energy with the same size blades.

Like a wind turbine, the power that a tidal stream device can convert can be estimated as follows:

$$P = C_p \frac{1}{2} \rho V^3 A \tag{25.9}$$

where V is the velocity of the tidal stream, C_p is the rotor's power coefficient (often also called the efficiency), and A is the swept area of the energy conversion device. With a tidal stream of 3 m/s, the power density is $P/A = \frac{1}{2}\rho V^3 = (0.5)(1000 \text{ kg/m}^3)(3 \text{ m/s})^3 = 13.5 \text{ kW/m}^2$. To achieve the same power density with a wind turbine, the wind velocity would need to be about 28 m/s.

Tidal-stream energy technologies are often feasible in areas with peak spring tidal currents greater than 1.5 m/s and water depths more than 25 m. Based on the characteristics of the energy conversion device, these technologies can generally be classified into six different groups, as depicted in Table 25.2.

Reimagine Our Future

Can you imagine elephants once again freely roaming the mesmerizing African plains? It seems that such issues are not first and foremost technological issues but rather begin with a change of "heart."

25.4 Wave Power

Wave energy conversion systems convert energy from the waves in oceans or large bodies of water and convert it to electricity via a generator. Wave power has a high availability, being available up to 90% of the time, while the availability of solar and wind power ranges from 20% to 30%. Wave power depends on wave speed, which is dependent on the depth of the water.

Wave power density (P') may be defined as the power of the wave per unit length of the wave front. This can be as high as 125 kW/m (Rusu and Onea 2017; Pörtner et al. 2022), as shown in Figure 25.6. Due to the nearly 800 times greater density of seawater to air, wave power is of a higher power density than wind power. For instance, for the Chicago area, the average annual wind power density is 0.58 kW/m², but it is 8.42 kW/m² for wave power. For the same region, the average power density of solar energy is about 0.17 kW/m².

The wave power global resource is estimated to be 2 TW (Europe: 240 GW, Australia: 280 GW, USA: 220 GW, Africa: 320 GW, Asia: 320 GW). On the west coast of the United States, wave energy has a potential of 250 TWh/yr, equivalent to 67% of the region's net electricity generation in 2019, which is enough to power 23 million homes.

For the effective conversion of energy from waves, such systems must cope with waves of varying heights and direction. They must also be able to withstand storms and operate with high availability for 30 years without excessive maintenance. Although wave energy is a renewable energy resource, large-scale wave-energy conversion systems have high construction costs and require undersea electricity cables. These systems also affect the marine ecosystem, as well as shipping and fishing.

25.4.1 Wave Energy Conversion

Most waves on the sea surface are caused by wind. Air streamlines are closer over wave crests, which, according to Bernoulli's theorem (Eq. (3.42)), means that the air pressure is lower at the crest of a wave than at its trough. This pressure differential causes the water surface to rise. This is what we call waves.

Table 25.2 Characteristics of the main types of tidal power turbines used in large and micro power schemes.

Type of tidal turbine		Rated power (MW$_e$)	Optimal tidal speed (m/s)	Rotor diameter (m)	Efficiency or power coefficient (%)	Major manufacturers
Horizontal axis turbines		0.1–2	1.5–4	8–20	40–45	Orbital Marine Power, Andritz, ABB, Sustainable Marine Energy, Simec Atlantis Energy, Nova Innovation, Voith Hydro
Vertical axis turbines		0.35–1.5	Not public information	5–8	40	Blue Energy Canada, Current2Current, Current Power AB, Norwegian Ocean Power
Oscillating hydrofoils			1.2–3.5	Not public information	Not public information	UEK Corp., Minesto
Ducted turbines		1.5	Not public information	16	Not public information	Tidal Energy System Corp., Tocardo, Lunar Energy, Rotech

					Developers
Archimedes screws		>1	Not public information	80–85	Greenheat Systems, Minesto, Norwegian Ocean Power, Vortex Hydro Energy
Tidal kites	0.5–1.2	1.2–3	1–3.5	Not public information	Minesto, BioPower Systems

Wave power density (kW/m of wave front)

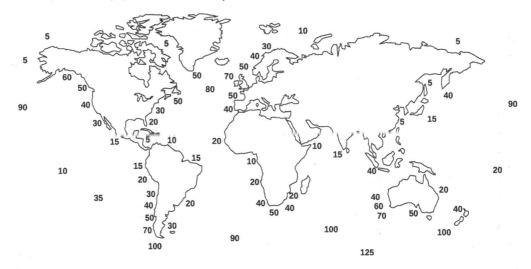

Figure 25.6 Approximate distribution of global wave power density (kW/m of wave front).

In general, the power that can be extracted from a wave front is related to the energy (E) and period (T) of the wave:

$$P = \frac{E}{T} \tag{25.10}$$

The potential energy contained in a wave (due to the surface being lifted from the troughs to the peaks) is (Jaffe and Taylor 2018):

$$E_{pot} = \int\limits_{0}^{L} dx \int dy \, \frac{1}{2}\rho g z^2 = \frac{1}{4}\rho g a^2 \cos^2 \omega T \tag{25.11}$$

with a being the amplitude of the wave (in m), T the period of the wave (in seconds), and ω the angular velocity (or "dispersion") of the wave (in rad/s):

$$\omega = \frac{2\pi}{T} \equiv \sqrt{gk} \tag{25.12}$$

with k being the wave number.

The kinetic energy contained in a wave is (Jaffe and Taylor 2018):

$$F_{kin} = \int\limits_{0}^{L} dx \int dy \int\limits_{-\infty}^{0} dz \, \frac{1}{2}\rho a^2 \omega^2 c^2 e^{2kz} \sin^2 \omega t = \frac{\rho a^2 \omega^2 A}{4k} \sin^2 \omega t \tag{25.13}$$

where $A = LW$ is the wave area in the xy-plane.

For *deep-water waves* (where the water depth $h > \lambda/2$), the wave speed or "celerity" (c) is related to the wavelength (λ) as follows (Elmore and Heald 1985):

$$c = \frac{\omega}{k} = \sqrt{\frac{g\lambda}{2\pi}} \tag{25.14}$$

For *shallow-water waves* (where the water depth $h < \lambda/20$):

$$c = \sqrt{gh} \tag{25.15}$$

where h is the depth of water. Figure 25.7a illustrates the salient variables in these equations, while Figure 25.7b shows how wave speed varies with the wavelength and depth of water.

(a)

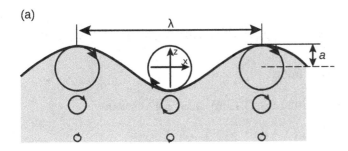

(b)

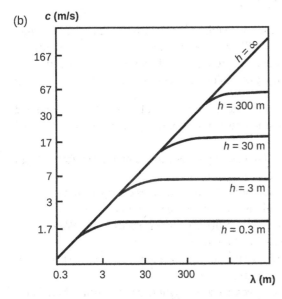

Figure 25.7 (a) Wave motion, showing wave length and amplitude. (b) Relationship of wavelength, velocity, and depth of water. *Source:* After Stoker (1992) and Elmore and Heald (1985).

Combining the potential and kinetic energy, Eqs. (25.11) and (25.13), the energy density (E') of a wave is:

$$E' = \frac{E}{A} = \frac{1}{2}\rho g a^2 \qquad \left[\text{kJ/m}^2\right] \tag{25.16}$$

The power contained in a wave per unit width of a wave front *in shallow- to intermediate-depth water* is:

$$P' = \frac{1}{4}\rho g a^2 \sqrt{g/k} = \frac{1}{4}\frac{\rho g^2 a^2}{\omega} = \frac{1}{8\pi}\rho g^2 a^2 T \qquad \left[\text{kW/m}\right] \tag{25.17}$$

Example 25.4   Wave Power

Consider an ocean wave with an amplitude of $a = 1$ m and a period of 12 s. Calculate the wave front's energy density and power density.

Given: $a = 1$ m, $T = 12$ s

Required: We must find E' and P'.

Solution strategy: We will use Eqs. (25.16) and (25.17) to calculate the energy density and power density of the wave front.

Analysis:

Energy density of wave front: $E' = \dfrac{E}{A} = \dfrac{1}{2}\rho g a^2 = (0.5)(1025 \text{ kg/m}^3)(9.81 \text{ m/s}^2)(1 \text{ m})^2 = 5.03 \text{ kJ/m}^2$

Power developed per unit width of wave front:

$$P' = \frac{1}{8\pi}\rho g^2 a^2 T = \frac{1}{8\pi}(1025 \text{ kg/m}^3)(9.81 \text{ m/s}^2)^2(1 \text{ m})^2(12 \text{ s}) = 47.1 \text{ kW/m}$$

25.4.2 Types of Wave Energy Conversion Devices

Utility-scale electricity generation from wave energy requires device arrays, rather than larger devices. As with wind turbine generators, devices are likely to be chosen for specific site conditions. Wave energy conversion devices may be classified according to their principle of operation, as shown in Table 25.3. The different types of energy conversion devices may be classified as follows:

- *Oscillating water column*: Passing waves raise the water level within a hollow, demi-submerged structure, causing the enclosed air to compress and flow to the atmosphere, driving a (Wells-type) turbine.
- *Submerged pressure differential*: The rise and fall of passing waves cause a pressure differential in the structure to trigger pressure pumps and generate electricity.
- *Point absorber (heaving buoy)*: This floating or submerged buoy generates energy from the buoy's movement caused by waves in all directions relative to the base connection.
- *Bulge wave*: A device is placed parallel to the waves, capturing energy from its surge. Water flows through the flexible device and passes through a turbine to exit.
- *Oscillating water surge converter*: This structure uses the surge movement of the wave (back-and-forth motion) to capture energy in an oscillating arm.
- *Rotating mass*: The heaving and swaying in the waves cause a weight to rotate within this device. This rotation drives an electric generator.
- *Attenuator*: The attenuator consists of multiple connected segments or a single long and flexible part that extracts energy from waves by following the parallel motion of the waves.
- *Overtopping device*: The water of passing waves is captured in a reservoir and released through a shaft. A turbine in the shaft generates energy when water passes.

Example 25.5 Using the Wave Clapper® in Chicago, on Lake Michigan

Eco Wave Power produces a "terminator"-type wave energy converter called the Wave Clapper® (see Table 25.3 and Figure 25.8). The box on pages 557 and 558 features an interview with the person behind this innovation. The Wave Clapper has an energy conversion efficiency of 50%. An energy entrepreneur wants to use this system to generate renewable energy for people living around Lake Michigan, close to Chicago. The average electricity energy usage is 32 kWh$_e$/day per customer (household) in Chicago. The entrepreneur has also determined that the average annual wave height of Lake Michigan is 0.64 m and the average wave period is 4.61 s (NDBC 2022). The wave energy has a capacity factor of 36%. The average density of the water in Lake Michigan is 999.3 kg/m³. (a) What is the power density per unit length of a wave front in Lake Michigan? (b) There is about 400 m of harbor wall available for mounting a Wave Clapper. How much power will such a Wave Clapper produce and how many households could receive electricity from this technology?

Table 25.3 Characteristics of the main types of wave energy conversion devices used in large and micro power schemes.

Type of wave energy conversion system		Rated power (MW$_e$)	Efficiency (%)	Major manufacturers
Oscillating water column	 Source: Wave Swell.	1.25–2.6	15	Voith Hydro, Ocean Energy Ltd, REWEC3, bioWAVE, AW-Energy, Wave Swell
Oscillating body	Pressure differentials Source: Bambora.	1.5	Not public information	Bombora Wave Power, Carnegie Clean Energy, Etymol Ocean Power
	Point absorber (heaving buoy) Source: Ocean Energy	0.05–1	30–51	Sinn Power, Eco Wave Power, Mitsui Engineering and Shipbuilding, Ocean Power Technologies, AWS Ocean Energy, Wave Star Energy, CorPower, Seatricity, C-Power, Ocean Energy, 40 South Energy
	Point absorber (terminator) Source: Eco Wave Power, used with permission.	0.1	50	Eco Wave Power
	Bulge wave Source: Fortum.	0.75	14–21	Fortum, Checkmate Seaenergy
	Oscillating water surge converter Source: AW-Energy, used with permission.	0.35	Not public information	AW-Energy, Atargis Energy

(Continued)

Table 25.3 (Continued)

Type of wave energy conversion system		Rated power (MW$_e$)	Efficiency (%)	Major manufacturers
	Rotating mass	0.75	Not public information	Wello
	Source: Wello.			
	Attenuator	0.75–2.25	Not public information	Weptos, Mocean Energy, Ecomerit Technologies
	Source: Weptos, used with permission.			
Overtopping		0.02–11	21–26	Wave Dragon, Naples Harbor
	Source: Wave Dragon, CC BY.			

Source: Boren (2021), IRENA (2020a), Lehmann et al. (2017), Lewis et al. (2011), and Zhang et al. (2022).

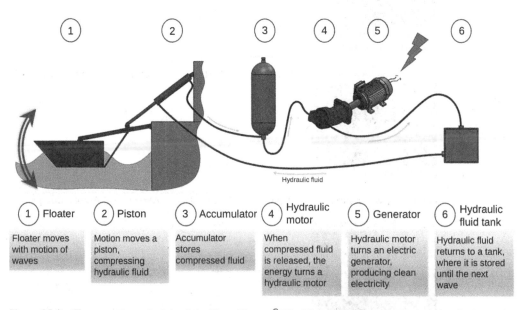

Figure 25.8 The operating principle of the Wave Clapper®. The Wave Clapper can be mounted on harbor walls where the energy conversion units capture wave power. *Source:* Eco Wave Power.

Given: $a = 0.64$ m, $T = 4.61$ s, $CF = 0.36$, $\rho = 999.3$ kg/m³, household energy consumption = 32 kWh$_e$/day, available harbor wall length = $L = 400$ m, Wave Clapper® efficiency = 50%.

Required: We must find (a) P', and (b) P and the number of households that could be served by a series of Wave Clappers mounted on the harbor wall.

Solution strategy: (a) The power developed per unit width of wave front can be calculated using Eq. (25.17). (b) Once we know that, we can calculate the total power that the series of Wave Clappers can produce and the number of households that will receive electricity from the plant.

Analysis:

a) P':

From Eq. (25.17): $P' = \dfrac{1}{8\pi}\rho g^2 a^2 T = \dfrac{1}{8\pi}\left(999.3 \text{ kg/m}^3\right)\left(9.81 \text{ m/s}^2\right)^2 (0.64 \text{ m})^2 (4.61 \text{ s}) = 7.2 \text{ kW/m}$

b) *P and number of households*:

For a harbor wall of 400 m, the total power available from waves is:

$P = P'L = (7.2 \text{ kW/m})(400 \text{ m}) = 2.9 \text{ MW}$

If each household in Chicago consumes 32 kWh of electricity per day, the average power consumption is (32 kWh$_e$/household)/(24 h/day) = 1.3 kW. A Wave Clapper has a 50% efficiency and the capacity factor of wave energy conversion in Lake Michigan is 36%. The number of households that can be supplied with wave-generated electricity 24 h a day is then:

$\dfrac{(2.9 \times 10^6 \text{ W})}{(1.3 \times 10^3 \text{ W}_e/\text{household})}(0.5)(0.36) = 402 \text{ households}$

Discussion: Chicago is known as "the windy city." This calculation shows that there is immense untapped energy in the waves of the surrounding Lake Michigan, produced by the wind. For approximately each 1-m length of the harbor wall fitted with a Wave Clapper device, one household's daily electricity needs are met. From the manufacturer's website, each Wave Clapper can produce a maximum of 5 kW$_e$, which means that the Wave Clapper will be able to meet a required demand of 4 kW$_e$ each (i.e., 1 m of harbor wall can facilitate the production of at least 1 kW$_e$ to provide electricity for at least one household).

Changing the World – One Wave at a Time

At a time when the world is depleting its non-renewable energy resources, scientists across the globe are searching for innovative renewable options. One such alternative is the power that lies within the waves of the ocean. However, wave energy conversion devices need to be robust and flexible if they are to cope with the widely varying operating conditions. A meter-deep ocean swell, with a 10-s period, for instance, carries a power density of 40 kW/m of swell width. An ocean swell with a 10-m amplitude and a 12-s period, on the other hand, has a power density of 5 MW/m of swell width.

Inna Braverman (37) took on the challenge of harnessing energy from ocean waves. She co-founded Eco Wave Power in 2011 to take advantage of this renewable energy resource that can theoretically produce double the amount of electricity needed on the planet today.

Inna explains that most wave energy developers have chosen to pursue the development of offshore wave energy technologies, which have their share of hurdles. These include the cost and complexity of installing offshore technologies and concerns that mooring devices to the ocean floor and transmission lines that transport the energy might harm marine life. In addition, there can be extreme wave heights offshore, and unfortunately, no man-made equipment can survive the loads of such extreme offshore waves. "There is also the challenge of designing buoys that can efficiently capture the full power of wave movement." To overcome the above-mentioned challenges, Eco Wave Power designed an onshore/nearshore solution. Eco Wave Power installs its wave-energy converters on harbor walls or breakwaters.

"Unlike offshore installations," says Inna, "we don't require ships, divers, underwater moorings, and cables, thus we significantly save costs, while not disrupting marine life. In addition, the floaters are tailor-made to the location of implementation for optimal energy generation and are equipped with a patented storm protection mechanism (see Figure 25.8). When the waves are too high for our system to handle, the floaters automatically rise above the water level, and lock in an upward position."

The Eco Wave Power team already has orders totaling more than 400 MW. These include 20 MW in Portugal, 77 MW in Turkey, 2 MW in Spain, and others. In 2021, Eco Wave Power was listed on the Nasdaq First North exchange (see Figure 25.9), a growth platform for hundreds of tech start-ups, becoming the first Israeli company to list on the Swedish Nasdaq. In July 2021, the company also listed on Nasdaq Capital Market in the USA under the stock symbol: WAVE.

Having just spent US$450,000 building its Gibraltar station, Inna says that Eco Wave Power's system can be scaled with relatively low cost and scalability risk. The company hopes to soon complete its second grid-connected wave energy array in Israel, which is done in collaboration with and co-funded by Israel's Energy Ministry, which recognized the Eco Wave Power technology as a "pioneering technology." Eco Wave Power is also planning a system for Los Angeles. According to the US Energy Information Agency, wave energy could supply 66% of all the USA's energy needs.

Inna Braverman was recognized by Wired Magazine as one of the "females changing the world," as well as by Fast Company as one of the world's "most creative people in business for 2020." She is also the winner of the United Nations' Global Climate Action Award.

Power, https://www.ecowavepower.com/

Figure 25.9   An elated Inna Braverman (37), center, celebrates the listing of her company on the Nasdaq stock exchange. *Source:* Nasdaq, Inc., Vanja Savic.

25.5 Outlook

The International Energy Agency predicts global hydropower capacity to increase to almost 2000 GW_e and that of pumped storage three- to fivefold to 400–700 GW_e by 2050. That means that, between 2022 and 2050, more than 1300 GW_e of new hydropower capacity is needed (IEA 2022).

Hydro energy: In 2020, the world had hydroelectric schemes with a combined capacity of around 1.2 TW_e. According to most projections to reach net-zero by 2050, the world will need to increase hydroelectric capacity to 1444 GW_e by 2030 and 1822 GW_e by 2050. For hydro-pumped storage, in 2020, the world had a combined capacity of 121 GW_e. To reach net-zero by 2050, the world will need to increase its hydro-pumped storage capacity to 225 GW_e by 2030 and 325 GW_e by 2050.

Tidal energy: The global market for ocean energy (tidal stream and wave energy conversion) was estimated at 58.7 MW_e in 2020 and is now projected to reach 348.1 MW_e by 2026. However, climate change is causing sea levels to rise, which, in turn, is changing the volume of water in an estuary or inlet between mean high tide and mean low tide. The rising sea levels are thus also affecting tidal current, tidal asymmetry, sediment transport patterns, and the feedback loops between them. A rise in the sea level is therefore bound to adversely affect tidal stream-energy conversion systems.

Wave energy: The harnessing of wave power is still an emerging technology. Owing to its optimal characteristics, such as the modular nature of devices, the wide availability of resources and low lifecycle emissions, it is attracting interest from both the energy industry and academia. However, due to its complexity, harsh operating environment, and immense capital and maintenance costs, no single wave-energy device design has yet emerged as optimal. The size density of wave energy conversion systems is a major challenge, especially for viable systems in the vicinity of 20 MW_e. Therefore, although wind technology boasts a sustained degree of research and development, and investment, wave power continues to lag.

In 2019, the levelized cost of electricity for water-based energy conversion systems was as follows: hydroelectric: US $0.067/kWh, tidal stream: US $0.20/kWh to US $0.45/kWh, and wave: US $0.30/kWh to US $0.55/kWh. Figure 25.10 illustrates the present costs for tidal and wave systems and shows that costs will reduce dramatically as the global installed capacity increases.

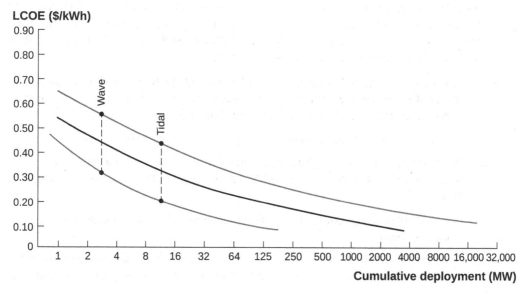

Figure 25.10 Estimated cost reduction curve and LCOE estimates for tidal and wave energy conversion systems. *Source:* Adapted from IRENA (2020a).

Innovation is not just a matter of developing new devices. It also relies on the development of new policies that allow the deployment of innovations to the market at a rapid enough speed. A clear system governing marine spaces and marine resource management rights allocation is therefore crucial to increase investor confidence and facilitate the eventual commercialization of this technology.

Problems

25.1 Consider a horizontal-axis tidal stream system that comprises two rotors. Each rotor has a diameter of 20 m and a power coefficient of 45%. Calculate the combined power output of the system with a tidal stream of 2.5 m/s.

[Answer: 2.2 MW]

25.2 A tidal barrage across an estuary has a surface area of 22 km^2. If the tidal range is 3.8 m, calculate (a) the maximum amount of energy that can be extracted per day, and (b) the maximum available power. You may assume that the water density is 1000 kg/m^3.

[Answers: (a) 1.56×10^{12} J/day; (b) 69.7 MW]

25.3 Ocean waves have a wave length of 40 m and an amplitude of 1.65 m. A large heaving buoy will be used to convert wave energy to electricity. The density of seawater is 1025 kg/m^3. Calculate (a) the maximum energy density (kJ/m), and (b) the power density (kW/m) of a wave front. What circumference of energy-absorbing buoy surface will be required to generate 500 kW if the buoy has an energy conversion efficiency of 38%?

[Answers: (a) 13.7 kJ/m^2; (b) 54.1 kW/m; (c) 24.3 m]

25.4 With the era of building big dams over in the United States, a growing number of existing dams are being modified to produce hydropower. According to proponents, these projects avoid the damaging impacts of new dams and could generate enough renewable electricity for several million homes. In 2016, the US Department of Energy found that retrofitting thousands of non-powered dams in the United States could realistically add as much as 4.8 GW$_e$ of generation capacity to the grid by 2050.

a) Calculate the required water flow rate of a 100 MW$_e$ hydropower plant for hydraulic heads of 10 , 30, 60, and 90 m. What conclusions can you draw from this, especially considering the effects of droughts initiated by climate change?

[Answers: 10 m: 1132 m^3/s; 30 m: 378 m^3/s; 60 m: 189 m^3/s; 90 m: 126 m^3/s]

b) What type(s) of turbine (e.g., Kaplan, Francis, Pelton) will be required to effectively produce 100 MW$_e$ in the above scenarios? Explain your choice(s) and show clear diagrams for each.

[Answers: 10 m: Cross-flow turbine; 30 m: Kaplan turbine; 60 m: Francis turbine; 90 m: Francis turbine]

c) Do an online search for manufacturers of the required turbine. Summarize the technical details of such a turbine and show a manufacturer's drawing or photo of the turbine.

What would such a turbine cost? Also, estimate the cost to install such a turbine into an existing dam structure.

[Answer: Open-ended question. However, hydroelectric-retrofit costs are around US $500/kW$_e$. A 100-MW$_e$ retrofit would therefore cost around US $50 million.]

d) Based on your research to retrofit a dam with a 100 MW$_e$ power plant, calculate the revenue generated over a 30-year lifetime. Assume a discount rate of 8%. The operator charges 6 ¢/kWh. Make realistic assumptions about the capacity. Clearly state all other assumptions and show all calculations.

[Answer: Assuming a 100% capacity factor, the revenue earned over 30 years is US $1.57 billion. The present value is US $157 million. The present value is therefore about three times the investment cost.]

e) What are the ramifications (social, technological, economic, environmental, and political) of retrofitting dams to produce electrical power instead of building new dams? Do independent research when addressing this question. Consider that a 100-MW$_e$ hydroelectric power plant requires a dam structure that contains around 50,000 t of cement. Also be sure to address how many greenhouse-gas emissions are avoided by retrofitting a dam rather than operating a 100-MW$_e$ natural gas-fired steam power plant. Clearly state your assumptions and show all your calculations. In addition, consider how such a small hydro storage facility might impact a future electricity grid that is fed by intermittent wind and solar power.

[Answer: Open-ended question. Among the STEEP aspects (social, technological, environmental, economic, and political), the reader should also calculate the CO_2 emissions avoided by using hydroelectric power rather than, say, burning natural gas for a steam-powered power plant. If a gas-fired steam power plant emits 0.5 kg CO_2/kW$_e$h, with the plant operating for 262,800 h (30 years), it means that $(0.5 \text{ kg } CO_2/\text{kWh}_e)(262,800 \text{ h})$ $(100 \times 10^3 \text{ kW}_e)(0.5 \text{ kg } CO_2/\text{kWh}_e) = 6.57 \times 10^9 \text{ kg } CO_2$ or 6.57 million tons of CO_2 emissions have been avoided. Also, the fact that a dam is being retrofitted means that around 50,000 t of cement is being saved. Concrete production releases around 1 kg CO_2 per kilogram cement. Therefore, as the dam has already been constructed, the following amount of CO_2 is not released into the atmosphere:

$$(1 \text{ kg } CO_2/\text{kg cement})(50,000 \text{ t cement})\left(\frac{1000 \text{ kg}}{t}\right) = 50 \times 10^6 \text{ kg } CO_2. \text{ So, retrofitting an}$$

existing dam for hydroelectric power instead of using a natural gas-fired steam power plant saves the emission of $(6.57 \times 10^9 \text{ kg } CO_2) + (50 \times 10^6 \text{ kg } CO_2) = 6.62 \times 10^9 \text{ kg}$ CO_2 over a 30-year period.]

f) Map the interactions between the United Nations' Sustainable Development Goals (SDGs) that pertain to the hydroelectric retrofitting of a dam. Your mapping should be part of a high-impact infographic of no more than 500 words. For inspiration, read the following article: Måns et al. (2016).

Hint: You may also want to use the following visualization tool: SDG Interlinkages Analysis and Visualization Tool, https://sdginterlinkages.iges.jp/visualisationtool.html

[Answer: Open-ended question, but the reader should at least discuss and show the linkages between SDG 3.9, 7.1, 7.2, 7.3, 9.3 and 12.]

References

Boren, B. (2021). Distributed embedded energy converters for ocean wave energy harvesting: Enabling a domain of transformative technologies (No. NREL/CP-5700-80484). Golden, CO: National Renewable Energy Laboratory.

DOE (2022). US hydropower market report – 2021. US Department of Energy. https://www.energy.gov/sites/prod/files/2021/01/f82/us-hydropower-market-report-full-2021.pdf

DOE (2020). A comparison of the environmental effects of open-loop and closed-loop pumped storage hydropower. US Department of Energy. https://www.energy.gov/sites/prod/files/2020/04/f73/comparison-of-environmental-effects-open-loop-closed-loop-psh-1.pdf

Elmore, W.C. and Heald, M.A. (1985). *Physics of Waves*. North Chelmsford, MA: Courier Corporation.

IEA (2022). World Energy Outlook 2022. International Energy Agency. https://www.iea.org/reports/world-energy-outlook-2022/

IHA (2021). 2021 Hydropower status report. International Hydropower Association.

IRENA (2020a). *Innovation Outlook: Ocean Energy Technologies*. Abu Dhabi: International Renewable Energy Agency.

IRENA (2020b). Innovative operation of pumped hydropower storage. Abu Dhabi: International Renewable Energy Agency. https://www.irena.org/-/media/Files/IRENA/Agency/Publication/2020/Jul/IRENA_Innovative_PHS_operation_2020.pdf

Jaffe, R.L. and Taylor, W. (2018). *The Physics of Energy*. Cambridge, UK: Cambridge University Press.

Keovilignavong, O., Nguyen, T.H. and Hirsch, P. (2023). Reviewing the causes of Mekong drought before and during 2019–20. *International Journal of Water Resources Development* 39 (1): 155–175.

Khojasteh, D., Lewis, M., Tavakoli, S., Farzadkhoo, M. et al. (2022). Sea level rise will change estuarine tidal energy: a review. *Renewable and Sustainable Energy Reviews* 156: 111855.

Lehmann, M., Karimpour, F., Goudey, C.A., Jacobson, P.T. et al. (2017). Ocean wave energy in the United States: Current status and future perspectives. *Renewable and Sustainable Energy Reviews* 74: 1300–1313.

Lewis, A., Estefen, S. Huckerby, J. Musial, W. et al. (2011). Ocean energy. In *IPCC Special Report on Renewable Energy Sources and Climate Change Mitigation* (eds O. Edenhofer, R. Pichs-Madruga, Y. Sokona, K. Seyboth, et al.). 497–534 Cambridge/New York, NY: Cambridge University Press.

Måns, N., Griggs, D. and Visbeck, M. (2016) Policy: Map the interactions between Sustainable Development Goals. *Nature* 534 (7607): 320–322

NDBC (2022). National Data Buoy Center, National Oceanic and Atmospheric Administration. www.ndbc.noaa.gov/

Pörtner, H.O., Roberts, D.C., Adams, H., Adler, C. et al. (2022). Climate change 2022: Impacts, adaptation and vulnerability. IPCC Sixth Assessment Report.

Prandle, D. (1984). Simple theory for designing tidal power schemes. *Advances in Water Resources* 7 (1): 21–27.

Rusu, L. and Onea, F. (2017). The performance of some state-of-the-art wave energy converters in locations with the worldwide highest wave power. *Renewable and Sustainable Energy Reviews* 75: 1348–1362.

Schmitt, R.J., Bizzi, S., Castelletti, A. and Kondolf, G.M. (2018). Improved trade-offs of hydropower and sand connectivity by strategic dam planning in the Mekong. *Nature Sustainability* 1 (2): 96–104.

Stoker, J.J. (1992). *Water Waves: The Mathematical Theory with Applications* (Vol. 36). New York: Wiley.

Waters, S. and Aggidis, G. (2016). Tidal range technologies and state of the art in review. *Renewable and Sustainable Energy Reviews* 59: 514–529.

Zhang, X., Zhang, H., Zhou, X. and Sun, Z. (2022). Recent advances in wave energy converters based on nonlinear stiffness mechanisms. *Applied Mathematics and Mechanics* 43 (7): 1081–1108.

26

Geothermal Energy and Storage

The Earth's deep heat is an energy resource everywhere beneath our feet. It can be accessed to provide heat and generate electricity at an adequate scale to meet the growth associated with economic development while aligning with the transitioning energy sector—including as a carbon-free energy source for industrial and district heat, and transportation.

The temperature at the center of the earth, about 6500 km deep, is about the same as the surface of the sun, nearly 6000 °C. The earth's interior is estimated to contain 44.2 TW of geothermal power (Pollack et al. 1993). This is more than twice the amount needed to supply the total global primary energy consumption in 2015. This heat is continually replenished by the decay of naturally occurring radioactive elements in the earth's interior and will remain available for billions of years, ensuring an essentially inexhaustible supply of energy (Blodgett and Stack 2009).

Geothermal heat flow is expressed visibly at the earth's surface as volcanoes, fumaroles, hot springs, and geysers. Although volcanoes represent the hottest and most visible form of geothermal energy, there is a range of such forms of energy in the subsurface, with temperatures from thousands of degrees to a few degrees above ground-surface temperatures. Much of this energy can be used for productive purposes.

Temperatures above 150 °C are widely—but not uniformly—distributed underground and become more common with increasing depth. Commercial electricity generation is generally economic from geothermal resources at temperatures above 150 °C. Geothermal resource temperatures at a depth of 7 km are accessible with existing drilling technology.

Both the energy conversion process and the end-use application of geothermal energy varies with resource quality, which is primarily a function of temperature. The *high capacity factor* (>90%) of geothermal energy means that the energy produced by geothermal power plants is nearly "always on." The high capacity factor also means that geothermal power plants can generate about two to four times as much electricity as a wind or solar-energy plant of the same installed capacity (Vimmerstedt et al. 2022).

In 2020, the USA remained the leader in installed geothermal capacity, with approximately 3.67 GW_{th}. This represents close to 25% of the world's total online capacity. More than 90% of this capacity is in California and Nevada (IEA 2022a).

Geothermal resources, in a range of temperatures, can be used economically for a variety of electric and non-electric applications. Three categories of geothermal resources can be identified (see Figure 26.1):

- *Geothermal (or ground-source) heat-pump resources:* The ubiquitous presence of shallow soil, rock, and/or aquifers—and, specifically, their thermal storage properties—presents a vast and

Energy Systems: A Project-Based Approach to Sustainability Thinking for Energy Conversion Systems,
First Edition. Leon Liebenberg.
© 2024 John Wiley & Sons, Inc. Published 2024 by John Wiley & Sons, Inc.
Companion website: www.wiley.com/go/liebenberg/energy_systems

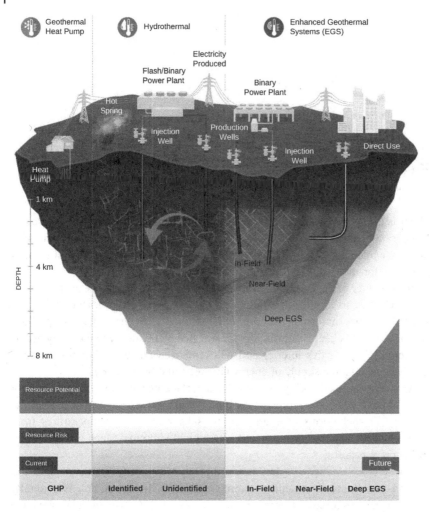

Figure 26.1 The diversity of geothermal resources and applications. *Source:* US Department of Energy, CC BY-SA 4.0.

important geothermal resource. The thermal storage capacity of the shallow earth enables its use as a heat-exchange medium for low-grade thermal energy. Geothermal heat pumps use this thermal storage to increase the efficiency and reduce the energy consumption of heating and cooling applications for residential and commercial buildings. Geothermal heat pumps (GHPs) can serve as effective systems for space cooling (in summer) and heating (in winter), as the ground temperature below 6-m depths in most regions remains stable throughout the year (i.e., between 10 and 15 °C).

(*Note:* Air-source heat pumps (ASHPs) are often the technology of choice for air-conditioning in regions where both space heating and cooling are required throughout the year. Air-conditioning and refrigeration systems, including heat pumps, were covered in Chapter 14.)

- *Hydrothermal resources:* Naturally occurring hydrothermal resources contain the basic elements of heat in the earth, along with groundwater and rock characteristics (i.e., open fractures that allow fluid flow) that are sufficient for the recovery of heat energy, usually through produced hot water or steam. Hydrothermal resources can range in temperature from a few degrees above ambient conditions to temperatures greater than 375 °C. Above this higher range, a new class of innovative subsurface and surface production technologies will likely be required to convert geothermal energy resources for beneficial use.

- *Enhanced geothermal systems:* Unconventional geothermal resources, often referred to as enhanced (or "engineered") geothermal systems (EGS), contain heat similar to conventional hydrothermal resources, but lack the necessary groundwater and/or rock characteristics to enable energy extraction without innovative subsurface engineering and transformation. Unconventional EGS resources can be found at any above-ambient temperature that supports energy conversion for a given end-use technology application. The resource has potential applications across the geothermal technology spectrum, although practical application will be limited by the costs of the required engineering. An emerging technology is to increase the temperature of productive saline aquifers by injecting high-temperature waste heat, e.g., injecting heat generated by compressing gases for underground storage (Stumpf 2023).

26.1 Geothermal (or "Ground-Source") Heat-Pump Resources

The rudiments of heat pumps were covered in Section 14.2. A typical residential heat pump has a *COP* of 4, which means that the heat pump needs just 1 unit of electricity input to provide 4 units of heat output. The heating cycle can be reversed to provide cooling services. *Geothermal heat pumps* are assumed to be economically viable if their *COP*, defined in Eqs. (26.3) and (26.7), are in the following ranges (NRC 2002; Stumpf 2023):

For heating (winter): $3.0 < COP_{heating} < 4.0$

For cooling (summer): $3.5 < COP_{cooling} < 6.7$

Geothermal heat pump resources refer to the shallow-earth environment, which is composed of rocks and soil at depths from a few centimeters below the ground to average depths of about 10 m. At these depths, which are typically below the *frost* (or "freezing") *line*, ground temperatures remain constant year-round (between about 10 and 15 °C). For most areas, this means that soil temperatures are usually warmer than the air in winter, and cooler than the air in summer.

Therefore, the thermal energy storage properties of the rocks and soil allow them to act as a heat sink—absorbing excess heat—during summer, when surface temperatures are relatively higher, and as a heat source during winter, when surface temperatures are lower. Geothermal heat pumps therefore take advantage of the thermal storage properties of ground, using thermal energy removed from buildings and seasonally stored in the ground during summer cooling operations to keep buildings warm in winter at reduced rates of electricity consumption. In addition, GHPs cool buildings at higher efficiencies than conventional air conditioners as the temperature of the shallow earth is generally cooler than the ambient air in summer.

Due to the analogy with electrochemical batteries, this geothermal resource is colloquially referred to as a "geothermal battery." If a GHP is operated with cooling only or heating only, the heat sink will either become saturated or depleted. To maintain the equilibrium of this "geothermal battery" (in summer), GHPs should return as much heat as is removed from the ground (in winter), similar to charging and discharging an electrochemical battery. This implies that GHPs operating in warm climates may require supplemental cooling (say, from a traditional air-conditioning system) during extremely hot days, while supplemental heating might be required in cold climates on extremely cold days.

A GHP system includes the following:

- A ground or borehole heat exchanger, which comprises a group of pipes buried or drilled in the ground, immersed in a surface water body, or exchanging heat directly with groundwater. However, instead of circulating groundwater, some companies are working on circulating another (benign) fluid, perhaps even liquefied carbon dioxide, in a completely closed-loop system (Eavor 2023).
- An energy-delivery system such as a heating, ventilation, and air-conditioning (HVAC) system with ductwork for forced air heating or cooling, and/or in-floor piping for radiant heating.
- A heat pump, which pumps thermal energy between the delivery system and the ground heat exchanger (see Section 14.2).

The geothermal heat exchanger transfers heat between the ground and a fluid, usually with a water/antifreeze mixture. There are several types and configurations of ground heat exchangers, the majority of which are closed-loop systems. Figure 26.2 illustrates closed- and open-loop systems using dry ground, groundwater, or surface water. Geothermal heat pumps are the most expensive type of geothermal technology in all countries, due to the earthworks or drilling needed to install the underground heat exchanger. This can represent more than half the total price of the system (although their far greater efficiency and durability—boreholes last for over 50 years—mean that they can be competitive on a levelized cost basis).

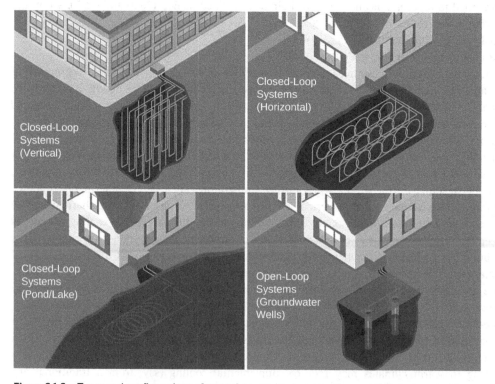

Figure 26.2 Types and configurations of ground-source heat pumps. *Source:* US Department of Energy, CC BY.

26.1.1 Energy Balance of Geothermal Heat Pump in Heating Mode

The heat exchanged between a GHP and groundwater is (Holm et al. 2015):

$$\dot{Q}_{gw} = \dot{m}_{gw} c_{p,gw} \Delta T \tag{26.1}$$

where $\dot{Q}_{gw}$ is the groundwater thermal potential (kW), $\dot{m}_{gw}$ is the mass flow rate of groundwater into the heat exchanger (kg/s), $c_{p,gw}$ is the specific heat of the ground water (4.18 kJ/kg°C), and ΔT is the temperature difference (°C) between the supply groundwater and return groundwater of the (groundwater) heat exchanger, as illustrated in Figure 26.3. In heating mode (Figure 26.3a):

$$\Delta T = T_{gw} - T_{out} \tag{26.2}$$

where T_{gw} is the groundwater temperature, i.e., supply water temperature to the heat exchanger, and T_{out} is the outlet groundwater temperature from the heat exchanger. In heating mode, the groundwater is returned in a cooled state.

The *COP* in heating mode is defined as follows (also see Section 14.1):

$$COP_{heating} = \frac{\text{Heating effect}}{\text{Power input}} = \frac{\dot{Q}_H}{\dot{W}_{in}} \tag{26.3}$$

where $\dot{Q}_H$ is the rate at which heat is delivered to a building (via a heat exchanger) for space or water heating, and $\dot{W}_{in}$ is the required electrical power to run the compressor.

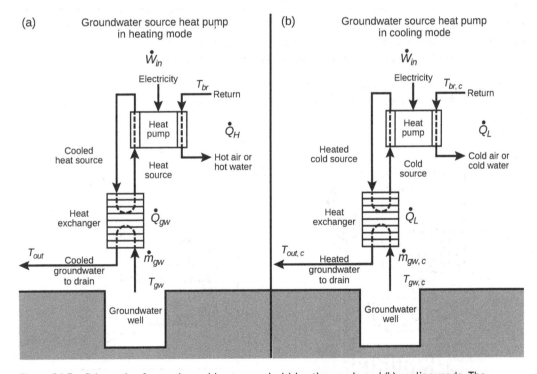

Figure 26.3 Schematic of a geothermal heat pump in (a) heating mode and (b) cooling mode. The ground heat-exchanger loop is represented as a "heat exchanger" and is connected to a heat pump and an indoor delivery system. Once groundwater has passed through a heat exchanger, it is discharged to a stream, pond, drainage ditch, or similar ground feature. *Source:* Adapted from Holm et al. (2015).

From Figure 14.1, it follows that

$$\dot{Q}_H = \dot{Q}_{gw} + \dot{W}_{in} \tag{26.4}$$

Combining Eqs. (26.3) and (26.4):

$$\dot{Q}_H = \left(\frac{COP_{heating}}{COP_{heating} - 1}\right)\dot{Q}_{gw} \tag{26.5}$$

26.1.2 Energy Balance of Geothermal Heat Pump in Cooling Mode

The heat exchanged between the GHP and groundwater in cooling mode is (Holm et al. 2015):

$$\dot{Q}_{gw,c} = \dot{m}_{gw,c}c_{p,gw}\Delta T_c \tag{26.6}$$

where $\dot{Q}_{gw,c}$ is the groundwater cooling potential (kW), $\dot{m}_{gw,c}$ is the mass flow rate of groundwater (kg/s) into the heat exchanger operating in cooling mode, as shown in Figure 26.4, and ΔT_c is the temperature difference between the return water of the groundwater heat exchanger and the supply water from the well.

In cooling mode (Figure 26.3b):

$$\Delta T_c = T_{out,c} - T_{gw,c} \tag{26.7}$$

where $T_{out,c}$ is the outlet groundwater temperature at the heat exchanger operating in cooling mode or the temperature of the heated groundwater to discharge, and $T_{gw,c}$ is the groundwater temperature supplied to the heat exchanger operating in cooling mode.

The *COP* in cooling mode is defined as follows (also see Section 14.1):

$$COP_{cooling} = \frac{\text{Cooling capacity}}{\text{Power input}} = \frac{\dot{Q}_L}{\dot{W}_{in}} \tag{26.8}$$

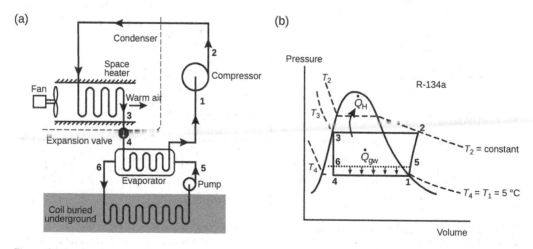

Figure 26.4 Geothermal heat pump used for space heating. (a) Layout and (b) representation of heat pump cycle on a pressure-volume diagram.

where $\dot{Q}_L$ is the cooling capacity (kW) of the heat pump to cool spaces or water, and $\dot{W}_{in}$ is the electrical power (kW) necessary to operate the compressor.

From Figure 14.1, it follows that:

$$\dot{Q}_L = \dot{Q}_{gw,c} - \dot{W}_{in} \tag{26.9}$$

Combining Eqs. (26.8) and (26.9):

$$\dot{Q}_L = \left(\frac{COP_c}{COP_c + 1} \right) \dot{Q}_{gw,c} \tag{26.10}$$

Reimagine Our Future

We are rapidly changing the ways in which we use and produce energy. How could these changes be accelerated to bring about massive decarbonization by 2035?

26.1.3 GHP Potential

The current geothermal contribution to US energy capacity is less than 1%. However, the potential is more than 8% by 2050 (IEA 2022a). Geothermal heat pumps maintain approximately 3% annual growth in the United States, with current installations exceeding 1.7 million units. About 40% of installations are residential. The other 60% are commercial or institutional. In 2020, the total installed capacity for direct geothermal use (i.e., for heating) in the United States was 485 MW_{th}. The total installed capacity for GHPs in the United States was 20.7 GW_{th}. This capacity is predicted to increase to 60 GW_{th} by 2050 (DOE 2019; IEA 2022a, 2022b).

The global installed geothermal heat pumps totaled around 108 GW_{th} in 2019. The global geo-thermal heat pump market is expected to grow from $3.56 billion in 2021 to $3.88 billion in 2022, and $5.29 billion in 2026 (Lund and Toth 2021).

Major players in the geothermal heat pump market are BDR Thermea, Bosch Thermotechnology, Carrier, Stiebel Eltron, Trane, Vaillant Group, Viessmann, NIBE Energy Systems, NIBE Industrier AB (Climate Master Inc., WaterFurnace, Enertech), Daikin, Glen Dimplex, and Maritime Geothermal.

Example 26.1 Comparison of a Geothermal Heat Pump and a Gas Furnace to Heat a Building

The air inside a building must be heated during winter. One can use a gas furnace to heat the space, or one could use a geothermal heat pump. Compare the hourly operating cost (¢/h) of these two systems to heat the space based on the following parameters, also summarized in Figure 26.4:

Building:

- The air inside a building must be maintained at 20 °C.
- The outdoor temperature is 0 °C.
- The building's heat loss is 75 MJ/h.

Heat pump:

- It operates on an ideal vapor-compression cycle.
- The refrigerant used is R-134a.

- The heat transfer fluid circulates inside a pipe loop placed 3 m underground, which provides heat to the heat pump's evaporator at an average temperature of 10 °C.
- The refrigerant enters the compressor as a saturated vapor at 5 °C and leaves the condenser as a saturated liquid at 55 °C.
- The refrigerant condenser coil is maintained at 55 °C (no losses are assumed) which heats air, that is circulated through the building using a fan.
- The cost of electricity is 9 ¢/kWh$_e$.

Gas furnace:

- The energy conversion efficiency is 100% (no losses are assumed).
- The cost of gas is 85 ¢/therm, which is approximately 2.9 ¢/kWh$_{th}$.

Required: The hourly operating cost (¢/h) of the geothermal heat pump and the gas furnace.

Solution strategy: We will use Eqs. (26.1)–(26.9) to analyze the heat pump, and we will use energy conservation principles to analyze the gas furnace. For the heat pump, we should find the enthalpies and entropies at the state points shown in Figure 26.5. Using those, we must calculate the compressor power, heat transfer rates in the condenser and water tank, the water-to-air heat exchanger, and the evaporator. We can then determine the *COP*.

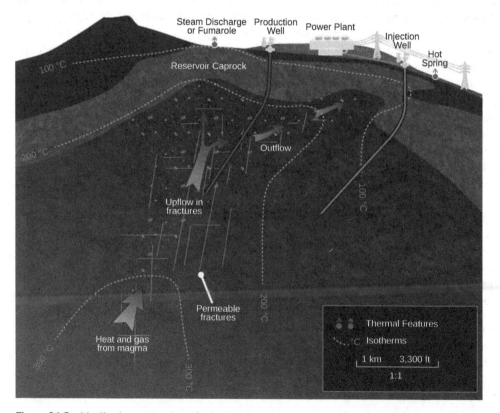

Figure 26.5 Idealized cross-section of a hydrothermal resource showing various conceptual elements of a high-temperature hydrothermal reservoir. *Source:* US Department of Energy, CC BY.

Assumptions: Assume that the compression process is adiabatic and reversible, i.e., it is isentropic. Assume that the condenser exit is at the saturated liquid condition. We will also assume that the pressure drops in the condenser, evaporator, and piping are negligible.

Analysis:

a) Geothermal heat pump

Using the process numbering of Figure 26.4, we may analyze the geothermal heat pump.

Enthalpies and entropies

The thermodynamic properties of R-134a can be read off tabulated values found in textbooks, or we could use Engineering Equation Solver (EES) or the widely used Refprop software, among others. A free version of Mini-Refprop is available from the National Institute of Standards and Technology (NIST): https://trc.nist.gov/refprop/MINIREF/MINIREF.HTM Using EES, the following values are found:

Process 1–2:

The refrigerant enters the compressor at 5 °C as a saturated vapor ($x_1 = 1.0$).
Therefore:

$$h_1 = h_{g,5 °C} = 250.1 \text{ kJ/kg}$$

$$s_1 = s_{g,5 °C} = 0.9164 \text{ kJ/kgK}$$

The refrigerant is superheated at state point 2, with $T_2 = 55 °C$ and $p_2 = p_3 = p_{sat, 55 °C}$. Also, process 1 to 2 is isentropic, so $s_2 = s_1 = 0.9164$ kJ/kgK, from which we find that $p_2 = p_3 = p_{sat, 55 °C} = 1.49$ MPa.
Therefore:

$$h_2 = 279.93 \text{ kJ/kg}$$

Compressor power

From Eq. (14.1): $\cancel{q_{12}} - w_{12} = \Delta h_{12} = h_2 - h_1$
So, $w_{12} = h_1 - h_2 = (250.1 - 279.92) \text{ kJ/kg} = -29.8 \text{ kJ/kg}$

Process 2–3:

$$h_3 = h_{f,55 °C} = 129.34 \text{ kJ/kg}$$
$$q_{23} = h_3 - h_2 = (129.34 - 279.92) \text{ kJ/kg} = -150.58 \text{ kJ/kg}$$

Heat transfer rates of the condenser

From Eq. (14.3): $q_{23} - \cancel{w_{23}} = \Delta h_{23} = h_3 - h_2 = (129.34 - 279.92) \text{ kJ/kg} = -150.6 \text{ kJ/kg}$

Process 3–4:

$$h_4 = h_3 = 129.34 \text{ kJ/kg}$$

Heat transfer rates of the evaporator

From Eq. (14.3): $q_{51} - \cancel{w_{51}} = \Delta h_{51} = h_1 - h_5 = (250.1 - 79.32) \text{ kJ/kg} = 170.8 \text{ kJ/kg}$

Coefficient of performance

From Eq. (26.3): $COP_{heating} = \dfrac{\text{Heating effect}}{\text{Work input}} = \dfrac{q_H}{w_{in}} = \dfrac{150.6 \text{ kJ/kg}}{29.8 \text{ kJ/kg}} \approx 5.0$

Operating cost of geothermal heat pump

Cost of electricity: 9 ¢/kWh$_e$

Compressor: $\dot{W}_{in} = \dot{m} w_{in}$

with the required mass flow rate of the refrigerant being:

$$\dot{m} = \frac{\dot{Q}_{loss}}{q_{heating}} = \frac{(75{,}000 \text{ kJ/h})(1 \text{ h/3600 s})}{150.6 \text{ kJ/kg}} = 0.14 \text{ kg/s (or 498 kg/h)}$$

Therefore, the required compressor power is $\dot{W}_{in} = \dot{m}w_{in} = (0.14 \text{ kg/s})(29.8 \text{ kJ/kg}) = 4.2 \text{ kW}_e$
The cost of operating the compressor of the geothermal heat pump is:

$$(4.2 \text{ kW}_e)(9 \text{ ¢/kWh}_e) = 37.8 \text{ ¢/h}.$$

b) **Gas furnace**

The cost of operating the gas furnace is $(75{,}000 \text{ kJ/h})(1 \text{ h/3600 s})(2.9 \text{ ¢/kWh}_{th}) = 60.9 \text{ ¢/h}$

Discussion: The operating cost of the gas furnace is about one-and-a-half times more than that of the geothermal heat pump. This situation improves further in favor of the geothermal heat pump as the price of natural gas increases during periods such as the energy crisis of 2022 (e.g., to 4 ¢/kWh$_{th}$), and if electricity is provided by photovoltaics, at around (3 ¢/kWh$_e$) rather than 9 ¢/kWh$_e$ from traditional grid supply. Other metrics that could be used to compare energy systems include environmental costs and benefits, and life-cycle costs (see Chapter 6).

Reimagine Our Future

Universities typically deliver highly skilled engineers who are either unaware or emotionally detached from the ethical, ecological, cultural, social, and international implications of their work. Engineering education seems to be very strong on inculcating *know-how* but not *know-why*. What changes can be incorporated in engineering education to remedy this situation?

26.2 Hydrothermal Resources

The natural formation of a hydrothermal resource typically requires three principal elements: heat, water, and permeability. When water is heated in the earth, hot water or steam can become trapped in porous and fractured rocks beneath a layer of relatively impermeable caprock, resulting in the formation of a hydrothermal reservoir. Geothermal water or steam may emanate naturally from the reservoir and manifest at the surface as hot springs or geysers. However, most stays trapped underground in rock, under pressure and only accessible through drilling (Figure 26.5).

The fluid in convective *hydrothermal resources* can be vapor- (steam), or water-dominated, with temperatures ranging from 100 °C to over 300 °C. High-temperature geothermal fields are most common near tectonic plate boundaries and are often associated with volcanoes and seismic activity, as the crust is highly fractured and thus permeable to fluids, resulting in heat sources being readily accessible.

The US Geological Survey (USGS)'s resource assessment estimates that the identified hydrothermal resources of >90 °C in the United States have the potential to provide a mean total of 9057 MW$_e$ of electric power generation (Williams et al. 2008). By the end of 2020, the global installed power of hydrothermal plants was about 16 GW$_{th}$. Those plants generated around 97 TWh$_{th}$ of heat (IHA 2021).

As shown in Figure 26.6, there are numerous viable sub-marine geothermal vents around the world. The global mean power density of these hydrothermal systems is around 60–100 mW/m^2

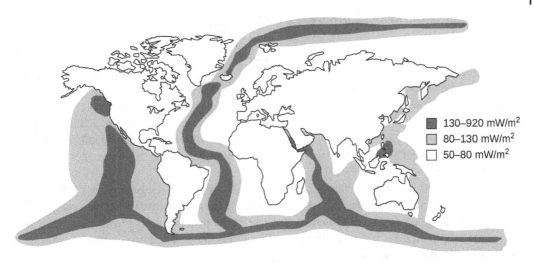

Figure 26.6 Global map of the Earth's surface heat flow, in mW/m². The global mean power density of hydrothermal systems is between 130 and 919 mW/m². *Source:* Adapted from Davies (2013).

(Davies 2013). Some sub-marine vents (called "black smokers") can produce seawater temperatures of 400 °C and realize thermal capacities ranging from 60 MW$_{th}$ to 5 GW$_{th}$ (IEA 2011; NOAA 2020). Such geothermal vents could be exploited directly without drilling and produce power by means of an encapsulated sub-marine plant. However, research and development is needed since there are no technologies available to commercially tap energy from offshore geothermal resources.

Geothermal heat can also be economically extracted from many *deep aquifer systems* all over the world. Many such locations can be reached within a depth of 3 km, with moderate heat flow in excess of 50–60 mW/m², and rock and fluid temperatures in excess of 60 °C (IEA 2011). Figure 26.7 shows the locations of global deep aquifers.

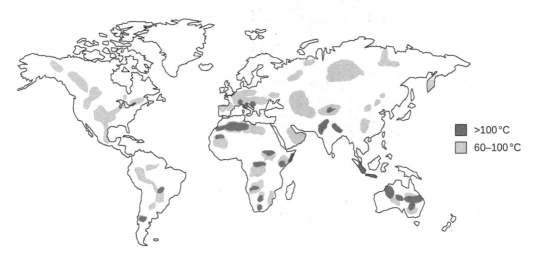

Figure 26.7 World map of deep (<3 km) aquifer systems. *Source:* Adapted from DOE (2019).

26.3 Enhanced (or "Engineered") Geothermal Systems

So far, the utilization of geothermal energy has been concentrated in areas of naturally occurring water or steam, and sufficient rock permeability. However, the vast majority of geothermal energy within drilling reach, which can be up to 5 km given current technology and economics, is in relatively dry and low-permeability rock. Heat stored in low-porosity and/or low-permeability rocks is commonly referred to as hot rock resources. These resources are characterized by limited pore space and/or minor fractures. They therefore contain insufficient water and are insufficiently permeable for natural exploitation. The way to address these difficulties is to use *engineered* (or "enhanced") *geothermal systems* (EGS).

With EGS, regions with a high thermal gradient are identified, and the hot dry rock is fractured with high-pressure water, similar to the technology used in "fracking" (hydraulic fracturing) in the oil and gas industry. Test wells are drilled to characterize the heat resource, and an energy conversion system is built on the surface. Ideally, EGS will inject cool water and extract heated water in a closed loop, as illustrated in Figure 26.8.

Enhanced geothermal systems have to contend with several challenges, including possible short-circuiting of the water flow between injection and withdrawal, environmental water losses, operating under immense injection pressures (e.g., 25 MPa) and high temperatures (400 °C), and induced seismicity.

In stable, "old" continental tectonic areas, where temperature gradients are low (7–15 °C/km), and where permeability is also low, but with a less favorable state of stress, depths will be

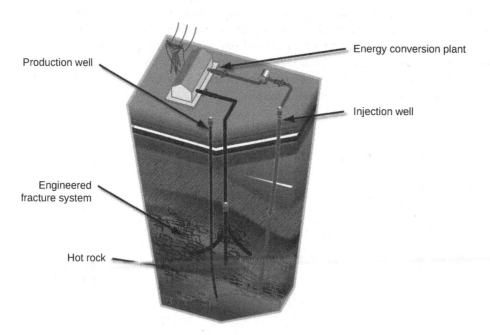

Figure 26.8 Illustration of an enhanced (or "engineered") geothermal system. *Source:* US Department of Energy, CC BY-SA 4.0.

significantly greater, and developing an EGS resource will be less economic. Technologies that allow energy to be tapped from hot-rock resources are still in the demonstration stage and require innovation and experience to become commercially viable.

Among current EGS projects worldwide, the Soultz-sous-Forêts in France has been operating the longest, since 2011. Although the plant has a maximum power output of only 1.65 MW$_e$, the project revealed how EGS projects could be conducted.

26.3.1 Superhot Rock Geothermal Systems

Superhot rock with temperatures of 400 °C can be found at depths of around 20 km. Commercializing and deploying superhot-rock geothermal energy broadly will require new technology, including rapid ultra-deep drilling methods, heat-resistant well materials and tools, and ways to develop deep-heat reservoirs in hot dry rock. The first such system may be commercialized before 2030 if adequate private investment can be secured (CATF 2022).

26.4 Generating Electricity and Heat with Geothermal Systems

Many existing geothermal power plants use steam produced by *flashing* (i.e., reducing the pressure of) the geothermal fluid produced from the reservoir. Today, geothermal power plants can use water in the vapor phase, in a combination of the vapor and liquid phases, or only in the liquid phase. The choice of plant depends on the depth of the reservoir, and the temperature, pressure, and nature of the entire geothermal resource. The three main types of plant are flash steam, dry steam, and binary plants. All current forms of accepted geothermal development use re-injection as a means of sustainable resource exploitation.

26.4.1 Flash Steam Plants

The most-found geothermal resources contain reservoir fluids with a mixture of hot liquid (water) and vapor (mostly steam). Flash steam plants make up about two-thirds of the geothermal installed capacity today. They are used where water-dominated reservoirs have temperatures above 180 °C (DOE 2019). In these high-temperature reservoirs, the liquid water component boils, or "flashes," as pressure drops. Flashing takes place in a *separator*, where liquid water is separated from (saturated) steam in an isenthalpic manner. Separated steam is piped to a turbine to generate electricity, and the remaining hot water may be flashed again twice (a double flash plant) or three times (a triple flash plant) at progressively lower pressures and temperatures to obtain more steam. The cooled brine and the condensate are usually sent back down into the reservoir through injection wells. Combined-cycle flash steam plants use the heat from the separated geothermal brine in binary plants to produce additional power before re-injection (Figure 26.9a). Flash steam plants range in capacity from 0.3 to 110 MW$_e$.

26.4.2 Dry Steam Plants

Dry steam plants, which make up about a quarter of geothermal capacity today, directly utilize dry steam (at around 150 °C) that is piped from production wells to the plant and then to a steam turbine. The control of steam flow to meet electricity demand fluctuations is easier than in flash steam

(a)

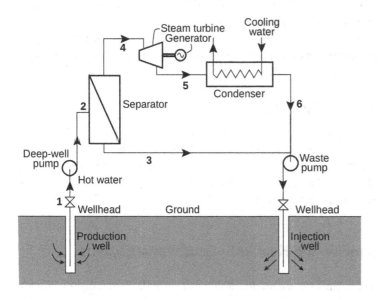

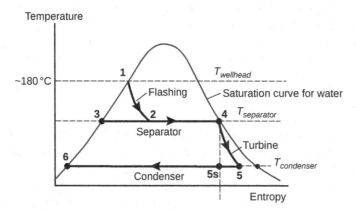

Figure 26.9 Types of electricity-generating power plants using geothermal energy: (a) a flash steam plant (a single-flash system is illustrated), (b) a dry steam plant, and (c) a binary (organic Rankine cycle) plant.

plants, where continuous up-flow in the wells is required to avoid gravity collapse of the liquid phase. In dry steam plants, the condensate is usually re-injected into the reservoir or used for cooling (Figure 26.9b). Dry steam plants range in capacity from 0.3 to 110 MW$_e$.

26.4.3 Binary Plants

Electrical power generation units that use binary cycles (i.e., with two fluid loops) constitute the fastest-growing group of geothermal plants, as they can use low- to medium-temperature resources, which are more prevalent. Binary plants that use an organic Rankine cycle or Kalina cycle (which uses a solution of two fluids with different boiling points for its working fluid), typically operate with temperatures varying from as low as 73 °C (at Chena Hot Springs, Alaska) to 180 °C (DOE

(b)

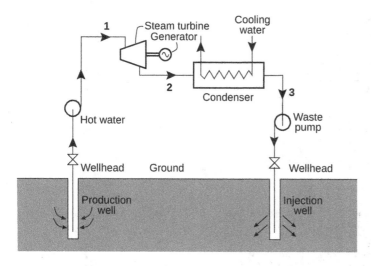

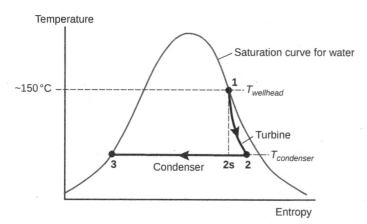

Figure 26.9 (Continued)

2019). In these plants, heat is recovered from the geothermal fluid using heat exchangers to vapor-ize an organic fluid with a low boiling point (e.g., butane or pentane in the ORC and an ammonia–water mixture in the Kalina cycle) and to drive a turbine. Figure 26.9c illustrates the two fluid loops, from which the name "binary plant" has been derived.

The ORC is the dominant technology for low-temperature geothermal resources. Under certain design conditions, the Kalina cycle can operate at a higher cycle efficiency than conventional ORC plants. The lower-temperature geothermal brine that leaves the heat exchanger is re-injected back into the reservoir in a closed loop, thus promoting sustainable resource exploitation. Today, binary plants have an 11% share of the installed global generating capacity and a 44% share in terms of the number of plants (DOE 2019). Binary plants range in capacity from 0.1 to 45 MW$_e$.

(c)

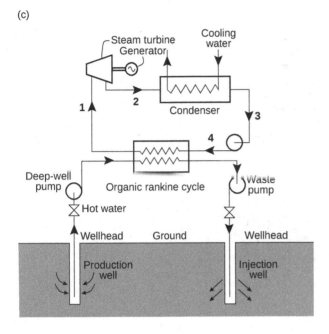

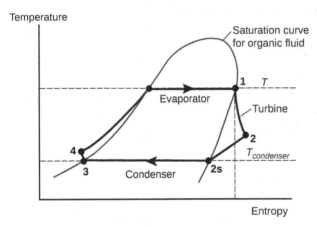

Figure 26.9 (Continued)

Example 26.2 Single-Flash Geothermal Power Plant

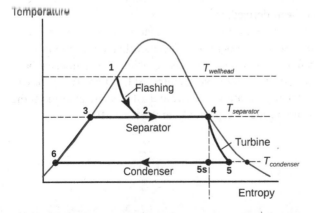

Consider the temperature-entropy diagram of a single-flash geothermal cycle (right). Saturated hot water exits the wellhead at 80 kg/s and is received at 275 °C and 5.9 MPa at state point 1. The condenser temperature is 50 °C (state points 5 and 6). The separator (between state points 3 and 4) should operate at the average of the maximum and minimum cycle temperatures. Estimate the thermal efficiency of this cycle.

Solution strategy: We can use the theory derived for ideal Rankine cycles in Section 13.1. The thermal efficiency of this cycle is $\eta_{th} = \dfrac{w_{net}}{q_{in}}$.

Analysis:

From Section 13.1 for Rankine cycles, we know the following:

with $w_{net} = w_{out} \cancel{-w_{in}} = x_2(h_4 - h_{5s})$

and $q_{in} = h_1 - h_6$

The enthalpies can be read off property tables of water, or using a resource such as EES:

$$h_1 = h_{f,275\,°C} = 1210.7 \text{ kJ/kg}$$

$$h_6 = h_{f,50\,°C} = 209.3 \text{ kJ/kg}$$

$$\therefore q_{in} = h_1 - h_6 = (1210.7 \text{ kJ/kg}) - (209.3 \text{ kJ/kg}) = 1001.4 \text{ kJ/kg}$$

$$T_4 = \frac{T_{well} + T_{cond}}{2} = \frac{275 + 50\,°C}{2} = 162.5\,°C$$

$$\therefore h_4 = h_{g,162.5\,°C} = 2760.2 \text{ kJ/kg}$$

Also: $s_4 = s_{5,s} = 6.7280 \text{ kJ/kgK}$

$$\therefore x_{5,s} = \frac{s_{5,s} - s_{f,50\,°C}}{s_{fg,50\,°C}} = \frac{6.7280 \text{ kJ/kgK} - 0.7038 \text{ kJ/kgK}}{7.3710 \text{ kJ/kgK}} = 0.82$$

$$\therefore h_{5,s} = x_{5,s}h_{fg,50\,°C} + h_{f,50\,°C} = 0.82(2382 \text{ kJ/kg}) + 209.34 \text{ kJ/kg} = 2162.6 \text{ kJ/kg}$$

To calculate the efficiency, we now only need to calculate the vapor quality at state point 2. We may assume that the process from 1 to 2 is isenthalpic: $h_1 = h_{2,275°\,C} = 1210.7 \text{ kJ/kg}$

$$x_2 = \frac{h_{2162.5\,°C} - h_{f,162.5\,°C}}{h_{fg,162.5\,°C}} = \frac{1210.7 \text{ kJ/kg} - 686.36 \text{ kJ/kg}}{2073.8 \text{ kJ/kg}} = 0.25$$

$$\therefore w_{net} = x_2(h_4 - h_{5s}) = 0.25(2760.2 \text{ kJ/kg} - 2162.6 \text{ kJ/kg}) = 149.4 \text{ kJ/kg}$$

$$\therefore \eta_{th} = \frac{w_{net}}{q_{in}} = \frac{149.4 \text{ kJ/kg}}{1001.4 \text{ kJ/kg}} = 0.15 \text{ or } 15\%$$

26.5 Sustainability of Geothermal Power Plants

26.5.1 Costs

The high initial investment cost and the uncertainty of the exploration of reservoirs due to a limited knowledge of the reservoir structure and long-term performance are major challenges associated with EGS. However, the proof of concept with the successful testing and installation of EGS in recent years, and the baseline understanding of potential resources indicate the potential viability

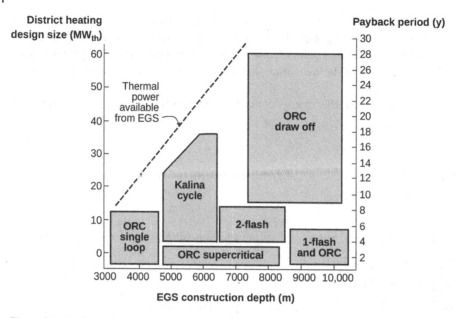

Figure 26.10 Typical life-cycle economic results of geothermal energy systems, considering the production of heat and electricity for an area of 24,000 inhabitants in Switzerland. *Source:* Gerber (2014) / Taylor & Francis.

of this technology. Figure 26.10 illustrates the typical life-cycle results (over a 25-year period) of geothermal power plants as a function of depth.

Geothermal project costs are highly site sensitive. Typical costs for geothermal power plants range from $1870 to $5050 per kilowatt (electric), noting that binary plants are normally more expensive than direct dry-steam and flash plants. The levelized cost of electricity (LCOE) of a geothermal power plant ranges from $0.04 to $0.14 per kilowatt-hour, assuming a 25-year economic life. By 2027, the LCOE of geothermal power is expected to drop to less than $0.037 per kilowatt-hour. The LCOE of geothermal power will continue to decrease if emerging technologies like EGS and using superhot dry rocks are successful (Aghahosseini and Breyer 2020; EIA 2022; IRENA 2022).

26.5.2 Environmental Effects

Deep pressurized hot water sources usually contain quantities of dissolved gases (CO_2, H_2S, NH_3, and CH_4). These are released during depressurization ("flashing") and cooling, and generate oxidation products such as SO_2 and NO_x. Metal salts in solution, including salts of mercury, boron, arsenic, and other metals, may either precipitate in geothermal brine ponds forming pipe scale that must be disposed of, or be released into the atmosphere from cooling towers as fine-grained particulate matter (Soltani et al. 2021). Methane is harmful to the ozone layer and has a high short-term (in decades) global warming potential (see Chapter 5). Mercury and arsenic that enter the food chain are well-known risks to human health.

Gas emissions are mainly related to dry-steam and flash-steam geothermal power plants. Dry-steam plants do not produce any mineral-laden brine, so their environmental impacts are lower than those of flash-steam plants. Binary plants also operate in a closed-loop mode with the direct return of fluids to depth, thus do not produce liquid or gaseous emissions. Exhaust emissions in geothermal power plants are caused by construction equipment, transportation, and wells. However, in many cases, these emissions are negligible compared to fugitive ones.

Table 26.1 compares the lifecycle emissions of geothermal power plants with other power plants. Geothermal power plants emit up to 10 times less CO_2 than natural gas-fired steam power plants. When compared to renewable resources, geothermal power plants also have fewer greenhouse-gas emissions than solar and biomass power plants. In addition, geothermal power plants emit far lower quantities of SO_2 and particulate matter than fossil-fired power plants. Figure 26.11 compares the avoided CO_2 emissions of various types of geothermal power plants compared to the best fossil-fired technology.

Geothermal power plants take up less space than many others as well, as is illustrated in the space taken up by the various types of power plants listed below (Soltani et al. 2021):

- 56 MW_e geothermal plant: 7460 m^2/MW_e
- 670 MW_e nuclear fission plant: 10,000 m^2/MW_e
- 25 MW_e wind farm: 16,000 m^2/MW_e
- 47 MW_e solar thermal plant (Mojave Desert): 28,000 m^2/MW_e
- 2.3 GW_e coal-fired steam power plant (including strip mining): 40,000 m^2/MW_e
- 10 MW_e photovoltaic plant (south-western USA): 66,000 m^2/MW_e

The pressure below the crust usually declines upon extraction of geofluids for geothermal activities. This could result in land subsidence, which can be as much as 15 m over a 50-year period

Table 26.1 Comparison of lifecycle emissions of some power plants.

Type of power plant	CO_2 (kg/MWh$_e$)	SO_2 (kg/MWh$_e$)	NO_x (kg/MWh$_e$)	Particulates (kg/MWh$_e$)
Coal-fired steam power	990	4.7	1.9	1.0
Gas-fired steam power	550	0.09	1.3	0.06
Photovoltaic plant	99	0.000228	0.00034	0.000119
Nuclear fission plant	22	—	—	—
Wind plant				
Onshore (1.5 MW_e)	10.2	0.00004	0.00003	0.00004
Offshore (1.5 MW_e)	8.9	0.00004	0.00002	0.00001
Geothermal plant				
EGS	16.9–49.8	—	—	—
Binary	42–62	0.0004–0.0005	—	—
ORC	80.5	0.0003	—	—
Single-flash	12	0.00006	—	—
Double-flash	3.9	0.00003	—	—
Hydrothermal geysers (open loop)	40.3	0.00009	0.0006	Negligible

Source: Adapted from Soltani et al. (2021).

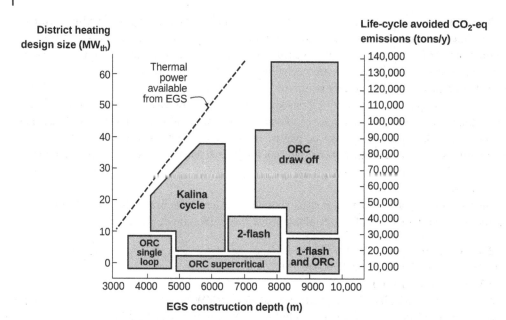

Figure 26.11 Avoided CO_2 emissions over a 25-year period compared to best-technology fossil-fired plants, considering the production of heat and electricity for an area of 24,000 inhabitants in Switzerland. *Source:* Gerber (2014) / Taylor & Francis.

(Narasimhan and Goyal 1984; Delleur 2006; DiPippo 2012). More dangerous, though, are the seismic effects that accompany the operation of an EGS reservoir. Between 1977 and 2017, seismic activity that can be related to the operation of geothermal power plants was reported around the world, with events measuring between 2 and 5.5 on the Richter scale.

In 2017, the Pohang EGS site in South Korea caused an earthquake with a magnitude of 5.5, which resulted in death, injury, and loss of property. In 2006, an EGS site in Basel, Switzerland, was suspected to have initiated an earthquake with a magnitude of 3.4. Following this event, all EGS operations in Switzerland were shut until 2020. Like many other countries, Switzerland is now actively pursuing geothermal power again (SwissInfo 2022).

Low-efficiency flash-steam geothermal systems require extensive water-cooling towers that consume more than 50% of the water used in the plant. (Dry-cooled binary cycles do not require excess water.) Figure 26.12a shows that the typical water consumption of a geothermal power plant can be double that of a combined-cycle power plant, but about two-thirds that of a conventional fossil-fired power plant. Figure 26.12b shows that the typical amount of heat discarded from geothermal power plants can be six times that of conventional fossil-fired power plants.

The pressure of geothermal reservoirs declines with the increasing extraction of fluids, which may result in the disappearance of hot springs and geysers. Failures in well casings can also cause the rapid depletion of groundwater (100–150 t/h). The water that is re-injected into wells may contain high concentrations of harmful elements, which may seep into the surrounding groundwater or surface water with harmful effects to fauna and flora. Further, land-based geothermal power plants are often located in protected areas, where the development and exploitation of resources can result in significant ecological damage (Delleur 2006).

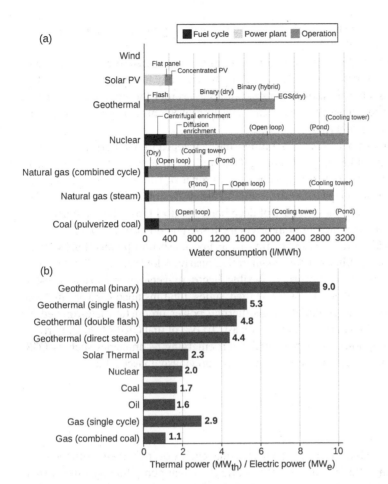

Figure 26.12 (a) Heat rejected per unit of power output from a geothermal power plant compared to others. (b) Comparison of the life-cycle water consumption for a geothermal power plant compared to others. *Source:* After Aghahosseini and Breyer (2020); DOE (2019); IRENA (2017); Soltani et al. (2021).

26.6 Outlook

Geothermal heat pumps are simpler and cheaper to deploy than their utility-scale counterparts. Not surprisingly, the installed capacity of GHPs increased by more than 9% between 2015 and 2022. GHPs now account for nearly 70% of geothermal direct-use capacity worldwide (IRENA 2022). However, to reach net-zero emissions by 2050, specialists deem that 50% of all space heating needs must be met by heat pumps by 2045 (IEA 2022a). Although the majority of heat pump installations are currently of the relatively cheap air-source variety, geothermal heat pumps are predicted to play a much more important role in the future, considering their year-round stable-temperature energy source and their superb *COP* (3–6). However, subsidies remain key to increase the competitiveness of geothermal heat pumps (IEA 2022c).

In 2022, geothermal energy only provided 0.33% (around 16 GW$_e$) of the world's electricity, with little change from its contribution in 1990 (IEA 2022c). In 2022, there were also only three EGS electrical power plants—all near the border of France and Germany—generating around 11 MW$_e$, which is enough to power around 9000 homes.

Even as EGS projects have struggled, new exploration techniques (borrowed from the oil and gas industry) have emerged, including the possibility of drilling horizontally instead of just vertically. Sophisticated steering systems allow these drills to negotiate complex rock formations to create viable underground reservoirs. There have also been technology improvements in exploring, discovering, developing, and managing geothermal resources.

For instance, a new startup company, Darcy Solutions, has developed in-borehole heat exchangers that can supply 35–140 kW of heating and cooling from aquifers (Darcy 2023). For larger geothermal energy systems, this would drastically reduce the number of boreholes needed (Stumpf 2023). The prospect of using closed-loop wells and circulating fluids other than groundwater seems promising (Eavor 2023). Also promising is the use of deep boreholes (>1 km), where temperature gradients of 17–24 °C/km have been measured in Sweden (Rosberg and Erlström 2021).

The prospects for geothermal energy are rapidly improving. Specialists now predict that by 2050, the United States could increase geothermal power generation nearly 26-fold compared to that generated in 2021, representing 60 GW$_e$ of always-on, flexible electricity-generation capacity (Tester et al. 2021). That would make up about 4% of the United States' total installed capacity in 2050 (DOE 2019). Globally, the technical potential for geothermal power production is estimated at 162–200 GW$_e$ by 2050, with EGS plants contributing at least 70 GW$_e$ of that (IRENA 2022; Pörtner et al. 2022). This potential will, however, only be realized if emerging technologies such as EGS continue to mature and enable access to resources that were previously inaccessible.

Problems

26.1 Consider a house that needs a total of 21 MWh$_{th}$ thermal energy over a three-month (90-day) period in winter. Assume that the house has a geothermal heat pump that operates with a *COP* of 3. The ground has a heat capacity of 3 MJ/m^{3}K and a thermal conductivity of 1.5 W/mK. Calculate (a) the volume of ground (V) for which the temperature must be reduced by 8 °C to provide thermal energy to supply the season's heating energy needs for the house, (b) the power requirement of the compressor, and (c) the area of a horizontal piping field needed for the rate of thermal energy replacement through ground conduction to match the average rate of power needed over winter, assuming a thermal gradient (dT/dz) near the pipes of 5 °C/m.

[Answers: (a) 2100 m^3; (b) 6.4 kW; (c) 427 m^2.]

Hint: (a) First calculate the heat (in MWh$_{th}$ or GJ) that must be removed from the ground using the definition of *COP*. That value of heat must be equal to $Q = c \, \Delta T V$, where c and V are the heat capacity and volume of the ground, respectively. (b) The required compressor power can be calculated by dividing the heat removed from the environment by the period of 90 days. (c) The heat flux surrounding the piping can be approximated by Fourier's Law as being $q = k(\mathrm{d}T/\mathrm{d}z)$, where k is the thermal conductivity of the ground. The fluxes above and below the ground-loop piping may be assumed to be equal, so that the total rate of heat transfer is $\dot{Q} = 2 \, q A$.

26.2 You are the lead mechanical engineer for a new campus instructional facility that will feature a ground-source ("geothermal") heat pump. The heat pump's evaporator (or condenser) will receive hot (or cool) water from several boreholes drilled in front of the building. At this stage in the design process, you need to select a GHP capable of heating and/or cooling the new building. Sophisticated energy modeling software informs you that the cooling load is 150 t, and the heating load is 135 t (Stumpf et al. 2021b). *Note:* 1 thermal ton is equivalent to 3.52 kW_{th}. Your selected heat pump manufacturer has a brand of heat pumps with a mean *COP* of 3.

a) What size heat pump (in kW_{th}) will you select and why?

[Answers: Using the *COP*, the cooling load is 529 kW_{th} and the heating load is 475 kW_{th}. To ensure a balanced ("geothermal battery") approach, select a heat pump that will provide 475 kW_{th}. The additional 529 kW_{th} minus 475 kW_{th} = 54 kW_{th} could be supplied by a separate air-conditioning system during peak cooling times.]

b) How much power will the GHP compressor require at the peak of the cooling season?

[Answer: 158 kW_e.]

c) If the ground can transfer 0.15 kW_{th} per meter (depth) in a well, and the Department of Health has instructed you that the maximum drill depth is 138 m, how many wells will you need to drill (assuming a vertical ground-loop system)?

[Answer: (475 kW_{th})/(20.7 kW_{th}/well) = 23 wells.]

d) You forgot to factor in the effect that ethylene glycol ("antifreeze") will have on the GHP system. If the closed-loop piping contains 40% ethylene glycol (and 60% water), how many additional wells should be drilled? (*Hint:* Glycol at this concentration transfers 10% less heat than pure water.)

[Answer: The heat that can be extracted changes to 20 kW_{th} per well, so that (475 kW_{th})/(20 kW_{th}/well) = 24 wells are required.]

26.3 Consider the Nesjavellir Geothermal Power Plant in Iceland, which produces 120 MW_e of electrical power. The geothermal well delivers around 1100 L of hot water (85 °C) per second, which will be used to provide heat for district heating. If the plant produces no electricity and all the hot water flows through heat exchangers, what is the heating capacity of this plant? The heat transfer fluid exits the heat exchanger at 20 °C, and the energy conversion efficiency of the district heating system is 95%.

[Answer: 285 MW_{th}.]

26.4 The Mahomet Aquifer in Illinois is composed of sand and gravel that was deposited in the valley of a river that was formed around 1.5 million years ago. This ancient river was later disrupted by floods that permanently altered the landscape. The last remnants of this valley were covered by sand, silt, gravel, and clay left behind by advancing glaciers over 130,000 years ago. The aquifer underlies 1.26 million acres (5100 km^2) of land in east-central Illinois and ranges from 15 to 60 m in thickness. It supplies over 380,000 m^3 per day of groundwater for public water use (for around 900,000 people), industrial supply, and irrigation (Holm et al. 2015; Ammons et al. 2018; Stumpf et al. 2021b).

The aquifer has an average depth of 45 m, where the average water temperature is 15 °C throughout the year (Holm et al. 2015; Stumpf et al. 2021b). Energy entrepreneurs are evaluating the prospects of using the Mahomet Aquifer as a geothermal energy resource to heat or cool the interior of some buildings in a business park located above the aquifer. The idea is to use water from the aquifer to provide heat to the evaporator units of water-source heat pumps located in the various buildings (see Figure 26.4). Evaluate the technical viability of this idea by calculating, among others, the volume flow rate of water that will be required to supply $1\,MW_{th}$ of thermal power to the buildings in winter, the *COP* of the heat pump in heating mode, and the electrical energy costs (¢/h) of these aquifer-source heat pumps compared to space heating provided by traditional gas furnaces. The average air temperature in winter is $0\,°C$.

[Answer: Open-ended question. The reader should, however, refer to Example 26.1 for a related problem.]

26.5 Read the following technical report: Robins et al. (2021)

Then, produce a 1000-word essay (with diagrams or photos) that clearly illustrates the environmental benefits and adverse effects of geothermal energy.

[Answer: Open-ended question. The reader should, however, reflect on the information in Chapter 26 when interpreting the knowledge shared in the technical report.]

26.6 Perform independent research about the social impacts of geothermal energy. Then, produce a high-impact fact sheet of no fewer than three pages, or about 400 words (with accompanying graphics), that addresses the following issues and how they might pertain to the use of geothermal energy: public understanding, poverty reduction, recruitment, health and safety, development of local business, and education. *Hint:* The following paper might be a good starting point for your research: Soltani et al. (2021).

[Answer: Open-ended question.]

References

Aghahosseini, A. and Breyer, C. (2020). From hot rock to useful energy: a global estimate of enhanced geothermal systems potential. *Applied Energy* 279 (December): 115769.

Ammons, C., Ballenger, E., Barnett, T., Bennett, S., et al. (2018). Mahomet aquifer protection task force: findings and recommendations. https://www.ilga.gov/reports/.

Anderson, A. and Rezaie, B. (2019). Geothermal technology: trends and potential role in a sustainable future. *Applied Energy* 248: 18–34.

Beckers, K.F. and McCabe, K. (2019). GEOPHIRES v2.0: updated geothermal techno-economic simulation tool. *Geothermal Energy* 7 (1): 1–28.

Beckers, K.F., Lukawski, M.Z., Anderson, B.J., Moore, M.C. et al. (2014). Levelized costs of electricity and direct-use heat from enhanced geothermal systems. *Journal of Renewable and Sustainable Energy* 6 (1): 013141.

Blodgett, L. and Stack, K. ed. (2009). *Geothermal 101: Basics of Geothermal Energy Production and Use.* Geothermal Energy Association.

Budiono, A., Suyitno, S., Rosyadi, I., Faishal, A. et al. (2022). A systematic review of the design and heat transfer performance of enhanced closed-loop geothermal systems. *Energies* 15 (3): 742.

CATF (2022). Superhot rock energy: a vision for firm, global zero-carbon energy. Clean Air Task Force. https://www.catf.us/resource/superhot-rock-energy-a-vision-for-firm-global-zero-carbon-energy/.

Cornwall, W. and Larson, E. (2022). Catching fire. *Science* 377 (6603): 252–255.

Darcy (2023). *Putting Groundwater to Work*. Darcy Solutions. https://darcysolutions.com/

Davies, J.H. (2013). Global map of solid Earth surface heat flow. *Geochemistry, Geophysics, Geosystems* 14 (10): 4608–4622.

Delleur, J.W. (2006). *The Handbook of Groundwater Engineering*. Boca Raton, FL: CRC Press.

DiPippo, R. (2012). *Geothermal Power Plants: Principles, Applications, Case Studies and Environmental Impact*. Oxford, UK: Butterworth-Heinemann.

DOE (2019). GeoVision. US Department of Energy. https://www.energy.gov/eere/geothermal/articles/geovision-harnessing-heat-beneath-our-feet.

Durga, S., Beckers, K.F., Taam, M., Horowitz, F. et al. (2021). Techno-economic analysis of decarbonizing building heating in Upstate New York using seasonal borehole thermal energy storage. *Energy and Buildings* 241: 110890.

Eavor (2023). Eavor Technologies. https://www.eavor.com/

EIA (2022). Levelized costs of new generation resources in the Annual Energy Outlook 2022. US Energy Information Administration (EIA).

Gallup, D.L. (2009). Production engineering in geothermal technology: a review. *Geothermics* 38 (3): 326–334.

Gerber, L. (2014). *Designing Renewable Energy Systems: A Life Cycle Assessment Approach*. Boca Raton, FL: CRC Press.

Holm, T.R., Lu, X. and Larson, D.R. (2015). *Feasibility of Groundwater Source Heat Pumps for Space Heating and Cooling in Mason County and the American Bottoms Area, Illinois*. Champaign, IL: Illinois Sustainable Technology Center.

IEA (2022a). Installation of about 600 million heat pumps covering 20% of buildings heating needs required by 2030. International Energy Agency (IEA). https://www.iea.org/reports/installation-of-about-600-million-heat-pumps-covering-20-of-buildings-heating-needs-required-by-2030.

IEA (2022b). World Energy Outlook 2022. International Energy Agency (IEA). https://www.iea.org/reports/world-energy-outlook-2022/.

IEA (2022c). The future of heat pumps. International Energy Agency (IEA). https://www.iea.org/reports/the-future-of-heat-pumps

IEA (2011). Technology roadmap. Geothermal heat and power. International Energy Agency (IEA). https://iea.blob.core.windows.net/assets/f108d75f-302d-42ca-9542-458eea569f5d/Geothermal_Roadmap.pdf

IHA (2021). 2021 hydropower status report. Sector trends and insights. International Hydropower Association (IHA). https://www.hydropower.org/publications/2021-hydropower-status-report

IRENA (2022). Renewable power generation costs in 2021. International Renewable Energy Agency (IRENA), Abu Dhabi.

IRENA (2017). Geothermal power. Technology brief. International Renewable Energy Agency (IRENA), Abu Dhabi.

Lee, I., Tester, J.W. and You, F. (2019). Systems analysis, design, and optimization of geothermal energy systems for power production and polygeneration: state-of-the-art and future challenges. *Renewable and Sustainable Energy Reviews* 109: 551–577.

Lu, S.M. (2018). A global review of enhanced geothermal system (EGS). *Renewable and Sustainable Energy Reviews* 81: 2902–2921.

Lund, J.W. and Toth, A.N. (2021). Direct utilization of geothermal energy 2020 worldwide review. *Geothermics* 90: 101915.

Lund, J.W., Bjelm, L., Bloomquist, G. and Mortensen, A.K. (2008). Characteristics, development and utilization of geothermal resources – a Nordic perspective. *Episodes Journal of International Geoscience* 31 (1): 140–147.

Narasimhan, T.N. and Goyal, K.P. (1984). Subsidence due to geothermal fluid withdrawal. Man-induced land subsidence. *Reviews in Engineering Geology* 6: 35–36.

NOAA (2020). Ocean floor features. National Oceanic and Atmospheric Administration (NOAA). https://www.noaa.gov/education/resource-collections/ocean-coasts/ocean-floor-features

NRC (2002). Commercial earth systems. A buyer's guide. Natural Resources Canada (NRC). https://publications.gc.ca/collections/Collection/M92-251-2002E.pdf.

NREL (2021). 2021 US geothermal power production and district heating market report. US National Renewable Energy Laboratory. https://www.nrel.gov/docs/fy21osti/78291.pdf

Olasolo, P., Juárez, M.C., Morales, M.P. and Liarte, I.A. (2016). Enhanced geothermal systems (EGS); a review. *Renewable and Sustainable Energy Reviews* 30. 133–144.

Pollack, H.N., Hurter, S.J. and Johnson, J.R. (1993). Heat flow from the earth's interior: analysis of the global data set. *Reviews of Geophysics* 31 (3): 267–280.

Pörtner, H.O., Roberts, D.C., Adams, H., Adler, C. et al. (2022). Climate change 2022: Impacts, adaptation and vulnerability. IPCC Sixth Assessment Report.

Ren, F., Wang, J., Zhu, S. and Chen, Y. (2019). Multi-objective optimization of combined cooling, heating and power system integrated with solar and geothermal energies. *Energy Conversion and Management* 197: 111866.

Robins, J.C., Kolker, A., Flores-Espino, F., et al. (2021). 2021 US geothermal power production and district heating market report (No. NREL/TP-5700-78291). National Renewable Energy Laboratory (NREL), Golden, CO, USA.https://www.nrel.gov/docs/fy21osti/78291.pdf

Rosberg, J.E. and Erlström, M. (2021). Evaluation of deep geothermal exploration drillings in the crystalline basement of the Fennoscandian Shield Border Zone in south Sweden. *Geothermal Energy* 9 (1): 1–25.

Schibuola, L. and Tambani, C. (2022). Environmental impact and energy performance of groundwater heat pumps in urban retrofit. *Energy and Buildings* 261: 111964.

Soltani, M., Kashkooli, F.M., Souri, M., Rafiei, B. et al. (2021). Environmental, economic, and social impacts of geothermal energy systems. *Renewable and Sustainable Energy Reviews* 140: 110750.

Sørensen, B. (2017). *Renewable Energy: Physics, Engineering, Environmental Impacts, Economics and Planning*. Cambridge, MA: Academic Press.

Stumpf, A.J. (2023). Personal communication with Dr Andrew Stumpf, Prairie Research Institute, University of Illinois at Urbana-Champaign.

Stumpf, A.J., Lin, Y-F., and Stark, T.D. (2021a). Subsurface characterization, monitoring, and modeling of a geothermal exchange borefield for the Campus Instructional Facility at the University of Illinois at Urbana-Champaign. Illinois State Geological Survey, Circular 606.

Stumpf, A.J., Keefer, D.A. and Turner, A.K. (2021b). Overview and history of 3-D modeling approaches. In A.K. Turner, H. Kessler M.J. van der Meulen (eds.), *Applied Multidimensional Geological Modeling: Informing Sustainable Human Interactions with the Shallow Subsurface*, pp.93–112. New York: Wiley Online Books.

Swissinfo (2022). Can deep geothermal projects help secure Swiss energy independence? https://www.swissinfo.ch/eng/sci-tech/can-deep-geothermal-projects-help-secure-swiss-energy-independence-/47606424

Tester, J.W., Beckers, K.F., Hawkins, A.J. and Lukawski, M.Z. (2021). The evolving role of geothermal energy for decarbonizing the United States. *Energy and Environmental Science* 14 (12): 6211–6241.

Tester, J.W., Drake, E.M., Driscoll, M.J., Golay, M.W. et al. (2012). *Sustainable Energy: Choosing Among Options*. Cambridge, MA: MIT Press.

Tester, J.W., Anderson, B.J., Batchelor, A.S., Blackwell, D.D. et al. (2006). *The Future of Geothermal Energy*. Boston, MA: Massachusetts Institute of Technology.

Vimmerstedt, L., Akar, S., Mirletz, B., Sekar, A. et al. (2022). Annual technology baseline: The 2022 electricity update (No. NREL/PR-6A20-83064). National Renewable Energy Laboratory (NREL), Golden, CO, USA.

Williams, C.F., Reed, M.J., Mariner, R.H., DeAngelo, J. et al. (2008). Assessment of moderate- and high-temperature geothermal resources of the United States. US geological survey fact sheet 2008–3082. US Geological Survey, Washington, DC, USA.

Zinsalo, J.M., Lamarche, L. and Raymond, J. (2022). Performance analysis and working fluid selection of an organic Rankine cycle power plant coupled to an enhanced geothermal system. *Energy* 245: 123259.

27

Storage of Electricity and Heat

How do you "bottle" renewable energy for when the sun does not shine, and the wind does not blow? This is a vexing question for engineers working to develop a greener electrical grid. Massive, utility-scale battery banks are perhaps one answer. However, they are still too expensive and can only store energy for a few hours, not for day-long stretches of cloudy or calm weather that hamper the operation of solar and wind energy conversion systems.

The fast-growing global energy demand calls for an increase in renewables (and nuclear power) to replace fossil fuels, with the aim of reducing the carbon footprint to help address climate change. Renewable energy consumption is expected to be close to the share of liquid fuels, leveling at $\sim$250 quads by 2050 (IEA 2022). However, that scenario will only materialize with effective energy storage to cope with periods of peak demand (of electricity and heat), avoiding the use of costly "peaking" power plants, like gas turbines. Energy storage is also required to deal with unexpected capacity losses (of electricity or heat), which is needed to smoothen fluctuations due to variable sources like wind and solar energy.

There are many strategies for storing energy. They are categorized as mechanical, thermal, electrochemical, or electrostatic (Figure 27.1). Chemical energy is stored in fuels such as gasoline or hydrogen. However, that will not be covered here. (Chapter 9 deals with energy storage in chemical fuels such as hydrogen.)

- *Mechanical energy storage* involves the storage of potential energy or kinetic energy, e.g., pumped hydro storage, gravity-based "falling weight" systems, compressed air energy storage (CAES), liquefied air energy storage (LAES), and liquefied CO_2 storage.
- *Thermal energy storage systems* use thermal energy to store and release heat or to generate electricity, e.g., sensible heat storage, latent heat storage, and thermochemical heat storage.
- *Electrochemical energy storage* refers to the use of chemical batteries which store electricity in electrodes or electrolytes, e.g., aqueous flow batteries, metal-anode batteries, and hybrid flow batteries.
- *Electrostatic energy storage* refers to the use of technology such as ultracapacitors (also called supercapacitors).

Of particular importance to our contemporary energy challenges is *long-duration energy storage* (LDES), which excludes electrostatic energy storage systems as it stores energy only for relatively short periods.

Energy Systems: A Project-Based Approach to Sustainability Thinking for Energy Conversion Systems,
First Edition. Leon Liebenberg.
© 2024 John Wiley & Sons, Inc. Published 2024 by John Wiley & Sons, Inc.
Companion website: www.wiley.com/go/liebenberg/energy_systems

Energy Storage Technologies

Mechanical	Thermal	Electrochemical	Electrostatic	Chemical
- Pumped hydro storage - Flywheel - Falling weight - Compressed air energy storage - Liquified air (or CO_2) energy storage	- Sensible heat storage - Latent heat storage - Thermochemical energy storage	- Metal anode batteries (lead-acid; nickel-based, sodium-based; lithium-ion) - Flow batteries (redox; hybrid)	- Ultracapacitors	- Hydrogen (fuel cell)

Figure 27.1 Salient types of long-duration electrical and heat storage systems.

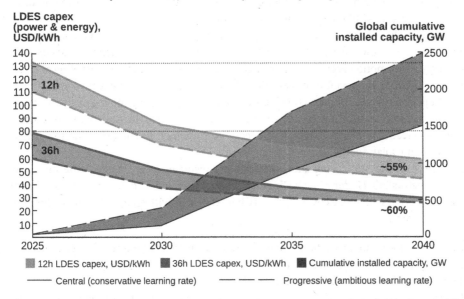

LDES capex evolution versus power capacity additions

Figure 27.2 Expected cost performance of long-duration energy storage by 2040. Costs in US dollars. *Source:* Adapted from McKinsey (2021).

Figure 27.2 illustrates that the cost performance of long-duration energy storage is expected to improve sharply (by around 60% by 2040), boosting capacity deployment.

27.1 Mechanical Energy Storage

27.1.1 Pumped Hydro Storage

Section 25.2 covers pumped hydro storage systems. These storage systems are the most widespread and account for 93% of large-scale energy storage in the United States and 10% of the total

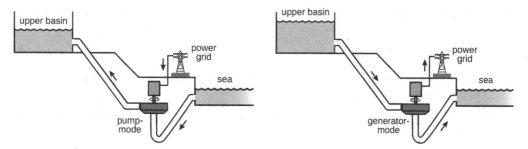

Figure 27.3 Operating principle of a pumped hydro storage facility using seawater. The decommissioned Okinawa Yanbaru pumped storage facility in Japan could store 564,000 m³ of seawater. The entire inner surface of the reservoir was covered with an impermeable liner to prevent seawater from leaking and damaging the surrounding vegetation. With an effective head of 136 m and maximum flow of 26 m³/s, the power plant could produce around $P = \eta \rho g \dot{V} h_{turbine} = (0.9)(1025 \text{ kg/m}^3)(9.81 \text{ m/s}^2)(26 \text{ m}^3/\text{s})(136 \text{ m}) = 32 \text{ MW}_e$ (also see Example 25.2).

hydropower capacity worldwide. When energy demand is low, water is pumped to a higher reservoir; and during peak-demand, the water flows downhill via an electrical generator. These systems' nickname of "water battery" is therefore apt. This mature technology is being revamped with the use of subsurface reservoirs rather than dams. Instead of using fresh water, seawater may be successfully used (Figure 27.3). However, large seawater (or fresh water) storage facilities must contend with revenue uncertainty, uncertainty around energy technology development, the reducing costs of grid-scale battery technology, and the long development approval time frames. In 2020, a 225-MW seawater storage facility was cancelled after years of development, mainly for these reasons (Arup 2020).

The amount of power (P) that can be converted in a pumped hydro storage system can be estimated from

$$P = \eta \rho g \dot{V} h_{turbine} \tag{27.1}$$

where $\dot{V}$ is the volumetric flow rate (m³/s), $h_{turbine}$ is the pressure head (m) above the turbine, ρ is the water density, and η is the mechanical energy conversion efficiency.

27.1.2 Compressed Air Energy Storage

Compressed air energy storage (CAES) uses surplus electricity from the grid or renewable sources to run an air compressor. The compressed air is stored in large underground tanks or caverns until energy is needed, at which point it is released through a gas turbine to generate electricity that is fed back into the grid. Large underground caverns provide impressive energy storage capabilities, but also significant heat losses ($\sim$50%).

We may analyze CAES systems by considering air to be an ideal gas, so that we could use the ideal gas equation of state, $pV = nRT$, where n is the number of moles of air. From Sections 2.4 and 11.1, if air is compressed isothermally ($T_1 = T_2$) from an initial volume (V_1) to a final volume (V_2), the work done is

$$dW = -p\,dV \equiv -(nRT/V)dV \tag{27.2}$$

Integrating Eq. (27.2) between V_1 and V_2, we find the work required to compress air from state point 1 to 2:

$$W = p_1 V_1 \, \ln(V_1/V_2) \tag{27.3}$$

Equation (27.3) also represents the maximum work that could be extracted by expanding the compressed air through a gas turbine.

Using the thermodynamic expressions developed in Section 11.1, the required power input $\dot{W}_{comp}$ of an air compressor is:

$$\dot{W}_{comp} = \dot{m} c_p T_1 \left[\left(\frac{p_2}{p_1} \right)^{\frac{\gamma-1}{\gamma}} - 1 \right] \tag{27.4}$$

where $\dot{m}$ is the mass flow rate of air (kg/s), c_p is the specific heat of air at constant pressure (around 1 kJ/kgK at an air temperature of 300 K), T_1 is the air temperature before compression, p_1 and p_2 are, respectively, the inlet and outlet pressures of the air, and $\gamma = c_p/c_v$.

The CAES systems have typical energy densities in the order of ~ 1 MJ/m^3. One such plant, developed by Hydrostor for the Californian market, provides 8–12 hours of energy storage as opposed to the 1–4 hours that current battery technologies can feasibly provide. This long-duration energy storage is essential for establishing the pathway to California's decarbonized electricity grid.

Rather than venting the heat generated as the air is compressed (as may be deduced from Eq. (7.2)), the planned Hydrostor system for California captures that heat and stores it in a separate thermal storage tank. It then uses that to reheat the air as it is fed into the turbine stage, which increases the efficiency of the system (Figure 27.4). The CAES efficiencies are between 40% and 52% (Tong et al. 2021). However, the new Hydrostor system has a reputed efficiency of 60%. The shown CAES plant also has an expected lifetime of 50 years, accompanied by low lifecycle

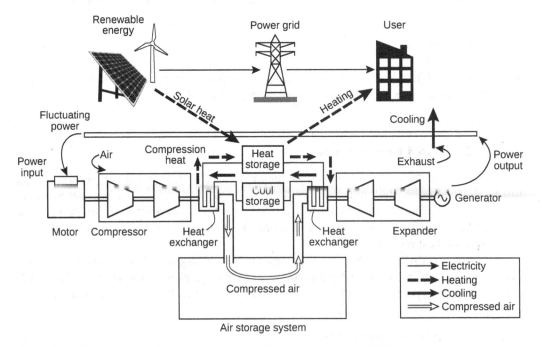

Figure 27.4 Illustration of a large-scale compressed air energy storage (CAES) plant.

emissions. Figure 27.4 illustrates such a large-scale CAES plant, which uses renewable energy (solar and wind power) to run the air compressors.

Example 27.1   Compressed Air Energy Storage System

For the CAES plant outlined in Figure 27.4, what storage volume is required to generate 4 GWh if the air is stored at 7 MPa?

Given: We are given the required energy output of compressed air in a CAES system.

Required: We must find the volume of compressed air that will provide an energy equivalent of 4 GWh.

Solution strategy: We will use Eq. (27.4) to calculate the work that expanding air can do: $W = p_1 \mathcal{V}_1 \ln(\mathcal{V}_1/\mathcal{V}_2)$.

Assumptions: Assume that the atmospheric pressure is 100 kPa and the air temperature is 300 K. We may also assume that air behaves as an ideal gas and that the compression process is isothermal.

Analysis:

Convert 4 GWh to joules: $(4 \times 10^9 \text{ Wh}) \left(\dfrac{1}{1 \text{ Ws}}\right) \left(\dfrac{3600 \text{ s}}{\text{h}}\right) = 14.4 \times 10^{12} \text{ J}$

From Eq. (27.4): $W = p_1 \mathcal{V}_1 \ln(\mathcal{V}_1/\mathcal{V}_2)$

We know from Chapter 3 that for an ideal gas changing isothermally from state 1 to state 2:

$\dfrac{p_1 \mathcal{V}_1}{T_1} = \dfrac{p_2 \mathcal{V}_2}{T_2}$. Therefore, $\dfrac{\mathcal{V}_1}{\mathcal{V}_2} = \dfrac{p_2}{p_1}$, and

$\therefore W = p_1 \mathcal{V}_1 \ln(p_2/p_1) \equiv p_2 \mathcal{V}_2 \ln(p_2/p_1)$

$W \equiv 14.4 \times 10^{12} \text{ J} = p_2 \, \mathcal{V}_2 \ln(p_2/p_1) = (7 \times 10^6 \text{ Pa}) \mathcal{V}_2 \ln(7 \times 10^3 \text{ kPa}/100 \text{ kPa})$

$\therefore \mathcal{V} = 484.2 \times 10^3 \text{ m}^3$

Discussion: If the storage volume were perfectly spherical: $\mathcal{V} \doteq \frac{4}{3}\pi R^3 = 484.2 \times 10^3 \text{ m}^3$, which provides a required container radius of $R = 48.7$ m. To construct such a massive storage vessel from steel would be exorbitantly expensive. It follows that the use of underground caverns or disused underground mines presents unique and possibly more economically feasible ways of storing compressed air. However, the underground cavern would first need to be sealed, and that is also costly.

27.1.3   Flywheels

Flywheels store kinetic energy contained in massive objects spinning at high velocities. The rotational masses are subjected to high centrifugal forces and material stresses, which means that strong and lightweight materials are typically used in their construction, including carbon fiber and plastics. Potentially, carbon nanotube technology could play a decisive role in the future of flywheel energy storage systems.

The kinetic energy (E_{kin}) contained in an object with angular velocity ω (rad/s) and with an inertia of I (kg m^2), is:

$$E_{kin} = \frac{1}{2}I\omega^2 \tag{27.5}$$

A solid disk with mass (m) and radius (r) has the following rotational inertia:

$$I = \frac{1}{2}mr^2 \tag{27.6}$$

Combining Eqs. (27.5) and (27.6), the kinetic energy per unit mass of a uniform disc of radius r, mass m, angular velocity ω is:

$$\frac{E_{kin}}{m} = \frac{1}{4}\omega^2 r^2 \tag{27.7}$$

Modern flywheel energy storage systems have energy densities ranging between 10 and 100 Wh/kg, and power densities ranging between 100 and 10,000 W/kg. It is not uncommon for contemporary flywheel energy storage facilities to have capacities in the order of 20 MW$_e$/80 MWh$_e$, i.e., with a four-hour discharge duration. Further, as may be deduced from Eq. (27.7), less-dense ("lighter") materials have a larger storage capacity.

The maximum angular velocity of these flywheels is determined by the maximum tensile stress of a material. The high velocities of the flywheels generate large frictional (or "windage") losses. To circumvent this, engineers have developed flywheels that spin in vacuums and shafts that are suspended in magnetic bearings. Figure 27.5 shows a modern hybrid energy storage plant, which features flywheels and chemical batteries used for electrical grid "smoothing."

(a)

(b)

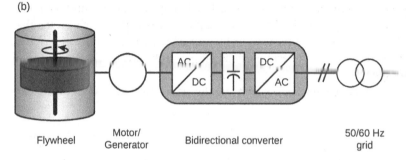

Flywheel   Motor/Generator   Bidirectional converter   50/60 Hz grid

Figure 27.5 (a) A hybrid energy storage pilot plant in Almelo, The Netherlands. The flywheels are in the foreground with the containerized lithium-ion battery storage systems behind them. The hybrid system combines 8.8 MW$_e$/7.12 MWh$_e$ of lithium-ion batteries with six flywheels, adding up to 3 MW$_e$ of power. The electrical energy from this plant is used for frequency-stabilizing primary control power to the transmission grid. *Source:* Leclanché; S4 Energy. (b) The typical energy conversion set-up of a flywheel, which generates alternating current via an electrical generator.

27.2 Thermal Energy Storage

Thermal energy storage systems use thermal energy to store and release heat or to generate electricity. These systems can be classified as either sensible heat storage (where the temperature of a liquid or solid is raised or lowered, e.g., hot or chilled water), latent heat storage (where energy conversion happens during an isothermal phase change, e.g., with ice or paraffin), or thermochemical heat storage (which uses endothermic or exothermic chemical reactions, e.g., the sorption of water).

These technologies use different mediums to store heat, including molten salts, concrete, aluminum alloy, rock material, or an array of heat transfer fluids. The charging equipment could range from electrical resistance heaters to heat engines and high-temperature heat pumps, among many other options.

In buildings, solar heat absorbed by materials during the day provides heating at night when it is emitted. This can be extended to seasonal stores. In the concentrated solar plants covered in Section 23.5, they use molten sodium and potassium nitrate salts with heat capacities of around 1.5–2 kJ/kgK, and densities one-and-a-half to two times that of water (Bonk et al. 2018). There is immense potential in using thermal energy storage at various scales, especially for district, commercial, and industrial applications.

27.2.1 Ice Storage ("Latent Heat Storage")

Latent heat storage is based on heat absorption or release when a phase-change material undergoes a phase change, like water becoming ice, or ice melting. Compared to sensible heat storage (where the temperature of the fluid changes), latent heat storage possesses a greater density of stored energy. This makes an ice storage system ideal for use in buildings air-handling units, for instance (Figure 27.6). Further, thermal storage is an environmentally friendly technology that aids in economically shaping end-use demand.

Many building owners or rate payers can attribute about half of their monthly electrical bills to peak electricity use. Switching a portion of cooling energy consumption to off-peak hours can help building owners reduce both energy costs and overall operating expenditures. Thermal energy storage tanks are a proven way of storing cooling produced at night when electricity costs are lower. *Melting ice* that is stored inside storage tanks provides cooling during periods of peak electrical costs or temporary power outages. The ice can be produced using an advanced vapor compression chiller system (Chapter 14).

27.2.2 Liquefied Gases

Liquefied air or liquefied CO_2 can be stored in well-insulated cryogenic containers. Energy is provided by pressurizing and vaporizing the air (or CO_2). The superheated air (or CO_2) passes through a gas turbine that drives an electrical generator. If waste heat is used to vaporize the liquid air (or CO_2), an overall energy conversion efficiency of 70% can be obtained. Liquefied air energy storage is also referred to as cryogenic energy storage (CES).

From Chapter 2, the maximum heat that is rejected at T_H when work W pumps heat from T_C to T_H, is:

$$Q = W \frac{T_H}{(T_H - T_C)} \tag{27.8}$$

Figure 27.6 The image shows an ice-enhanced air-cooled chiller plant used to cool school buildings. The Thermal Battery™ uses a bank of ice storage units (made by Calmac/Trane) and an advanced vapor-compression chiller (produced by Trane). An ice storage system, which uses 6300 L of water, can produce 600 kWh$_{th}$ of cooling. This technology is commonly called "Power-to-X," where "X" could be thermal storage, hydrogen storage or chemical storage. *Source:* Calmac/Trane.

while the maximum work from a gas turbine with Q flowing in at temperature T_H and with heat rejected at temperature T_C is:

$$W = Q \, \frac{(T_H - T_C)}{T_H} \tag{27.9}$$

The energy density of a liquefied air energy storage system could be as high as 40 kWh/m^3, but these systems are very expensive and have the disadvantage of long payback periods.

27.2.3 Thermochemical Energy Storage

Thermochemical energy storage (TCES) uses reversible gas-solid reactions to store thermal energy in a heat transfer fluid (HTF) in the form of chemical energy. Thermal energy, say, from a concentrated solar power (CSP) plant, is typically stored in a heat transfer fluid comprising hydroxides and carbonates (mainly calcium-based), but also metal oxides and perovskites that allow operating in an open loop with air as both the heat transfer fluid and the gaseous reactant (oxygen) during the heat charge/discharge steps.

TCES plants have a higher energy density than other types of thermal energy storage. These unique systems also have very long-duration storage capability and are able to operate at high temperatures of 400–1000 °C. Box 27.1 explains the rudiments of thermochemical energy storage at a concentrated solar plant.

Box 27.1   Thermochemical Energy Storage at a Concentrated Solar Plant

As a reaction system for thermochemical energy storage, a calcium oxide/calcium hydroxide/water ($CaO/Ca(OH)_2/H_2O$) reaction system has received attention from energy engineers because of the low cost and environmentally friendly reaction materials. The TCES system is used in conjunction with a concentrated solar plant, as illustrated in Figure 27.7.

The gas-solid reaction of the system is:

$$Ca(OH)_2 \ (s) \rightleftharpoons CaO(s) + H_2O(g) \qquad \Delta H = 104 \text{ kJ/mol} \qquad (27.10)$$

The forward endothermic reaction is known as *dehydration* and the reverse exothermic reaction is known as *hydration*.

The heat absorbed during dehydration is stored, with an energy storage density of 1.4 MJ/kg of $Ca(OH)_2$, as long as calcium oxide and water are separated from each other. The heat output occurs during hydration. A TCES system based on the $CaO/Ca(OH)_2/H_2O$ reaction can be utilized at high operating temperatures between 400 and 600 °C, depending on the water vapor pressure.

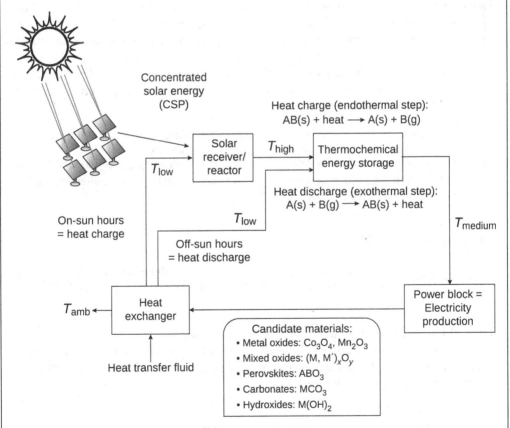

Figure 27.7   During on-sun hours, the heat transfer fluid flows inside the solar receiver and is used to store heat in the TCES system (heat charge, endothermal). The exiting heat transfer fluid is used to run a turbine. During off-sun hours, the heat transfer fluid flows directly through the TCES system for heat recovery (discharge step, exothermal) to increase its temperature and provide heat to the downstream process, thereby enabling continuous operation. *Source:* André and Abanades (2020) / CCBY 4.0 / MDPI / Public domain.

During on-sun hours, the heat transfer fluid flows inside the solar receiver and is used to store heat in the TCES system (known as "heat charging" or the endothermic stage). The exiting heat transfer fluid is used to run a turbine. During off-sun hours, the heat transfer fluid flows directly through the TCES system for heat recovery (known as the "discharge" or exothermic phase) to increase its temperature and provide heat to the downstream process, thereby enabling continuous operation.

Example 27.2 Thermochemical Energy Storage at a University Campus

Imagine a scenario where a university decides to completely stop using fossil fuels. The university's power plant, which produces electricity and heat for the campus, has decided to replace fossil fuels with a calcium oxide (CaO)-based thermochemical energy storage system to produce energy. CaO can react with steam, producing heat just as burning coal and natural gas do. The energy density of CaO is ∼1.8 MJ/kg.

a) Find the rate of reaction (kg/s) needed to produce a power output of 100 MW, which is sufficient to supply the entire campus with heat and electricity. Assume 60% net efficiency.

b) In another scenario, assume there is a heavy snowstorm in the university town and that the supply of natural gas, which is the main fuel source for the university's power plant, is cut off. It would take approximately four days for the natural gas pipeline to be fixed. How much CaO (in tons) would be needed to keep supplying the campus with energy during this period? Assume the same net efficiency of 60%.

Given: We are given the energy density of CaO: 1.8 MJ/kg. Using this stored CaO, 100 MW must be produced in a system with an overall energy conversion efficiency of 60%.

Required: (a) First, we must find the rate of reaction (kg/s) needed to produce 100 MW. (b) Then, for a different scenario, we need to calculate how many tons of CaO will be required to supply the university's energy needs for four days.

Analysis:

a) **Reaction rate for 100 MW**

$$\text{Reaction rate} = \left(\frac{\text{Power output}}{\text{Energy density}}\right)\left(\frac{1}{\text{Net efficiency}}\right) = \left(\frac{100 \text{ MJ/s}}{1.8 \text{ MJ/kg}}\right)\left(\frac{1}{0.6}\right) = 92.6 \text{ kg/s}$$

b) **Mass of CaO required to supply 100 MW for four days**

We should first convert CaO's energy density of 1.8 MJ/kg to its equivalent in MWh/ton, seeing that we are required to find the number of tons of CaO required to produce 100 MW (at 60% efficiency) continuously for four days:

$$\left(\frac{1.8 \text{ MJ}}{\text{kg}}\right)\left(\frac{1 \text{ Ws}}{1 \text{ J}}\right)\left(\frac{1 \text{ h}}{3600 \text{ s}}\right)\left(\frac{1000 \text{ kg}}{1 \text{ ton}}\right) = 0.5 \frac{\text{MWh}}{\text{ton}}$$

$$\text{Total material required} = \frac{\text{Required energy}}{\text{Energy density} \times \text{Net efficiency}}$$

$$= \frac{(100 \text{ MW})(4 \text{ days})(24 \text{ h/day})}{(0.5 \text{ MWh/ton})(0.6)} = 32,000 \text{ tons CaO}$$

Thank you to Dr. Arpit Dwivedi (Cache Energy) for providing this question.

Table 27.1 illustrates the advantages of operating a CSP plant with a TCES system rather than with a sensible heat storage system or a latent heat storage system.

27.2.4 Electrothermal Energy Storage

In 2019, Siemens Gamesa began the operation of its revolutionary hot-rock thermal energy storage system, which has the potential to cost-effectively store gigawatt hours of renewable energy for up to two weeks with minimal losses.

An electrothermal energy storage (ETES) pilot plant in Hamburg, Germany, converts electrical energy obtained from wind turbines into hot air using a resistance heater and a blower to heat about 1000 tonnes of volcanic rock to 750 °C. When required, the plant converts the stored thermal energy back into electricity using a 30 MW_e steam turbine-generator combination. Due to efficient insulation, the heat can be stored for a week or longer – at a fraction of the cost of battery storage. According to Siemens Gamesa, the pilot plant can store up to 130 MWh for a week, which the local utility sells on the market.

Systems such as this may eventually resolve the problem of the variability of wind and solar, and enable near-baseload renewable energy output. The emphasis is on developing cost-effective, efficient, and scalable energy storage systems, such as ETES. This presents an elementary building block for the further expansion of renewable energy and the success of transitioning to no-carbon energy.

Table 27.1 Comparison of a thermochemical energy storage facility with sensible and latent heat storage facilities.

Storage type	Sensible heat storage	Latent heat storage	Thermochemical energy storage
Energy density (kWh/kg)	0.02–0.03	0.05–0.1	0.5–1.0
Volumetric energy density (kWh/m³)	50	100	500
Energy storage period	Limited, due to thermal losses	Limited, due to thermal losses	Theoretically unlimited
Theoretical energy transport (km)	<5 km	<5 km	>100 km
Technological complexity	Simple	Medium	Complex
Cost	Medium	Medium	High

Saving the Planet Through Salient, Clean Alternatives

Following worldwide concern about climate change and global warming, the search is on for a more environmentally conscious, ethical option to replace lithium-ion batteries. These batteries are the current industry standard. However, they contain metals such as cobalt, nickel and manganese, which are toxic and can contaminate water supplies and ecosystems if they leach out of landfills.

Ryan Brown (29) was convinced that alternative battery technology could be developed. In 2017, he started working on developing a zinc-based battery with a PhD student, Brian Adams. Together, they founded the company Salient Energy, through which they are commercializing this invention. "Not only is the zinc-based battery a cleaner option, it is also cheaper and safer than its widely used lithium-ion counterpart," explains Ryan.

Embarking on a venture with such scientific risk requires one to think big, says Ryan. "Innovating in deep tech is driven by the unforgiving nature of financial mathematics: any idea that requires a few years and millions of dollars before it can be deemed feasible needs to have the potential to result in a massive payoff." This requires targeting a huge market opportunity and keeping the big goal in mind. But it also requires fine attention to the details that matter in achieving the opportunity.

Ryan points out that Salient Energy's zinc-ion battery is based on the same production processes as lithium-ion batteries. "This was done to simplify manufacturing and create a product that could be used as a drop-in replacement for lithium-ion batteries," he explains.

When the team started developing the zinc-based alternative, the consensus view in industry was that lithium-ion was on an unstoppable growth curve and that the only viable applications for new batteries were in areas where lithium-ion could not compete. "Instead of accepting this to be true, my curiosity drove me to investigate the drivers of lithium-ion cost reduction. I noticed that manufacturing costs were falling rapidly, but material costs were essentially flat." This inspired the belief that they could make a battery that paired low material costs with standard lithium-ion manufacturing to achieve a durable cost advantage over lithium-ion.

This was, however, not without its challenges. "Developing a new battery technology is a long and expensive process that is wholly dependent on third-party financing," explains Ryan. This means that he needs to identify opportunities to bolster the company's credibility or communicate its progress in a way that is easy for a non-expert audience to understand. He is clearly doing a great job, as Salient Energy has already raised US $10 million to develop its technology.

"The company's goal is to keep the clean energy transition moving at the pace that Mother Nature demands by providing a scalable and sustainable alternative to lithium-ion batteries for energy storage." Its ambition is to have a zinc-ion battery in-market by 2025 to replace lithium-ion as the industry standard before the end of the decade.

Ryan Brown, co-founder of Salient Energy, https://salientenergy.ca/

27.3 Electrochemical Energy Storage

27.3.1 Lead-Acid Battery

The first practicable rechargeable battery was the lead-acid battery, which is relatively simple to produce. Lead-acid batteries are used in a variety of electrical power applications, with "sizes" ranging from 1 to 1000 Ah. Lead-acid batteries cover a range of power sizes, from a few kilowatts to tens of megawatts. These batteries have moderate energy conversion efficiencies of around 70%, last around 10 years, and are relatively cheap. A typical lead-acid battery can be recharged about 300 times. In the United States and Europe, about 95% of lead-acid batteries are recycled.

Lead-acid batteries typically comprise a metallic-lead anode and lead-oxide cathode in a sulfuric acid electrolyte. The electrolyte physically segregates the two electrodes from direct electron transfer while allowing ions to transport both charges and masses across the cell. The open-circuit potential (V^0) of this battery is 2.04 V (0.356 V at the anode, and 1.685 V at the cathode). Its energy density is 20–30 Wh_e/kg and its power density is approximately 150 W_e/kg.

The chemical reactions of a lead-acid battery may be written as follows:

$$\text{Anode:} \qquad Pb + H_2SO_4 \rightarrow PbSO_4 + 2H^+ + 2e^- \tag{27.11}$$

$$\text{Cathode:} \qquad PbO_2 + 2H^+ + 2e^- + H_2SO_4 \rightarrow PbSO_4 + 2H_2O \tag{27.12}$$

$$\text{Net reaction:} \qquad PbO_2 + Pb + 2H_2SO_4 \rightarrow 2PbSO_4 + 2H_2O \; \Delta E^0 = 2.04\,V \tag{27.13}$$

In Eq. (27.13), the reaction from left to right is for discharging, and from right to left is for charging the battery, while ΔE^0 depicts a cell's open-circuit voltage. Refer to Table 27.2 for the standard potentials of a few half-reactions commonly encountered with chemical batteries.

Engineers are revisiting the grid storage applications of the lead-acid battery that is more than a century old. Lead-acid batteries have a recycling rate of nearly 95%, compared to the dismal rates for recycling other batteries. Further, today's lead-acid cells discharge only 20–30% of their theoretical potential. Research on basic material properties is addressing how much more of that capacity could be used.

Table 27.2 Half-reactions of a few commonly used materials and their standard potentials.

Half-reaction	Standard potential, E^0 (V)
$Li^+ + e^- \rightarrow Li$	−3.04
$K^+ + e^- \rightarrow K$	−2.93
$Ca^{2+} + 2e^- \rightarrow Ca$	−2.87
$Na^+ + e^- \rightarrow Na$	−2.71
$Mg^{2+} + 2e^- \rightarrow Mg$	−2.37
$Zn^{2+} + 2e^- \rightarrow Zn$	−0.76
$Co^{2+} + 2e^- \rightarrow Co$	−0.27
$V^{3+} + e^- \rightarrow V^{2+}$	−0.26
$2H^+ + 2e^- \rightarrow H_2$	0.00
$Cu^{2+} + 2e^- \rightarrow Cu$	+0.34
$Ag^+ + e^- \rightarrow Ag$	+0.80
$2H^+ + VO_2^+ + e^- \rightarrow VO^{2+} + H_2O$	+1.00
$O_2 + 4H^+ + 4e^- \rightarrow 2H_2O$	+1.23
$Cl_2 + 2e^- \rightarrow 2Cl^-$	+1.36
$F_2 + 2e^- \rightarrow 2F^-$	+2.87

27.3.2 Lithium-Ion Battery

The rechargeable lithium-ion battery (LIB) is deemed to be an ideal power source for portable electronics, military and medical applications, power tools, electric vehicles, and grid-scale energy storage. Lithium-ion batteries have high energy densities of around 100–300 Wh_e/kg. Costs have fallen, and lithium-ion batteries are now being used for load-balancing on the grid. Capacities of 100 MW_e/400 MWh_e are not uncommon, such as the Tesla Megapack project in California's Ventura County. Modern lithium-ion batteries that are used in electric vehicles can sustain around 1500 charge-discharge cycles, about five times more than lead-acid batteries.

A disadvantage of lithium-ion batteries is that lithium is chemically very reactive. This was overcome in 1980 through *intercalation* – a reversible process that moves lithium ions into or out of graphite without damaging it. The development of electric vehicles is increasing the demand for lithium-ion batteries, and costs are falling through "learning." Ongoing research and development is seeking to improve safety and decrease the charging time of these batteries to less than 10 minutes for electric vehicles. Research is also active pertaining to recycling of lithium-ion batteries (see Box 27.2).

Lithium-ion batteries usually use carbon-based materials at the anodes, due to their low cost and high reversibility during charge and discharge cycling (Figure 27.8). The cathodes of these batteries usually feature a lithiated metal oxide such as lithium cobalt oxide ($LiCoO_2$), lithium manganese oxide ($LiMn_2O_4$), lithium nickel manganese cobalt oxide ($LiNiMnCoO_2$), lithium iron phosphate ($LiFePO_4$), or lithium titanate ($Li_4Ti_5O_{12}$). The electrolyte in lithium-ion batteries is usually a mixture of a lithium salt such as lithium hexafluorophosphate ($LiPF_6$), which is dissolved in an organic carbonate solvent. The Li^+ migrates to the anode during charging and to the cathode during discharging through intercalation.

The electrolyte physically segregates the two electrodes from direct electron transfer, while allowing ions to transport both charges and masses across the cell. With lithium-polymer batteries, the liquid electrolyte in a lithium-ion battery is replaced with a solid electrolyte, which improves the

---

Box 27.2 Recycling of Lithium-Ion Batteries

The global electric vehicle (EV) stock grew to 10 million in 2020, and 160 GWh_e lithium-ion batteries were produced to power these electric cars. With deeper EV penetration, the global lithium demand has reached a new record (345,000 metric tons of lithium carbonate equivalent in 2020). There could soon be serious shortages of lithium, often labeled "white gold."

A re-examination of recycling strategies is crucial. Recycling also presents an opportunity for batteries to reduce socio-economical risks in relation to non-domestic supply chains in different countries (Bauer et al. 2022). Better recycling methods will not only prevent toxic pollution, but will also help governments boost their economic and national security by increasing supplies of key battery metals (such as lithium, cobalt, and nickel) that are controlled by one or two nations (Morse 2021; Burton et al. 2022).

Looking at the batteries' lifecycle, challenges will emerge because the raw materials that are required for the LIB chemistry are scarce and rarely sourced sustainably. Further, present battery manufacturing and end-of-life disposal practices are far from sustainable. Clearly, it is essential to design, manufacture, use, dispose of, and recycle LIBs in a sustainable way.

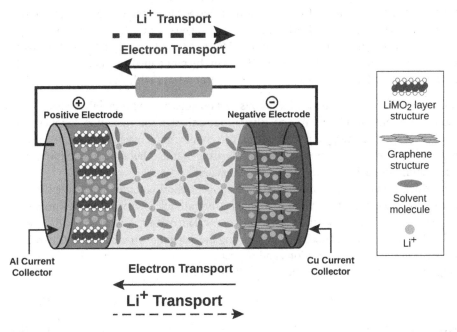

Figure 27.8 Schematic of a lithium-ion battery cell. The cathode comprises an aluminum-based collector while the anode comprises a copper-based collector. *Source:* Adapted from MIT Energy Initiative (2022).

battery's safety and makes it lighter. As the polymer itself is extremely thin, it also enables greater flexibility in terms of shape and design – it need not be contained in a rigid case and can be made to be extremely compact.

The electrochemistry of a lithium-ion battery may be simplified as follows:

$$\text{Anode:} \qquad LiC_6 \rightleftharpoons C_6 + Li^+ + e^- \tag{27.14}$$

$$\text{Cathode:} \qquad CoO_2 + Li^+ + e^- \rightleftharpoons LiCoO_2 \tag{27.15}$$

$$\text{Net reaction:} \qquad LiC_6 + CoO_2 \rightleftharpoons C_6 + LiCoO_2 \qquad \Delta E^0 = 3.6\,V \tag{27.16}$$

In Eq. (27.16), the reaction from left to right is for discharging, and from right to left is for charging the battery.

27.3.3 Lithium-Air Batteries

Lithium-air batteries (LABs) have the highest energy density of commercial rechargeable batteries. Lithium-air batteries have a specific theoretical energy density of 11.5 kWh/kg, although in practice they "only" attain around 5.2 kWh/kg. It is not uncommon for lithium-air batteries to be able to sustain around 700 charge-discharge cycles, nearly double that of lead-acid batteries, but half that of lithium-ion batteries.

There are two types of lithium-air batteries, one based on aqueous electrolytes and the other using nonaqueous electrolytes (Liu et al. 2020):

- *Nonaqueous lithium-air batteries* operate under acidic conditions and have varied specific theoretical energies, depending on the type of lithium-oxygen product formed during discharge. For instance, the formation of Li_2O_2 ($2Li + O_2 \rightleftharpoons Li_2O_2$), Li_2O ($4Li + O_2 \rightleftharpoons 2Li_2O$), and LiOH ($4Li + 2H_2O + O_2 \rightleftharpoons 4LiOH$) leads to specific energies of 3495, 5361, and 3350 Wh_e/kg, respectively. These are several times higher than those of typical lithium-ion batteries (100–300 Wh_e/kg).
- *Aqueous lithium-air batteries* operate under alkaline conditions and typically have the following reaction: $4Li + 6H_2O + O_2 \rightleftharpoons 4LiOH \cdot H_2O$, the forward reaction characterizing discharge and the backward one characterizing charge. During discharge, the generated LiOH is dissolved in the electrolyte until it reaches its saturation solubility; $LiOH \cdot H_2O$ then precipitates and deposits on the cathode. Since $LiOH \cdot H_2O$, rather than LiOH (as in the non-aqueous system), is the product, the specific energy of the aqueous system is lower, around 2170 Wh_e/kg.

A lithium-air battery's cathode is made of a porous carbon substrate with metal catalysts (e.g., manganese, cobalt, ruthenium, platinum, and silver) that enhance the reduction kinetics and increase the cathode's specific energy density. During battery charging, oxygen donates electrons to the lithium via reduction at the cathode. Atmospheric oxygen must therefore be present at the cathode, but contaminants such as water vapor can damage it.

At the anode, which is made of lithium metal, the electrochemical potential forces the lithium metal to release electrons via oxidation (without involving the cathodic oxygen). The lithium metal reacts with atmospheric oxygen during the discharge process. The oxygen is reduced by Li^+ to create lithium oxide (Li_2O) or lithium peroxide (Li_2O_2) during the discharge process. During charging, the lithium oxides decompose to lithium ions and oxygen.

The following chemical reactions are typical of a lithium-air battery:

$$\text{Anode:} \qquad Li \rightleftharpoons Li^+ + e^- \tag{27.17}$$

$$\text{Cathode:} \qquad Li^+ + e^- + O_2 \rightleftharpoons LiO_2 \tag{27.18}$$
$$Li^+ + e^- + LiO_2 \rightleftharpoons Li_2O_2$$

$$\text{Net reaction:} \qquad 2Li + O_2 \rightleftharpoons Li_2O_2 \qquad \Delta E^0 = 2.96\,V \tag{27.19}$$

27.3.4 Flow Batteries

Unlike other chemical batteries, flow batteries (or "redox flow batteries") store energy in electrolytes rather than in the electrodes (Figure 27.9), so capacity is only limited by the volume of the electrolytes. Flow batteries are characterized by their high efficiencies and many discharge cycles, typically on the order of 20,000 charge-discharge cycles. However, these batteries have lower energy densities than metal-ion batteries.

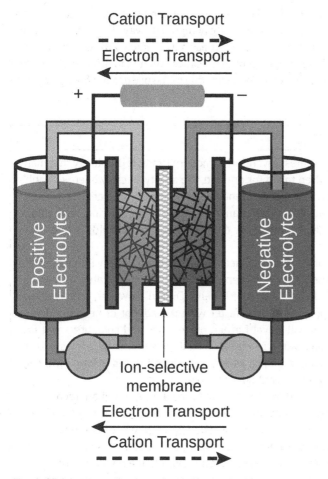

Cation Transport

Electron Transport

$+$ $-$

Positive Electrolyte

Negative Electrolyte

Ion-selective membrane

Electron Transport

Cation Transport

Figure 27.9 Schematic of a vanadium redox flow battery. *Source:* Adapted from MIT Energy Initiative (2022).

The most common type of flow battery is the *vanadium redox flow battery*. However, vanadium is exceedingly expensive, so cheaper materials are being researched. Vanadium redox batteries store energy by employing vanadium redox couples (V^{2+}/V^{3+} in the negative and V^{4+}/V^{5+} in the positive half-cells). These active chemical species are always fully dissolved in sulfuric acid electrolyte solutions (Zhang et al. 2022) (see Figure 27.9).

Flow batteries are alternatives for grid-scale energy storage, offering the promising characteristics of high scalability, design flexibility, and decoupled energy and power. "Decoupling" refers to the ability to have independent control of energy storage and power output by scaling up the electrolyte storage tanks and the electrodes, respectively. This is difficult to implement in rechargeable metal-ion batteries. Flow batteries are therefore attracting renewed interest with the significant advances that are being made in the relevant materials chemistry, performance metrics and characterization.

However, the widespread application of flow batteries is restricted in their utilization as they make use of scarce and expensive materials such as vanadium and are accompanied by environmental and health concerns. In particular, the production of vanadium-redox flow batteries has a high impact on global warming potential (GWP), ozone depletion potential (ODP), the emission of particulates, acidification potential, and cumulative energy demand. This contribution is almost exclusively driven by emissions associated with the production of vanadium pentoxide, which is used in battery electrolyte. Selecting more environmentally benign pathways to produce the same electrolyte can significantly decrease the environmental impacts associated with this technology. Depending on the type of flow battery, electrolytes have both cancer-related toxicity (such as when using a bromine electrolyte) and non-cancer-related toxicity (due to the use of glass fiber-reinforced polyester resins) (CEC 2021).

The Future of Batteries

If you were to ask anyone about the greatest invention of modern times, it is unlikely that many would say the lithium-ion battery. Often taken for granted, the ability to carry energy in one's pocket has been the foundation upon which our current lifestyle is built. These are the thoughts of BS (Mechanical Engineering and Nuclear, Plasma, and Radiological Engineering) sophomore, **Dimitri Kalinichenko** (20). Dimitri is upbeat about our energy future, and he holds strong views on the subject.

Dimitri believes that the success of our future society will depend on battery innovation. He admits that advances in battery technology are currently limited by issues related to flammability and the environmentally friendly extraction of lithium. Dimitri, however, enthuses that "batteries will only get safer, cheaper, and charge faster with the advent of new aluminum-ion technologies. As a result, our cars will travel further, our phones and computers will last longer on the go, our homes will remain warm in winter and cold in summer, regardless of grid failures, lights will remain on during the strongest winds, and our connected lifestyle will continue when the sun doesn't shine, and the wind doesn't blow."

This seems appealing. We will no longer have to conform to the limitations of our technology, waiting for hours for our cars to charge, closing our laptops because they have run out of battery energy, or being tied to outlets for our electrical needs. Such is the power of energy storage.

"The energy transition has long been presented as sacrificing on our current lifestyle," Dimitri says. "Nothing could be further from the truth; we stand on the brink of a revolution of what is possible. You can't carry oil around to burn when you need energy; you can carry a battery around."

Dimitri Kalinichenko is a sophomore doing a dual degree in Mechanical Engineering, and Nuclear, Plasma and Radiological Engineering at the University of Illinois at Urbana-Champaign.
Source: Dimitri Kalinichenko.

27.3.5 Standard Electrochemical Potentials

Table 27.2 shows the standard electrochemical potential of a few salient materials used in contemporary batteries. All reduction potentials are measured against that of hydrogen, i.e., $2H^+ + 2e^- \rightarrow H_2$, which is assigned a value of $E^0 = 0$ V. Some half-reactions oxidize H_2 (e.g., $Cu^{2+} + 2e^- \rightarrow Cu$), while others (e.g., $Fe^{2+} + 2e^- \rightarrow Fe$) reduce H^+. Any compound that reduces H^+ has a negative reduction potential.

Box 27.3 The Lithium-Ion Battery Pack Used in a Tesla Electric Vehicle

Electric vehicles like the Tesla Model 3 (Figure 27.10a) can recharge their lithium-ion batteries (Figure 27.10b) to 80% of full capacity at quick-charging stations within 30 min. The lithium-ion battery pack used in the Tesla Model 3 electric vehicle has a maximum energy capacity of 74 kWh$_e$ (Figure 27.10b), which provides the vehicle with a range of around 570 km. The battery pack consists of 4416 cells, with each cell producing a nominal voltage of around 3.5 V (see Eq. (27.16) and Table 27.1). The 4416 cells are connected in groups of 46 cells per "brick," with the same brick distribution in the four modules. That configuration delivers peak power of 258 kW$_e$ at 355 V-DC.

(a) (b)

Figure 27.10 (a) A 2023 Tesla Model 3 electric vehicle, and (b) the 74-kWh$_e$ Li-ion battery pack for a 2021 Tesla Model 3 EV. *Source:* Tesla.

Reimagine Our Future

The transition to sustainable energy requires a reform in global politics and economics, among others. Lewis Mumford suggested a long time ago that the political economy can be altered not by a massive uprising or movement, but rather by "a steady withdrawal of interest." This suggests that we would find what previously captivated our imagination was now neither interesting nor important. Imagine discovering that gasoline was no longer interesting or important because you now had an electric vehicle with batteries charged by photovoltaic panels? Can you imagine not even needing your own electric vehicle as you used your bicycle or electrified communal transport to commute to work?

27.4 Electrostatic Energy Storage

An *ultracapacitor*, sometimes called a supercapacitor, stores energy electrostatically rather than electrochemically. Energy storage is achieved without a chemical reaction in the form of an electric field between two electrodes. The main difference between ultracapacitors and conventional capacitors is that ultracapacitors have a very high energy density as they have a larger electrode surface area, coupled with a much thinner electrical layer between the electrode and the electrolyte. Ultracapacitors feature an electrolyte ionic conductor instead of an insulating material (Yu et al. 2013).

Ultracapacitors have high capacitances (in the order of kilofarads) compared to capacitors used in conventional electronics that operate in the micro- or picofarad range. Ultracapacitors bridge the gap between electrolytic capacitors and rechargeable batteries. They have high-power densities (500–5000 W/kg), can accept and deliver a charge much faster than rechargeable batteries (e.g., within 5 ms), tolerate many charge and discharge cycles (around 500,000 charge–discharge cycles are common), and have a high energy conversion efficiency (85–98%). Ultracapacitors are deemed to be environmentally friendly, as no heat or hazardous substances are released during their discharge. They also typically last for 20 years. However, ultracapacitors have a low energy density (0.1–5 Wh/kg), high self-discharge rates (typically 14% per month) and come at a high cost (US $300–2000/kWh).

Although ultracapacitors can only store around 10% of the energy of chemical batteries, they can charge (and discharge) much quicker than lithium-ion batteries, for instance. This makes ultracapacitors ineffective as a general energy storage medium for passenger vehicles, for example. However, as ultracapacitors can charge much faster than metal-ion batteries, they are viable for use in vehicles such as buses that must stop frequently at known points (see Figure 27.11).

Ultracapacitors are therefore used in applications that require many rapid charge–discharge cycles, rather than long-term compact energy storage – in automobiles, buses, trains, cranes and elevators, where they are used for regenerative braking, short-term energy storage, or burst-mode power delivery (Grbovic 2013).

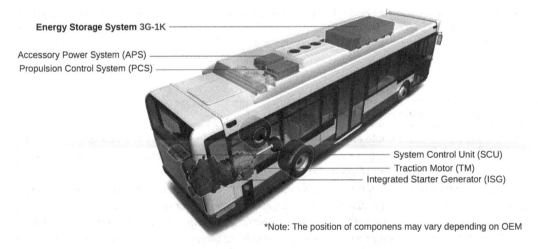

Figure 27.11 Image of a citybus with electrical propulsion featuring an ultracapacitor ("Energy Storage System 3G-1K") mounted on its roof. The electrostatic double-layer capacitor can deliver 200 kW and 1 kWh at around 700 V. The ultracapacitor consists of many cells in series and in parallel, much as is the case for photovoltaic arrays and chemical batteries. *Source:* BAE Systems.

Unlike ordinary capacitors, ultracapacitors do not use a conventional solid dielectric. Rather, they use electrostatic double-layer capacitance and electrochemical pseudocapacitance, both of which contribute to the total capacitance of the capacitor (Zhu et al. 2020). Electrostatic double-layer capacitors (EDLCs) use carbon electrodes or derivatives with much higher electrostatic double-layer capacitance than electrochemical pseudocapacitance. The separation of charge is around 0.3–0.8 nm, much smaller than in a conventional capacitor.

Hybrid capacitors, such as the lithium-ion capacitor, use electrodes with differing characteristics: one exhibiting mostly electrostatic capacitance and the other exhibiting mostly electrochemical capacitance.

When a discharged capacitor is connected to a direct-current voltage source, it is charged. The capacitor voltage rises over time from zero as it tends toward the charging voltage. A charging current flows, which assumes a peak value at the start and decreases to zero with time. The greater the electrical resistance and the greater the capacitance (C), the longer the charging process takes. As an ultracapacitor is charged or discharged, its voltage (V) and current (I) will vary with time. Therefore, the capacitor's power is also time-dependent.

The electrostatic energy stored in an ultracapacitor cell is given by:

$$W = \frac{1}{2}CV^2 \qquad [\text{J}] \tag{27.20}$$

The power that the ultracapacitor releases is:

$$P = VI \qquad [\text{W}] \tag{27.21}$$

When capacitors are *connected in parallel* (Figure 27.12a), their equivalent capacitances can be calculated as follows:

$$\Delta V_1 = \Delta V_2 = \Delta V$$
$$Q_{tot} = Q_1 + Q_2 \equiv C_{eq}\,\Delta V = C_1\Delta V_1 + C_2\Delta V_2 \tag{27.22}$$
$$\therefore C_{eq} = C_1 + C_2$$

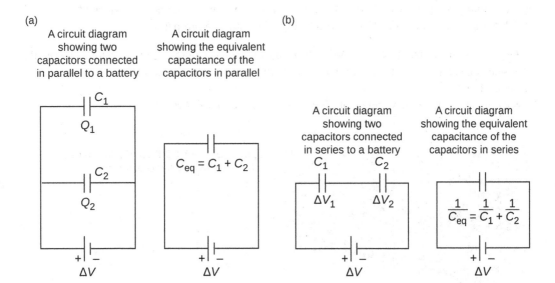

(a) A circuit diagram showing two capacitors connected in parallel to a battery   A circuit diagram showing the equivalent capacitance of the capacitors in parallel

(b) A circuit diagram showing two capacitors connected in series to a battery   A circuit diagram showing the equivalent capacitance of the capacitors in series

Figure 27.12 (a) Two capacitors connected in parallel, and (b) in series.

For n number of capacitors in parallel, we have:

$$C_{eq} = C_1 + C_2 + \dots + C_n \qquad \text{(parallel connection)} \tag{27.23}$$

When capacitors are *connected in series* (Figure 27.12b), their equivalent capacitances can be calculated as follows:

$$Q_1 = Q_2 = Q$$
$$\Delta V_{tot} = \Delta V_1 + \Delta V_2 \equiv \frac{Q}{C_{eq}} = \frac{Q}{C_1} + \frac{Q}{C_2} \tag{27.24}$$
$$\therefore \frac{1}{C_{eq}} = \frac{1}{C_1} + \frac{1}{C_2}$$

For n number of capacitors in series, we have

$$\frac{1}{C_{eq}} = \frac{1}{C_1} + \frac{1}{C_2} + \dots + \frac{1}{C_n} \qquad \text{(series connection)} \tag{27.25}$$

Example 27.3 Ultracapacitor Used in a Citybus

An electric citybus is equipped with a lithium-ion battery pack and an ultracapacitor module. When braking, the electric motors change over to generating mode and send current to the ultracapacitor and the batteries (this is called "regenerative braking"). The ultracapacitor is used to supply energy when rapid acceleration is required and to store energy when rapid deceleration occurs.

The ultracapacitor bank comprises several Maxwell BoostCap 3000 cells (Figure 27.13). Each of the cells is rated at 2.7 V, 3000 F, and 3 Wh_e. The ultracapacitor bank is required to produce 80 kW for 30 seconds at 350 V.

a) Determine the total energy requirement (in joules) to cope with this demand. Disregard energy losses.
b) Determine the total capacitance required.
c) How many Maxwell BoostCap 3000 cells must be connected in series to produce 350 V?
d) How many Maxwell BoostCap 3000 cells must be connected in parallel?

Required: Using a 3000-F, 2.7-V, 3-Wh_e ultracapacitor cell, we must calculate the total energy, W, the total capacitance, C, and the number of cells that must be connected in series and parallel to produce 350 V.

Solution strategy: We have adequate information to use Eqs. (27.20) to (27.25).

Figure 27.13 A 3000-F, 2.7-V, 3-Wh_e ultracapacitor cell. *Source:* Maxwell Technologies.

Analysis:

a) **Total energy requirement**

$$W = Pt = (80 \times 10^3 \text{ J/s})(30 \text{ s}) = 2400 \text{ kJ}$$

b) **Total required capacitance**

$$W = 2400 \times 10^3 \text{ J} \equiv \frac{1}{2}CV^2 = \frac{1}{2}C(350 \text{ V})^2$$

$$\therefore C = \frac{2(2400 \times 10^3 \text{ J})}{(350 \text{ V})^2} \equiv 39 \text{ F}$$

c) **Number of cells in series**

The maximum cell potential is 2.7 V and the maximum system voltage is 350 V, so we need $\frac{350 \text{ V}}{2.7\text{V/cell}} = 130$ cells connected in series.

Note: These 130 capacitors in series have an effective capacitance of:

$$\frac{1}{C'} = \frac{1}{C_1} + \frac{1}{C_2} + \dots + \frac{1}{C_{130}} = \frac{130 \text{ F}}{3000 \text{ F}} = 0.043$$
$$\therefore C' = 23.3 \text{ F}$$

d) **Number of cells in parallel**

The cell-level capacitance is $C_{system} \times$ (number of cells in series) = (39 F)(130 cells) = 5070 F

Therefore, the number of cells that must be connected in parallel is $\frac{5070 \text{ F}}{3000 \text{ F/cell}} = 1.7$.

Note: Alternatively, one could state (from b) that the total capacitance should be 39 F. In a parallel circuit, capacitances are additive. Therefore, the required number of groups would be $\frac{39 \text{ F}}{23 \text{ F}} = 1.7$.

We need to round up 1.7 to a whole number, 2. Two 3000-F modules must therefore be connected in parallel, with each module comprising 130 cells connected in series. A total of $2 \times 130 = 260$ Maxwell BoostCap 3000 cells will be required to build up this ultracapacitor.

27.5 Comparison of Energy Storage Technologies

Table 27.3 compares a few salient energy storage technologies. Figure 27.14 does the same, but in the form of a Ragone plot, which plots the rated power against the energy storage capacities as a function of discharge time.

Flywheels can boost power nearly instantaneously to provide load-leveling services, such as voltage and frequency regulation. Although flywheels cannot provide long-term storage yet, that will change in the not-too-distant future with the development of higher-strength carbon-fiber

Table 27.3 Comparison of some salient grid-scale energy storage technologies, including capital (or levelized) cost of storage.

Storage type	System	Power density (kW/m³) or W/kg	Volumetric energy density (kWh/m³)	Gravimetric energy density (Wh/kg)	Lifetime (cycles) or (years)	Energy efficiency (%)	Maximum storage duration (h)	Response time	LCOS (US $/kWh)
Mechanical	Pumped hydro storage	0.1–4	0.2–1.5	0.5–1.5	$10^3 - 10^5$ (40–50 yr)	65–85	0–15	Minutes	1000–5000
	Compressed air energy storage	5–15	2–20	30–100	5×10^4 (30 yr)	55–80	6–24	Seconds to minutes	40–150
	Flywheel	1000–5000	10–400	10–100	$10^4 - 10^6$	80–95	N/A	Seconds	7000–8000
Thermal	Ice storage, or "latent heat storage"	—	100–400	150–250	>1000	75–90	25–100	Hours	60–200
	Liquefied gases	—	25–200	10–120	30 yr	80	4–25	Minutes	250–650
	Thermochemical energy storage	—	600–900	500–600	40 yr	85–90	Days to months	Seconds	100–500
	Electrothermal energy storage or "sensible heat storage"	10–30 W/kg	50–80	10–200	5–40 yr	30–60	200	Hours	5–30
Electro-chemical	Lead-acid battery	90–700	50–90	25–60	$250 - 1800$ (12 yr)	65–85	25–100	<5 ms	150–370
	Lithium-ion battery	300–700	200–750	80–300	$500 - 5 \times 10^5$ (10 yr)	85–98	50–200	<5 ms	130–450
	Vanadium redox flow battery	150–200 W/kg	—	10–70	15 yr	68–85	25–50	<5 ms	400–800
Electrostatic	Ultracapacitor	40,000–120,000	10–35	2–10	$10^4 - 10^6$ (10–15 yr)	85–98	Days to months	<5 ms	3600–10,000
Chemical	Hydrogen	>500	100–3000	300–2000	> 20,000 (35 yr)	20–50	500–1000	<5 ms	8–10

Source: Aziz et al. (2022), Bloomberg NEF (2022), Cole et al. (2021), IEA (2022), Koohi-Fayegh and Rosen (2020), Lazard (2022), McKinsey (2021), and Dwivedi (2023).

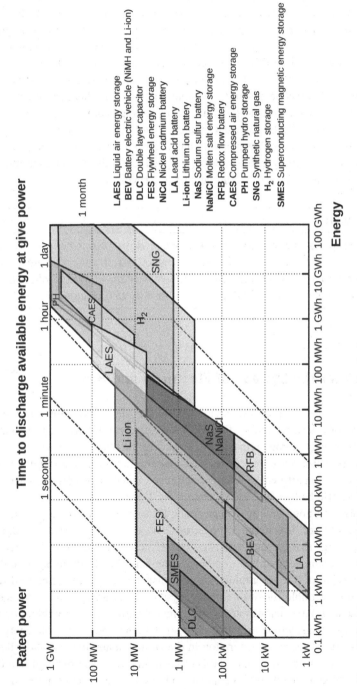

Figure 27.14 Ragone plot of a few salient energy storage technologies.

materials and magnetic bearings made with high-temperature superconductors, both of which will dramatically increase the rate of spin and mitigate energy lost by friction.

For storage of two to eight hours, lithium-ion batteries dominate. However, lithium is less abundant, and the cobalt used in some batteries is even more scarce. The battery's solvent electrolyte is flammable. The solvents could also create a hazard in the event of a leak.

The most promising battery technologies for long-term applications can decouple power from the energy capacity. In such battery types, energy storage capacity can be increased by adding to the volume of electrolytes stored in the tanks, such as with vanadium redox ("flow") batteries, but they are exceedingly expensive and have many issues linked to environmental unsustainability. Organic electrolytes could, however, diminish these concerns and reduce flow batteries' cost.

Non-chemical storage options include hydrogen, gravity- and thermal-based systems, and compressed air. There are many concepts and combinations that store heat in fluids, or in concrete blocks, or carbon-fiber blocks, with various ways of converting heat back to electricity. These systems are attractive, in part because their base materials are generally cheap.

As electric vehicles become more prevalent, their collective batteries can become a grid-scale storage medium. For instance, every grid-enabled (equipped for bidirectional charging) Ford F-150 Lightning electric truck and Tesla EV could be part of a roving storage fleet.

Reimagine Our Future

There is no more time to contemplate *delusional optimism* on the one hand and *despair* on the other. Our best hope appears to have the courage to hear the truth about climate destabilization and the urgent need to decarbonize our actions. But we must then have the will to *act with courage*. Leading by example, what will you do to help inspire such change?

27.6 Outlook for the Storage of Electricity and Heat

Energy storage enables the deep decarbonization of electric power systems that rely heavily on intermittent sources of renewable energy, like wind and solar, without sacrificing system reliability. However, grid-scale energy storage remains too expensive for massive amounts of renewables to compete against legacy electricity sources such as coal and natural gas. Recognizing that reality, in January 2020, the US Department of Energy announced a goal to lower the levelized cost of storage on an electricity grid by 90% by 2030 to around 5 ¢/kWh.

Innovations in long-duration energy storage systems are now attracting unprecedented interest from governments, utilities, transmission operators, and energy entrepreneurs. Unsurprisingly, investment in the energy storage sector is rising fast. However, the outlook for energy storage is complicated as energy storage is linked to other elements of the electricity system: demand-side management, expansion of the geographic extent of dispatch areas, and enhanced transmission capacity. Importantly, variable renewable energy systems (and associated energy storage) must compete with new energy conversion systems that combust natural gas in combination with carbon capture and storage systems, which are often cheaper than variable renewable energy systems.

To achieve zero-carbon electricity generation in the United States and elsewhere before 2040, it will be necessary to accelerate innovation. If this innovation is to be successful, it requires the integration of social, economic, regulatory, technical, and environmental considerations.

Problems

27.1 Redo the problem in Example 27.2. Then, assume that CaO costs US \$100/ton and has an energy density of 1.8 MJ/kg or 500 kWh/ton. (a) How many cycles of energy storage and discharge are needed before the material cost of CaO can be cheaper than coal and natural gas? Assume coal and natural gas cost 2 ¢/kWh. (b) If CaO replaces coal as an energy source, how many CO_2 emissions would be prevented per ton of CaO for the number of cycles calculated above? Assume that 1 kg of CO_2 is generated for every kWh of energy produced from coal. Neglect the cost of storing energy in CaO.

Hint: The cost of CaO will be $\dfrac{\$100/\text{ton}}{500\ \text{kWh/ton}} = 20\ ¢/\text{kWh}$

[Answers: (a) $\dfrac{20\ ¢/\text{kWh}}{2\ ¢/\text{kWh}} = 10$ cycles; (b) The amount of CO_2 prevented is 5 tons CO_2 per ton of CaO used.]

27.2 Concentrated solar reflectors focus the sun's rays on a collector, through which molten salt flows to capture heat.

The salt, also called "solar salt," typically comprises a mixture of sodium nitrate and potassium nitrate, which cannot explode under high pressure (like water can), has good thermal capacitance (2620 kJ/kgK, about 0.63 times the heat capacity of water), good fluidity, and low vapor pressure. This type of thermal energy storage is also relatively cheap (US \$55/kW$_{\text{th}}$), which is attractive for long-duration storage compared to technologies such as lithium-ion batteries.

The hot salt tanks are made of 347H stainless steel, with dimensions of approximately 40-m diameter by 10-m height, operating at 565 °C. The "cold" salt storage tanks are similar-sized carbon steel vessels operating at about 300 °C. While the cold salt tanks are reliable and cost-effective, the hot tanks are significantly more expensive, and suffer from reliability issues.

The Greater Los Angeles has an electrical energy demand of about 14.5 GW$_e$. How many hot and cold salt storage tanks would be required to provide the Greater Los Angeles area with electrical power for four days? Assume an ambient temperature of 30 °C.

[Answer: Around 285 hot-fluid tanks and 562 cold-fluid tanks are required.]

27.3 The energy source of a battery is the same as that of combusting gasoline or burning wood – the rearrangement of chemical bonds. The binding forces of atoms are electric, and batteries cleverly exploit this fact to produce electricity directly instead of producing heat first. In theory, batteries should be better storage media than gasoline, not worse, because they are optimally efficient and generate little waste heat.

However, they lose this efficiency advantage through the large extra weight they must carry to function. Conventional batteries do not use oxygen from the air, but instead carry all their reactants on board.

a) Using the information from Table 27.3 and Figure 27.14, estimate the mass of a battery pack for a 2010 Model 3 Tesla EV.

b) Do some online research and provide a breakdown of the type and mass of materials used in the Tesla Model 3 EV's Li-ion batteries. State your reference(s).

c) Briefly comment on the social, technological, environmental, economic, and political ramifications of the extensive use of lithium-ion batteries.

[Answers: (a) 600 kg; (b) Tesla uses different lithium-ion batteries in its current 3-series electric vehicles. Some EVs have lithium manganese oxide (LMO) batteries, others have lithium nickel manganese cobalt oxide (NMC), and others may also feature lithium iron phosphate (LFP) batteries. In general, though, an 80-kWh battery would use around 70 kg of graphite, 30 kg of aluminum, 30 kg of copper, 500 g of lithium polymer, 4 kg of plastics, and 35 kg of a carbonate electrolyte; (c) Open-ended question.]

27.4 Comment, in no fewer than 300 words, on the technical feasibility and economic viability of using lead-acid batteries, not lithium-ion batteries, for grid-scale energy storage. Be sure to highlight the advantages and disadvantages of using this mature technology. You might find the following article most helpful in your research: Lopes and Stamenkovic (2020).

[Answer: Open-ended question. However, the cited article contains an excellent summary of the advantages of using the well-established lead-acid battery technology, not least of which being the more than 99% recyclability rate of lead-acid batteries.]

References

André, L. and Abanades, S. (2020). Recent advances in thermochemical energy storage via solid-gas reversible reactions at high temperature. *Energies* 13 (22): 5859.

Arup (2020). Cultana Pumped Hydro Energy Storage Project. Arup. https://arena.gov.au/assets/2020/10/cultana-pumped-hydro-energy-storage-project-phase-2.pdf

Aziz, M.J., Gayme, D.F., Johnson, K., Knox-Hayes, J. et al. (2022). A co-design framework for wind energy integrated with storage. *Joule* 6 (9): 1995–2015.

Bauer, C., Burkhardt, S., Dasgupta, N.P., Ellingsen, L.A.W. et al. (2022) Charging sustainable batteries. *Nature Sustainability* 5 (3): 176–178.

Bloomberg NEF (2022). *Lithium-Ion Battery Price Survey*. New York: Bloomberg NEF.

Bonk, A., Sau, S., Uranga, N., Hernaiz, M. et al. (2018). Advanced heat transfer fluids for direct molten salt line-focusing CSP plants. *Progress in Energy and Combustion Science* 67: 69–87.

Burton, M., Farchy, J. and Cang, A. (2022). Behind the nickel mess on the London Metal Exchange. *Bloomberg UK*. https://www.bloomberg.com/news/articles/2022-03-18/behind-the-nickel-mess-on-the-london-metal-exchange-quicktake#xj4y7vzkg.

CEC (2021). Life cycle assessment of environmental and human health impacts of flow battery energy storage production and use. CEC-500-2021-051. California Energy Commission.

Cole, W., Frazier, W. and Augustine, C. (2021). Cost projections for utility-scale battery storage: 2021 update. NREL/TP-6A20-79236. National Renewable Energy Laboratory.

Dwivedi, A. (2023). Personal communication with Dr Arpit Dwivedi, Cache Energy, 29 January. https://cache-energy.com/

Grbovic, P.J. (2013). *Ultra-Capacitors in Power Conversion Systems: Applications, Analysis, and Design from Theory to Practice*. New York: Wiley.

IEA (2022). World Energy Outlook 2022. International Energy Agency (IEA). https://www.iea.org/reports/world-energy-outlook-2022/

Koohi-Fayegh, S. and Rosen, M.A. (2020). A review of energy storage types, applications and recent developments. *Journal of Energy Storage* 27: 101047.

Lazard (2022). Levelized cost of storage. Lazard. https://www.lazard.com/perspective/levelized-cost-of-energy-levelized-cost-of-storage-and-levelized-cost-of-hydrogen/

Liu, T., Vivek, J.P., Zhao, E.W., Lei, J. et al. (2020). Current challenges and routes forward for nonaqueous lithium-air batteries. *Chemical Reviews* 20 (14): 6558–6625.

Lopes, P.P. and Stamenkovic, V.R. (2020). Past, present, and future of lead-acid batteries. *Science* 369 (6506): 923–924.

McKinsey (2021). Net-zero Power. Long Duration Energy Storage for a Renewable Grid. New York: McKinsey & Company.

MIT Energy Initiative (2022). The Future of Energy Storage. Cambridge, MA: MIT Energy Initiative. https://energy.mit.edu/futureofenergystorage

Morse, I. (2021). A dead battery dilemma. *Science* 372 (6544): 780.

Tong, Z., Cheng, Z. and Tong, S. (2021). A review on the development of compressed air energy storage in China: technical and economic challenges to commercialization. *Renewable and Sustainable Energy Reviews* 135: 110178.

Yu, A., Chabot, V. and Zhang, J. (2013). Electrochemical Supercapacitors for Energy Storage and Delivery: Fundamentals and Applications. Abingdon, UK: Taylor & Francis.

Zhang, L., Feng, R., Wang, W. and Yu, G. (2022). Emerging chemistries and molecular designs for flow batteries. *Nature Reviews Chemistry* 6 (8): 524–543.

Zhu, T., Lot, R., Wills, R.G. and Yan, X. (2020). Sizing a battery-supercapacitor energy storage system with battery degradation consideration for high-performance electric vehicles. *Energy* 208: 118336.

Mini Project 9

Grid-Storage Batteries

Driven by technological advances, facilities are being built with storage systems that can hold enough renewable energy to power hundreds of thousands of homes. The advent of "big battery" technology addresses a key challenge for green energy – the intermittency of wind and solar.

In Morro Bay, California, energy entrepreneurs are developing one of the world's largest utility-scale storage batteries. The storage facility will comprise 600 MW$_e$ of lithium-ion batteries, which could be produced by Tesla's Gigafactory facility in Nevada. The 600 MW$_e$ battery would store enough energy to power around 450,000 homes. The battery will store renewable energy from the grid when it is plentiful like at the middle of the day when the sun is shining. The battery will release its energy during periods of peak energy usage, thus negating the need for gas turbine "peaker" plants with their large emissions.

The DeCordova Energy Storage Facility in Granbury, Texas. The 260 MW$_e$/260 MWh$_e$ battery energy storage project is the largest of its kind in the Lone Star State. The proposed Morro Bay project in California will however be one of the largest in the world, when implemented. *Source:* Vistra, https://vistracorp.com/

Energy Systems: A Project-Based Approach to Sustainability Thinking for Energy Conversion Systems,
First Edition. Leon Liebenberg.
© 2024 John Wiley & Sons, Inc. Published 2024 by John Wiley & Sons, Inc.
Companion website: www.wiley.com/go/liebenberg/energy_systems

1) What is the range of energy densities and costs of modern grid-storage batteries (chemical, mechanical, and other batteries)? Tabulate your answers (type of battery, manufacturer, energy density in MWh_e/kg, cost in $/MWh_e$), and include a URL to the website of the company or product. Based on your research, what type of battery will dominate the grid storage market in five years' time?

2) What policies could be instituted to help promote the adoption of lithium-ion grid battery storage facilities?

3) Summarize the technical specifications of the Tesla Megapack batteries that are suitable for providing $600 MW_e$ of storage. How much do such batteries cost? How long do such batteries last? What warranties are provided? What are the maintenance costs of these batteries?

4) How much energy will the Morro Bay storage battery feed into the grid in summer and winter? How much revenue will the operating company earn for selling its energy to the grid? How much new tax revenue will that generate for the city? Clearly state your assumptions and show all calculations.

5) What type of materials and in what quantities will a $600 MW_e$ lithium-ion battery need? What is the materials resource for each of the salient materials? Where will these materials be sourced? What are the social, economic, and environmental implications of extracting these materials from the identified areas? Is this a sustainable scenario? Explain.

6) Summarize grid-storage battery technologies in a high-impact manner by means of a 500-word fact sheet.

The fact sheet must use a combination of text and graphics. The text should be for a general audience and concise. The fact sheet must also cite the references that you used in your research. The graphics should help communicate the challenge addressed, your proposal to use grid-storage batteries, and what it would achieve, if implemented. Be sure to specifically address the following aspects in your fact sheet:

- What bearing does the use of grid-storage batteries have on one or more of the United Nations' Sustainable Development Goals (SDGs), paying attention to the relevant SDGs and keeping in mind the connections between people and nature. Some questions to consider: How widespread is the challenge? Are many people, ecosystems or other forms of life affected? Does the use of grid-storage batteries deal with a specific location or community? If so, explain.

- Why do you believe that the use of grid-storage batteries will be successful? Who or what will benefit from this action? How would your proposal to use grid-storage batteries, if implemented, promote, or impede the achievement of the mentioned SDGs?

- Describe your plans to achieve your intended result. Some questions to consider: How long will it take to implement your solution? How long will it stay in effect? What resources (financial, human, technological, physical, other) do they require? Would your proposal require or benefit from partnerships with any companies, non-profit organizations, governments, or other organizations? What are the main obstacles, and have you considered how they can be overcome? How will progress be monitored and evaluated?

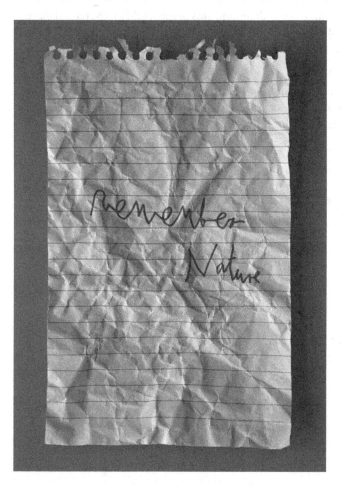

Remember nature

Remember Nature was a project initiated by artist and political activist Gustav Metzger, co-curated by artists Jo Joelson and Andrea Gregson with Hans Ulrich Obrist and the Serpentine Gallery on 4 November 2015. Metzger's call to action urged arts, professionals, and students from all disciplines to create new work to "Remember Nature," addressing global issues such as extinction, climate change, and environmental pollution. A launch event took place at Central Saint Martins College of Art and Design, and at other art schools on the day. The project was reenacted with a "day of action" in 2022.

"Remember Nature," digital collage on notepaper. *Source:* Image © London Fieldworks, courtesy London Fieldworks.

Energy Systems: A Project-Based Approach to Sustainability Thinking for Energy Conversion Systems, First Edition. Leon Liebenberg.
© 2024 John Wiley & Sons, Inc. Published 2024 by John Wiley & Sons, Inc.
Companion website: www.wiley.com/go/liebenberg/energy_systems

Week 10 – Decarbonizing Transportation, Buildings, Heavy Industry, and Power Generation

Global emissions have soared by two-thirds in the three decades since international climate talks began. What is apparently needed to reduce emissions to the required extent is a new approach that creates incentives for leading countries and industries to spark transformative technological revolutions. These strategies are required, in particular, for energy systems used in transportation, buildings, and heavy industry, which together emit around 56% of the world's CO_2. Global electricity production accounts for another 40% of the world's CO_2 emissions. Therefore, if we can successfully decarbonize transportation, buildings, heavy industry, and power generation, we would be on the road to energy sustainability.

Chapter 28 – Decarbonizing transportation, buildings, industry, and electrical power: Transportation; buildings; industry; electric power generation; outlook
Mini Project 10 — Decarbonizing the steel industry

28

Decarbonizing Transportation, Buildings, Industry, and Electrical Power

The world gets more energy efficient every year, but levels of emissions keep rising (Figure 28.1). That is the reason why *deep decarbonization* ideas rely on the replacement of fossil fuels rather than reducing fossil-fuel consumption per person or per unit of performance.

The aggregate impact of nuclear, hydroelectric, and solar/wind generation reduced the global reliance on fossil fuels from around 95% of primary energy in 1975 to around 85% in 2020. Energy transitions apparently take a long time and require massive investments. The International Energy Agency (IEA) nonetheless expects reliance on fossil fuel to decline at a more rapid pace from 2022, fueled in part by "Big Oil" companies becoming "Big Energy" companies, and by a faster global electric vehicle (EV) transition (IEA 2022b). Despite this move in the right direction, the IEA projects that the world may still be 66% reliant on fossil fuels by 2050 (IEA 2022a).

Reimagine Our Future
Would large corporations act for the public good if they were not forced to do so by the state?

As can be inferred from Figure 28.1, global emissions have soared by about two-thirds in the 30 years since international climate talks began. Carbon emissions are *rising* at around 2% per year, while climate scientists inform us that we should be *reducing* carbon emissions by more than 8% per year to be consistent with keeping global warming at 1.5 °C compared to pre-industrial times. It now appears to be impractical to make such rapid cuts globally with an industrial complex that does not usually change quickly (Victor 2020; Cembalest 2021). At the current rate of development, we should pass the 1.5 °C goal well before 2030 and the 2 °C goal soon after that. We will be very lucky if we can contain global warming at 3 °C by 2050.

There is little sense of urgency to decarbonize society. We appear to be waiting for progressive countries and companies to lead the way by demonstrating practical, scalable ways to achieve deep decarbonization. However, those key countries and companies are not doing what they should, and certainly not at the rapid rate that is required.

Will countries, industries, and individuals perhaps be more motivated to bring about rapid decarbonization if there were adequate incentives (such as rebates when using low-carbon technologies) and disincentives (such as carbon taxes and feebates when using high-emission technologies)? These attempts have not been successful so far. Global agreements on climate change have not yet delivered their promises (Thunberg 2023). It is apparent that rapid progress now requires

Energy Systems: A Project-Based Approach to Sustainability Thinking for Energy Conversion Systems,
First Edition. Leon Liebenberg.
© 2024 John Wiley & Sons, Inc. Published 2024 by John Wiley & Sons, Inc.
Companion website: www.wiley.com/go/liebenberg/energy_systems

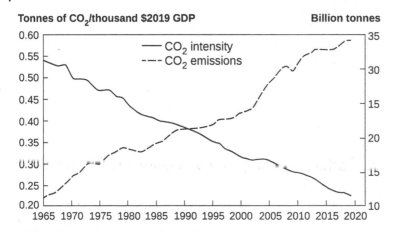

Figure 28.1 Global CO_2 intensity is declining, but CO_2 emissions are rising unabated. *Source:* After BP (2022), Cembalest (2022), and IEA (2022a).

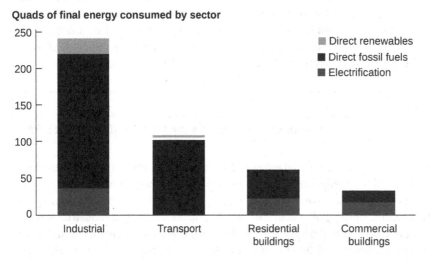

Figure 28.2 Low levels of electrification in global industry and transport. *Source:* After BP (2022), Cembalest (2022), EIA (2022), and IEA (2022a).

changing the facts on the ground. We need to urgently employ new methods of energy conversion in manufacturing, transportation, buildings, and electricity generation (Figure 28.2). This means that we need new technology and business models, and immense resolve. And we need to make better use of the technology that we have.

While much of the focus to eliminate carbon emission by 2050 has been on the main culprits – power plants, buildings, and cars – more than one-third of emissions come from heavy transport such as trucks and planes, and the heat-intensive manufacture of materials such as steel and cement (Figure 28.3). We will not reach our goal without also decarbonizing those sectors. However, those sectors are widely considered hard to abate – stubbornly resistant to decarbonization, which many believe would be slow, costly, and unprofitable (Lovins 2021).

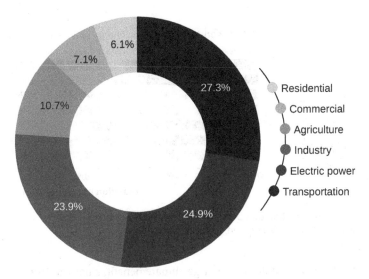

Figure 28.3 Global CO_2 emissions in 2022, by sector. *Source:* IEA (2022a) / CCBY 4.0 / Public domain.

Decarbonization requires a series of technological revolutions in each of the major emitting sectors. Each sector will require customized strategies. Those would be focused on providing increasingly competitive and abundant renewable electricity to displace fossil fuels (Zehner 2012). Fossil-fired (coal, oil, gas) power plants would eventually be starved of revenue, while their levelized cost of electricity would rise dramatically. Electrified transport, along with heat-intensive manufacturing processes using renewable energy, would also devalue and strand their fossil-fueled counterparts.

Reimagine Our Future

Between now and 2050, an estimated US \$120 trillion needs to be invested globally in low-carbon technologies. This requires an average investment of 2% of global GDP per year in decarbonization solutions, including renewable energy, energy efficiency, and other low-carbon technologies. This transformation requires well-designed policies and regulations, customized business models, and dedicated platforms for sharing best practices and lessons learned, and more. How would you share best energy practices with your friends? When we reimagine the future, we stretch both our individual imaginations and our collective imagination. So, invite your friends or family to describe their own hopes and dreams for a clean energy future.

28.1 Transportation

Today, transportation accounts for about 25% of global CO_2 emissions. Passenger cars and light vehicles account for 40–50% of global transport energy use (Figure 28.4). Although passenger cars and light vehicles are being successfully electrified, other transportation categories (buses and heavy trucks) are lagging, and some categories (shipping and aircraft) are more difficult to electrify (Figure 28.4a).

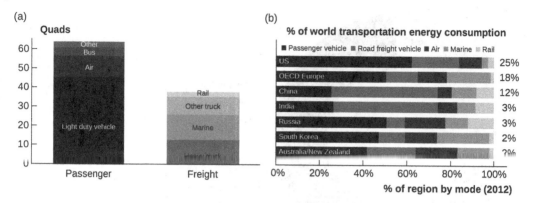

Figure 28.4 (a) World transportation energy consumption by mode, and (b) transportation energy consumption by region and mode. *Source:* After EIA (2022) and IEA (2022b).

A faster EV revolution in the United States could have a large climate benefit, since the United States accounts for 25% of global transport energy consumption, and since light vehicles represent 60% of this amount. Both figures are the highest in the world (Cembalest 2021). Globally, the EV share in 2020 was 4.5%, up from ~2.5% in 2018 and 2019. Note how this compares to IEA scenarios of 20–40% of EV shares in 2030 (Figure 28.4b).

28.1.1 Electric Vehicles

In this review of electric vehicles, only battery electric vehicles (BEV) and plug-in hybrid electric vehicles (PHEV) are considered. It should be clear from Figure 28.5 why hybrid electric vehicles (HEV) are not considered. HEVs usually feature an internal combustion engine as the prime mover, with limited battery back-up. However, in this book we are concerned with electric vehicles that will make a massive impact in decarbonization, namely EVs and BEVs, which have large battery capacities. If those large-capacity batteries are charged with electricity obtained from renewable energy sources, those EVs will make a great impact in lowering CO_2 emissions.

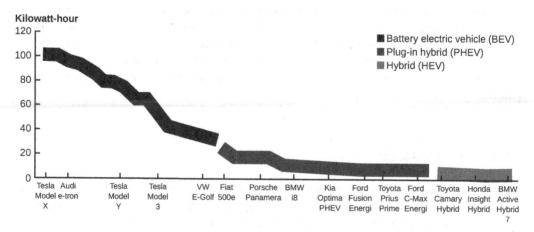

Figure 28.5 Electric vehicle battery capacity. *Source:* Adapted from Car and Driver (2022), IEA (2022b), and Inside EVs (2022).

Reimagine Our Future

What if highways were electric? Germany is testing this idea to help make the trucking industry emission free. In Frankfurt, such a pilot system feeds electricity to trucks as they drive, using wires strung above the roadway and a pantograph mounted on the cab. This idea seems very clever, delivering power directly from the electricity grid to electrical motors on trucks. What challenges might need to be faced if this idea were to be widely adopted?

Electric truck with pantograph driving along a highway in Frankfurt. *Source:* Scania.

28.1.2 Batteries

Figure 28.6 shows the enormous drop in cost of lithium-ion batteries between 2013 and 2022. Lithium-ion battery technology is associated with a learning rate of around 18%. (Also see Section 6.3 and Table 27.3.) However, battery prices (for EVs, buses, and stationary storage) have been rising steadily since December 2022 due to rising prices for raw material and battery components, and soaring inflation. The volume-weighted average prices for lithium-ion battery packs across all sectors have increased to US $151/kWh$_e$ in 2023; a 7% rise in real terms from 2021/2022. Market specialists expect battery prices to start dropping again in 2024, when lithium prices are expected to ease as more extraction and refining capacity comes online. Based on the updated observed learning rate, the average price for lithium-ion battery packs should fall below US $100/kWh$_e$ by 2026 (BNEF 2022a).

The expected reduction in battery cost will help boost the pace of EV sales. Some analysts project that EVs will reach cost parity with conventional cars by 2026 (BNEF 2022b). It will not be a

Battery cost $ per kWh

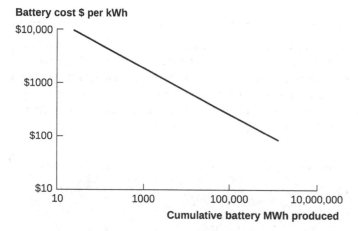

Cumulative battery MWh produced

Figure 28.6 The learning curve of lithium-ion batteries between 2013 and 2022. *Source:* After BNEF (2022b) and ARK (2022).

surprise if this situation changes soon, as some core materials used in batteries are in short supply, especially cobalt, nickel, and aluminum. Analysts think that we are currently producing only 10% of the battery materials that will be needed in 10 years' time, which implies that 90% of the battery supply chain does not yet exist (WSJ 2022).

28.1.3 Policies

In cars, policies aimed at boosting sales and lowering the costs of electric vehicles – such as subsidies that decline as technology improves, as well as investments in battery charging infrastructure – can play a role, as they already have in California, USA, as well as in China and some European countries (Victor 2020). Even more experimentation is needed, however, in realms such as battery charging, so that EVs become a viable business and move beyond niche applications.

28.1.4 Reimagining the Future of Transportation

In the United States, 40% EV penetration would increase electricity consumption by around 400 TWh$_e$ compared to the current generation of 3930 TWh$_e$, and by more in the future, depending on the growth rate of the total vehicle stock (IEA 2022b). Such an increase in electricity consumption will require vast investments in load management to prevent surges in demand that could overwhelm the electric grid.

The US grid is geared to cope with household energy consumption of 1–2 kWh$_e$ (per hour), while a Level 2 EV battery charger can consume 8–9 kWh$_e$ per hour (EIA 2020; FR 2022). It should be clear that the US electricity grid requires urgent updating to enable it to cope with the expected mass electrification of transport. Between 2004 and 2020, the US transmission "grid miles" only grew by 1.2% per year. It would have to increase to around 6% to cope with the expected increase in electrical load due to the mass electrification of vehicles, among other factors (Larson 2020).

Globally, the uptake of EVs has been slow, despite falling battery costs. This slow uptake could be ascribed to one or more of the following (BNEF 2022a, 2022b; Cembalest 2021):

- The lifecycle of conventional (gasoline or diesel) light vehicles has increased from 9 years in 2000 to 13 years in 2022, according to the US Department of Transport (DOT 2022). This delays the replacement of such cars with EVs.

- EVs cost more than comparable gasoline or diesel vehicles.
- *"Range anxiety"*: Prospective EV owners might be afraid that an EV might not have enough battery charge to reach its destination, leaving its occupants stranded. In many European countries, owners are not replacing their diesel or gasoline cars, but merely buying an additional EV for short-distance commuting (Burlig et al. 2021; Muehlegger and Rapson 2023).

Globally, the share of EVs in the transportation sector is small, only 1% (IEA 2022b). The use of renewables in transportation is dominated by biofuels, mostly bioethanol and biodiesel, in certain countries. Shipping and aviation have also made comparatively little progress to transition to renewable energy.

Analysts nonetheless predict that advanced technologies in the transportation sector could cut transport emissions to only 3 Gt of CO_2 annually by 2050 (Figure 28.7). That would represent a 70% reduction compared to the case in 2020. If this could be achieved, the transportation sector would be responsible for around 30% of emission cuts by 2050 (IRENA 2020).

Best-case scenarios of several analysts predict that the transportation sector can significantly increase the electrification of passenger transport, as well as the use of biofuels (Dorr and Seba 2020; IRENA 2020; IEA 2022a, 2022b). Also, green hydrogen will increasingly be used as a transport fuel. Such a combination would lead to a drop of nearly 70% in oil consumption by 2050 compared to 2015. The share of electricity in the transportation energy sector could realistically rise from just above 1% in 2015 to 33% in 2050, 85% of which is renewable. Biofuels would increase their share from just below 3% to 22% in the same period.

Best-case scenarios of several analysts also predict that total liquid biofuel production could grow from 129 billion liters in 2015 to just over 900 billion liters in 2050. Nearly half of this total would be from conventional biofuels, the production of which would more than triple, requiring significant upscaling. The other half would be from advanced biofuels, which can be produced from a wider variety of feedstocks than conventional biofuels, but which supply just 1% of biofuels today. The steep increase in biofuel production requires careful planning that fully considers the sustainability of biomass supply.

New energy sources, in combination with information and communication technologies, are changing the transportation industry. As performance improves and battery costs fall, sales of

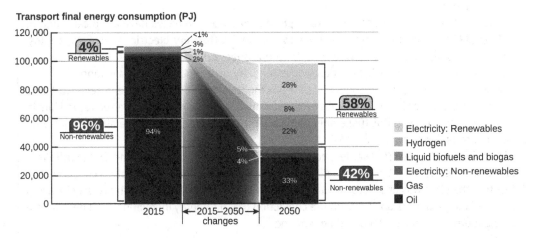

Figure 28.7 Final energy consumption in the energy sector, between 2015 and 2050, to reach net-zero targets by 2050. *Source:* Reproduced from IRENA (2020) with permission from the International Renewable Energy Agency.

electric vehicles, electric buses and electric two- and three-wheelers are growing. In 2017, around three million electric vehicles were on the road. However, analysts predict that the number could realistically increase to over one billion by 2050. To achieve this, most of the passenger vehicles sold from about 2040 would need to be electric. While about half the stock of passenger vehicles is projected to be electric by 2050, closer to 75% of passenger car activity (passenger-kilometers) would also be provided by electric vehicles.

Hydrogen will increasingly be used as a transport fuel, albeit limited to large vehicles powered by fuel cells, including city buses, locomotives, and ferries. The production of hydrogen from renewable power may provide an important option in efforts to meet demand flexibly and to expand renewable power generation. Although the technology is not yet ready for widespread commercialization, some countries believe that hydrogen will be a viable transport fuel.

To meet these objectives, an estimated US $14 trillion of total investment would be required in the transportation sector by 2050. Around US $3.4 trillion would be needed to develop the biofuel (predominantly advanced biofuels) and hydrogen industries. The balance would be needed to develop electrification and energy efficiency.

Rethinking the Electric Vehicle Business with Curiosity and Courage

A young entrepreneur's desire to ensure a cleaner, more equitable future for all has led to the establishment of a company that is leading the era of clean energy transportation and is driving positive change in communities through the development and commercialization of an array of specialized electric vehicles.

John Walsh (29) had always wanted to be an entrepreneur. He was fascinated by the possibilities of providing scalable, economically viable clean-tech solutions. As founder and CEO of Endera, his company has already built and delivered more than 50 specialized electric vehicles and has orders for 2000 more to the value of US $300 million.

John admits that curiosity is the driving force behind his company's innovations. "I have always been curious about humanity and the impact it has on the environment. Our goal is to make electric vehicles the preferred solution in our commercial markets." Endera builds its electric shuttles, school buses and delivery trucks in a 250,000 square-foot factory. Its clients include airports, corporate campuses, and hotels, who find this kind of turnkey value proposition very attractive.

Clients also like the fact that Endera's specialized electric vehicles have a typical range of 150 mi (241 km) and take under an hour to charge. "With every electric vehicle we deploy, we are reducing emissions, as well as cost to our operators. By making our vehicles cheaper and cleaner, we hope to achieve energy sustainability on a global scale," says John.

John is not new to start-up companies. He has founded three successful clean-tech start-ups. No doubt his degrees in economics, finance and management from MIT and Caltech stand him in good stead for success. But these only complement John's passion to make a difference. And people are taking notice.

As a *Forbes 30 under 30* awardee, he has participated in an honorary round table with President Joe Biden. He has also been on ABC's *Shark Tank*, where he pitched one of his concepts. The fact that the pitch was unsuccessful clearly did not bring John down. He is relentlessly focusing on something that he is good at with the conviction that his energy innovations will be valuable in the future. This recipe seems to be working.

John Walsh, Founder and CEO of Endera, in front of one the company's commercial electric vehicles. *Source:* Endera Marketing, https://www.enderamotors.com/

28.2 Buildings

Today's building sector accounts for about one-third of global energy consumption, and 15% of direct end-use CO_2 emissions. This figure rises to 30% if indirect emissions from the electricity and heat used in buildings are included. Energy use in the building sector accounts for around 3 Gt of direct CO_2 emissions in 2021. Space and water heating accounted for 80% of direct CO_2 emission in the buildings sector, equivalent to around 2.5 Gt CO_2 (IEA 2022a).

In buildings, there are two distinct challenges. One concerns new buildings – where, for example, advanced heat pump technology can allow for the electrification of heating systems, but fewer than 10% of buildings globally use heat pumps (IEA 2022c). The other, much bigger problem is how to cut emissions from existing buildings, which will account for most emissions from the building sector. Here the actions are a blend of technology (e.g., it is easier to retrofit efficient heating and air-conditioning systems) and regulation, since in most countries, the big barriers to changing buildings are not just cost, but also building codes (Victor 2020).

28.2.1 Heat Pumps

Between 2018 and 2021, per unit of energy, electricity was two to six times more expensive than natural gas in many countries over the last three winters (Figure 28.8). Therefore, a heat pump would need a coefficient of performance, *COP*, of at least 2 to 6 in those places for fuel cost expenses to break even.

It is important to consider the seasonal performance of heat pumps, as most air-source heat pumps cannot operate effectively at ambient air temperatures below $-10\,°C$, where *COP*s typically dip below 2 (Figure 28.9).

In the United States, the primary approach to deep decarbonization is "all-electric," whereby all existing fossil fuel-based heating systems should be replaced by high-efficiency electric heat pumps (HPs), combined with the expansion of renewable electricity supply. If all fossil fuel-based heating

Cost per megajoule of energy, electricity price divided by natural gas price, for industrial users

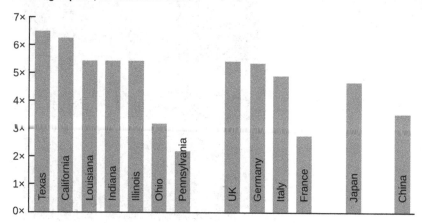

Figure 28.8 Electricity is three to six times more expensive than natural gas (BP 2022; Cembalest 2022; EIA 2022; IEA 2022b).

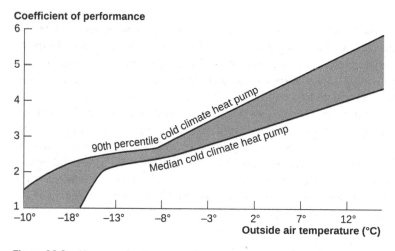

Figure 28.9 Air-source heat pump performance versus outdoor air temperature. *Source:* Adapted from Waite and Modi (2020).

could be replaced with heat pumps in the United States, it would result in an estimated 70% (equivalent to approximately 500 GW$_e$) increase in the nationwide peak load (Waite and Modi 2020).

Detailed energy simulations also show that, without increasing peak loads, currently available electric heat pumps could reduce fossil fuels to 43% of total heating energy supply (currently 70%). Future advances in heat pump technology could reduce this further to 23%. However, several challenging regions would remain (Waite and Modi 2020). The strategic use of legacy (fossil fuel-based) energy infrastructure could facilitate a more flexible transition to low-carbon heating.

As discussed in Section 26.2, water-source or ground-source heat pumps could be deployed instead of air-source heat pumps that are limited by cold ambient air. However, these "geothermal" heat pumps are more expensive than their air-source counterparts, which limits their adoption.

28.2.2 Policies

Several policies are currently being developed to support the heat pump market. This is prompted by record-high energy prices and the anticipation of upcoming challenges following the war in Ukraine and unrest in the Middle East. Several European countries provide subsidies to incentivize the use of heat pumps. These are aided by the application of high fossil fuel taxes. For example, the carbon tax in Norway is around US \$130 per ton of fuel oil compared to only US \$11 in the United States (Yeung 2022).

The European Union aims to phase out fossil fuels for residential heating by 2040. To achieve that goal, 40% of residential and 65% of commercial buildings should be electrified with 35 million new heat pumps. The EU countries will probably adopt strict combustion regulations, which will hopefully be adopted by the rest of the world (IAE 2021; Yeung 2022). The United States indeed signaled the critical importance of heat pumps in the recent Defense Production Act. Accordingly, a variety of financial instruments will be used to accelerate the adoption of heat pumps, including grants, rebates, and subsidies.

If electricity becomes cheaper than gas, users will enjoy a major incentive to switch to electrical heat pumps for space and water heating (and cooling). Rebalancing taxes and levies on electricity and gas could also help improve heat pump competitiveness in many markets.

Building codes and standards should also be updated to better reflect the urgent need for heat pumps. For example, in California, heat pumps were chosen as a baseline building heating and cooling technology in 2021. This means that new buildings must meet energy-efficiency standards equal to or above those of high-efficiency heat pumps. In China, a new building regulation came into force in 2022, setting requirements for installed heating, ventilation, and air-conditioning (HVAC) and efficiency improvements. (See also page 246 for the discussion on weatherizing of buildings to reduce energy costs.)

A ban should also be placed on replacing oil and coal boilers in buildings. For instance, in 2022, France banned the replacement of oil and coal boilers under that country's new building code. From 2023, New York City has banned the use of natural gas in new buildings up to seven storeys high, and from 2027 in those over seven storeys (IEA 2022a).

28.2.3 Reimagining the Future of Buildings

The building sector currently covers a residential and commercial floor area of 150 billion m^2. This is projected to increase to 270 billion m^2 by 2050. Buildings make a significant contribution to global emissions and need to play a central role in efforts to reduce them. The building sector has done little so far to promote the energy transition. In 2015, globally, only an estimated 36% (including traditional biomass) of the energy used in buildings was renewable (Figure 28.10).

To reach net-zero carbon emissions by 2050, many analysts predict that renewables should, and could, provide for 77% of the global energy demand (Dorr and Seba 2020; IRENA 2020; IEA 2022a, 2022b). Electricity demand in the buildings sector is projected to increase by 70% by 2050, despite improvements in appliance efficiency. This will be due to strong growth in electricity demand (mainly in emerging economies) and increases in the electrification of heating using heat pumps and seasonal energy storage.

This best-case scenario includes the use of highly efficient appliances, including smart home systems with advanced controls for lighting and heating, improved heating and cooling systems, better

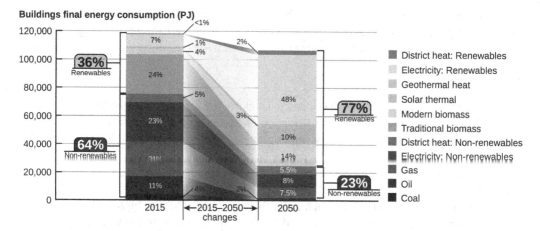

Figure 28.10 The increasing use of electricity in buildings and the decline of fossil fuels. Breakdown of final energy consumption in the building sector, by source. *Source:* Reproduced from IRENA (2020) with permission from the International Renewable Energy Agency.

insulation, the replacement of gas boilers by heat pumps and other efficient boilers, and retrofitting old and new buildings to make them energy efficient.

To reach this projected best-case scenario by 2050, a cumulative investment of US $38 trillion is required between now and 2050. An additional US $1.6 trillion would be required for the deployment of renewable energy systems in buildings.

A significant increase in the share of modern renewables (excluding traditional uses of biomass) for heat and other direct use must take place. The largest increase will be in solar thermal systems, which could increase the total solar collector area ten-fold, from around 600 million m² to over 6000 million m² (IRENA 2020).

Heat pumps are also poised to play a critical role in reaching net zero by 2050. Their use to heat buildings must be significantly expanded. Heat pumps achieve energy efficiencies three to six times higher than boilers and can be powered by renewable electricity. Realistically, it is projected that the number of heat pumps in operation would increase from around 190 million in 2021 to over 600 million units by 2050 (IEA 2022c). They could supply 27% of the heat demand in the building sector by 2050 (IRENA 2020; IEA 2022c).

Efficient and clean district energy systems would provide 16% of building heat demand, more than double the level in 2021. The projected shift in cooking technologies from fuel to electricity will further promote renewable energy sources, due to the expected high share of renewable power by 2050. Electric stoves, such as induction cookstoves, can cut the energy demand of cooking by three to five times (IRENA 2020). More renewable-based stoves that use modern biofuels and solar energy could also be deployed.

By 2050, most new and renovated buildings will be more energy efficient and will rely largely on renewable technology. Most efficiency investments will be spent on making buildings more energy efficient. Early action is, however, required to avoid stranded assets and meet future re-investment needs.

Bioenergy is expected to remain the largest renewable fuel source in buildings. It could meet about 30% of heating and cooking demand by 2050. This implies a three-fold increase relative to 2021 levels (IRENA 2020). The use of biogas is also expected to increase.

28.3 Industry

Plastic, cement, steel, ammonia, and other industrial materials form the building blocks of the modern world. Of the 150 EJ that the industrial sector consumed in 2021, only 20% consisted of electricity. Most of these industries use high-temperature heat for the various manufacturing processes (Figure 28.11). Most of that heat is provided by burning coal, natural gas, and oil (Figure 28.2). Unsurprisingly, the industrial sector is the largest fossil fuel end-user on a global basis (Figure 28.3). In 2020, industrial processes were responsible for producing around 14 Gt of CO_2. Table 28.1 illustrates the typical distribution of thermal loads in heavy gas-emitting industries in the United States, where more than 50% of industrial processes have heat loads more than 1 MW_{th}.

Clearly, if we were to decarbonize industry, we must decarbonize heat. The process heat could be provided by renewable electricity, zero-carbon fuels, or zero-carbon heat.

28.3.1 Electrification of Industrial Processes

Some industrial processes (e.g., primary metals, secondary or "recycled" steel, machinery, wood products, plastic and rubber) can be electrified to eventually use more renewable energy as the grid is decarbonized (Table 28.2). Certain mining activities also have high electrification potential, especially activities related to transport, excavation, pit crushing, and belt conveying systems (Griffith 2021; Du Plessis et al. 2013). However, it will be more difficult to decarbonize industry sectors such as chemicals, pulp and paper, and food, but waste heat could be captured in combined heat and

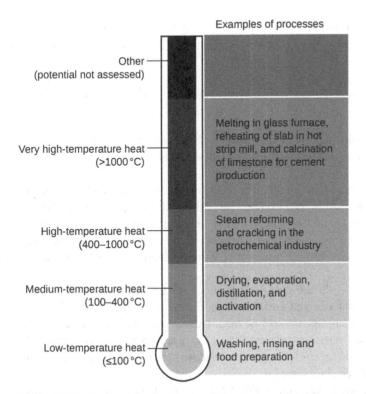

Figure 28.11 Share of total fuel consumption in the industrial sector. *Source:* IEA (2022a) / CCBY 4.0 / Public domain.

Table 28.1 Distribution of thermal loads as a fraction of facilities in 14 top greenhouse gas-emitting industries in the United States.

Facility annual average heat load (MW$_{th}$)	Fraction of facilities (%)
0–1	39.6
1–10	26.5
10–100	18.9
100–1000	2.1
Unreported	13

Source: After McMillan and Ruth (2018).

Table 28.2 Industrial sectors with high electrification potential. (HVAC: heating, ventilation, and air-conditioning; CHP: combined heat and power.)

Industrial sector	Heat requirement (°C)	Fuel consumption shares (%)		
		HVAC	Heat	CHP
Primary metals (e.g., steel)	1200	6	75	7
Fabricated metal	430–680	20	61	7
Machinery	730	46	39	4
Secondary steel	1425–1540	4	87	0
Wood products	180	10	50	14
Vehicle parts (drying)	150	31	33	12
Plastics and rubber	260	20	33	24

Source: Adapted from McKinsey (2020).

power (CHP) systems. CHP-intensive sectors are harder to electrify since producers would need to purchase energy previously obtained at little to no cost and redesign their heat-related manufacturing processes (Cembalest 2022).

Other hard-to-electrify sectors include non-metallic minerals such as glass, brick, and cement (Swanepoel et al. 2014), which require process heat at temperatures greater than 1400 °C (Table 28.3). These products do not conduct electricity and are therefore more difficult to electrify, compared to, say, steel, which does conduct electricity. Another hard-to-abate industry is oil refining, which has high levels of "own-use" fuel consumption, a source of energy that will be lost when switching to electricity.

The big challenge will be to electrify the industrial sectors with low/medium electrification potential as that sector uses more than two-and-a-half times the energy of the high-potential sectors (Cembalest 2021). The upfront switching costs will be massive, and industrial companies would have to contend with the fact that electricity is around three to six times more expensive than natural gas, for instance (Figure 28.8). The efficiency gains of electric heating versus chemical combustion might offset some of the cost, but not a lot.

Like in the building sector, electrical heat pumps will play a critical role in decarbonizing industry. Despite the possibility of using green electricity to power them, heat pumps are perfectly suited to capture waste heat from industrial processes (Liebenberg and Meyer 1998). To meet industry's high-temperature demands, heat pumps are now being developed that can deliver heat at

Table 28.3 Industrial sectors with medium-to-low electrification potential.

Industrial sector	Heat requirement (°C)	Fuel consumption shares (%)		
		HVAC	Heat	CHP
Food and beverages	120–500	4	25	40
Chemicals	100–850	1	32	43
Pulp and paper	650	2	21	63
Non-metallic minerals	870–1600	3	90	1
Oil/coal products	220–540	0	58	22

Source: Adapted from McKinsey (2020).

temperatures of more than 180 °C. These heat pumps employ refrigerants such as the hydro-fluoro-olefin HFO-1336mzz-Z that is chemically stable to around 250 °C (DryFiciency 2023). Soon, refrigerants will be able to operate stably at around 300 °C, thus further expanding the scope of industrial heat pump applications (Obrist et al. 2023).

However, these industrial heat pumps have high upfront capital costs relative to gas boilers, which translate to payback periods of five years or more in typical industrial applications. These heat pumps also hardly exceed a heating capacity of 1 MW_{th}, whereas around 45% of industrial heat is required in the range of 1–100 MW_{th} and beyond. Further, if industrial waste heat is to be successfully used, heat pumps should be situated within a few hundred meters of the industrial site to avoid thermal losses during the transport of the waste heat fluid.

28.3.2 Zero-Carbon Fuels for Industrial Processes

Another approach to decarbonize industrial heat is to use zero-carbon fuels rather than fossil fuels. This is relatively simple to achieve where these fuels are combusted only for heat production, such as with boilers to generate steam. Zero-carbon substitutes such as hydrogen, ammonia, biofuels, and synthetic hydrocarbons can be employed.

But direct fuel substitution is not straightforward in many other industrial processes where hydrocarbon fuels provide not only thermal energy, but serve as a reactant as well. This is the case in steel production, where iron ore is reduced to atomic iron by reacting it with metallurgical coke (Kim et al. 2022). In a blast furnace, the coke (C) is partially oxidized by oxygen to produce heat and carbon monoxide (CO), which, in turn, functions as the main reducing agent, converting iron oxide to iron:

$$2C(s) + O_2(g) \rightarrow 2CO(g) + heat \tag{28.1}$$

$$Fe_2O_3(s) + 3CO(g) \xrightarrow{\; heat \;} 2Fe(s) + 3CO_2(g) \tag{28.2}$$

For such processes, where a carbon-containing fuel is intrinsic to the process chemistry, either the emitted carbon dioxide can be captured and geologically sequestered, or new process chemistries must be developed using carbon-free reactants. For example, iron can be directly reduced with hydrogen instead of coke (Thiel and Stark 2021):

$$Fe_2O_3(s) + 3H_2(g) \xrightarrow{\; heat \;} 2Fe(s) + 3H_2O(g) \tag{28.3}$$

In this case, hydrogen provides both the heat to drive the reaction and serves as the reductive reactant. If "green hydrogen" can be used (see Section 9.3), this approach would yield decarbonized steel manufacturing (Bhaskar et al. 2020). Section 9.3 expounds on why hydrogen (H_2) and ammonia (NH_3) will be critical in the transition to net-zero emissions. Section 9.4 explains why biofuels and synthetic hydrocarbon fuels will also play an important role.

28.3.3 Zero-Carbon Heat for Industrial Processes

Instead of using combustion to convert chemical energy into a zero-carbon fuel and then to heat, zero-carbon heat can also be converted directly from environmental sources, like solar radiation (Chapter 23) and geothermal energy (Chapter 26). Zero-carbon heat can also be generated from nuclear fission, where micro-nuclear power plants could play an important role in future (Chapter 17).

To achieve cost and technical parity with carbon-rich heat, research and development in all three technologies (electrification, zero-carbon fuels, and zero-carbon heat) is needed to increase attainable top temperatures, reduce capital costs, and, in the case of solar thermal, increase resource dispatchability (Thiel and Stark 2021). Table 28.4 summarizes the challenges and opportunities of each of these technologies.

Table 28.4 Challenges for decarbonizing heat processes in industry when using renewable electricity, zero-carbon fuels, or zero-carbon heat.

Technology		Key challenges	Research and development needs
Electrification of heat	Industrial heat pumps	Limited maximum temperature, high upfront cost, scaling	Novel refrigerants, low-cost manufacturing techniques
	Electrical resistive heating	Scaling	—
Zero-carbon heat	Solar-thermal	Limited maximum temperature, intermittency, low spatial density	Low-cost collectors, high-temperature materials, long-duration thermal storage
	Geothermal	High upfront cost, high-resolution understanding of geothermal resource, access to kilometers-deep geothermal sources	Resource characterization, deep drilling technology, working fluids
	Nuclear	Limited maximum temperature, safety and security, scaling	High-temperature designs, development of micro and small nuclear reactors (10–300 MW$_{th}$), and very-high-temperature (>900 °C) gas reactors
Low-carbon fuels	Hydrogen	Combustion stability, production cost, storage and transportation	Combustion modeling and control, advanced burner (combustion chamber) design, advanced materials
	Ammonia	Combustion stability, production cost, storage and transportation	Combustion modeling and control, advanced burner (combustion chamber) design, electrosynthesis
	Biofuels	Lifecycle emissions, production cost, storage and transportation	Drop-in substitute fuels, economically viable retrofitting
	CO_2-derived hydrocarbon fuels	Direct-air carbon capture and process costs, synthesis technology, scaling	Direct-air carbon capture processes, improved CO_2 reduction

Source: Adapted from Kim et al. (2022) and Thiel and Stark (2021).

Waste Heat: Energy Entrepreneur Turns to an Overlooked Renewable Resource

Nearly a third of energy consumed in industry is lost as heat to the environment. This represents a massive loss of energy, with concomitant increased emissions of carbon dioxide, as fossil fuels are usually burnt to make up for the loss of heat.

Dr. Antoine Meffre (39) decided to use his engineering skills and knowledge to address this unacceptable situation. He founded a company in France, Eco-Tech Ceram, which uses ceramic technology to capture and reuse high-temperature waste heat from industrial processes.

In France, thermal losses from industrial processes translate to around 100 terawatt-hours. This wasted energy is equivalent to a loss of about US $5 billion a year. Antoine realized the opportunity to help industry recover the waste heat while building a profitable company.

The Eco-Stock® heat recovery unit, which Antoine and his team developed, takes waste heat from industrial furnaces and stores it in ceramic at temperatures up to 1000 °C. The ceramic can be made from coal fly ash that is recovered from coal combustion or from cofalit that is recovered from asbestos plasma inertization, construction waste or municipal solid waste, further reducing the carbon footprint. The stored heat is returned to the furnace or boiler when it is needed, increasing the thermal efficiency of the process.

"I have surrounded myself with a team of men and women who are experts and committed to a decarbonized, profitable and sustainable industry that will participate in the development of a virtuous and resilient economy," Antoine enthuses. He is an action-oriented problem solver, who successfully launched his company during adverse economic conditions. "With minimal human resources, we had to master several disciplines: thermal engineering, materials science, industrial process engineering, finance, contracts, project management and commissioning." His team clearly mastered all these business aspects.

So far, Antoine has raised more than US $50 million in start-up funding, and his company has already attracted some big-name clients. These include a world leader in the field of tiles, Wienerberger; the bathroom and tableware company, Villeroy & Boch; and one of the world's largest steelmakers, ArcelorMittal. Antoine's innovations have also earned him several awards, including one from the World Innovation Contest, the ArcelorMittal Innovator's Trophy, and a European Union Horizon 2020 award.

Antoine laments that "worldwide, the recovery of waste heat from industry must be prioritized. For instance, the annual wasted heat from industry in France corresponds to the energy consumption of 2.5 million people. This is unacceptable!" Every bit of high-quality thermal energy must be retrieved and used. But, as Antoine philosophizes, "the best energy is that which we do not consume."

Dr. Antoine Meffre, founder and CEO of Eco-Tech Ceram in France.
Source: Eco-Tech Ceram, © Nathalie Oundjian (https://ecotechceram.com/en/energy-storage/).

The Eco-Stock® waste heat exchanger can recover 1 MW of power and 2 MWh of energy when operating at 600 °C. The energy storage unit loses only 5–7% of heat per day. For efficient operation, the heat should be discharged within one to three days, depending on the price of competing energy. The Eco-Stock® has a typical energy payback period of only three months and sports an energy return on investment (EROI) ratio of 45. *Source:* Eco-Tech Ceram.

28.3.4 Policies

The penetration of low-carbon technologies into markets follows a familiar S-shaped curve, with the emergence of a new technological system, its diffusion into widespread use, and the reconfiguration of whole markets around the new system.

Today, only a subset of political jurisdictions – mainly in Europe and parts of the USA, and in a few other countries – have demonstrated that they are highly motivated to act. The silver lining in this disturbing fact is that these leaders can get a lot done – if they have the right strategy. Much of what is needed to improve technologies and markets in the initial phase can happen in small groups of countries where incentives for change are strongest (Victor 2020).

The steel industry has progressed the least with decarbonization efforts. Because steel is a globally traded commodity, no company will make this shift on its own. Active industrial policy is therefore required, which implies direct subsidies for companies that are testing green steel production methods, as is happening in Sweden, Austria, Australia, and a few other countries. Beyond subsidies, government policy can link new supplies of green steel with users who are most willing to pay higher initial costs – including governments, which can create guaranteed markets through their purchasing policies. The engineering of decarbonized systems may prove relatively easy once enough companies, governments, and consumers focus on the need.

With this industrial policy approach, a new green steel industry is slowly emerging. Industrial-scale pilot plants are slated for the early 2030s. Those projects are, however, concentrated in the European Union and Japan as the various governments have offered attractive industrial policies. The problem is that this rate of innovation and testing is much too slow and indecisive. Governments and companies must test a wider array of green steel techniques and build a larger market for green steel faster. The same applies to cement and other high-polluting industries.

While the details vary across each sector, the pictures are similar. Leaders need to channel political energy from the growing public concern about climate change into policies aimed at changing the incentives to test and deploy new technologies (Victor 2020). As the process gains momentum, the political strategies must shift away from small groups of highly motivated leaders toward broader diffusion. This kind of shift in policy strategy is most evident in the electric power sector, which is farthest along in the process of decarbonization. Success with renewables has focused governments on the next frontier: the integration of renewables while keeping grid power reliable and affordable.

28.3.5 Reimagining the Future of Industry

To date, the industry sector has been the biggest laggard with respect to the energy transition. In 2015, renewables (mainly biofuels) provided only around 7% of industry's direct energy use (i.e., excluding electricity) (see Figure 28.12). Electricity supplied almost 27% of the energy consumed by the sector.

As shown in Figure 28.3, the industrial sector is the second-largest emitter of energy-related CO_2. Industry is responsible for a third of the emissions worldwide. Figure 28.12 summarizes the findings of many analysts who project that the industry sector would need to cut its carbon emissions by more than half by 2050 so that we could achieve a net-zero carbon society, globally. Even if industry were to achieve this lofty goal, it would still emit 5.1 Gt of CO_2 in 2050 (IRENA 2020).

With the best-case scenarios, industry will become the largest source of CO_2 emissions by 2050. Its share is projected to rise from 29% in 2015 to 46% of annual emissions by 2050. Within the sector,

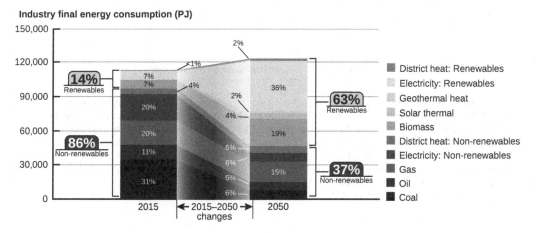

Figure 28.12 Industrial final energy consumption, now and projected to 2050. *Source:* Reproduced from IRENA (2020) with permission from the International Renewable Energy Agency.

chemical, petrochemical, and steel will still be among the largest emitters, as they employ energy-intensive and high-temperature processes that are difficult to decarbonize (Figure 28.12).

To achieve the level of decarbonization proposed under best-case scenarios, analysts predict that cumulative investments of US $5 trillion will be required by 2050.

Reimagine Our Future

Demand response could play an important role in decarbonization. With demand response a building or industrial process can contribute to balancing the electric grid if its consumption is reduced in response to the grid's current needs. For instance, electrical loads could be automatically modified during peak demand, even without the consumer being aware of it. In a residential home, hot-water temperatures could automatically be set lower than normal and refrigerator temperatures could be set higher. And, in industry, non-critical but energy-intensive processes (such as air compression) could be automatically switched to run during cheaper (non-peak demand) periods. This could save the owner a lot of money, while also benefiting the environment. Why is demand response not implemented in the entire built environment and in every industry, in every country?

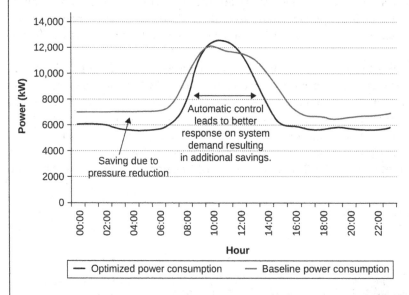

A compressed-air system at a South African gold mine was retrofitted with an automatic compressor control system featuring compressor cascading and pressure bandwidth control. The goal was to implement a simple demand reduction strategy to afford meaningful electrical energy savings. The automatic control strategy realized a saving of 1.25 MW$_e$ (on a baseline of 7.22 MW$_e$) during the electricity utility's evening peak demand window. This represents a reduction of 17.3% in electrical power consumption during the evening peak period, and savings of nearly US $360,000 per year. *Source:* Liebenberg et al. (2012). Republished with permission of JESA.

28.4   Electric Power Generation

To achieve net-zero emissions by 2050, the electricity sector needs to decarbonize almost completely. As illustrated in several chapters, this can be achieved by using renewable energy systems, increasing energy efficiency (i.e., by doing more with less), and making power systems more flexible.

It is, however, crucial that government policies also look beyond just renewables – for example, to flexible gas-fired power plants that capture carbon pollution before it is released into the atmosphere, and to advanced nuclear plants with zero local emissions. Such plants can help keep grids reliable as they shift to lots of renewable power with the associated intermittency of power delivery.

As solar and wind energy systems expand, their learning rates will improve. The lower costs imply that every dollar invested will go further. Up to 2022, total investment in renewable energy has remained roughly flat, but the installed capacity of wind and solar is soaring (BNEF 2022c).

Figure 28.13 summarizes the predictions of some analysts. Global electricity consumption is projected to double by 2050 (to 42,000 TWh$_e$) compared to 2015 levels. By 2050, the share of renewable energy in generation would be 85%, up from 29% in 2021 (IRENA 2022). Solar and wind capacity will lead the way, rising from 800 GW$_e$ in 2021 to 13,000 GW$_e$ by 2050. The output of geothermal, bioenergy, and hydropower should increase by 800 GW$_e$ over the same period. Decentralized renewable power generation is expected to grow from just 2% of total generation in 2021 to 21% by 2050, a 10-fold increase. To achieve these goals, however, requires no new coal plants to be built, and 95% of coal plants in operation today to be phased out. That might be doable in the long term, but the energy crisis initiated by the recent war in Ukraine has required many countries to recommission their coal-fired power plants.

Electricity generation (TWh/y)

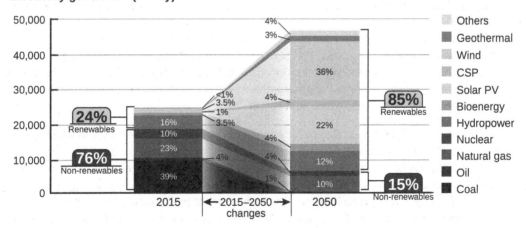

Figure 28.13   The rising importance of solar and wind energy in the power sector. *Source:* Reproduced from IRENA (2020) with permission from the International Renewable Energy Agency.

To create a power system with 85% renewable power will require investments to deploy variable renewable energy systems such as onshore-wind (33%), photovoltaic (31%), and concentrated solar power (12%). As the share of renewable energy in electricity generation rises, investments will also be needed for storage, transmission and distribution capacity, and for flexible generation and demand response. Between 2015 and 2050, investments in these areas could add an estimated US $9 trillion in this best-case scenario. Cumulatively, the investment needs of the power sector (beyond generation capacity) could reach US $24.6 trillion by 2050.

28.5 Outlook

Demand-side energy options pertain to doing more with less, using smarter technologies. Energy efficiency, therefore, refers to getting more energy services per unit of energy consumed. For instance, LED light bulbs use 75% less energy than incandescent bulbs, saving money and reducing greenhouse gas emissions. Other examples of energy efficiency include weather stripping around doors and windows to avoid leaking heat, electric vehicles instead of gas guzzlers, properly inflated vehicle tires to minimize rolling resistance, more efficient industrial electric motors and compressors, or the recovery of waste heat in industrial plants. Supply-side options to energy efficiency could include adding energy storage systems and converting to combined heat and power plants, among many others. According to the International Energy Agency, improving energy efficiency could deliver over 40% of the greenhouse gas emissions reductions needed to meet global climate goals.

Decarbonization requires a series of *technological revolutions* in each of the major emitting sectors. Consolidating the various scenarios for decarbonizing transportation, industry, buildings, and the power sector, and incorporating current developments surrounding the energy crisis spurred on by geopolitical events such as the recent war in Ukraine, two broad-based scenarios are highly probable (Figure 28.14):

- The baseline economic transition scenario shows how the energy transition might evolve from today because of no further policy action, economic forces, and technology tipping points (BNEF 2022c).
- The net-zero scenario describes an economics-led transformation of the energy economy to achieve net-zero emission by 2050. This scenario combines the faster and greater deployment of renewables, nuclear power, and other low-carbon dispatchable technologies (e.g., hydrogen and biofuels). Taking a sector-led approach, this scenario describes a credible pathway to meet the net-zero goals of the Paris Agreement (BNEF 2022c).

High global oil and gas prices offer an incentive for households and businesses to adopt renewable energy solutions. In Europe, policymakers are fast-tracking permits for installing renewables and simplifying regulations for retrofitting buildings to be more energy efficient. European and Australian steel makers are investing in *green hydrogen* and teaming up with energy companies to build wind-to-hydrogen plants or even using micro-nuclear power plants. Car makers are turning to green steel which is produced using renewable energy. *Power-to-X technologies* are being developed to harness sunlight and wind. In 2022, the United States adopted the Inflation Reduction Act, which includes subsidies for the local manufacture of clean technologies. Western countries

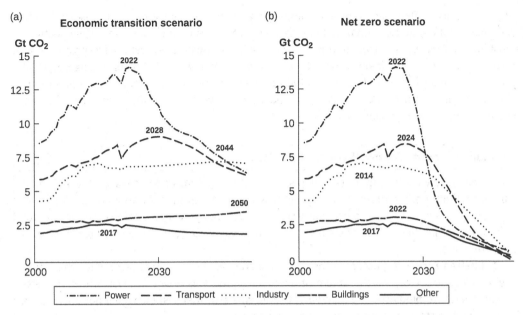

Figure 28.14 Two probable scenarios: (a) business-as-usual ("Economic Transition Scenario"), or (b) going all out to achieve net-zero by 2050 ("Net Zero Scenario"). *Source:* Adapted from BNEF (2022c) and IEA (2022a).

are indeed ramping up domestic manufacturing of green technologies to improve energy security. But these decarbonization efforts are too few and happening too slowly.

To ensure that the world decarbonizes by 2050 (Figure 28.14b), *we need immediate and sustained change on a massive scale.* Current and planned policies to limit global warming to 1.5 °C or even 2 °C are however inadequate to ensure that the world will not exceed its energy-related "carbon budget" already by 2030. The deployment of renewable energy systems must increase at least six-fold if we are to achieve "net zero" by 2050 (IRENA 2018, 2020). The share of electricity in total energy use must double, with substantial electrification of transport and industry, and of heat. Renewable energy systems could then make up nearly two-thirds of energy consumption and 85% of power generation (IRENA 2022). Combined with energy efficiency, this scenario could provide more than 90% of the climate mitigation needed to limit global warming to 2 °C.

Carbon abatement is not only feasible and urgently needed – it could provide ample rewards if implemented strategically. Some of the world's most power-intensive (and dirty) industries could be displaced by new technologies and advances in materials and manufacturing technologies, novel business models, and smart policies with lucrative incentives, along with aggressive investments. If this transition is managed strategically, it will be able to decouple transportation, industry, buildings, and power generation from climate. We do not seem to lack engineering capabilities, but rather a strategic approach to creating the incentives and markets required for sparking new technology and businesses (Victor 2020; Lovins 2021; Smil 2022; Costanza 2023).

Our ingenuity is often marked by us coming together to tackle what appear to be insurmountable goals. Now is the time for humanity to do that again. Success requires *less moralizing* about sustainability and *more strategizing and real action.* There is however not yet a widespread sense of urgency to accelerate our actions to decarbonize our way of living. We do not like change, especially if it is messy and uncomfortable. We should however embrace the challenges and opportunities when reimagining our future. It is going to take all of us working together with laser-like focus, and with honesty, compassion, respect, and grit.

Innovation Converts Organic Waste to Diesel Fuel

With the worldwide energy crisis, coupled with the problem of food waste, the potential of successfully deriving fuel from organic waste is just waiting to be exploited. A further benefit of such a system, particularly in remote, energy-scarce regions, would be for off-grid communities to use their own unwanted waste to generate energy for their own requirements.

Dr Chelsea Tucker (25) accepted the challenge and has co-developed technology that could help address the worldwide problem of energy security. This novel catalyst, which converts organic waste to biodiesel, is a modular, off-grid, waste-to-fuel system, which not only makes use of food waste, but also farm waste and sewage.

Dr. Chelsea Tucker (25), a chemical engineer who co-developed a novel catalyst to convert waste to biodiesel. Chelsea is the research coordinator for the international Waste-to-Fuel project, which includes researchers from South Africa, India, and Brazil. *Source:* Chelsea Tucker.

"It is a profound experience to have a project you care so much about spark the interest of a global audience," Chelsea recently said. Her project is part of the international waste-to-fuel initiative. Chelsea has won numerous international awards for her research breakthroughs, which include the simplification of a waste-to-biogas process. This was a precursor to producing biodiesel and developing the novel catalyst. In doing this, Chelsea's innovation offers a solution to two of the world's most pressing issues: a glut of waste and a lack of energy in emerging economies.

Chelsea's innovation has another big advantage. "Organic waste sent to landfills decomposes to produce greenhouse gases, which exacerbates global warming. By using organic waste to create energy, we are preventing tons of carbon dioxide and methane from entering the atmosphere."

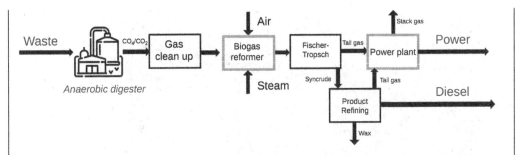

Chelsea Tucker's innovative waste-to-fuel system, which features a simplification of the widely used Fischer–Tropsch process. Chelsea's design uses biogas obtained from the anaerobic digestion of municipal solid waste. Biogas is converted by an air-fed reformer to syngas, a mixture of CO and H_2. Uniquely, the syngas is then fed to a single-pass Fischer–Tropsch reactor, which employs a manganese catalyst that ensures high conversion efficiencies. The liquid products are refined to diesel and wax, and the gases are fed to an electrical power plant.

Chelsea constantly questions things around her. She is therefore keen for not only remote communities to use this technology. As the waste-to-fuel initiative works on an entirely circular economy (everything that is thrown away gets either recycled or reused), she hopes that the public will get behind waste recycling and separation, putting their organic waste aside for municipal pick-up. The idea is that organic waste would go to a local depot, where it can be dropped into a local decentralized diesel system and converted to electricity that will be used throughout the community, as well as diesel.

"The diesel will have two functions. It can be sold to community members for use in their cars (at a competitive price in comparison to imported fuels), or it could enable backup electricity generation. In this way, the load shedding of electricity will become a distant memory."

Chelsea's work has been widely recognized. She is a L'Oreal Women in Science awardee, a Queen Elizabeth Diamond Jubilee scholar and an Allan Gray Orbis Fellow. She was also a top-10 finalist in the international Falling Walls competition.

Problems

28.1   More than 20 million electric vehicles could be sharing US roads in the next five years. Updating battery-charging infrastructure is key to scaling up the industry. Building owners will play an important role in providing charging solutions at their buildings. If you were a building manager, what novel EV battery charging system would you implement to not only generate an income for the building's maintenance, but to also service the building's occupants who drive EVs?

[Answer: This is an open-ended question.]

28.2   Some people think that we should reconsider some of the traditional practices of indigenous peoples from around the world. After all, many of those peoples have successfully handled climatological disasters and other unforeseen events. What do you think? How might we successfully integrate ancient practices or those used by indigenous peoples into modern buildings to help bring about sustainable energy? (See Figure 28.15.)

[Answer: This is an open-ended question.]

Figure 28.15 Students from the University of Utah design and build single-family homes in collaboration with the Navajo Nation in southeastern Utah, emphasizing sustainability and a respect for the unique social and cultural needs of its remote desert location. Students study indigenous architecture and southwestern vernacular traditional building methods. Then they design the houses, work on construction documentation, and prepare project management documents. Students spend a full semester at the site overseeing the construction, https://soa.cap.utah.edu/designbuild/.
Source: Image used with permission from the University of Utah.

28.3 Factors such as fuel economy, electricity grid mix, vehicle choice, and temperature affect EV greenhouse-gas emissions compared to vehicles with internal combustion engines. Successfully decarbonizing the transportation sector depends on understanding their combined effects (Bullard 2022). Read the following paper, which performs lifecycle assessments to compare the EV and internal combustion EV "well-to-wheel" greenhouse-gas emissions in the USA from 2018 to 2030: Challa et al. (2022).

Summarize your findings and produce a simple diagram or chart that compares the various vehicle drives and energy sources in emissions of CO_2-equivalent per kilometer driven. (The chart could look like the one produced by the IEA, https://www.iea.org/data-and-statistics/charts/well-to-wheels-greenhouse-gas-emissions-for-cars-by-powertrains.)

[Answer: Open-ended question, but the final diagram should look like the cited one.]

28.4 Study Figure 28.16, which summarizes the energy conversion efficiency of a 60-foot citybus using either a hydrogen fuel-cell drive or battery-electric drive. A bus that uses a fuel-cell drive and green hydrogen (obtained from solar-electric hydrolysis) has a range of around 190 km. The hydrogen fuel-cell bus costs around US $1.2 million. A bus with a battery-electric drive that uses solar electricity has a range of more than 445 km. This bus costs around US $500,000. Why do so many mass transit districts still deploy fuel-cell electric buses rather than the clearly more efficient, and cheaper, battery-electric buses? Share your insights in a 300-word essay with energy calculations.

[Answer: Open-ended question.]

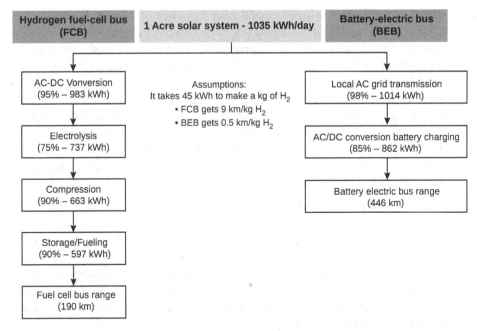

Figure 28.16 Comparison of the energy efficiencies of a hydrogen fuel-cell citybus and the same type of bus propelled by a battery-electric drive. Both drives obtain their electricity from the same photovoltaic array. *Source:* After Fournier (2023).

28.5 There are five well-known strategies for decarbonizing our toughest sectors (buildings, transport, industry, and power):

- *Replace:* Rapidly scale green technologies to outcompete legacy (fossil fuel) rivals and supplant obsolete technology assets.
- *Transform:* Create novel incentives and business models that reward innovative competitors, challenging incumbent industries with breakthrough technologies.
- *Redesign:* Integrate new design methods, technologies, materials, and manufacturing techniques to disrupt legacy industrial ecosystems.
- *Migrate:* Relocate basic materials industries using cheaper production unlocked by clean energy.
- *Align:* Harmonize customers' and providers' incentives by rewarding frugal structural design and "servitizing" basic materials.

Read Amory Lovins's original paper and then produce a 500-word narrative memo that provides a clear example for each of the five strategies on how to decarbonize energy in buildings, transport, industry, or by producing electric power: Lovins (2021).

[Answer: Open-ended question.]

28.6 The US electricity grid has been called the "largest machine in the world", comprising 7700 power plants, 3300 utilities and 4.3 million km of power lines. In the process of *electrifying everything*, policymakers will need to ensure the stability of this "machine." US electricity customers experienced eight hours of power interruptions in 2020 (EIA 2021). Is it possible to electrify large parts of a modern economy? Explain in no fewer than 300 words.

[Answer: Open-ended question.]

28.7 While hydrogen (H_2) has high energy density by weight, it has very low energy density by volume (see Table 7.5). This suggests that long-haul ships might be ideally suited to using hydrogen fuel. The size of hydrogen storage tanks on ships might need to be very large, and if ships instead use liquefied hydrogen (with a larger volumetric energy density, according to Table 7.5), the refrigeration could be prohibitively high (liquid hydrogen must be stored at cryogenic conditions of $-253\,°C$). One must also consider the safety aspects of liquid hydrogen, leakage and detonation risks, and the need for novel bunkering infrastructure. Instead of using hydrogen, do research into using ammonia (NH_3) produced from green hydrogen via the Haber Bosch process (also see Section 9.3). Could green ammonia fueled internal combustion engines be the answer to sustainable long-haul shipping? Discuss your findings in no fewer than 500 words.

[Answer: Open-ended question.]

28.8 Cement manufacturing uses 0.5% of all the USA's energy, but produces 2% of CO_2. Globally, this figure is 7% of CO_2 emissions. The CO_2 is produced from both the burning of a hydrocarbon fuel and the basic chemical reaction that forms cement from limestone (CaO). Perform combustion calculations and show that producing 1 kg of cement results in the emission of about 0.9 kg of CO_2. Studying the following papers might stand you in good stead when answering this question:

- Worrell et al. (2001).
- Ali et al. (2011).

[Answer: In the two cited papers.]

28.9 German manufacturer GEA Heating & Refrigeration Technologies has installed a 1.4-MW_{th} Grasso V heat pump that runs on ammonia (NH_3 or R717) at the Mars Wrigley Confectionery chocolate factory in Veghel, The Netherlands. The heat pump has an average *COP* of 5.9. The heat pump uses low-temperature heat recovered from the facility's refrigeration system to heat up water to $63\,°C$. This water is then used in various processes within the facility, for example chocolate and syrup storage and air-handling units. The heat pump has led to a reduction of 1,000,000 m^3 of natural gas per year, equivalent to about 1000 tons of CO_2-equivalent per year, or to about 1800 tons of CO_2 per year if green electricity is applied.

Provide your own case study illustrating how a heat pump could be integrated in a large food and beverage plant. Select a heat pump from GEA's online heat pump catalog to meet the heating duties of your selected food or beverage plant. Clearly state the operating conditions of the heat pump, its yearly-averaged *COP*, the capacity factor of the plant, and CO_2 emissions saved compared to performing the same operation with a gas boiler.

[Answer: Open-ended question.]

References

Ali, M.B., Saidur, R. and Hossain, M.S. (2011). A review on emission analysis in cement industries. *Renewable and Sustainable Energy Reviews* 15 (5): 2252–2261.

Bhaskar, A., Assadi, M. and Nikpey Somehsaraei, H. (2020). Decarbonization of the iron and steel industry with direct reduction of iron ore with green hydrogen. *Energies* 13 (3): 758.

BNEF (2022a). Lithium-ion battery pack prices rise for first time to an average of $151/kWh. Bloomberg New Energy Finance. https://about.bnef.com/blog/lithium-ion-battery-pack-prices-rise-for-first-time-to-an-average-of-151-kwh/

BNEF (2022b). Electric Vehicle Outlook 2022. Bloomberg New Energy Finance. https://about.bnef.com/electric-vehicle-outlook/

BNEF (2022c). New Energy Outlook 2022. Bloomberg New Energy Finance. https://about.bnef.com/new-energy-outlook/

BP (2022). BP Statistical Review of World Energy 2022, 71st edition.

Bullard, C. (2022). Will the cars of future run on biofuels or solar electricity? Personal communication with Professor Clark Bullard, University of Illinois at Urbana-Champaign, June 2022.

Burlig, F., Bushnell, J., Rapson, D. and Wolfram, C. (2021). Low energy: estimating electric vehicle electricity use. *AEA Papers and Proceedings* 111 (May): 430–435.

Car and Driver (2022). Longest range electric cars, ranked. Car and Driver, 7 June 2022. https://www.caranddriver.com/shopping-advice/g32634624/ev-longest-driving-range/

Cembalest, M. (2021). *2021 Annual Energy Paper. Eye on the Market*, 11th edition. New York: JP Morgan Asset & Wealth Management.

Cembalest, M. (2022). *2022 Annual Energy Paper. Eye on the Market*, 12th edition. New York: JP Morgan Asset & Wealth Management.

Challa, R., Kamath, D. and Anctil, A. (2022). Well-to-wheel greenhouse gas emissions of electric versus combustion vehicles from 2018 to 2030 in the US. *Journal of Environmental Management* 308 (April): 114592.

Costanza, R. (2023). *Addicted to Growth: Societal Therapy for a Sustainable Wellbeing Future*. Abingdon-on-Thames, Oxfordshire: Routledge.

Dorr, A. and Seba, T. (2020). RethinkX. Disruption, implications, and choices. Rethinking energy 2020–2030. A RethinkX sector disruption report.

DOT (2022). Average age of automobiles and trucks in operation in the United States. US Department of Transport. https://www.bts.gov/content/average-age-automobiles-and-trucks-operation-united-states

DryFiciency (2023). DryFiciency – Industrial heat pump for a climate-neutral European industry. Wienerberger. https://heatpumpingtechnologies.org/annex58/wp-content/uploads/sites/70/2022/07/hthpannex58dryfwbgfinal-1.pdf

Du Plessis, G.E., Liebenberg, L. and Mathews, E.H. (2013). The use of variable speed drives for cost-effective energy savings in South African mine cooling systems. *Applied Energy* 111: 16–27.

EIA (2020). Hourly electricity consumption varies throughout the day and across seasons. US Energy Information Agency. https://www.eia.gov/todayinenergy/detail.php?id=42915

EIA (2021). US electricity customers experienced eight hours of power interruptions in 2020. US Energy Information Agency. https://www.eia.gov/todayinenergy/detail.php?id=50316

EIA (2022). Annual Energy Outlook 2022. US Energy Information Agency. https://www.eia.gov/outlooks/aeo/

Fournier, D. (2023). *Personal Communication with Don Fournier, 7 February*. Champaign, IL: Energy Consultant.

Griffith, S. (2021). *Electrify: An Optimist's Playbook for Our Clean Energy Future*. MIT Press.

IAE (2021). Phase-out regulations for fossil fuel boilers at EU and national level. Institute for Applied Ecology, Freiburg, Germany. https://www.oeko.de/fileadmin/oekodoc/Phase-out_fossil_heating.pdf

IEA (2022a). World Energy Outlook 2022. International Energy Agency. https://www.iea.org/reports/world-energy-outlook-2022/

IEA (2022b). Global EV Outlook 2022. Trends in electric light-duty vehicles. International Energy Agency https://www.iea.org/reports/global-ev-outlook-2022/trends-in-electric-light-duty-vehicles

IEA (2022c). The future of heat pumps. International Energy Agency. https://www.iea.org/reports/the-future-of-heat-pumps

Inside EVs (2022). Which electric car is the cheapest? What sort of range should I expect from a certain car? Inside EVs, July 2022. https://insideevs.com/reviews/344001/compare-evs/

IRENA (2018). *Global Energy Transformation. A Roadmap to 2050*. Abu Dhabi: International Renewable Energy Agency.

IRENA (2020). *Energy Transformation 2050*. Abu Dhabi: International Renewable Energy Agency.

IRENA (2022). *Renewable Capacity Statistics 2022*. Abu Dhabi: International Renewable Energy Agency.

Kim, J., Sovacool, B.K., Bazilian, M., Griffiths, S., et al. (2022). Decarbonizing the iron and steel industry: a systematic review of sociotechnical systems, technological innovations, and policy options. *Energy Research and Social Science* 89 (July): 102565.

Larson, E. (2020). *Net-zero America: Potential Pathways, Infrastructure, and Impacts*. Princeton University.

Liebenberg, L. and Meyer, J.P. (1998). Potential of the zeotropic mixture R-22/R-142b in high-temperature heat pump water heaters with capacity modulation. *ASHRAE Transactions* 104: 418.

Liebenberg, L., Velleman, D. and Booysen, W. (2012). A simple demand-side management solution for a typical compressed-air system at a South African gold mine. *Journal of Energy in Southern Africa* 23(2): 20–29.

Lovins, A. (2021). Decarbonizing our toughest sectors-profitably. *MIT Sloan Management Review* 63 (1): 46–55.

McKinsey (2020). *Plugging In: What Electrification Can Do for Industry*. McKinsey Global Institute.

McMillan, C. and Ruth, M. (2018). *Industrial Process Heat Demand Characterization (No. 91)*. Golden, CO: National Renewable Energy Laboratory-Data (NREL-DATA).

Muehlegger, E.J. and Rapson, D.S. (2023). Correcting estimates of electric vehicle emissions abatement: Implications for climate policy. *Journal of the Association of Environmental and Resource Economists* 10 (1): 263–282.

Obrist, M.D., Kannan, R., McKenna, R., Schmidt, T.J. et al. (2023). High-temperature heat pumps in climate pathways for selected industry sectors in Switzerland. *Energy Policy* 173: 113383.

Smil, V. (2022). Beyond magical thinking: Time to get real on climate change, 19 May 2022. *Yale Environment* 360.

Swanepoel, J.A., Mathews, E.H., Vosloo, J. and Liebenberg, L. (2014). Integrated energy optimisation for the cement industry: A case study perspective. *Energy Conversion and Management* 78: 765–775.

Thiel, G.P. and Stark, A.K. (2021). To decarbonize industry, we must decarbonize heat. *Joule* 5 (3): 531–550.

Thunberg, G. (2023). *The Climate Book: The Facts and the Solutions*. Penguin Press.

Victor, D. (2020). Deep decarbonization: A realistic way forward on climate change, 28 January 2020. *Yale Environment* 360.

Waite, M. and Modi, V. (2020). Electricity load implications of space heating decarbonization pathways. *Joule* 4 (2): 376–394.

Worrell, E., Price, L., Martin, N., Hendriks, C., et al. (2001). Carbon dioxide emissions from the global cement industry. *Annual Review of Energy and the Environment* 26(1): 303–329.

WSJ (2022). Rivian CEO warns of looming electric-vehicle battery shortage, 18 April 2022. *Wall Street Journal*.

Yeung, P. (2022). How Norway popularized an ultra-sustainable heating method. Reasons to be cheerful, 17 January 2022. https://reasonstobecheerful.world/heat-pumps-norway-efficiency-emissions/

Zehner, O. (2012). *Green Illusions. The Dirty Secrets of Clean Energy and the Future of Environmentalism*. University of Nebraska Press.

Mini Project 10

Decarbonizing the Steel Industry

Steel manufacture currently produces 8% of global anthropogenic CO_2 emissions. Most steel productions begin by burning coal with iron ore in a blast furnace. Coal combustion generates heat but is also part of the chemical process in the furnace where oxygen is stripped from the ore to make pure iron (also known as pig iron). This process is called coking. The pure iron is then turned into steel in an electric arc furnace. The waste product (from combining the carbon in the coal with the oxygen in the ore) is large quantities of CO_2. The entire process emits around 1.9 tons of CO_2 for every ton of steel.

Molten iron is poured into a furnace for purification and alloying to become steel at the ThyssenKrupp steelworks in Duisburg, Germany. *Source:* Katpatuka / Wikimedia Commons / CC BY 4.0.

Energy Systems: A Project-Based Approach to Sustainability Thinking for Energy Conversion Systems,
First Edition. Leon Liebenberg.
© 2024 John Wiley & Sons, Inc. Published 2024 by John Wiley & Sons, Inc.
Companion website: www.wiley.com/go/liebenberg/energy_systems

1) Do independent research about steel manufacturing. Provide a detailed process map that explains the traditional steel manufacturing process that employs a blast furnace and an electric arc furnace, followed by a continuous casting machine or rolling machine, forming machines, and coating machines. Summarize each of the salient steps in the process and provide a breakdown of the typical energy (heat and electricity) requirement for each of these steps.

2) Carbon dioxide emissions could be significantly reduced by abandoning the blast furnace process altogether. One alternative approach for making pure iron would be to run electrical current through molten iron oxide in a process known as iron ore pyro-electrolysis. Provide an explanatory diagram showing this process and its typical current and voltage requirements.

3) The electrical energy demands for pyro-electrolysis are huge, but without the need for coal as part of the process itself, that energy could come from a no-carbon source such as nuclear power or green hydrogen. Consider only green hydrogen. How could such green hydrogen be produced on-site? Clearly describe the process, provide explanatory diagrams, and summarize the salient chemistry for producing green hydrogen. How much green hydrogen would be required to power a traditional arc furnace?

4) How many megatons of CO_2 would be saved when using iron ore pyro-electrolysis compared to the traditional process?

5) How does the use of green hydrogen to produce green steel (via iron ore pyro-electrolysis) relate to the urgent needs facing planetary and social systems addressed in the United Nations' Sustainable Development Goals (SDGs)? You should interpret and report the findings of at least five peer-reviewed journal papers published in the last five years.

6) Based on your research to retrofit a steel manufacturing facility with a pyro-electrolysis plant and based on the projected energy savings compared to a traditional steel-manufacturing plant, determine the monthly repayments if the developers were to borrow all the money with a 20-year, 6%-interest loan. Assume a realistic capacity factor. State your assumptions and show all calculations. What would the payback period be?

Appendix A

Table A.1 Units of Energy and Power.

Unit	SI Equivalent
1 electron volt [eV]	1.602×10^{-19} J
1 eV per molecule	96.49 kJ/mol
1 calorie (thermochemical)	4.184 J
1 British thermal unit [BTU]	1055 J
1 kilowatt-hour [kWh]	3.6 MJ
1 horsepower [hp]	746 W
1 BTU/lb	2.32 kJ/kg
1 cubic meter of natural gas	36 MJ
1 kilogram of methane	61.3 MJ
1 cubic meter [m^3] of biogas	23 MJ
1 liter of gasoline	32.9 MJ
1 kilogram of gasoline	43.8 MJ
1 liter of diesel oil	35.9 MJ
1 kilogram of diesel oil	42.7 MJ
1 cubic meter of hydrogen at 1 atm	10 MJ
1 kilogram of hydrogen	120 MJ
1 therm (US)	105.5 MJ
1 barrel of oil equivalent [BOE]	6.118 GJ
1 ton of coal equivalent [TCE]	29.308 GJ
1 tonne of oil equivalent [toe]	41.868 GJ
1 quadrillion BTU [quad] = 10^{15} Btu	1.055 EJ
1 terawatt-year [TWy]	31.56 EJ
1 watt [W]	1 J/s
1 horsepower [hp] (electric)	746 W
1 ton of air conditioning	3.517 kW
1 BTU/h ft^2	3.154 W/m^2

Energy Systems: A Project-Based Approach to Sustainability Thinking for Energy Conversion Systems,
First Edition. Leon Liebenberg.
© 2024 John Wiley & Sons, Inc. Published 2024 by John Wiley & Sons, Inc.
Companion website: www.wiley.com/go/liebenberg/energy_systems

Table A.2 Conversion Factors.

Quantity	Colloquial Unit	SI Unit
Mass	1 ounce [oz]	0.0284 kg
	1 pound [lb]	0.454 kg
	1 ton (US)	907.2 kg
	1 metric tonne [t]	1000 kg
Length	1 foot [ft]	0.305 m
	1 mile [mi]	1609 m
	1 inch [in]	0.0254 m
Time	1 year [y]	3.156×10^7 s
	1 hour [h] (= 60 min)	3600 s
	1 minute [min]	60 s
Force	1 pound [lb]	4.448 N
Energy	1 British thermal unit [BTU]	1055 J
	1 foot-pound [ft-lb]	1.355 J
	1 kilowatt-hour [kWh]	3,6000,000 J
Power	1 watt [W]	1 J/s
	1 horsepower [hp]	746 J/s
Area	1 acre	4047 m^2
	1 hectare [ha]	10,000 m^2
Volume	1 liter [l]	0.001 m^3 (= 1000 cm^3)
	1 fluid ounce [fl oz] (US)	0.029
	1 gallon [gal]	$3.785 \times 10^{-3} / 4.546 \times 10^{-3}$ m^3 (US/Imperial)
	1 oil barrel [bbl] (=159 l)	0.159 m^3
Density	1 pound per cubic foot [lb/ft^3]	16.018 kg/m^3
Speed	1 mile per hour [mph]	0.447 m/s = 1.609 km/h
	1 knot [kn]	0.514 m/s
	1 foot per second [ft/s]	0.305 m/s
Rotational speed	1 revolution per minute [rpm]	0.1047 rad/s
Pressure	1 atmosphere [atm]	101,325 Pa
	1 pound per square inch [psi]	6896 Pa
	1 bar	100,000 Pa
Temperature	° celsius [°C]	K − 273.15
	° fahrenheit [°F]	32 + 1.8 °C
Temperature difference	1°F	0.56 °C (and 0.56 K)
Thermal resistance	1 ft^2 °F h/BTU	0.176 Km^2/W
Magnetic field	1 gauss [G]	0.0001 T
Radiation	1 rad	0.01 Gy
	1 rem	0.01 Sv
	1 curie [Ci]	3.7×10^{10} Bq
Viscosity	1 poise [P]	0.1 Pa s
Permeability	1 darcy [D]	9.869×10^{-13} m^2

Table A.3 SI Units for Various Physical Quantities.

Quantity	SI unit	Symbol	Definition of Derived Unit
Mass	kilogram	kg	
Length	meter	m	
Time	second	s	
Frequency	hertz	Hz	/s
Force	newton	N	$kg\,m/s^2$
Energy	joule	J	$kg\,m^2/s^2$
Power	watt	W	J/s
Pressure	pascal	Pa	N/m^2
Charge	coulomb	C	A s
Current	ampere	A	
Electromagnetic potential	volt	V	J/C (or J/As)
Resistance	ohm	Ω	V/A
Capacitance	farad	F	C/V (or As/V)
Inductance	henry	H	V s/A
Magnetic field	tesla	T	$V\,s/m^2$
Amount	gram-mole	mol	
Temperature	kelvin	K	
Activity	becquerel	Bq	s^{-1}
Absorbed dose	gray	Gy	J/kg
Dose equivalent	sievert	Sv	J/kg
Luminous intensity	candela	cd	
Luminous flux	lumen	lm	cd sr
Illumination	lux	lx	$Cd\,sr/m^2$
Plane angle	radian	rad	
Solid angle	steradian	sr	

Table A.4 Prefixes, Symbols, and Factors.

yocto (y)	10^{-24}	kilo (k)	10^3
zepto (z)	10^{-21}	mega (M)	10^6
atto (a)	10^{-18}	giga (G)	10^9
femto (f)	10^{-15}	tera (T)	10^{12}
pico (p)	10^{-12}	peta (P)	10^{15}
nano (n)	10^{-9}	exa (E)	10^{18}
micro (μ)	10^{-6}	zetta (Z)	10^{21}
milli (m)	10^{-3}	yotta (Y)	10^{24}

Table A.5 Fundamental Constants and Useful Physical Quantities (Approximated).

π	3.142
e	2.718
Planck's constant (h)	6.626×10^{-34} J s
Planck's (reduced) constant ($\hbar = h/2\pi$)	1.055×10^{-34} J s
Speed of light (c)	2.998×10^{8} m/s
Vacuum permeability (μ_0)	$4\pi \times 10^{-7}$ N A^2
Vacuum permittivity (ϵ_0)	$(\mu_0 c^2)^{-1} = 0.054$ F/m $= 8.854 \times 10^{-12}$ C/V m
Avogadro constant (N_A)	6.022×10^{23} mol^{-1}
Boltzmann constant (k_B)	1.381×10^{-23} J/K
Stefan–Boltzmann constant (σ)	5.67×10^{-8} W/m^2 K^4
Electron charge (e)	1.602×10^{-19} C
Electron mass (m_e)	9.109×10^{-31} kg $\cong 0.511$ MeV/c^2
Proton mass (m_p)	1.673×10^{-27} kg $\cong 938.272$ MeV/c^2
Atomic mass unit, or dalton (u)	1.661×10^{-27} kg $\cong 931.494$ MeV/c^2
Mean radius of Earth's orbit (1 AU)	1.496×10^{11} m
Earth's mean equatorial radius	6.378×10^{6} m
Mass of Earth	5.972×10^{24} kg
Average solar constant above atmosphere	1366 W/m^2
Global geothermal flux / average gradient	44.2 TW / 25 K/km
Gravitational acceleration	9.807 m/s^2
Standard temperature and pressure (STP)	273.15 K, 100 kPa
Gas constant ($R = N_A k_B$)	8.314 kJ/kmol K
Water – latent heat of melting / vaporization at STP	334 kJ/kg / 2.26 MJ/kg
Specific heat capacity of water (15 °C) / air (STP)	4.186 kJ/kg K / 1.0035 kJ/kg K
Mass density of water (15 °C) / air (STP)	999.1 kg/m^3 / 1.275 kg/m^3
Half-life of ^{235}U / ^{238}U / ^{239}Pu	7.0×10^{8} / 4.5×10^{9} / 2.4×10^{4} yr
Half-life of ^{232}Th / ^{40}K / ^{222}Rn	1.4×10^{10} yr / 1.25×10^{9} yr / 3.8 d
Average annual environmental radiation exposure	3×10^{-3} Sv/yr

Index

Energy Systems: A Project-Based Approach to Sustainability Thinking for Energy Conversion Systems, First Edition. Leon Liebenberg.
© 2024 John Wiley & Sons, Inc. Published 2024 by John Wiley & Sons, Inc.
Companion website: www.wiley.com/go/liebenberg/energy_systems